Mammalogy

 SAUNDERS COLLEGE PUBLISHING

Philadelphia Fort Worth Chicago
San Francisco Montreal Toronto London Sydney
Tokyo

Mammalogy
Third Edition

Terry A. Vaughan

Northern Arizona University
Flagstaff, Arizona

Text Typeface: Garamond Book
Compositor: University Graphics
Acquisitions Editors: Michael Brown and Ed Murphy
Project Editor: Carol Field
Copy Editor: Irene Nunes
Art Director: Carol Bleistine
Cover Art + Design: Lawrence R. Didona
Production Manager: Tim Frelick
Assistant Production Manager: JoAnn Melody

Library of Congress Cataloging in Publication Data

Vaughan, Terry A.
  Mammalogy.

  Bibliography: p.
  Includes index.

  1. Mammals.   I. Title.
QL703.V38   1985   599   85-10754
ISBN 0-03-058474-4

MAMMALOGY 3rd edition   ISBN 0-03-058474-4

Printed in the United States of America.
Library of Congress catalog card number 85-10754.

6   032   9876543

Saunders College Publishing
Holt, Rinehart and Winston
The Dryden Press

Preface

This text is intended for use by upper division or graduate college or university students majoring in zoology. Because no two instructors cover the same material in presenting a course on mammalogy, I have tried to treat the biology of mammals broadly enough to make this book useful to different instructors with contrasting approaches to the subject.

In this third edition most of the organization of the previous editions has been retained. Many of the chapters dealing with the biology of mammals have been largely rewritten or supplemented with recently published material, whereas the chapters on the orders of mammals have been changed relatively little. I have not attempted to present an exhaustive review of the world literature on mammals, but instead have dealt with subjects that I regard as important, interesting, and basic to an understanding of mammals. New knowledge is being published at such a rate that any text in science is to a large extent a progress report that is already partly out of date when it is published.

If this text is useful, holds the interest of students, and is respected by members of the far-flung community of mammalogists, it is primarily due to the work of the researchers on whose studies I have based this book. The organization, choice of topics, and selection of illustrative material is mine, but the book should be credited largely to those whose names appear in the bibliography.

The overall layout of this book still reflects the influence of Carl W. May, former Biology Editor of Saunders College Publishing, whose firm editorial hand guided the preparation of the first edition. Others from whom I sought help on the earlier editions, and whose contributions are still evident in the present edition of this book, include Edwin H. Colbert, E. Raymond Hall, Edalee Harwell, Jason A. Lillegraven, Richard E. MacMillen, Larry G. Marshall, Thomas J. O'Shea, W. Leslie Robinette, Jane Stanton and the late Hugh Stanton, and the late Hobart M. Van Deusen.

I have had a great deal of assistance on this third edition from many people. I would like to thank the following friends and colleagues for their especially important help: William A. Clemens, Nicholas Czaplewski, Mary R. Dawson, William R. Downs, John F. Eisenberg, George E. Goslow, Jr., Jennifer U. M. Jarvis, Farish A. Jenkins, Jr., Karl F. Koopman, Gerald L. Kooyman, Thomas H. Kunz, Jason A. Lillegraven, Richard E. MacMillen, Larry G. Marshall, Kenneth S. Norris, Thomas J. O'Shea, Leonard Radinsky, Nancy Siepel-Hyatt, Jack S. States, Jill E. Varnum, and Thomas G. Whitham.

I also gratefully acknowledge the help of the staff of Saunders College Publishing. Special thanks are extended to Carol Field, Ed Murphy, and Irene Nunes, each of whom contributed important advice, critical comments, and forbearance.

During part of 1983, my wife Rosemary and I were on sabbatical leave in Kenya, East Africa. Information gained during this trip and some photos taken in national parks and reserves were included in this edition. In Kenya, Professor Festo A. Mutere and his wife Melsa were our gracious hosts. They helped in many ways to make our stay both productive and pleasant. Much of our time was spent camping at Lake Baringo, where we studied bats and observed hippos. At Lake Baringo we received many courtesies and essential help and advice from Betty Roberts, Elizabeth Roberts, Murray Roberts, and Terry Stevenson.

Finally, I offer my sincerest gratitude to my wife, Rosemary, who helped with the preparation of this third edition in many ways. In addition to doing all of the typing for this project and helping with the photographic work, the bibliography, the index, and the drawings, she critically read all of the newly written material. Her insistence on clarity and directness salvaged many of my abstruse and tortuous sentences. I appreciate especially that her enthusiasm and genuine enjoyment of the project made it a pleasure for me.

Terry A. Vaughan

Contents

Chapter 1

Introduction

The Domain of Mammalogy

Mammalogy—the division of zoology dealing with mammals—has occupied the efforts of scientists of many kinds. Vertebrate zoologists have studied such aspects as the structure, taxonomy, distribution, and life histories of mammals; physiologists have considered mammalian hibernation and water metabolism; physicists and engineers have studied mammalian echolocation and locomotion; geologists and vertebrate paleontologists have outlined the patterns of mammalian evolution; and psychologists and ecologists have considered mammalian behavior. In addition, a wealth of information has been contributed by perceptive observers lacking zoological training. Prehistoric drawings and carefully molded figures of mammals attest to our long-standing fascination with the beauty and grace of mammalian forms.

Mammals have been regarded as worthy of study by this wide variety of workers for many reasons. The practical aspects have attracted some. Through the study of various kinds of laboratory mammals, much has been learned about mammalian histology and about the effects of diseases and drugs; work on domestic breeds of mammals has improved meat production; and research on game species has shown how sustained yields of these animals can be achieved through appropriate management techniques. To most students and researchers, however, practicality is not foremost. To this group, mammals are beautiful and fascinating creatures with remarkable physiological, structural, and behavioral adaptations to an amazing array of life styles, and thus living mammals in their natural settings are the focal point of interest. The adaptations themselves, how they enable mammals to efficiently exploit demanding environmental conditions, how they evolved, and the interaction between mammals and their environment are all fascinating lines of inquiry. The most productive studies have been the result of the intense interest of researchers in a biological relationship rather than their preoccupation with solving a practical problem. In this book I deal primarily with the impressive literature on mammals that has resulted from such basic research. This research cannot be regarded as "impractical" since, ideally, perspective gained from such work must guide our decisions affecting the survival of threatened species of mammals (and, indeed, entire ecosystems).

During the last 30 years, our knowledge of mammalian biology has expanded tremendously: echolocation (natural sonar) has been studied intensively in both bats and marine mammals; the remarkable ability of some mammals to live under conditions of extreme aridity with no drinking water has been explained; the circulatory and metabolic adaptations associated with temperature regulation and metabolic economy have been studied; adaptations to deep diving in marine mammals have been investigated; hibernation and migration and the mechanisms influencing them have received attention; important contributions have been made to our knowledge of mammalian population cycles and the factors that may control them; studies of functional morphology have enlarged our understanding of mammalian terrestrial, aquatic, and aerial locomotion. Probably no field has been more tardy in developing than that of animal behavior. This field has developed rapidly in the last few years, however—perhaps in response to our belated realization that the time for studying some species under natural conditions is growing short—and has contributed tremendously to our appreciation of how finely tuned to their environment mammals are. Broad and vital gaps in our basic knowledge of mammals remain, but work continues.

Classification

In any careful study, one of the vital early steps is the organization and naming of objects. As stated by Simpson (1945:1) with reference to animals, "It is impossible to examine their relationships to each other and their places among the vast, incredibly complex phenomena of the universe, in short to treat them scientifically, without putting them into some sort of formal arrangement." The arrangement of organisms is the substance of taxonomy, but modern taxonomists, perhaps better termed systematists, are less interested in identifying and classifying animals than in studying their evolution. These systematists bring information from such fields as genetics, ecology, behavior, and paleontology to bear on the subjects of their research. They attempt to base their classifications on the most reliable evidence of animal phylogenetic relationships. Excellent discussions of the importance of systematics to our knowledge of animal evolution are given by Simpson (1945) and Mayr (1963).

Because of the difficulties that arise when a single kind of animal or plant is recognized by different common names by people in different areas, or by many common names by people in one area, scientists more than 200 years ago adopted a system of naming organisms that would be recognized by biologists throughout the world. Each known kind of organism has been given a binomial (two part) scientific name. The first, the *generic name,* may be applied to a number of kinds, but the second name refers to a specific kind, a *species.* As an example, the blacktail jackrabbit of the western United States is *Lepus californicus.* To the genus *Lepus* belong a number of similar, but distinct, long-legged species of hares, such as *L. othos* of Alaska, *L. europaeus* of Europe, and *L. capensis* of Africa. Because considerable geographic variation frequently occurs within a species, a third name is often added; this designates a particular *subspecies.* Thus, the large-eared and pale-colored subspecies of *L. californicus* that lives in the deserts of the western United States is *L. c. deserticola;* the smaller-eared and dark-colored subspecies from coastal California is *L. c. californicus.*

The species is the basic unit of classification. A modern and widely accepted definition of species is given by Mayr (1942:120): "Species are groups of actually or potentially interbreeding natural populations, which are reproductively isolated from other such groups." Each species is generally separated from all other species by a "reproductive gap," but within each species there is the possibility for gene exchange. As put by Dobzhansky (1950), all members of a species "share a common gene pool."

Clearly, however, not all species resemble each other to the same degree or are equally closely related. The hierarchy of classification, based on the starting point of the species, has been developed to express degrees of structural similarity and, ideally, phylogenetic relationships between species and groups of species. The taxonomic scheme includes a series of categories, each higher category more inclusive than the one below. Using our example of the hares, a number of long-legged species are included in the genus *Lepus.* This genus, and other genera containing rabbit-like mammals, form the *family* Leporidae; this family, and the family Ochotonidae (the pikas), share certain structural features not possessed by the other mammals and belong to the *order* Lagomorpha; this order, and all other mammalian orders, form the *class* Mammalia, members of which differ from all other animals in the possession of hair, mammary glands, and many other features. Mammals, birds, reptiles, amphibians, and fish all possess an endoskeleton, and these groups (in addition to some others) form the *phylum* Chordata. All of the phyla of animals (Protozoa, Porifera, Coelenterata, and so forth) are united in the Animal Kingdom. The classification of our jackrabbit can be outlined as follows:

> Kingdom Animal
> Phylum Chordata
> Class Mammalia
> Order Lagomorpha
> Family Leporidae
> Genus *Lepus*
> Species *Lepus californicus*

Further subdivision of this classification scheme may result from the recognition of additional categories, such as subclass, superorder, or subfamily.

Most ordinal names end in *-a,* as in Carnivora; all family names end in *-idae,* and all subfamily names end in *-inae.* In the following discussions, contractions of the names of orders, families, or subfamilies will often be used as adjectives for the sake of convenience: leporid will refer to Leporidae, leporine to Leporinae, lagomorph to Lagomorpha, and so on.

Some similarities between different kinds of animals are due to parallelism or to convergence. *Parallelism* occurs when two closely related kinds of animals pursue similar modes of life for which similar structural adaptations have evolved. The similar specializations of the skull and dentition (elongate snouts and reduced number of teeth) that occur in a number of genera of nectar-feeding Neotropical bats are examples of parallelism. *Convergence* involves the development of similar adaptations to similar (or occasionally nearly identical) styles of life by distantly related species. The golden moles of Africa (p. 86) and the marsupial "moles" of Australia (p. 65) are examples of convergence. These animals belong to different

mammalian infraclasses (Eutheria and Metatheria, respectively; p. 45), and their lineages have been separate for over 70 million years. Their habits are much the same, however, and structurally they resemble each other in many ways.

An outline of the classification of mammals used in this book is given in Chapter 4. It is not based on any single published classification, but in some ways it reflects current taxonomic thought. Although it departs from his system in many minor ways, the classification used here is partially that of Simpson (1945). It should be stressed that no universal agreement has been reached on the classification of mammals. Our knowledge of many groups of mammals is incomplete, and future study may demonstrate that some of the families listed here can be discarded because they contain animals best included in another family; perhaps a family or two are yet to be described. The present classification, then, is not used by all mammalogists, and it is by no means immutable.

Describing Specialization

The terms "primitive," "specialized," and "advanced" are used repeatedly in the chapters on orders and families. A primitive mammal is one that has not departed far from the ancestral type, or at least has retained many structural characters typical of the ancestral type. A monotreme (egg-laying mammal, such as the duck-billed platypus or spiny anteater) is more primitive than a house cat because it lays eggs and has a small brain and bones in the pectoral girdle that are found in reptiles but not in other mammals. In these ways the monotreme resembles reptiles, the ancestral group of mammals, whereas the cat does not.

A specialized mammal is one that, in becoming adapted to a particular mode of life, has departed strongly from the ancestral structural plan. A horse is specialized because, in becoming adapted to a life in which speed afoot and the ability to feed on grasses are important, the limbs became elongate, all digits but the third were essentially lost, and the cheek teeth developed complicated occlusal surfaces. In these ways (and many others), the specialized horse has departed strongly from the structure of the primitive mammal.

The term "advanced" is frequently used in comparisons of different members of an evolutionary line or taxonomic group. An advanced species is one in which the particular structural features that characterize the group are highly developed. As an example, the present-day big brown bat is more advanced than the early Eocene bat *Icaronycteris index.* Both are specialized for flight, but because of many refinements in the flight apparatus of the big brown bat, it is the far more perfectly adapted flier.

Plan of the Book

The first part of this book covers preliminary material. Chapter 2 deals with the characteristics of mammals, that is, the structural features that characterize the group. Chapter 3 briefly covers the evolution of mammals from reptilian ancestry, and Chapter 4 is an introduction to the classification of mammals. In roughly the first half of the remainder of the book (Chapters 5 through 17), the orders and families of mammals are discussed; the second half (Chapters 18 through 25) treats selected aspects of the biology of mammals. My main hope is that from this coverage of the subject students can gain a general understanding of, and an appreciation for, the form and function of mammals.

The anatomical drawings, which I regard as essential to the ordinal chapters, should help students understand the discussions of structure and function. I have made liberal use of these drawings, most of which illustrate skulls, teeth, or feet. The profiles of skulls are from the left side, and all occlusal views of teeth show the right upper or the left lower tooth row.

The bibliography in the appendix includes both papers that are cited in the text and additional publications by authors whose papers are cited. The latter are included because I feel they contain material students may find useful.

Chapter 2

Mammalian Characteristics

Mammals owe their spectacular success to many features. Many of the most important and most diagnostic mammalian characteristics serve to further intelligence and sensory ability, to promote endothermy, or to increase the efficiency of reproduction or of securing and processing food. The senses of sight and smell are highly developed, and the sense of hearing has undergone greater specialization in mammals than in any other vertebrates. Efficient gathering and utilization of a tremendous variety of foods are aided by specializations of the dentition and the digestive system. The perfection of endothermy has allowed mammals to remain active under a wide array of environmental conditions, and specializations of the postcranial anatomy, particularly the limbs and feet, have enabled them to make effective use of this activity. In some species, extended periods of parental care have enabled parents to train their young in demanding foraging patterns and in complex social behavior.

The basic structural plan of the mammalian body was inherited from a relatively unspectacular reptilian group, the mammal-like reptiles of the synapsid order Therapsida (p. 28). Members of this ancient order followed an evolutionary path that diverged strongly from the reptilian path from which arose the vastly more spectacular and more successful dinosaurs and other ruling reptiles (subclass Archosauria). The key to the marginal persistence of the therapsids through the Triassic (Table 3–1) was perhaps their ability to move and to think more quickly than their archosaurian contemporaries. These same abilities probably enabled the descendants of the therapsids, the mammals, to

survive through the Jurassic and Cretaceous periods, when the dinosaurs completely dominated the terrestrial scene. Also of major importance to early mammals was their highly specialized dentition, which probably allowed them to utilize certain foods more efficiently than could reptiles.

An important morphological trend in the therapsid-mammalian line was toward skeletal simplification. An engineer may redesign a machine and increase its efficiency by reducing the number of parts to the minimum consistent with the effective performance of a particular function; a similar type of simplification occurred in the therapsid skeleton. In the skull and lower jaw, which in primitive reptiles consisted of many bones, a number of bones were lost or reduced in size. The limbs and limb girdles were also simplified to some extent, and their massiveness was reduced. As a result, the advanced therapsid skeleton roughly resembles that of the egg-laying mammals (order Monotremata), but the limbs of some therapsids were less laterally splayed than are those of today's specialized monotremes.

When mammals first appeared in the Triassic, then, they represented no radical structural departure from the therapsid plan, but had simply attained a level of development (involving a dentary/squamosal jaw articulation; p. 29) that is interpreted by most vertebrate paleontologists as indicating that the animals had crossed the mammalian-reptilian boundary. Many of the mammalian characters discussed in this chapter resulted from evolutionary trends clearly characteristic of therapsid reptiles. Unfortunately, the fossil record cannot indicate when various important features of the soft anatomy became established, and one can only speculate whether or not advanced therapsid reptiles had such features as mammae, hair, or a four-chambered heart.

Although the fossil record yields no clear evidence as to when endothermy became established in therapsids or early mammals, we know that today the metabolic "machinery" of mammals differs markedly from that of reptiles. Mammals have a three- to sixfold greater capacity for energy production than do reptiles and a standard metabolic rate some eight times higher (Else and Hulbert, 1981). Relative to total body size, the internal organs of mammals are larger than those of reptiles, and mitochondrial membrane surface areas for the heart, kidney, and brain are far greater in mammals, as are mitochondrial enzyme activity and thyroid activity. These differences are clearly related to the mammalian capacity for high energy production and were probably part of a suite of anatomical and physiological features that allowed mammals (perhaps even the earliest ones) to be nocturnal.

Crompton, Taylor, and Jagger (1978) hypothesize that homeothermy in mammals developed in two steps. Early mammals developed homeothermy as an adaptation to nocturnal foraging for insects, but maintained body temperatures some 10 C° lower than the 38 to 40°C of most mammals of today. The lower body temperature was maintained without an increase in the resting metabolic rate above the reptilian level. The second step involved an elevation of body temperature and metabolic rate in some mammals, and these mammals could be active in the heat of the day and largely avoid prolonged evaporative cooling.

The following sources discuss the characteristics of mammals: Romer and Parsons, 1977; H. M. Smith, 1960; Weichert, 1965; J. Z. Young, 1957. Especially useful is the treatment by Romer in *Vertebrate Paleontology* (1966:187–196).

Soft Anatomy

Skin Glands.   The skin of mammals contains several kinds of glands not found in other vertebrates; the most important of these are the

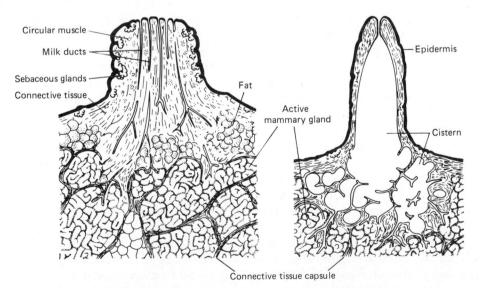

Figure 2-1. A section through the nipple and associated tissues of a primate (left) and the nipple of an artiodactyl (right). (From Hildebrand, 1974)

mammary glands. These glands provide nourishment for the young during their postnatal period of rapid growth. Mammary glands consist of a complex system of ducts which reach the surface of the skin through a prominence called a nipple or teat (Fig. 2–1). During late pregnancy, the epithelium of the ducts is stimulated by endocrine secretions (estrogens and progesterone) and divides rapidly to produce secretory alveoli, which after birth produce milk. The production of milk is stimulated by secretions (prolactin and growth hormone) of the anterior lobe of the pituitary. Nursing and emptying of milk from the mammary glands result in nervous stimulation of the anterior lobe of the pituitary and continued production of prolactin and milk.

The composition of milk differs from species to species. Cow's milk contains about 85 percent water; the dry weight includes approximately 20 percent protein, 20 percent fat, and 60 percent sugars (largely lactose), as well as vitamins and salts in roughly the proportion in which they are found in blood. The milk of mammals whose young grow unusually rapidly contains high levels of protein and fat. Seals, for example, produce milk with roughly 12 times as much fat and five times as much protein as cow's milk.

The period of association between the mother and its young during lactation and suckling is one of close mother-young social bonds. For many mammals, especially those with complex foraging and social behavior patterns, this is an extremely important time of training and preparing the young for adult life.

In most mammals, the young suck milk from the projecting nipples. In monotremes, however, nipples are lacking and the young suck from tufts of hair on the mammary areas (Burrell, 1927; Ewer, 1968:236). Whales, dolphins, and porpoises have muscles that force milk into the mouth of the young, a seemingly necessary adaptation in animals that have no lips and are therefore unable to suck. The number of nipples varies from two in a number of mammals to about 19 in the mouse-opossum (*Marmosa;* Tate, 1933:36).

Other types of skin glands are also important in mammals. The watery secretion of the *sweat glands* functions primarily to promote evaporative cooling but also eliminates some waste materials. In humans and some ungulates, sweat glands are broadly distributed over the body surface, but in most mammals they are more restricted. In some insectivores, rodents, and carnivores, sweat glands occur only on the feet or on the venter, and they are completely lacking in the Cetacea and in some bats and rodents. Hair follicles are supplied with *sebaceous glands* (Fig. 2–2), the oil secretion of which lubricates the hair and skin. A variety of *scent glands* and *musk glands* are found in mammals. These glands are variously used as attractants, for marking territories, for communication during social interactions, or for protection. The smell of skunk is familiar to all but the most city-bound and has caused the temporary banishment of many a farm dog. A musk gland marked by a patch of dark hairs occurs on the top of the tail of wolves and coyotes, as well as on the tail of many domestic dogs. The functions of some mammalian scent glands in connection with social behavior are discussed in Chapter 20.

Hair. The bodies of mammals are typically covered with hair, a unique mammalian feature that has no structural homolog among other vertebrates. Hair was perhaps developed by therapsid reptiles before a scaly covering was lost. In modern mammals that possess scaly tails or bony plates (such as armadillos), hairs project from beneath the scales in a regular pattern. A similar pattern of hair distribution, perhaps reflecting the ancestral condition, is also seen in mammals without scales. (Note the pattern of hair projection on your own arm, for example.)

A hair consists of dead epidermal cells that are strengthened by keratin, a tough, horny tissue made of proteins. A hair grows from living cells in the hair root. Each hair consists of an outer layer of cells arranged in a scalelike pat-

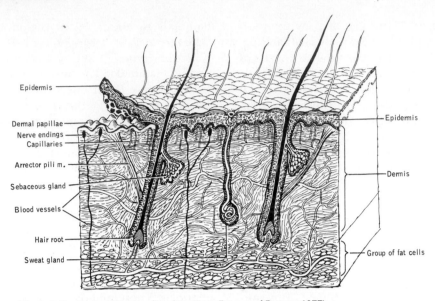

Figure 2-2. Section of mammalian skin. (From Romer and Parsons, 1977)

tern, the *cuticular scale,* a deeper layer of highly packed cells, the *cortex,* and in some cases a central core of cuboidal cells, the *medulla* (Fig. 2–3). The color of hair depends on pigment in either the medulla or the cortical cells; the cuticular scale is usually transparent.

The coat of hair, termed the *pelage,* functions primarily as insulation. The dissipation of heat from the skin surface to the environment and the absorption of heat from the environment are retarded by the pelage. Seals, sea lions, and walruses, many of which live in extremely cold water, are insulated by both hair and subcutaneous blubber. Some mammals are hairless,

or nearly so. These either live in warm areas or have specialized means of insulation other than hair. The essentially hairless whales and porpoises have thick layers of blubber that provide insulation. Hair is sparse on elephants, rhinoceroses, and hippopotami; these animals live in warm areas, have thick skins that offer some insulation, and have such favorable mass/surface ratios because of their large size (p. 449) that retention of body heat is no problem.

Hair, being nonliving material, is subject to considerable wear and bleaching of pigments. During periodic molts, usually once or twice a year, old hairs are lost and new ones replace them. This often occurs in a regular pattern of replacement (Fig. 2–4). In many north-temperate species, the molts are in the spring and fall and the summer pelage is generally shorter and has less insulating ability than the winter pelage. In some species that live in areas with continuous snow-cover in the winter, the summer pelage is brown and the winter coat is white. The arctic fox, several species of hares, and some weasels follow this pattern.

The color of most small terrestrial mammals closely resembles the color of the soil on which they live. In his careful study of conceal-

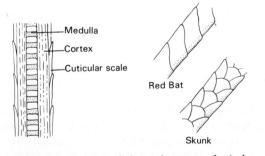

Figure 2-3. Structure of a hair and two types of reticular scale patterns. (After T. I. Storer and Usinger, 1965)

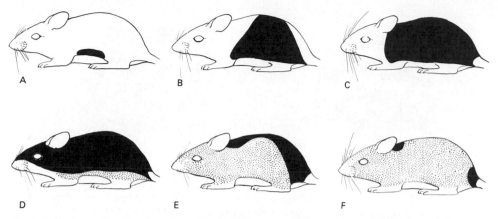

Figure 2–4. The pattern of postjuvenile molt in the golden mouse *(Ochrotomys nuttalli)*. Black areas indicate portions where adult hair is replacing juvenile hair. Stippled areas indicate new adult pelage. (After Linzey and Linzey, 1967)

ing coloration in desert rodents of the Tularosa Basin of New Mexico, Benson (1933) found that white sands were inhabited by nearly white rodents and adjoining stretches of black lava were home to black rodents. Broadly speaking, nocturnal mammals that are active against dark substrates, such as dark forest soils, are dark-colored, whereas those foraging over light-colored soils, such as the soil found in deserts, arc relatively pale. In large, diurnal, open-country dwellers, such as African antelope, however, coloration may be related to temperature control. The pale bodies of the Arabian oryx *(Oryx leucoryx)* and the addax *(Addax nasomaculatus),* both desert dwellers, may be important in reflecting light and reducing the intake of solar radiation.

Countershading is a color pattern common to mammals and many other vertebrates. Under most lighting conditions, the back of an animal is more brightly illuminated than is the underside. If a mammal were all of a single color, the underside would appear very dark relative to the back and the form of the animal would be obvious. When the back and sides are darkly colored and the underside and insides of the legs are white, however—an almost universal color pattern among terrestrial mammals—the well-lighted back reflects little light and the shaded white venter tends to strongly reflect

light. The result is that the form of the animal becomes obscured to some extent and the animal becomes less conspicuous.

The color patterns of mammals serve a variety of purposes. The pelages of some ungulates and some rodents are marked by white stripes that tend to obliterate the shapes of the animals when they are against broken patterns of light and shade. The eye is one of the most conspicuous and unmistakable vertebrate features; some facial markings in mammals may serve importantly to obviate the bold pattern of the eye by superimposing a more dominant and disruptive pattern (Fig. 2–5). If these markings only occasionally allow an animal to go unnoticed by a predator, or if they cause a predator to be indecisive in its attack for but a fraction of a second, they have adaptive value.

Even in broad daylight, the stripes of zebras cause distant herds to fade into their background, but another adaptive importance has been considered by Cott (1966). The stripes are so patterned as to create an optical illusion: the animal's apparent size is increased (Fig. 2–6). In the dim light in which most predators hunt, this illusion may cause a slight miscalculation of range and an occasional inaccurate leap.

What about the glaringly white rump patch of the pronghorn antelope (Fig. 13–19), however, or the conspicuous white-next-to-black

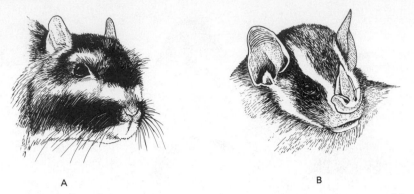

Figure 2-5. The faces of two mammals in which facial masks reduce the conspicuousness of the eye: (A) South American viscacha *(Lagostomus maximus)*, (B) tent-building bat *(Uroderma bilobatum)*.

markings of some African antelope, the bold white eye ring of many carnivores (Fig. 10–3), and the white-on-black pattern of skunks? No single explanation can be applied to these diverse patterns, but each makes an animal, or at least part of it, more obvious rather than less so. The antelope's markings seemingly function as warning signals to other herd members when an individual begins to run from danger; the bounding gait of some antelope shows off these markings. The eye rings of carnivores are perhaps important as accents for the eyes and face and may accentuate facial expressions used during intraspecific social interactions. The black-and-white coloration of skunks, on the other hand, makes these defensively well-endowed animals conspicuous and unmistakable to their would-be predators.

Fat and Energy Storage. Although fat (adipose tissue) is by no means a feature unique to mammals, in these animals it is of particularly vital importance. Fat serves three major functions in mammals: (1) energy storage, (2) a source of heat and water, and (3) thermal insulation. The lives of many species of mammals are punctuated by times of crisis when food is in short supply or energy demands are unusually high. Those mammals that hibernate must store enough energy to sustain life through periods when no food is available, for example, and some huge filter-feeding whales spend their winters in plankton-poor tropical waters where they do little or no feeding. Such mammals survive by metabolizing stored fat. During times when males are competing for mates or defending territories or when females are lactating, stored fat is often the key to survival, and those individuals with the greatest amounts of stored fat have the highest reproductive success. An example of the severity of this energy crisis in a

Figure 2–6. (A) Horizontal and (B) actual markings of a zebra, showing the optical illusion of increased size. (Partly after Cott, 1966)

A

B

female gray seal is given by R. Young (1976): during 15 days of lactation, the mother lost 45 kilograms of body weight while its young gained 27 kilograms. Desert dwellers and mammals of temperate areas often have localized fat storage (in the tail or in the inguinal or abdominal region, for example), whereas boreal and aquatic species typically store fat subcutaneously over much of the body, and this layer is important as insulation as well as for food storage.

Circulatory System. In keeping with their active life and their endothermic ability, mammals have a highly efficient circulatory system. A complete separation of the systemic circulation and the pulmonary circulation has been achieved in mammals. The four-chambered heart functions as a double pump: the right side of the heart receives venous blood from the body and pumps it to the lungs; the left side receives oxygenated blood from the lungs and pumps it to the body. The fascinating evolution of the mammalian heart and circulatory pattern is described in detail by Hildebrand (1974:277).

As might be expected because of the great size difference between the smallest and the largest mammal (the weights of the 2-gram shrew and the 160,000-kilogram whale differ by a factor of 80,000,000), the heart rate is highly variable from species to species. The rate in nonhibernating mammals varies from under 20 beats per minute in seals to over 1300 in shrews (Table 2–1). Especially remarkable is the ability of some mammals to alter their heart rates rapidly. As an extreme example, a resting big brown bat *(Eptesicus)* has a rate of about 400 beats per minute; this rate increases almost instantly to about 1000 when the bat takes flight and generally returns to the resting rate within 1 second after flight stops (Fig. 2–7).

The erythrocytes (red blood cells) of mammals are biconcave disks. They extrude their nuclei when they mature, apparently as a means of increasing oxygen-carrying capacity.

Respiratory System. In mammals, the lungs are large and, together with the heart, virtually fill the thoracic cavity. Air passes down the trachea, into the bronchi, and through a series of branches of diminishing size into the bronchioles, from which branch alveolar ducts. Clustered around each alveolar duct is a series of tiny terminal chambers, the alveoli. Exchange of gases between inspired air and the bloodstream occurs in the alveoli; the thin alveolar membranes are surrounded by dense capillary beds. In humans, the lungs contain about 300 million alveoli, which provide a total respiratory surface of about 70 square meters— some 40 times the surface area of the body.

Table 2-1: Heart Rates of Selected Mammals

Species	Common Name	Weight	Heart Rate (beats/min)
Erinaceus europaeus	European hedgehog	500–900 g	246 (234–264)
Sorex cinereus	Gray shrew	3–4 g	782 (588–1320)
Eutamias minimus	Least chipmunk	40 g	684 (660–702)
Sciurus carolinensis	Gray squirrel	500–600 g	390
Phocaena phocaena	Harbor porpoise	170 kg	40–110
Mustela vison	Mink	0.7–1.4 kg	272–414
Phoca vitulina	Harbor seal	20–25 kg	18–25
Elephas maximus	Asiatic elephant	2000–3000 kg	25–50
Equus caballus	Horse	380–450 kg	34–55
Sus scrofa	Swine	100 kg	60–80
Ovis aries	Sheep	50 kg	70–80

Data from Altman and Dittmer, 1964:235.

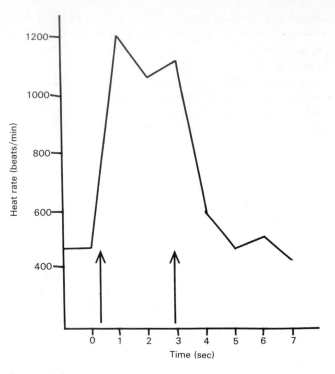

Figure 2-7. Heart rate of the big brown bat *(Eptesicus fuscus)* at rest and during flight. The arrows indicate the beginning and end of flight. (After Studier and Howell, 1969)

Air is forced into the lungs by muscular action that increases the volume of the thoracic cavity and decreases the pressure within the cavity. Some volume increase is gained by the forward and outward movement of the ribs under the control of intercostal muscles, but of greater importance is the muscular diaphragm (a structure unique to mammals). When relaxed, the diaphragm is bowed forward, but when contracted its central part moves back-

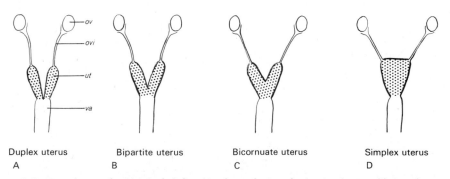

Duplex uterus
A

Bipartite uterus
B

Bicornuate uterus
C

Simplex uterus
D

Figure 2-8. Several types of uteri (stippled) found in placental mammals, showing degrees of fusion of the two "horns" of the uterus. (A) Duplex uterus occurs in the orders Lagomorpha, Rodentia, Tubulidentata, and Hyracoidea. (B) Bipartite uterus is known in the order Cetacea. (C) Bicornuate uterus is found in the order Insectivora, in some members of the orders Chiroptera and Primates, and in the orders Pholidota, Carnivora, Proboscidea, Sirenia, Perissodactyla, and Artiodactyla. (D) Simplex uterus is typical of some members of the order Chiroptera and Primates and of the order Edentata. (From H. M. Smith, 1960)

ward toward the celomic cavity, thus increasing the volume of the thoracic cavity.

Reproductive System.   In mammals, both ovaries are functional and the ova are fertilized in the oviducts. The embryo develops in the uterus and lies within a fluid-filled amniotic sac. Nourishment for the embryo comes from the maternal bloodstream by way of the placenta. (The female reproductive cycle and the establishment of the placenta are discussed in Chapter 21.) The structure of the uterus is variable (Fig. 2–8).

The male copulatory organ, the penis, contains erectile tissue and is surrounded by a sheath of skin, the *prepuce.* In many species the penis contains a bone, the *os penis,* or baculum, which may differ markedly even between closely related species (Fig. 2–9) and may

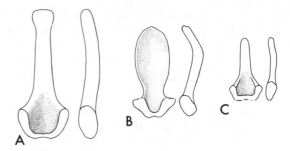

Figure 2-9. The bacula of several species of New Guinean murid rodents, showing differences in the structure of this bone in closely related mammals. (From Lidicker, 1968)

therefore be of considerable use in taxonomic studies. The tip of the penis has an extremely complicated form in some species (Fig. 2–10). The testes of mammals, instead of lying in the celomic cavity as in other vertebrates, are typi-

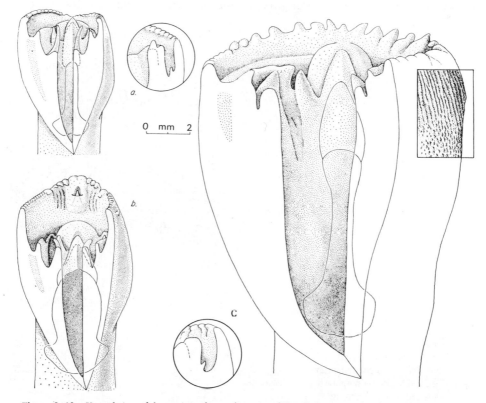

Figure 2-10. Ventral view of the penises of several species of New Guinean murid rodents, showing the complex structure of the organ in these mammals. (From Lidicker, 1968)

cally contained in the *scrotum,* a saclike structure that lies outside the body cavity but is an extension of the celomic cavity. The testes either descend permanently from the celomic cavity into the scrotum when the male reaches reproductive maturity or are withdrawn into the body cavity between breeding seasons and descend when the animal again becomes fertile. In most mammals, the maturation of sperm cannot proceed normally at the usual deep-body temperature and the scrotum functions as a "cooler" for the testes and developing sperm.

Brain. Compared with the brains of other vertebrates, that of the mammal is unusually large. This greater size is due largely to a tremendous increase in the size of the cerebral hemispheres. These structures were ultimately derived from a part of the brain important in lower vertebrates in receiving and relaying olfactory stimuli.

Most characteristic of the brain of higher mammals is the great development of the *neopallium,* a mantle of gray matter that first appeared as a small area in the front part of the cerebral hemispheres in some reptiles. In mammals the neopallium has expanded over the surface of the deeper, primitive vertebrate brain. The surface area of the neopallium is vastly increased in many mammals by a complex pattern of folding (Fig. 2–11). A new development in placental mammals is the *corpus callosum,* a large concentration of nerve fibers that passes between the two halves of the neopallium and provides communication between them.

The unique behavior of mammals is largely a result of the development of the neopallium, which functions as a control center that has come to dominate the original brain centers. Sensory stimuli are relayed to the neopallium, where much motor activity originates. Present actions are influenced by past experience;

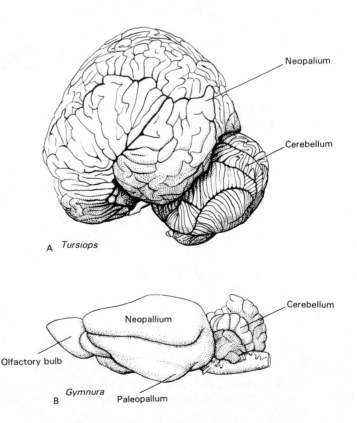

Figure 2–11. The brain of (A) a porpoise and (B) a hedgehog. The neopallium is greatly enlarged and highly convoluted in the specialized and intelligent porpoise but relatively small and smooth-surfaced in the primitive hedgehog. (*Tursiops* after Kruger, 1966; *Gymnura* after Romer and Parsons, 1977.)

learning and "intelligence" are important. The size of the brain relative to total body size seems not always to be a reliable guide to intelligence, for brain size apparently need not increase in proportion to increases in body size to maintain intelligence. The degree of development of convolutions on the surface of the neopallium is perhaps a better indication of intelligence.

Sense Organs. The sense of smell is acute in many mammals, probably in part as a result of the development of turbinal bones in the nasal cavities (Fig. 2–12). The olfactory bulbs and olfactory lobes form a great part of the brain in some insectivores and are reasonably large in carnivores and rodents. The sense of smell is poorly developed and the olfactory part of the brain is strongly reduced in whales and the higher primates; the olfactory system is absent in porpoises and dolpins (Kruger, 1966:247).

The sense of hearing is highly developed in mammals, and in some 20 percent of mammalian species hearing provides an important substitute for vision (Chapter 24). Mammals alone have an external structure (the *pinna*) to intercept sound waves; the pinnae may be extremely large and elaborate in some mammals, particularly in bats (Fig. 8–13). Pinnae are missing (presumably secondarily lost) in some insectivores, phocid seals, and cetaceans. The *external auditory meatus,* the tube leading from the pinna to the tympanic membrane, is typically long in mammals and extremely long in cetaceans. The middle ear is an air-filled chamber that houses the three *ossicles* and is typically enclosed by a bony *bulla* (Fig. 2–13). The mammalian *cochlea* is more or less coiled. (Some variations in the structure of the mammalian ear are discussed on p. 510).

The mammalian eye resembles that of most amniote vertebrates. In most nocturnal mammals, the *tapetum lucidum* is well developed. This is a reflective structure within the choroid that improves night vision by reflecting light back to the retina. (This reflection accounts for the shine when an animal's eyes are picked up by headlight beams at night.) Although in most

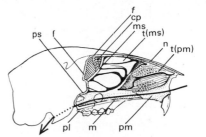

Figure 2–12. Cutaway view of the nasal chamber of the Abert Squirrel *(Sciurus aberti),* showing the complicated arrangement of turbinal bones. The entire right half of the nasal part of the skull is removed, exposing the left side of the nasal chamber. The arrow shows the main air path from the external to the internal nares, but some air circulates through the upper part of the chamber and over the turbinal bones. Abbreviations: *cp,* cribriform plate of the mesethmoid bone (through which the branches of the olfactory nerve pass out of the braincase); *f,* frontal; *m,* maxillary; *ms,* mesethmoid; *n,* nasal; *pl,* palatine; *pm,* premaxillary; *ps,* presphenoid; *t(ms),* turbinals connected to the mesethmoid; *t(pm),* turbinals connected to the premaxillary.

mammals the eyes are well developed, in some insectivores and some cetaceans they are strongly reduced in both size and function. In such species the eyes, able to differentiate only between light and dark, and may serve primarily to aid the animal in maintaining the appropriate nocturnal or diurnal activity cycles (Herald, Brownell, et al., 1969; Lund and Lund, 1965).

Most mammals have *vibrissae.* These are the whiskers on the muzzle and the long, stiff hairs that are present on the lower legs of some mammals. The vibrissae are tactile organs, and those on the face probably enable nocturnal species to detect obstacles near the face. The vibrissae on the muzzle generally arise from a structure termed the *mystacial pad* and are controlled by a complex of muscles (Fig. 2–14).

Digestive System. Salivary glands are present in mammals, and in some ant-eating species

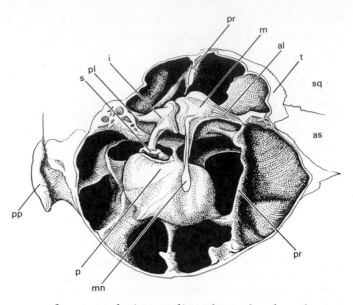

Figure 2-13. Lateral view of the right middle ear chamber (anterior is to the right) of the Abert squirrel *(Sciurus aberti),* with the auditory bulla largely removed. The complex partitioning of the bulla, the position of the ear bones and the inner ear, and the ligamentous bracing of the malleus and incus are shown. In life the manubrium of the malleus rests against the tympanic membrane. Abbreviations: *al,* anterior ligament (of the malleus); *as,* alisphenoid; *i,* incus; *m,* malleus; *mn,* manubrium; *p,* periotic; *pl,* posterior ligament; *pp,* paroccipital process; *pr* partitions of bulla; *s,* stapes; *sq,* squamosal; *t,* tympanic.

they are specialized for the production of a mucilaginous material that makes the tongue sticky. The stomach is a single saclike compartment in most species but is complexly subdivided in ruminant artiodactyls, in cetaceans, and in sirenians (Fig. 2–15). In herbivorous species, digestion is frequently accomplished partly by microorganisms that inhibit the stomach or the *cecum,* a blind sac that opens into the posterior end of the small intestine.

Muscular System. The mammalian limb and trunk musculature has been highly plastic. Dif-

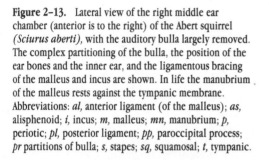

Figure 2-14. The superficial facial muscles of the cotton rat *(Sigmodon bispidus);* these muscles partly control facial expression. (After Rinker, 1954)

ferent evolutionary lines have developed muscular patterns beautifully adapted to diverse modes of locomotion. Cetaceans are the fastest marine animals, certain carnivores and ungulates are the most rapid runners, and bats as fliers are more maneuverable than birds. Some muscular specializations favoring specific types of locomotion are described in the ordinal chapters. Especially notable in mammals is the great development of dermal musculature. In many mammals these muscles form a sheath over most of the body and allow the skin to move. These dermal muscles have differentiated and have moved over much of the head (Fig. 2–14), where they control many essential actions. In mammals there are no more vital voluntary muscles than those that encircle the mouth; these function during suckling and are among the first voluntary muscles to be subjected to heavy use. Facial muscles move the ears, close the eyes, and control the subtle changes in expression that are so important in the social lives of many mammals.

The Skeleton

General Features. The mammalian skeleton differs from that of the reptile in several basic

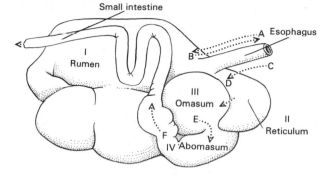

Figure 2–15. The four-chambered stomach of a ruminant artiodactyl. As the animal feeds, it swallows the vegetation, which is then stored in the rumen. While the animal rests, it regurgitates the food from the rumen and "chews its cud" (remasticates the food). The food then goes to the reticulum, omasum, and abomasum, where digestion is aided by a diverse microbiota. (After T. I. Storer and Usinger, 1965)

ways, all of which may well be related to the active style of life of mammals. The mammalian skeleton is more completely ossified (Fig. 2–16), a feature perhaps associated with the need for well-braced attachments for muscles. Considerable fusion of bones has also occurred, as, for example, in the pelvic girdle. The skeleton has become simplified in mammals; this development seemingly increases the flexibility of the axial skeleton and allows the limbs greater speed and range of movement. The greater range of movement is of particular advantage to arboreal creatures, which many early mammals may have been. The simplification of the skeleton may have also been advantageous in terms of metabolic economy—the less bone, the less energy invested in its development and maintenance. Further, selection may have favored a light skeleton in the interest of quick movement with relatively little expenditure of energy.

To an animal as active as a mammal, well-formed articular surfaces on limb bones and

solid points of attachment for muscles are highly advantageous during the period of skeleton growth as well as during adult life. Mammals have abandoned the pattern of bone growth typical of reptiles. In many reptiles, skeletal growth may continue throughout much of life. Growth in reptiles occurs at the ends of limb bones by ossification of the deep parts of a persistently growing cartilaginous cap; such a pattern clearly limits the establishment of a well-formed joint. In mammals, however, skeletal growth is generally restricted to the early part of life. The articular surfaces and some points of attachment of large muscles become well formed and ossified early, while rapid growth is still under way. Growth continues at a cartilaginous zone where the end of the bone and its articular surface, the *epiphysis,* join the shaft of the bone, the *diaphysis* (Fig. 2–17). When full growth is attained, this cartilaginous zone of growth becomes ossified, fusing the epiphysis and diaphysis. Because within a given

Figure 2–16. Lateral view of the right side of the pectoral girdle of (A) a lizard and (B) a mammal, showing the greater ossification and simplification of the structure in the mammal. Abbreviations: *ac,* anterior coracoid; *acr,* acromion process; *cl,* clavicle; *cor,* coracoid process; *icl,* interclavicle; *sc,* scapula; *sp,* spine of scapula.

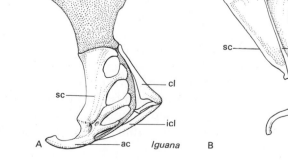

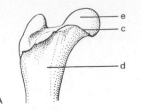

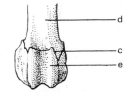

Figure 2-17. The (A) proximal and (B) distal ends of the right femur of a young hedgehog *(Erinaceus europaeus),* showing the epiphyses *(e),* diaphysis *(d),* and intervening cartilaginous zone *(c).*

species this fusion usually occurs at a certain age, the degree of closure of the "epiphyseal line" is useful in estimating the age of a mammal.

The Skull. The braincase of the mammalian skull (Fig. 2–18) is large. In addition to its primary function of protecting the brain, the braincase provides a surface from which the temporal muscles originate. (In many mammals, these are the most powerful muscles that close the jaws.) A *sagittal crest* increases the area of origin for the temporal muscles in many mammals; the *lambdoidal crest* gives origin to the temporal muscles and some cervical muscles. The *zygomatic arch* is usually present as a structure that flares outward from the skull. It protects the eyes, provides origin for the masseter muscles, and forms the surface with which the condyle of the dentary (lower jaw) bone articulates. The zygomatic arch may be reduced or lost, as in some insectivores and cetaceans, or may be enlarged, as in those groups in which the masseter muscles largely supplant the tem-

poral muscles as the major jaw muscles (Fig. 16–3). The mammalian skull has a secondary palate (p. 31 and Fig. 3–4), and there are usually *turbinal bones* within the nasal cavities (Fig. 2–12).

A number of *foramina* (openings) perforate the braincase and allow passage of the cranial nerves (Fig. 2–18). In some rodents, the infraorbital foramen, through which blood vessels and a branch of the fifth cranial nerve (the trigeminal) pass, is enormously enlarged in association with specializations of the masseter muscles (Fig. 16–2). The incisive foramina, present in the palates of many mammals, house an olfactory organ (Jacobson's organ) that allows the mammal to "smell" the contents of its mouth. These olfactory organs are widespread among vertebrates; a snake puts the tips of its forked tongue against this part of the palate after "testing" its immediate environment.

Sounds that cause vibration of the tympanic membrane are mechanically transmitted by the three ear ossicles (Fig. 2–13) through the air-filled chamber to the inner ear. The footplate of the stapes fills an opening into the inner ear and, acting like a piston, transforms the movements of the ossicles to vibrations of the fluid in the cochlea. The inner ear, with the cochlea and semicircular canals, is contained by the periotic bone, which is generally covered by the squamosal bone but is exposed as the mastoid bone in some mammals (Fig. 2–18). The auditory bulla is formed by the expanded tympanic bone or by the tympanic bone plus the entotympanic bone, a bone found only in mammals. The bulbous tympanic bullae in many mammals look like structures remembered too late and hurriedly stuck to the skull, but they are highly modified in some species in connection with specialized modes of life.

The lower jaw is formed by the *dentary bone,* which typically has a coronoid process, on which the temporalis muscle inserts, a coronoid fossa, in which the masseter muscles insert, and an angular process, to which a jaw-opening muscle (the digastricus) attaches. In

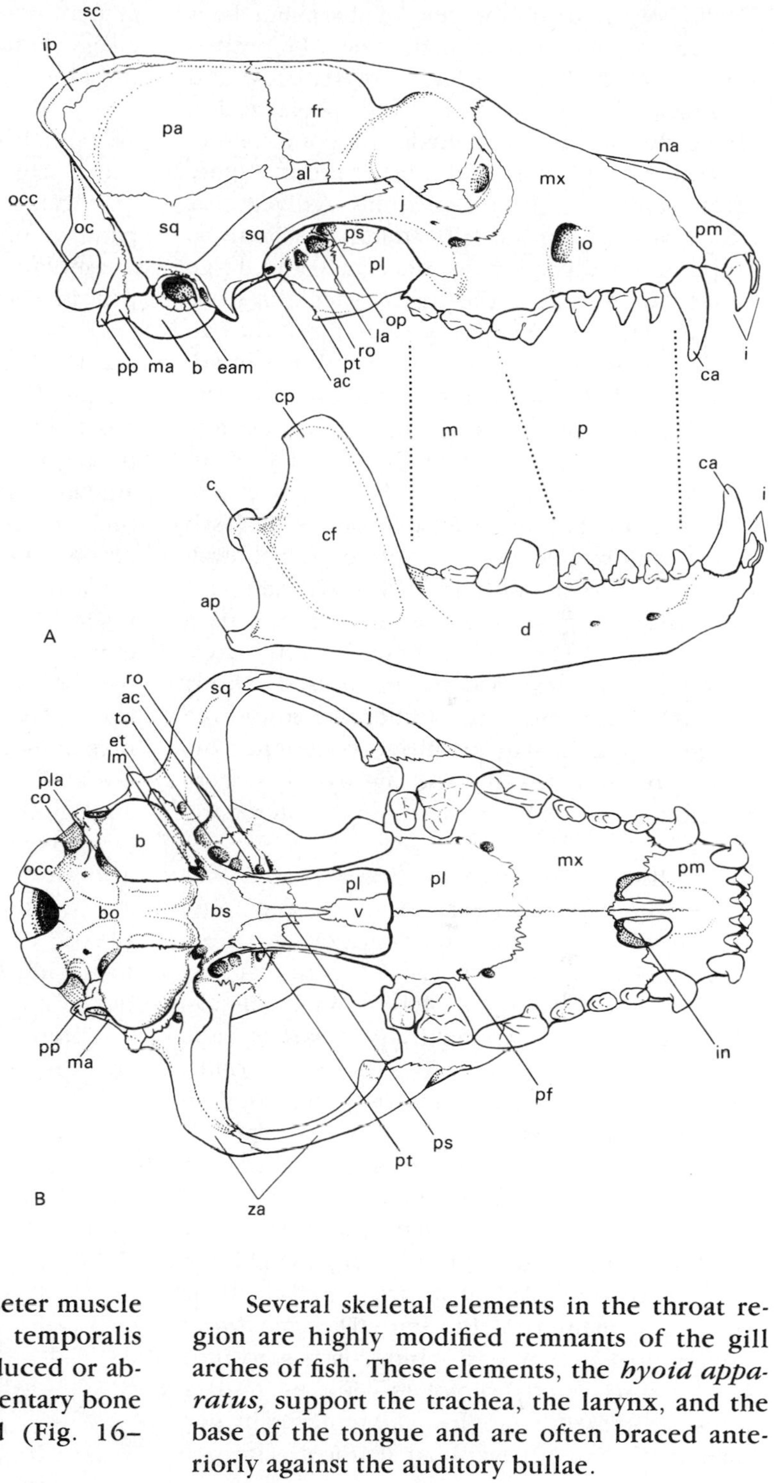

Figure 2–18. (A) Side and (B) top views of the skull of the African hunting dog *(Lycaon pictus),* showing bones, foramina, and teeth. Abbreviations: *ac,* alisphenoid canal; *al,* alisphenoid; *ap,* angular process; *b,* auditory bulla; *bo,* basioccipital; *bs,* basisphenoid; *c,* condyle; *ca,* canines; *cf,* coronoid fossa; *co,* anterior condyloid foramen; *cp,* coronoid process; *d,* dentary; *eam,* external auditory meatus; *et,* eustachian tube; *fo,* foramen ovale; *fr,* frontal; *i,* incisors; *in,* incisive foramen; *io,* infraorbital foramen; *ip,* interparietal; *j,* jugal; *l,* lacrimal; *la,* anterior lacerate foramen; *lm,* medial lacerate foramen; *m,* molars; *mx,* maxillary; *na,* nasal; *oc,* occipital; *occ,* occipital condyle; *op,* optic foramen; *p,* premolars; *pa,* parietal; *pf,* posterior palatine foramen; *pl,* palatine; *pla,* posterior lacerate foramen; *pm,* premaxillary; *pp,* paroccipital process; *ps,* presphenoid; *pt,* pterygoid; *ro,* foramen rotundum; *sc,* sagittal crest; *sq,* squamosal; *v,* vomer; *za,* zygomatic arch.

some herbivores, in which the masseter muscle is enlarged at the expense of the temporalis muscle, the coronoid process is reduced or absent and the posterior part of the dentary bone becomes dorsoventrally broadened (Fig. 16–9A).

Several skeletal elements in the throat region are highly modified remnants of the gill arches of fish. These elements, the *hyoid apparatus,* support the trachea, the larynx, and the base of the tongue and are often braced anteriorly against the auditory bullae.

Teeth. Without doubt, one of the major keys to the success of mammals has been the possession of teeth. Fish, amphibians, reptiles, and mammals all have teeth, but the specialization of the dentition in mammals has gone far beyond anything found in the other groups. Only among mammals have dentitions evolved that are capable of coping with items so difficult to prepare for digestion as dry grass and large bones. So varied are the dental specializations of mammals and so closely related are they to specific styles of feeding and to patterns of adaptations of the skull, jaws, and jaw musculature, that to know in detail the dentition of any mammal is to understand much about its way of life. Most of our knowledge of the early evolution of mammals is based on studies of fossil teeth, which, because of their extreme hardness, are often the only parts of early mammals that are preserved. Teeth alone can tell us a great deal. The earliest known vertebrates, which were jawless, had bodies encased in bony plates. When some of the arches that supported the gill apparatus in primitive vertebrates became modified into jaws and jaw supports, teeth developed on the bony plates that bordered the mouth.

Although the teeth in different dentitions differ widely in number, structure, and function, in most mammals the dentition is *heterodont;* that is to say, it consists of teeth that vary in both structure and function. In mammals, teeth occur on the premaxillary, maxillary, and dentary bones (Fig. 2–18). The anteriormost teeth, the *incisors* and *canines,* are used to gather or kill food, whereas the more specialized cheek teeth, the *premolars* and *molars,* are used to grind or slice food in preparation for digestion. In many mammals the canines are used in stereotyped displays during social interactions. Characteristically, two sets of teeth appear in a mammal's lifetime. The *deciduous* dentition develops early and consists of incisors, canines, and premolars—but no molars. These "milk teeth" are lost and replaced by permanent teeth as the animal matures. The permanent dentition consists of a second set of incisors, canines, and premolars but also includes the molars, which have no deciduous counterparts. The deciduous dentition of some species bears little resemblance to the permanent dentition (Fig. 2–19 and 8–32A).

The form, function, and origin of the cusp patterns of the cheek teeth, and especially of the molars, are of particular interest. The basic primitive molar, termed *tribosphenic* in reference to the basically three-cusped pattern of the occlusal surface, is found in a number of fossil mammals. (The occlusal surfaces of teeth are those that contact their counterparts of the opposing jaw—the surfaces the dentist generally attacks when putting in a filling.) A stroke of luck for functional morphologists is that the American opossum *(Didelphis virginiana)* and some other living marsupials have molars that resemble those of primitive mammals that coexisted with dinosaurs. Opossums are omnivorous, eating insects and other small animals and soft plant material; probably many Mesozoic mammals had similar diets. Careful studies of jaw action in the opossum, therefore, can probably indicate how the molars functioned in primitive mammals over 70 million years ago. The studies of Crompton (1971), Crompton and Jenkins (1968), and Crompton and Hiiemae (1969, 1970) provide much of the basis for the following discussion of the functional morphology of molars.

In the opossum, the molars serve two func-

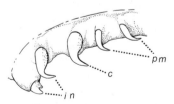

Figure 2-19.   The deciduous upper dentition (of the left side) of a Neotropical fruit-eating bat *(Artibeus lituratus)*. Abbreviations: *c,* canine; *in,* incisors; *pm,* premolars. (From Vaughan, 1970b)

tions. For up to 60 percent of the time involved
in chewing and throughout the initial stages of
chewing, the high cusps of the upper and lower
cheek teeth crush and puncture the food without coming together. After the food is thus
pulped, it is sliced by the six matching shearing
surfaces shown in Fig. 2–20. This shearing is facilitated by the way in which food is trapped
and steadied by the opposing molars (Fig. 2–
21). Chewing occurs on one side of the jaw at a
time. During cutting strokes, the jaw action is
not one of simple up and down movements. Instead, precise lateral adjustments of the jaw during mastication enable opposing molars to slide
against each other. As shown in Fig. 2–21, this
movement involves a transverse as well as an
upward component as the lower molars shear
against the uppers. Attrition facets on the molars
of Mesozoic mammals indicate that occlusion of
the teeth during chewing has always had this
transverse component (Butler, 1972). Major
shearing surfaces are those designated in Fig. 2–
20 as *1* and *2,* but additional cutting occurs
when the surfaces on the sides of the cusps of
the quadrate posterior part (the talonid) of the
lower molars shear against their counterparts on
the upper molars. As a result of this complex
pattern of occlusion, each time the three pairs
of opposing molars of one side of the jaw come
together, 18 cuts are made in the food, and thus
food already pulped is rapidly sectioned. Natural selection has favored the evolution of "efficient" dentitions because time spent in masti-

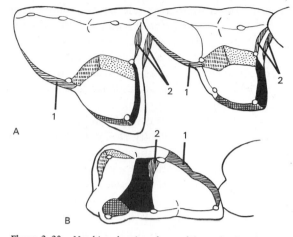

Figure 2–20. Matching shearing planes of the occlusal surfaces of
the (A) upper and (B) lower molars of the Virginia opposum
(Didelphis virginiana). (After Crompton and Hiiemae, 1969)

cating food means greater energy expenditure
and less time available for food gathering. In
some species, this also means a greater period
of vulnerability to predators.

Two major evolutionary trends in molar
structure appeared in the Mesozoic and became
pronounced in the Cenozoic. In carnivores,
some of the cheek teeth became bladelike and
the shearing function was elaborated. In the interest of powerful sectioning of flesh, transverse
jaw action was reduced in these animals and a
variety of associated changes in the skull, jaws,
and jaw musculature occurred. In herbivores,
however, which must finely macerate plant material in preparation for digestion, the molars

Figure 2–21. (A) Opposing molars of the
Virginia opossum, showing the opposing
cusps that steady, puncture, and crush food.
(B) Movement of the lower teeth as they
shear against the uppers. These teeth are
viewed from the front. See text for
description of action. (B after Crompton and
Hiiemae, 1969)

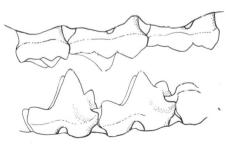

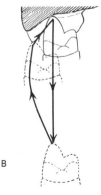

A   B

became quadrate, transverse jaw action came to be of primary importance, and distinctive features of the skull, jaws, and jaw musculature favoring this action developed. The dentitions of a modern carnivore, a modern herbivore, and a primitive Mesozoic mammal are compared in Fig. 2–22.

The number of teeth of each type in the dentition is designated by the *dental formula*. This is written as the number of teeth of each kind on one side of the upper jaw over the corresponding number in the lower jaw. Such a formula is incisors 3/3, canines 1/1, premolars 4/4, molars 2/3. Because the teeth are always listed in this order, the formula may be shortened to 3/3, 1/1, 4/4, 2/3. (The skull in Fig. 2–18 has this dental formula.) The dental formula lists the teeth of only one side; therefore, the total number of teeth in the formula must be doubled to give the total number of teeth in the

dentition. As an example, the arrangement for humans is 2/2, 1/1, 2/2, 3/3 × 2 = 32. The basic maximum number of teeth in placental mammals is 44 (3/3, 1/1, 4/4, 3/3), but marsupials commonly have more than this number. The number of teeth is frequently reduced, and a few placentals completely lack teeth; some specialized placentals, most notably odontocete cetaceans, have more than 44 teeth and have *homodont* dentitions (those in which all teeth are alike).

The mammalian tooth typically consists of an inner material, *dentine,* covered by a layer of *enamel* (Fig. 2–23). Dentine consists largely of hydroxyapatite $[3(Ca_3PO_4)_2 \cdot Ca(OH)_2]$, has an organic fiber content of about 30 percent, and is harder than bone. Enamel also consists almost entirely of hydroxyapatite, which in the enamel of all living mammals but monotremes is arranged in a prismatic crystalline pattern.

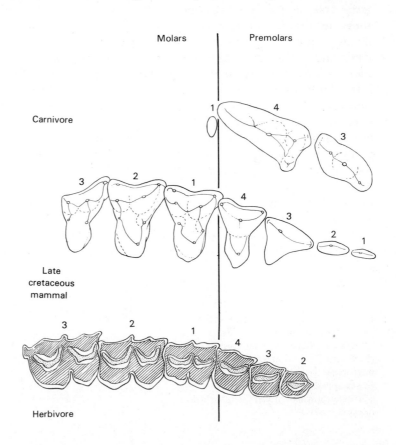

Figure 2–22. Comparisons of the occlusal surfaces of the cheek teeth of a primitive mammal (*Kennalestes,* from Late Cretaceous), a carnivore *(Lynx),* and a herbivore *(Cervus).* (Partly after Crompton and Hiiemae, 1969)

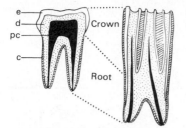

Figure 2-23. Generalized sections of mammalian teeth, showing the structure and materials. The black area is the pulp cavity *(pc)*, the stippled part is dentine *(d)*, the cross-hatched part is cement *(c)*, and the unshaded areas are enamel *(e)*. The molar on the left is similar to that of primates and is low-crowned; the molar on the right is similar to that of a horse and is high-crowned.

Enamel, the hardest mammalian (or vertebrate) tissue, is only 3 percent organic. The tooth is bound to the jaw by *cement,* a relatively soft material that may also form part of the crown of the tooth. Most mammals have teeth that are *brachyodont,* or short-crowned, and growth ceases after the tooth is fully grown. Many herbivores, because their teeth are subject to rapid wear—resulting from abrasion by silica in grass and by soil particles that adhere to plants—have *hypsodont,* or high-crowned, teeth (Fig. 2-23). As a further adaptation to abrasive food, in some mammals some teeth (and in some rodents all teeth) grow continuously and are termed *ever-growing teeth.* The roots of mammalian teeth are often divided; primitively, the upper molars have three roots and the lower molars two. Incisors and canines are single-rooted, and premolars may have one or two roots. The dentitions of herbivores usually serve

only two functions: the incisors, or the incisors and the canines, clip vegetation and the cheek teeth grind the food. Between these teeth there is usually a space called a *diastema.* This is typical of rodents, for example.

The shape of the molar crown varies in response to the demands of different diets (Fig. 2-24). In pigs and in some rodents, carnivores, and primates, the molars are *bunodont,* which means that the cusps form separate, rounded hillocks that crush and grind food. In herbivores the molars may be *lophodont,* with cusps forming ridges, or *selenodont,* with cusps forming crescents; in these cases the teeth finely section and grind vegetation. In the dental batteries of many insectivores, bats, and carnivores are *sectorial* teeth. These have bladelike cutting edges that section food by shearing against the edges of their counterparts in the opposing jaw (as do the fourth upper premolar and the first lower molar in the dentition in Fig. 2-18).

The triangular primitive upper molar (Fig. 2-25A) is marked by three major cusps, the *protocone,* the *paracone,* and the *metacone,* and the apex of the triangle (the protocone) points inward. The lower molar (Fig. 2-25B) has two sections, an anterior *trigonid* and a posterior *talonid.* The trigonid is triangular; the apex of the triangle (the *protoconid*) points outward, and the *paraconid* and *metaconid* form the inner edge. The talonid has two major cusps, the *hypoconid* and *entoconid.*

Axial Skeleton. Compared with that of the quadrupedal reptile, the mammalian vertebral column allows far greater freedom of head movement and powerful dorsoventral, rather than lateral, flexion of the spine. Most of the distinctive structural features of the mammalian vertebral column are related to these functional

Figure 2-24. Three major types of right upper molariform teeth. Anterior is to the right; the outer edge of each tooth is toward the top. The cross-hatched parts are dentine. The lophodont tooth shows an advanced stage of wear.

Bunodont

Lophodont   Selenodont

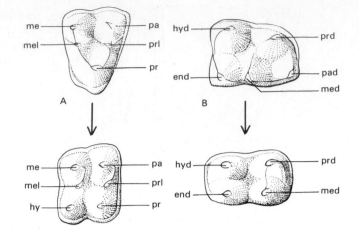

Figure 2-25. Basic cusp pattern of mammalian molars: (A) right upper molars; (B) left lower molars. The upper pair of teeth represent the primitive cusp pattern; this was modified in some evolutionary lines by the addition of a cusp (hypocone) in the upper tooth and the loss of a cusp (paraconid) in the lower tooth, yielding more or less quadrate teeth (lower pair) adapted to omnivorous or herbivorous diets. Abbreviations: *end*, entoconid; *hy*, hypocone; *hyd*, hypoconid; *me*, metacone; *med*, metaconid; *mel*, metaconule; *pa*, paracone; *pad*, paraconid; *pr*, protocone; *prd*, protoconid; *prl*, protoconule. (After Romer, 1966)

contrasts (pp. 33). The mammalian vertebral column has five well-differentiated sections: *cervical, thoracic, lumbar, sacral,* and *caudal.* Only the anteroposteriorly compressed thoracic vertebrae bear ribs. In some groups, such as edentates and bats, the rigidity of the rib cage is greatly enhanced by a broadening of the ribs. The first two cervicals are highly modified, the sacral vertebrae are more or less fused to support the pelvic girdle, and differentiation of the vertebrae of each region is typical (Fig. 2–26). Usually from 25 to 35 presacral vertebrae are present. All mammals, with the exception of several edentates and the manatee (order Sirenia), have seven cervical vertebrae.

The sternum is well developed and solidly anchors the ventral ends of the ribs, helping to form a fairly rigid rib cage. The sternum is not highly variable, but in some bats departs strongly from the typical mammalian plan.

Limbs and Girdles. In most terrestrial mammals, the main propulsive movements of the limbs are fore and aft; the toes point forward and the limbs are roughly perpendicular to the ground. In the most highly cursorial species (*cursorial* mammals are those adapted for running), the joints distal to the hip and shoulder tend to limit movement to a single plane. This allows reduction of whatever musculature does not control flexion and extension, and results in lighter limbs. The mammalian pelvic girdle has a characteristic shape, with the *ilium* projecting forward and the *ischium* and *pubis* extending backward (Fig. 3–5); these bones are solidly

Figure 2-26. Vertebrae of the gray fox *(Urocyon cinereoargenteus),* showing the great structural variation in the parts of the vertebral column. The vertibrae are viewed from the right side; anterior is to the right. (A) Fifth cervical vertebra, (B) axis (second cervical), (C) atlas (first cervical). Abbreviations: *af,* articular facet for the capitulum of the rib; *c,* centrum; *ns,* neural spine; *prz,* prezygapophysis; *ptz,* postzygapophysis; *tp,* transverse process.

A

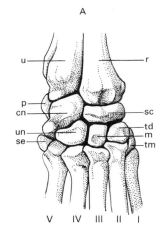

B

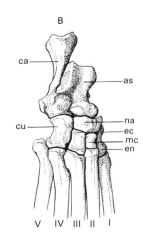

Figure 2-27. Primitive patterns of the podials (foot bones) of mammals: (A) carpus of a hedghog *(Erinaceus europaeus),* (B) tarsus of the wolverine *(Gulo luscus).* The centrale, a carpal element that in some mammals with primitive limbs lies proximal to the trapezoid and magnum, is missing in the hedgehog. Abbreviations: *as,* astragalus; *ca,* calcaneum; *cn,* cuneiform; *cu,* cuboid; *ec,* ectocuneiform; *en,* entocuneiform; *m,* magnum; *mc,* mesocuneiform; *na,* navicular; *p,* pisiform; *r,* radius; *sc,* scapholunar (fused scaphoid and lunar); *se,* accessory sesamoid; *td,* trapezoid; *tm,* trapezium; *u,* ulna; *un,* unciform. The metacarpals and metatarsals are numbered.

fused. In the shoulder girdle of placentals, the *coracoid* and *acromion* are usually reduced to small processes on the scapula and the reptilian interclavicle is gone (Fig. 2–16); the clavicle is reduced or absent in some cursorial species.

In the *manus* (hand) and *pes* (foot) of mammals, there is a standard pattern of bones (Fig. 2–27), but many variations on this basic theme (some of which are described in the chapters on orders) occur among mammals with specialized types of locomotion, such as flight (bats), swimming (cetaceans and pinnipeds), or rapid running (ungulates, rabbits, some carnivores). The primitive mammalian number of digits (five) and the basic phalangeal formula of two phalanges in the thumb *(pollex)* and first digit of the hind limb *(hallux)* and three phalanges in each of the remaining four digits (2-3-3-3-3) are retained by many mammals. Common specializations involve loss of digits, reduction in the numbers of phalanges, or, occasionally, addition of phalanges *(hyperphalangy),* as in whales and porpoises (Fig. 15–3).

Chapter 3

Mammalian Origins

Mammals arose from an ancient reptilian lineage, the mammal-like reptiles (subclass Synapsida), that appeared at the start of the reptilian radiation 300 million years ago. From a late Carboniferous divergence from other reptilian evolutionary lines, mammal-like reptiles became the most important reptile group in the Permian and Early Triassic periods. (Table 3–1 lists the geologic periods.) In the Triassic, however, before the appearance of dinosaurs, synapsids dwindled in importance, and they were nearly extinct by the end of this period. The earliest mammals, from the Late Triassic, were the unspectacular descendants of the declining synapsid line.

The first two thirds of mammalian history, during the 140 million years from the Late Triassic to the Late Cretaceous, has been regarded as the dark ages of mammalian history. (This segment of mammalian history is discussed in fascinating detail by Lillegraven, Kielan-Jaworowska, and Clemens, 1979.) During this vast span of time, the evolutionarily adventuresome dinosaurs dominated the terrestrial scene, and their radiation resulted in a diverse array of herbivorous and predatory types, some of which were highly specialized for bipedal locomotion. Dinosaurs were the largest land animals of all time: nearly all species were very large, and some species reached weights of 86 tons (Colbert, 1962). Only a few species were smaller than 10 kilograms as adults. By comparison, Mesozoic mammals were insignificant. Most were mouse-sized, and the occasional "giant" was probably not as large as the occasional species of relatively tiny dinosaur. Mesozoic mammals probably hid by day and foraged at night; the poor fossil record of the postcranial skeleton suggests that they adhered conservatively to a mouselike body form and to quadrupedal locomotion. Why Mesozoic mammals were so uniformly small will perhaps never be known. Lillegraven (1979a) speculated that predation by juvenile dinosaurs was the selective force making large size disadvantageous. Only tiny mammals would have been able to find concealment from these predators.

Minor mammalian adaptive radiations occurred in the Mesozoic, and the order Multituberculata, which appeared in the Jurassic, had a long history that reached far into the Cenozoic. Only after the extinction of the dinosaurs in the Late Cretaceous, however, did mammals begin the dramatic adaptive burst that led to their dominant position in the Cenozoic Era.

Mammal-Like Reptiles

Skull Characteristics

The synapsid skull is characterized by a single opening low in the temporal part, with the postorbital and squamosal bones meeting above the opening. From a primitive reptilian skull with no temporal openings, a skull type still retained by turtles (Fig. 3–1), various patterns of perforation of the temporal part of the skull developed among early reptiles (Fig. 3–2). The openings are thought by some researchers to have developed originally to increase the freedom for expansion of the adductor muscles of the jaw; these muscles primitively attached inside the solid temporal part of the skull. Others have held that a selective advantage was gained by

Table 3-1: Geologic Time Scale Since Life Became Abundant

Era	Period	Estimated Time Since Beginning of Each Period (millions of years)	Epoch	Typical Mammals and Mammalian Ancestors
Cenozoic	Quaternary	2+	Recent	Modern species and subspecies; extirpation of some mammals by humans
			Pleistocene	Appearance of modern species or their antecedents; widespread extinction of large mammals
	Tertiary	65	Pliocene	Appearance of modern genera
			Miocene	Appearance of modern subfamilies
			Oligocene	Appearance of modern families
			Eocene	Appearance of modern orders
			Paleocene	Adaptive radiation of marsupials and placentals
Mesozoic	Cretaceous	130		Appearance of marsupials and placentals
	Jurassic	180		Archaic mammals
	Triassic	230		Therapsids; appearance of mammals
Paleozoic	Permian	280		Appearance of therapsids (from which mammals evolved)
	Carboniferous	350		
	Devonian	400		
	Silurian	450		
	Ordovician	500		
	Cambrian	570		

After Romer and Parsons, 1977

the reduction in the weight of the skull due to the temporal openings. According to another explanation, the stresses on the skull were greatest around the periphery of the temporal area, whereas the middle of the area was not needed for bracing of the jaw area or for muscle

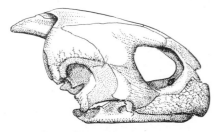

Figure 3-1. An anapsid skull (sea turtle, *Chelonia*). Note the unbroken shield of bone in the temporal region.

attachment. Therefore, the loss of bone in the middle of the temporal "shield," when solid bone in this area was no longer of adaptive importance, contributed importantly to metabolic economy by reducing the mass of bone in the skull that had to be developed and maintained (R. C. Fox, 1964).

In any case, the general trend in progressive mammal-like reptiles was clearly toward the enlargement of the temporal opening (Fig. 3–3) and toward the movement of the jaw muscle origins from the inner surface of the temporal shield to the braincase and to the zygomatic arch, the remnant of the lower part of the original temporal shield. In some advanced mammal-like reptiles, the postorbital bar, a vestige of the anterior part of the temporal shield, was lost.

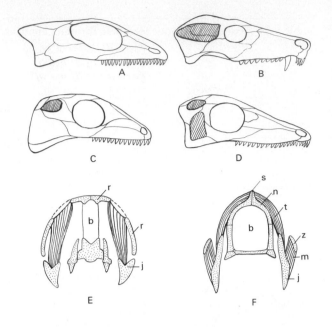

Figure 3-2. (A–D) Diagrammatic views of skulls of reptiles showing the different arrangements of temporal openings. (A) Anapsid skull (no temporal opening), (B) synapsid skull (postorbital and squamosal bones meeting above opening), (C) euryapsid skull (postorbital and squamosal bones meeting below opening), (D) diapsid skull (two temporal openings). (E, F) Cross sections of skulls showing attachments of jaw muscles. (E) Pelycosaur, with the jaw muscles originating within the remaining parts of the temporal shield (*r*); the sides of the braincase are cartilaginous. (F) Mammal, with the jaw muscles originating on the new and completely ossified braincase (*n*), on the saggital crest (*s*), and on the zygomatic arch (*z*), a remnant of the original skull roof. Abbreviations: *b,* brain cavity; *j,* lower jaw; *m,* masseter muscle; *n,* new braincase formed partly by extensions of bones that originally formed the skull roof; *r,* skull roof; *t,* temporal muscles; *z,* zygomatic arch.

Evolution

The subclass Synapsida includes the order Pelycosauria, a primitive, largely Permian group, and the order Therapsida, a more advanced assemblage with a long record extending from the Middle Permian to the Early Jurassic. The pelycosaurs, the dominant Early Permian reptiles, were carnivores that probably fed largely on fish and amphibians at the water's edge. This mode of life may have been responsible for their lack of structural diversity and rather primitive limbs. In contrast, therapsids were terrestrial and were both herbivores and carnivores. Selection for more efficient terrestrial locomotion and masticatory apparatus in these animals led to structural change and diversity, and it was from this progressive group that mammals arose.

Therapsids span the morphological gap from fairly primitive reptiles to animals of nearly mammalian grade, and they display many

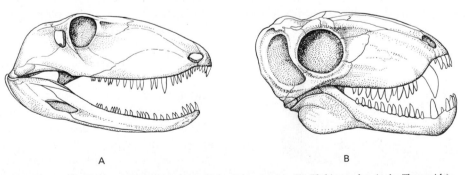

Figure 3-3. Synapsid skulls: (A) *Ophiacodon* (order Pelycosauria), (B) *Phthinosuchus* (order Therapsida). (After Romer, 1966) Note the enlargement of the temporal opening in *Phthinosuchus,* the more advanced synapsid.

morphological trends leading to the basic mammalian anatomical plan. Major trends include (1) reduction of the temporal shield as the temporal opening enlarged, (2) movement of the jaw muscle origins to the braincase and zygomatic arch, (3) development of two occipital condyles, (4) expansion of the maxillary and palatine bones backward and toward the midline to form a secondary palate (Fig. 3–4), (5) development of a strongly heterodont dentition, and (6) expansion of the dentary bone at the expense of the other jaw elements and establishment of a dentary-squamosal jaw joint. The postcranial skeleton was changed by fusion of the cervical and lumbar ribs with the vertebrae as pleurapophyses, by alteration of the joint between the atlas and axis (p. 33), by partial abandonment of the primitive, spraddle-legged limb posture (p. 32), by modification of the limb girdles (Fig. 3–5), and by simplification (by fusion of elements) of the tarsus and carpus (Fig. 2–26).

Cynodonts, Therapsid Ancestors of Mammals

One division of the order Therapsida, the suborder Theriodonta, includes an array of advanced, predaceous, mammal-like reptiles. One infraorder of the suborder Theriodonta, the Cynodontia, is generally regarded as the ancestral group of mammals. Only in cynodonts do advanced stages of many of the typically mammalian trends occur, and a broad picture of the reptile-mammal transition emerges from a study of cynodont structure.

A handy and widely used landmark in reptilian-mammalian evolution is the structure of the jaw articulation. In reptiles this joint is typically between the quadrate bone of the skull and the articular bone of the lower jaw, but in mammals the squamosal and dentary bones form this joint. The earliest mammals are regarded as mammals and not reptiles because of the presence of the dentary-squamosal joint, but the situation is far from simple. In cynodont therapsids, there are several stages in the transformation of the jaw joint. An intermediate stage, developed by several cynodont families, is the development of a secondary jaw joint, in addition to the quadrate-articular joint, between the surangular bone and the squamosal bone. This secondary joint probably braced the quadrate-articular joint against backward displacement of the jaw during chewing. A further step toward the mammalian jaw was taken by one family of cynodonts and is illustrated by *Probainognathus*. In this cynodont, an articular depression

Figure 3–4. Palatal views of synapsid skulls. (A) *Scymnognathus* (order Therapsida); note that the internal nares (*o*) open into the anterior part of the mouth. (B) *Cynognathus* (order Therapsida); note that the maxillaries (*m*) and palatines (*pa*) have extended medially, forming a shelf that shunts air from the external nares to near the back of the mouth. Abbreviations: *m*, maxillary; *o*, internal narial opening; *p*, premaxillary; *pa*, palatine; *pt*, pterygoid. (After Romer, 1966)

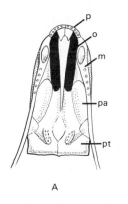

A

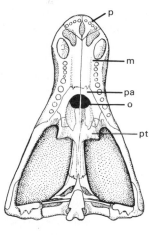

B

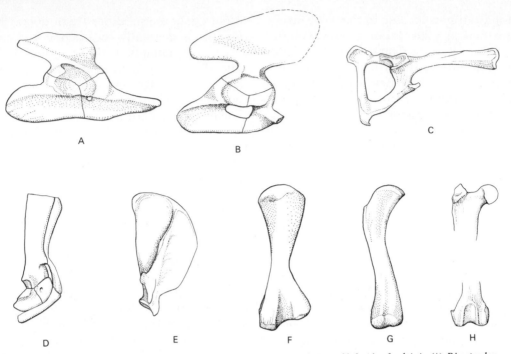

Figure 3-5. Bones of limbs and girdles of reptiles and mammals. Lateral view of left side of pelvis in (A) *Dimetrodon* (order Pelycosauria), (B) *Cynognathus* (order Therapsida), (C) *Erinaceus* (order Insectivora). Lateral views of the right scapulae of (D) *Kannemeyeria* (order Therapsida), (E) *Lynx rufus* (order Carnivora). Anterior views of the left femora of (F) *Ophiacodon* (order Pelycosauria), (G) a cynodont (order Therapsida), (H) *Lynx rufus* (order Carnivora). (Reptiles after Romer, 1966)

(glenoid fossa) developed in the squamosal bone, and into it fit the surangular bone of the lower jaw, braced by an "articular" process of the dentary bone (Fig. 3–6). In the Late Triassic, some cynodonts demonstrated the dentary-squamosal jaw joint for the first time. This involves a posterior extension of the dentary bone into the glenoid fossa of the squamosal bone, with the quadrate-articular joint present medial to this "mammalian" jaw joint. As a later development, the quadrate and articular bones no longer formed a jaw joint but functioned together with the stapes to transmit vibrations from the tympanic membrane to the oval window of the inner ear.

The cynodont-mammalian trend toward enlargement of the dentary bone and reduction of the postdentary bones has generally been regarded as having developed under chronic selection for improvement in the bracing of the jaw joint and greater chewing efficiency. Allin (1975), however, regarded the postdentary bones of cynodonts as important transmitters of sound to the inner ear. In his view, a reflected part of the angular bone supported a tympanic membrane, and the articular and quadrate bones transmitted vibrations from this membrane (via the stapes) to the inner ear (Fig. 3–7). Accordingly, reduction of the size of the articular and quadrate bones would improve their sensitivity to vibrations and would enhance the sense of hearing. Finally, when freed from the lower jaw in mammals, the articular, quadrate, and angular bones became part of the ear apparatus (the malleus, incus, and tympanic ring, respectively).

The development of masseter muscles with essentially the same attachments as those of

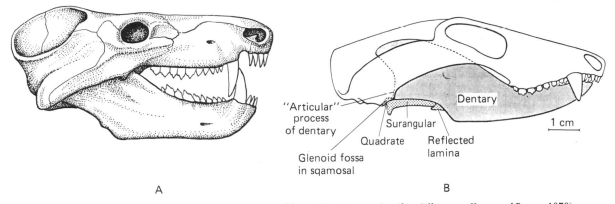

"Articular" process of dentary

Glenoid fossa in sqamosal

Quadrate

Surangular

Reflected lamina

Dentary

1 cm

A

B

Figure 3-6. Therapsid skulls: (A) *Cynognathus* (after Romer, 1966), (B) *Probainognathus* (from Lillegraven, Kraus, and Brown, 1979).

mammals was a cynodont innovation that occurred in no other therapsid line. These muscles originate on the zygomatic arch and insert on the lateral surface of the dentary bone and are powerful adductors of the jaws (they close the jaws). The development of these muscles resulted in several important functional refinements (Crompton and F. A. Jenkins, 1979). First, the masseter muscles formed part of a muscular sling that suspended the jaw and enhanced the precise control of transverse jaw movements. Second, these muscles increased the force of the bite. Third, the forces produced by the bite were "focused" through the point of the bite, and thus the stress on the jaw joint was reduced.

A shift in the structure and function of the dentition can be traced through the cynodont–

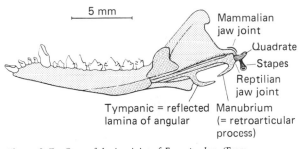

5 mm

Mammalian jaw joint

Quadrate

Stapes

Reptilian jaw joint

Tympanic = reflected lamina of angular

Manubrium (= retroarticular process)

Figure 3-7. Parts of the jaw joint of *Eozostrodon*. (From Crompton and Jenkins, 1979)

early mammal evolutionary line. The front teeth were of primary importance in primitive therapsid reptiles; the incisors and canines were robust and the cheek teeth relatively weak. The reverse was true in advanced cynodonts and early mammals, in which the cheek teeth were the more robust series. The development in cynodonts of masseter muscles and the concentration of jaw action power through the postcanine teeth attended these changes in function. Progressively greater precision and breadth of movement of the lower jaw set the stage for the later evolution in mammals of a complex molar cusp pattern (Crompton, 1974:434).

Certain other features that evolved only in cynodonts foreshadow important mammalian structures. The postcanine teeth of some cynodonts were tricuspid and resembled those of some of the earliest known mammals. In other cynodonts, the cheek teeth were complex and double-rooted, and precise patterns of occlusion developed.

Several other structural features that became well developed in cynodonts are typical of both ancient and modern mammals. One such is the secondary palate, a structure formed by an inward and backward extension of the premaxillary, maxillary, and palatine bones. This bony plate lies beneath the original roof of the mouth and forms a passage that shunts air from the external nares at the front of the snout to the internal narial openings at the back of the mouth (Fig. 3–4). Such a bypass allows mammals to breathe while food is being chewed. The incisive foramina (p. 18) were large in cynodonts, a reflection perhaps of the importance of the sense of smell to these animals. The reduction in size of lumbar ribs and the retention of a thoracic rib cage in cynodonts may have been associated with the development of a muscular diaphragm and the respiratory movements typical of mammals.

The limbs and girdles of cynodonts were modified as the spraddle-legged reptilian limb posture was partially abandoned in favor of a more vertical limb orientation. The ilium shifted forward, and the pubis and ischium moved backward as fore-and-aft limb movement became more important than lateral movement (Fig. 3–5).

The limbs of some cynodonts were slim and adapted to rapid running. An overly simplistic approach has often been taken in describing the differences between reptilian and mammalian limb postures. The reptilian posture has been characterized as spraddle-legged, with the humerus and femur directed horizontally, whereas mammalian limbs have been described as moving directly fore and aft and being positioned nearly vertically beneath the body. Actually, this latter posture is typical only of cursorial mammals. F. A. Jenkins (1971) found that, during locomotion in a group of noncursorial species, the humerus and femur function in postures more horizontal than vertical and at oblique angles relative to the parasagittal plane. The studies of Jenkins further demonstrate that the limb postures of terrestrial mammals are extremely diverse. Certainly a trend toward a vertical limb posture can be detected in cynodont reptiles and in early mammals, but the stereotypical picture of the vertical limb posture shared by all terrestrial mammals should be abandoned.

Cynodonts, and therapsids in general, were clearly active terrestrial reptiles with well-developed senses of hearing and smell, and later species were probably endotherms (Feder,

1981). Why, after such a long period of dominance, did the progressive cynodonts and other therapsids become extinct? A major cause may have been competition from dinosaurs (Colbert, 1982). Early theriodont therapsids were often the size of large dogs, whereas the last surviving cynodonts were squirrel-sized and the earliest mammals were no larger than mice. The therapsid–early mammal evolutionary line was apparently under intense selection for small size. As mentioned earlier in relation to Mesozoic mammals, the smaller size of later therapsids may have made available to them more retreats secure from dinosaurs.

Early Mammals

As mentioned above, the first two thirds of mammalian history is documented by a spotty fossil record that leaves many geographic areas, time periods, and anatomical changes unrepresented. The following discussions, based largely on the work of Crompton and F. A. Jenkins (1979) and Jenkins and Crompton (1979), summarize our present knowledge of this phase of mammalian evolution. A more complete (and perhaps different) story will be told as future discoveries fill the gaps in our knowledge.

Several kinds of mammals are known from the Late Triassic, but only members of the family Morganucodontidae (order Triconodonta, Fig. 3–8) are well enough represented in the fossil record to provide evidence of both cranial

Figure 3–8. Reconstruction of *Eozostrodon,* a Triassic mammal of the family Morganucodontidae. The length of this animal was about 107 mm. (After Crompton and Jenkins, 1968)

and postcranial skeletal anatomy. Although these early mammals clearly evolved from cynodont ancestry, they displayed the following structural features that distinguish them from even the most advanced cynodonts:

1. In species of like body size, the morganucodontid brain was three or four times larger than that of even the most advanced therapsids, a reflection perhaps of greater neuromuscular coordination and improved auditory and olfactory acuity.

2. The condyle of the dentary bone fit into the glenoid fossa of the squamosal bone.

3. The cheek teeth were differentiated into premolars and molars, and the premolars were probably preceded by deciduous teeth.

4. Chewing was on one side of the jaw at a time, and the lower jaw on the side involved in chewing followed a triangular orbit as viewed from the front (Fig. 2–21).

5. During chewing, the inner surface of the upper molars sheared against the outer surface of the lower molars (Figs. 3–9 and 2–21).

6. The cochlear region of the skull was far larger and more conspicuous ventrally than in cynodonts.

7. Body weight, probably 20 to 30 grams, was an order of magnitude smaller than in any Middle Triassic cynodont.

8. The pelvis was essentially mammalian, with a rodlike ilium and a small pubis.

9. As part of a series of specializations allowing rotary head movement, the dens of the axis was large and protuberant and fit into the atlas (Fig. 2–26B).

10. The thoracic and lumbar vertebrae arched dorsally, the thoracic vertebrae had narrow, posteriorly directed neural spines, and the lumbar vertebrae bore dorsally directed neural spines and anterolaterally inclined transverse processes (Fig. 3–10).

Whereas in reptiles horizontal undulations of the body accompany quadrupedal locomotion, in mammals dorsoventral flexion and extension of the vertebral column increase the length of the stride and the speed of locomotion. The arched vertebral column and the differentiation of the thoracic and lumbar vertebrae in quadrupedal mammals are associated with muscular specializations that control such movements of the vertebral column. These features are unique to mammals.

What was the style of life of these earliest mammals? F. A. Jenkins and Parrington (1976) regarded morganucodonts and their Triassic triconodont relatives as insectivores with considerable climbing ability. The apparent ability of the hallux to move independently of the other digits indicates grasping ability, and enlargement of the foramina of the cervical vertebrae through which nerves contributing to the bra-

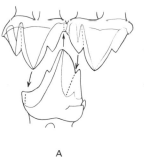

Figure 3-9. (A) Shearing planes of opposing molars of a primitive, Late Triassic mammal; the shearing surfaces are outlined (after Crompton, 1974). (B) Occlusal view of the lower molars of an opposum *(Didelphis virginiana),* showing the tongue-in-groove fit of the anterior and posterior surfaces of adjacent teeth.

A

B

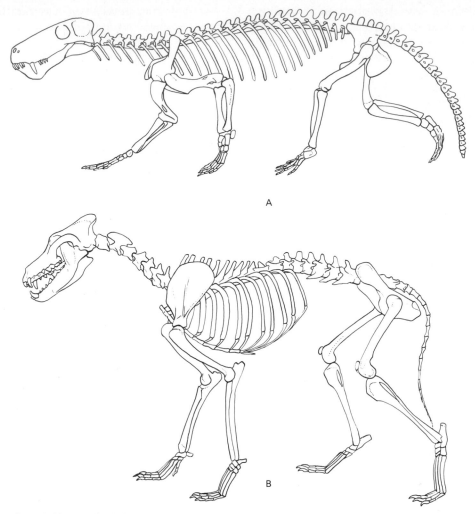

Figure 3-10. (A) The skeleton of *Lycaenops,* a Permian mammal-like reptile (after Colbert, 1948). (B) The skeleton of the dire wolf *(Canis dirus),* a modern mammal (after Stock, 1949, courtesy Los Angeles County Museum of Natural History). Note the dorsal arching of the spine, the semivertical posture of the neck, and the markedly different forms of the thoracic and lumbar vertebrae in the dire wolf.

chial plexus passed is suggestive of refined neuromuscular control of the forelimbs. These mammals were likely secretive, nocturnal creatures that depended heavily on their well-developed senses of hearing and smell. Endothermy probably favored nocturnal activity, and the animals must have been covered with hair, but it seems unlikely that they had developed the myriad adaptations necessary for coping with

the high temperatures encountered during diurnal activity.

These earliest mammals possessed a suite of skeletal features that marked them clearly as mammals, but what was their reproductive pattern and did they have mammary glands? Lillegraven (1979b) states that "the development of lactation was probably a key feature in the origin and later success of mammals in adapting to the changing environments of the Mesozoic and Cenozoic, and was unquestionably fully functional well before the end of the Triassic." One compelling line of histologic evidence supports this view. The mammary tissue of all living mammals is essentially identical, despite the fact that the "nontherian" and therian evolutionary lines diverged before the end of the Triassic. In all probability, then, the histologic similarities in mammary tissue are due to inheritance by both divisions of mammals from a common Late Triassic ancestor that possessed mammary glands. Further, mammary glands and deciduous dentition, which allowed the delay of the growth of the complex adult dentition in the juvenile mammal, probably occurred together. A dentition capable of masticating food can be delayed in a young mammal that is nourished by its mother's milk. During the nursing period, however, a tight social bond between mother and young is essential. Therefore, deciduous teeth, delayed adult dentition, mammary glands, lactation, maternal care, and a tight bond between mother and nursing young must have evolved in concert. When therian mammals abandoned egg-laying and began bearing living young (viviparity) is unknown; living prototherian mammals (order Monotremata) still lay eggs, and therian mammals may have kept this pattern long after the Triassic.

Although of great interest in connection with the story of mammalian evolution, the early mammals we have been considering were but insignificant members of a Late Triassic terrestrial fauna that was becoming increasingly dominated by dinosaurs, the most spectacular land vertebrates of all time. The tiny Late Triassic mammals were innovative in unspectacular ways that furthered their survival in the shadow of the dinosaurs. How very different would have been the vast sweep of post-Triassic vertebrate evolution, and how altered would be the face of the earth today, if the little Late Triassic mammals had proven vulnerable to some contemporary reptilian predator and had relinquished the scene completely to the reptiles.

Mesozoic Mammalian Radiations

Current evidence indicates that mammals originated monophyletically from cynodont reptiles (Hopson and Crompton, 1969), but by the Late Triassic mammals had diverged into several stocks. For some time, paleontologists held that the phylogeny of early mammals was best depicted as a dichotomy between two of the early groups, the Kuehneotheriidae and the Morganucodontidae. From the Morganucodontidae, which had basically triconodont molars (Fig. 3–11A), several taxa of nontherian mammals (triconodonts, docodonts, and monotremes) may have evolved. On the other hand, the Kuehneotheriidae, with triangular molars (Fig. 3–

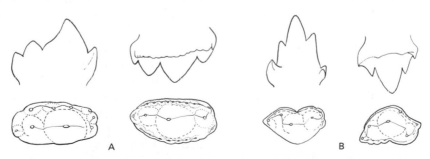

Figure 3–11. Diagrams of the molars of Triassic mammals. (A) *Eozostrodon* (family Morganucodontidae), (B) *Kuehneotherium* (family Kuehneotheriidae). In each case, the lower molar is on the left and the upper molar is on the right. (Modified from Crompton, 1974)

11B), may have given rise to therians (symmetrodonts, pantotheres, marsupials, and eutherians). It was an additional group, however, the Haramiyidae, that cast doubt on this rather simple view of early mammalian evolution. Haramiyids appeared before the other two families and are of unknown affinities (although they have been proposed as the ancestors of multituberculates). At present, the relationships among these and other taxa of early mammals remain equivocal. A species from Arizona *(Dinnetherium nezorum),* recently described by F. A. Jenkins, Crompton, and Downs (1983), lived contemporaneously with kuehneotheriids and morganucodontids but represents yet another early line of descent. This species provides further evidence that a simple dichotomy between morganucodontids (nontherians) and kuehneotheriids (therians) is no longer an accurate representation of the complex early evolution of mammals. Most of the Late Triassic and Jurassic groups of mammals were seemingly "experimental" lines that disappeared before the end of the Mesozoic. Only the monotremes (or at least vestiges of this group), marsupials, and eutherians survive today.

Among the prototherians, one of the oldest and most primitive groups is the order Triconodonta, known from the Late Triassic to the Early Cretaceous. Triconodonts were predaceous; the largest genus was nearly the size of a house cat. The dentition was heterodont, with as many as 14 teeth in a dentary bone. The canines were large, and typically the molars had three cusps arranged in a front-to-back row (Fig. 3–12A, B).

The order Docodonta is represented by several primitive genera known from the Late Jurassic. Members of this group have roughly quadrate teeth, with the cusps not aligned anteroposteriorly (Fig. 3–12C, D). The braincase and postcranial skeleton seem to be on a reptilian level of development.

Among nontherian mammals, the order Multituberculata is especially remarkable. These were the first mammalian herbivores, and

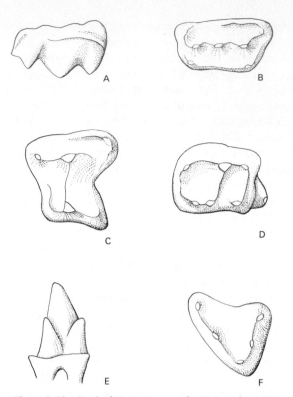

Figure 3–12.   Teeth of Mesozoic mammals. (A) Lateral and (B) occlusal view of right upper molar of a triconodont (order Triconodonta). Occlusal views of (C) right upper and (D) left lower molar of *Docodon* (order Docodonta). (E) Lateral view of lower left molar and (F) occlusal view of upper right molar of a symmetrodont (order Symmetrodonta). (From Romer, 1966)

although they disappeared in the early Tertiary and left no descendants, they were highly successful. Multituberculates appear first in the Late Jurassic, and their fossil record spans 100 million years. These animals were widespread in both the Old World and New World and were the ecological equivalents of rodents in some ways. The strongly built lower jaw provided attachment for powerful jaw muscles; there were usually two (but sometimes three) incisors above and two below, and a diastema was present in front of the premolars (Fig. 3–13A). Typical of some advanced multituberculates were upper molars with three parallel rows of cuspules and remarkably specialized bladelike posterior lower premolars (Fig. 3–13B, C).

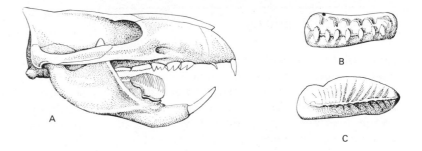

Figure 3-13. (A) *Ptilodus* (order Multituberculata) skull. Occlusal view of (B) upper and (C) lower cheek teeth of *Ptilodus*. (From Romer, 1966, after Simpson, 1937)

Multitubcrculates persistently retained several primitive features. The olfactory lobes of the brain were large, the cerebrum was smooth, the incisive foramina were large, and the cochlea of the ear was very small relative to that of contemporary placentals of comparable size. Considered together, these features suggest a rather primitive mammal that could not remain long in competition with eutherians, but the fossil record indicates otherwise. For over 70 million years, multituberculates and eutherians coexisted. The decline of the multituberculates began in the late Paleocene and spanned 20 million years. The competition probably began with condylarths (ancestors of ungulates) in the Late Cretaceous, intensified when primates became common in the Paleocene, and became overwhelming in the Eocene, when rodents became ubiquitous (Van Valen and Sloan, 1966). Multituberculates appear last in the early Oligocene fossil record of Wyoming and South Dakota.

The therian order Symmetrodonta is known from the Late Triassic to the Late Cretaceous. Symmetrodonts are among the oldest known mammals and were probably predaceous. The molar crown pattern is marked by three fairly symmetrically situated cusps (Fig. 3-12E, F).

Because it is generally accepted that eutherian and metatherian mammals evolved from the order Pantotheria, this group is of particular interest. Pantotheres have been found mainly in Late Jurassic rocks, and the following pantothere features are traceable to eutherian and metatherian mammals. The profile of the ventral border of the dentary bone is interrupted by an angular process; the lower molar has a posterior "heel" (Fig. 3-14B), which is represented in eutherians and metatherians by the talonid, the posterior section of the lower molar (Fig. 3-14D). The talonid has a basin into which the protocone of the upper molar fits. The shape of the anterior trigonid section of the pantothere lower molar resembles the comparable part of this tooth in some eutherians and metatherians; also, the pantothere triangular upper molar resembles the corresponding tooth of some primitive eutherians and metatherians (Fig. 3-14A, C).

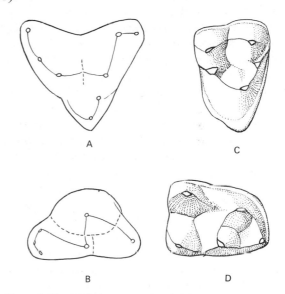

Figure 3-14. (A) Right upper and (B) left lower molar of *Aegialodon*, an early Cretaceous eupantothere (order Pantotheria). The upper molar is a hypothetical reconstruction. (C and D) Comparable teeth of the primitive Eocene eutherian mammal *Omomys*, a tarsier-like primate. (C and D after Romer, 1966)

Cretaceous Mammals

A broad view of the Cretaceous—a period of great biotic change—provides a background against which the late Mesozoic evolution of mammals can be viewed. The Late Jurassic was a time of considerable interchange of biotas between continents, as indicated by the occurrence in western Europe, East Africa, and western North America of identical or closely related species of reptiles (Colbert, 1973:175) and by intercontinental similarities among floras (Vachrameev and Akhmet'yev, 1972:419). After the earliest part of the Cretaceous, however, dispersal between continents became sharply restricted. In the New World, a series of transgressions of a seaway from the Arctic Ocean to the Gulf of Mexico divided North America for much of the Cretaceous into two separate centers for the evolution of terrestrial plants and animals. In Eurasia also the dispersal of land animals was restricted: Europe was essentially an archipelago of islands during the Cretaceous, and the "Turgai Strait" Seaway separated the land faunas of Europe and Siberia. Angiosperms probably evolved in the Early Cretaceous (Krassilov, 1973; J. Muller, 1970) and certainly underwent a major adaptive radiation then, an event that strongly affected the evolution of land faunas. Coadaptive evolution between angiosperm flowers and insects, for example, fostered a Cretaceous insect radiation.

Most dramatic were Cretaceous changes in the fortunes of the dinosaurs. Throughout the Jurassic and Early Cretaceous they were diverse and abundant and dominated the terrestrial scene. In the Late Cretaceous, several herbivorous groups—the ankylosaurs, ceratopsians, and hadrosaurians—diversified in association with the increasing importance of angiosperms and decline of gymnosperms. By the close of the Cretaceous, dinosaurs were gone. Their extinction was perhaps caused in part by climatic shifts tending toward reduced equability, that is, greater seasonal and daily temperature extremes (Axelrod and Bailey, 1968). There is at present no general agreement among scientists on the factors influencing these climatic shifts.

From the basal Late Triassic mammalian family Kuehneotheriidae, several families evolved in the Jurassic, and *Aegialodon,* a descendant of one of these families, is known from the Early Cretaceous fossil record of England. This genus has been described on the basis of a single, worn, lower molar (Fig. 3–14B), the structure of which strongly suggests that *Aegialodon* belonged to the lineage that gave rise to marsupials and eutherians (Table 3–2). These "modern" mammals appear as two divergent evolutionary lines in the Cretaceous.

Through much of the Early Cretaceous, land dwellers were barred from intercontinental movement by oceans and seaways; thus, populations of mammals on different continents evolved in isolation under different environmental pressures. This isolation of premarsupial and preplacental stocks may well have favored

---

Table 3-2: Partial Classification of Mammals

Class Mammalia
  Subclass Prototheria
    Infraclass Eotheria
      Order Triconodonta
        Family Morganucodontidae
      Order Docodonta
        Family Docodontidae
    Infraclass Ornithodelphia
      Order Monotremata
        Family Tachyglossidae
        Family Ornithorhynchidae
    Infraclass Allotheria
      Order Multituberculata
        (?) Family Haramiyidae
  Subclass Theria
    Infraclass Trituberculata
      Order Symmetrodonta
        Family Kuehneotheriidae
      Order Pantotheria
    Infraclass Metatheria
      Order Marsupialia
    Infraclass Eutheria (including all orders of placental mammals)

Partly after Crompton and F. A. Jenkins, 1973, and Hopson, 1970

their differentiation. Each group seemingly faced some comparable adaptive problems, but, as in the case of reproduction, each group developed unique solutions to these problems.

The fragmentary Cretaceous mammalian fossil record (with many geographic areas and time periods unrepresented) can be interpreted as indicating different places of origin for marsupials and eutherians. The earliest known undoubted marsupials are from Late Cretaceous rocks in Canada. Cretaceous marsupials are known only from western North America and from one locality in Peru, suggesting a western North American origin (Clemens, 1979a). Mammals from the Early Cretaceous fossil record of Mongolia are probably the earliest known eutherians, and they appeared somewhat later in North America. The earliest North American eutherians were perhaps derived from an Asian ancestral stock. Kielan-Jaworowska (1975) hypothesized that marsupials were unable to move from North America to Asia in the Late Cretaceous, but that eutherians reached North America from Asia at this time. By the latest Cretaceous, the previous diversity of marsupials was drastically reduced by a major episode of extinction, whereas eutherians at this time experienced an important radiation.

An important addendum to the story of Cretaceous mammals, and a further indication that evolutionary patterns are seldom simple, centers on the doubtful systematic positions of some fossils from the Cretaceous of North America and Asia. *Deltatheridium* (Fig. 3–15) from Mongolia, for example, has a distinctive complex of structural features and is accordingly regarded by some (Butler and Kielan-Jaworowska, 1973) as neither eutherian nor marsupial but as a "therian of metatherian-eutherian grade." Lillegraven (1974:265) postulates that the marsupial-eutherian dichotomy was not yet entirely clear-cut in the Cretaceous; equivocal mammals such as *Deltatheridium* were perhaps representative of evolutionary lines that were replaced by marsupials or eutherians before the Cenozoic.

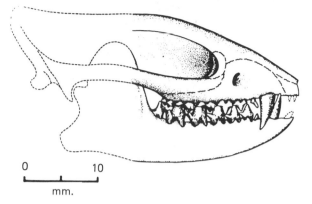

Figure 3–15. Skull of *Deltatheridium* (from Lillegraven, Kraus, and Bown, 1979).

Modern mammals (eutherians, marsupials, and therians of metatherian-eutherian grade) underwent an adaptive radiation in the Cretaceous. This is indicated in part by the diversification of dentitions after the earliest Cretaceous. Similar significant radiations occurred among dinosaurs (Colbert, 1973) and multituberculates at this time. These radiations probably reflect the availability of some important new food source. Considerable literature (Clemens, 1970:375; Lillegraven, 1974:277; Lillegraven, Kraus, and Brown, 1979) points to the overriding importance of the Early Cretaceous appearance and adaptive burst of flowering plants (class Angiospermae). The seeds of some angiosperms develop within an edible and nutritious fruit. Angiosperm fruits and seeds are eaten today by many mammals and were probably important to Cretaceous mammals. Mammals probably have a long history of contributing to angiosperm seed dispersal: Krassilov (1973) reported that the seed coats of some of the earliest angiosperms bore hooklets capable of tangling in the fur of foraging mammals. The Lepidoptera (moths and butterflies) appeared in the Cretaceous (M. R. MacKay, 1970), probably in response to the food offered by the flower nectar and leaves of angiosperms. The Isoptera (termites) also appeared at this time, and the Coleoptera (beetles) underwent an adaptive radiation. These insect groups are important foods for mammals today and were perhaps similarly important in the Cretaceous. The diversification of dentitions may well have enabled Cretaceous mammals not only to exploit plant foods but also to profit tremendously from the expanding diversity and growing populations of insects.

For mammals, then, the Mesozoic, and especially the Cretaceous, was a time of experimentation. Natural selection, partly in the form of predation by an imposing array of reptilian carnivores and probably some avian predators and partly in the form of competition from reptiles, birds, and other mammals, affected many changes in mammalian structure and function.

Behavioral, physiological, and anatomical changes that increased the efficiency of feeding, reproduction, and thermoregulation may have been critical. Various structural plans evolved and were workable for different lengths of time; some evolutionary side branches proved sterile. During the Mesozoic time of evolutionary trial and error, however, the basic mammalian structural plan was tested, retested, and perfected and the major taxa were established. The extinction of the dinosaurs at the end of the Cretaceous, the diversity of flowering plants, and perhaps the cooling climatic trend set the stage for the dramatic adaptive burst of mammals at the start of the Cenozoic.

Chapter 4

Classification of Mammals

Despite their remarkable success, mammals are much less diverse than are most invertebrate groups. This is probably due in part to their far greater size and to the high energy requirements of endothermy, and thus to the inability of mammals to exploit great numbers of restricted ecological niches. Roughly 1000 genera and 4060 species of mammals are currently recognized. These figures are insignificant by comparison with those for invertebrates. There are, for example, an estimated 750,000 species of insects, 30,000 of protozoans, and 107,000 of mollusks.

Chapters 6 through 17 consider the orders and families of mammals listed in Table 4-1. In these chapters, such features as group size, present geographic distribution, time of appearance in the fossil record, structural characters, and brief life histories are given for each order and family. Whenever appropriate, morphology is related to function so that the remarkable structural and functional diversity displayed by mammals can be appreciated.

I devote considerable attention to the or-

ders and families of mammals not because I wish to put primary stress on the taxonomic aspect of mammalogy, but rather as an attempt to provide students with sufficient information on the various kinds of mammals to make the subsequent discussions of mammalian biology meaningful. The edge of students' interest is often dulled if they must deal with information about completely unfamiliar kinds of animals. It seems pointless to me to discuss population cycles of arvicolines, for example, if students have only a vague idea of what an arvicoline is. The chapters on orders, then, should serve as a background for the chapters on selected aspects of the biology of mammals.

The classification that follows is partly that of McKenna (1975b), but departs in several places to accommodate my preferences. The classification of rodents is that of Carleton (1984). No classification system yet proposed has gained universal acceptance, but Simpson's (1945) has been the standard for many years and is still widely used.

Table 4–1:   A Classification of Living Mammals

Classification	Common Name
Subclass Prototheria	
Order Monotremata (3 species)[a]	
Family Tachyglossidae (Aust)[b]	Spiny anteaters
Ornithorhynchidae (Aust)	Duck-billed platypus
Subclass Theria	
Infraclass Metatheria (marsupials)	
Order Marsupialia (242 species)	
Family Didelphidae (NA)	Opossums
Caenolestidae (SA)	Rat opossum
Dasyuridae (Aust)	Marsupial "mice," "rats," "carnivores"
Peramelidae (Aust)	Bandicoots
Notoryctidae (?)	Marsupial "mole"
Phalangeridae (Aust)	Cuscuses, phalangers
Petauridae (Aust)	Gliders
Burramyidae (Aust)	Pigmy possums
Macropodidae (Aust)	Kangaroos, wallabies
Tarsipedidae (Aust)	Honey possum
Vombatidae (Aust)	Wombats
Phascolarctidae (Aust)	Koala
Infraclass Eutheria (placentals)	
Order Insectivora (406 species)	
Family Erinaceidae (NA)	Hedgehogs
Talpidae (Eur)	Moles
Tenrecidae (Af)	Tenrecs
Chrysochloridae (Af)	Golden moles
Solenodontidae (?)	Solenodons
Soricidae (Eur)	Shrews
Order Macroscelidea	
Family Macroscelididae (Af)	Elephant shrews
Order Scandentia	
Family Tupaiidae (SE Asia)	Tree shrews
Order Dermoptera (2 species)	
Family Cynocephalidae (?)	Flying lemurs
Order Chiroptera (853 species)	
Family Pteropodidae (Eur)	Old World fruit-eating bats
Rhinopomatidae (?)	Mouse-tail bats
Emballonuridae (Eur)	Sac-winged bats
Craseonycteridae (SE Asia)	Bumble-bee bat
Noctilionidae (?)	Bulldog bats
Nycteridae (?)	Hollow-faced bats
Megadermatidae (Eur)	False vampire bats
Rhinolophidae (Eur)	Horseshoe bats
Phyllostomidae (?)	Leaf-nosed bats
Mormoopidae (?)	Moustached bats
Desmodontidae (SA)	Vampire bats
Natalidae (SA)	Funnel-eared bats
Furipteridae (?)	Smoky bats
Thyropteridae (?)	Disk-winged bats
Myzapodidae (?)	Sucker-footed bat
Vespertilionidae (Eur or NA)	Common bats
Mystacinidae (?)	Short-tailed bat
Molossidae (Eur)	Free-tail bats

Table 4-1:   A Classification of Living Mammals

Classification	Common Name
Order Primates (166 species)	
Family Lemuridae (Mad)	Lemurs
Indridae (Mad)	Indrid lemurs
Daubentoniidae (Mad)	Aye-aye
Lorisidae (Asia)	Lorises
Galagidae (Af)	Galagos
Tarsiidae (Eur)	Tarsiers
Cebidae (SA)	New World monkeys
	Marmosets
Cercopithecidae (Af)	Old World monkeys
Pongidae (Af)	Great apes, gibbons
Hominidae (Af)	Humans
Order Carnivora (284 species)	
Family Canidae (NA or Eur)	Wolves, foxes, jackals
Ursidae (Eur)	Bears
Procyonidae (NA)	Raccoons, ring-tail cats, etc.
Mustelidae (?)	Skunks, badgers, weasels, otters, wolverines
Viverridae (Eur)	Civets, genets, mongooses
Hyaenidae (Asia)	Hyenas, aardwolf
Felidae (NA or Eur)	Cats
Otariidae	Sea lions, fur seals
Odobènidae	Walrus
Phocidae	Earless seals
Order Hyracoidea (11 species)	
Family Procaviidae (Af)	Hyraxes
Order Proboscidea (2 species)	
Family Elephantidae (Asia)	Elephants
Order Sirenia (5 species)	
Family Dugongidae	Dugongs, sea cows
Trichechidae	Manatees
Order Perissodactyla (16 species)	
Family Equidae (NA)	Horses, asses, zebras
Tapiridae (Eur)	Tapirs
Rhinocerotidae (Eur)	Rhinoceroses
Order Artiodactyla (171 species)	
Family Suidae (Eur)	Swine
Tayassuidae (Eur)	Javelinas, peccaries
Hippopotomidae (Asia)	Hippopotami
Camelidae (NA)	Camels, llamas
Tragulidae (Eur)	Chevrotains
Cervidae (Asia)	Deer, elk, moose, caribou, etc.
Giraffidae (Asia)	Giraffe, okapi
Antilocapridae (NA)	Bison, antelope, gazelles, sheep, goats,
Bovidae (Eur)	cattle, etc.
Order Xenarthra (31 species)	
Family Myrmecophagidae (SA)	Anteaters
Bradypodidae (?)	Tree sloths
Dasypodidae (SA)	Armadillos
Order Pholidota (8 species)	
Family Manidae (Eur)	Scaly anteaters
Order Tubulidentata (1 species)	
Family Orycteropodidae (Eur or Asia)	Aardvarks

Table 4–1:   A Classification of Living Mammals

Classification	Common Name
Order Cetacea (78 species)	
Family Balaenidae (?)	Right whales
Eschrichtiidae (?)	Gray whale
Balaenopteridae (?)	Rorquals
Physeteridae (?)	Sperm whales
Monodontidae (?)	Narwhals, belugas, Irrawaddy dolphins
Ziphiidae (?)	Beaked whales
Delphinidae (?)	Ocean dolphins
Phocoenidae (?)	Porpoises
Platanistidae (?)	River dolphins
Order Rodentia (1750 species)	
Family Aplodontidae (NA)	Mountain beaver
Sciuridae (Eur or NA)	Squirrels, marmots
Castoridae (NA)	Beavers
Geomyidae (NA)	Pocket gophers
Heteromyidae (NA)	Kangaroo rats, pocket mice
Dipodidae (Asia)	Jumping mice, jerboas
Muridae (Eurasia)	Muroid rodents
Anomaluridae (?)	Scaly-tailed flying squirrels
Pedetidae (Af)	Spring hare
Ctenodactylidae (Eurasia)	Gundis
Gliridae (Eur)	Dormice
Selviniidae (Eur)	Desert dormouse
Bathyergidae (Asia)	Mole rats
Hystricidae (Asia)	Old World porcupines
Petromuridae (Af)	Rock rat
Thryonomyidae (Af)	Cane rats
Erethizontidae (SA)	New World porcupines
Chinchillidae (SA)	Chinchillas, viscachas
Dinomyidae (SA)	Pacarana
Caviidae (SA)	Cavies, guinea pigs
Hyrochoeridae (SA)	Capybara
Dasyproctidae (SA)	Agoutis
Agoutidae (SA)	Pacas
Ctenomyidae (SA)	Tuco-tucos
Octodontidae (SA)	Degus, rock rats
Abrocomidae (SA)	Chinchilla rats
Echimyidae (SA)	Spiny rats
Capromyidae (SA)	Hutias, cavies
Myocastoridae (SA)	Nutria
Heptaxodontidae (SA)	Giant hutias
Order Lagomorpha (65 species)	
Family Ochotonidae	Pikas
Leporidae	Rabbits, hares

[a]Numbers of species are approximate.

[b]Geographic origins: Af, Africa; Aust, Australia; Eur, Europe; Mad, Madagascar; NA, North America; SA, South America; SE Asia, Southeast Asia.

Chapter 5

Noneutherian Mammals: Monotremes and Marsupials

Monotremes and marsupials can conveniently be considered apart from the rest of the mammals. Both are primitive in a variety of ways, and both have a reproductive pattern different from that of other mammals: monotremes lay eggs; marsupials bear tiny and poorly developed young, and most have a choriovitelline placenta that differs from the chorioallantoic placenta of placental mammals (p. 424). The classification of the major groups of mammals reflects the phylogenetic isolation of the monotremes from the marsupials and placentals. Monotremes belong to the subclass Prototheria. The evolutionary line that gave rise to prototherians has been separate from that of therians for 190 million years. Both marsupials and placentals belong to the subclass Theria, but the marsupials are in one infraclass (Metatheria) and the placentals in another (Eutheria). The evolutionary paths of marsupials and placental mammals diverged from a common ancestor over 100 million years ago, in the Early Cretaceous.

Order Monotremata

Although represented today by but three genera, each with a single species, and therefore constituting a minor segment of the Recent mammalian fauna, monotremes are of great interest for several reasons. Morphologically, they closely resemble no other living mammals, and they possess some features more typical of reptiles than of mammals. Monotremes lay eggs and incubate them in birdlike fashion, and yet they have hair and suckle their young. The order Monotremata includes the family Tachyglossidae (echidnas, or spiny anteaters), which are found in Australia, Tasmania, and New Guinea, and the family Ornithorhynchidae (duck-billed platypuses), restricted to eastern Australia and Tasmania.

Morphology

Many structural features distinguish monotremes from other mammals. The monotreme skull is uniquely birdlike in appearance (Fig. 5–1). It is toothless except in young platypuses, the sutures disappear early in life, and the elongate and beaklike rostrum (Fig. 5–1) is covered by a leathery sheath (this sheath is horny in birds). The lacrimal bones are absent and the jugal bones are small or absent, whereas these bones are present in most therian mammals. Evidences of prefrontal and postfrontal bones, typically reptilian elements that are missing in therian mammals, occur on the frontal bones of monotremes. There is no auditory bulla, but the chamber of the middle ear is partially surrounded by oval tympanic rings.

Figure 5-1. Skull of the spiny anteater *(Tachyglossus aculeatus)*. Length of skull 111 mm.

Monotreme appendages represent excellent examples of "mosaic evolution" (Crompton and F. A. Jenkins, 1973). The shoulder girdle retains a bone pattern typical of therapsids, the forelimb has a rather reptilian posture resulting in part from fossorial specializations, and the pelvis and posture of the hind limbs are essentially therian.

The monotreme pectoral girdle contains an interclavicle, clavicles, precoracoids, and coracoids (Fig. 5–2) and provides a far more rigid connection between the shoulder joint and the sternum than does the girdle characteristic of therian mammals. Large epipubic bones extend forward from the pubes in both sexes. Cervical ribs are present, and the thoracic ribs lack tubercles, processes that occur on the ribs of most other mammals and are braced against the transverse processes of the vertebrae.

As put by A. B. Howell (1944:26), no monotremes "by any strength of the imagination might be considered cursorial." Monotremes have retained a limb posture that is similar in some ways to that of reptiles, and in monotremes this posture is associated with limited running ability. In the Australian echidna *(Tachyglossus aculeatus),* the humerus remains roughly horizontal to the substrate during walking (F. A. Jenkins, 1970). Rotational movement of the humerus, rather than fore-and-aft movement as in most mammals, is largely responsible for propulsion. Because in reptiles the limb posture is splayed, the fore and hind feet touch the ground well to the side of the shoulder and hip joints, respectively. The limb

posture in the echidna partially departs from this pattern because the forearm angles medially and the manus (forepaw) is roughly ventral to the shoulder joint; in the hind limb the foot is roughly ventral to the knee. The posture of the hind limb of the echidna resembles that of many generalized therian mammals. When the echidna is in motion, its body is elevated well above the ground in nonreptilian fashion. Despite the advances in limb posture in the echidna, locomotion is slow and appears labored and awkward.

Reproduction

The monotreme reproductive system and pattern are completely unique among mammals. Eggs are laid, and these are telolecithal (the yolk is concentrated toward the vegetal pole of the ovum) and meroblastic (early cleavages are restricted to a small disk at the animal pole of the ovum), as are birds' eggs. Only the left ovary is functional in the platypus (Asdell, 1964:2), as in most birds, but both ovaries are functional in the echidna. Shell glands are present in the oviducts. There is a cloaca, and in males the penis is attached to the ventral wall of this cavity. The testes are abdominal, and seminal vesicles and prostate glands are absent. The female echidna temporarily develops a pouch-like structure (not homologous to the pouch developed by some marsupials) when young are being incubated or cared for, but the platypus never develops one. The mammae lack nipples,

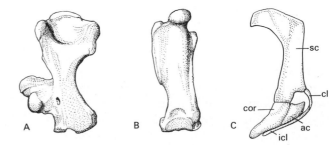

Figure 5–2. Bones of monotremes. (A) Left humerus and (B) right femur of the spiny anteater *(Tachyglossus aculeatus);* (C) pectoral girdle of the duck-billed platypus *(Ornithorhynchus anatinus).* Abbreviations: *ac,* acromion; *cl,* clavicle; *cor,* coracoid; *icl,* interclavicle; *sc,* scapula. (C after Romer 1966)

and the young suck milk from two lobules in the pouch in the echidna (Fig. 5–3) or from the abdominal fur in the platypus.

A long period of maternal care of the young is typical of monotremes. The platypus lays a single egg in a leaf nest in a burrow, where incubation lasts for 12 days. The newly hatched young is tiny (11 mm in length) and nearly embryonic in appearance. The mother suckles the young and broods it (keeps it warm) for nearly 16 weeks, and it develops slowly. The first growth of hair appears seven weeks after hatching, and the eyes do not open until about nine weeks after hatching (Fleay, 1944). Development of the single young echidna is similarly slow. The young is kept in its mother's pouch until it is about 12 weeks old, at which time its eyes open and it leaves the pouch to live in its mother's burrow. Weaning is at about 20 weeks. Both platypuses and echidnas have a low reproductive rate, apparently no more than one young a year.

Paleontology

The fossil record of monotremes consists almost entirely of Australian Pleistocene species referable to recent genera. The single earlier species *(Obdurodon insignis),* assigned to the family Ornithorhynchidae, is known from the middle Miocene of Australia (M. E. Woodburne and

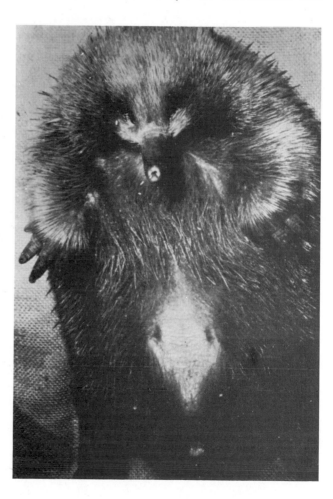

Figure 5-3. Ventral view of a live echidna *(Tachyglossus aculeatus),* showing the beaklike rostrum, the poorly developed pouch (typical of the nonbreeding season), and the tufts of hair at the mammary lobules. (M. L. Augee)

Tedford, 1975). Possible relationships between monotremes and various Mesozoic mammalian lineages have been discussed (Clemens, 1979b; Kermack, 1963; Simpson, 1959), and a relationship between monotremes and multituberculates has been proposed (Kermack and Kielan-Jaworowska, 1971; Kielan-Jaworowska, 1974). Monotremes seem certainly to have descended from some Mesozoic group within the nontherian division of mammalian ancestry, but the identity of this group remains unknown.

Family Tachyglossidae. Members of this group have a robust body covered with short, sturdy spines (Fig. 5–4) that are controlled by unusually powerful panniculus carnosus muscles. *Zaglossus,* the New Guinea echidna,

A

B

Figure 5–4. Two species of monotremes: (A) Australian spiny anteater *(Tachyglossus aculeatus),* (B) New Guinea spiny anteater *(Zaglossus bruijni). (Tachyglossus* by M. L. Augee; *Zaglossus* by H. M. Van Deusen)

weighs from 5 to 10 kilograms, and the Australian spiny anteater *(Tachyglossus)* from about 2.5 to 6 kilograms. The rostrum is slender and beaklike, the dentary bones are slender and delicate, and the long tongue is protrusile and covered with viscous mucous secreted mostly by the enlarged submaxillary salivary glands. Food is ground between spines at the base of the tongue and adjacent transverse spiny ridges on the palate. The limbs are powerfully built and are adapted for digging. The humerus is highly modified by broad extensions of the medial and lateral epicondyles that provide unusually large surfaces for the origins of some of the powerful muscles of the forearm (Fig. 5–2). In *Zaglossus,* the number of claws is variable; some animals have only three claws front and rear whereas others have a full complement of five (Van Deusen, 1969.) In *Tachyglossus,* all digits have stout claws. The ankles of male echidnas (and of some females) bear medially directed spurs, the function of which is not known.

These animals have highly specialized modes of life. They are powerful diggers and can rapidly escape predators by burrowing. Food consists largely of termites and ants *(Tachyglossus)* or earthworms and soil arthropods *(Zaglossus).* Foraging involves turning over stones and digging into termite and ant nests, and the prey is captured by the sticky tongue.

The Australian spiny anteater is a true hibernator and becomes torpid in response to cold and lack of food. During periods of torpor in experimental animals, the body temperature (5.5°C) was close to ambient temperature (5.0°C) and the heart rate dropped to 7 beats per minute (Augee and Ealey, 1968). Experimental animals were able to arouse spontaneously. One could dig slowly when its body temperature was only 10.5°C and its heart rate 7 beats per minute.

Family Ornithorhynchidae. The duck-billed platypus is smaller than the echidna, weighing from 0.5 to 2.0 kilograms. Some structural features of the platypus are associated with its semiaquatic mode of life. The pelage is dense,

and, as in the muskrat *(Ondatra),* the underfur is woolly. The external auditory meatus is tubular, as in the beaver *(Castor).* The eye and ear openings (pinnae are absent) lie in a furrow that is closed by folds of skin when the animal is submerged. The feet are webbed, but the digits retain claws that are used for burrowing. The web of the forefoot extends beyond the tips of the claws and is folded back against the palm when the animal is digging or when it is on land. The ankles of the male platypus have grooved and medially directed spurs that are connected to venom glands.

Although the young have teeth, the gums of adults are toothless and covered by persistently growing, horny plates. Anteriorly, the occlusal surfaces of the plates form ridges that are used to chop food; posteriorly, the plates are flattened crushing surfaces. Some additional mastication is accomplished by the flattened tongue, which acts against the palate. The elongate rostrum bears a flattened, leathery bill that has remarkable tactile ability.

The platypus inhabits a variety of waters, including mountain streams, slow-moving and turbid rivers, lakes, and ponds, and is primarily a bottom feeder. Along with some plants, aquatic crustaceans, insect larvae, and a wide variety of other animal material are taken during dives that last for roughly 1 minute. The platypus takes refuge in burrows dug into banks adjacent to water. Seasonal torpor occurs in some parts of its range.

Order Marsupialia

Marsupials and placentals represent two evolutionary lines that have been separate since the Early Cretaceous or the Late Jurassic (Clemens, 1968; Lillegraven, 1974; Slaughter, 1968). As a result of their long, independent history, marsupials differ structurally from placentals in many ways. In their semiarboreal habits and omnivorous-insectivorous diet, the didelphids are seemingly the present counterparts of early therians. Today, only two important strongholds for marsupials remain: the Australian region (Australia, Tasmania, New Guinea, and nearby islands) and the Neotropics (southern Mexico, Central America, and most of South America). Where isolated from placentals for long periods of time, marsupials have undergone remarkable radiation. Most marsupials have functional counterparts among placentals.

Morphology

The marsupial skull frequently has a small, narrow braincase housing small cerebral hemispheres with simple convolutions. Ossified auditory bullae, when present, are usually formed largely by the alisphenoid bone rather than by the tympanic (both ectotympanic and entotympanic), petrosal, and/or basisphenoid bones, as occurs in most placentals. The marsupial palate characteristically has large vacuities (Fig. 5–5), and the angular process of the dentary bone is inflected medially (except in *Tarsipes* and only

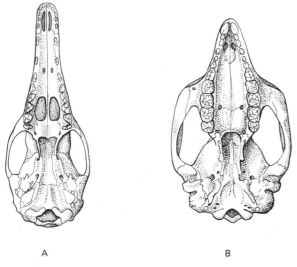

A   B

Figure 5-5. Ventral views of marsupial skulls. (A) New Guinea bandicoot (*Peroryctes raffrayanus,* family Peramelidae; length of skull 82 mm); (B) ring-tail possum (*Pseudocheirus corinnae,* family Petauridae; length of skull 97 mm). (After Tate and Archbold, 1937)

weakly so in *Phascolarctos* and *Myrmecobius*). The dentition is unique in that there are never equal numbers of incisors above and below, except in the family Vombatidae; the cheek teeth primitively include 3/3 premolars and 4/4 molars.

Marsupials often have highly specialized feet associated with specialized types of locomotion (Fig. 5–6). The unusual patterns of specialization of the hind feet are probably a result of an arboreal heritage and the early development of an opposable first digit and an enlarged and powerfully clutching fourth digit. As in the monotremes and multituberculates, epipubic bones extend forward from the pubic bones in both sexes (except in *Thylacinus,* in which the epipubic bones are vestigial, and in the extinct Borhyaenidae, in which they are absent).

Reproduction

The marsupial reproductive pattern differs sharply from that of placentals. The females of about 50 percent of the marsupials of today have a marsupium (an abdominal pouch) or abdominal folds within which the nipples occur. The number of nipples varies from two in the family Notoryctidae and some members of the Dasyuridae to 27 in some members of the family Didelphidae. Individual variation in the number of nipples often occurs within a species. The female reproductive tract is bifid, i.e., the vagina and uterus are double (Fig. 5–7). In all but the family Notoryctidae, which is adapted for digging, the testes are contained in a scrotum anterior to the penis.

The gestation period is characteristically short (8 to 43 days), and the young are tiny and rudimentary at birth. Newborn marsupials probably possess the minimal anatomical development allowing survival outside the uterus. Organogenesis has just begun, the separation of the ventricles of the heart is incomplete, the lungs are vascularized sacs lacking alveoli, and the kidneys lack glomeruli. Also lacking are cra-

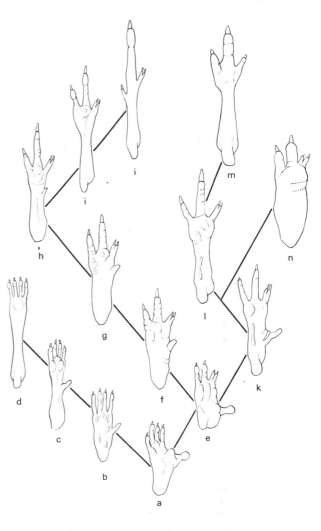

Figure 5-6. Ventral views of marsupial right hind feet, showing patterns of specialization associated with various styles of locomotion. (A) The presumed basic arboreal type, represented by the foot of a didelphid. The lines indicate possible evolutionary pathways leading to greater specialization. Feet of the following kinds are shown. Dasyuridae: (B) *Phascogale,* (C) *Sminthopsis,* (D) *Antechinomys.* Petauridae: (E) *Pseudocheirus.* Peramelidae: (F) *Perameles* sp., (G) *Peroryctes* sp., (H) *Peroryctes* sp., (I) *Macrotis,* (J) *Chaeropus.* Macropodidae: (K) *Hypsiprymnodon,* (L) *Potorous,* (M) *Macropus,* (N) *Denrolagus.* (After A. B. Howell, 1944)

nial nerves II to IV and VI, eye pigments, eyelids, and cerebral commissures (nerve fiber bundles connecting the cerebral hemispheres). Despite this minimal development, however, the naked, blind, and delicate newborn is able to make its way at birth from the vulva to the marsupium. Here it attaches to a nipple and remains there for a period of time greatly exceeding the gestation period. The weight of the young marsupial when it leaves the pouch and that of the newborn placental are roughly the same in species of comparable full-grown size (Sharman, 1970).

Most marsupials, with the exception of the Peramelidae, have a choriovitelline placenta that lacks villi (see p. 429 for further material on marsupial reproduction).

Paleontology

The earliest undoubted marsupial fossils are from the Late Cretaceous of North America. At that time, the dinosaurs were still dominant and the only surviving groups of primitive Mesozoic mammals were the Multituberculata and the Triconodonta. The niches previously filled by primitive Mesozoic mammals were seemingly

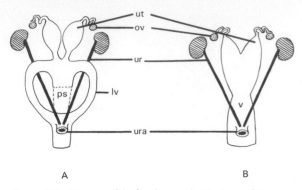

Figure 5–7. Diagrams of the female reproductive tracts of (A) marsupials and (B) placentals. Abbreviations: *lv,* lateral vagina; *ov,* ovary; *ps,* pseudovaginal canal; *ura,* urethra; *ur,* ureter; *ut,* uterus; *v,* vagina. (After Sharman, 1970)

being occupied in the Late Cretaceous by the dominant mammalian groups of today, the marsupials and the placentals, with the persistent multituberculates remaining important.

The most primitive marsupial family, and the stem group from which all other marsupials evolved, is Didelphidae. Members of this family were present in North America in the Late Cretaceous and have a nearly continuous fossil record there through the middle Miocene. The European record shows several genera from the early Eocene through the middle Miocene. During most of the Cenozoic, the Australian region and South America have been the two centers of marsupial diversification. In both regions marsupial radiations occurred under partial or complete isolation from competition with eutherians.

Many experts believe that marsupials arose in South America and moved northward into North America in the Cretaceous and southward to Australia, via Antarctica, in the Late Cretaceous or earliest Teritary. The presence of a marsupial in Eocene rocks in Antarctica strongly supports a southern route of marsupial dispersal (M. O. Woodburne and Zinsmeister, 1982). Marsupials reached Australia well before placentals did and underwent a spectacular Cenozoic radiation. Twelve living marsupial families, including 75 genera, and five known extinct families resulted from this radiation.

To one meeting the present Australian marsupial fauna for the first time, the numbers of species and the structural extremes are impressive. By comparison with the fauna of the late Pleistocene, however, the present fauna is severely depleted. Many species of large marsupials became extinct between the late Pleistocene and historic time, and further reductions occurred in historic times. The extinct families Thylacoleonidae and Diprotodontidae are especially noteworthy. *Thylacoleo* was a predaceous Pleistocene form roughly the size of an African lion. The third premolars were greatly elongated shearing blades, and the strongly built front limbs had retractile claws (Keast, 1972:227). The Diprotodontidae were represented in the Pleistocene Epoch by several genera. One *(Diprotodon)* was roughly the size of a rhinoceros and is the largest marsupial known. A number of very large kangaroos (family Macropodidae) also became extinct before historic times. A giant among them is *Procoptodon goliah,* an extremely short-faced macropodid (Fig. 5–8) that stood about 3 meters high.

Why did these imposing marsupials disappear? This question has been considered carefully by Merrilees (1968), who stressed the possible influence of humans. Aborigines probably entered Australia 35,000 years ago, in the late Pleistocene, at a time when the marsupial fauna included the large species just mentioned. Aborigines used fire extensively and so perhaps used grass fires to hunt the large marsupials. Post-Pleistocene aridity and human-made fires may together have tipped the balance against the large marsupials. In any case, only modern faunas are found in archaeological sites spanning the last 20,000 years.

Further reductions in the number of species of marsupials began with the coming of European peoples to Australia. The combined effects of heavy grazing by livestock, clearing of the land for agriculture, and the introduction of

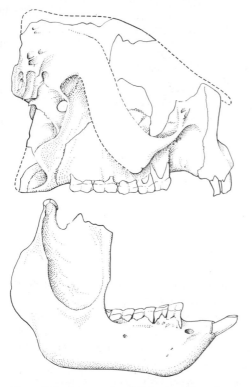

Figure 5-8. *Procoptodon goliah,* a huge, browsing, macropodid marsupial from the Pleistocene fossil record of Australia. Length of skull 218 mm. (After Tedford, 1967)

the Old World rabbit *(Oryctolagus cuniculus)* caused widespread decline in the abundance of many species and a number of extinctions (Calaby, 1971). As an example, of the some 45 species of kangaroos that occupied Australia just prior to the entry of Europeans, three are extinct and the populations of roughly an additional dozen seem to have declined drastically, although some species may have become rare before historic times.

The marsupial radiation in South America rivaled that in Australia. There are 93 living and fossil genera of marsupials known from South America and 105 from Australasia (L. G. Marshall, 1981). Accompanying the striking structural diversification of the marsupials in re-

sponse to a wealth of available habitats was a convergence by many types toward various placentals. By the Middle Tertiary, there were marsupials that structurally (and undoubtedly functionally) resembled shrews, moles, rodents, and carnivores. In the Early Tertiary, several groups of placentals were in South America. Edentates and various ungulate groups (p. 362) seemingly "owned" the large herbivore adaptive zone throughout much of the Tertiary, to the complete exclusion of marsupials.

Order Marsupialia

All of the South American marsupials presumably evolved from a basal, insectivorous-omnivorous didelphid stock, and the occurrence of at least 14 genera of didelphids at a late Paleocene fossil locality near Rio de Janeiro documents an Early Tertiary radiation. Perhaps the most spectacular didelphid descendants were the family Borhyaenidae and the saber-tooth family Thylacosmilidae (L. G. Marshall, 1977a).

The Borhyaenidae includes a number of marsupials with dentitions that suggest styles of feeding ranging from omnivorous to carnivorous. One borhyaenid, *Stylocynus,* was roughly the size of a bear and presumably omnivorous, and a number of small and medium-size species were also omnivores. In some borhyaenids, the first digit of the hind foot is partly opposable, a feature indicative of semiarboreal habits. All known borhyaenids are rather short-legged, and the terrestrial types lack marked cursorial specializations. Members of the genus *Borhyaena* have a skull rather like that of a wolf (Fig. 5–9).

The Pliocene family Thylacosmilidae includes five species, all of which have recurved, saber-like upper canines. The roots of these teeth extend nearly to the occipital part of the skull, and their tips are protected by a flange on the dentary bone (Fig. 5–9). Although basically well adapted to a carnivorous life, both borhyaenids and thylacosmilids had canines that

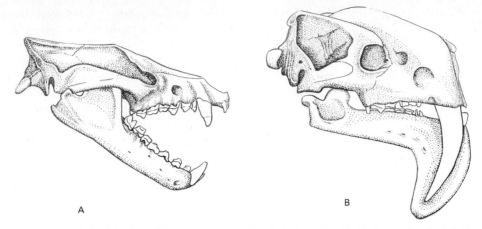

Figure 5-9. Skulls of members of the extinct marsupial family Borhyaenidae. (A) *Borhyaena* (length of skull 230 mm); (B) *Thylacosmilus* (length of skull 232 mm). (After Romer, 1966)

wore rapidly because they had only a thin layer of enamel. In compensation, the roots in some forms remained open through much of life, permitting continued growth (Patterson and Pascual, 1972:262). The shape of the sabers of *Thylacosmilus* and skeletal features usually associated with powerful neck musculature indicate that the sabers were used as slashing or stabbing weapons as the predators clung to large prey with their powerful forelimbs (Marshall, 1976), a predatory style also used by eutherian saber-tooth cats (p. 169). The limbs of thylacosmilids are short, and most prey was probably captured by a surprise attack rather than a chase.

The complex story of the decline and eventual extinction of the Borhyaenidae is told by Marshall (1977a). Many early Eocene borhyaenids resembled didelphid marsupials and were of moderate size; by the early Oligocene, however, several very large forms had evolved. The largest of all *Proborhyaena gigantea,* which had a skull roughly 1 meter long, disappeared by the end of the early Oligocene, and with its extinction a trend began toward reduced size in carnivorous borhyaenids. This evolutionary step moved counter to the trend toward large size that is common to many mammalian lineages. Further, the lack of cursorial adaptations in bor-

hyaenids is unusual for a group of large carnivorous mammals and may have resulted from competition with large carnivorous birds.

During the Early Tertiary in South America, the adaptive zones available to carnivores were probably not controlled or dominated by any one group, and, from early Oligocene through Pliocene times, three families of large carnivorous birds shared with borhyaenids a predatory mode of life. *Phororhacos,* a predaceous member of the family Phororhacidae (order Gruiformes), stood 1.5 to 3 meters tall and had a heavily built, hooked bill mounted on a skull about the size of that of a horse. The wings were tiny and the bird was flightless, but the hind legs were long and slim and the bird must have been a swift runner. Probably the phororhacoid birds and the borhyaenid carnivores partitioned the resources available to carnivores: the phororhacoids developed and maintained large size, occupied open savanna areas, and used their cursorial ability in pursuing swift prey; the borhyaenids were of more moderate size, occupied wooded country, and largely killed slow-moving prey. The Late Tertiary reduction in the size of borhyaenids may well have been part of an adaptive trend toward reduction of competition with the imposing phororhacoids (Patterson and Pascual, 1972:262).

Further declines in the fortunes of borhyaenids were apparently associated with the mid-Miocene arrival in South America of members of the theretofore entirely North American raccoon family (Procyonidae). The fossil record indicates that replacement of the larger omnivorous borhyaenids by the omnivorous procyonids was complete by the late Pliocene. Extinction of other borhyaenids is thought by Marshall (1977a) to be related to possible competition with marsupials of the family Didelphidae. The small to medium-size borhyaenid omnivores declined in the middle Pliocene and became extinct by the late Pliocene, whereas didelphids with similar adaptations appeared in the middle Pliocene and underwent a striking adaptive radiation in the late Pliocene.

The history of the Borhyaenidae was thus intertwined with that of the phororhacoid birds, the immigrant South American procyonids, and the didelphid marsupials. By the late Pliocene, the borhyaenids were gone and the three latter groups prevailed.

An event of major importance to the South American mammalian fauna occurred in the late Pliocene. At that time the land bridge connecting North and South America was established and a flood of northern placentals moved southward while some southern mammals moved northward (p. 360). Although the Borhyaenidae were already extinct by this time and were thus not affected by this collision of faunas, the saber-tooth thylacosmilids were still present and were perhaps decisively affected. The occurrence of placental saber-tooth cats in early Pleistocene strata immediately above late Pliocene beds bearing thylacosmilids suggests the possibility of competitive replacement.

Carnivory was just one of a number of feeding patterns of South American marsupials. *Necrolestes* from the middle Miocene was probably insectivorous and fossorial and may have had a mode of life similar to that of the Australian marsupial ''moles'' (Notoryctidae).

The family Caenolestidae, which still persists in relict populations along the Andes Cordillera, appears first in the Eocene of South America. Some caenolestids were convergent toward multituberculates in a series of features, including general skull form, enlargement of the anterior incisors, and structure of the serrate, lower pair of cheek teeth (Fig. 5–10). Both multituberculates and the multituberculate-like caenolestids (subfamily Abderitinae) probably resembled rodents in feeding habits.

The most rodent-like marsupials yet known are two South American species from the Early Tertiary. These species compose the family Groeberiidae and are remarkable in having such features as enlarged incisors with enamel only on the anterior surfaces, a sharp reduction in the number of cheek teeth, and a broad diastema.

Of special interest is the family Argyrolagidae, a supreme example of evolutionary convergence. This unique family, considered in detail by Simpson (1970b), is known from the Pliocene and Pleistocene of South America and probably diverged early from primitive didelphid ancestry. Argyrolagids did not resemble closely any other group of marsupials, but they possessed a series of morphological characters that are found today in such specialized rodents as kangaroo rats (Heteromyidae) and jerboas (Dipodidae). These rodents occupy mainly sparsely vegetated desert or semiarid areas; all are saltatorial, and all share certain distinctive morphological features. The hind limbs are long, and the hind feet are modified by the loss of digits and, in some cases, by the fusion of metatarsal bones (Fig. 16–14). The long hind limbs of argyrolagids are highly specialized along similar lines: only the third and fourth

Figure 5-10. Jaw of a Miocene caenolestid, showing the highly specialized, trenchant cheek tooth. (After Romer, 1966)

Figure 5-11. Hind foot of *Argyrolagus* (Argyrolagidae), an extinct marsupial that resembled the kangaroo rat. Note that the appressed metatarsals of digits three and four form a structure resembling the cannon bone of artiodactyls. (After Simpson, 1970b)

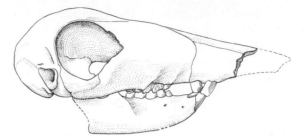

Figure 5-12. Skull of *Argyrolagus*. Note the similarity between the form of this skull and that of the kangaroo rat shown in Figure 16-21. Length of skull 55 mm. (After Simpson, 1970b)

digits of the foot are retained, and the metatarsals are closely appressed and resemble to some extent the cannon bone of an artiodactyl (Fig. 5–11).

All of these animals have cheek teeth adapted to grinding and incisors suited for gnawing, and in both kangaroo rats and argyrolagids the cheek teeth are rootless and have a simple occlusal surface. (Among rodents, however, such an occlusal surface is not unique to kangaroo rats.)

Condensation of moisture on the cool nasal mucosa during exhalation is a means of reducing pulmonary water loss in kangaroo rats, and the unusual tubular extension of the nasal cavity anterior to the incisors in these animals (Fig. 16–13) is associated with improved water conservation (p. 484). In the argyrolagids, there is an even more elongate extension of the nasal cavity (Fig. 5–12). Here, too, this specialization may have facilitated the maintenance of water balance in an arid environment.

The entire form of the skull in kangaroo rats and jerboas is modified by the enormous auditory bullae, and in argyrolagids the bullae are also inflated. This enlargement of the bullae has been shown to be one of a remarkable series of specializations that allow kangaroo rats to detect faint, low-frequency sounds made by their predators. Very probably the enlarged bullae of the argyrolagids served a similar end.

It seems, then, that in two lineages which have been separate since at least Early Cretaceous, but which occupy (or occupied) similar dry habitats, nearly identical suites of characteristics have evolved.

One cannot help but wonder why argyrolagids became extinct, for "by all rules of analogy and theories of extinction, they should have survived, as did their close ecological analogs in North America, Asia, Africa, and Australia" (Simpson, 1970b). At present, the mystery of the extinction of the argyrolagids remains unsolved. As the fossil record of South American mammals becomes more complete, information bearing on the extinction of the argyrolagids and other marsupial groups will probably come to light.

Marsupials Versus Eutherians: Relative Competitive Abilities

When one views the course of mammalian evolution since marsupials and eutherians diverged from a common ancestory in Early Cretaceous times, the general impression is that the two groups are not of equal adaptive ability. Several lines of evidence can be cited.

1. Marsupials have not equaled the remarkable functional radiation of eutherians. There are no flying or marine marsupials, and some extremely productive food sources have never been tapped. Marine plankton, utilized by two orders of eutherians (Cetacea and Carnivora), and flying insects, fed on by bats (Chiroptera), have never been marsupial fare.

2. Marsupials have been far more conservative in structural plan. None have modified limbs into fins or wings as eutherians have.

3. Marsupials have not been able to exploit great size. Although there were several large marsupials in the Pleistocene, the largest living marsupial (the red kangaroo) is only 1/1300 the size of the largest eutherian (the blue whale).

4. Marsupials have never realized the advantages of highly social behavior.

5. Marsupials have not developed the systematic diversity of eutherians. Only about 6 percent (250) of the total number of species of living mammals (4060) are marsupials.

Although sadly incomplete and equivocal, the fossil record does suggest a competitive edge for eutherians. In North America, where marsupials appeared before eutherians, by the latest Cretaceous marsupials had declined seriously but eutherians had radiated. In South America, similarly, where a Tertiary marsupial radiation occurred in isolation from a balanced assemblage of eutherians, the diversity of marsupials declined late in the Tertiary and in the Pleistocene, perhaps at first because of the entry into South America of only a few eutherians and then (with the emergence of the land bridge connecting North and South America) because of an invasion of eutherians from the north. Es-

pecially impressive is the total extinction of the South American marsupial carnivores (of the families Borhyaenidae and Thylacosmilidae) and their ultimate replacement by eutherian carnivores. (These examples lose force if considered in light of the hypothesis of Matthew (1915) that mammals that evolved on a large continent are competitively superior to those that originated on a smaller one.)

Although there is no general consensus on the matter, the view of many scholars is that marsupials are adaptively and competitively inferior to eutherians. Perhaps the best course here is to catalog briefly some basic differences between the two groups, to provide an introduction to the conflicting views appearing in the literature, and to avoid offering a resolution for an unresolved problem.

Each of the following may be associated with adaptive-competitive differences between the two groups:

1. Marsupials have a brief gestation period and bear almost embryonic young that undergo most of their basic development while attached to the mother's nipple and nourished by milk. Eutherians, however, have long gestation periods and the young are much more developed at birth.

2. The cerebral cortex develops more rapidly and attains greater volume in eutherians than in marsupials (F. Muller, 1969).

3. Behavioral plasticity is greater in eutherians: social groups with long-term dominance hierarchies and cooperative rearing of young occur only among eutherians, and territoriality, an important aspect of eutherian behavior, is uncommon in marsupials.

4. Antipredator behavior is more highly developed in eutherians: unified herd action, cooperative defense of young, complex vocal and visual communication, and sustained high-speed running are known only among eutherians.

5. The relatively low diploid number of marsupials has been mentioned (Hayman, 1977; Lillegraven, 1975) as possibly being related to their lack of evolutionary flexibility.

6. The investment of energy by the mother is probably lower in marsupials than in eutherians (Parker, 1977), but eutherians are seemingly able to reproduce more rapidly.

The view that the marsupial style of reproduction, and marsupial biology in general, represent an alternate but not inferior solution to survival problems has been discussed by a number of workers (including Kirsch, 1977; Parker, 1977; and Pond, 1977). For considerations of the comparative adaptiveness of the marsupial and eutherian patterns of reproduction, see Lillegraven (1975, 1979b).

South American Marsupials

Family Didelphidae.   Didelphids are the most generalized marsupials and constitute the oldest known family, dating from the early part of Late Cretaceous. As mentioned, this is the basal family of the marsupial radiation. The Didelphidae includes 14 Recent genera comprising 70 species and occurs from southeastern Canada, with the American opossum *(Didelphis marsupialis;* Fig. 5–13), to southern Argentina, with the Patagonian opossum *(Lestodelphis halli).* One species *(Dromiciops australis)* that inhabits cool, southern Andean forests is regarded by Marshall (1984) as the single living member of the family Microbiotheriidae.

In these New World opossums, the rostrum is long (Fig. 5–14), the braincase is usually narrow, and the sagittal crest is prominent. The dental formula is 5/4, 1/1, 3/3, 4/4 = 50. The incisors are small and unspecialized, and the canines are large. The upper molars are basically tritubercular with sharp cusps, and the lower molars have a trigonid and a talonid (Fig. 5–15).

Except for the opposable and clawless hallux in all species, a feature probably inherited from arboreal ancestral stock, and the webbed hind feet in the water opossum *(Chironectes minimus),* the feet are unspecialized, with no loss of digits or syndactyly (the condition in which two digits are attached by skin, as shown by a number of examples in Fig. 5–6.) The foot posture is plantigrade. A marsupium is present in some didelphids but is represented by folds of skin protecting the nipples in others and is absent in some. The tail is long and usually prehensile.

Figure 5-13. A didelphid marsupial *(Didelphis marsupialis).* In the United States, this animal is common in the southeast and in many areas along the Pacific coast. (San Diego Zoo)

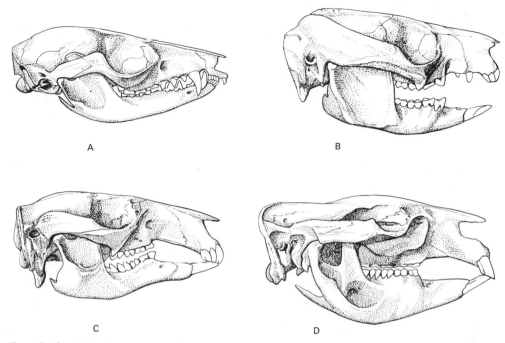

Figure 5–14. Skulls of marsupials. (A) Mouse opossum (*Marmosa canescens,* Didelphidae); length of skull 35 mm. (B) Brush-tail possum (*Trichosurus vulpecula,* Phalangeridae); length of skull 87 mm. (C) Wallaby (*Wallabia bicolor,* Macropodidae); length of skull 135 mm. (D) Wombat (*Vombatus ursinus,* Vombatidae); length of skull 180 mm.

Although they occupy a wide range of habitats, didelphids are primarily inhabitants of tropical or subtropical areas, where they are often locally abundant. Most didelphids are partly arboreal and are omnivorous. The water opossum, however, is aquatic and carnivorous, and the woolly opossum (*Caluromys* spp.) is largely herbivorous.

The small mouse opossum (*Marmosa),* a widespread Neotropical didelphid, is one of the most abundant small mammals in some parts of Mexico. Although mouselike in general appearance, it seems to be largely insectivorous in some areas, at least during the summer (R. B. Smith, 1971). Poorly defined folds of skin protect the nipples of *Marmosa,* and the young simply hang on to the nipples and the mother's venter as best they can.

Family Caenolestidae. Recent members of this family (three genera and seven species) bear the common name of rat opossum. Three relict, disjunct populations occupy forested areas of the Andes Mountains of northern and western South America. The earliest known caenolestids are found in the Eocene fossil record of South America, and in the Oligocene and Miocene a diverse group of caenolestids, including some highly specialized types (Fig. 5–10), appeared.

Caenolestids resemble shrews because of their elongate heads and small eyes (Fig. 5–16). The skull is elongate and the brain primitive; the olfactory bulbs are large and the cerebrum lacks fissures. The dental formula is 4/3-4, 1/1, 3/3, 4/4 = 46 or 48; the first lower incisors are large and procumbent, and the remaining lower incisors, the canine, and the first premolar are unicuspid. Kirsch (1977) found that the lower incisors are used like rapiers to stab prey. The atlas bears a movable cervical rib. The feet are unspecialized, the tail is long but not prehensile, and there is no marsupium.

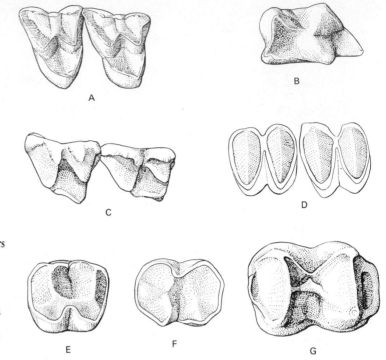

Figure 5-15. Occlusal views of marsupial molars.
(A) Second and third right upper and (B) third lower
left molar of a mouse opossum (*Marmosa canescens,*
Didelphidae). (C) Second and third upper right molars
of the New Guinea bandicoot (*Peroryctes
raffrayanus,* Peramelidae; after Tate and Archbold,
1937). (D) Second and third right upper molars of a
wombat (*Vombatus ursinus,* Vombatidae). (E)
Second upper right and (F) third lower left molar of a
phalanger (*Trichosurus vulpecula,* Phalangeridae).
(G) Second upper right molar of a wallaby (*Wallabia
bicolor,* Macropodidae).

Australian Marsupials

Family Dasyuridae. Dasyurids (Fig. 5–17)
are more progressive than didelphids, both dentally and with regard to limb structure. Although the earliest known dasyurid is from the
Australian Middle Tertiary, the family must have
arisen at a far earlier time. Recent members of
this family include 16 genera and 53 species,
and the geographic range includes Australia,
New Guinea, Tasmania, the Aru Islands, and
Normanby Island (Van Deusen and J. K. Jones,
1967:61).

Many of the major characters of dasyurids
are shared by other marsupials, but several features are diagnostic of the former. The dental
formula is 4/3, 1/1, 2-3/2-3, 4/4 = 42-46. The
incisors are usually small and either pointed or
bladelike, the canines are large and have a sharp
edge, and the upper molars have three sharp
cusps adapted to an insectivorous and carnivorous diet. The skulls of some dasyurids resemble

Figure 5-16. A caenolestid marsupial *(Lestoros inca).* This
individual came from Peru, at an elevation of 3530 meters. (J. A.
W. Kirsch)

Figure 5–17. Four members of the family Dasyuridae. (A) A marsupial "rat" *(Dasyuroides byrnei);* (B) a marsupial "mouse" *(Sminthopsis crassicaudata);* (C) a marsupial "mouse" *(Antechinus stuartii);* (D) the native "cat" *(Dasyurus viverrinus).* (*Dasyuroides* and *Antechinus* by J. Hudson; *Sminthopsis* and *Dasyurus* by A. Robinson)

rather closely those of didelphids (Fig. 5–18). The forefoot has five digits, and the hind foot has four or five digits. The hallux is clawless and usually vesitigial and is absent in some cursorial genera (Figs. 5–6 and 5–19). There is no syndactyly. The foot posture is plantigrade in many species, but the long-limbed jumping marsupials, such as *Antechinomys* and the cursorial and carnivorous native "cat" (*Dasyurus viverrinus;* Fig. 5–18) and Tasmanian "wolf" (*Thylacinus cynocephalus;* Fig. 5–20), are digiti-

grade. The marsupium is often absent; when present, it is often poorly developed. The tail is long and well furred, conspicuously tufted in some species, and never prehensile. The size of dasyurids ranges from that of a shrew *(Planigale)* to that of a medium-size dog *(Thylacinus).*

A wide variety of terrestrial habitats are occupied by dasyurids, and a few species are arboreal. There are a remarkably diverse array of marsupials in the family Dasyuridae. The

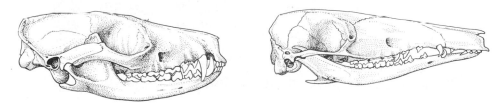

Figure 5–18. Skulls of marsupials. (A) Native "cat" (*Dasyurus viverrinus;* Dasyuridae); length of skull 72 mm. (B) Bandicoot (*Perameles* sp., Peramelidae); length of skull 81 mm.

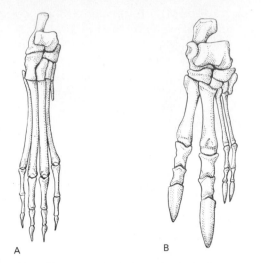

Figure 5-19. Feet of two marsupials. (A) A terrestrial species, the Australian native "cat" (*Dasyurus viverrinus,* Dasyuridae). (B) An arboreal species, a tree kangaroo (*Dendrolagus* sp., Macropodidae). (After Marshall, 1972)

smaller species fill the feeding niche occupied in Eurasia and North America by shrews (family Soricidae) and resemble these animals in the possession of long-snouted heads and unspecialized limbs. A group of rat-size dasyurids seems adapted to preying on insects and small vertebrates, and the desert-dwelling genus *Antechinomys* has long, slender limbs and a long, tufted tail and uses a rapid, bounding, quadrupedal gait. Another group, the native "cats," consists of somewhat civet-like dasyurids that weigh from roughly 0.5 to 3 kilograms and prey on a variety of small vertebrates. These cats are agile and effective predators, and, although primarily terrestrial, they are capable climbers. The largest marsupial carnivores are the Tasmanian devil *(Sarcophilus harrisii)* and the Tasmanian "wolf" *(Thylacinus cynocephalus).* The Tasmanian devil is a stocky, short-

Figure 5-20. The Tasmanian "wolf" (*Thylacinus cynocephalus,* Dasyuridae), which may now be extinct. (E. H. Colbert)

limbed dasyurid, weighing from 4.5 to 9.5 kilograms; it is now restricted to Tasmania. It is a persistent scavenger but will also kill a wide variety of small vertebrates. The Tasmanian wolf is doglike in both size and general build (Fig. 5–20); it has long limbs and a digitigrade foot posture.

Although it is treated here as a dasyurid, *Thylacinus* has been separated from the dasyurids and put in the family Thylacinidae by Ride (1970:226). Now extinct in New Guinea, on the Australian mainland, and very possibly in Tasmania, *Thylacinus* is able to prey on such large animals as the larger species of wallabies.

The most divergent dasyurid is the numbat, or banded anteater *(Myrmecobius fasciatus)*—a small, long-snouted animal often regarded as the sole representative of the family Myrmecobiidae. The teeth are small and widely spaced in the long tooth row, and the long, protrusile tongue is used in capturing termites. This animal was formerly widespread in eucalyptus forests, in which fallen branches and logs provided lush populations of termites, but the commercial clearing of these forests has severely restricted the range of the numbat.

Family Peramelidae.   Members of this family are called bandicoots and are characterized in general by an insectivore-like dentition and a trend toward specialization of the hind limb for running or hopping. Eight Recent genera, represented by 19 Recent species, are known, mainly from Australia, Tasmania, and New Guinea. Some species of bandicoots have been extirpated or have become uncommon over parts of their former range owing, apparently, to the grazing of livestock, to brush fires, and to the introduction of various placental mammals.

The smaller bandicoots are the size of a rat, with the largest species weighing roughly 7 kilograms (Fig. 5–21). The dental formula is 4-5/3, 1/1, 3/3, 4/4 = 46 or 48. The incisors are small, and the molars are tritubercular (Fig. 5–15) or quadritubercular. The rostrum is slender (Fig. 5–18), and the ears of some species resemble those of rabbits. The marsupium is present and opens to the rear, and bandicoots, alone among marsupials, have a chorioallantoic placenta. Although often long, the tail is not prehensile. The fourth digit of the hind foot is always the largest, and the remaining digits are variously reduced (Fig. 5–22). The hind foot posture is usually digitigrade, and the hind limbs are elongate. The opposable hallux, probably inherited by peramelids from an arboreal ancestral stock, is rudimentary or may be lost. The second and third digits of the hind foot are joined (syndactylous) as far as the distal phalanges by an interdigital membrane, and the muscles of these digits are partially fused, allowing them to act only in unison (F. W. Jones, 1924). An extreme degree of cursorial specialization occurs in the pig-footed bandicoot *(Chaeropus ecaudatus):* the forelimb is functionally didactyl and the hind foot is functionally monodactyl during running. The second and third digits of the forelimb are large and clawed; the first and fifth are absent, and the fourth is vestigial. The two species of rabbit-eared bandicoots (*Macrotis lagotis* and *M. leucura*) are placed by Archer and Kirsch (1977) in a separate family (Thylacomyidae).

The structure and function of the specialized peramelid hind foot are unique and have been described in detail by L. G. Marshall (1972). In mammals, extreme reduction in the number of digits is usually associated with good running ability. Most highly cursorial ungulates have retained only the third digit (in the case of the horse) or digits three and four (in the case of some antelope). A similar trend occurs in peramelids. Probably partly due to an early development of syndactyly involving the second and third digits and the use of these digits for grooming, the general trend in the cursorial peramelids is toward the reduction of all digits but the fourth, with a great enlargement of this digit (Fig. 5–22). These specializations are accompanied by an alteration in the structure and function of the tarsal bones. The ectocuneiform bone makes broad contact with the proximal end of the fourth metatarsal and partially sup-

Figure 5-21.   Two peramelid marsupials. (A) New Guinea bandicoot *(Peroryctes raffrayanus);* (B) "rabbit" bandicoot *(Macrotis lagotis).* (A by S. O. Grierson; B by A. Robinson)

ports this digit, a character unique to peramelids. The mesocuneiform is lost, and the weight of the body is borne mainly by the cuboid, ectocuneiform, navicular, and astragalus bones. The calcaneum does not serve a major weight-bearing function, but of course serves as a point of insertion for the extensors of the foot, muscles of great importance in locomotion.

Horses, antelope, and peramelids provide beautiful examples of different structural means of solving a similar functional problem, in this case, refining running ability. In the horse only

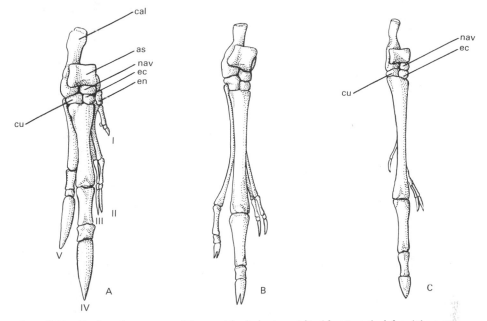

Figure 5-22. The feet of some peramelid marsupials; the least specialized foot is on the left and the most specialized is on the right. (A) Long-nosed bandicoot *(Perameles* sp.); (B) rabbit bandicoot *(Macrotis* sp.); (C) pig-footed bandicoot *(Chaeropus ecaudatus).* Abbreviations: *as,* astragalus; *cal,* calcaneum; *cu,* cuboid; *ec,* ectocuneiform; *en,* entocuneiform; *nav,* navicular. The digits are numbered in A. (After L. G. Marshall, 1972)

the third digit is retained, and it is supported largely by the ectocuneiform, the navicular, and the astragalus bones (Fig. 12-3). In the pronghorn antelope, only two digits are retained and the cannon bone (the fused third and fourth metatarsals) is supported largely by the fused cuboid and navicular bones and the fused mesocuneiform and ectocuneiform bones; the calcaneum is no longer a weight-bearing element. In the most cursorial peramelid *(Chaeropus),* the fourth digit is greatly enlarged and is supported, as outlined above, by the cuboid, navicular, ectocuneiform, and astragalus bones (Fig. 5-22). Marshall (1972) points out, however, that the structure of the hind limb of peramelids is not entirely modified for running, perhaps because of the burrowing tendencies of these animals. The fibula is large and movement at the ankle joint is not restricted to a single plane, as it is in most cursorial mammals. (Cursorial adaptations are discussed in more detail on p. 184.)

Bandicoots are largely insectivorous but also eat small vertebrates, a variety of invertebrates, and some vegetable material. Some species take refuge in nests that they build of plant debris; both species of *Macrotis* dig burrows in which they hide during the day. *Chaeropus* is jackrabbit-like: it squats beneath saltbushes *(Atriplex)* in semiopen areas and depends on speed to escape its enemies (F. W. Jones, 1924:171). This highly cursorial bandicoot may be extinct, not having been seen since about 1926.

Family Notoryctidae. This remarkable family is represented by a single species of marsupial "mole" that inhabits sandy soils in arid parts of northwestern and south central Australia. Many of the diagnostic characters of these mouse-size animals are adaptations for fossorial life. The eyes are vestigial, covered by skin, lensless, and, as indicated by the specific name *Notoryctes typhlops* (*typhlops* means "blind" in greek), nonfunctional. The ears lack pinnae. The nose bears a broad cornified shield, and the nostrils are narrow slits. The dental formula is usually 4-3/3, 1/1, 2/3, 4/4 = 44-42 but the incisors vary in number. The incisors, canines, and all but the last upper premolar are unicuspid; the paracone and metacone of the upper molars form a prominent single cusp, and the lower molars lack a talonid. As an adaptation serving to brace the neck when the animal forces its way through the soil, the five posterior cervical vertebrae are fused. The forelimbs are robust, and the claws of digits three and four are remarkably enlarged and function together as a spade; the other digits are reduced. The central three digits of the hind feet have enlarged claws, the small first digit has a nail, and the fifth digit is vestigial. The marsupium is partially divided into two compartments, each with a single nipple. The fur is long and fine-textured, and varies in color from silvery white to yellowish red.

These animals use their powerful forelimbs and armored rostrum to force their way through soft, sandy soil. When the animal forages near the surface, the soil is pushed behind it and no permanent burrow is formed. The food is predominantly invertebrate larvae.

Family Phalangeridae.   In this family are the possums and cuscuses, a group of primarily arboreal animals. Three genera and about 15 species are known. The brush-tailed possum *(Trichosurus vulpecula)* is one of the most familiar of Australian mammals, for it frequently maintains resident populations in suburban areas, where it often seeks shelter in roofs of houses and feeds on cultivated plants.

These marsupials are of moderate size, ranging in weight from 1 to 6 kilograms. The skull is broad and has deep zygomatic arches (Fig. 5–14B). The molars are bilobed with rounded cusps (Fig. 5–15E, F). As adaptations to arboreal life, the hands and feet are large and have a powerful grasp, and the tail is prehensile. The cuscuses have short ears and woolly fur and look much like a teddy-bear (Figs. 5–23 and 5–24).

Members of this family mostly inhabit

Figure 5-23.   Two views of a New Guinea cuscus (*Phalanger maculatus,* Phalangeridae). Note the prehensile tail with the traction-producing ridges on the bare distal part of the ventral surface. (S. O. Grierson)

Figure 5-24. Two New Guinea cuscuses (Phalangeridae). (A) *Phalanger vestitus;* (B) *P. orientalis.* (A by S. O. Grierson; B by H. M. Van Deusen)

wooded areas, but the adaptable brush-tail possum also occupies treeless areas, where it takes refuge in rocks or in the burrows of other mammals. This animal is locally destructive to plantations of introduced pines. Phalangerids are omnivorous and are known to eat a wide variety of plant material as well as insects, young birds, and birds' eggs. The brush-tail possum is solitary and has a sternal scent gland, considerably larger in males, which produces a musky smell that is used in the scent marking of objects within the animal's territory. This marsupial is one of Australia's most valuable fur bearers.

During 1959, there were 107,500 brush-tail possum skins marketed in Victoria.

Family Petauridae. This family includes the ring-tail possums, so named because of their prehensile tails, and the greater and lesser gliding possums, some of the handsomest and most remarkable of all marsupials (Fig. 5–25). Five genera and 23 species are currently recognized.

Most members of this family are fairly small; weights range from about 100 grams to 1.5 kilograms. The skull is broad, and the four-cusped molars have fairly sharp outer cusps forming a roughly W-shaped ectoloph. The tail

Figure 5-25. Three members of the family Petauridae. (A) Ring-tail possum *(Pseudocheirus forbesi)*; (B) gliding possum *(Petaurus breviceps)*; (C) striped possum *(Dactylopsila trivirgata)*. (S. O. Grierson)

is prehensile in some petaurids and long and bushy in others. Some species are strikingly marked (Fig. 5–25B, C). The gliders (*Petaurus* and *Schoinobates*) have furred membranes that extend between the limbs and function as lifting surfaces for gliding. In these gliders, the claws are sharp and recurved, like those of a cat, and increase the ability of the animal to cling to the smooth trunks and large branches of trees.

The petaurids are nocturnal and arboreal creatures and inhabit wooded areas. (Figure 5–26 shows a habitat of the gliders.) Ring-tail possums are nocturnal and strictly herbivorous, eating both leaves and fruit. They make conspicuous nests (dreys) of leaves and twigs in the dense scrub of eastern and southeastern Australia. Curious specializations, similar to those of the primate *Daubentonia,* occur in two petaurid genera of striped possums. In *Dactylopsila,* and to a more advanced degree in *Dactylonax,* the fourth digit of the hand is elongate and slender, and its claw is recurved. In addition, the incisors are robust and function roughly as do those of rodents. Striped possums tear away tree bark with their incisors and extract insects from crevices and holes in the wood with the specialized fourth finger and the tongue. The conspicuous striped color pattern of *Dactylopsila* (Fig. 5–25C) is of interest because it is associated, as in skunks, with a powerful, musky scent.

The gliders are strikingly similar to flying squirrels *(Glaucomys)* in gliding style and ability and some can glide over 100 meters. *Schoinobates,* the greater glider, is remarkable in having perhaps the most specialized marsupial diet: its food is entirely leaves and blossoms, chiefly those of eucalyptus trees. Sugar gliders *(Petaurus breviceps)* live in family groups, and scent marking plays an important role in the so-

Figure 5–26.   Rain forest in eastern Victoria, Australia. This community is inhabited by two species of lesser gliding possums *(Petaurus)* and by greater gliding possums *(Schoinobates volans)*. (D. Harrison)

cial organization of the group. Each individual has a particular odor recognized by the others. The cohesion of the group is also aided by mutual scent marking, for all members of the group become permeated with the scent of the group's dominant males (Schultze-Westrum, 1965).

Family Burramyidae. The type genus of this family was known for many years only from Pleistocene fossil material; finally, in 1966, at a ski lodge on Mt. Hotham in Victoria, a representative of the genus was found alive. More recently, it has been found at other localities. This family contains four genera and seven species of small, mouselike marsupials, called pigmy possums.

These diminutive marsupials are from 120 to 295 millimeters in total length, are delicately built, and have large eyes and mouselike ears (Fig. 5–27). The tail is long and prehensile in all species and has a lateral fringe of hairs in the pigmy glider *(Acrobates pygmaeus).* This species has a narrow gliding membrane bordered by a fringe of long hairs. Traction between the digits and the trunks and branches of trees is increased in the pigmy glider by expanded pads at the tips of the fingers and toes; the surfaces of the pads have ridges that further increase the clinging ability of these animals.

Members of this family are restricted to wooded areas. They are apparently insectivorous-omnivorous, but the feeding habits of some members of the group are not known. As in some small placental mammals, the ability to become torpid during cold weather is well developed in some burramyid marsupials. Pigmy gliders become torpid in their nests on cold days, and the tails of the members of two genera of pigmy possums (*Cercartetus* and *Eudromicia*) become greatly enlarged with fat as winter approaches and these animals undergo periods of torpor.

Family Macropodidae. Members of this familiar marsupial group, which includes the kangaroos, euros, and wallabies (Fig. 5–28), are the ecological equivalents of such ungulates as antelope. Both macropodids and ungulates are cursorial, and both have highly specialized limbs. Rather than using quadrupedal locomotion, however, as ungulates do, the macropodids are primarily bipedal. Further, both groups are herbivorous and have skulls and dentitions specialized for this mode of feeding. Even some specializations of the digestive system are similar in these two groups. Ride (1970:50) cites studies by Waring and Main on digestion in the tammar *((Macropus eugenii),* euro *(M. robustus)* and quokka *(Setonix brachyurus).* These animals have intestinal bacteria that digest the cell walls of plants; these (and probably other) macropodids can thus utilize the digestible inner parts of the plant cells and the byproducts of the bacterial digestion and are able to digest

Figure 5-27. A pygmy possum (*Cercartetus concinnus,* Burramyidae). (A. Robinson)

Figure 5–28. Three kinds of macropodid marsupials. (A) Great gray kangaroo *(Macropus giganteus)*; (B) pademelon *(Thylogale billardierii)*; (C) red-necked wallaby *(Macropus rufogriseus)*. (D. Harrison)

the bacteria. The ruminant ungulates also depend on bacteria to increase the efficiency of their utilization of vegetation (p. 189).

The present distribution of the 17 Recent genera and approximately 60 Recent species of macropodids includes New Guinea, Bismarck Archipelago, the D'Entrecasteaux Group, Australia, and, by introduction, some islands near New Guinea and New Zealand. The family Macropodidae appears first in the Middle Tertiary (late Oligocene or early Miocene) of Australia. Wallabies and a kangaroo are known from the Late Tertiary, and unusually large macropodids occured in the Pleistocene.

Living macropodids vary tremendously in size and structure. The musky rat kangaroo *(Hypsiprymnodon moschatus)* weighs only 500 grams, whereas the great gray kangaroo *(Macropus giganteus),* the largest living marsupial, reaches 2 meters in height and approximately 90 kilograms in weight. The marsupium is usually large and opens anteriorly. The macropodid skull is moderately long and slender, and the rostrum is usually fairly long (Fig. 5–14C). The dental formula is 3/1, 1-0/0, 2/2, 4/4 = 34 or 32. The upper incisors have sharp crowns with their long axes oriented more or less front to back. The tips of the procumbent lower incisors are held against a leathery pad just behind the upper incisors when the animals gather vegetation. This specialized arrangement serves a cropping function similar to that of the lower incisors and the premaxillary pad in the artiodactylan ungulates lacking upper incisors. There is a broad diastema between the macropodid incisors and the premolars. The molars are quadritubercular and bilophodont (Fig. 5-15G). In many macropodids, the last molar does not erupt until well after the animal becomes adult. A unique situation occurs in the little rock wallaby *(Peradorcas concinna),* in which nine molars may erupt in succession. Usually four or five molars are functional at one time, and replacement is from the rear as the molars are successively lost from the front.

Macropodids are highly specialized for jumping. The forelimbs are five-toed and usually small; they are used for slow movement on all fours or for food handling (Frith and Calaby, 1969). The hind limbs are elongate, especially the fourth metatarsal. The hallux is missing in all but *Hypsiprymnodon.* Digits two and three are small and syndactylous and used for grooming; the fourth is the largest digit; and the fifth is often also robust (Fig. 5-29). The unusual pattern of digital reduction and the dominance of the fourth digit in the most highly cursorial Australian marsupials are perhaps due to the arboreal ancestry of these animals. In these ancestors, the foot was five-toed and the hallux opposable; the fourth was the longest remaining digit, and the foot was adapted to grasping branches. With specialization of the foot for running or hopping, the hallux was lost and the longest toe, the fourth, became the most important digit. In most macropodids, the foot is functionally two-toed during rapid locomotion, which is characteristically bipedal, but in *Megaleia* the foot is functionally one-toed (Fig. 5-29C). In the macropodid tarsus, there is no contact between the ectocuneiform and the fourth metatarsal (in contrast to the arrangement in the

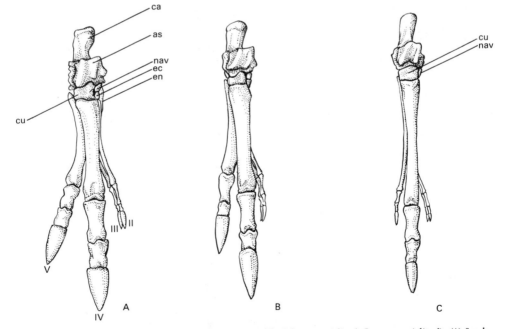

Figure 5-29. The foot bones of some macropodid marsupials (A least specialized, C most specialized). (A) Scrub wallaby (*Thylogale* sp.); (B) kangaroo (*Macropus* sp.); (C) red kangaroo *(Megaleia rufa).* Abbreviations: *as,* astragalus; *ca,* calcaneum; *cu,* cuboid; *ec,* ectocuneiform; *en,* entocuneiform; *nav,* navicular. The digits are numbered in A.

Peramelidae; Fig. 5–22). Because the hind limb posture of macropodids is basically plantigrade, the calcaneum is an important weight-bearing element of the tarsus (which it is not in the digitigrade Peramelidae). The macropodid tail is usually long and robust, and functions in the more specialized species as a balancing organ and as the posterior "foot" of the tripod formed by the plantigrade hind feet and tail, on which the animal can sit when not in motion.

Several macropodid genera depart from the familiar structural pattern of kangaroos and from the grazing or browsing habit. *Hypsiprymnodon,* a muskrat-size inhabitant of rain forests and riparian situations, has a tail of modest length and retains all the digits of the hind foot. The hind limbs are not greatly elongate, and the animal uses quadrupedal rather than saltatorial locomotion. Animal material forms a large share of the food of this seemingly primitive macropodid. The tree kangaroos (*Dendrolagus,* Macropodinae) spend considerable time on the ground but frequently use their arboreal ability to escape from danger. This mode of life is reflected by the large and robust forelimbs with strong recurved claws, by the hind limbs, which are not strongly elongate, and by the short, broad hind foot (Fig. 5–19B). Saltation, typical of terrestrial kangaroos, has not been completely abandoned by tree kangaroos; not only are these animals agile climbers, but they also leap from one tree to another and from tree to ground. Their food is large fruit and leaves.

The running ability of the larger kangaroos (*Macropus*) is impressive. Speeds on level terrain of roughly 88 kilometers per hour are attained, and leaps covering distances of 13.5 meters and height of 3.3 meters have been reported (Troughton, 1947:213). The solitary hare wallabies (*Lagorchestes*), which are roughly the size of a large rabbit, also are renowned for their great speed, These animals have a jackrabbit-like style of escape. They hide beneath bushes or clumps of grass, burst out suddenly when frightened, and run away at high speed. The highly developed jumping ability of

macropodids allows these animals to move easily for long distances between scattered sources of water or forage and to escape enemies by erratic leaps. These abilities, rather than the capacity for great speed, are perhaps of primary adaptive importance. Saltation may have been developed by small forms ancestral to kangaroos as a means of erratic escape in open areas. This ricochetal style of locomotion is known in a number of desert-dwelling rodents. According to A. B. Howell (1944:247), the kangaroo's "method of traveling by saltation was hardly begun for the purpose of ultimate speed. Rather has it built speed into the locomotor pattern that was already established, probably for some other purpose." The locomotion of rock wallabies (*Petrogale*) is adapted to the rocky country they inhabit. According to F. W. Jones (1924:231), their movements are spectacular: "There seems to be no leap it will not take, no chink between boulders into which it will not hurl itself."

Family Tarsipedidae. This family contains but one species, the highly specialized, slender-nosed honey possum, or noolbender (*Tarsipes spencerae*). This remarkable animal's many specializations obscure its relationships to other marsupials, and its taxonomic position has long been uncertain.

Tarsipes is small, only about 15 to 20 grams in weight, and has a long, prehensile tail. The pelage is marked by three longitudinal stripes on the back. The rostrum is long and fairly slim, and the dentary bones are extremely slender and delicate. The cheek teeth are small and degenerate, and only the upper canines and two medial lower incisors are well developed. The snout is long and slender, and the long tongue has bristles at its tip (these specializations are similar to those of some nectar-feeding bats of the family Phyllostomidae). All digits but the syndactylous second and third digits of the hind feet have expanded terminal pads resembling to some extent those of the primate *Tarsius*.

Honey possums occur in forested and

shrubby areas of southwestern Australia. Like the hummingbirds and nectar-feeding bats, honey possums feed on nectar, pollens, and to some extent small insects that live in flowers. The long, protrusile tongue is used to probe into flowers. *Tarsipes* can climb delicately over even the insecure footing of clusters of flowers at the ends of branches and often clings upside down to flowers while feeding. Although the animal is still common in some areas today, the expansion of agriculture in southwestern Western Australia is restricting the honey possum's range.

Family Vombatidae. This family is represented by two living genera and three species. Known as wombats, these animals are completely herbivorous and show remarkable structural convergence toward rodents. Because of the efforts of humans, wombats have become scarce or absent over much of their former range and now are restricted to parts of eastern and southern Australia, Tasmania, and the islands between Australia and Tasmania.

Wombats are stocky animals with small eyes and rodent-like faces (Fig. 5–31); their body weight can exceed 35 kilograms (Troughton, 1947:144). The skull and dentition bear a striking resemblance to those of some rodents (Figs. 5–14D and 5–15D). The skull is flattened, the rostrum is relatively short, and the heavily built zygomatic arches flare strongly to the sides. The area of origin of the anterior part of the masseter muscle is marked by a conspicuous depression in the maxillary and jugal bones that is similar to the comparable depression in the maxillary and premaxillary bones of the beaver (*Castor;* Fig. 16–9B). The dental formula is 1/1, 0/0,1/1, 4/4 = 24; all teeth are rootless and ever-growing. Only the anterior surfaces of the incisors bear enamel, and the incisors and the first premolars are separated by a wide diastema. The molars are bilophodont (Fig. 5–15D). As in rodents, the coronoid process of the dentary bone is reduced and the masseter muscle, rather than the temporalis, is the major muscle of mastication.

The limbs are short and powerful, and the foot posture is plantigrade. The forefeet have five toes; all digits have broad, long claws. The hallux is small and clawless, but the other digits have claws. Digits two and three of the hind feet are syndactylous. The tail is vestigial. The marsupium opens posteriorly and contains one pair of mammae.

This family is first known from the middle Miocene Epoch (Gill, 1957), and both Recent

Figure 5–30. A wombat (*Vombatus ursinus,* Vombatidae). (A. Robinson)

genera have fossil species. The Pleistocene trend toward large size apparent in many other mammalian groups also occurred in the Vombatidae, as evidenced by the huge Pleistocene "wombat" *Phascolonus.*

Wombats are powerful burrowers. Their burrows are extensive networks of tunnels and are sufficiently wide to admit a small person. As a boy, Nicholson (1963) studied wombats in Victoria by crawling into their burrows. He found that, almost without exception, they were amiable and inquisitive, and that they seemed sociable and visited each other's burrows. Young wombats learned to burrow in their mother's burrow system by digging small subsystems but abandoned the maternal burrows about four months after leaving their mother's pouches. Wombats dig burrows in the open or beneath rock piles, as do marmots *(Marmota),* their North American rodent counterparts. They live in level or mountainous terrain supporting dry or moist sclerophyll forests or grassland and are able to go for long periods without drinking water. Their food is largely herbs and grass but includes bark, roots, and fungi (Troughton, 1947:139, 140).

Family Phascolarctidae. The familiar koala, or native "bear" (*Phascolarctos cinereus;* Fig. 5–31) is the sole member of this family. This highly specialized herbivore is restricted to some wooded parts of southeastern Australia.

The tufted ears, naked nose, and chunky, tail-less form make the koala one of the most distinctive of Australian marsupials. It is a fairly

Figure 5-31. The koala (*Phascolarctos cinereus,* Phascolarctidae). (D. Harrison)

large marsupial; the adult ranges from 8 to 10 kilograms in weight. The skull is broad and sturdily built, and the dentary bones are deep and robust (Fig. 5–32A). The dental formula is 3/1, 1/0, 1/1, 4/4 = 30. The roughly quadrate molars have crescentic ridges (Fig. 5–32B), and there is a diastema in both the upper and lower

Figure 5-32. (A) Skull of the koala (*Phascolarctos cinereus,* Phascolarctidae; length of skull 132 mm). (B) Occlusal view of the second molar, upper right tooth row, showing the crescentic areas of dentine, exposed by wear, and the complex pattern of furrows.

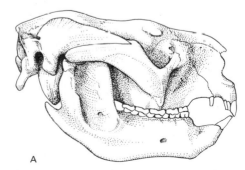

A

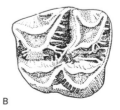

B

tooth rows between the cheek teeth and the anterior teeth. Tree branches are grasped between the first two and the last three fingers of the hand and between the clawless first digit and the remaining digits of the foot; the long, curved claws aid in maintaining purchase on smooth branches.

Koalas are fairly sedentary and feed on only a few species of smooth-barked eucalyptus trees. Maturation of a koala takes considerable time. A single young is born and is carried in the pouch for six months, after which it rides on its mother's back for a few more months. The young koala is dependent on its mother for one year, and sexual maturity is not reached until three or four years of age. According to Ride (1970:88), koalas grunt in piglike fashion when feeding at night and wail continuously when alarmed. Koalas lived in southwestern Western Australia during the Late Pleistocene, but no longer occur there even though suitable habitat is present.

Chapter 6

Introduction to Eutherian Mammals

The story of eutherian mammals begins with Early Cretaceous fossils from Khovboor, Mongolia. The record is sketchy, however, and leaves unrepresented vast geographic areas, great sweeps of Cretaceous time, and critical anatomical changes. Most of the fossils are isolated teeth or fragments of jaws with partial tooth rows, and they represent only parts of Asia, western North America, and a single locality each in France and Peru. The entire first half of the Cretaceous has yielded very little eutherian material, but a better record is available from the Late Cretaceous.

In spite of the paucity of fossils, some conclusions regarding Cretaceous eutherians can be made. They were small, from the size of a shrew to the size of a marmot, and their structural diversity in the Late Cretaceous reflects radiations during the 70 million years of Cretaceous time. Their skulls lacked an auditory bulla and were typically long, with a narrow braincase and long, narrow snout. The incisors varied in number from 5/4 to 3/3. Early Cretaceous eutherians had five premolars, whereas Late Cretaceous eutherians had four. There were three molars which were typically tribosphenic (bearing three major cusps) with high, sharp cusps (Fig. 6–1). The hands and feet of several genera are known, and these lack opposable first digits, indicating that Cretaceous eutherians were not basically arboreal.

Late Cretaceous eutherians probably played a variety of ecological roles. The diverse dentitions were seemingly adapted to insectivory, carnivory, and herbivory, and a reasonable speculation would be that Cretaceous eutherians

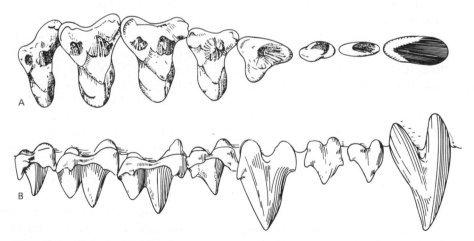

Figure 6–1. (A) Occlusal and (B) labial view (view from the inside) of the right upper canine and cheek teeth of *Kennalestes gobiensis;* the dentition of this eutherian is the most primitive of any yet known. (From Kielan-Jaworowska, 1969)

Table 6-1:   Comparison of Marsupials and Eutherians

Metatheria	Eutheria
Braincase relatively small	Braincase relatively large
Posterior palatal vacuities usually present	Posterior palatal vacuities developed in only a few forms
Usually, a poorly developed auditory bulla derived from alisphenoid	Auditory bulla commonly well developed and of various origins
Angle of lower jaw usually inflected medially	Angle of lower jaw usually not inflected medially
Dental formula derived from 5/4, 1/1, 3/3, 4/4	Dental formula derived from 3/3, 1/1, 4/4, 3/3
Essentially monophyodont teeth; only third premolar replaced	Diphyodont teeth; replacement of most antemolar teeth
Epipubic bones in both sexes of most forms	No epipubic bones
None possess baculum or os clitoridis	Baculum or os clitoridis commonly present
Cerebral hemispheres usually small	Cerebral hemispheres tend to be large
Young carried externally on nipples and commonly enclosed in pouch	Young in uterus or nest; rarely carried on nipples and never enclosed in true pouch
No true embryonic trophoblast	Embryonic trophoblast present, as distinguished from the inner cell mass
Erosion of uterine wall by extraembryonic membranes of developing young uncommon	Erosion common
Choriovitelline placenta in most forms; chorioallantoic placenta in only a few genera	"Chorioallantoic" placenta largely replaces choriovitelline placenta
Altricial young	Precocial young common

From Kielan-Jaworowska, Bown, and Lillegraven, 1979.

functioned variously as climbers, jumpers, diggers, and locomotor generalists. Interpretation of fossil postcranial material is difficult, for we know that some living mammals (squirrels, *Sciurus* sp., for example) are jacks of several locomotor and dietary trades.

Although sparse, the fossil record documents a latest Cretaceous establishment of the evolutionary lines leading to the orders Insectivora, Carnivora, Perissodactyla, Artiodactyla, and Primates, and suggests a very early Cenozoic differentiation of the orders Dermoptera, Chiroptera, Rodentia, and Cetacea. The tracing back of many eutherian evolutionary lines through the Early Cenozoic and Late Mesozoic awaits a more complete fossil record.

The morphological and physiological features listed in Table 6–1 distinguish the marsupials (infraclass Metatheria) and the eutherians (infraclass Eutheria) and serve as an excellent diagnosis of these two major groups of living mammals.

The accounts of eutherian mammals in the following chapters include brief comments on when orders and families appeared and, in some cases, sketches of the evolutionary histories of taxa. The student should regard such comments as summaries of our current knowledge but should bear in mind that the fossil record is sadly incomplete for some groups. Evolutionary lines that have been restricted to tropical areas, small and delicate animals, and arboreal types had relatively little chance of leaving a fossil record. The first appearance of the Chiroptera (bats) in the early Eocene, for example, tells us only that by this time all the basic chiropteran adaptations had been perfected and that bats were part of the North American fauna. The history of these small, fragile, basically tropical creatures must extend far back, however, perhaps into latest Cretaceous times. We simply lack documentation of this history and doubtless of the early histories of many other groups as well.

Chapter 7

Insectivores and Their Relatives

This chapter deals with the order Insectivora, which includes the hedgehogs, moles, tenrecs, golden moles, solenodonts, and shrews, along with three seemingly closely related groups, the elephant shrews (Macroscelidea), tree shrews (Scandentia), and flying "lemurs" (Dermoptera). Whereas the order Insectivora is an important group today, occupying a wide diversity of habitats in both the Old and New Worlds, the elephant shrews, tree shrews, and dermopterans are restricted geographically and the dermopterans, at least, are probably a relict group and are limited to tropical refugia.

Because the most primitive eutherian mammals were insectivorous and because their descendants have often retained dentitions that remain adapted to an insect diet (although frequently highly specialized), the tendency has been to include some primitive types together with all modern descendants in the order Insectivora. This order has thus long been used as a convenient catchall repository for taxa of doubtful affinities. As an example, the tree shrews and elephant shrews have often been in-cluded in the order, although they strongly depart structurally and behaviorally from the taxa grouped here in Insectivora. Taxonomic assignment of various equivocal fossil types has also been difficult and controversial. Members of the Cretaceous Asiatic family Deltatheriidae (Fig. 7–1A), for example, have been regarded by some workers as insectivores important in the evolution of such groups as the Carnivora; on the basis of recent evidence, however, these animals appear to be members of a sterile evolutionary line, separate from both marsupials and eutherians, that died out in the Late Cretaceous (Lillegraven, 1974:265). The problem of the classification of "Insectivora" and their relatives remains complex, and the system used here is provisional at best.

An insectivorous style of life was common to many Cretaceous eutherians, and, although their phylogenetic relationships are still a source of controversy, the three major groups of Cretaceous mammals (Lepictidae, Palaeoryctidae, and Zalambdalestidae) have often been put in the order Insectivora. In any case, the lin-

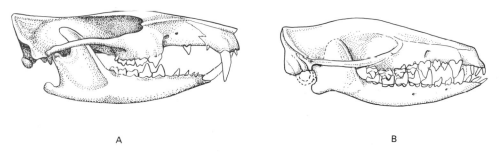

A   B

Figure 7-1. Skulls of insectivore-like mammals from the Cretaceous. (A) *Zalambdalestes* (length of skull 50 mm; after Romer, 1966); (B) *Kennalestes* (length of skull 27 mm; after Kielan-Jaworowska, 1975).

eages we are discussing are ancient: the roots of some of the modern families included here in the order Insectivora probably reach back into Late Cretaceous or Early Cenozoic times.

Morphology of the Insectivora

Although considerable morphological diversity is represented among the mammals grouped in the order Insectivora, for all Recent members the following characters are reasonably diagnostic: the tympanic bone is annular, no auditory bulla is present, and the entotympanic bone is absent; the tympanic cavity is often partially covered by processes from adjacent bones; the olfactory bulbs are longer than the rest of the brain and are largely interorbital; the eyes and optic foramina are usually small; the jugal is reduced or absent, and the zygomatic arch is incomplete in some groups; the orbitosphenoid bone is mainly anterior to the braincase; the teeth have sharp cusps, and usually the crown pattern of primitive placentals is recognizable; the anterior dentition is often modified by the enlargement and specialization of the incisors and the reduction of the canines; the limbs (except those of the fossorial groups) are usually unspecialized and are never adapted to saltation.

Venomous Insectivores

Except for the duck-billed platypus (order Monotremata), the only mammals known to be venomous are shrews. Over 350 years ago, there were reports on the symptoms that developed when humans were bitten by the short-tailed shrew of North America *(Blarina brevicauda),* and work in the present century has confirmed that this animal has venomous saliva. It has also been demonstrated that the European water shrew *(Neomys fodiens)* and the Haitian solenodon *(Solenodon paradoxus)* are venomous. The salivas of other insectivores closely related to these two shrews have been studied but are apparently not toxic. In some people, the bites

of *Suncus murinus* cause minor aches and hypersensitivity and reddening of the skin, especially at the finger joints (G. L. Dryden, personal communication).

Both *N. fodiens* and *B. brevicauda* have similar adaptions for delivering venom, and the effects of the venoms are similar. (Venomous insectivores and their venoms are discussed by Pournelle, 1968, and Pucek, 1968.) In these shrews, the first lower incisors have concave medial surfaces, forming a crude channel, and the ducts from the venom-producing submaxillary salivary glands open near the base of these teeth. *Neomys fodiens* salivates copiously during attacks on prey, and saliva is seemingly channeled to wounds via the two first lower incisors. Pearson (1942) showed that mice injected with extracts of the submaxillary glands of this shrew were strongly affected; the activity of the mice was reduced rapidly by what seemed to be a neurotoxic action (a neurotoxin impairs normal function of the nervous system). Frogs bitten by *N. fodiens* were partially immobilized and, when forced to move, were uncoordinated. Laboratory mice injected with a homogenate of these salivary glands immediately developed paralysis of the hind limbs.

The toxic saliva of *Solenodon* is produced by the submaxillary salivary glands, is carried by ducts to the base of the large and deeply channeled posterior surfaces of the second lower incisors, and presumably enters a wound by capillary action. The saliva of this insectivore is similar in effect to the saliva of the venomous shrews.

Perhaps of greatest interest is the functional importance of venom to insectivores. *Blarina brevicauda* can kill mice considerably larger than itself, and Eadie (1952) reported that meadow voles *(Microtus pennsylvanicus)* were an important fall and winter food of this shrew. Frogs and small fish are known to be preferred foods of *N. fodiens*. Both of these shrews attack prey from behind and direct bites at the neck and base of the skull, an area where neurotoxic venom might be readily introduced into the central nervous system. The adaptive importance to a very small predator of making its relatively large prey helpless would seem to be great, and one wonders why more shrews are not venomous.

Order Insectivora

Family Erinaceidae. Members of the family Erinaceidae, the hedgehogs, are morphologically primitive but remain successful even in areas highly modified by humans. The family is represented today by eight genera and 16 species; they occur in Africa, Eurasia, southeastern Asia, and the island of Borneo. Erinaceids are first known from the Oligocene, and fossil material is known from the Oligocene to the Pliocene in North America, and from the Oligocene to the Recent in the Old World. The family Adapisoricidae, containing primitive relatives of the hedgehogs, is represented from the Cretaceous to the Oligocene in both hemispheres.

Erinaceids vary from the size of a mouse to the size of a small rabbit (1.4 kilograms). The eyes and pinnae are moderately large, and the snout is usually long. The zygomatic arches are complete. The dental formula is 2-3/3, 1/1, 3-4/2-4, 3/3 = 36-44. The first upper and, in some species, the first lower incisors are enlarged, but the front teeth (Fig. 7–2) never reach the degree of specialization typical of shrews. In hedgehogs, the upper molars have simple nonsectorial cusps, with the paracone

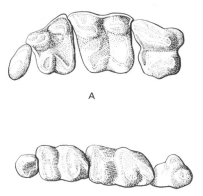

Figure 7-3. Cheek teeth of a hedgehog (*Erinaceus* sp., Erinaceidae). Fourth premolar and three molars of the (A) upper right and (B) lower left tooth rows.

and metacone near the outer edge; the hypocone completes the quadrate form of the upper tooth (Fig. 7–3A). Both the trigonid and talonid of the lower molars are well developed (Fig. 7–3B). The molars are thus better adapted to an omnivorous than to an insectivorous diet. The feet retain five digits in all but one genus, and the foot posture is plantigrade. An obvious specialization is the possession of spines in members of the subfamily Erinaceinae (Fig. 7–4). In

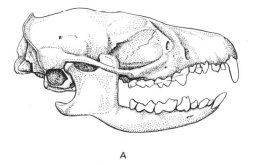

Figure 7-2. Skull of an insectivore, the hedgehog (*Erinaceus* sp., Erinaceidae; length of skull 32 mm).

Figure 7-4. European hedgehog *(Erinaceus europaeus)* with young. (G. L. Dryden)

these animals, the sheet of muscle beneath the skin (panniculus carnosus) is greatly enlarged and controls the pulling of the skin around the body and the erection of the spines.

In various parts of their wide range, hedgehogs occupy deciduous woodlands, cultivated lands, and tropical and desert areas. They are onmivorous, but animal food seems to be preferred and a wide variety of invertebrates are eaten.

Kingdon (1974a:32) reports that hedgehogs attack and kill small snakes and, during the attack, direct their spines forward, leaving only a small part of their body exposed to the strikes of the snake. Hedgehogs seem remarkably resistant to snake venom. Some members of this family protect themselves by rolling into a tight ball with the spines erect.

Members of the subfamily Erinaceinae are probably heterothermic. (Heterothermic animals can regulate their body temperature physiologically, but temperature is not regulated precisely or at the same level at all times.) Hibernation occurs in the widespread genus *Erinaceus,* and estivation is practiced by the desert species *Parachinus aethiopicus.* A related species from India, *P. micropus,* has survived in captivity for periods of from four to six weeks without food or water (Walker, 1968:133), and this species and *Hemiechinus auritus* are known to have winter periods of dormancy in India. In Kenya, *Erinaceus albiventris* disappears and apparently hibernates through the long dry season from May to September or October. In this instance, the animals are probably responding primarily to food shortages, for temperatures remain moderate through this period.

Family Talpidae. This family includes a group of small rat- or mouse-size animals usually referred to as moles. These predominantly burrowing insectivores (17 genera and 31 species) occur in parts of North America, Europe, and Asia. The European fossil record of talpids begins in the late Eocene; talpids are known first in the New World from the Oligocene. Apparently, the anatomical modifications typical of

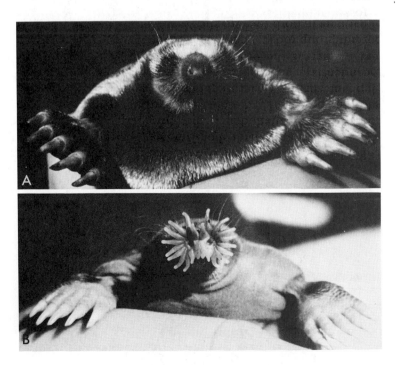

Figure 7-5. Heads and forelimbs of two moles (Talpidae). (A) Hairy-tailed mole *(Parascalops breweri).* (B) Star-nosed mole *(Condylura cristata);* the unique "star" of finger-like structures is found in no other mammal. (G. L. Dryden)

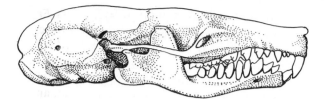

Figure 7-6. Skull of the eastern mole (*Scalopus aquaticus,* Talpidae; length of skull 37 mm).

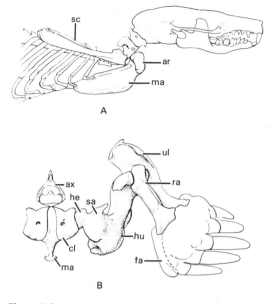

Figure 7-8. Details of the pectoral girdle and forelimb of the eastern male (*Scalopus aquaticus,* Talpidae). (A) Side view of the pectoral girdle. (B) Anterior view of part of the pectoral girdle and the forelimb, with the shoulder joint slightly disarticulated to show the head of the humerus and its secondary articular surface. The head of the humerus articulates with the glenoid fossa of the scapula, and a second articulation (involving considerably larger surfaces) occurs between the secondary articular surface of the humerus and an articular surface of the clavicle. Abbreviations: *ar,* articular surface of the clavicle; *ax,* axis; *cl,* clavicle; *fa,* position of the falciform bone; *he,* head of the humerus; *hu,* humerus; *ma,* manubrium of the sternum; *ra,* radius; *sa,* secondary articular surface of the humerus; *sc,* scapula; *ul,* ulna.

Recent fossorial genera were attained early, for the Recent European genus *Talpa* is first known from the Miocene.

The head and forelimbs of most talpids are modified for fossorial life (Fig. 7–5). The zygomatic arch is complete, the tympanic cavity is not fully enclosed by bone, and the eyes are small and often lie beneath the skin. The snout is long and slender, the ears usually lack pinnae, and the fur is characteristically lustrous and velvety. The dental formula is 2-3/1-3, 1/0-1, 3-4/3-4, 3/3 = 32-44. The first upper incisors are inclined backward (Fig. 7–6), and the upper molars have W-shaped ectolophs (Fig. 7–7). In the fossorial species, the forelimbs are more or less rotated from the usual orientation typical of terrestrial mammals in such a way that the digits point to the side, the palms face backward, and the elbows point upward (Fig. 7–8). In addition, the phalanges are short, the claws are long,

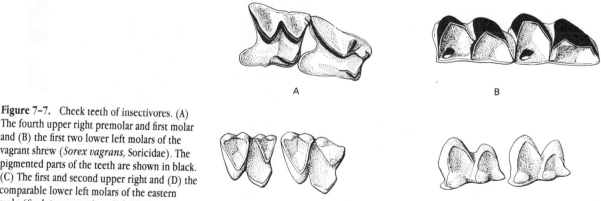

Figure 7-7. Cheek teeth of insectivores. (A) The fourth upper right premolar and first molar and (B) the first two lower left molars of the vagrant shrew (*Sorex vagrans,* Soricidae). The pigmented parts of the teeth are shown in black. (C) The first and second upper right and (D) the comparable lower left molars of the eastern mole (*Scalopus aquaticus,* Talpidae).

and the clavicle and humerus are unusually short and robust. The scapula is long and slender and serves both to anchor the forelimb solidly against the axial skeleton and to provide advantageous attachments for some of the powerful muscles that pull the forelimb backward. The anteriormost segment of the sternum (the manubrium) is greatly enlarged and extends forward to beneath the base of the skull. These specializations both increase the area for attachment of the large pectoralis muscles and move the shoulder joint forward and allow the forepaws to remove or loosen soil beside the snout.

The clavicle is short and broad and provides a large secondary articular surface for the humerus. The double articulation of the shoulder joint, with articular contacts between the humerus and the scapula and clavicle, provides an unusually strong bracing for this joint during the powerful rotation of the humerus that accompanies the digging stroke of the forelimb. In some genera, the falciform bone is large and increases the breadth of the forepaw and braces the first digit (Fig. 7–8).

One talpid, the Asiatic shrew mole *(Uropsilus)*, lacks fossorial or aquatic specializations and resembles a shrew in general form.

The remarkable subfamily Desmaninae inhabits the Pyrenees Mountains and some mountains in Portugal, southeastern Europe, and parts of Russia. These animals are adapted to semiaquatic life. They have webbed forefeet, and the greatly enlarged hind feet are webbed and bear a fringe of stiff hairs that increase their effectiveness as paddles. These animals also have flexible snouts that have an extremely highly developed sense of touch and smell and enable the animals to detect underwater prey.

Fossorial talpids occur typically in moist and friable soils in forested, meadow, or streamside areas and feed largely on animal material. A species that occurs in the eastern United States *(Scalopus aquaticus),* however, locally penetrates the moderately dry sandhill prairies of eastern Colorado, where the characteristic ridges of soil made by the animals appear only during wet weather. In most areas, these ridges are a common evidence of the presence of moles and are made by the animals as they travel just beneath the surface by forcing their way through the soil. Soil from deep burrows is deposited on the surface in more or less conical mole hills. The semiaquatic Old World desmanines live along the banks of lakes, ponds, or streams and feed largely on aquatic inverte-

Figure 7-9. A tenrec (*Tenrec ecaudatus,* Tenrecidae). (J. F. Eisenberg and E. Gould)

Figure 7-10. A Madagascar "hedgehog" (*Echinops telfairi*, Tenrecidae). (J. F. Eisenberg and E. Gould)

brates. Their burrows open beneath the surface of the water and extend upward to a nest chamber above the water level. Moles are not known to hibernate or estivate.

Family Tenrecidae. The tenrecs are a distinctive group of primitive insectivores that vary widely in structure and in habits. Various tenrecs bear a general resemblance to such diverse mammals as shrews, hedgehogs, muskrats, mice, and otters. The family includes 12 genera and 34 species and inhabits Madagascar, the Comoro Islands, and western central Africa. The meager fossil record of tenrecs indicates little about their evolution. The oldest fossil records are the Miocene record of East Africa, and Pleistocene fossils are known from Madagascar.

Probably the ancestral stock of the tenrecs on Madagascar dispersed there from Africa in the Eocene Epoch.

Tenrecs vary from roughly the size of a shrew to the size of a cottontail rabbit. The snout is frequently long and slender. The jugal bone is absent, the eye is usually small, and the pinnae are conspicuous. The tympanic bone is annular, and the squamosal bone forms part of the roof of the tympanic cavity. The anterior dentition varies from species to species. The first upper premolars are never present, and the molars are 3/3 in all but *Tenrec* (4/3) and *Echinops* (2/2). The upper molars have crowns that are triangular in occlusal view, and only in one genus *(Potamogale)* is a W-shaped ectoloph present in these teeth.

An unusually broad array of adaptive types occurs within the *Tenrecidae.* *Tenrec* roughly resembles a tail-less, coarse-pelaged, long-snouted opossum (Fig. 7–9) and has spines interspersed with soft hairs. It is omnivorous. *Echinops* (Fig. 7–10), *Hemicentetes* (Fig. 7–11), and *Setifer* are also spiny, and the latter two genera resemble hedgehogs (Erinaceidae) closely. In these two genera, the panniculus carnosus muscle is powerfully developed and enables the animals to erect the spines. It also contributes to the ability of these animals to roll into a ball. The feet and head are tucked beneath the body during this protective movement, and "the sphincter muscles running

Figure 7-11. A streaked tenrec (*Hemicentetes semispinosus*, Tenrecidae). (J. F. Eisenberg and E. Gould)

around the body at the junction of the spiny dorsum and the hairy venter permit the spiny dorsal skin to be drawn together, thus enclosing the animal in an impregnable shield of spines" (E. Gould and Eisenberg, 1966). These authors found that newborn *Echinops* and *Setifer* reacted to being disturbed by rolling into a ball. *Hemicentetes* has a group of 14 to 16 specialized quills on the middle of the back, the "stridulating organ," that rub together when underlying dermal muscles are twitched to produce sounds in a variety of repetitive patterns. Differences in these sounds depend on differences in associated behavior of the animals (E. Gould, 1965) and may be used in intraspecific communication.

The subfamily Potamogalinae, the otter shrews, includes animals that in many ways are the most remarkable members of the Tenrecidae. Members of this relict group live in western central Africa; they are the only living members of a primitive lineage and have probably survived because of their highly specialized, semiaquatic style of life. Although the giant otter shrew *(Potamogale velox)* has been known to scientists since 1860, the genus to which the dwarf otter shrews belong *(Micropotamogale)* was not described until 1954 (Heim de Balsac, 1954).

The giant otter shrew is quite large for an insectivore, measuring 600 millimeters in length and weighing about 1 kilogram, and is highly specialized for the life of a miniature otter. The body is long and streamlined, the limbs are rather short and stocky, and the large tail is laterally compressed. Propulsion beneath the water is controlled by lateral movements of the flattened tail, and a number of features are associated with this locomotor style. The caudal vertebrae have high neural spines and transverse processes, and most have chevron bones. These unusual caudal vertebrae provide attachment points for the powerful tail musculature, which is aided by the greatly enlarged gluteal muscles. The posterior parts of the gluteal muscles, which in quadrupedal mammals move the

hind limbs, attach to the muscles overlying roughly the first five caudal vertebrae and move the tail. In the sinuous motion of the back and tail, and even in overall body form, *Potamogale* resembles a large salamander. Giant otter shrews live in permanent streams and rivers and in coastal swamps, and although they rely partly on fish, they seem to prefer fresh-water crabs to other food (Kingdon, 1974a:17). The habits of potamogales remain poorly known and provide fascinating opportunities for the resourceful biologist.

Some tenrecs are known to become torpid under natural conditions during seasons of food shortage (J. F. Eisenberg and Gould, 1970) or in the laboratory for unknown reasons (G. L. Dryden, personal communication).

Family Chrysochloridae.   Another type of fossorial insectivore is typified by chrysochlorids, the golden moles. These animals resemble "true" moles (Talpidae) but even more closely resemble, in fossorial adaptations and in function, the marsupial "moles" (Notoryctidae). The five genera and roughly 17 species constituting the family Chrysochloridae occur widely in southern Africa, where they occupy forested areas, savannas, and sandy deserts. The earliest fossil chrysochlorids from the Miocene of East Africa resemble Recent species, and these and Pleistocene fossil material give no firm evidence of the derivation of the group. Butler (1969) suggests that the Tenrecidae and the Chrysochloridae may be related; and J. F. Eisenberg (1981) raises each to ordinal rank.

Golden moles have modes of life similar to those of the fossorial members of the Talpidae and possess some parallel adaptations as well as some contrasting structural features. The ears of golden moles lack pinnae, and the small eyes are covered with skin. The pointed snout has a leathery pad at its tip (Fig. 7–12A). The zygomatic arches are formed by elongate processes of the maxilla, and the occipital area includes bones, the tabulars, not typically found in mammals. The skull is abruptly conical instead of being flattened and elongate as it is in many in-

Figure 7–12. The cape golden mole (*Chrysochloris asiatica;* Chrysochloridae). (A) Note the leathery nose pad and the greatly enlarged claws of digits two and three. (B) Note the enormously enlarged malleus. This specialization may aid in conduction of sound through the bones of the skull by increasing the inertia of the malleus and may improve hearing under ground. The "third bone" of the forelimb replaces the flexor muscle of the third digit. (A by J. U. M. Jarvis; B by E. N. Keen)

sectivores. An auditory bulla is present and is formed largely by the tympanic bones; the malleus is enormously enlarged (Fig. 7–12B). The dental formula is usually 3/3, 1/1, 3/3, 3/3 = 40. The first upper incisor is enlarged, and the molars are basically tritubercular and lack the stylar cusps and W-shaped ectoloph typical of talpids. The permanent dentition of golden moles emerges fairly late in life. The forelimbs are powerfully built, and the forearm rests against a concavity in the rib cage. The fifth digit of the hand is absent, and digits two and three usually have huge picklike claws. A "third bone" is present in the forelimb (Fig. 7–12B). The forelimbs are not rotated as are those of talpids, but more or less retain the usual mammalian posture, with the palmar surfaces downward.

Golden moles are adept burrowers. J. A. Bateman (1959) studied golden moles in the laboratory and found that a 60-gram golden mole could push up a 9-kilogram weight covering its cage; this amounts to exerting a force equal to 150 times the animal's weight. During digging, the wedge-shaped head butts upward as the claws of the powerful forelimbs sweep downward and backward (Kingdon, 1974a:24, 25), and when the animal is close to the surface a ridge marks the course of its progress. Both deep and shallow burrows are constructed; the depth of the burrows may depend on the amount of soil moisture. The roofs of shallow burrows in sandy soil frequently collapse, leaving a furrow in the sand as a trace of the former burrow. The diet of golden moles consists mostly of invertebrates; two desert-dwelling genera (*Cryptochloris* and *Eremitalpa*) also eat legless lizards. In sandy deserts, *Eremitalpa* occasionally forages for insects blown by the wind into furrows in the sand (F. C. Eloff, 1967).

Family Solenodontidae. Represented today by but one genus and two species, the solenodons are relict types that are unable to survive in competition with other placentals recently introduced into their ranges. Solenodons occurred in sub-Recent and Recent times in Cuba,

Haiti, and Puerto Rico but are now restricted to Haiti (*Solenodon paradoxus*) and to Cuba, where a declining and endangered population of *Solenodon cubanus* occurs. The extinct genus *Nesophontes* occupied the West Indies at least until the arrival of the Spaniards. The introduction by humans of the house rat (*Rattus*), the mongoose (*Herpestes*), and dogs and cats into the West Indies and the extensive clearing of land for agriculture combined to cause the rapid decline of the solenodons. These animals are now rare in most areas, and hopes for their survival under natural conditions are dim.

Solenodons are roughly the size of a muskrat and have the form of an unusually large and big-footed shrew. The five-toed feet and the moderately long tail are nearly hairless. The snout is long and slender, the eyes are small, and the pinnae are prominent. The zygomatic arch is incomplete, no auditory bulla is present, and the dorsal profile of the skull is nearly flat. The dentition is 3/3, 1/1, 3/3, 3/3 = 40. The first upper incisor is greatly enlarged and points backward slightly; the second lower incisor has a deep lingual groove that may function to transport the toxic saliva that empties from a duct at the base of this tooth. The upper molars lack a W-shaped ectoloph and are basically tritubercular. A sharp and bladelike (trenchant) ridge is formed by a high crest at the outer edge of each molar.

Solenodons are generalized omnivorous feeders that prefer animal material. They often find food by rooting with their snouts or by uncovering animals with their large claws. Being rather archaic creatures that seem to have little competitive ability, their distribution on islands is probably the key to their continued, if tenuous, survival.

Family Soricidae. Members of this family, the shrews, are among the smallest and least conspicuous of mammals. In many areas, they are the most numerous insectivores; they have the widest distribution of any of the Insectivora and are the most familiar. The family Soricidae is represented today by 20 genera and some 290 species and occurs throughout the world except in the Australian area, most of South America, and the polar areas. Soricids appear in the Oligocene in both Europe and North America. Because soricids are rare as fossils, their early evolution is obscure; they may have evolved from an early erinaceid ancestral stock (M. R. Dawson, 1967a:20).

Shrews are small: the smallest weighs 2.5 grams (the smallest living mammal), and the largest weighs roughly 180 grams, the weight of a rat. The snout is long and slim, the eyes are small, and the pinnae are usually visible. The feet are five-toed and, except for fringes of stiff hairs on the digits in semiaquatic species and enlarged claws in semifossorial forms, unspecialized. The foot posture is plantigrade or semiplantigrade. The narrow and elongate skull usually has a flat dorsal profile (Fig. 7–13B); there is no zygomatic arch or tympanic bulla, and the tympanic bone is annular (Fig. 7–13A). The specialized dentition consists of from 26 to 32 teeth; the dental formula of *Sorex* is 3/1, 1/1, 3/1, 3/3 = 32. In the subfamily Soricinae, the teeth are pigmented; the first upper incisor is large and hooked and bears a notch and projection resembling those on the upper mandible of a falcon (Fig. 7–13B). Behind the first upper incisor is a series of small unicuspid teeth (presumably incisors, a canine, and premolars); the fourth upper premolar is large and has a trenchant ridge; and the upper molars have W-shaped ectolophs (Fig. 7–7A). Both the trigonid and talonid of the lower molars are well developed (Fig. 7–7B), and the first lower incisor is greatly enlarged and procumbent (leaning forward).

Because they are unusually small, shrews can exploit a unique mode of foraging. Many shrews patrol for insects beneath logs, fallen leaves, and other plant debris and in the narrow spaces and crevices beneath rocks. Rodent surface runways and burrows may also be used as feeding routes. Because of their style of foraging, shrews are seldom observed, even in areas where they are common. Although the shrew is typically associated with moist conditions, some species, such as the gray shrew (*Notiosorex crawfordi*) of the southwestern United States and the piebald shrew (*Diplomesodon pulchellum*) of southern Russia, inhabit desert areas. Aquatic adaptations in some species allow them to dive and swim and feed mainly on aquatic invertebrates. One of the most aquatic species is the Tibetan water shrew (*Nectogale*

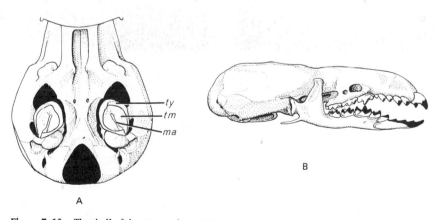

Figure 7–13. The skull of the vagrant shrew (*Sorex vagrans,* Soricidae; length of skull 17 mm). (A) Ventral view of the basicranial region, showing the annular (ringlike) tympanic bone (*ty*), the tympanic membrane (*tm*), and the malleus (*ma*). (B) Side view, showing the pincer-like anterior incisors. The pigmented parts of the teeth are shown in black.

elegans), which inhabits mountain streams and feeds primarily on fish. In this species, the streamlined shape is enhanced by the strong reduction of the pinnae, and the digits and feet have fringes of stiff hairs that greatly increase their effectiveness as paddles. The distal part of the tail is laterally compressed, and the edges bear lines of stiff hairs.

Innate following behavior, unique to some shrews of the subfamily Crocidurinae, results in the formation of "caravans." When a female and her litter are moving, the first young grabs her tail with its mouth, the next young takes the first young's tail in its mouth, and so on, forming a chain of young that under some conditions is dragged by the mother.

Order Macroscelidea: Elephant Shrews

The relationships between the elephant shrews and other groups of mammals are unclear, due in part to a poor fossil record. Elephant shrews have often been regarded as members of the het-

Figure 7-14. Two species of elephant shrews (Macroscelididae): (A) spectacled elephant shrew *(Elephantulus rufescens);* (B) four-toed elephant shrew *(Petrodromus tetradactylus).* (G. Rathbun)

erogeneous and inclusive order Insectivora but differ from the hedgehogs, shrews, moles, and tenrecs in having complete auditory bullae, entotympanic bones, a large jugal bone and complete zygoma, and relatively small olfactory lobes. In an effort to stress their uniqueness, I treat the elephant shrews as a separate order which comprises a single family (Macroscelididae).

The elephant shrews are distinctive African mammals specialized for rapid bounding. Their present distribution includes Morocco, Algeria, and the part of Africa south of the Red Sea. There are four genera and 19 living species. Fossils are known only from Africa, where an extinct species of the still-living *Rhynchocyon* is known from the Miocene and an extinct genus is present in the Pliocene.

The size of elephant shrews varies, with some being as small as a mouse and others as large as a large rat. They have large eyes and ears and remarkably long snouts (Figs. 7–14 and 7–15). The limbs are long and slender, and the tail is moderately long. The forelimbs and hind limbs have four or five digits; in *Rhynchocyon* the forefeet are functionally tridactyl, with the first digit absent and the fifth much reduced,

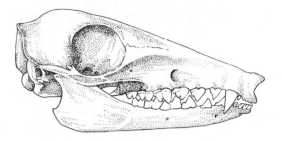

Figure 7-16. Skull of golden-rumped elephant shrew (*Rhynchocyon chrysopygus;* length of skull 67 mm).

whereas the hind feet have four digits. This more extreme loss of digits in the forelimb than in the hind limb is most unusual in mammals. The very large orbits of the skull are never bordered by a complete postorbital bar (Fig. 7–16). The dental formula is 1-3/3, 1/1, 4/4, 2/2-3 = 36-42; the last upper premolar is the largest molariform tooth, and the upper canine is double-rooted. The upper molars are quadrate and have four major cusps.

A variety of habitats are occupied by elephant shrews, including open plains, savannas, thornbush, and tropical forests. *Rhynchocyon,* the giant elephant shrew, is a forest dweller and in some places, such as along the coast of Kenya, occupies relict strips of evergreen forest. *Rhynchocyon chrysopygus* (Fig. 7–15) has been studied by Rathbun (1979) in the coastal forests of Kenya, where these diurnal animals feed on a wide variety of invertebrates, many of which they dig from the leaf litter with the long claws of their front feet. This species is strikingly colored: a dark chestnut brown, with almost a purplish cast, covers the back and flanks, against which the bright yellow rump patch stands out in sharp contrast. Rathbun has found that this species typically lives in pairs occupying stable territories and that scent marking by both sexes is an important territorial behavior. A resident pair will chase conspecific intruders, with males chasing males and females chasing females. Rathbun postulates that the conspicu-

Figure 7-15. Golden-rumped elephant shrew *(Rhynchocyon chrysopygus).* (G. Rathbun)

ous yellow rump functions as a target during aggressive encounters. He has found that the skin beneath this patch is thicker than skin over the rest of the body and that scars and cuts are concentrated beneath the patch.

Studies by Rathbun (1976, 1979) on the diurnal spectacled elephant shrew (Fig. 7–14A) have shown that the territories of this brush dweller contain intricate patterns of trails and that 24 percent of the daylight behavior of a territorial male is devoted to cleaning these trails. The front feet are used to meticulously sweep away leaf and twig litter. This species spends its life above ground, never seeking refuge in burrows, and usually forages within 1 meter of trails. These elephant shrews escape predators by bounding along the trails at amazing speed. Scent marking by feces and sternal glands is concentrated on territorial borders. Rathbun found that the bulk of the diet is termites and ants.

Figure 7–17. A tree shrew (*Tupaia longipes,* Tupaiidae). (M. W. Sorenson)

Order Scandentia: Tree Shrews

This order is represented by a single family, Tupaiidae. Members of this family resemble small, long-snouted squirrels (Fig. 7–17) and occur from India through Burma to the islands of Sumatra, Borneo, and the Philippines. The order is represented entirely by its Recent members (five genera and 16 species).

The dental formula is 2/3, 1/1, 3/3, 3/3 = 38; the upper incisors are caniniform, and the upper canine is reduced. The upper molars have trenchant, W-shaped ectolophs (Fig. 7–18A), and the lower molars retain the basic insectivore pattern (Fig. 7–18B). There is no loss of digits, and all digits have strongly recurved claws. The long tail is usually heavily furred. Tree shrews have well-developed postorbital processes that join the zygoma (Fig. 7–19).

Tree shrews occupy deciduous forests and forage both in the trees and on the ground. They are oppotunistic feeders and utilize a variety of foods, but animal material and fruit are preferred. The feather-tailed tree shrew *(Ptilocercus lowei)* of the Malay Peninsula is arboreal and nocturnal, but other tree shrews are diurnal. Tree shrews are characteristically highly vocal. The mountain tree shrew lives in social groups in which a rigid dominance hierarchy is apparent, whereas in parts of Borneo other species occupying lowland areas do not form social groups (Sorenson and Conaway, 1968).

Order Dermoptera

Members of this order are generally called flying lemurs because of their lemur-like faces and their ability to glide between trees (Fig. 7–20A). One family, Cynocephalidae, with but one genus *(Cynocephalus)* and two species, represents the order. The distribution includes tropical forests from southern Burma and southern Indochina, Malaya, Sumatra, Java, Borneo,

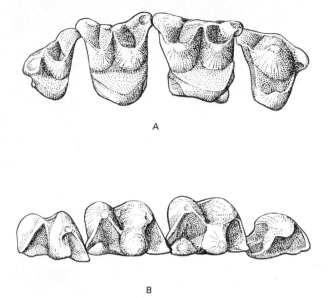

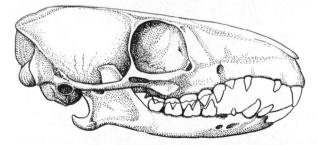

Figure 7–18. Cheek teeth of a tree shrew (*Tupaia* sp., Tupaiidae). Fourth premolar and molars of (A) upper right and (B) lower left tooth rows.

and nearby islands to southern Mindanao and some of the other southern islands of the Philippine group. An extinct dermopteran family (Plagiomenidae) appears in the late Paleocene and early Eocene records of North America, but the family Cynocephalidae is known only from the Recent.

Members of the family Cynocephalidae are of modest size (roughly 1 to 1.75 kilograms) and have large eyes and faces that resemble those of Old World fruit bats or some prosimian

Figure 7–19. Skull of a three shrew (*Tupaia* sp., Tupaiidae; length of skull 52 mm).

primates (Fig. 7–20B). The brownish, chestnut, or gray pelage is irregularly blotched with white. Although the molars have retained a basically three-cusped insectivore pattern, the broad cheek teeth have a shearing action that includes a large transverse component. This action, and the highly crenulated enamel of the molars, provide for efficient mastication of plant material (Rose and Simons, 1977). The anterior dentition is specialized: the lateral upper incisor is caniniform, and the first two lower incisors are broad and pectinate (comblike). The unusual lower incisors are used to groom the fur but may also be used to scrape leaves when the animal feeds. The dental formula is 2/3, 2/2, 2/2, 3/3 = 38.

A broad, furred membrane extends from the neck to near the ends of the fingers, between the limbs, and from the hind foot to the tip of the tail (Fig. 7–20B). The hands and feet retain five digits which bear needle-sharp, curved claws for clutching branches. As in bats, the neural spines of the thoracic vertebrae are short, the sternum is keeled, the ribs are broad, the radius is long, and the distal part of the ulna

Figure 7-20. (A) Flying lemur *(Cynocephalus volans)* gliding between trees in a tropical forest in Mindanao (Philippine Islands). (B) Flying lemur with a young animal clinging with needle-sharp claws to the bare skin where the gliding membrane joins the body. (C. H. Wharton, © 1948 National Geographic Society)

is strongly reduced. The great lengthening of the intestine typical of herbivorous mammals is well illustrated by flying lemurs. *Cynocephalus,* which has a head plus body length of only about 410 millimeters, has an intestinal tract approaching 4 meters in length, nine times the head-and-body length (Wharton, 1950:272). The cecum, a blind diverticulum at the proximal end of the colon, is greatly enlarged (to about 48 centimeters in length) and is divided into compartments. This chamber harbors microorganisms that help break down cellulose and other relatively indigestible carbohydrates. Cecal enlargement is usually associated with an herbivorous diet (as in many rodents).

Flying lemurs are nocturnal and are slow but skillful climbers, but they are nearly helpless on the ground. They can glide distances well over 100 meters in traveling to and from feeding places.

The diet includes leaves, buds, flowers, and fruit. Winge (1941:145) reports that the enlarged tongue and specialized lower incisors are used in cowlike fashion in picking leaves. According to Wharton (1950), flying lemurs seek refuge during the day in holes in trees, and several individuals may occupy the same den. He reports that these animals invariably remain upside down while traveling along branches and feeding. The distribution of flying lemurs is being restricted in some areas by the clearing of forests for agriculture, and in some regions the animals are hunted for their meat and their fur.

Chapter 8

Chiroptera: Bats

Because of their secretiveness, their appearance in gathering darkness during Shakespeare's "very witching time of night," their ability to fly, and their unusual form, which for the casual observer sets them apart from the more familiar groups of mammals, bats have long been central figures in supersitition and have provided shadowy forms for poetry:

> Bats, and an uneasy creeping in one's scalp
> As the bats swoop overhead!
> Flying madly.
>
> Pipistrello!
> Black piper on an infinitesimal pipe.
> Little lumps that fly in air and have voices
>   indefinite, wildly vindictive;
>
> Wings like bits of umbrella.
>
> Bats!
>
> Creatures that hang themselves up like an
>   old rag, to sleep;
> And disgustingly upside down.
> Hanging upside down like rows of disgust-
>   ing old rags
> And grinning in their sleep.
> Bats!

This is the bat of D. H. Lawrence, but the biologist's bat has taken on substance. Accelerated research on bats in recent years has revealed some of the fascinating aspects of the lives of these animals: extraordinarily complex social behavior, involving harems maintained by males and the use of an array of vocal communication signals; a coordinated assemblage of neuromuscular and behavioral adaptations allowing bats to perceive in detail their prey and their environment by the use of sound; and an ability unsurpassed in other mammals to conserve energy daily or to survive through periods of stress by drastic reductions in the metabolic rate. Biologists have come to realize not only that bats deserve respect as remarkably specialized products of at least 55 million years of evolution but also that they merit our protection for their importance in terrestrial ecosystems as efficient predators of insects.

Bats are a remarkably successful group today and constitute the second largest mammalian order (behind Rodentia); approximately 168 genera and 853 species of living bats are known. Bats are nearly cosmopolitan in distribution, being absent only from arctic and polar regions and from some isolated oceanic islands. Although bats are frequently abundant members of temperate faunas, they reach their highest densities and greatest diversity in tropical and subtropical areas. In certain Neotropical localities, for example, there are more species of bats than of all other kinds of mammals combined. Bats occupy a number of terrestrial environments, including temperate, boreal, and tropical forests; grasslands; chaparral; and deserts. Because man-made structures often afford excellent roosting sites and agricultural areas provide high insect populations, bats are doubtless more abundant in some areas now than they were before these areas were occupied by Europeans. In some areas, however, the molesting of colonies of bats by humans has caused alarming declines in populations, and the poorly controlled use of insecticides presents another threat.

Two sharply differentiated suborders of bats are recognized. The suborder Megachiroptera includes the family Pteropodidae, the Old

World fruit bats, and the suborder Michrochiroptera includes the other 17 families of bats. Microchiropterans are nearly cosmopolitan in distribution and are largely insectivorous. Two functional contrasts between the megachiropterans and the microchiropterans are of particular importance. Megachiropterans are not known to hibernate and maintain body temperature within fairly narrow limits by physiological and behavioral means, whereas many microchiropterans are heterothermic, and some hibernate for long periods. In addition, whereas microchiropterans use echolocation as their primary means of orientation and can fly and capture insects in total darkenss, most megachiropterans use vision and therefore are helpless in total darkness. One exception is the megachiropteran *Rousettus,* in which the ability to echolocate perhaps evolved independently. *Rousettus* uses clicks made by the tongue as the basis for its acoustical orientation (Novick, 1958a, 1958b). All microchiropterans use ultrasonic pulses produced by the larynx.

Echolocation, a means of perceiving the environment even in total darkness (Chapter 24), and flight, allowing great motility, have been two major keys to the success of bats. These abilities enable bats to occupy at night many of the niches filled by birds during the day. In addition, the remarkably maneuverable flight of bats facilitates a mode of foraging for insects that birds have never exploited. Heterothermy, allowing bats to hibernate or to operate at a lowered metabolic output during part of the diel cycle, has enabled these animals to occupy areas only seasonally productive of adequate food and to utilize an activity cycle involving only nocturnal or crepuscular foraging periods. The metabolic economy resulting from hibernation and from lowered metabolism during part of the diel cycle has affected the longevity of some bats. For their size, some michrochiropteran bats are remarkably long-lived. *Myotis lucifugus,* a small bat weighing roughly 10 grams, may live as long as 30 years (Keen and Hitchcock, 1980).

Morphology

Many of the most important diagnostic features of bats are adaptations for flight. The bones of the arm and hand (with the exception of the thumb) are elongate and slender (Fig. 8–1),

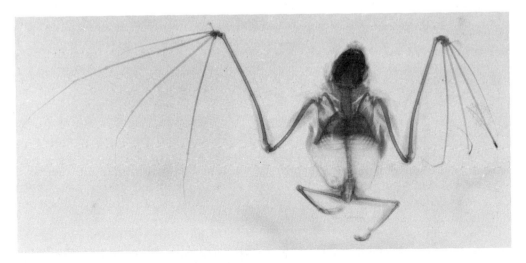

Figure 8–1. An X-ray photograph of the big fruit-eating bat (*Artibeus lituratus,* Phyllostomidae), showing the great elongation of the bones of the arm and hand. (From Vaughan, 1970b)

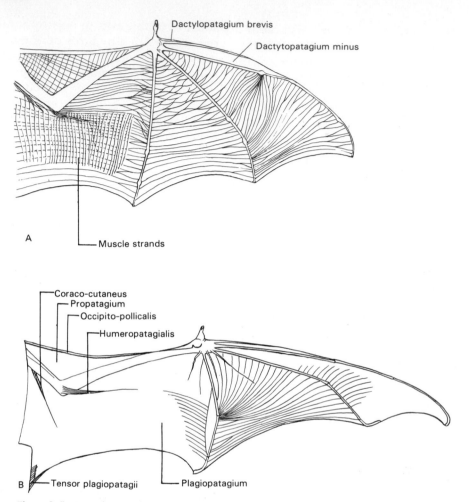

Dactylopatagium brevis

Dactytopatagium minus

A

Muscle strands

Coraco-cutaneus
Propatagium
Occipito-pollicalis
Humeropatagialis

B   Tensor plagiopatagii   Plagiopatagium

Figure 8-2. Ventral views of the wings of two bats, showing the parts of the wing and muscles and the elastic fibers that brace the membranes. (A) The big fruit-eating bat (*Artibeus lituratus,* Phyllostomidae); note the muscle strands that reinforce the plagiopatagium and the system of elastic fibers. This broad-winged bat does not remain on the wing for long periods. (B) The western mastiff bat (*Eumops perotis,* Molossidae). This narrow-winged bat is a fast and enduring flier. (A from Vaughan, 1970a; B from Vaughan, 1970b)

and flight membranes extend from the body and the hind limbs to the arm and the fifth digit (plagiopatagium), between the fingers (chiro-patagium), from the hind limbs to the tail (uro-patagium), and from the arm to the occipitopollicalis muscle (propatagium, Fig. 8–2B). In some species, the uropatagium is present even when the tail is absent. The muscles bracing the wing membranes are often well

developed and anchor a complex network of elastic fibers (Fig. 8–2A). Rigidity of the out-stretched wing during flight is partly controlled by the specialized elbow and wrist joints, at which movement is limited to the anteroposterior plane.

In most microchiropteran species, the en-larged greater tuberosity of the humerus locks against the scapula at the top of the upstroke

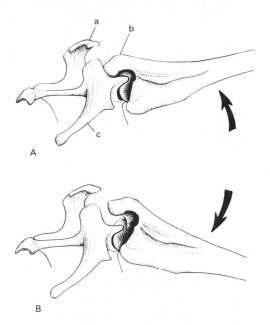

Figure 8-3. Anterior view of the left shoulder joint of a free-tail bat *(Molossus ater)* at (A) the top of the upstroke and (B) during the downstroke. The greater tuberosity of the humerus (*b*) locks against the scapula at the top of the upstroke, transferring the responsibility for stopping this stroke to the muscles binding the scapula to the axial skeleton. During the downstroke, the greater tuberosity of the humerus moves away from its locked position. This type of action and this type of shoulder joint also occur in the Vespertilionidae and other advanced families of bats. The acromion process (*a*) and the coracoid process (*c*) of the scapula are shown. (From Vaughan 1970a)

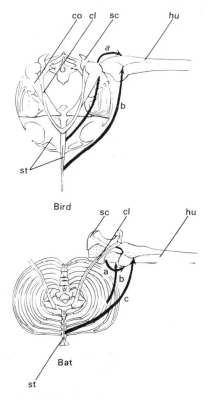

Figure 8-4. Anterior views of the thorax and part of the left forelimb of a bird and a bat, with some of the major muscles controlling the wing-beat cycle shown diagramatically. In the bird, the supracoracoideus muscle (*a*) raises the wing and the pectoralis muscle (*b*) powers the downstroke; both muscles originate on the sternum. In the bat, the downstroke is primarily controlled by three muscles: the subscapularis (*a*), the serratus anterior (*b*), and the pectoralis (*c*). Only the pectoralis originates on the sternum. Many muscles power the upstroke in bats. Abbreviations: *cl,* clavicle; *co,* coracoid; *hu,* humerus; *sc,* scapula; *st,* sternum. (From Vaughan, 1970a)

(Fig. 8–3), allowing the posterior division of the serratus anterior muscle, which tips the lateral border of the scapula downward to help power the downstroke of the wing (Fig. 8–4). The adductor and abductor muscles of the forelimb raise and lower the wings and are therefore the major muscles of locomotion. (In the contrasting arrangement found in terrestrial mammals, the flexors and extensors provide most of the power for locomotion.) The distal part of the ulna is reduced in bats, and the proximal section usually forms an important part of the articular surface of the elbow joint (Fig. 8–5). The clavicle is present and articulates proxi-

mally with the enlarged manubrium and distally with the enlarged acromion process and enlarged base of the coracoid process (Fig. 8–4). The hind limbs are either rotated to the side 90 degrees from the typical mammalian position and have a reptilian posture during quadrupe-

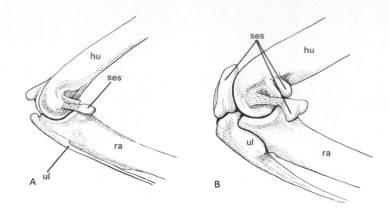

Figure 8–5. Lateral view of the right elbow of (A) a myotis (*Myotis volans,* Vespertilionidae) and (B) a free-tail bat (*Molossus ater,* Molossidae). Abbreviations: *hu,* humerus; *ra,* radius; *ses,* sesamoid; *ul,* ulna. (From Vaughan, 1970a)

dal locomotion, or they are rotated 180 degrees, have a spider-like posture, and are used primarily to suspend the animal upside down from a horizontal support. The fibula is usually reduced, and support for the uropatagium, in the form of the calcar bone (Fig. 8–6), is usually present.

The evolution of the muscular control pattern of the wing-beat cycle typical of microchiropteran bats has seemingly been strongly influenced by their use of echolocation. Highly maneuverable flight is essential for these bats because objects are perceived in detail only at fairly close range. In contrast, birds use vision for more long-range perception of their environment and thus have relatively little need for extremely maneuverable flight. In both groups, similar trends toward rigidity of the axial skeleton and lightening of the wings occur, but many of the muscular and skeletal specializations that enable these animals to control their wings differ in the two groups. The pectoral girdle in birds is braced solidly by a tripod formed by the clavicula and coracoids anchored to the sternum and by the nearly bladelike scapula, which rests almost immovably against the rib cage. The pectoralis and supracoracoideus muscles, both of which originate on the sternum, supply nearly all the power for the wing beat.

In bats, nearly the reverse mechanical arrangement occurs: the scapula is braced against the axial skeleton by the clavicle alone, movements of the clavicle during flight increase flight efficiency (Hermanson, 1981), and the job of powering the wing beat is shared by many muscles (Fig. 8–4). This division of labor is made possible partly because the scapula is free to rotate on its long axis. The pectoralis, the posterior division of the serratus anterior, and the clavodeltoideus muscles control the downstroke of the wings; only the pectoralis originates on the sternum. The muscles of the deltoideus and trapezius groups and the supraspinatus and infraspinatus muscles largely power the upstroke. The subscapularis is responsible for fine control of the wings during the entire wing-beat cycle (Hermanson and Altenbach, 1981).

A morphological trend of critical importance to bats and all other flying animals is toward the reduction of wing weight. Propulsion is obtained in all flying animals by movement of the wings, and the kinetic energy produced by such movement depends upon the speed and weight of the wing. The amplitude of a stroke and its speed are progressively greater toward the wing tip. Consequently, reduction of the weight of the distal parts of the wing results in a reduction of the kinetic energy developed during a wing stroke. A considerable advantage in metabolic economy is thus gained, for as less kinetic energy is developed during each stroke,

Figure 8-6. A fishing bat *(Noctilio labialis)*, showing several stages in the wing-beat cycle. A bone (the calcar) braces the uropatagium next to the wrist. (A) Top of the upstroke; (B) midway through the downstroke; (C) end of the downstroke; (D) midway through the upstroke. (By J. S. Altenbach, courtesy C. Brandon)

less energy is necessary to control the wings. In addition, light wings can be controlled with speed and precision during the extremely rapid maneuvers used when bats chase flying insects. Reduction of the weight of the wings has been furthered in bats by many specializations. Movement at the elbow and wrist joints is limited to one plane, thus eliminating musculature involved in rotation and bracing at these joints. In addition, the work of extending and flexing the wings is transferred from distal muscles (of the forearm and hand) to large proximal muscles (pectoralis, biceps, and triceps), thereby allowing a reduction in the size of the distal musculature. Certain forearm muscles are made nearly inelastic by investing connective tissue, and because of this modification and specializations of their attachments, these muscles "automati-

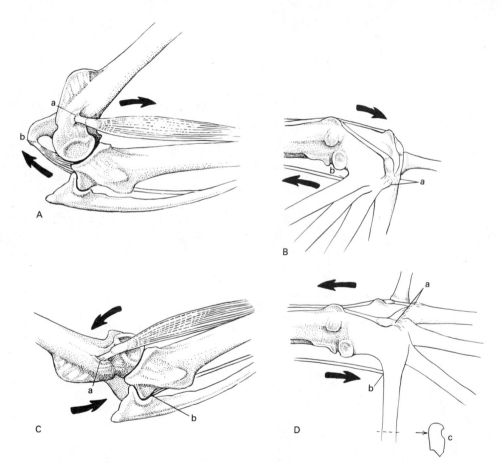

Figure 8-7. Lateral views of the elbow joint of the leaf-chinned bat *(Mormoops megalophylla),* showing the "automatic" flexion and extension of the fingers caused by certain forearm muscles in many advanced bats. (A) Flexion of the elbow joint moves the origins of the extensor muscles *(a) toward* the wrist and the origin of one flexor muscle *(b) away* from the wrist. (B) Because the flexor muscle is largely inelastic, with flexion at the elbow, the distal tendon of the flexor *(b)* pulls on the fifth digit and tends to flex the fingers. (C) With extension at the elbow joint, the origin of the extensor muscles *(a)* is moved away from the wrist and the origin of the flexor muscle *(b)* moves toward the wrist. (D) This action pulls the extensor tendon *(a)* toward the elbow and releases tension on the flexor tendon *(b),* thus extending the fingers. In D the complex cross-sectional shape of the fifth digit *(c)* is shown. (From Vaughan and G. C. Bateman, 1970)

cally" extend the chiropatagium with extension at the elbow joint or flex the chiropatagium with flexion at this joint (Fig. 8–7).

The hind limbs of bats are generally quite thin but are not drastically reduced in length because of their importance in supporting the trailing edge of the plagiopatagium and the lateral edge of the uropatagium. According to D. J. Howell and Pylka (1976), the fact that bats generally hang upside down is related to the thinness of the hind legs, which probably evolved under selective pressures favoring reductions in weight. The hanging position was perhaps necessitated by the inability of the delicate femur to tolerate compression stresses associated with other roosting postures.

Flight

The three modern groups of flying animals—insects, birds, and bats—are all highly successful. Viewing the terrestrial scene, there are more flying than nonflying species of animals, but each flying group has evolved a different type of wing: that of birds is formed of feathers braced by a simplified forelimb skeleton along the leading edge; insect wings are membranous sheets of chitin braced by intricate patterns of chitinous veins; and bat wings are sheets of skin braced by the five-digited forelimb and elastic connective tissue. Flight styles also differ. Birds usually depend on relatively fast and not especially maneuverable flight. Insects usually use extremely rapid wing beats, a variety of flight speeds, and often a remarkable ability to hover. Most bats, however, utilize slow, highly maneuverable flight. As might be expected, diverse and complex mechanical and aerodynamic problems are faced by these groups of fliers, and animal flight remains incompletely understood. Inasmuch as 22 percent of terrestrial mammals are bats, we must come to grips with flight in bats.

Most students have been introduced at least once to the basic aspects of aerodynamics; this

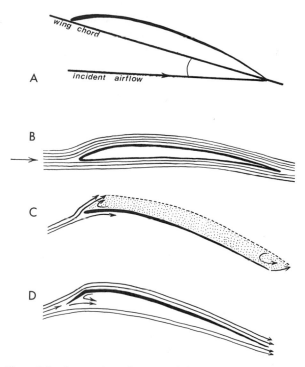

Figure 8–8. Cross sections of wings (airfoils) and air flow over the wings. (A) Thin airfoil, showing the *angle of attack,* the angle between the wing chord and the incident air flow; (B) flow of air over a thick airfoil; (C) turbulence and separated air flow over a wing at a high angle of attack; (D) the addition of a *leading-edge flap* keeps the air flowing smoothly over the surface. (After Norberg, 1972)

topic can therefore be treated briefly. Because the wings of animals usually provide both the thrust and the lift necessary for sustained flight, whereas in aircraft the wings provide only the lift, flight in animals presents special problems.

Lift is generated when an airstream sweeps over a wing with an asymmetrical cross section. The profile of the cross section of a wing (the airfoil) varies widely from one species of flying animals to another, but characteristically in birds and bats it has an arched dorsal surface and a concave ventral surface (Fig. 8–8B). The tendency is for the parts of the airstream flowing over the opposite surfaces of the wing to arrive at the trailing edge simultaneously; this necessitates faster movement of air over the dorsal surface than over the ventral surface. The more rapidly the air moves over a surface, the less pressure it exerts, a relationship described by Bernoulli in 1738 and exploited by flying vertebrates for over 150 million years. The unequal forces on opposing wing surfaces creates lift, a force opposing the force of gravity on a flying animal.

Lift is also created when a surface is presented at an angle of attack to the airstream (angle of attack is the angle that the chord line of the airfoil makes with the plane of motion of the wing: Fig. 8–8A). Within limits, lift can be increased by raising the angle of attack. When, however, the angle of attack becomes so great that the air moving over the upper surface breaks away from the wing and forms turbulent eddies, the lift produced by the wing abruptly falls as the drag sharply rises and stalling occurs. (Drag is the force exerted by air on an object in motion and in a direction opposite that of the motion.)

Wing performance is influenced by a series of variables. Lift increases directly as the surface area of the wing, but so does drag; lift increases (within limits) as the camber of an airfoil increases (camber is the curvature, or arching, of an airfoil), but this also increases drag; lift increases as the square of the speed, as does drag. Intuitively, then, one might expect that some of the constraints forcing modifications of wing design on fast fliers are of relatively little importance in slow-flying bats. This seems to be true and leads us to a consideration of the unique structure and function of the chiropteran wing.

The wings of bats form very thin airfoils of high camber. Several important features enhance the performance of these wings in the low-speed flight typical of most bats (Norberg, 1969, 1972, 1981). Thin airfoils, essentially cambered plates, are more effective in producing high lift at low speeds than are conventional airfoils with some thickness. Of further importance is the ability of the bat to vary the camber of the wing in the interest of producing high lift

at low speeds. Camber of the bat wing is largely under the control of the occipitopollicalis muscles, the flexors of the thumb, the inclination of the dactylopatagium minus (Fig. 8–6), the fifth digit, and the hind limbs.

Compared with birds, bats have low wing loadings (Table 8–1; Norberg, 1981). Wing loading is the ratio of body weight to wing area (W/S); in general, the lower the wing loading, the slower an animal can fly and still maintain adequate lift to remain airborne. Most bats also have broad wings with a low aspect ratio, which is the relationship of the length of a wing to its mean breadth and for wings of irregular shape is expressed as the ratio of the span squared to the wing area (b^2/S). Some bats that fly rapidly and remain in flight for long periods have long, narrow, high-aspect-ratio wings (Fig. 8–2B). Narrow wings suffer less loss of lift than do broad wings, owing to air spillage from the high-pressure area on the ventral surface to the low-pressure area on the dorsal surface of the wing tip. Wings that are strongly tapered toward the tip minimize this spillage and loss of lift and are typical of fast-flying bats.

Another feature that differs strongly in bats with contrasting flight styles is the breadth of the membranes that form the leading edge of the wing: the propatagium, the dactylopatagium brevis, and the dactylopatagium minus (Fig. 8–2). These membranes can be canted downward to form "leading-edge flaps," structures that greatly improve the efficiency of cambered-plate, low-speed airfoils (Fig. 8–8C). The leading-edge flap is larger in low-speed wings than in high-speed wings. Several sorts of flaps are used in aircraft to avoid stalling at low speeds, and in birds the alula (the small "bastard wing" formed by feathers attached to the thumb) and the slotting between feathers at the wing tip serve this function. By tending to avoid the separation of air flow from flight surfaces behind them, these slots improve low-speed performance at high angles of attack. Leading-edge flaps are especially effective in airfoils with thin sections and a thin leading edge (J. H. Abbott

Table 8-1:   Comparison of Wing Loadings (Grams of Body Weight/Square Centimeter of Wing Area) of Bats and Birds

Species	Weight (g)	Wing Area (cm²)	Wing Loading
Phyllostomid bats			
Artibeus phaeotis	9.9	101.74	0.098
A. lituratus	59.6	380.36	0.180
Phyllostomus discolor	42.2	261.60	0.162
Glossophaga soricina	10.6	99.29	0.106
Desmodontid bats			
Desmodus rotundus	27.8	198.94	0.161
Vespertilionid bats			
Myotis nigricans	4.2	67.58	0.062
Rhogeessa tumida	3.9	55.91	0.070
Molossid bats			
Molossus sinaloae	23.8	133.33	0.179
M. molossus	16.1	95.15	0.169
Birds			
Ruby-throated Hummingbird	3.0	12.4	0.242
House Wren	11.0	48.4	0.227
Chimney Swift	17.0	104.0	0.163
Redwinged Blackbird	70.0	245.0	0.286
Mourning Dove	130.0	257.0	0.364
Peregrine Falcon	1222.5	1342.0	0.911
Common Loon	2425.0	1358.0	1.786
Golden Eagle	4664.0	6520.0	0.715
Canada Goose	5662.0	2820.0	2.007

From Lawlor, 1973; Poole, 1936.

and von Doenhoff, 1949), precisely the type of airfoil formed by the wings of bats.

To produce lift, an airfoil must move through the air, and this requires a means of propulsion. In animals, propulsion is created by movements of the wings, and photographs of the wing-beat cycle in bats in level flight indicate that the downstroke is the power stroke and the upstroke is largely a recovery stroke (Fig. 8–6). During the downstroke, the wings are fully extended and the powerfully braced fifth digit and the hind limbs maintain the plagiopatagium at a fairly constant angle of attack, but the air pressure against the membranes becomes progressively greater toward the wing tip as the speed of the wing increases. This increase in pressure, coupled with the elasticity of the membranes between the digits, causes the trailing edges of the chiropatagium to lag behind the well-braced leading edge—in effect, the wing tip is twisted into a propeller-like shape and serves a propeller-like function. As the wing tip sweeps rapidly downward, it tends to force air backward, resulting in forward thrust of the animal. The membrane between the third and fourth digits (dactylopatagium longus) is probably of primary importance in producing thrust. During the upstroke, or recovery stroke, the wing is partly flexed, the stroke is directed upward and to some extent backward, and the force of the air stream partially aids the movement. Judging in part from the large muscles that power the downstroke and the relatively small muscles that control the upstroke, one would expect that the latter demands relatively little power and energy.

What about the function of the proximal segment of the wing, the plagiopatagium, during the wing-beat cycle? This segment retains an angle of attack appropriate to the production of lift through much of both the downstroke and the upstroke, and it seems to function primarily as a lifting surface. In many bats, a series of muscular and skeletal specializations insure that the fifth digit is braced against forces tending to cause extension beyond the point at which the angle of attack is proper for the development of lift.

Some bats can fly very slowly and some can hover, and during these types of flight the action of the wings is different from that used in level flight. When the nectar-feeding bat *Leptonycteris sanborni* hovers the downstroke is directed largely forward and the upstroke is directed backward (Fig. 8–9). The posture of the wings during the downstroke is similar to that in level flight, but because the stroke is largely horizontal, vertical thrust is developed. The upstroke, however, is complicated by a reversal of the usual posture of the wing tip: the tip turns over in such a way that the dorsal surface of the chiropatagium faces downward and the leading edge of the wing still leads in this stroke but is posterior to the trailing edge (Fig. 8–9). Toward the end of the upstroke, the reversed wing tip is flipped rapidly backward and produces considerable upward thrust, and at the start of the downstroke the wing tip swings into its normal posture. This powerful flip probably demands considerable energy, but the vertical thrust that it develops strongly augments the thrust resulting from the downstroke and enables the bat to remain nearly stationary in the air. Probably because of the high energy cost of

Figure 8–9.   A nectar-feeding bat *(Leptonycteris sanborni)* in flight, showing some positions of the wings during slow or hovering flight. (J. S. Altenbach)

hovering, it is generally used only briefly by bats.

Although they appear to interrupt awkwardly the otherwise smooth chiropatagium, the arm and fingers enhance low-speed flight performance by serving as turbulence generators. At the low speeds and high angles of attack common to the wings of many bats, a turbulent boundary layer (the thin layer of air very near the surface of the wing that is affected by friction with the surface) is better able to remain against the wing and continue to allow the development of lift than is a laminar boundary layer, which tends to separate completely from the wing surface with a consequent loss of lift and rise in drag (Fig. 8–10). The turbulent boundary layer formed behind turbulence generators (such as the arm or digits of bats) serves as a transition layer within which there is exchange of momentum between the fast-flowing outer layers of air and the inner layers, which are slowed by friction with the surface of the wing. The turbulent boundary layer can thus transfer kinetic energy to the surface layers from the rapid outer layers and can retain laminar flow in the lift-producing outer layers. In small, slow fliers, which most bats are, the type of air flow induced by turbulence generators is especially important in maintaining lift at high angles of attack.

During the early evolution of the bat wing, selection seemingly favored refinements in design that allowed the development of high lift at low speeds. Later, however, perhaps in the Eocene and Oligocene epochs, bats underwent an adaptive radiation involving, in part, exploitation of various styles of flight. The wings of some bats (members of the family Molossidae and some members of the family Emballonuridae, for example) developed characteristics advantageous during rapid flight. Because lift varies as the square of the speed of an airfoil, it would seem that rapid-flying bats could afford the luxury of higher wing loadings because of the greater lift developed per unit of wing area at higher speeds. Because drag also increases as

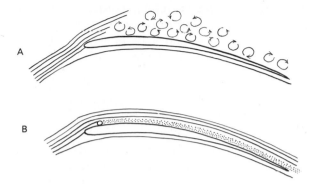

Figure 8–10. Effect of a turbulence generator. (A) Separation of air flow over a wing and loss of lift; (B) the boundary layer of air flowing over the wing changes from laminar to turbulent flow after encountering a protruding structure (such as the digit of a bat), and the lift produced by the airfoil is retained. (After Norberg, 1972)

the square of the speed, however, a reduction in wing surface area, angle of attack, and camber during rapid flight would be highly advantageous. Wing design in rapid-flying bats is clearly the result of a series of evolutionary compromises, and not all these animals have wings that are alike; nonetheless, a number of bats have evolved roughly the same type of high-speed wing. (High speed is used here only in a relative sense, for probably few bats achieve speeds in level flight above 80 kilometers per hour.) This wing is narrow and often has an elongate and strongly tapered tip, the membranes that form the leading edge flap are narrow, and the camber is low. Together these features reduce drag and favor efficient (in terms of energy outlay), rapid, and enduring flight. In the case of free-tail bats (Molossidae), the wing membranes contain an unusually large number of elastic fibers that brace the membranes against the force of the airstream at high speeds.

The muscles controlling the wing-beat cycle in bats are composed of several kinds of fibers. In small bats (such as *Myotis*), all muscle fibers are highly oxidative and fast-acting in association with the rapid wing beats and high metabolic rates of these bats. In broad terms, as body size in bats increases, so does the relative proportion of slow-twitch fibers to fast-acting fibers. This shift seems to reflect the lower wing beat and metabolic rates in larger bats (Armstrong, Ianuzzo, and Kunz, 1977; Strickler, 1980).

Paleontology

Because of their small size, ability to fly, delicate structure, and greatest abundance in tropical areas where fossilization seldom occurs, bats are rare as fossils. Consequently, the evolution of bats is poorly known. The earliest undoubted fossil bat (*Icaronycteris index,* Fig. 8–11) is from early Eocene beds in Wyoming. On the basis of this beautifully preserved specimen, Jepsen (1966) described the extinct family Icaronycteridae. Although this bat has several primitive features, such as claws on the first two digits of the hand and fairly short, broad wings, its basic limb structure is that of modern bats. The upper molars of *Icaronycteris* have the W-shaped ectoloph typical of most insectivorous bats, and this bat has been put in the suborder Microchiroptera. An Eocene bat was found to have moth scales in its gut, indicating that the moth-eating habit of bats is of great antiquity (J. D. Smith, Richter, and Storch, 1979). Late Eocene and early Oligocene deposits in France have yielded the earliest records of the modern microchiropteran families Emballonuridae, Megadermatidae, Rhinolophidae, and Vespertilionidae. Megachiropterans appear first in the Oligocene record of Italy.

The Eocene appearance of a bat clearly beautifully adapted for flight, the Oligocene appearance of many modern families, and the assignment by paleontologists of fossils from the late Eocene to the still-living genus *Rhinolophus* indicate an early origin of bats. This antiquity is further indicated by the fact that some genera were essentially as they are now in the Oligocene, a time when horses were three-toed and no bigger than sheep and bears and antelope had not yet appeared on the scene. The

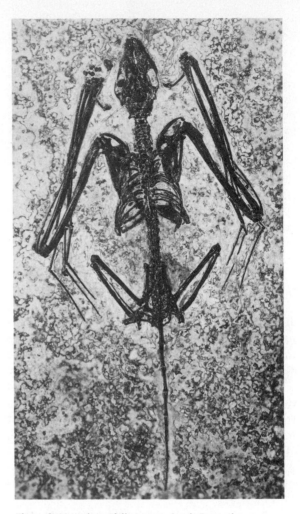

Figure 8–11. A beautifully preserved early Eocene bat *(Icaronycteris index).* (From Jepsen, 1970)

wings of an Oligocene molossid bat of the still-living genus *Tadarida,* for example, were found by Sige (1971) to be nearly identical to those of present-day members of this family. Although no clear fossil evidence bears on the matter, a Paleocene origin of bats seems probable, and a late Cretaceous divergence of chiropteran ancestors from primitive insectivore stock seems possible. Jepsen, in his excellent discussion of the evolution of bats (1970:22), summed up the status of our knowledge: "At present bat history has a completely open end, in the distant past, that only more fossils can close."

Suborder Megachiroptera

Inasmuch as only one family represents this suborder today, the descriptions given below for the family Pteropodidae characterize the suborder Megachiroptera.

Family Pteropodidae. Most pteropodids are fruit eaters, and many species are called flying foxes because of their foxlike faces and large size. These bats are abundant and often conspicuous members of many tropical biotas in the Old World. This family is represented by 41 Recent genera and 165 Recent species. Pteropodids occur widely in tropical and subtropical regions from Africa and southern Eurasia to Australia, and on many South Pacific islands as far east as Samoa and the Carolines.

Members of this family are often large, up to 1.4 kilograms in weight and 1.2 meters in wing span, but some are small (16 grams in weight and 245 millimeters in wing span). The Megachiroptera have the following features which set them apart from the Microchiroptera.

Figure 8-12. A tube-nosed bat (*Nyctimene* sp., Pteropidae) from New Guinea. (M. B. Fenton)

The face is usually foxlike, with large eyes, usually a moderately long snout with a simple, unspecialized nose pad, and simple ears lacking a tragus (Fig. 8–12). (The tragus, a fleshy projection of the anterior border of the ear opening, may be seen on a member of the Vespertilionidae in Fig. 8–13E.) The orbits are large and are bordered posteriorly by well-developed postorbital processes which may meet to form a postorbital bar. The rostrum is never highly modified (Fig. 8–14). The dental formula is 1-2/0-2, 1/1, 3/3, 1-2/2-3, = 24-34. The molars are never tuberculosectorial with W-shaped ectolophs, as in most microchiropterans, but are low, moderately flat-crowned, more or less quadrate, and lacking in stylar cusps (Fig. 8–15). The teeth are adapted basically to crushing fruit. The wing is primitive in having two clawed digits, and the greater tuberosity of the humerus is not enlarged (Fig. 8–16) to make contact with the scapula at the top of the upstroke. The tail is typically short or rudimentary.

Broadly speaking, pteropodids eat two types of food. Most members of the subfamily Pteropodinae are fruit eaters, whereas members of the subfamily Macroglossinae eat mostly nectar and pollen. The fruit eaters as a rule are large bats with fairly robust or moderately reduced dentitions. The jaws in these species are usually fairly long, or, in some species that presumably eat hard fruit, the jaws are shorter and the teeth and dentary bones are unusually robust. The fruit bats often roost in trees in large colonies (Fig. 8–17); in the case of the African genus *Eidolon,* as many as 10,000 have been observed roosting together. Fruit bats occasionally travel long distances during their nocturnal foraging, and *Pteropus* regularly flies at least 15 kilometers from roosting sites to feeding areas (Breadon, 1932; Nelson, 1965a, 1965b; F. N. Ratcliffe, 1932). The fruit eaters are usually not particularly maneuverable fliers but have a steady, direct style of flight. They are adroit at clambering in vegetation, where the clawed first and second digits of the wing come into play.

Figure 8–13. Faces of some microchiropteran bats: (A) Neotropical fruit-eating bat (*Artibeus anderseni;* Phyllostomidae); (B) long-nosed bat (*Leptonycteris sanborni,* Phyllostomidae), a nectar and pollen feeder. The following bats are insectivorous: (C) leaf-chinned bat (*Mormoops megalophylla,* Mormoopidae); (D) hoary bat (*Lasiurus cinereus,* Vespertilionidae); (E) pallid bat (*Antrozous pallidus,* Vespertilionidae); (F) African free-tail bat (*Tadarida pumila;* Molossidae). (A by R. M. Warner; D and E by P. Brown; F by R. P. Vaughan)

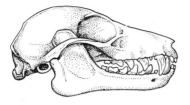

Figure 8-14. The skull of a megachiropteran bat (*Pteropus* sp., Pteropodidae); length of skull 62 mm.

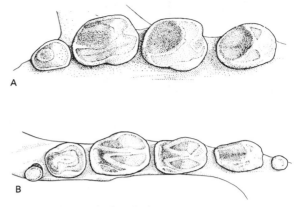

Figure 8-15. The cheek teeth of a megachiropteran bat (*Pteropus* sp.). (A) Right upper tooth row, showing two molars and two premolars. (B) Lower left tooth row, showing three premolars and three molars. (From Vaughan, 1970b)

Hypsignathus monstrosus, the hammerhead bat, is unique among mammals in the fantastic degree to which the vocal apparatus is specialized in the males. This large, frugivorous pteropodid, with a wing spread approaching 1 meter, occupies tropical forests in much of central Africa. Communal displays by males in courtship areas enter importantly into the breeding cycle of *Hypsignathus* (p. 396). The males on the courtship arena, which is called a lek, use a penetrating call, described by Kingdon (1974a:170) as "gutteral, explosive, and blaring," to attract females. The remarkable specializations of the vocal apparatus clearly

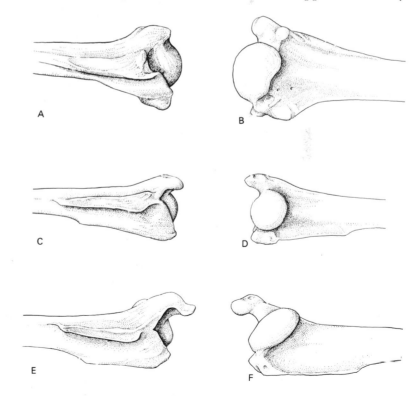

Figure 8-16. The proximal end of the right humerus in three bats. Anterior views are on the left and posterior views on the right. (A, B) *Pteropus* sp. (Pteropodidae); (C, D) *Myotis lucifugus* (Vespertilionidae); (E, F) *Molossus ater* (Molossidae). (From Vaughan, 1970b)

Figure 8-17. Part of a flying fox "camp" in northeasern Australia. (R. E. Carpenter)

evolved in association with the importance of loud vocalizations during breeding displays.

Externally, the most striking feature of the male is the strange hammer-head appearance (Fig. 8–18). This is due in part to the enlarged and elevated nasal bones but is accentuated by a large pouch that encloses the rostrum and extends back over the cranium. These features enhance the resonance of the calls, and pharyngeal sacs in the throat are probably also resonators. Equally impressive are internal features attending the massive enlargement of the larynx. This structure, which contains huge vocal chords, has moved into the thorax, where it occupies most of the space filled by the heart and lungs in other mammals. As a result of this migration, the large trachea lies against the diaphragm and curves sharply craniad to the lungs, which are also forced against the diaphragm. The thoracic cavity thus serves largely as a container for the huge larynx in male *Hypsignathus,* with a drastic sacrifice in lung capacity. Kingdon has called this animal a flying loudspeaker; this characterization seems especially

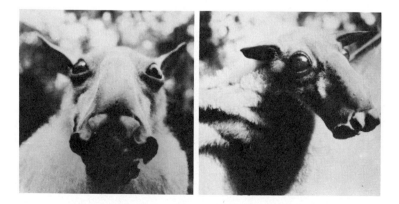

Figure 8-18. Two views of the head of a male hammer-head bat *(Hypsignathus monstrosus),* showing the highly modified lips and inflated rostrum. These specializations are associated with the ability to produce very loud, resonant sounds. (R. L. Peterson)

The pteropodids that eat nectar and pollen are smaller than their fruit-eating relatives and have long, slender rostra, strongly reduced cheek teeth, and delicate dentary bones. The tongue is long and protrusile and has hairlike structures at its tip to which pollen and nectar adhere. Pollen, which adheres to the fur (Fig. 8–19) and is ingested when the bats groom themselves, is probably an essential source of protein to nectar-feeding bats (p. 318). Some species roost in groups in caves, and some roost solitarily in vegetation. Flight is slow and maneuverable.

Suborder Microchiroptera

Recent members of this suborder are usually small. The eyes are often small, the rostrum is usually specialized, and the nose pad and lower lips may be modified in a variety of ways (Figs. 8–13 and 24–2). The ears have a tragus in all but members of the family Rhinolophidae, are usually complex, and are frequently large. The postorbital process is usually small. Dentitions vary tremendously, but most microchiropterans (except the desmodontids and some members of the Phyllostomidae) have tuberculosectorial molars; the upper molars have a W-shaped ectoloph with strongly developed stylar cusps, and in the lower molars the trigonid and talonid are roughly equal in size (Fig. 8–20A). In many insectivorous species and in some frugivorous members of the family Phyllostomaidae, one or more premolars above and below are caniniform, and in some insectivorous species the premaxillae are separate (Fig. 8–21).

The flight apparatus of the microchiropterans is more progressive than that of the megachiropterans. In microchiropterans, the second digit does not bear a claw and lacks a full complement of phalanges, and its tip is connected by a ligament to the joint between the first and second phalanges of the third digit. During flight, this connection allows the second digit to brace the third digit, which forms much of the

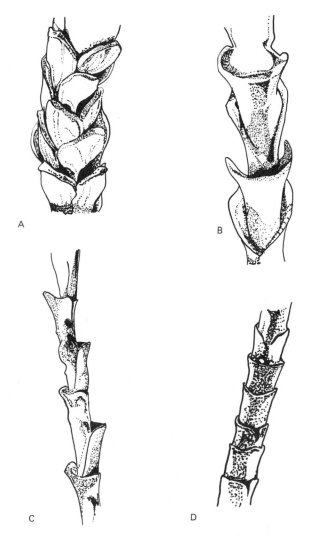

Figure 8-19. Hair of (A) *Epomophorus* sp., (B) *Glossophaga* sp., and (C) *Musonycteris harrisoni*, all nectar-eating bats. Note that these hairs are all adapted to catching pollen. (D) Hair of *Pteronotus davyi*, an insectivorous bat. (After D. J. Howell)

apt when one considers the enlarged lips of the males, which can be formed into almost perfect megaphones. The anatomy of the larynx is described by R. Schneider, Jurg Kugn, and Kelemen (1967), and early studies on the morphology of the vocal apparatus of this bat were carried out by Matschie (1899) and H. Lang and Chapin (1917).

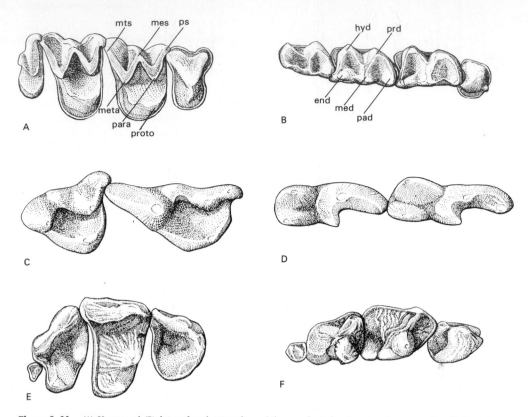

Figure 8–20. (A) Upper and (B) lower fourth premolar and three molars of an insect-eating vespertilionid *(Lasiurus cinereus)*. (C) Upper and (B) lower second and third molars of a nectar-eating phyllostomid *(Leptonycteris sanborni)*. (E) Upper and (F) lower fourth premolar and three molars of a fruit-eating phyllostomid *(Artibeus jamaicensis)*. Abbreviations: *end,* entoconid; *hyd,* hypoconid; *med,* metaconid; *mes,* mesostyle; *meta,* metacone; *mts,* metastyle; *pad,* paraconid; *para,* paracone; *prd,* protoconid; *proto,* protocone; *ps,* parastyle.

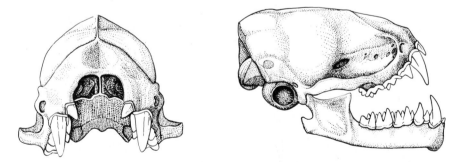

Figure 8–21. The skull of the hoary bat *(Lasiurus cinereus,* Vespertilionidae): (A) anterior view, showing the emarginate front of the palate; (B) side view, showing the shortened rostrum typical of some insect-feeding bats; length of skull 17 mm.

leading edge of the distal part of the wing against the force of the airstream (Norberg, 1969). The greater tuberosity of the humerus is usually enlarged and locks against a facet on the scapula at the top of the upstroke of the wings (Fig. 8–3). The size of the tail and uropatagium varies (Figs. 8–6 and 8–9). The shape of the wing varies according to foraging pattern and style of flight. In general, slow, maneuverable fliers have short, broad wings and rapid, enduring fliers have long, narrow wings (Fig. 8–2B).

Since their divergence from primitive insectivore stock, perhaps in the Cretaceous or early Paleocene, microchiropteran bats have undergone a remarkable adaptive radiation. There are 17 Recent families of microchiropterans and approximately 128 genera now recognized. This large number of families and genera reflects the great structural diversity and widely contrasting modes of life within this suborder.

Family Rhinopomatidae. Members of this small family, containing but one genus with three species, occur in northern Africa and southern Asia east to Sumatra. These animals are called mouse-tail bats because of the long tail that is largely free from the uropatagium. No fossil representatives of the family are known.

These small bats (10 to 15 grams in body weight) are considered the most primitive members of the Microchiroptera. The premaxillaries resemble those of megachiropterans in being separate from one another, and their palatal portions are much reduced. The second digit of the hand, in contrast to the arrangement in all other microchiropterans, retains two well-developed phalanges. Perhaps the clearest indication of the primitiveness of these bats is the structure of the shoulder joint. In contrast to the situation in most microchiropterans, the greater tuberosity of the humerus is small and does not lock against the scapula at any point in the wing-beat cycle. Other rhinopomatid features include laterally expanded nasal chambers; no fusion of cervical, thoracic, or lumbar vertebrae; and a complete fibula.

The dentition is adapted to an insectivorous diet. The molars are tuberculosectorial; the upper molars have W-shaped ectolophs of the usual microchiropteran type. The dental formula is 1/2, 1/1, 1/2, 3/3 = 28. These are fairly small bats (the length of the head and body is up to 80 millimeters) with slender tails whose length approaches that of the head and body. The eyes are large, and the anterior bases of the large ears are joined by a fold of skin across the forehead.

Mouse-tail bats are insectivorous and typically occupy hot, arid areas. They roost in a wide variety of situations, including fissures in rocks, houses, ruins and caves; one species roosts in large colonies in some Egyptian pyramids. Although locally common, mouse-tail bats are outnumbered by other types of bats over much of their range and are not as important today as other microchiropteran families. Rhinopomatids perhaps hibernate in some areas. Large deposits of subcutaneous fat occur in the abdominal area and around the base of the tail in individuals from some localities. These bats tolerate body temperatures as low as 22°C and can spontaneously rewarm themselves (Kulzer, 1965).

Family Emballonuridae. This family contains a variety of bats that are frequently called sac-wing or sheath-tail bats. They range in size from small (about 4 grams) to large (up to 105 grams); *Taphozous peli,* an African emballonurid, is among the largest of the insectivorous microchiropterans, with a wing spread of nearly 70 centimeters. The small emballonurids have a wing span of about 240 millimeters. Twelve genera and about 48 species are currently recognized, and the wide geographic range of emballonurids includes the Neotropics (much of southern Mexico, Central America, and northern South America), most of Africa, southern Asia, most of Australia, and the Pacific Islands east to Samoa. The earliest fossil emballonurid is from the Eocene or Oligocene of Europe.

These small bats combine a number of

primitive features with several noteworthy specializations. In the possession of postorbital processes and reduced premaxillaries that are not in contact with one another, emballonurids resemble pteropodids. In addition, the shoulder and elbow joints are primitive. An advancement over the rhinopomatids is the retention of only the metacarpal in the second digit; the flexion of the proximal phalanges of the third digit onto the dorsal surface of the third metacarpal is a specialization also found in some advanced families of bats. External obvious specializations include a glandular sac in the propatagium in some genera and the emergence of the tail from the dorsal surface of the uropatagium. The nose is simple; that is to say, it lacks leaflike structures or complex patterns of ridges and depressions. In addition to the more common gray and dark brown species of emballonurids, some species of one genus *(Saccopteryx)* have whitish stripes on the back, and members of the genus *Diclidurus* are white.

These insectivorous bats typically inhabit tropical or subtropical areas, where they use a great variety of roosting sites. Emballonurids occupy houses, caves, culverts, rock fissures, hollow trees, vegetation, or the undersides of rocks and dead trees for daytime retreats and usually roost in colonies. They are often fairly tolerant of well-lighted situations. In East Africa, *Taphozous mauritianus* often roosts on the trunks of large trees such as baobab trees *(Adansonia digitata;* Fig. 8–22). In some areas, emballonurids probably forage mainly over water. Some members of the genus *Taphozous* have long, narrow wings, are swift and dashing fliers, and often forage in clearings and above the canopies of tropical forests.

A distinctive feature of some emballonurids is the glandular sac in the propatagium. Recent work on one Neotropical species *(Saccopteryx bilineata)* has shown that this sac, especially well developed in males, is used in ritualized displays during the breeding season (p. 394). **Family Craseonycteridae.** This family, described by J. E. Hill (1974), is found only in

Figure 8–22. A baobab tree *(Adansonia digitata)* in Kenya, East Africa. Cavities in these trees are used as daytime retreats by various kinds of mammals, including some species of bats.

Thailand. As far as is known, only one species *(Craseonycteris thonglongyai)* represents the family. The common name is bumble-bee bat.

Craseonycteris is delicately built and is one of the smallest living mammals. It has small eyes and large ears. The premaxillae are not fused to adjacent bones, a feature that may increase the mobility of the upper lip, and the much-reduced coronoid process of the dentary bone probably allows a wide gape of the jaws.

The dental formula is 1/2, 1/1, 1/2, 3/3 = 28 and is of the usual insectivorous type with W-shaped ectolophs on the upper molars. The greater tuberosity of the humerus extends beyond the head of the humerus and may serve as a locking device; the second digit of the wing has only one very short phalanx; and the wing is broad. The pelvis and axial skeleton are highly specialized: the last three thoracic vertebrae and all but the last two lumbars are fused, and the sacral vertebrae are fused, whereas the pelvis is delicately built. The hind limbs are slender, and the fibula is threadlike. These bats resemble members of the families Rhinopomatidae and Emballonuridae in several ways, and Hill (1974) considers the Craseonycteridae to be rather closely related to these families.

This tiny bat roosts by day in caves and eats small arthropods (J. E. Hill and Smith, 1981), which are probably gleaned from leaves.

Family Nycteridae. Members of this small family (14 species of one genus) are called hollow-faced bats. These bats occur in Madagascar, Africa, the western Arabian peninsula, the Malay Peninsula, and parts of Indonesia, including Sumatra, Java, and Borneo. No fossil nycterids are known.

These fairly small bats weigh from about 6 to 30 grams and have wing spans ranging from 250 to 350 millimeters. They can be recognized by their large ears, very small eyes, and distinctive "hollow" face. The skull has a conspicuous interorbital concavity (Fig. 8–23) that is probably associated with the beaming of the ultrasonic pulses used in echolocation. This concavity is connected to the outside by a slit in the facial skin. The dental formula is 2/3, 1/1, 1/2, 3/3 = 32, and the molars are tuberculosectorial. Postcranially, these bats combine primitive and specialized features. The shoulder and elbow joints are fairly primitive, but the retention of only the metacarpal of the second digit of the hand and the reduction of the number of phalanges of the third digit to two are obvious specializations. The pectoral girdle is modified in the direction of enlargement and strengthening of the bracing of the sternum, a pattern parallel to the trend in birds toward the strengthening of the pectoral girdle. The sternum in nycterids is robust, and the mesosternum is strongly keeled. The manubrium is broad, the first rib is unusually strongly built, and the seventh cervical and first thoracic vertebrae are fused.

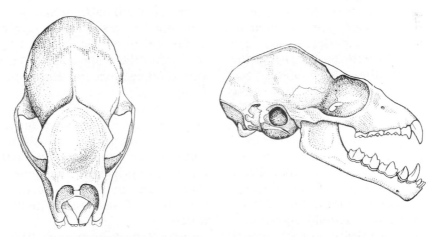

Figure 8-23. The skull of a slit-faced bat (*Nycteris thebaica*, Nycteridae): (A) dorsal view, showing the depression in the forehead; (B) side view, showing the flattened profile; length of skull 19 mm. (After J. E. Hill and Carter, 1941)

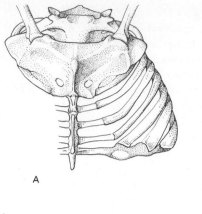

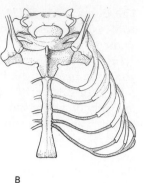

Figure 8-24. Ventral views of the thorax and lateral views of the sternum of (A, C) a rhinolophid *(Hipposideros commersoni)* and (B, D) a vespertilionid *(Myotis yumanensis).* Note the highly specialized sternum of *Hipposideros,* to which the first two ribs are fused. (From Vaughan, 1970b)

This general pattern also occurs in the family Megadermatidae (a family to which the nycterids are closely related) and reaches its most extreme development in the family Rhinolophidae (Fig. 8–24). Because the specializations of the pectoral girdle in these bats parallel to some extent the roughly similar modifications in birds, they could be associated with a progressive structural trend in bats. Actually, it is doubtful that this is the case. Some of the most advanced and successful families of bats have less birdlike pectoral girdles than those in the microchiropteran families listed above, but have modifications of the shoulder and elbow joints and forelimb musculature that provide for efficient flight. Perhaps the nycterid-megadermatid-rhinolophid pectoral girdle is associated with a foraging style typified by short intervals of flight. In any case, this style of pectoral girdle seems to be a divergent type and does not represent a progressive morphological trend common to most "advanced" microchiropterans.

Hollow-faced bats inhabit tropical forests and savanna areas and seem to feed largely on arthropods picked from vegetation or from the ground. Nycterids are amazingly delicate and maneuverable fliers, and when foraging they often seem to drift effortlessly around the trunks of large trees and near foliage. Flying insects form part of the diet, but orthopterans and flightless arachnids, such as spiders and scorpions, are also important food items. *Nycteris grandis* of Africa eats a remarkably varied assortment of animals, including orthopteran and lepidopteran insects, fish, amphibians, birds, and bats (Fenton and Fullard, 1981). These bats remain on the wing only for fairly short intervals, for they retire to a resting place to eat their larger prey.

Nycterids roost in a variety of situations, and some are even known to occupy burrows made by porcupines and aardvarks. In Kenya, in some remote safari camps, the pits dug as essential parts of privies are occasionally used as daytime retreats by *Nycteris thebaica,* to the consternation of the uninitiated users of these toilets.

Family Megadermatidae. This is not a large family, consisting of but four genera and five species. These bats are known as false vampires,

Figure 8–25. The African yellow-winged bat (*Lavia frons;* Megadermatidae). Note the extremely large nose leaf and huge ears. (R. P. Vaughan)

ing 1 meter and weighs up to nearly 200 grams; smaller species have wing spreads of about 320 millimeters and weigh about 25 grams. The ears are large and are connected across the forehead by a ridge of skin. The snout bears a conspicuous "nose leaf," and the eyes are large and prominent (Fig. 8–25). The premaxillae and upper incisors are absent; the upper canines project forward and have a large secondary cusp (Fig. 8–26B). The molars are tuberculosectorial; the dental formula is 0/2, 1/1, 1-2/2, 3/3 = 26-28. In *Megaderma,* and to a still greater extent in *Macroderma,* the W-shaped ectoloph of the upper molars is modified by the partial loss of the commissures connecting the mesostyle to the paracone and metacone. This trend is toward the development of an anteroposteriorly aligned cutting blade and may be associated with the carnivorous habits of these genera. The shoulder and elbow joints are primitive; the second digit of the hand has one phalanx, and the third has two phalanges. The pectoral girdle has specializations similar to those of the nycterids, but the strengthening of the pectoral girdle is carried further in megadermatids. The manubrium of the sternum is broader in megadermatids than in the nycterids and is fused with the first rib and the last cervical and first thoracic vertebrae into a robust ring of bone. The megadermatid sternum is moderately keeled. The tail is very short or absent.

These bats occur in tropical forests and sa-

an inappropriate title as they neither resemble vampires nor feed on blood. They occur in tropical areas in East Africa, southeastern Asia including Indonesia, the Phillippines, and Australia. The fossil record of megadermatids is scanty; the earliest fossil is from the Eocene or Oligocene of Europe.

These are fairly large, broad-winged bats. The largest species has a wing spread approach-

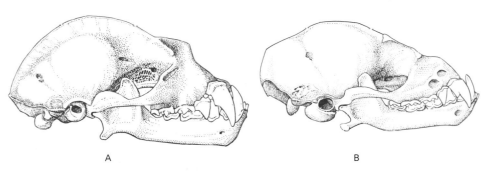

A   B

Figure 8–26. Skulls of two African bats that eat large beetles. (A) Giant leaf-nosed bat (*Hipposideros commersoni,* Rhinolophidae); length of skull 32 mm; (B) African false vampire bat (*Cardioderma cor,* Megadermatidae); length of skull 27 mm.

vannas, often near water, and eat a variety of foods. Of the five species of megadermatids, three are known to be carnivorous and two are mostly insectivorous. The Australian ghost bat *(Macroderma gigas),* an unusually large, pale-colored megadermatid, feeds on a variety of small vertebrates. In some areas, it seems to feed largely on other bats. The ghost bat and related species in southeastren Asia frequently consume their prey while hanging from the ceilings of spacious covered porches or verandas of large homes and detract from the gracious atmosphere by littering the floors with feet, tails, and other discarded fragments of frogs, birds, lizards, fish, bats, and rodents. Prakash (1959) observed *Megaderma lyra* of India eating bats of the genera *Rhinopoma* and *Taphozous,* a gecko, and a large insect. In the stomach contents of these bats, he found bones of amphibians and fishes. The carnivorous species of megadermatids may hunt partly by sight. Another megadermatid, the partially diurnal, insectivorous African *Lavia frons,* uses a style of foraging similar to that of a flycatcher, hanging from a branch and making short flights to capture passing insects.

Cardioderma cor of Africa has the most sedentary style of foraging known for any insectivorous bat (Vaughan, 1976). This entirely nocturnal bat perches in low vegetation when foraging and at some seasons regularly gives loud, humanly audible calls that seem to be territorial announcements. The body revolves through approximately 360 degrees as the hanging bat meticulously "scans" the ground, listening for sounds made by terrestrial invertebrates, such as large beetles and centipedes. When prey is detected, the bat flies directly to the ground, snatches up the food, and returns to the same perch to consume it. When insects are abundant during the wet seasons, *Cardioderma* spends very little time in flight: on some nights it perches for periods averaging nearly 11 minutes and spends less than 1 percent of its foraging time in flight. In the dry season, however, when prey abundance declines, flights from perch to

perch are more frequent and more time is spent in flight, although flights after prey average only 3 seconds. Considering all seasons, flights after prey average only 5 seconds. In *Cardioderma,* and perhaps in all megadermatids, the technique of searching for prey has departed markedly from that of most insectivorous bats, whereas the style of flight and the morphology of the forelimb have remained generalized.

Megadermatids roost in many types of places, from hollow trees, caves, and buildings in the case of most species, to sparse, occasionally sunlit vegetation in the case of *Lavia.* This bat has been found by Wickler and Uhrig (1969) to occupy fairly small foraging territories and to have several humanly audible calls during social interactions.

Family Rhinolophidae. This is a large and successful Old World family, with ten genera and approximately 125 species. Its members are often called horseshoe bats because of the complex and basically horseshoe-shaped cutaneous ridges and depressions on the nose (Fig. 8–27). Some rhinolophids are quite small, with body weights of nearly 6 grams and wing spans of 250 millimeters, but *Hipposideros commersoni* of Africa, at the other extreme, is the largest insectivorous bat, some individuals reaching body weights of over 100 grams and wing spans of over 500 millimeters. The geographic distribution includes much of the Old World from western Europe and Africa to Japan, the Philippines, Indonesia, Melanesia, and Australia. This may be an extremely ancient family, for some late Eocene fossils from Europe have been assigned to the living genus *Rhinolophus.*

Because of the unique and complex face, rhinolophids are one of the most unmistakable groups of bats. The ears are usually large but lack a tragus, and the eyes are small and inconspicuous. The tail is of moderate length in some species but is small or rudimentary in others. The pectoral girdle is remarkable because it represents the extreme development of the trend (that occurs also in the Nycteridae and Megadermatidae) toward powerful bracing and

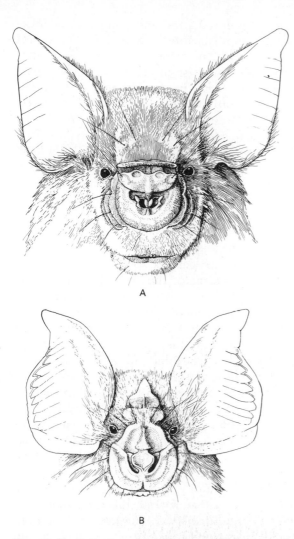

A

B

Figure 8-27. Faces of bats of the family Rhinolophidae: (A) giant leaf-nosed bat *(Hipposideros commersoni)*; (B) *Rhinolophus landeri.*

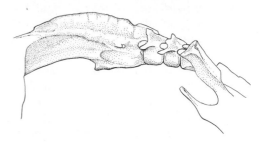

Figure 8-28. Lateral view of the left side of the fused lumbar vertebrae of a funnel-eared bat *(Natalus stramineus,* Natalidae). (From Vaughan, 1970b)

enlargement of the sternum. In the most extreme manifestation of this trend, the seventh cervical vertebra, the first and second thoracic vertebrae, the first and most of the second rib, and the enormously enlarged and shieldlike manubrium of the sternum are fused into a powerfully braced ring of bone (Fig. 8–24). The shoulder joint has a moderately well-developed locking device. In some rhinolophids, all but the last two lumbar vertebrae are fused; a simi-

lar specialization occurs in the Natalidae (Fig. 8–28). The pelvis is uniquely modified by enlargement of the anterior parts and an accessory connection between the ischium and the pubis. These unusual pelvic specializations may be responses to the mechanical stresses imposed on the hind limbs and pelvis by the repeated take-offs and landings that occur during foraging in some of these bats. When they roost, these bats often hang upside down, and the hind limbs are rotated 180 degrees from the usual mammalian posture so that the plantar surfaces of the feet face forward. The extreme adaptations for strengthening the pectoral and pelvic girdles that are typical of rhinolophids occur to a comparable degree in no other family of bats.

Horseshoe bats are common in many areas, and in Germany the "Hufeisennase" is a familiar inhabitant of attics and church steeples. These bats have wide environmental tolerances; various species inhabit temperate, subtropical, tropical, and desert regions. Rhinolophids hibernate in some parts of their range and characteristically rest or hibernate with the body enshrouded by the wing membranes. Several species in East Africa are migratory. The food is largely arthropods, and the style of foraging resembles that of the nycterids and some megadermatids. Horseshoe bats pick spiders and insects from vegetation or capture flying insects in midair, and *Rhinolophus ferrumequinum* was observed to alight on the ground and capture flightless arthropods (Southern, 1964). When foraging, *Hipposideros commersoni* of Africa hangs fairly high in trees, uses echolocation to detect large and straight-flying beetles at distances up to 20 meters, and makes brief and precise interception flights that last an average of but 5.1 seconds (Vaughan, 1976). This bat returns to the perch to consume prey, which consists of very large beetles (up to 60 millimeters in length). Like *H. commersoni,* a number of rhinolophids make short foraging flights and do not remain continuously on the wing while foraging. Perhaps the wing membranes are important in some species in aiding in the capture of insects. F. A. Webster and Griffin (1962) demonstrated photographically that one species of rhinolophid is able to capture insects in the chiropatagium.

In contrast to many bats that emit pulses from the open mouth in echolocation, rhinolophids keep the mouth closed during flight; the ultrasonic pulses used in echolocation are emitted through the nostrils and are beamed by the complex nasal apparatus (Mohres, 1953). A remarkable series of coordinated behaviors is associated with the highly specialized rhinolophid style of echolocation (p. 503).

Most horseshoe bats are colonial, but some are solitary. Many kinds of roosting sites are used; caves, buildings, and hollow trees are generally preferred, but foliage and the burrows of large rodents are used by some species.

Family Noctilionidae.   Although this family is not important in terms of numbers of species (it contains but two species of one genus), it is of special interest because one species is structurally and behaviorally highly specialized for eating fish. Noctilionid bats are often referred to as bulldog bats or fishing bats. They occupy the Neotropics from Sinaloa, Mexico, and the West Indies to northern Argentina. There is no fossil record of noctilionids.

Both structurally and in appearance, noctilionids are distinctive. They are fairly large (from roughly 30 to 60 grams in weight and up to 585 millimeters in wing spread), and the heavy lips, somewhat resembling those of a bulldog, pointed ears, and simple nose make the face unmistakable (Fig. 8–29). The dorsal pelage varies in color from orange to dull brown, and a whitish or yellowish stripe is usually present from the interscapular area to the base of the tail. The hind limbs and feet are remarkably large, especially in *Noctilio leporinus,* and the feet have sharp, recurved claws. The premaxillae are complete, and in adults the two maxillae are fused together and are fused with the premaxillae, forming a strongly braced support for the enlarged upper medial incisors. The dental formula is 2/1, 1/1, 1/2, 3/3 = 28. The teeth are robust, and the molars are tuberculosectorial.

The seventh cervical vertebra is not fused to the first thoracic, the shoulder and elbow joints are primitive, and the second digit of the hand has a long metacarpal and a tiny vestigial phalanx. The pelvis is powerfully built, with the ischia strongly fused together and fused to the posterior part of the laterally compressed, keel-like sacrum. The tibia and hind foot of *N. leporinus* have a series of unusual specializations, which will be considered below.

The feeding habits of the two species of *Noctilio* differ (Fleming, Hooper, and Wilson, 1972; Hooper and Brown, 1968). *Noctilio labialis* eats largely insects, which it seems to catch over water. D. J. Howell and Burch (1974) report that several individuals taken in June in Costa Rica had fed largely on the pollen of a tropical tree *(Brosimum);* perhaps the animals scooped floating pollen from the surface of the water. *Noctilio leporinus,* however, is a markedly atypical microchiropteran in that it eats largely fish. The style of foraging of this species involves the use of the hind claws as gaffs (Bloedel, 1955). This bat recognizes concentrations of small fish or single fish immediately beneath the surface of the water by detecting (by means of echolocation) the ripples or breaks in the surface that these fish create (Suthers, 1965, 1967). The bat skims low and drags its feet in the water, with the limbs rotated so that the hooklike claws are directed forward. (This in-

Figure 8-29. The face of a fishing bat (*Noctilio labialis,* Noctilionidae). (N. Smythe and F. Bonaccorso)

volves rotation of the hind limbs 180 degrees from the typical mammalian position.) When a small fish is "gaffed," it is brought quickly from the water and grasped by the teeth. From 30 to 40 small fish were captured in this fashion per night by a *N. leporinus* under laboratory conditions. A series of modifications of the hind limb are clearly advantageous in allowing this animal to pursue efficiently its specialized style of foraging. The long calcar, which is roughly as long as the tibia, the calcaneum, the digits and claws, and the distal part of the tibia are all strongly compressed so that they are streamlined with respect to their direction of movement when they are dragged through the water. During foraging sweeps, the short tail is raised and the bladelike calcar is pulled craniad and clamped against the flattened side of the tibia. In this way, the large uropatagium is brought clear of the water and the streamlined calcar and tibia knife through the water, producing a minimum of drag.

Noctilionids roost during the day in groups in hollow trees and rock fissures, caves, and occasionally buildings. *Noctilio leporinus* is seemingly most common in tropical lowland areas, frequently occurring along coasts, where it forages along rivers or streams, over mangrove-lined marshes and ponds, or over the sea. In western Mexico in the dry season, I have taken individuals as they foraged over small disconnected ponds in a nearly dry stream bed. These ponds supported large numbers of small fish.

Family Mormoopidae. The three genera and eight species that make up this family were traditionally considered members of the family Phyllostomidae (G. S. Miller, 1907:118; Simpson, 1945:57), but recent studies have shown that they differ so markedly from the phyllostomids that they merit recognition as a separate family (Forman, 1971; J. D. Smith, 1972; Vaughan and Bateman, 1970). All mormoopids can appropriately be called leaf-chinned bats because in all species a conspicuous, leaflike flap of skin occurs on the lower lip. These bats are largely tropical in distribution and occur from the southwestern United States and the West Indies south to Brazil.

Leaf-chinned bats are fairly small, weighing between 7 and 20 grams, and have several distinctive external specializations. The snout and chin always have cutaneous flaps or ridges (that reach their most extreme form in *Mormoops,* Fig. 8–13C), but a nose leaf is never present. The ears are moderately large, have a tragus, and vary in shape but always have large ventral extensions that curve beneath the fairly small eyes. The tail is short and protrudes from the dorsal surface of the fairly large uropatagium. The rostrum is tilted more or less upward (this feature is most extremely developed in *Mormoops*), and the floor of the braincase is elevated. The coronoid process of the dentary bone is reduced, allowing the jaws to gape widely. The teeth are of the basic insectivorous type; the dental formula is 2/2, 1/1, 2/3, 3/3 = 34.

The number of phalanges in the hand is as in the phyllostomids, but the shoulder and elbow joints differ markedly from the phyllostomid pattern. The greater tuberosity of the humerus in mormoopids does not form a well-developed locking device with the scapula; the head of the humerus is more or less elliptical, perhaps favoring a specialized wing-beat cycle. The elbow joint is specialized in all species, and, in *Mormoops,* modifications of the distal end of the humerus and the forearm musculature provide for a highly efficient "automatic" flexion and extension of the hand. The musculature of the hand is reduced and simplified; this lightens the hand and probably favors maneuverability and endurance. The hind limbs do not have the spider-like posture typical of phyllostomids, but instead have a reptilian posture that allows the animal to crawl on the walls of caves with considerable agility.

Leaf-chinned bats are among the most abundant bats in many tropical localities, where they are seemingly the major chiropteran insectivores. They are most common in tropical forests but occur also in some desert areas. Some species appear early in the evening; their insect-catching maneuvers resemble those of their temperate zone counterparts, the vespertilionids. Leaf-chinned bats usually roost in caves or deserted mine shafts and may concentrate in large numbers. A colony of *Mormoops* observed by Villa-R. (1966:187) in Neuvo Leon, Mexico, contained more than 50,000 bats, and a colony of four species of mormoopids in Sinaloa, Mexico, was estimated to contain between 400,000 and 800,000 bats (G. C. Bateman and Vaughan, 1974). When the bats from the latter colony emerged in the evening, they swept down the nearby arroyos and trails in such numbers and at such speeds that one hesitated to move across their path. When they form large colonies, these bats seem to disperse many kilometers from their roosting site to forage at night and remain continuously on the wing for several hours. Their impact on tropical ecosystems must be great, for the bats in the Sinaloan colony probably consume over 1400 kilograms of insects per night. It is not surprising that the bats must disperse over a wide area to forage.

Family Phyllostomidae. This is the most diverse family of bats with respect to structural variation and contains more genera than any other chiropteran family. Forty-six genera and 138 Recent species are included in the family. These Neotropical "leaf-nosed bats" are so named because of the conspicuous leaflike structure that is nearly always present on the nose (Figs. 8–13A and 8–13B; 24–2). These bats have exploited the widest variety of foods used by any family of bats. Some leaf-nosed bats have retained insectivorous feeding habits, but some are carnivorous and eat small vertebrates, including rodents, birds, and lizards. Some eat nectar and pollen, and some are frugivorous. Phyllostomids are the most important bats in the Neotropics and occur from the southwestern United States and the West Indies south to northern Argentina. These bats can be traced back to the Miocene of Colombia.

The great structural variation that occurs in the Phyllostomids is largely associated with an adaptive radiation into a wide variety of feeding niches. Within the family are some fairly small bats (*Choeroniscus* has a wing spread of roughly 220 millimeters and weighs 8 grams) as well as the largest New World bat (*Vampyrum,* with a wing spread of over 1 meter and a body weight of up to 190 grams). In most species, the nose leaf is conspicuous and spear-shaped, but in a few species the nose leaf is rudimentary or highly modified (Fig. 8–13B). The ears vary from extremely large to small, and a tragus is present. The tail and uropatagium are long in some species, with many stages of reduction and the absence of the tail and uropatagium being represented by various species. Some species have a uropatagium but lack a tail; only *Sturnira* has completely lost the uropatagium. The wings are typically broad; the second digit has one phalanx and the third has three phalanges. The shoulder joint has a moderately well-developed locking device formed between the greater tuberosity of the humerus and the scapula, but the elbow joint and forearm musculature are primitive, and the forelimb is, for a bat, generalized. Probably all phyllostomids, whatever their feeding habits, remain on the wing only for short periods during foraging.

The forelimbs are not used only for flight but are important in many species in handling food as well as in climbing over and clinging to vegetation (as in the fruit-eating species). The importance of such use of the forelimbs has probably favored the retention in phyllostomids of limbs more generalized than those of many strictly insectivorous bats. The seventh cervical and first thoracic vertebrae are not fused, and no fusion of elements to form the sturdy pectoral ring characteristic of rhinolophoid bats occurs in phyllostomids. In some leaf-nosed bats, however, the sternum is strongly keeled. The ventral parts of the pelvis are lightly built in most species, but the ilia are robust and more or less fused to the sacral vertebrae. These vertebrae are fused into a solid mass that becomes laterally compressed posteriorly. The acetabulum is characteristically directed dorsolaterally; the hind limbs are rotated 180 degrees from the usual mammalian orientation and have a spider-like posture. Because of this position of the hind limbs, some phyllostomids are unable to walk on a horizontal surface and use the hind limbs only for hanging upside down.

All of the Recent leaf-nosed bats probably evolved from an ancestral type that had tuberculosectorial teeth adapted to a diet of insects. Of the six Recent subfamilies, however, only one has retained this dentition, and in some species there is no trace of the ancestral pattern. The noteworthy adaptive radiation of phyllostomids will be traced by considering the dentitions and foraging habits of each subfamily.

The subfamily Phyllostominae deviates least from the ancestral structural plan, and some species retain insectivorous feeding habits. This subfamily contains all of the leaf-nosed bats with tuberculosectorial teeth of the ancestral type; however, in some species (*Chrotopterus* and *Vampyrum,* for example) the W-shaped ectoloph of the upper molars is distorted by the reduction of the stylar cusps and the closeness of the protocone, paracone, and metacone. Most members of this subfamily are insectivorous, and some species are known to pick insects either from vegetation or from the ground. On the other hand, a few of the largest phyllostomines resemble their Old World look-alike ecological counterparts, the megadermatids, in their carnivorous habits. The large phyllostomine species—*Phyllostomus hastatus, Trachops cirrhosus, Chrotopterus auritus,* and *Vampyrum spectrum*—are known to feed on small vertebrates. *Trachops* specializes on frogs (Fig. 8–30), which it locates by their calls (Tuttle, 1981). Beneath the roosts of *V. spectrum,*

Figure 8-30. A frog-eating bat *(Trachops cirrhosus)* about to capture a frog. (Taken on Barro Colorado Island, Panama, by M. D. Tuttle, Milwaukee Public Museum, Milwaukee, Wisconsin; reprinted from M. D. Tuttle: Frog-eating bat, *Science,* 1981, vol. 214, cover photograph; © 1981 The American Association for the Advancement of Science)

feathers and the tails of rodents and geckos frequently give indications of feeding preferences. The means by which these carnivorous-omnivorous bats perceive small vertebrates is not known. They may well hear the faint sounds made as their prey moves, and the large eyes of these bats indicate that hunting may also involve the use of vision. Bats of this type generally have large ears, however, suggesting highly discriminatory echolocation.

Nectar feeding is popular among tropical vertebrates (as indicated by the presence of over 300 species of hummingbirds in the American tropics) and has also been adopted by bats of the subfamily Glossophaginae of Mexico and Central and South America and by bats of the subfamily Phyllonycterinae of the West Indies. These bats feed on the nectar and pollen of a great variety of plants and have many structural features associated with this mode of life. The tongue is long and protrusile and has a brush-like tip (Fig. 8–31); the rostrum is elongate and

the dentaries are slender (Fig. 8–32B). The cheek teeth have largely lost the tuberculosectorial pattern (Fig. 8–20C, D). The hairs of at least some of these nectar feeders have divergent scales (Fig. 8–19) that catch pollen as the bat feeds on nectar. This pollen is swallowed when the bats groom their fur and provides a protein supplement without which the animals could not survive (p. 318).

In nectar-feeding bats, the wings are usually broad and the uropatagium is reduced. Although they probably are not able to remain on the wing for long periods of time, as can some insectivorous bats, nectar feeders can maneuver delicately through dense tropical vegetation and can hover. Flowers are seemingly located by the sense of smell, and these bats feed by hovering briefly and thrusting their long tongue into the flowers. The pollination of many night-blooming Neotropical plants is accomplished by nectar-feeding phyllostomids, just as many plants in the Old World tropics are pollinated

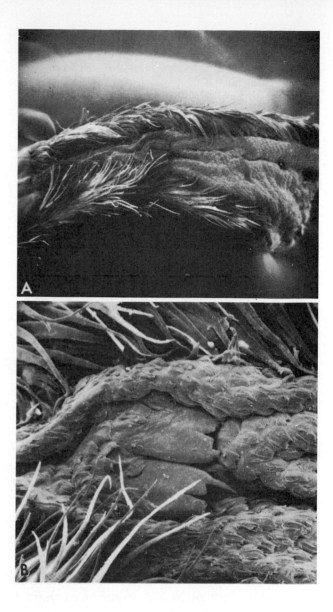

Figure 8-31. Scanning electron micrograph of the tongue of a nectar-feeding bat *(Leptonycteris sanborni)*, magnified (A) ×20 and (B) ×100. The tip of the tongue is to the left. (D. J. Howell and N. Hodgkin)

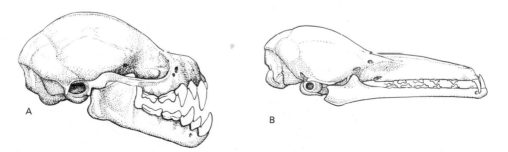

Figure 8-32. Skulls of leaf-nosed bats (Phyllostomidae): (A) a fruit eater *(Artibeus phaeotis)*; length of skull 19 mm; (B) a nectar feeder *(Choeronycteris mexicana)*; length of skull 30 mm.

by nectar-feeding pteropodids. Selective forces determined by this method of pollination have probably been important in the evolution of flower structure and in the timing of pollen and nectar production in these plants (p. 319). Glossophagine bats are clearly not so tightly restricted to a nectar and pollen diet as they were formerly thought to be (D. J. Howell and Burch, 1974). A dietary continuum from heavy reliance on insects (in *Glossophaga*) to virtually total dependence on nectar and pollen (in *Leptonycteris*) occurs in the subfamily.

The members of three subfamilies of phyllostomids—Carollinae, Sturnirinae, and Stenoderminae—are frugivorous. The success of these groups and the richness of this food source in the Neotropics is indicated by the fact that within the family Phyllostomidae, the largest and most abundant group of Neotropical bats, half of the species (approximately 60 out of 121) are basically fruit eaters. Several variations on this fruit-eating theme can be recognized. Members of the subfamily Carollinae have reduced molars with the original tuberculosectorial pattern largely obliterated. These bats apparently prefer ripe, soft fruit and are known to eat a great variety of it. The second frugivorous subfamily, Sturnirinae, is composed of fairly small, often brightly colored bats that have robust molars with no trace of the basic tuberculosectorial pattern. Indeed, their molars strongly resemble those of New World monkeys (Cebidae). Sturnirines eat small and often hard fruit, such as the fruits of low-growing species of nightshade (*Solanum* sp.).

The third frugivorous subfamily, Stenoderminae, contains bats with robust teeth that are highly modified for crushing fruit. The upper molars have lost the stylar cusps, and the inner portion is much enlarged and marked by complex rugosities (Fig. 8–20E,F). The rostrum is short, and the coronoid process of the dentary bone is fairly high in many species, conferring considerable mechanical advantage for powerful jaw action to the large temporal muscles. Large species of the stenodermine genus *Artibeus* are remarkably abundant in some Neotrop-

ical areas, and their piercing calls are characteristic sounds of the tropical nights. Often many *Artibeus* of several species concentrate on a single fig tree (*Ficus* sp.) with abundant fruit. In central Sinaloa, Mexico, two students and I camped beneath such a fig tree—but only for one night. The activities of dozens of *A. lituratus, A. hirsutus,* and *A. jamaicensis* caused a nearly continuous rain of fruit and bat excrement throughout much of the night, and with sunrise came herds of aggressive local pigs to gather from the ground the night's fallout of figs. Stenodermines often eat unripe and extremely hard fruit, and it is perhaps as an adaptation to this type of food that the robust teeth and powerful jaws evolved.

Family Desmodontidae. This family contains the vampire bats, the only mammals that feed solely on blood. Only two genera and three species constitute this group, but they are widely distributed from northern Mexico southward to northern Argentina, Uruguay, and central Chile. No extinct genera have been found, but *Desmodus* is known from the Pleistocene Epoch.

Vampire bats are fairly small. Mexican specimens of the common vampire bat *(Desmodus rotundus)* usually weigh about 30 grams and have an average wing span of 365 millimeters. The skull and dentition are highly specialized. The rostrum is short, the braincase is high (Fig. 8–33), and in all species the cheek teeth are reduced in both size and number. In *D. rotundus,* the most specialized species, the dental formula is 1/2, 1/1, 2/3, 0/0 = 20. The upper incisors are unusually large and are compressed and bladelike, as are the upper canines. These teeth have remarkably sharp cutting edges. The cheek

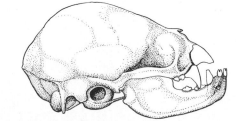

Figure 8-33. The skull of a vampire bat (*Desmodus rotundus,* Desmodontidae); length of skull 24 mm.

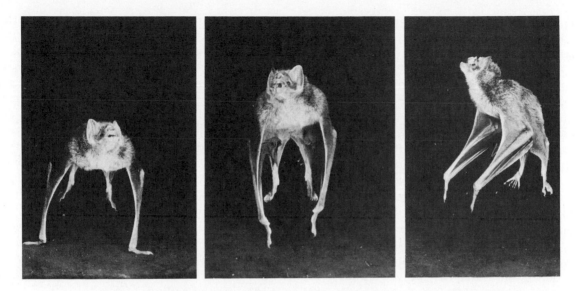

Figure 8–34. A vampire bat (*Desmodus rotundus*) leaping. Note the use of the long, robust thumbs. (J. S. Altenbach)

teeth are tiny. Except for the canine, the lower teeth are small. The thumb in *Desmodus* is unusually long and sturdy and contributes an additional segment with three joints to the forelimb during quadrupedal locomotion. The hind limbs are large and robust, and the fibula is not reduced. The proximal part of the femur and the tibia and fibula are flattened and ridged; these irregularities provide large surfaces for the attachment of the powerful hind limb musculature. Vampire bats can run rapidly and easily and can even jump short distances (Fig. 8–34). Their flight is strong and direct and not highly maneuverable.

Vampire bats can detect temperature differences, an ability probably of importance when they are seeking a place to bite. Temperature stimuli are probably detected in three pits in the hairless skin surrounding the nose (L. Kurten and Schmidt, 1982). The surface temperature of this skin is up to 9 C° lower than that of nearby parts of the face.

The feeding habits of this family are of particular interst. Vampire bats begin foraging after complete darkness and have one foraging period per night. *Diaemus* and *Diphylla* prefer the blood of birds, but *Desmodus* feeds on mammals. *Desmodus* alights on the ground near its chosen host, usually a cow, horse, or mule, and climbs up the foreleg to the shoulder or neck. The bat uses its upper incisors and canines to make an incision several millimeters deep from which it "laps" blood with its tongue. Vampire bats occasionally feed on the feet of cattle, at which time their ability to jump quickly may enable them to avoid injury when the host animal moves its feet. In *Desmodus,* the ingestion of blood is facilitated by an anticoagulant in the saliva that retards clotting. It has been estimated that each bat takes a meal of blood each night that amounts to over 50 percent of the fasting weight of the bat; a vampire bat weighing 34 grams, then, takes roughly 18 grams of blood per night (Wimsatt, 1969a). Because of this nightly drain of blood from cattle in certain localities and because vampire bats transmit rabies and other diseases, they are of great economic importance in many Neotropical areas. Occasionally vampire bats feed on humans.

Renal function in vampire bats is of great interest. These animals begin urinating soon after they begin to feed and rapidly lose much of the water taken in with the blood meal. This enables the bats to fly back to their roosts with less expenditure of energy and at less risk from predation than if they were burdened by the full weight of the ingested blood. Back in the roost, the bats continue digesting the now partially dehydrated blood and are faced with a problem of excreting large amounts of nitrogenous wastes without losing excessive amounts of water. At this time, rather than freely excreting water, as was done earlier in the foraging-digestion cycle, the kidney exhibits a remarkable ability to concentrate wastes, and highly concentrated urine is excreted. In this notable instance, a tropical mammal that lives in environments that seldom lack accessible water has evolved a kidney surpassing that of many desert mammals in its ability to concentrate urine and conserve water (Horst, 1969).

Family Natalidae. This small Neotropical family includes a single genus with five species. These bats are commonly referred to as funnel-eared bats and occur from Baja California, northern Mexico, and the West Indies southward to Colombia, Venezuela, and Brazil. The only fossils are those of the living genus from the Pleistocene and Recent records.

These small bats weigh from roughly 5 to 10 grams and have slender, delicate-looking limbs, broad wing membranes, and a large uropatagium that encloses the long tail. The funnel-shaped ears with a tragus, the simple nose lacking any sort of nose leaf, and the long, soft pelage that is frequently yellowish or reddish are characteristic. The skull has a long, wide rostrum with complete premaxillaries, and the braincase is high. The teeth are tuberculosectorial; the dental formula is 2/3, 1/1, 3/3, 3/3 = 38. The humeroscapular locking device is well developed, reduction of the phalanges of the hand is well advanced (the second digit lacks a phalanx and the third has two), and the manubrium of the sternum is unusually broad and has a well-developed keel. Some of the most distinctive natalid features, however, are those

of the axial skeleton that reduce its flexibility: the thoracic vertebrae are anteroposteriorly compressed and fit tightly together; the ribs are broad, and the narrow intercostal spaces are largely spanned by sheets of bone; all except the last two lumbar vertebrae are fused into a solid, laterally compressed, dorsally and ventrally keeled mass (Fig. 8–28); and the sacral vertebrae are mostly fused. As a result of these specializations, the strongly arched thoracolumbar section of the vertebral column is nearly rigid, with movement between this and the sacral section of the column allowed only by the "joint" formed by the last two lumbar vertebrae. These specializations seem to brace and cushion the verbebral column against shock transmitted to it by the hind limbs when this bat alights on the ceilings of caves.

Funnel-eared bats are insectivorous, and their foraging flight is slow, delicate, and maneuverable. Individuals released in dense vegetation are amazingly adroit at flying slowly through small openings between the interlacing branches of trees and shrubs. These bats inhabit tropical and semitropical lowlands and foothills and typically roost in groups in warm, moist, and deep caves or mines. These are handsome little bats; groups of *Natalus stramineus* scattered over the ceiling of a cave look like bright orange jewels in the beam of a flashlight.

Family Furipteridae. This small family contains but two genera, each with one species. These bats occur in northern South America south to southern Brazil and northern Chile, and in Trinidad. Furipterids, known as smoky bats because of their grey pelage, are seemingly closely related to the Natalidae, Thyropteridae, and Myzopodidae. All these groups share certain structural similarities. No fossil furipterids are known.

Externally, furipterids resemble natalids in the structure of the ears and in the slender build. The shoulder joint and the fused lumbar vertebrae are also similar in these families. Furipterids differ from natalids in minor features of the skull and dentition, such as partially cartilaginous premaxillaries and reduced canines. The furipterid dental formula is 2/3, 1/1, 2/3, 3/3 = 36. The thumb of smoky bats is greatly reduced and functionless.

These bats apparently are not common, and their habits are poorly known. They are insectivorous and have been found in caves and buildings. Most of the area inhabited by smoky bats is tropical, but *Amorphochilus* occurs in arid coastal sections of northwestern South America.

Family Thyropteridae. This family includes two small Neotropical species of bats, which are known as disk-winged bats because of the remarkable sucker disks that occur on the thumbs and feet. These animals, and the one member of the family Myzopodidae, are the only bats and the only mammals that have true suction cups. Disk-winged bats occur in southern Mexico, Central America, and South America as far as Peru and southern Brazil. No fossil thyropterids have been recorded.

In general appearance and in many skeletal details, these small (weight about 4 grams; wing span 225 millimeters), delicately formed bats resemble natalids, but the lumbar vertebrae are not fused as in the latter. The skulls of natalids and thyropterids are similar, and the dental formulas are the same. The thumb is reduced but retains a small claw, and its first phalanx has a sucker disk. The second digit is short, being represented by only a rudimentary metacarpal, and as a result the membrane between digits two and three is unusually small. The third digit has three bony phalanges. The digits of the feet have only two phalanges each, the third and fourth digits are fused, and the metatarsals bear suction disks which have a complex structure that allows them to act as suction cups. The bats can cling to smooth surfaces and can even climb a vertical glass surface. A fibrocartilaginous framework braces each disk; the rim of the disk consists of 60 to 80 chambers, each supplied by a sudoriferous (sweat) gland, which improves the tightness of contact with the substrate by insuring that the face of the disk is constantly

moistened. The disk itself lacks muscles, but specialized forearm muscles produce suction by cupping the middle of the face of the disk and release suction by lifting a section of the disk rim (Wimsatt and Villa-R., 1970).

Disk-winged bats are insectivorus and restricted to tropical forests. Their roosting habits are highly specialized. They roost only in the young, slightly unfurled leaves of certain tropical plants that are partially or completely shaded by larger trees. Such a roosting site is provided by the "platanillo" (*Heliconia* sp.), which resembles the banana plant. When a young leaf of this plant is beginning to unroll, it forms a tube roughly 1.3 meters long and 25 millimeters in diameter with a small opening at its tip. Several disk-winged bats may occupy such a tubular leaf in a head-to-tail row, heads upward, with the sucker disks anchoring them to the slippery surface of the smooth leaf. Because of the leaf soon unfurls, it is suitable for occupancy for only about 24 hours, and the bats move periodically to new and more suitable leaves. Findley and Wilson (1974) found that these bats usually roost in social groups of six or seven, that the bats of a given group always roost together, and that each group occupies an exclusive area within which it roosts in the daytime.

Family Myzopodidae. The only species representing this family is *Myzopoda aurita,* the sucker-footed bat, a species restricted to Madagascar. No fossils of this family are known.

The structure of its shoulder joint indicates that this bat is probably related to the Natalidae, Furipteridae, and Thyropteridae. The lumbar vertebrae are not fused as they are in natalids. The cheek teeth are of the standard tuberculo-sectorial-insectivorous type, and the dental formula is 2/3, 1/1, 3/3, 3/3 = 38. The ears are very large, and the ear opening is partly covered by an unusual mushroom-shaped structure found in no other bat. The claw of the thumb is rudimentary, and the thumb bears a sucker disk. Only the metacarpal of the second digit is bony; the third digit has three ossified phalanges. The foot bears a sucker disk on its sole, and as in thyropterids, each digit has only two phalanges. In *Myzopoda,* the metatarsals are fused and all the toes fit tightly against one another.

Myzopoda appears to be rare, and its life history is unknown. Its dentition indicates insectivorous feeding habits.

Family Vespertilionidae. This is the largest family of bats in terms of numbers of species and the most widely distributed. Thirty-three genera and approximately 310 species are included in this family, and in temperate parts of the world these are usually by far the most common bats. In the New World, vespertilionids occur from the tree line in Alaska and Canada southward throughout the United States, Mexico, and Central and South America. All of the Old World is inhabited north to the tree line in northern Europe and Asia. Most islands, with the exception of some that are remote from large land masses, support vespertilionids. As can be inferred from their geographic distribution, these bats occupy a wide variety of habitats, from boreal coniferous forests to barren sandy deserts. In the Neotropics, however, they are greatly outnumbered by bats of other families, particularly by leaf-nosed bats (Phyllostomidae).

Perhaps because of the diversity of habits and structure represented within the Vespertilionidae, no common name for this group is in general use; they are usually simply called vespertilionid bats. This family can be traced back to the middle Eocene in both Europe and North America, but it apparently did not reach Africa and South America until the Pleistocene. The genus *Myotis* is remarkable for its broad geographic distribution, which includes roughly the entire area occupied by the Vespertilionidae, and for its long fossil record, which begins in the middle Oligocene of Europe.

Vespertilionids are rather plain-looking bats that lack the distinctive facial features characteristic of many families. A nose leaf is rarely present, nor do complex flaps or pads occur on the lower lips. The eyes are usually small. The ears are of moderate or large size, and the tragus is present but differs in shape markedly from one species to another. These bats are usually small, weighing from 4 to 45 grams, with wing spans ranging from 200 to 400 millimeters. The wings are typically broad, and the uropatagium is large and encloses the tail. The shoulder joint is of an advanced type and provides for a locking of the large greater tuberosity of the humerus (Fig. 8–16C,D) against the scapula at the top of the upstroke of the wing. The elbow joint is also advanced, and the spinous process of the medial epicondyle, which is well developed in many species, enables certain forearm muscles to "automatically" extend and flex the hand (Fig. 8–7). The shaft of the ulna is vestigial, but the proximal portion forms an essential part of the elbow joint. The second digit of the hand has two bony phalanges, and the third digit has three. The fibula is rudimentary. The manubrium of the sternum has a keel, but the body of the sternum has at best a slight ridge. Except in one genus *(Tomopeas),* all presacral vertebrae are unfused.

The teeth are tuberculosectorial, and the W-shaped ectoloph of the upper molars is always well developed. The dental formula is 1-2/2-3, 1/1, 1-3/2-3, 3/3 = 28-38. The skull lacks postorbital processes; the palatal parts of the premaxillaries are missing, and the front of the palate is emarginate (Fig. 8–21). In general, vespertilionids are mostly small, plain bats that are characterized by refinements of the flight apparatus that make them efficient, maneuverable fliers.

Most vespertilionids are insectivorous, and in their ability to capture flying insects they are unexcelled. Most children in Europe and North America gain their first experience with bats by watching vespertilionid bats, silhouetted against the twilight sky, making abrupt turns and sudden dives while pursuing insects. The most commonly used vespertilionid foraging technique is probably also the most demanding: it involves the pursuit and capture of flying insects, which means that the bats must remain on the wing throughout most of their foraging periods. The insects, detected by echolocation, are usually followed in their erratic flight by a series of intricate maneuvers by the bat and are either captured in the mouth or, in the case of some species of bats, trapped by a wingtip or by the uropatagium (F. A. Webster and Griffin, 1962). This type of foraging demands highly maneuverable flight, and this is the type of flight to which vespertilionids seem best adapted.

Styles of foraging vary from one type of vespertilionids to another. The tree-roosting bats *(Lasiurus)* remain on the wing throughout their foraging, while others alight to eat large prey. Some species snatch insects or arachnids from leaves or pounce on them on the ground. The pallid bat *(Antrozous pallidus),* a common species in the southwestern United States, feeds on such large terrestrial arthropods as scorpions, Jerusalem crickets *(Stenopelmatus),* and sphinx moths (Sphingidae). This bat often uses porches, shallow caves, or abandoned buildings to rest and eat its prey. Often there are accumulations of discarded legs, fragments of exoskeletons, and wings beneath these roosts. Some vespertilionids capture insects from the surface of the water, and several species of *Myotis* capture fish or crustaceans from the water, probably by gaffing the prey with the claws of the feet. Some vespertilionids are known to have an early evening and a predawn foraging period. Kunz (1973b) found that in Iowa several temporal patterns of foraging occurred among six species of vespertilionid bats. These species had major periods of foraging within five hours after sunset, and some of them had a well-marked second foraging period while others had only a minor second period or none at all. The nocturnal activity patterns of most vespertilionids remain unknown, but in all probability the times of peak activity of some bats coincide with times of peak activity of their insect prey.

A wide variety of roosting places are utilized by vespertilionids. They adapt well to urban life and frequently roost during the day in attics, in spaces between rafters, or behind shutters or loose boards. Crevices in rocks, spaces beneath rocks or behind loose bark, caves, mines, holes in trees, and foliage are also utilized. Often these bats are colonial, frequently with maternity colonies of females with young occupying one roost and adult males using another. Many species, however, such as the foliage-roosting bats, roost singly or in small groups. Some species rest for part of the night beneath bridges or in porches or buildings, often in places never used as daytime retreats.

In temperate regions, many vespertilionids hibernate. Although the hibernation sites of some species are not known, some well-known species hibernate in caves and mines or buildings, and some species migrate fairly long distances to reach favorite hibernacula. Excellent reviews of hibernation (W. H. Davis, 1970; McNab, 1982) and migration in bats (Griffin, 1970) are available. Two small European species are known to migrate over 1600 kilometers from Russia to Bulgaria (Krzanowski, 1964). Some temperate-zone vespertilionids hibernate for short periods, and may be at least intermittently active in the winter, but others in colder areas hibernate throughout the winter except for occasional short arousals. In tropical areas, vespertilionids may respond to seasonal changes in insect abundance by local migrations. This is the case in southern Kenya. The weight of much observational evidence points toward migration in foliage-roosting bats in the United States. However, some red bats *(Lasiurus borealis)* remain in cold regions in the central United States throughout the winter and may be active on warm days (W. H. Davis and Lidicker, 1956).

Family Mystacinidae. One rather aberrant species, *Mystacina tuberculata,* the short-tailed bat, is the sole member of this family. This species occurs only in New Zealand, and the family has no fossil record.

Mystacina has some characteristics of vespertilionids and some of free-tail bats (Molossidae) but in general is sufficiently distinct from either group to merit recognition as a member of a separate family. Vespertilionid characteristics of *Mystacina* include the advanced locking shoulder joint, one phalanx in the second digit and two in the third, and the lack of fusion of presacral vertebrae. The skull is roughly like that of vespertilionids, but there is no anterior palatal emargination. The teeth are tuberculosectorial, and the dental formula is 1/1, 1/1, 2/2, 3/3 = 28. The tongue is partly protrusile and bears a brush of fine papillae on its tip. The limbs resemble in some ways those of molossids: the wing membranes and uropatagium are tough and leathery, the first phalanx of the third digit folds back on the dorsal surface of the metacarpal, and the hind foot is unusually broad; the fibula is complete, and the hind limb is robust. Unlike vespertilionids or molossids, the tail of *Mystacina* is short and protrudes from the dorsal surface of the uropatagium. Each of the claws of the thumb and foot has a secondary talon at its ventral base.

This unusual bat has a generalized diet that includes fruit, nectar, pollen, and insects (M. J. Daniel, 1979). The wing can be folded compactly, owing to the unique flexion pattern of the third digit, and during quadrupedal locomotion it is partially protected by the leathery proximal part of the plagiopatagium. The limbs are seemingly well adapted to quadrupedal locomotion, and the talons are used in excavating roosting tunnels in decaying trees. This bat roosts in a wide variety of places, including abandoned seabird burrows.

Family Molossidae.   Members of this family, the free-tail bats, are important components of tropical and subtropical chiropteran faunas throughout much of the world. Twelve genera and 84 species are included in the family. Molossids occupy the warmer parts of the Old World, from southern Europe and southern Asia southward, as well as Australia and the Fiji Islands. In the New World, they occasionally occur as far north as Canada, but the main range begins in the southern and southwestern United States and the West Indies and extends southward through all but the southern half of Chile and Argentina.

Structurally, this is a peripheral group of bats; the most extreme manifestations of many of the typically chiropteran adaptations for flight occur in the Molossidae. These bats weigh from 8 to 54 grams, and wing spans range from 240 to 450 millimeters. The greater tuberosity of the humerus is large (Fig. 8–16E,F), and the locking device between it and the scapula is highly developed. The origins of the extensor carpi radialis longus and brevis and flexor carpi ulnaris muscles are well away from the center of rotation of the elbow joint and probably act more effectively than in any other bats as "automatic" extensors and flexors of the hand. The wing is typically long and narrow (Fig. 8–2B), with the fifth digit no longer than the radius, and the membranes are leathery because they are reinforced by numerous bundles of elastic fibers. In many species, there are structural refinements that favor high-speed flight. In many molossids, for example, the radius and forearm muscles are flattened and the arrangement of specialized hairs (Fig. 8–35) is such that the forearm is streamlined with respect to the airstream during flight (Vaughan and M. M. Bateman, 1980). In the interest of rigidity of the outstretched wing during flight, movement at the wrist and elbow joint is strictly limited to one plane. The muscles that brace the fifth digit and maintain an advantageous plagiopatagium attack angle during the downstroke of the wings are large and unusually highly specialized. Except for fusion of the last cervical and first thoracic vertebrae, the presacral vertebrae are unfused. The body of the sternum is not keeled.

The general appearance of molossids is distinctive. The tail extends well beyond the posterior border of the uropatagium when the bats are not in flight, and the fur is usually short and velvety. (In one genus, *Cheiromeles,* the fur is so short and sparse that the animal appears naked.) The muzzle is broad and truncate, and the thick lips are wrinkled in some species (Fig. 8–13F). Typically, the ears are broad, project to

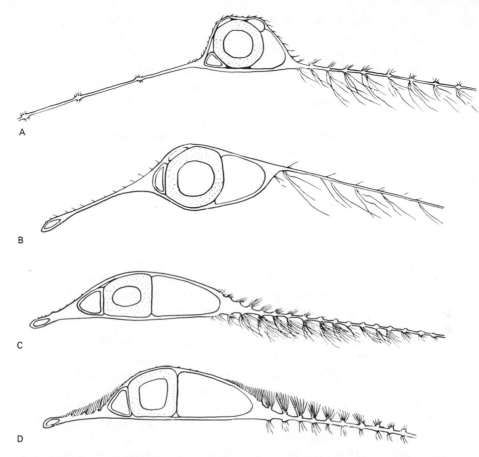

Figure 8–35. Cross-sectional views of the forearms and wing membranes of four species of bats, showing in C and D the pronounced streamlining of the wing by hair tracts and the flattening of the forearm in molossids. Two slow fliers: (A) *Nycteris thebaica* and (B) *Eptesicus fuscus;* two fast fliers: (C) *Molossus ater* and (D) *Tadarida condylura.* (Reprinted from Vaughan and M. M. Bateman, 1980, p. 74, by permission of Texas Tech Press)

the side, and are like short wings. As viewed from the side, the pinnae are arched and resemble an airfoil of high camber. The ears are frequently braced by thickened borders and are connected by a fold of skin across the forehead. Because of their unique design, the ears in most species do not directly face the force of the airstream during flight, an adaptation probably of considerable importance to these fast-flying bats.

The skull is broad, the teeth are tuberculosectorial, and the dental formula is 1/1-3, 1/1, 1-2/2, 3/3 = 26-32. Several characteristically molossid features are associated with the well-developed quadrupedal locomotion typical of these bats. The first phalanges of digits three and four flex against the posterodorsal surfaces of their respective metacarpals, allowing the chiropatagium to be folded into a compact bundle, no longer than the forearm, that is manageable when the animals run. The feet are broad and have sensory hairs along the outer edges of the first and fifth toes. The fibula is not reduced, and the short hind limbs are stoutly built. Within the structural limits of the basic chiropteran plan, these bats have seemingly made the best of two types of locomotion. The highly specialized wings are clearly adapted to fast, efficient flight, and the primitive hind limbs have not lost their ability to serve in rapid quadrupedal locomotion.

These insectivorous bats are remarkable for their speedy and enduring flight. Whereas most bats fly fairly close to the ground or to vegetation when foraging, many molossids fly high and may move long distances during their nightly foraging. Some populations of Brazilian free-tail bats *(Tadarida brasiliensis),* the bats that occur in great numbers in Carlsbad Caverns and other large caves in the southwestern United States, fly at least 90 kilometers to their foraging areas each night (R. B. Davis, Herreid, and Short, 1962). These bats were observed in Texas with radar and helicopters, and dispersal flights were tracked (T.C. Williams, Ireland, and J. M. Williams, 1973). Dispersing bats were recorded at elevations of over 3000 meters, and masses of bats moved at an average speed of 40 kilometers per hour. The western mastiff bat *(Eumops perotis)* forages over broad areas and in southern California may on occasion fly more than 650 meters above the ground (Vaughan, 1959:22). Because of the temperature inversions that frequently prevail for many nights in this area, in the winter these high-flying bats may be surrounded by air warmer than that at the ground and may be catching insects that are flying in the warm air strata. Some molossids remain in flight for much of the night; foraging periods of at least six hours have been recorded for some species.

The flight of many molossids is unusually fast, and some species rival swifts and swallows in aerial ability. In some areas of Mexico, early-flying mastiff bats *(Molossus ater)* mingle with late-flying flocks of migrating swallows, and the bats seem at least the equals of the swallows in speed and maneuverability. One gets the impression, in fact, that the swallows are hastened to their roosts by the sudden appearance in abundance of their chiropteran counterparts.

Some molossids make spectacular dives when returning to their roosts. The western mastiff bat, for example, often makes repeated high-speed dives and half loops past the roosting site. It returns to its roost in a cliff by diving toward the cliff base, pulling sharply upward at the last instant, and entering the crevice with momentum to spare. Several other molossids are known to return to their roosting places by similar maneuvers. Because the wings of many molossids are narrow and have relatively small sur-

Figure 8-36. The roosting place of mastiff bats (*Eumops perotis,* Molossidae) in a granite cliff. The animals occupy the space beneath the tongue-shaped slab of rock at the upper right. (From Vaughan, 1959)

face areas relative to the weights of the bats, these animals must attain considerable speed before they can sustain level flight. As a result, some species roost high above the ground in cliffs (Fig. 8–36), buildings, or palm trees, in situations where they can dive steeply downward for some distance in order to gain appropriate flight speed. These species are unable to take flight from the ground.

At least some species of molossids seem to forage in groups, the cohesion of which is insured in part by loud vocalizations audible to humans. It is common knowledge among field workers that often when molossids are heard overhead or captured in nets over water, single individuals are the exception and groups are the rule. Group foraging with communication between members would be advantageous to widely dispersing bats that take daytime shelter in communal roosts and seek concentrations of insects (perhaps the molossid strategy).

Most molossids inhabit warm areas. Migration, therefore, is not generally characteristic of these bats. The Brazilian free-tail bat, however, is known to make extensive migrations from the United States to as far south as southern Mexico (Villa-R. and Cockrum, 1962). The tremendous deposits of guano in some large caves inhabited by molossids attest to the effect that large colonies of these bats must have on insect populations in some areas.

Chapter 9

Order Primates

The order Primates is of particular interest not only because it includes our closest relatives but because its members display a fascinating breadth of structural and behavioral adaptations to their environments. The primate radiation can be viewed as an exploitation of arboreal herbivory, arboreal locomotion, manual dexterity, stereoscopic vision, and complex social behavior and communication. The ancestors of humans are among the relatively few primates that adopted a largely terrestrial mode of life.

Primates other than humans have been most successful in tropical and subtropical areas, where today they pursue mostly arboreal modes of life. Some primates, such as baboons and chimpanzees, have become partly or mostly terrestrial, but only humans have become fully bipedal. Approximately 51 genera and 168 species of primates are living today, of which 16 genera and about 50 species are in the New World.

Although primates are basically herbivorous, many are omnivorous and seem to be largely opportunistic feeders. The molars of primates are largely bunodont and brachyodont and have the quadrate form typical of molars of many herbivores or generalized feeders. Early in the evolution of primates, a hypocone was added to the upper molar and the paraconid of the lower molar disappeared, leaving a basically four-cusped pattern (Fig. 9–1). The primate trend toward shortening of the rostrum is probably related to the importance of stereoscopic vision.

Primates constitute one of the oldest eutherian orders, dating from the late Cretaceous Period of North America. Primates underwent an early radiation from the Late Cretaceous to the late Eocene in both Europe and North America, and the families that resulted from that radiation are grouped together within the suborder Plesiadapiformes (Fig. 9–2). The more primitive families among living primates (Lemuridae, Indridae, Daubentoniidae, Lorisidae and Galagidae) are included here in the Suborder Strepsirhini, whereas the more advanced families (Tarsiidae, Cebidae, Cercopithecidae, Pongidae, and Hominidae) are placed in the Suborder Haplorhini. This classification largely follows that of Thorington and Anderson (1984), and departs from the time-honored recognition of the Suborders Prosimii and Anthropoidea.

A tremendous burst of primate literature has appeared within the last two decades, and at least five journals are devoted exclusively to primatology. So vast and evergrowing has the literature on primates become, that the Regional Primate Research Center of the University of

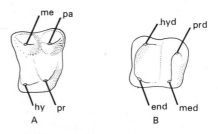

Figure 9–1. A diagrammatic representation of the basic four-cusped crown pattern of primates: (A) right upper molar; (B) left lower molar. Abbreviations; *end*, entoconid; *hy*, hypocone; *hyd*, hypoconid, *me*, metacone; *med*, metaconid; *pa*, paracone; *pr*, protocone; *prd*, protoconid.

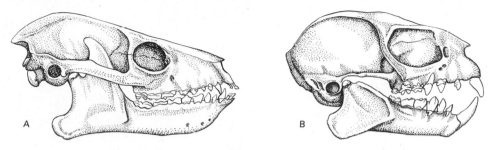

Figure 9-2. Skulls of fossil primates: (A) an Eocene lemuroid (*Notharctus,* Adapidae); length of skull 75 mm; (B) an Eocene tarsier-like primate (*Tetonius,* Anaptomorphidae); length of skull 46 mm. (After Romer, 1966)

Washington (Seattle, Washington, 98195) maintains a computerized primate bibliographic service.

Suborder Strepsirhini

The five families included in this suborder contain an assemblage of mostly arboreal mammals that in some cases bear only a marginal resemblance to the more "standard" primates (monkeys, great apes, and humans). The strepsirhines are relatively primitive primates and, in the case of several families, occupy restricted geographic areas and pursue specialized modes of life. Even among strepsirhines, however, the importance of vision, manual dexterity, and vocal communication is apparent.

Family Lemuridae. The lemurs inhabit Madagascar and the nearby Comoro Islands. Among the approximately 15 Recent species, belonging to eight genera, some are arboreal, some are semiarboreal, and some are largely terrestrial. These are the most primitive living primates. The fossil record of lemurids is from the Eocene, Pleistocene, and sub-Recent deposits in Madagascar. One extinct giant of presumably arboreal habits had an elongate skull 30 centimeters in length. The survival of lemurs is perhaps related to their insular distribution; they have been isolated on Madagascar since the early Cenozoic and have never been in competition with progressive primates (except hu-

mans). The four genera of dwarf and mouse lemurs are often placed in a separate family (Cheirogaleidae).

In contrast to most primates, the cranium of lemurs is elongate and the rostrum is usually of moderate length, giving the faces of some lemurs a foxlike appearance (Fig. 9–3). In more typical primate fashion, the lemurid braincase is large, crests for the origin of the temporal muscles are inconspicuous, and the foramen magnum is directed somewhat downward. The largest lemurs are roughly the size of a house cat; the smallest are the size of a mouse. The dental formula is 0-2/2, 1/1, 3/3, 3/3 = 32-36. The upper incisors are usually reduced or absent, and between those of the two sides is a broad diastema; the lower canine is incisiform, and the first lower premolar is caniniform. The molars are basically tritubercular. The pollex and hallux are more or less enlarged and are opposable in all genera (Fig. 9–4A). The pelage is woolly, the tail is long and heavily furred, the limbs are usually slim, and the tarsal bones are not greatly elongated (Fig. 9–5A). Conspicuous color patterns occur in some species.

Lemurs are variously omnivorous, insectivorous, or herbivorous-frugivorous and, depending on species, diurnal or nocturnal. They are agile climbers, and the hands are used both for climbing and for handling food. Some species make great leaps from branch to branch. Some lemurs store fat in preparation for estivation during the dry season. Large social groups occur

Figure 9–3. A lemur (*Lemur fulvus,* Lemuridae). (D. Schmidt, San Diego Zoo)

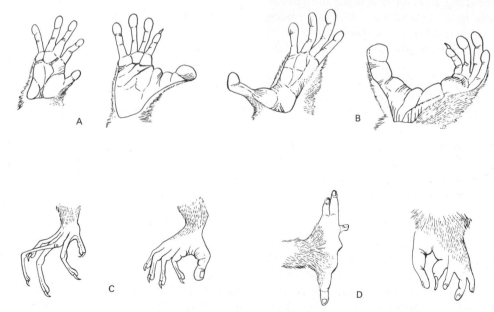

Figure 9–4. Hands and feet of some prosimian primates (the hand is on the left in each pair): (A) lemur (*Lemur mongoz,* Lemuridae); (B) indrid (*Propithecus diadema,* Indridae); (C) aye-aye (*Daubentonia madagascariensis,* Daubentoniidae); (D) potto (*Arctocebus calabarensis,* Lorisidae).

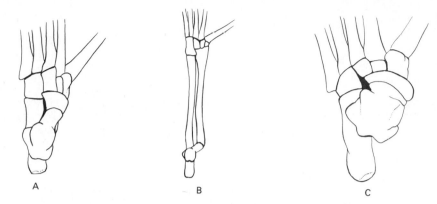

Figure 9-5. Dorsal views of the left feet of several primates, showing the tarsal bones: (A) lemur (*Lemur* sp., Lemuridae); (B) tarsier (*Tarsius,* Tarsiidae); (C) gorilla (*Gorilla gorilla,* Pongidae). Note the remarkable elongation of the calcaneum and the navicular in the tarsier. (From W. E. L. Clark, 1971)

in some species, and, as in higher primates, vocalization seems important in maintaining contact between members of a group. (Lemur social behavior is discussed on p. 396).

Family Indridae. The indrids, often called woolly lemurs, are not a diverse group, including but three genera and four species, and are restricted to Madagascar. There are Pleistocene and sub-Recent records of indrids from Madagascar, and matching the extinct huge lemurids are extinct indrids that are also of great size.

These animals are fairly large (up to 900 millimeters in head-and-body length). Two genera have a shortened rostrum and a monkey-like face; the snout is fairly long in the other genus. The dental formula is 2/2, 1/0, 2/2, 3/3 = 30. The upper incisors are enlarged, and the first lower premolar is caniniform. The hands and feet are highly modified for grasping branches during climbing (Fig. 9–4B). The pelage is conspicuously marked in some species, and the tail is long in three species.

These primates are largely herbivorous. Their leaf-eating habits resemble those of the Neotropical howler monkey (*Alouatta;* Cebidae) and the African colobus monkey (*Colobus;* Cercopithecidae). Indrids are typically fairly slow, deliberate climbers. The hind limbs are long relative to the front limbs, and when trav-

eling on the ground these primarily arboreal and diurnal animals proceed by a series of hops. The hands are used for climbing and for handling food, but manual dexterity seems limited and food is often picked up in the mouth. One genus is solitary or occurs in pairs; the other two genera typically live in small bands. A specialized laryngeal apparatus enables *Indri* to produce loud, resonant calls. These howls are given with greatest frequency in the morning and evening, as are the calls of the howler monkey, and perhaps function in maintaining territorial boundaries between neighboring bands. All indrid species are vocal to some extent.

Family Daubentoniidae. This family is represented by only one highly specialized living species *(Daubentonia madagascariensis)* with the common name of aye-aye. This nocturnal animal occurs locally in northern Madagascar, where it is restricted to dense forests and stands of bamboo. The fossil record of the Daubentoniidae consists of sub-Recent fossils from Madagascar of an extinct species that was larger than the surviving aye-aye.

Aye-ayes weigh approximately 2 kilograms. They have prominent ears and a long, bushy tail. The skull and dentition are remarkably specialized and depart strongly from the usual primate plan. The skull is short and moderately high.

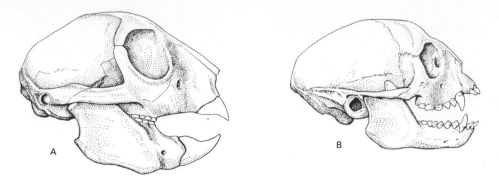

Figure 9-6. Skulls of primates: (A) aye-aye (*Daubentonia madagascariensis,* Daubentoniidae); length of skull 90 mm; (B) marmoset (*Saguinus geoffroyi,* Cebidae); length of skull 51 mm. (B after E. R. Hall and Kelson, 1959)

The orbit is prominent and faces largely forward; the postorbital bar and zygomatic arch are robust, and the rostrum is short and deep (Fig. 9–6A). The dentition differs from the basic primate type both in the extensive loss of teeth and in the strong specialization of the teeth that are retained. The dental formula is 1/1, 0-1/0, 1/0, 3/3 = 18 or 20. The canine is often absent, and the cheek teeth have flattened crowns with no clear cusp pattern. The laterally compressed incisors are greatly enlarged, wear to a sharply beveled edge because only the anterior surfaces are covered with enamel (as in rodents), and are ever-growing. Because of the shape of the teeth and the presence of a diastema between the incisors and the cheek teeth, *D. madagascariensis* was first described as a rodent. The hand is unique among primates. The digits are clawed, and all but the nonopposable pollex are long and slender; the third digit is remarkably slender (Fig. 9–4C). In the hind paw, the hallux is opposable and bears a nail, but the other digits are clawed.

Aye-ayes are arboreal, nocturnal, and mainly insectivorous. Their foraging technique is noteworthy. The elongate third finger is used to tap on wood harboring wood-boring insects; the aye-aye then listens carefully for insects within the wood, and the remarkable third digit is used for removing adult and larval insects from holes or fissures in the wood. When necessary, the powerful incisors tear away wood to enable the third digit to reach insects in deep burrows. Surprisingly, this strange mode of foraging is shared by two Australian genera of marsupials of the family Petauridae. In *Dactylonax,* the more specialized of these marsupials, the front incisors are modified and the manus is specialized along lines parallel to those in the hand of *Daubentonia,* except that the fourth rather than the third digit is the probing finger.

As is the case with many mammals that occupy limited areas, the future of the aye-aye seems dim, and its survival depends on the preservation of natural forests in northern Madagascar.

Family Lorisidae. The lorises are more widely distributed than are the primitive primates of Madagascar and are locally common. Lorises occur in Africa south of the Sahara, in India, Ceylon, and southeastern Asia, and in the East Indies. The fossil record of lorisids is scanty, but it suggests that these animals evolved in the Old World and have never occurred elsewhere. A Pliocene form is known from Asia. There are four genera and five species.

The eyes face forward in the lorisids (Fig. 9–7), rather than more or less to the side as in the lemurids, and the rostrum is short. Lorisids

Figure 9-7. A loris (*Loris tardigradus*, Lorisidae). (R. Garrison; San Diego Zoo)

Circulatory adaptations in the appendages provide for an increased blood supply to the digital flexor muscles, which are used in gripping branches during extended periods of contraction. Similar circulatory modifications, involving the formation of a *rete mirabile,* are also important in this and many other mammals in conserving body heat. (A *rete mirabile* is a complex meshwork of small arteries and veins that are intertwined so that the warm blood passing to an appendage is cooled by the cooler blood coming from the appendage and the cooler blood from the appendage is warmed by the arterial blood. One result of this system is the avoidance of much of the energy loss that would accompany the warming of drastically cooled blood from a poorly insulated limb as the blood entered the general bloodstream.)

These nocturnal primates are insectivorous and carnivorous, and prey is usually captured by the hands after a stealthy approach. The specialized lorisine genus *Arctocebus* spends considerable time upside down and is reported to sleep in this position (E. P. Walker, 1975: 419). ***Family Galagidae.*** Galagos are handsome animals that reach the size of a large squirrel and have very large eyes, expressive ears that resemble those of some bats, and a remarkable ability in some species to make prodigious arboreal leaps. Seven species of galagos within a single genus are recognized by W. C. O. Hill and Meester (1971). The tail is long and well furred and is used as a balancing organ during leaping. In contrast to lorises, galagos have unusually long hind limbs with powerful thigh muscles. The skull has a long rostrum (relative to those of other primates), and the dental formula is 2/2, 1/1, 3/3, 3/3 = 36; specialized lower incisors and canines are procumbent (Fig. 9-8A) and form a comblike structure (Fig. 9-9A) used in grooming the fur and in feeding on resin. The specialized hands and feet are well adapted to grasping: the thumb and hallux are both large and opposable, the fourth digits are unusually long, and the distalmost pads of the digits have well-developed traction ridges (Fig. 9-10) of

are arboreal, and their locomotion usually involves methodical hand-over-hand climbing. Lorises vary from the size of a rat to the size of a large squirrel. The braincase is globular, the facial part of the skull is often short and ventrally placed, and the anteriorly directed orbits are separated by a thin interorbital septum. The dental formula is 1-2/2, 1/1, 3/3, 3/3 = 34 or 36. The upper incisors are small, the lower canine is incisiform, and the molars are basically quadritubercular.

The manus and pes are specialized in a variety of ways for clutching branches. In the genus *Arctocebus,* a pincer-like hand has been developed by the reduction of digits two and three and a change in the posture of the remaining digits; the first digit of the pes is opposable and frequently greatly enlarged (Fig. 9-4D).

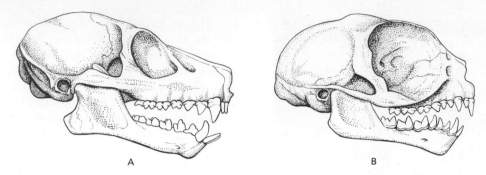

Figure 9-8. Skulls of prosimian primates: (A) galago (*Galago* sp., Galagidae); length of skull 65 mm; (B) tarsier (*Tarsius spectrum*, Tarsiidae); length of skull 36 mm. (From W. E. L. Clark, 1971)

importance during climbing. The second digit of the hind foot is short and bears a claw used for grooming, but all other digits bear flattened nails. The foot segment of the hind limb is long in association with the leaping ability; the elon-

gation involves the tarsus but is not as extreme as that in *Tarsius* (Fig. 9–5B).

Galagos are common in many sections of Africa, and the arboreal leaps of *Galago senegalensis* are a fascinating part of the twilight scene

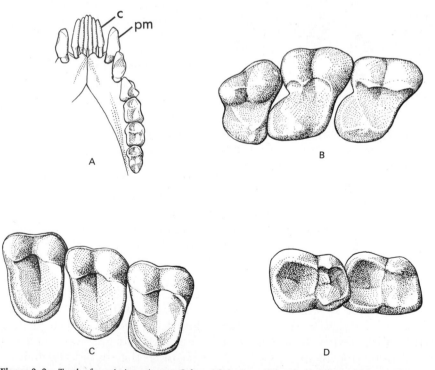

Figure 9-9. Teeth of prosimian primates. Galago (*Galago* sp., Galagidae): (A) lower right tooth row, showing the incisiform canine (*c*) and the caniniform premolar (*pm*); (B) upper right molars. Tarsier (*Tarsius spectrum*, Tarsiidae): (C) upper right molars; (D) first and second lower left molars. (From W. E. L. Clark, 1971)

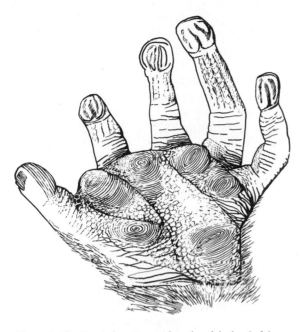

Figure 9-10. Traction patterns on the palm of the hand of the greater galago *(Galago crassicaudatus).*

in some parts of East Africa. This species (Fig. 9–11) usually takes refuge by day in family groups in holes in such trees as the baobab (Fig. 8–22), and the evening dispersal from these retreats often involves the use of repeatedly followed pathways through the trees. In Kenya, I often observed a galago that used the roof of our cabin and adjacent trees as part of its pathway. Many evenings after sunset, this animal could be heard landing on our thatched roof; it then took

six or so leaps to travel through a large umbrella tree *(Acacia tortilis),* after which it made an enormous leap, spanning about 4 meters horizontally, to a bush. It then dropped to the ground and hopped kangaroo-fashion some 20 meters to a group of trees, where it resumed its arboreal travel. A leap of 7 meters by a galago is mentioned by Kingdon (1971:318).

Galagos have an extremely varied diet but seem to prefer insects. When insects are abundant during the rainy season, galagos depend on this food source. A precisely judged leap terminating in a quick grab with one hand is a common style of capturing insects. Kingdon (1971:318) reports that *G. senegalensis* feeds also on seeds and small vertebrates and that it takes nectar from the large flowers of baobab trees. Of particular interest is the habit of eating the resin (gum) of trees such as acacias. In some areas with long dry seasons and periodic food shortages, this food, which contains polymers of pentose sugars, is the key to survival (R. D. Martin and Bearder, 1979).

Galagos have a variety of vocalizations that serve as warnings, as communication signals between mother and young, and perhaps as appeasement during intraspecific encounters. *Galago senegalensis* has a "vocabulary" of about ten basic sounds (Andersson, 1969), and the loud, raucous calls of the greater galago *(G. crassicaudatus)* are an impressive addition to the chorus of night sounds in wooded parts of East Africa.

Figure 9-11. Bushbabies, or lesser galagos *(Galago senegalensis).* (A) A group of five that slept together on a *Commiphora* branch during the day; (B) an individual preparing to leap from a *Commiphora* branch. (Taken in southern Kenya, East Africa. A by T. R. Huels, B by R. G. Bowker)

Suborder Haplorhini

The five families within this suborder are "higher" primates and have many progressive features not typical of members of the suborder Strepsirhini. Haplorhines—tarsiers, monkeys, marmosets, apes, and humans—are the most familiar primates and are vastly more important in terms of taxonomic diversity and adaptability than are the relatively primitive strepsirhines.

Family Tarsiidae. This family is represented today by three species of the genus *Tarsius* and occurs in jungles and secondary growth in Borneo, southern Sumatra, some East Indian islands, and some of the Philippine Islands. Tarsiids are known from the Eocene record of Europe, but there are no fossils representing the remainder of the Cenozoic.

The tarsier is roughly the size of a small rat and, with its large head, huge eyes, long limbs, and long tail, has a distinctive appearance (Fig. 9–12). The most conspicuous cranial features are the enormous orbits, which face forward and have expanded rims and a thin interorbital septum (Fig. 9–8B). The eye of the tarsier is apparently adapted entirely to night vision, for it lacks cones in the retina. The dental formula is 2/1, 1/1, 3/3, 3/3 = 34. The medial upper incisors are enlarged, the premolars are simple, the crowns of the upper molars are roughly triangular, and the lower molars have large talonids (Fig. 9–9D). The neck is short, a characteristic of many saltatorial vertebrates. All but the clawed second and third pedal digits have flat nails, and all digits have disklike pads (Fig. 9–13A). The limbs, especially the hind ones, are elongate; the tibia and fibula are fused.

The trend toward jumping ability that is apparent in galagos is developed to an extreme degree in the family Tarsiidae. As in all highly specialized jumpers, the hind foot is elongate, but in the tarsier the elongation is unique. It involves two tarsal bones (hence the name *Tarsius*) rather than metatarsals, as in such jumpers as jerboas and kangaroo rats (Fig. 16–14D,E). In *Tarsius,* the calcaneum and navicular are greatly elongate (Fig. 9–5B), whereas the metatarsals are not unusually long in relation to the phalanges (Fig. 9–13A). An important functional end is achieved by this unusual system of foot elongation: because the elongation has occurred in the tarsus, the dexterity and grasping

Figure 9-12. Tarsiers (*Tarsius* sp., Tarsiidae). (R. Garrison; San Diego Zoo)

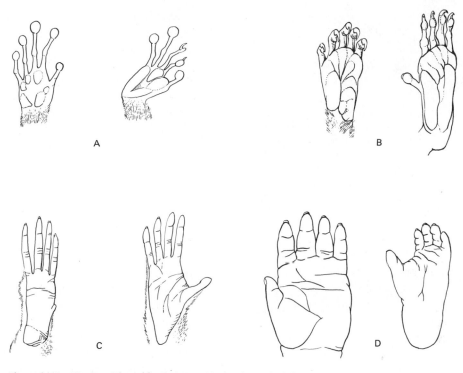

Figure 9–13. Hands and feet of four primates (the hand is on the left in each pair). (A) Tarsier (*Tarsius spectrum,* Tarsiidae); (B) marmoset (*Callithrix* sp., Cebidae); (C) woolly spider monkey (*Brachyteles arachnoides,* Cebidae); (D) gorilla (*Gorilla gorilla,* Pongidae). (A, B, D from W. E. L. Clark, 1971)

ability of the digits themselves (the metatarsals and phalanges) have not been sacrificed. A reduction of dexterity would have accompanied an elongation of the metatarsals and the resulting functional reduction of the number of digit segments. In elephant shrews and kangaroo rats, dexterity and gripping ability of the hind foot are not important and a more "direct" means of elongation—lengthening of the already somewhat elongate metatarsals—occurred.

Tarsiers are primarily arboreal and nocturnal and feed largely on insects, which they pounce upon and grasp with the hands. Fogden (1974:161) observed tarsiers quietly watching and waiting on a low perch and leaping down to the ground to capture insects. Although more highly adapted to leaping than any other primate, tarsiers can walk and climb quadrupedally, hop or run on their hind legs on the ground, and slide down branches (Sprankel, 1965). Tarsiers and some species of galagos share the ability to leap long distances with great precision, and in both of these types of primates the landing from a leap is largely bipedal. In association with jumping ability, much of the weight of the tarsier is concentrated in the hind limbs, which together constitute 21 percent of the total weight of the animal; the musculature of the thighs alone equals 12 percent of the body weight, largely owing to great enlargement of the quadriceps femoris (Grand and Lorenz, 1968), a powerful extensor of the shank. Some work has shown that tarsiers usually live in pairs, but Fogden (1974:162) observed single animals most frequently, suggesting that only large, dominant males associate with females. Perhaps the relative silence of tarsiers reflects their limited social life.

Figure 9-14. A howler monkey (*Alouatta palliata*, Cebidae). (San Diego Zoo)

Family Cebidae (Fig. 9-14). The New World monkeys belong to this family, which includes some 45 Recent species of 16 genera. Cebids range from southern Mexico through Central America to southern Brazil. They first appear in the early Oligocene record of South America. Primitive primates ancestral to the cebids may have entered South America from Central America on logs or debris that floated across the stretch of water that separated these land masses during much of the Tertiary Period.

Cebids, exclusive of the marmosets and tamarins, have the characters outlined in this paragraph. The limbs are elongate, the digits bear curved nails, and the pollex is not opposable and is small or absent in some species (Fig. 9-13C). The hallux is strongly opposable. The tail is typically long and is prehensile in many species. Size ranges from about 275 grams to 9 kilograms. The skull is more or less globular, with a high braincase and short rostrum. The orbits face forward, and the nostrils are separated by a broad internarial pad and face to the side. The dental formula is 2/2, 1/1, 3/3, 3/3 = 36, and the lateral pair and medial pair of cusps of the quadrate molars are separated by a central, anteroposteriorly aligned depression (Fig. 9-15B). Brightly colored and bare patches of skin on the rump (ischial callosities) do not occur in cebids, as they often do in Old World monkeys (family Cercopithecidae).

The subfamily Callitrichinae, the marmosets and tamarins, differ from other cebids in having triangular molars, in lacking the third molar (2/2, 1/1, 3/3, 2/2 = 32), in having chisel-shaped medial incisors, and in having claws on all digits but the hallux, which bears a nail. Marmosets and tamarins are small, weighing from some 70 to 1000 grams, and the heads of several species are adorned by manes or conspicuous tufts of fur. The hand resembles that of a squirrel (Fig. 9-13B).

Cebids typically occur in tropical forests, and most are diurnal. They are basically vegetarians; fruit is often preferred, but a wide variety of plant and animal material is eaten. The night monkey *(Aotes trivirgatus),* an inhabitant of Central and South America, is apparently insectivorous and carnivorous and may feed to some extent on bats. Cebids are active, intelligent animals and adroit climbers; some species move with amazing speed through the trees. For dazzling arboreal ability, the Neotropical spider monkey *(Ateles)* is probably surpassed only by the Old World gibbons (*Hylobates,* Pongidae).

Most cebids are vocal to some extent, and several species have loud, penetrating calls. Outstanding among these is the howler monkey, in which the hyoid apparatus is enlarged into a resonating chamber. The males emit loud roaring sounds that carry for long distances through the tropical rain forest. These sounds are seemingly important in keeping together the members of a troop, which may include up to 40 individuals. The territories of troops are probably announced and partly maintained by the loud vocalizations. Most cebids are gregarious, the most common social aggregation consisting of a

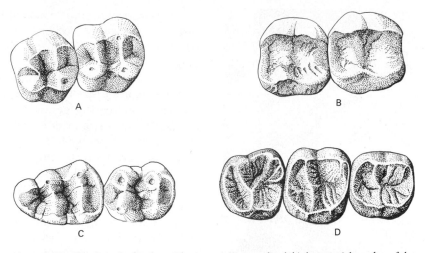

Figure 9-15. Cheek teeth of anthropoid primates: (A) second and third upper right molars of the orange-crown mangabey (*Cercocebus torquatus,* Cercopithecidae); (B) first and second upper right molars of a saki monkey (*Pithecia monachus,* Cebidae); (C) second and third lower left molars of a baboon (*Papio* sp., Cercopithecidae); (D) upper right molars of the orangutan (*Pongo pygmaeus,* Pongidae). Note the cross lophs on the teeth of the mangabey and the baboon, and the extra posterior cusp (hypoconulid) on the third lower molar of the baboon. (D from W. E. L. Clark, 1971)

family group. Some species form unusually large troops; the squirrel monkey *(Saimiri)* occurs in bands of up to 100 animals.

Marmosets and tamarins are omnivorous. The diet consists mostly of fruit and insects; lizards and small birds and their eggs may be important foods for some species. Marmosets are social and typically live in small family groups. In some Neotropical areas, they are the most common primates. These animals are often extremely vocal and emit a variety of high-pitched sounds and piercing alarm calls, some of which resemble those of birds that live in the same areas.

Family Cercopithecidae. These are the Old World monkeys and are the most successful primates in terms of numbers of species (some 70 Recent species of 12 genera). They occupy a wide range, including Gibraltar, northwestern Africa, Africa south of the Sahara, southern Arabian Peninsula, much of southeastern Asia east to Japan, Indonesia east to Timor, and the Philippine Islands. Among nonhominid primates,

cercopithecids have the greatest tolerance for cold climates; some occupy high forests in Tibet, and others live in northern Honshu, Japan, where winter snows occur. Cercopithecids first appear in the Oligocene record of Egypt and, like the living species, fossil forms are known only from the Old World. Just as in many other groups, some Pleistocene cercopithecids reached large sizes; an extinct South African baboon of this epoch reached the size of a gorilla.

In weight, cercopithecids range from 1.5 to over 50 kilograms, and some species are stocky in build, quite unlike most cebids. The nostrils are close together and face downward (Fig. 9–16), a condition termed catarrhine. The skull is often robust and heavily ridged, and, relative to that of cebids, the rostrum is long (particularly in the baboons). The dental formula is 2/2, 1/1, 2/2, 3/3 = 32, as in the apes (Pongidae) and humans (Hominidae). The medial upper incisors are often broad and roughly spoon-shaped; the upper canines are usually large and in some

Figure 9-16. Two cercopithecid monkeys: (A) Japanese macaques *(Macaca fuscata)* in northern Honshu, Japan; (B) African blue monkey *(Cercopithecus mitis)* in Kenya, East Africa. (A by C. B. Koford)

inner pair by two transverse ridges producing a bilophodont tooth. The last lower molar has an additional posterior cusp, the hypoconulid (Fig. 9–15C).

All of the digits have nails, and the pollex and hallux are opposable except in the strongly arboreal, leaf-eating genus *Colobus,* in which the pollex is vestigial or absent. The tail is vestigial in some species but long in others. Ischial callosities are well developed in many species, and the bare rump skin is frequently bright red. The conspicuous patch is used in conjunction with ritualized postures as a means of communication between members of a social group. Bare facial skin may also be red but is bright blue in the mandrill *(Papio sphinx)*. These patches of skin are more brightly colored in the male than in the female in some species. The olfactory epithelium is greatly reduced in cercopithecids, and apparently their sense of smell is rudimentary. The facial muscles are well developed and produce a wide variety of facial expressions. Some cercopithecids are brightly or conspicuously marked. For example, the variegated langur of Indochina *(Pygathrix nemaeus)* has a bright yellow face, a chestnut strip beneath the ears, black and chestnut limbs, a gray body, and a white rump and tail.

Although most cercopithecids are probably largely omnivorous, some are adapted to an herbivorous diet. Members of the subfamily Colobinae (the arboreal langurs and colobus monkeys) are herbivorous and frugivorous, and some species seem to feed primarily on leaves. The baboons *(Papio, Theropithecus)* are the most successful terrestrial cercopithecids, and one species, in areas where suitable trees are not available, assembles in large groups (up to 750 individuals) on cliffs (Kummer, 1968a:310). The remarkably complicated social behavior of baboons and of certain other Old World monkeys is reasonably well known (De Vore, 1975; Jay, 1968) and is discussed on page 397.

Interesting contrasts between the behavior of the baboons and that of the equally terrestrial

species are tusklike. When the jaws are closed, the lower canine rests in a diastema between the upper canine and the last incisor. The first lower premolar is enlarged and forms a shearing blade that rides against the sharp posterior edge of the upper canine (Fig. 9–17A). Most of the molars have four cusps, the outer pair connected to the

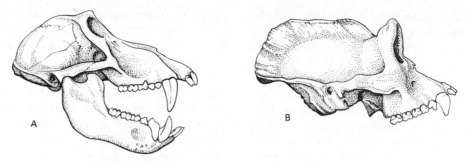

Figure 9-17. Skulls of haplorhine primates: (A) baboon (*Papio* sp., Cercopithecidae); length of skull 200 mm; (B) gorilla (*Gorilla gorilla,* Pongidae); length of skull 320 mm.

patas monkey *(Cercopithecus patas)* are discussed by K. R. L. Hall (1968:114). Savanna baboons are highly vocal, live in fairly large troops controlled by several dominant males, and are prone to noisy, rough, aggressive interactions. In contrast, the patas monkey usually maintains "adaptive silence" but has a repertoire of soft calls, lives in small troops, each with a single adult male that serves as a watchdog, is rarely aggressive, and never fights. The patas monkey has a slim greyhound-like build and is the fastest runner of all primates, having been timed at a speed of 55 kilometers per hour. Adaptations for speed in this animal include elongation of the limbs, carpals, and tarsals; shortening of the digits; reduction of the pollex and hallux; and the development of palmar and plantar pads. This remarkably cursorial primate has a quiet mode of life, usually attempts to escape detection, and depends on its speed to escape danger. In these respects, the patas monkey is the primate counterpart of the small antelope. The noisy baboon troop, however, frequently depends on its aggressive dominant males to confront and discourage a predator, and terrestrial locomotion is relatively unimportant as a means of escaping enemies.

Sexual dimorphism is pronounced in both the baboons and the patas monkey, as it is in many primates. The male baboon of South and East Africa weighs roughly 33 kilograms, the female 16.5 kilograms; the male patas monkey averages 13 kilograms, and the female 6.5 kilograms (K R L Hall, 1968:114). Probably all cercopithecids are basically social, and vocalizations and facial expressions play central roles in social interactions. The life span of these monkeys is long: a Chacma baboon *(Papio ursinus)* lived in captivity for 45 years, and life spans of 20 or 25 years in the wild may be common (Walker, 1968:447).

Family Pongidae. The gibbons and great apes are included in this family, members of which occur in equatorial Africa, southeastern Asia, Java, Borneo, Sumatra, and the Mentawi Islands. Eight Recent species of four genera are known; all species are restricted to tropical forests. The pongid record begins in the Oligocene of Egypt, when the gibbon-like genus *Propliopithecus* appears. Pongids evolved in Africa, reached Europe in the Miocene and Asia in the Pliocene and never occurred in the New World.

The family Pongidae includes two subfamilies that differ markedly in structure and mode of life. The gibbons, subfamily Hylobatinae, are arboreal and are the most rapid and spectacular climbers and brachiators of all mammals. Gibbons are relatively small, weighing from 5 to 13 kilograms, have remarkably long arms, and, during brachiation, use the hands like hooks rather than as grasping structures.

The great apes, subfamily Ponginae, vary from 48 to 270 kilograms in weight and have robust bodies and powerful arms. The hands and feet are similar to those of humans, but the hallux is opposable (Fig. 9–13D).

The pongid skull is typically robust and in older animals is marked by bony crests and ridges; it is long relative to its width (Fig. 9–17B). The teeth are large. The dental formula is 2/2, 1/1, 2/2, 3/3 = 32, as in cercopithecids and humans. The incisors are broad, and the premaxillae and anterior parts of the dentary bones are broadened to accommodate them. The canines are large and stoutly built but never tusklike. The upper molars are quadrangular and basically four-cusped, and the lower molars have an additional posterior cusp (hypoconulid). In contrast to cercopithecids, a trend toward elongation of the molars does not occur in pongids, and the molars lack well-defined cross ridges (Fig. 9–15D). The tooth rows are parallel, and the mandibular symphysis is braced by a bony shelf (the "simian shelf").

The forelimbs are longer than the hind limbs, and the hands are longer than the feet; all digits bear nails. Pongids have no tails. The thorax is wide, and the scapula has an elongate vertebral border. Adaptations allowing advantageous muscle attachments during erect or semi-erect stances include lengthening of the pelvis and enlargement and lateral flaring of the ilium. Regarding structural details, locomotor ability, brain size, and level of intelligence, the great apes are closer to humans than are any other mammals.

Pongids are largely vegetarians, but some are occasionally carnivorous. The chimpanzee (*Chimpansee*), for example, occasionally catches and eats the colobus monkey. Arboreal locomotion in pongids involves brachiation in some species. The gorilla (*Gorilla*) and chimpanzee are mostly terrestrial, but both are able climbers. Although capable of bipedal stance and limited bipedal locomotion, they are mostly quadrupedal. The behavior of pongids and cercopithecids has been studied intensively in recent years. Eisenberg (1981:160) lists some important literature on primate ecology and behavior. Owing primarily to the efforts of humans, who act almost as if their survival and prosperity were threatened by their next of kin, some of the great apes are dangerously close to extinction. Destruction of habitat and killing of the animals, fostered by our anachronistic feeling that our position as the dominant form of life justifies any form of exploitation of our environment, has led to serious reduction of some primate populations. Populations of wild orangutans (*Pongo pygmaeus*) are small and these are restricted to parts of the islands of Sumatra and Borneo. As a sad stroke of irony, an important drain on the declining populations of orangutans resulted from their capture and exportation to European and American zoos, institutions dedicated in part to the preservation of vanishing species.

Family Hominidae.   Humans are the only living members of the family Hominidae. In humans, the skull has a greatly inflated cranium, housing a large cerebrum, and the rostral part is virtually absent. The foramen magnum is beneath the skull, a feature associated with an upright stance. The dentition is not as robust as in the pongids: the incisors are less broad, the canines typically rise but slightly above adjacent teeth, and the cheek teeth are less heavily built. The premolars are usually bicuspid. The upper molars have four cusps; the first lower molar has five cusps, the second has four, and the third has five. The dental formula 2/2, 1/1, 2/2, 3/3 = 32 occurs in most individuals, but one or more of the posterior molars (the "wisdom teeth") may not appear. The tooth rows are not parallel, as they are in pongids, nor is the simian shelf present in the mandible. The pollex, but not the hallux, is opposable. With a change in posture and use of the forelimbs, the thorax has become broad and the scapulae have come to lie dorsal to the rib cage, as in bats, rather than lateral to the rib cage, as in most other mammals. As in the case of many primates, in humans the males are considerably larger than the females.

If the same standards used in the classification of other mammalian orders were applied to the primates, humans would be included with the gibbons and great apes in a single family. "Differences of no greater magnitude than those separating the hominids and the pongids characterize subfamilies in some other orders of mammals" (S. Anderson, 1967a:177). The ancestry of hominids can seemingly be traced back to the Miocene Epoch.

The astounding growth of human populations throughout much of the world is leading to progressively more acute problems that threaten, if not our very survival, at least our present style of life. It is obviously essential that people of different races learn to accept and appreciate living and working together as equals and that effective birth control measures be taken in most parts of the world in order to halt population growth. Our adaptive ability is being put to more critical tests today than ever before; our future may well be determined by choices we make within the next decade or two.

Chapter 10

Carnivores

Predation in mammals is an ancient and profitable, if not entirely honorable, occupation. Primitive carnivorous mammals (creodonts) appear in the early Paleocene before most of the Recent mammalian orders. Mammalian carnivores probably evolved in response to the food source offered by an expanding array of herbivorous mammals and underwent adaptive radiation as herbivores diversified.

The classification of carnivorous mammals used here recognizes the order Creodonta (primitive carnivorous mammals) and the order Carnivora. Included in the latter are the terrestrial carnivores (such as dogs, weasels, and bears) and Pinnipedia aquatic carnivores (seals, sea lions, and walruses).

Most Recent terrestrial carnivores are predaceous and have a remarkable sense of smell. Cursorial ability may be limited, as in the Ursidae and Procyonidae, or may be strongly devel-

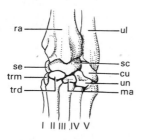

Figure 10–1. Anterior view of the left carpus of the grey fox *(Urocyon cinereoargenteus).* Abbreviations: *cu*, cuneiform; *ma*, magnum; *ra*, radius; *se*, sesamoid; *sc*, scapholunar; *trd*, trapezoid; *trm*, trapezium; *ul*, ulna; *un*, unciform.

oped, as in the cheetah and some canids. The braincase is large; the orbit is usually confluent with the temporal fossa; the turbinal bones are usually large, and their complex form provides a large surface area for olfactory epithelium. There are usually 3/3 incisors (3/2 in the sea otter, *Enhydra lutris*), and the canines are large and usually conical; the cheek teeth vary from 4/4 premolars and 2/3 molars in long-faced carnivores, such as the Canidae and Ursidae, to 2/2 premolars and 1/1 molars, as in some cats. The fourth upper premolar and the first lower molar are carnassials (specialized shearing blades). The teeth are rooted. The condyle of the dentary bone and the glenoid fossa of the squamosal bone are transversely elongate and allow no rotary jaw action and only limited transverse movement.

Cursorial adaptations evident in the carpus include the fusion of the scaphoid and lunar bones and the loss of the centrale (Fig. 10–1). The foot posture is plantigrade, as in ursids and procyonids, or digitigrade, as in canids, hyaenids, and felids. Little reduction of digits has occurred; the greatest reduction occurs in the hyenas and in the African hunting dog *(Lycaon pictus),* in which the manus and pes have four toes.

Order Creodonta

The oldest carnivorous mammals, order Creodonta, appeared in the early Paleocene, were the typical carnivores of the Paleocene and Eocene, and persisted in Old World tropical refugia into the early Pliocene. Some creodonts appear to have retained the insectivorous food

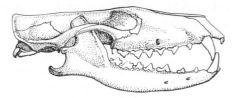

Figure 10-2. The skull of *Sinopa,* an Eocene creodont in which the carnassials are M^2 and M_3; length of skull approximately 150 mm. (After Romer, 1966)

habits of their ancestors, whereas some were carnivorous and some were omnivorous. Two Eocene genera had saber-like upper canines and probably killed relatively large prey. The creodont skull (Fig. 10–2) differs from that of the Carnivora in lacking an ossified auditory bulla and in having either no carnassial pair or having M^1 and M_2 or M^2 and M_3 forming the carnassials. (In these abbreviations for teeth, the capital letter represents the type of tooth, the number is the number of the tooth, and the superscript and subscript indicate upper and lower, respectively. For example, upper premolar four is PM^4 and lower molar one is M_1.)

The limbs of creodonts were primitive. The feet were usually five-toed and plantigrade, and the limbs were often short; the scaphoid and lunar of the carpus were not fused, and the centrale was present; the distal phalanges were fissured and in some species bore flattened, rather than clawlike, nails. Creodonts diversified in the Paleocene and Eocene epochs, and the Late Eocene genus *Andrewsarchus* is the largest known land carnivore. Its skull is roughly 1 meter in length.

Order Carnivora

Terrestrial Carnivores

Modern terrestrial carnivores are basically predatory types. The Carnivora largely replaced the Creodonta early in the Tertiary. The basal family of the Carnivora is the Miacidae, which first ap-

pears in middle Paleocene beds of North America and is not known after the Eocene. Resemblances between early miacids and *Cimolestes,* a Late Cretaceous insectivore of the extinct family Palaeoryctidae, suggest that the order Carnivora evolved from an ancestral stock within the order Insectivora. Miacids were small and perhaps mostly arboreal carnivores. They had the modern carnassial arrangement (PM^4 over M_1) but lacked ossified bullae. In contrast to the creodonts, the distal phalanges of miacids were not fissured.

Family Canidae. Twelve genera and about 34 Recent species make up this familiar family. Canids—wolves (Fig. 10–3), jackals (Fig. 10–4A), foxes (Fig. 10–4B), and dogs—occupy a great array of environments from arctic to trop-

Figure 10-3. An Alaskan wolf. (From the classic study *The Wolves of Mount McKinley,* by Adolf Murie, 1944)

Figure 10–4. (A) Black-backed jackal *(Canis mesomelas)* and (B) bat-eared fox *(Otocyon megalotis)*. These canids occur together in East Africa; the jackal eats mostly small mammals, the young of small antelope, and carrion, whereas the fox eats mostly insects, rodents, and lizards. (Taken in Kenya, by K. P. Dial)

ical. Prior to their dispersal with humans, canids occurred nearly worldwide except on most oceanic islands. The dingo *(Canis dingo)* was probably brought to Australia by early humans. The canids appeared in the late Eocene in Europe and North America and have occupied these areas continuously to Recent times.

Canids are broadly adapted carnivores; this is reflected in their morphology. The canid skull typically has a long rostrum (Fig. 10–5A) that houses a large nasal chamber with complex turbinal bones, a feature associated with a remarkable sense of smell. Most canids have a nearly complete placental complement of teeth (3/3, 1/1, 4/4, 2/3 = 42); the canines are generally long and strongly built, and the carnassials retain the shearing blades (Fig. 10–6A,B). The postcarnassial teeth have crushing surfaces, indicating a more flexible diet than that of the more strictly carnivorous cat family (Felidae). The limbs in the more cursorial species are long, and rotation at the joints distal to the shoulder and hip joints is reduced in the interest of cursorial ability. The clavicle is absent. The feet are digitigrade, and the well-developed but blunt claws are nonretractile. The forepaw usually has five toes, and the hind paw has four. The weight of canids ranges from 1 to 75 kilograms.

Some kinds of canids forage tirelessly over large areas, and lengthy pursuit is frequently part of the hunting technique. The coyote *(Canis latrans),* probably one of the swiftest canids, can run at speeds of up to about 65 ki-

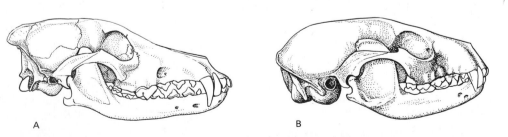

A   B

Figure 10–5. Skulls of two carnivores: (A) coyote *(Canis latrans);* length of skull 202 mm; (B) raccoon *(Procyon lotor);* length of skull 115 mm.

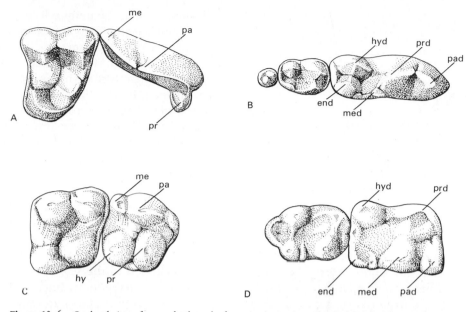

Figure 10-6. Occlusal view of some cheek teeth of two carnivores; in each pair the right upper teeth are on the left and the left lower teeth are on the right. The cusps of the carnassials (P^4 and M_1) are labeled. (A, B) Coyote *(Canis latrans)*; the cheek teeth both shear and crush. (C, D) Raccoon *(Procyon lotor)*; note that the upper carnassial has a hypocone and that all the teeth are adapted to crushing. Abbreviations; *end,* entoconid; *hy,* hypocone; *hyd,* hypoconid; *me,* metacone; *med,* metaconid; *pa,* paracone; *pad,* paraconid; *pas,* parastyle; *pr,* protocone; *prd,* protoconid.

lometers per hour. The fact that coyotes in many areas depend partly on jackrabbits for food is an impressive testimonial to this carnivore's speed. Canids often hunt in open country, and wolves *(Canis lupus)* and the African hunting dog *(Lycaon pictus)* seem to rely more on endurance than on speed when hunting. These canids, and the eastern Asian dholes *(Cuon alpinus),* habitually hunt in packs and kill larger prey than could be overcome by a solitary hunter. The gray fox *(Urocyon cinereoargenteus)* does not generally forage in open areas but is amazingly agile and can run rapidly through the maze of stems beneath a canopy of chaparral. The food of canids includes vertebrates, arthropods, mollusks, carrion, and many types of plant material. Blackbacked jackals *(Canis mesomelas)* have become a problem in parts of South Africa because of their extensive feeding on pineapples (Ewer, 1968:30), and coyotes in parts of the western United States feed heavily on such cultivated crops as melons and such uncultivated plant material as juniper berries and prickly-pear cactus fruit. The average canid is clearly an opportunist; this may in large part account for the great success of this family.

Family Ursidae. The bears are notable for their large size and their departure from a strictly carnivorous mode of life. This family contains six genera and eight species. Morphological and physiological evidence suggests that the giant panda, *Ailuropoda melanoleuca,* is a bear, and it is included here in the Ursidae. The distribution of ursids includes most of North America and Eurasia, the Malay Peninsula, the South American Andes, and the Atlas Mountains of extreme northwestern Africa. Bears inhabit diverse habitats, from drifting ice in the Arctic to the tropics, but are most important in boreal and temperate areas.

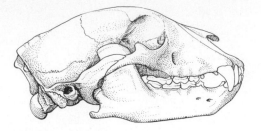

Figure 10–7. Skull of a black bear *(Ursus americanus)*; length of skull 289 mm. Note the diastema behind the anteriormost cheek tooth and the small upper carnassial (the third tooth in back of the canine).

Bears first appeared in Europe in the early Oligocene but did not enter North America until Miocene times. They probably reached South America and northwestern Africa in the Pleistocene and Recent respectively.

The bear skull retains the long rostrum typical of the canids, but the orbits are generally smaller and the dentition is very different (Fig. 10–7). The postcarnassial teeth are greatly enlarged, and the occlusal surfaces are "wrinkled" and adapted to crushing (Fig. 10–8B). On the other hand, the first three premolars are usually rudimentary or may be lost, and a diastema usually occurs between premolars. The upper carnassial is roughly triangular because of the posterior migration of the protocone and is

much smaller than the neighboring molars (Fig. 10–8B); both upper and lower carnassials no longer have a shearing function. The dental formula is usually 3/3, 1/1, 4/4, 2/3 = 42, but premolars may be lost with advancing age. The limbs, especially the forelimbs, are strongly built; the plantigrade feet have long, nonretractile claws. There are five toes on each foot. The ears are small, and the tail is extremely short. In size, bears range from that of a large dog to over 760 kilograms.

The abandonment of cursorial ability in favor of power in the limbs and the loss of the shearing function of the cheek teeth in favor of a crushing battery have accompanied the adoption of omnivorous feeding habits. The strong forelimbs can aid in the search for food by rolling stones or tearing apart logs, and the crushing surfaces of the molars can cope with many kinds of food, from insects and small vertebrates to berries, grass, and pine nuts. Carrion is also avidly sought. The polar bear *(Ursus maritimus)* has a more restricted diet, consisting largely of seals, and the giant panda eats mostly bamboo shoots. In areas with cold winters, bears hibernate for much of the winter in caves or other retreats protected from drastic temperature fluctuations.

Family Procyonidae. This family includes the raccoons *(Procyon)* and ring-tailed "cats" *(Bassariscus)* and their relatives. As with the

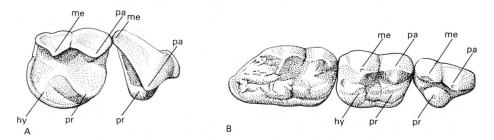

Figure 10–8. Occlusal views of some upper cheek teeth of two omnivorous carnivores. (A) Right upper carnassial (P^4) and first molar of the hog-nosed skunk *(Conepatus mesoleucus)*; note the blade on the carnassial and the broadened crushing surface on M^1. (B) Right upper carnassial and two molars of the black bear *(Ursus americanus)*; note the small, nontrenchant carnassial and the greatly lengthened molars. Abbreviations are as in Figure 10-6.

bears, omnivorous feeding habits have become predominant in procyonids. Seven genera and 19 species are known. The taxonomic position of the lesser panda *(Ailurus)* is controversial; it is here regarded as a procyonid. Procyonids occupy much of the temperate and tropical parts of the New World, from southern Canada through much of South America. The lesser panda occurs in south-central China, northern Burma, Sikkim, and Nepal. Procyonids chiefly inhabit forested areas, but the range of one species of ring-tailed cat *(Bassariscus astutus)* includes arid desert mountains and foothills. Procyonids are known from the late Oligocene to the Recent in North America, from the late Miocene to the late Pliocene in Europe, and from the Pliocene to the Recent in Asia. They reached South America from North America in the Pliocene.

The structural and functional departure of procyonids from the carnivorous norm has included adaptations favoring both omnivorous feeding habits and climbing ability. Associated with the omnivorous trend has been a specialization of the cheek teeth. The premolars are not reduced (Fig. 10–5B), as in the bears, but the shearing action of the carnassials is nearly lost. Instead, the carnassials are low-cusped crushing teeth; a hypocone was added to the upper, and in the lower the talonid was enlarged and broadened (Fig. 10–6C,D). In contrast to the elongate upper molars of bears, those of procyonids are usually broader than they are long. The coati *(Nasua narica)* has flattened, bladelike canines that are formidable defensive weapons. The dental formula is usually 3/3, 1/1, 4/4, 2/2 = 40. There are five toes on each foot; the foot posture is usually plantigrade, and the claws are nonretractile or semiretractile. The limbs are fairly long. The toes are separate, and the fore-paw has considerable dexterity in some species and is used in handling food. Tracks left by the human-like hand of the raccoon are familiar to many people. The tail is long, generally marked by dark rings, and prehensile in the arboreal kinkajou *(Potos flavus)*. Procyonids are of modest size, weighing from less than 1 to about 20 kilograms.

The familiar raccoon often takes advantage of cultivated crops. Corn is a staple food item for Midwestern raccoons, and they eat grapes, figs, and melons in parts of California (J. Grinnell, Dixon, and Linsdale, 1937:159). In addition, they prey on a variety of small vertebrates and some invertebrates. Some tropical procyonids are largely vegetarians. E. R. Hall and Dalquest (1963) report that in Veracruz the coati eats corn, bananas, and the fruit of the coyol palm and that kinkajous eat mostly fruit. The ring-tailed cat, on the other hand, is known to feed mostly on small rodents in some areas. Procyonids reach their greatest diversity and greatest densities in the Neotropics, where they are largely arboreal; in some tropical forests several species may occur together. In such areas, the nocturnal, quavering cries of kinkajous can be heard regularly.

Coatis are social animals and assemble in female-young tribes of from five to 20 or so animals. These animals are highly vocal and have a varied repertoire of communication calls.

Family Mustelidae. This large family, with 23 genera and some 63 Recent species, includes the weasels, badgers, skunks, and otters (Fig. 10–9). Mustelids occupy virtually every type of terrestrial habitat, from arctic tundra to tropical rain forests, and live in rivers, lakes, and the sea. The distribution is nearly cosmopolitan, but they do not inhabit Madagascar, Australia, or oceanic islands. Mustelids appear in the fossil records of North America and Eurasia in the early Oligocene, but they did not reach South America and Africa until the Pliocene.

These are typically fairly small, long-bodied carnivores with short limbs and a pushed-in face. The skull generally has a long braincase and a short rostrum (Figs. 10–10B and 10–11), and the postglenoid process partially encloses the glenoid fossa so that in some species the condyle of the dentary bone is difficult to disengage from the fossa. Obviously, little lateral and no rotary jaw action is possible. The denti-

Figure 10-11. The skull of the sea otter *(Enhydra lutris)*; length of skull 152 mm. The heavy cheek teeth are adapted to crushing marine invertebrates.

Figure 10-9. (A) California sea otter *(Enhydra lutris)* feeding. (B) Closeup of a sea otter's face. (A by N. Siepel-Hyatt; B courtesy California Game and Fish Department)

tion is quite variable but is generally 3/3, 1/1, 3/3, 1/2 = 34. The carnassials are trenchant in many species (Fig. 10–12C,D) but have been modified into crushing teeth in others; in the sea otter *(Enhydra lutris),* for example, none of the cheek teeth are trenchant, the carnassials have rounded cusps adapted to crushing, and the postcarnassial teeth, M^1 and M_2, are broader than they are long (Fig. 10–12E,F). The first upper molar is frequently hourglass-shaped in occlusal view (Fig. 10–12C) or may be expanded into a large crushing tooth, as in skunks (Fig. 10–8A). The limbs are usually short, the five-toed feet are either plantigrade or digitigrade, and the claws are never completely retractile. Anal scent glands are usually well developed; they are extraordinarily large in skunks and are used for defense. The tail is generally long, and the pelage may be conspicuously marked, as in skunks and badgers. Some mustelids have beautiful, glossy fur that has considerable value in the fur trade. In size, mustelids range from the smallest member of the order Carnivora, a circumboreal weasel *(Mustela nivalis)* which weighs 35 to 50 grams, to the fairly large sea otter (about 35 kilograms).

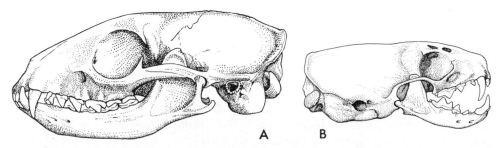

Figure 10-10. Small carnivore skulls: (A) white-tail mongoose *(Ichneumia albicauda)*; length of skull 106 mm; (B) least weasel *(Mustela nivalis)*; length of skull 31 mm.

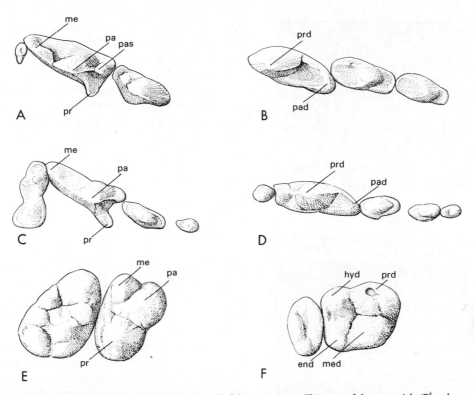

Figure 10-12. Occlusal views of the cheek teeth of three carnivores. The cusps of the carnassials (P^4 and M_1) are labeled, and abbreviations are as in Figure 10-6. In A the parastyle *(pas)* is also shown. (A) Upper and (B) lower entire sets of cheek teeth of the bobcat *(Lynx rufus)*. Note the lack of crushing teeth. Only sectorial teeth are present; the parastyle of P^4 increases the length of its shearing blade, and the loss of the talonid of M_1 makes this tooth entirely bladelike. (C) Upper and (D) lower entire sets of cheek teeth of the least weasel *(Mustela nivalis)*; the crushing teeth, M^1, and M_2 are reduced, and the shearing function of the cheek teeth is of major importance. (E) P^4 and M^1 and (F) M_1 and M_2 of the sea otter *(Enhydra lutris)*. Note that all the teeth have rounded cusps and are adapted to crushing; the carnassials retain no shearing function.

Mustelids, although basically carnivorous, pursue many styles of feeding. Most aggressively search for prey in burrows, crevices, or dense cover, and many are able killers. In Colorado, I observed a male long-tailed weasel *(M. frenata)* killing a young cottontail rabbit *(Sylvilagus audubonii)* roughly twice its own weight. The weasel killed by biting the back of the rabbit's skull repeatedly. According to Errington (1967:24), a mink *(Mustela vison)* "hugs its victim with the forelegs while it scratches violently with the hind legs and bites vital parts, especially about head and neck." Some mustelids, such as the beautiful and graceful marten *(Martes americana),* are swift and agile climbers and feed partly on arboreal squirrels. Otters (subfamily Lutrinae) are semiaquatic, or almost completely aquatic in the case of the sea otter, and feed on a wide array of vertebrates and invertebrates. The skunks, with no claim to remarkable agility or killing ability, seem to feed on whatever animal material is most readily available, which during the summer is generally insects.

Family Viverridae. This is the largest family in the order Carnivora with regard to numbers

Figure 10-13. (A) African civet *(Viverra civetta)*, a large, dog-like viverrid; (B) large spotted genets *(Genetta tigrina)*, adroit climbers that often feed on birds. (Taken in Kenya; A by R. G. Bowker, B by T. R. Huels)

of species and includes the civets, genets, and mongooses. Thirty-six genera and roughly 70 Recent species are recognized, but the taxonomy of this family is still uncertain. Viverrids inhabit much of the Old World, but the center of their distribution is in tropical and southern temperate areas, and they are absent from northern Europe and all but southern Asia, as well as from New Guinea and Australia. This is an old group: it appeared in Europe in the early Oligocene but did not reach Africa until the Pleistocene, and reached Madagascar only in the Recent.

Viverrids are usually small, fairly short-legged, long-tailed carnivores (Figs. 10–13 and 10–14). Like mustelids, some viverrids have well-developed scent glands. The viverrid skull frequently has a moderately long rostrum (Fig. 10–10A). The premolars are large, and the carnassials are usually trenchant. The upper molars are tritubercular and are wider than they are long; the lower molars have well-developed talonids. The dental formula is generally 3/3, 1/1, 3-4/3-4, 2/2 = 36-40. The five toes on each foot include a much reduced pollex or hallux. The foot posture is plantigrade or digitigrade, and the claws are partly retractile. The ears are generally small and rounded. Some species are banded, others are spotted, and still others are striped. The smallest viverrid weighs less than 1 kilogram, and the largest weighs 14 kilograms. Because of the great morphological variation within this family, it is difficult to frame a definitive structural diagnosis applicable to all species.

Viverrids have a variety of feeding habits. Most are carnivorous and eat small vertebrates or insects. Some feed on the ground, and some in trees. In some parts of East Africa, viverrids are remarkably common and diverse for carnivores, with half a dozen or more species occurring in the same area. Some diurnal species are highly social and easily seen (Fig. 10–14B). Foraging parties of the banded mongoose *(Mungos mungo)* feed heavily on the dung beetles attracted to the feces of large ungulates. The palm civets (subfamily Paradoxurinae) are omnivorous and often feed primarily on fruit or other plant material. Some viverrids are semi-aquatic and feed largely on aquatic animals, and some mongooses *(Herpestes)* kill and eat snakes, including certain highly venomous species. The habits of several viverrids are almost unknown; the Congo water civet *(Osbornictis piscivora),* for example, is represented in museums by only a few specimens and has never been observed in the wild by a biologist. Probably no family of carnivores is so poorly known as the viverrids; clearly, much field study could profitably be concentrated on this group.

Family Hyaenidae. Many carnivores will eat carrion if the opportunity arises, but most mem-

Figure 10-14. Two African viverrids: (A) white-tail mongoose *(Ichneumia albicauda)*, a solitary, nocturnal species; (B) dwarf mongoose *(Helogale parvula)*, a highly social, diurnal species. (Taken in Kenya, by R. G. Bowker)

bers of the family Hyaenidae have become specialized for carrion feeding. This is a small family, with but three genera and four Recent species. The distribution includes Africa, southwestern Asia, and parts of India. The Hyaenidae, probably derived from viverrid stock, appeared in Eurasia in the late Miocene and, except for *Chasmaporthetes* (which probably crossed the Bering Strait land bridge and is known from the Pleistocene of North America), has been an entirely Old World family.

Except for the aardwolf *(Proteles cristatus),* hyenas are characterized by rather heavy builds, forelimbs longer than hind limbs,

strongly built skulls, and powerful dentitions (Fig. 10-15A). The carnassials are well developed, and all of the cheek teeth have heavily built crowns adapted to crushing bone. The dental formula is 3/3, 1/1, 4/3, 1/1 = 34. The feet are digitigrade, and both the forepaws and the hind paws have four toes that bear blunt, nonretractile claws. The pelage is either spotted *(Crocuta;* Fig. 10-16) or variously striped *(Hyaena).* Hyenas weigh up to 80 kilograms.

The aardwolf (subfamily Protelinae) is more lightly built than the hyenas (Hyaeninae), and has a more delicate skull and smaller teeth (Fig. 10-15B). All teeth except the canines are

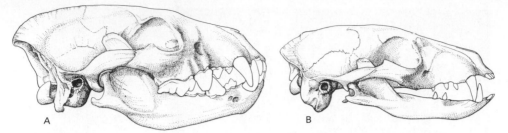

Figure 10-15. Hyaenid skulls: (A) spotted hyena *(Crocuta crocuta);* length of skull 248 mm; (B) aardwolf *(Proteles cristatus);* length of skull 148 mm.

small, and the cheek teeth are simple and conical. The dental formula is generally 3/3, 1/1, 3/2-1, 1/1-2 = 28-32, but frequently some of these teeth are lost (as in the skull shown in Fig. 10–15B). The forefeet have five toes, and the hind feet have four. The animal is striped and has a mane of long hair from neck to rump. The tail is quite bushy. When the animal is threatened and adopts a defensive posture, the hair of the mane and tail is erected but the mouth remains closed. *Proteles* has abandoned the open-mouthed threat used by most carnivores, which in its case would merely advertise the weakness of the dentition, in favor of extensive erection of the long hair. The aardwolf also releases fluid from the well-developed anal glands when attacked.

Hyenas in some areas specialize in scavenging on the kills of lions and other large carnivores and are able to drive cheetahs *(Acinonyx)* from their kills. They may also forage in villages at night for edible refuse. In the ability to crush large bones, they are unsurpassed, but studies by Kruuk (1966) and others have shown that spotted hyenas are also powerful predators. Often hunting in packs of up to 30 animals, these nocturnal hunters can bring down even zebras. Indeed, in the Ngorongoro Crater in Tanzania, a reversal of the usual pattern of interactions between lions and hyenas has occurred: spotted hyenas are better able than the lions to make regular kills, and lions live by driving hyenas from their kills and eating the carrion (Ewer, 1968:103).

The aardwolf eats mostly termites, to which it is directed largely by the sounds they make (Kruuk and Sands, 1972), and its unusually large auditory bullae may be associated with an enhancement of the sense of hearing. In contrast to many termite feeders, the aardwolf does not dig for termites, but laps them from the surface of the ground.

Family Felidae. Of all the carnivores, the cats are the most proficient predators. Some species regularly kill prey as large as themselves and may occasionally overcome prey several times their own weight (as in the case of the African lion and the giraffe). Throughout the history of the order Carnivora, since its earliest radiation, the cats have been the carnivores most highly specialized morphologically for a predaceous style of life.

Included in the family of living cats (Felidae) are four or five genera (depending on the taxonomic scheme followed) with about 37 species. The cats are quite a uniform group structurally; all cats, from the pampered tabby to the tiger, bear a strong family resemblance (Figs. 10–17 and 10–18). This family occurs nearly worldwide, with the exception of Antarctica, Australia, Madagascar, and various isolated islands.

In members of the family Felidae, the rostrum is short, an adaptation furthering a powerful bite, and the orbits in most species are large (Fig. 10–19). The number of teeth is reduced. The typical dental formula is 3/3, 1/1, 3/2, 1/1 = 30, and the anteriormost upper pre-

Figure 10-16. (A) Spotted hyena, in company with a black-backed jackal and white-backed vultures, feeding on the remains of a rhinoceros; (B) aardwolf foraging for termites. Note the proximity of the apparently unconcerned Thompson's gazelle. (A by M. P. Vaughan, B from Kruuk and Sands, 1972)

Figure 10-17. Two small New World cats that occur largely in tropical and subtropical areas: (A) ocelot *(Felis pardalis);* (B) jaguarundi *(Felis yagouaroundi).*

165

Figure 10-18. The cheetah *(Acinonyx jubatus),* the fastest cursorial mammal. Note the nonretractile claws. This animal is in a semicrouch and is stalking some Thompson's gazelles. (Taken in Kenya, by M. P. Vaughan)

molar is strongly reduced or lost (as in *Lynx).* The carnassials are well developed and have specializations that enhance their shearing ability (Fig. 10–12A,B). The foot posture is digitigrade. The forelimbs are strongly built, and the manus can be supinated; the claws are sharp and recurved and are completely retractile, except in the cheetah *(Acinonyx).* These features of the forelimbs allow cats to clutch and grapple with prey. Some species are spotted or striped; these color patterns enable the animals to conceal themselves effectively (Fig. 10–20). The weight of cats varies from about that of a domestic cat (3 kilograms) to 275 kilograms in the tiger.

The groundwork for an understanding of the phylogeny of cats was laid by Matthew (1910), and fossil material described in the last decade or so has provided important new evidence. Until the fossil record of the cats is more complete, any classification or outline of their phylogeny must be provisional.

The earliest cats, which appeared in Eurasia and North America in the Early Oligocene, were already strongly differentiated from other carnivores. All had retractile claws, sectorial carnassials, reduced cheek teeth anterior and posterior to the carnassials, and some had saber-like upper canines (Fig. 10–21). Early in their history, in the Oligocene, these cats were repre-

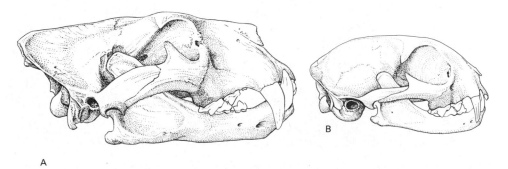

Figure 10-19. Felid skulls: (A) African lion *(Panthera leo);* length of skull 366 mm; (B) bobcat *(Lynx rufus);* length of skull 120 mm.

Figure 10-20. A leopard *(Panthera pardus)* well concealed against a background of mottled light and shadow. (W. L. Robinette)

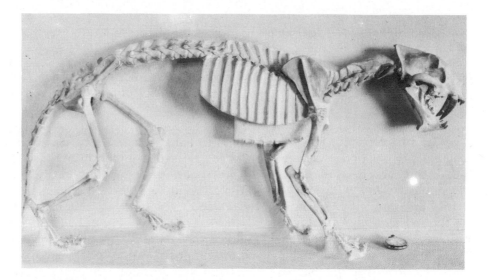

Figure 10-21. The skeleton of *Hoplophoneus primaevus* (Machairodontinae), an Oligocene sabertooth cat. (R. W. Wilson; South Dakota School of Mines)

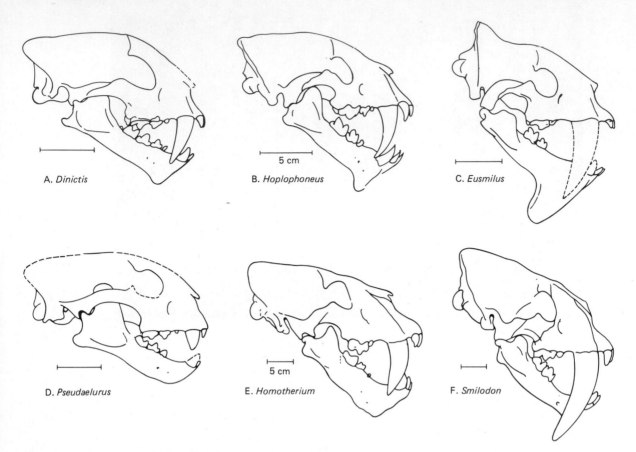

Figure 10-22. Some sabertooth cat skulls, showing (A, B, C) three grades of sabertooth specialization in the Oligocene and (D, E, F) three genera of sabertooth felids resulting from a Pliocene radiation. (Reprinted from Emerson and Radinsky, 1980, Paleobiology, vol. 6, p. 296)

sented by three grades of sabertooth specialization (Fig. 10–22). All members of this early sabertooth radiation are here included in the family Nimravidae (following L. G. Martin, 1980). This radiation culminated in the Pliocene with the most highly specialized of all sabertooth cats, *Barbourofelis fricki* (Fig. 10–23). The nimravids are not known after the Pliocene. Features that distinguish nimravids from felids include the lack of an auditory bulla in most species, the lack of a *septum bullae,* and the absence of a cruciate sulcus (a conspicuous, deep groove) on the brain.

The family to which all living cats belong, Felidae, appeared in the Miocene. The basal members of this family had short upper canines, but a Miocene and Pliocene radiation resulted in both sabertooth types and cats with short upper canines (Fig. 10–24). The Pleistocene felid *Smilodon,* although not as extreme in sabertooth specialization as *Barbourofelis,* was an imposing predator that survived until the end of the Pleistocene and may have coexisted with humans (G. J. Miller, 1969). All living felids have short, conical upper canines.

The spectacular sabertooth cats have received considerable attention from scientists,

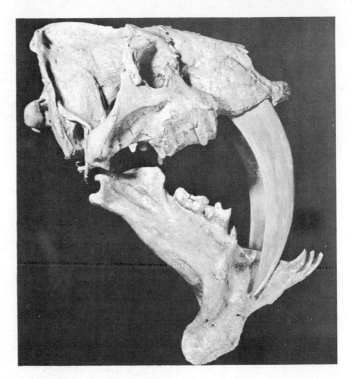

Figure 10–23. Skull of *Barbourofelis fricki,* an extinct Pliocene felid from Nebraska that represents the extreme in sabertooth specialization. Length of skull about 290 mm. (From C. B. Schultz, M. R. Schultz, and L. D. Martin, 1970)

but there is no consensus as to how these predators used their sabers in killing prey. The most probable explanation, in my view, is that of Emerson and Radinsky (1980), who believe that the sabers were used to inflict shallow slash wounds to the throat or side of the neck. Most species of sabertooth cats must have ambushed prey. The powerful forelimbs possessed by some sabertooth cats may have been used together with retractile claws to grasp and partially immobilize prey while the sabers slashed the throat. Presumably sabertooth cats had the edge over felids with short, conical upper canines in being able to kill somewhat larger prey. There is no question that the sabers were imposing weapons. Several fossil skulls have stab wounds made by sabers: one skull of a nimravid *(Nimravus)* was stabbed by another *(Eusmilus),* and one fossil dire wolf skull has a part of *Smilodon* saber imbedded in its forehead.

Although the precise mode of attack of sabertooth cats is uncertain, the functional importance of some of the sabertooth morphological features can be recognized. The reduced coronoid process of the dentary bone and the ventral shift in the position of the jaw joint, along with other features, suggest an unusually wide gape of the jaws (90 to 95 degrees for sabertooth cats versus 65 to 70 degrees for modern felids) to free the sabers for action. The more vertically oriented temporalis muscle and the closeness of the carnassials to the jaw joint allowed for a bite comparable in power to that of modern felids (Emerson and Radinsky, 1980). The sabers were laterally compressed and bladelike, and their edges bore serrations that must have facilitated slicing. The upper carnassials of the most specialized species had an additional cusp, anterior to the paracone and parastyle, that lengthened the carnassial shear. The heavy, ventrally extended mastoid process and the enlarged and posteriorly elongate transverse processes of the

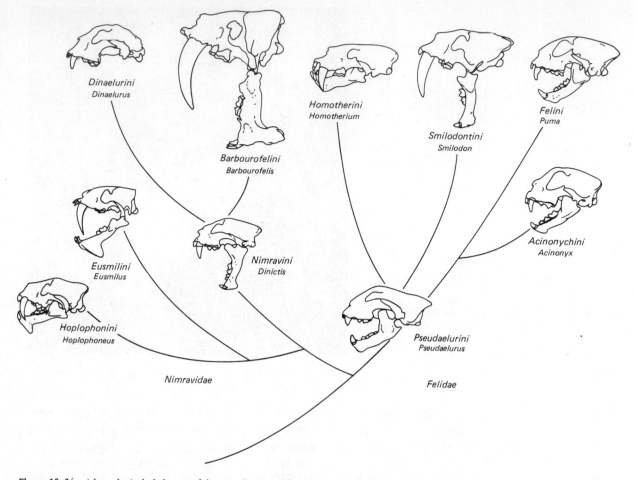

Figure 10–24. A hypothetical phylogeny of the cats. (Reprinted from L. D. Martin, 1980, p. 151, by permission of the Nebraska Academy of Sciences)

atlas and axis provided attachment for powerful muscles that pulled the head downward, probably during slashing.

The extinction of the sabertooth cats is difficult to explain. Sabertooth upper canines evolved independently four times among mammals (in Eocene creodonts, nimravids, felids, and the Pliocene marsupial family Thylacosmilidae), each time, perhaps, in response to the abundance of large herbivores. Probably in each case the extinction of the sabertooth carnivores was linked to the decline or extinction of their preferred prey.

As described on p. 377, cats usually catch prey by a stealthy stalk followed by a brief burst of speed. They are typically sight hunters, and some species spend considerable time watching for prey and waiting for it to move into striking distance. Many kinds of animals are eaten, from fish, mollusks, and small rodents to ungulates as large as buffalo. The African lion *(Panthera leo)* is unusual among felids in being highly so-

Figure 10-25. The face of a California sea lion *(Zalophus californianus)*. Note the valvular nostrils and the external ear. (U.S. Navy)

cial. *Felis planiceps* of southeastern Asia seems to be the only cat with noncarnivorous tendencies; this species prefers fruit (Goodwin, 1954:570).

Aquatic Carnivores

These are earless seals (Phocidae), the eared seals (Otariidae), and the walruses (Odobenidae). Seemingly Phocids evolved from mustelid ancestry, whereas otariids and odobenids derived from ursids.

Many distinctive morphological features of aquatic carnivores are adaptations to marine life. They are larger than land carnivores (ranging from 90 to 3600 kilograms in body weight). Large size saves energy in cold environments because of the favorable mass-to-surface-area ratio of large animals (p. 199). According to Scheffer

(1958:8), large size in pinnipeds is primarily an adaptation to a cold environment. The body is insulated by thick layers of blubber. The pinnae are either small or absent, the external genitalia and mammary nipples are withdrawn beneath the body surface, the tail is rudimentary, and only the parts of the limbs distal to the elbow and knee protrude from the body surface. As a result, the torpedo-shaped body has smooth contours and creates little drag during swimming. The slitlike nostrils (Fig. 10-25) are normally closed and are opened by voluntary effort.

The skull is partially telescoped, with the supraoccipital partially overlapping the parietals; the rostrum is usually shortened, and the orbits are usually large and encroach on the narrow interorbital area. Either one or two pairs of lower incisors are present. The canines are conical; the cheek teeth are homodont (none are modified as carnassials), two-rooted, and usu-

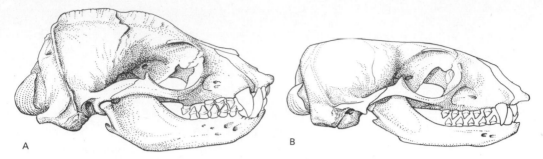

Figure 10–26. Sexual dimorphism in the skulls of the California sea lion *(Zalophus californianus):* (A) adult male; length of skull 290 mm; (B) adult female; length of skull 214 mm.

ally simple and conical (Fig. 10–26); they vary in total number from 12 to 24. In some pinnipeds, cheek teeth are characteristically lost with advancing age. The limbs and girdles are highly specialized. The clavicle is absent, and the humerus, radius, and ulna are short and heavily built. The pollex is the longest and most robust of the five digits and forms the leading edge of the winglike fore flipper. The pelvic girdle is small and nearly parallel to the vertebral column. The femur is broad and flattened. The first and fifth are the longest digits of the pes, and both the manus and pes are fully webbed.

The reduction of the vertebral zygapophyses and the absence of the clavicle allow the vertebral column and the forelimbs considerable flexibility and freedom of movement; these features may favor rapid maneuvering during the pursuit of prey. Although terrestrial locomotion is characteristically slow and laborious in most species, the importance of terrestrial locomotion when the animals haul themselves out on rocks or ice, or are on breeding grounds, has probably limited the extent to which the limbs of aquatic carnivores have become specialized for swimming.

Aquatic carnivores are extremely capable divers, with some species surpassing most cetaceans in this skill. The diving performance of the eared seals (Otariidae) is probably similar to that of many dolphins. The northern fur seal *(Callorhinus ursinus)* usually dives no deeper than 50 meters, but a dive to a depth of 190 me-

ters was recorded by Kooyman, Gentry, and Urquhant (1976). The duration of the dives was usually 2 to 5 minutes; the longest dive was 5.4 minutes. The sea lions probably have similar diving ability. The most spectacular diving occurs among the earless seals (Phocidae). The harp seal *(Phoca groenlandica)* of the North Atlantic reaches depths of at least 250 meters and can stay beneath the surface for many minutes. The Weddell seal *(Leptonychotes weddelli),* however, is the most impressive diver among aquatic carnivores. It commonly reaches depths of 300 to 400 meters and is known to dive to 600 meters (Kooyman, 1976b). Dives often last more than 40 minutes, and a dive of 1.17 hours was recorded (Kooyman, Kerem, et al., 1971). This duration surpasses that recorded for any marine mammal except the sperm whale *(Physeter catodon),* bowhead whale *(Balaena mysticetus),* and bottle-nosed whale *(Hyperoodon ampullatus).* Distances traveled during dives by Weddell seals are also remarkable. Kooyman (1968) found that during a single dive they can swim 5 kilometers from their breathing holes in the ice and return.

An integrated array of specializations is associated with the diving ability of seals. (This discussion is based largely on the fine work of Kooyman and his associates on the Weddell seal; their studies are cited in the bibliography.) The lungs of aquatic carnivores tend to be larger than those of terrestrial mammals of comparable size, a character important primarily in contributing to bouyancy and allowing the animals to rest at sea. The respiratory airways, even the terminal segments supplying the alveoli, are made rigid by cartilage and muscle. This feature, shared by cetaceans, allows for free passage of air from the alveoli when the lungs collapse under the great pressures to which the body is subjected during deep dives (p. 231). In addition, this rigidity contributes to the animal's ability to expire air from the lungs extremely rapidly and provides for very quick exchanges of large gas volumes. During extended dives, there is a redistribution of blood: blood flow to the major muscle masses is restricted in the interest of maintaining adequate oxygen supply to the brain and heart. It has been estimated that the metabolic rate of a seal during a long dive is less than 20 percent of the basal rate.

Especially dramatic is the precipitous drop in heart rate and cardiac output of blood during extended dives. Of interest here is the seal's psychological, or voluntary, control over blood flow. During brief dives, the heart rate declines modestly and little redistribution of blood occurs, but immediately after submersion for a long dive the heart rate drops drastically (Fig. 10–27) and there is redistribution of blood, indicating that the seal is anticipating a long dive and adjusting its blood flow accordingly. The heart rate during these prolonged dives drops from 85 to 16 beats per minute. The amazing

diving ability of seals, then, is not due to one or two revolutionary physiological or morphological changes but is the result of the reciprocal fine-tuning of many morphological, physiological, and psychological adaptations.

Family Otariidae.   This family, containing six genera and 12 Recent species, includes the eared seals and the sea lions (Figs. 10–25 and 10–28). These animals inhabit many of the coastlines of the Pacific Ocean and parts of the South Atlantic and Indian oceans. They are common along the Pacific coast of North America. The earliest otariids are known from the middle Miocene, and may have originated in the food-rich kelp reefs of the North Pacific (Scheffer, 1958:33).

Otariids differ from phocids in being less highly modified for aquatic life and better able to move on land. The hind flippers can be brought beneath the body and used in terrestrial locomotion, and well-developed nails occur on the three middle digits. A small external ear is present (Fig. 10–25). Males are much larger than females; in the northern fur seal *(Callorhinus ursinus)*, males weigh four and a half times as much as females. Considerable sexual dimorphism in the shape of the skull occurs in some species (Fig. 10–26), and in males the skull becomes larger and more heavily ridged with advancing age. The dental formula is 3/2, 1/1, 4/4, 1-3/1 = 34-38. The body is covered

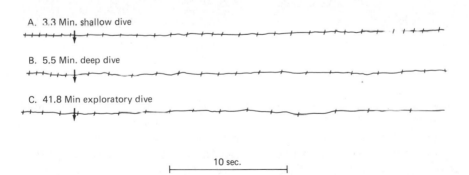

Figure 10–27.   Heart rate of a Weddell seal *(Leptonychotes weddelli)* at the beginning of three types of dives. Arrows indicate times of submersion. (Reprinted from Kooyman, 1972, by permission of Comparative Biochemistry and Physiology)

Figure 10–28. Steller sea lions *(Eumetopias jubata)* on Año Nuevo Island, off the coast of California: (A) bull with cows and young; (B) cows with young. (R. T. Orr)

with uniformly dark fur. Weights of otariids range from roughly 60 to 1000 kilograms.

These seals are generally highly vocal and utter a great variety of sounds. They tend to be gregarious all year round and are social during the breeding season, when they assemble in large breeding rookeries (p. 406). Propulsion in the water is accomplished by powerful downward and backward strokes of the forelimbs; speeds up to 27 kilometers per hour have been recorded (Scheffer, 1958:13). Otariids eat mostly squid and small fish that occur in schools, and they maneuver rapidly in pursuit of prey.

Family Odobenidae. This family contains only one species, *Odobenus rosmarus,* the walrus. This species occurs near shorelines in arctic waters of the Atlantic and Pacific oceans but may

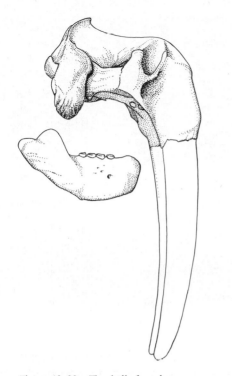

Figure 10-29. The skull of a walrus *(Odobenus rosmarus)*; length of skull approximately 355 mm. The tusks are enlarged upper canines.

stray southward to some extent along the coastlines. Odobenids first appear in the late Miocene and, like the otariids, may have had a North Pacific origin.

The walrus is a large pinniped (up to 1270 kilograms), with a robust build, a nearly hairless skin, and no external ears. The hind flippers can be brought beneath the body and are used for terrestrial locomotion, which is ponderous and slow. In both sexes, the upper canines are modified into long tusks (Fig. 10-29), which in the adult lack enamel. There are no lower incisors in adults, and 12 cheek teeth are usually present. The dental formula is 1-2/0, 1/1, 3-4/3-4, 0/0 = 18-24. On the huge mastoid processes attach the powerful neck muscles that pull the head downward.

Walruses feed on mollusks, which they rake from the sea floor by means of their lips and huge tusks. They are gregarious and polygynous and frequently assemble in large groups of more than 1000 individuals. They are migratory to some extent, moving southward in winter. Walruses make a variety of loud noises when out of the water and make a church-bell sound and rasps and clicks under water (Schevill, Lawrence, and Ray, 1963, 1966). The fact that the rasps and clicks are made during swimming suggests their use in echolocation.

Family Phocidae. There are 18 species and 13 genera of earless seals, the most abundant aquatic carnivores. They occur along most northern (above 30 degrees northern latitude) and most southern (below 50 degrees southern latitude) coastlines and in some intermediate areas. They appear in the middle Miocene, and presumably originated in the Northern Hemisphere.

The earless seals are more highly specialized for aquatic life than are other aquatic carnivores. As the vernacular name implies, there is no external ear. The hind flippers are useless on land but, as a result of lateral undulatory movements of the body, are the primary propulsive organs in the water. The fore flippers are short and well furred. The structure of the cheek teeth is highly variable, but is usually fairly simple. In the crab-eater seal *(Lobodon)*, however, the cheek teeth have complex cusps (Fig. 10-30). The pelage of most phocids is spotted, banded, or mottled (Fig. 10-31). These seals frequently have extremely heavy layers of subcutaneous blubber that give the

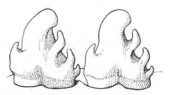

Figure 10-30. Medial view of two right lower cheek teeth of the crab-eater seal *(Lobodon carcinophagus)*. These complex teeth enable this animal to depend on filter feeding. (After Walker, 1968)

Figure 10–31. Harbor seals *(Phoca vitulina)* on Año Nuevo Island, off the coast of California. (R. T. Orr)

bodies smooth contours and, in some cases, a nearly perfect fusiform shape. Most species weigh from about 80 to 450 kilograms, but male elephant seals occasionally weigh as much as 3600 kilograms.

Many phocids are monogamous and form small, loose groups in which no social hierarchy is evident, but some, such as the elephant seal, are gregarious and polygynous and have a dominance hierarchy. The monogamous species are quiet, whereas the polygynous species are highly vocal (W. E. Evans and Bastian,

Figure 10–32. Northern elephant seal *(Mirounga angustirostris)* on Año Nuevo Island, off the coast of California. (R. T. Orr)

Figure 10-33.   A Weddell seal *(Leptonychotes weddelli)* swimming beneath the Antarctic ice. (G. L. Kooyman)

1969:437). The sole function of the proboscis of the male elephant seal (Fig. 10–32) is the production of vocal threats (Bartholomew and Collias, 1962).

The usual food of phocids is fish, cephalopods, and other mollusks. Large prey may be taken: Sterling (1969) reports a fish weighing 29.5 kilograms removed from the stomach of a Weddell seal *(Leptonychotes weddelli;* Fig. 10–33), and the powerful leopard seal *(Hydrurga leptonyx)* eats penguins and small seals. Two species of phocids are filter feeders and use the complex cheek teeth (Fig. 10–30) to filter crustaceans and other plankton from the water. So abundant is the filter-feeding Antarctic crab-eater seal *(Lobodon carcinophagus)* that it constitutes the major share of the world population of phocid seals.

Many Weddell seals of the Antarctic are year-round residents as far south as 79 degrees, where broad areas are covered all year with stationary sea ice and the seals must depend on scattered breathing holes. The seals use the upper canines and incisors to ream these holes open by using violent side-to-side thrashing of the head (Kooyman, 1975b). The ice is about 1 meter thick away from the holes, and the survival of seals depends on their ability at the end of a dive to find their way back to the original breathing hole or to locate a new one. These animals must be skilled navigators in darkness, for where many spend the winter the sun does not appear above the horizon for up to three and a half months.

Chapter 11

Subungulates: Hyraxes, Elephants, Sirenians

If general appearance were used as the single criterion for evaluating phylogenetic relationships, the rodent-like hyraxes, massive elephants, and ungainly aquatic sirenians would be judged to be three very distantly related mammals. In this case, however, appearances are deceptive, for the fossil record suggests that these groups evolved in Africa from a common ancestral stock related to the ungulates. Because of this presumed near-ungulate ancestry, hyraxes, elephants, and sirenians have long been called subungulates.

Hyraxes: Order Hyracoidea

Members of this unusual order are small, rodent-like creatures commonly called hyraxes or dassies, and their appearance (Fig. 22-13) gives little indication of their relationship to the ungulates. This is a small order, with a single Recent family, Procaviidae, and two fossil families of doubtful validity. Recent members include three genera with seven species. Hyracoids, which today occupy nearly all of Africa except the arid northwestern part, appeared first in lower Oligocene beds in Egypt, from which members of an extinct family and of the living family Procaviidae are known. The relationships of the hyracoids are uncertain, but they perhaps descended from an early "ungulate" stock that was also ancestral to the elephants and sirenians. Some early members of an extinct hyracoid family (Geniohyidae) reached the size of a tapir. The distribution of the structurally more conservative surviving family extended north of its present limits during the Pliocene, when a

gigantic procaviid occurred in western Europe as far north as France.

The roughly rabbit-size procaviids of today have a short skull with a deep lower jaw (Fig. 11-1). The dental formula is 1/2, 0/0, 4/4, 3/3 = 34. The incisors are specialized: the pointed, ever-growing uppers are broadly separated and triangular in cross section, and the flattened posterior surfaces lack enamel; the lowers are chisel-shaped and generally tricuspid. Behind the incisors is a broad diastema, and the cheek teeth are either brachyodont or hypsodont. The molars resemble those of a rhinoceros: the uppers have an ectoloph and two cross lophs, and the lowers have a pair of V-shaped lophs. The body is fairly compact, and the tail is tiny. The forefoot has four toes and the hind foot has three, and the feet are mesaxonic (the plane of symmetry goes through the third digit). The digits are joined to the bases of the last phalanges (Fig. 11-2), and except for the clawed second digit of the pes, all digits bear flattened nails. The plantigrade feet have specialized elas-

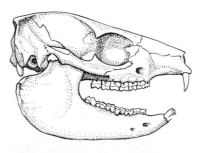

Figure 11-1. Skull of a rock hyrax (*Heterohyrax* sp.,); length of skull 98 mm. (After Hatt, 1936)

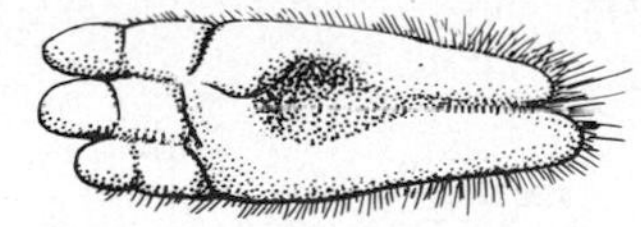

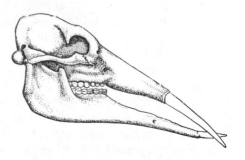

Figure 11-2. The sole of the right hind foot of a rock hyrax (*Dendrohyrax dorsalis*).

tic pads on the soles that are kept moist by abundant skin glands; in addition, the soles may be "cupped" by specialized muscles and provide remarkable traction. Although the clavicle is absent, as in cursorial mammals, the centrale of the carpus is present, a feature decidedly not characteristic of runners. The stomach is simple, but digestion is aided by microbiota in the pair of caecae of the colon and in the single iliocolic caecum.

Hyraxes are mainly herbivorous and are nimble climbers and jumpers. They occur in a variety of habitats, from forests and scrub country to rock outcrops and lava beds in grassland, and at elevations up to 5000 meters. The bush hyrax (*Heterohyrax brucei*) lives in cliffs, ledges, and talus and is adroit at climbing rapidly over steep rock faces and in trees.

Bush hyraxes and rock hyraxes (*Procavia johnstoni*) often form polygynous family groups, with one territorial adult male and several genetically related females and their young (H. N. Hoeck, Klein, and P. Hoeck, 1982). Hyraxes are relatively long-lived for small mammals, some individuals surviving more than 10 years in the wild (H. N. Hoeck, 1982).

The body temperature of the bush hyrax was found by Bartholomew and Rainy (1971) to be quite variable. This animal is known to make wide use of behavioral thermoregulation (p. 466).

Elephants: Order Proboscidea

Throughout much of the Cenozoic, some of the largest and most spectacular herbivores were proboscideans, and in the late Tertiary Period, a varied array of these animals occurred widely in North America, Europe, and Africa. The diversity of proboscideans was reduced in the Pleistocene Epoch, and today only two species represent this remarkable group. Because elephants now often threaten the interests of humans and because of the great value of their tusks, they are being extirpated over wide areas as human populations increase. Elephants occur today only in Africa south of the Sahara Desert (*Loxodonta*) and in parts of southeastern Asia (*Elephas*). Regrettably, we may be witnessing the final stages in the history of one of the most interesting mammalian orders.

The fossil record of proboscideans begins in the late Eocene Epoch of Egypt with *Moeritherium*. This tapir-size animal had a moderately primitive complement of teeth (3/3, 1/0, 3/3, 3/3 = 38), but the second incisors above and below were enlarged into short tusks. Proboscideans apparently reached North America in the late Miocene Epoch. These rather primitive proboscideans, of the Family Mastodontidae, had brachyodont teeth with few ridges; most or all of the cheek teeth were in place at one time, and both an upper and a lower set of tusks were usually present (Fig. 11–3).

Family Elephantidae. This family, to which both living species belong, is represented first by the Pliocene and Pleistocene genus *Stegodon*. Although the skull was not as short as in

Figure 11-3. The skull of *Gomphotherium*, a Miocene proboscidean. Length of skull and tusks roughly 1 m. (After Romer, 1966)

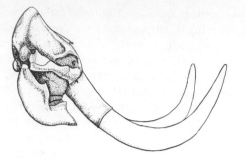

Figure 11-4. The skull of *Mammuthus,* a Pleistocene elephantid. Length of skull and tusks roughly 2.8 m. (After Romer, 1966)

Figure 11-5. A group of adult female and young African elephants *(Loxodonta africana)* in Buffalo Springs Game Reserve, Kenya, East Africa. (K. P. Dial)

more advanced elephants and the teeth were brachyodont, the occlusal surfaces of the cheek teeth had laminae (cross ridges) and no more than 2/2 cheek teeth were functional at one time. The lower tusks were vestigial, whereas those on the upper jaw were long and curved. The Pleistocene woolly mammoth *(Mammuthus primigenius)* was in some ways more specialized than the living elephants. It had a remarkably short, high skull (Fig. 11–4) and long tusks that occasionally crossed; the last molar had up to 30 laminae, more than occur in the living elephants. Entire frozen woolly mammoths have been found in Siberia and Alaska, and many graceful drawings made by Paleolithic peoples on the walls of caves depict these animals.

The two living proboscideans—the African elephant, *Loxodonta africana* (Fig. 11–5), and the Indian elephant, *Elephas maximus*—are the largest land mammals, reaching weights of 5900 kilograms. They have a long, dextrous proboscis (trunk) with one or two finger-like structures at its tip, large ears, and graviportal limbs. The limb bones are heavy, and the proximal segments of the limbs are relatively long; the ulna and tibia are unreduced, and the bones of the five-toed manus and pes are short and robust and have an unusual spreading digitigrade posture (Fig. 11–6). A heel pad of dense connective tissue braces the toes and largely supports the weight of the animal. As an adaptation allowing the efficient support of great weight, the long axis of the pelvic girdle is nearly at right angles to the vertebral column and the acetabulum faces ventrally. In addition, when the weight of the body is supported by the limbs, there is little angulation between limb segments; that is to say, each segment is roughly in line with other segments. The gait is unusual. As described by A. B. Howell (1944:53), an elephant "relies exclusively upon the walk or its more speedy equivalent, the running walk, which permits it to keep at least two feet always upon the ground. Not only does the weight make it advisable that this be distributed among each of the four feet when the animal is in mo-

Figure 11-6. The right hind foot of *Mastodon,* a late Tertiary and Pleistocene proboscidean. (After Romer, 1966)

tion, but the bulk doubtless requires that the equilibrial stresses be shifted as gradually as possible to each foot, rather than more abruptly as in the trot or gallop.''

The skull is unusually short and high, perhaps in response to a need for great mechanical advantage for the muscles that attach to the lambdoidal crest and raise the front of the head and the tusks. The skull contains numerous large air cells, particularly in the cranial roof.

The highly specialized dentition consists of the tusks (each a second upper incisor) and six cheek teeth in each half of each jaw. The pattern of cheek tooth replacement is remarkable. The cheek teeth erupt in sequence from front to rear, but only a single tooth or one tooth and a fragment of another are functional in each half of each jaw at one time. As a tooth becomes seriously worn, it is replaced by the next posterior tooth. The usage intervals for the cheek teeth in the African elephant are: 1, birth to second year; 2, 1.5 to fifth year; 3, second to eleventh year; 4, fifth to nineteenth year; 5, fifteenth to about sixtieth year; 6, twenty-third to 60+ years (Krumrey and Buss, 1968). The hypsodont cheek teeth are formed of thin laminae, each consisting of an enamel band surrounding dentine, with cement filling the spaces between the ridges (Fig. 11–7). The last molar, the tooth that must serve for much of the animal's adult life, has the greatest number of laminae. The premolars are considerably smaller, simpler, and less durable than the molars. In old elephants, some of the anterior laminae of the third molar may be lost while the remainder of the tooth is still functional. The unique pattern of tooth replacement and the complex occlusal surface provide for an enduring dentition and a long life. The life span of elephants is thought by Perry (1954) to be about 70 years.

Elephants occupy forests, semiopen or dense scrub, and savanna and are restricted to areas near water. They feed on a variety of trees, shrubs, grasses, and aquatic plants and characteristically strongly influence their environment (p. 297). Each individual eats up to 410 kilograms of forage daily. Their great size and strength enable them to ''ride down'' fairly large trees in order to feed on the leaves. Female elephants are highly social. They live largely in kinship groups from which adult males are excluded; these groups are held together by close social ties between adult females and between mothers and their young.

The greatest danger to elephants is encroachment on their habitat by humans. In Africa, national parks, where settlement by people is prohibited, offer some hope for the survival of wild populations.

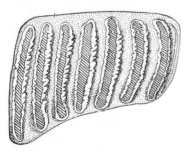

Figure 11–7. The occlusal surface of a molar of the Asiatic elephant *(Elephas maximus)*. The ridges of the lamellae are enamel, the cross-hatched areas are dentine, and the stippled areas are cement.

Sirenians: Order Sirenia

The sirenians, or seacows—dugongs and manatees—are the only completely aquatic mammals that are herbivorous and are one of the most anomalous mammalian orders. There are four living species of two genera (*Dugong,* Dugongidae; *Trichechus,* Trichechidae). According to J. K. Jones and Johnson (1967:367), sirenians occur in ''coastal waters from eastern Africa to Riu Kiu Islands, Indo-Australian Archipelago, western Pacific and Indian oceans; tropical western Africa; coastlines of Western Hemisphere from 30°N to 20°S, Caribbean region, and Amazon and Orinoco drainages in South

Figure 11–8. A female Florida manatee *(Trichechus manatus)* and her young. (U.S. Fish and Wildlife Service)

America; formerly also in Bering Sea.'' Sirenians probably share a common ancestry with the proboscideans and are known from Eocene deposits at such scattered points as Europe, Africa, and the West Indies. Nearly all fossil sirenians were tropical and marine.

Sirenians are large, reaching weights in excess of 1150 kilograms. They are nearly hairless except for bristles on the snout and have thick, rough, or finely wrinkled skin (Fig. 11–8). The nostrils are valvular, the nasal opening extends posterior to the anterior borders of the orbits, and the nasals are either reduced or absent. The skull is highly specialized, and the dentary is deep (Fig. 11–9). The tympanic bone is semicircular, and the external auditory meatus is small. The periotic bone has no bony attachment to the skull but is attached by ligaments. The lungs are unlobed, unusually long, oriented horizontally, and separated from the massive gut by a long, horizontal diaphragm. The orientation of the lungs and the dense and heavy bone allow the animal to use minor adjustments in lung volume to maintain a horizontal attitude while feeding at various depths. The heavy bone may counterbalance the added bouyancy due to excessive gas production in the gut. Postcranially, sirenians somewhat resemble cetaceans. The five-toed manus is enclosed by skin and forms a flipper-like structure, the pelvis is vestigial, and the tail is a horizontal fluke. There is no clavicle, and, unlike the case with cetaceans, the scapula is narrow and bladelike.

The teeth are unusual. The teeth of dugongs *(Dugong)* are large and columnar, lacking in enamel, and cement-covered. They have open roots, and the occlusal surfaces are wrinkled and bunodont. In manatees *(Trichechus),* by contrast, there is an indefinite large number of enamel-covered, cementless teeth, each with two cross ridges and closed roots. As teeth at the front of the tooth row wear out, they are replaced by the posterior teeth pushing forward. Five to eight teeth in each side of each jaw are functional at one time. Manatees evolved from dugongid ancestral stock in South America, and perhaps because of their more specialized dentition, completely replaced the dugongs in the New World. Horny plates cover the front of the palate and the adjacent surface of the mandible in all genera. The skull of *Trichechus* is modified by elongation of the nasal cavity, and this animal, alone among mammals, has only six cervical vertebrae. Some differences between the

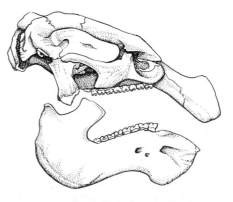

Figure 11–9. The skull of a manatee *(Trichechus manatus);* length of skull 360 mm. (After Hall and Kelson, 1959)

Table 11-1: Comparison of Sirenian Characteristics

Dugongidae	Trichechidae
Functional dentition: 1/0, 0/0, 0/0, 2-3/2-3 = 10-14	No functional incisors; numerous cheek teeth
Cheek teeth columnar, no enamel, cement-covered; roots single	Teeth with cross ridges, covered with enamel; cement absent; roots double; continuous tooth replacement
Premaxillaries large; nasals absent; nasal cavity short	Premaxillaries small; nasals present; nasal cavity long
Slender neural spines and ribs	Robust neural spines and ribs
Flippers lacking nails	Flippers with nails in two of the three species
Tail notched, like that of whales	Tail not notched but spoon-shaped

two families of sirenians are shown in Table 11–1.

Sirenians are heavy-bodied, slow-moving animals that inhabit coastal seas, large rivers, and lakes and graze while submerged for periods of up to about 15 minutes. Some individuals inhabiting the coasts of Florida move into rivers, springs, and industrial warm-water effluents in winter to avoid cold water (D. S. Hart-man, 1979). Manatees are known to make sounds under water (Schevill and Watkins, 1965). O'Shea (1982), observing free-ranging West Indian manatees *(Trichechus manatus),* found that these calls form a series of complex graded signals used to maintain contact between individuals and to communicate basic "motivational states."

Humans have been responsible for a great restriction of the range of sirenians and exterminated the Steller sea cow *(Hydrodamalis)* in about 1769, only 27 years after its discovery (Walker, 1968:1334). This gigantic sirenian was nearly 8 meters in length, probably reached over 6000 kilograms in weight, and inhabited shallow parts of the Bering Sea, where it fed on seaweed. Present serious declines in the populations of dugongs and manatees in some areas are due to persistent hunting by humans. In Florida, where they are stringently protected from hunters, most manatees bear scars inflicted by boat propellers, and collisions with boats and other human-caused factors are responsible for a large proportion of the observed mortality. Because of the rapidly increasing human population in Florida and the low reproductive rate of manatees, serious population declines of manatees due to habitat loss and accidental deaths seem likely in the future.

Chapter 12

Order Perissodactyla: Horses, Tapirs, Rhinoceroses

Since Eocene times, some of the most specialized and spectacular cursorial mammals have been perissodactyls. In the early Tertiary, these were the most abundant ungulates, but their diversity was reduced in the Oligocene, and with the diversification and "modernization" of the artiodactyls (pigs, camels, antelope, and their relatives) in the Miocene, the fortunes of perissodactyls further declined. The surviving perissodactylan fauna (consisting of five genera and 18 species) is but an insignificant remnant of this once important and diverse group and is vastly overshadowed by an impressive living artiodactylan assemblage (consisting of 185 species). Perissodactyls occur today largely in southern areas—Africa, parts of central and southern Asia, and tropical parts of southern North America and northern South America.

The term "ungulate" has no taxonomic status but refers to all hoofed mammals, both the Perissodactyla and the Artiodactyla. Ungulates are typically herbivorous and are adapted to rapid cursorial locomotion. Among the ungulates are some of the most graceful and handsome mammals and some that are in serious danger of extinction. An important key to understanding the biology of ungulates is knowledge of their cursorial and feeding adaptations.

Cursorial Specialization

Exceptional running ability has evolved independently in a number of mammalian groups. It has provided a means of escaping predators (as in ungulates, rabbits, and some rodents) or of capturing prey (as in carnivores). Cursorial adaptations appear early in the Cenozoic history of mammals. The early Eocene genus *Diacodexis,* the earliest known artiodactyl, had slim, elongate limbs and was highly cursorial (Rose, 1982), and the earliest known perissodactyl, *Hyracotherium,* of the late Paleocene Epoch, was similarly, but not as extremely, specialized. The refinement of cursorial adaptations in ungulates was favored by their Miocene occupation of expanding grasslands. Not only was speed the primary means of escaping predators in this open country, but long daily or seasonal movement to seek water or nutritious food probably became an important part of the ungulate mode of life.

Running speed is determined basically by two factors: stride length and stride rate (the number of strides per unit of time). Most important cursorial specializations lengthen the stride or increase its rate. Perhaps the most universal cursorial adaptation that lengthens the stride is lengthening of the limbs. In generalized mammals, or in many powerful diggers, the limbs are fairly short and the segments are all roughly the same length (Fig. 12–1). In cursorial species, however, the limbs are long and, in the most specialized runners, the metacarpals and metatarsals have become greatly elongate and the manus and pes are the longest segments (Fig. 12–1). The loss or reduction of the clavicle contributes further to the length of the stride. This occurs in carnivores, leporids, and

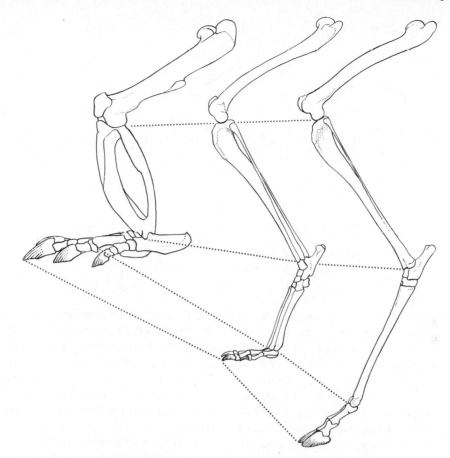

Figure 12-1. The hind limbs of three mammals: (A) armadillo *(Dasypus novemcinctus)*, a powerful digger with plantigrade feet; (B) coyote *(Canis latrans)*, a good runner with digitigrade feet; (C) pronghorn antelope *(Antilocapra americana)*, an extremely speedy runner with unguligrade feet. Note the lengthening of the shank and foot in B and C especially; the metatarsals have undergone the greatest lengthening. The limbs are not drawn to scale, but the femur is the same length in each drawing.

ungulates. With the loss of the clavicle, the scapula and shoulder joint are freed from a bony connection with the sternum and the scapula can change position to some extent and can rotate about a pivot point roughly at its center. Because the scapula is not anchored to the axial skeleton, the shoulder joint pivots upward and forward as the forelimb reaches forward during the stride and swings downward and backward when the forelimb moves back. Hildebrand (1960) estimated that such movements of the scapula added roughly 115 millimeters to the stride length of the running cheetah. As an additional advantage, when the forefeet strike the ground at the end of a forward bound, the impact is cushioned by the muscles that bind the scapula to the body, rather than the shock being transferred directly from the shoulder joint to the axial skeleton via the clavicle.

Substantial lengthening of the stride also results from an inchworm-like flexion and extension of the spine (Fig. 12–2). In small or

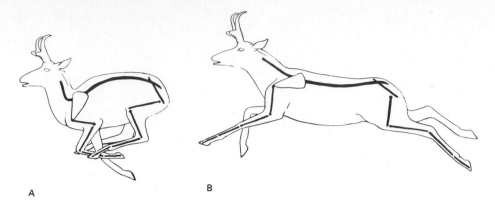

Figure 12–2. Two positions of a running pronghorn antelope, showing the flexion and extension of the vertebral column and the changing position of the scapula. (A) The forelimbs have just left the ground; the hind limbs are reaching forward and will touch the ground as the forelimbs swing forward. (B) The animal is bounding ahead after the limbs have driven against the ground; the forelimbs are reaching forward.

moderate-size runners, the flexors and extensors of the vertebral column are powerfully developed; the vertebral column extends as the forelimbs reach forward and the hind limbs are driving against the ground, and it flexes when the front feet move backward while braced against the ground as the hind limbs swing forward. Hildebrand estimated that such movements of the vertebral column could propel the cheetah at nearly 10 kilometers per hour if the animal had no legs!

The speed of limb movements and thus stride rate are similarly increased by a combination of structural modifications. The total speed of the foot, which drives against the ground and propels the animal, depends on the speed of movement at each joint of the limb. If another movable joint is added to a limb, the speed of the limb will be increased by the speed of movement at the new joint. The greater the number of joints that move in the same direction simultaneously, the greater the speed of the limb. For this reason, nearly all cursorial mammals have abandoned a plantigrade foot posture in favor of one that is a digitigrade or unguligrade (a foot posture in which the hoof-bearing tips of the digits contact the ground). This lifting of the heel from the ground allows another limb joint, that between the metapodials and the phalanges, to contribute to limb speed. In addition, the movable scapula and vertebral column, which help to increase stride length, also contribute their motion to total foot speed.

Specializations of the musculature also add importantly to limb speed and hence to running speed. The trend in many cursorial mammals is toward a lengthening of the tendons of some limb muscles (in association with the elongation of the distal segments of the limbs), and in some cases there has been a proximal migration of the insertion points of these muscles. Generally, the nearer the insertion point of a muscle approaches the joint it spans and at which it causes motion, the greater the advantage for speed; such specialized muscles are primarily geared for speed (but not for power). In these animals, a complex division of labor among muscles probably allows other muscles, having attachments that confer considerable mechanical advantage for power, to control certain relatively slow movements that get the animal in motion; the less powerful "high-gear" muscles are brought into play during high-speed running.

Speedy limb movements are further facilitated by a reduction of the weight of the distal parts of the limbs and the resultant reduction in the amount of kinetic energy that must be overcome at the end of one limb movement and the start of another. Inasmuch as the distal part of the limb moves more rapidly than the proximal part during a stride, reduction of weight of the distal parts is especially advantageous. Several specializations commonly serve this end. The most obvious is the loss of digits. In the most extreme mammalian cases—the front and hind limbs of the horse, and the hind limbs of the pig-footed bandicoot (Fig. 5–22C) and of the red kangaroo (Fig. 5–29C)—only the one digit that functions during running is retained. The strengthening of the distal joints of the limbs by modifications of bones and ligaments, and the limiting of movements at these joints to a fore and aft plane, obviate the need for muscular bracing and for muscles that produce rotary motions, further reducing distal weight. Also, the heaviest muscles are mostly in the proximal segment of the limb, thus keeping the center of gravity of the limb near the body. The combined effect of these modifications that reduce and redistribute weight is to favor rapid limb movement and to reduce the outlay of energy associated with that movement. (For excellent discussions of cursorial adaptations in mammals, see Hildebrand, 1959, 1960, 1965, 1974:487.)

The graceful legs of an antelope, with slim distal segments and the largest muscles bunched near the body (Fig. 13–19), display beautifully many of the cursorial modifications discussed here.

Living ungulates typically have many of the cursorial specializations just discussed. The feet are modified by the loss of some toes, by the alteration of the foot posture so that only the tips of the toes touch the ground, and by the development of hoofs. The limbs are usually slender, and the tendency in the most rapid runners is toward great elongation of the second segments (the forearm of the forelimb and the shank of the hind limb) and the distalmost segment (the manus and pes). The distal parts of the ulna and the fibula are strongly reduced in advanced ungulates, and the joints distal to the shoulder and hip joints tend to limit movement to the anteroposterior plane.

In the ungulate ankle joint, the calcaneum appears to be pushed aside, so to speak. In mammals in which no drastic reduction of digit number has occurred, the distal surface of the astragalus articulates with the navicular, the calcaneum articulates with the cuboid (Fig. 2–26), and the weight of the body is transferred through the digits, the distal carpals, both the astragalus and the calcaneum, and the tibia and fibula. In ungulates, a different arrangement occurs in association with the reduction of digit number: the astragalus rests more or less directly on the distal tarsal bones, which may be highly modified by fusion and loss of elements (Figs. 12–3 and 13–3) and the weight of the body is borne by the central digits (or digit, in the case of the horse), the distal tarsals, and the astragalus. The astragalus thus becomes the main weight-bearing bone of the two proximal

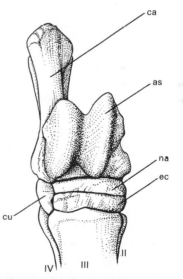

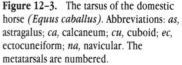

Figure 12–3. The tarsus of the domestic horse *(Equus caballus)*. Abbreviations: *as,* astragalus; *ca,* calcaneum; *cu,* cuboid; *ec,* ectocuneiform; *na,* navicular. The metatarsals are numbered.

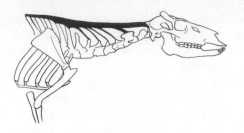

Figure 12–4. A schematic drawing of the nuchal ligament (in black) of an ungulate.

ligament begins to rebound, and when the foot is leaving the ground, the phalanges snap toward the flexed position (Fig. 13–2C). The familiar backward flip of the horse's foot just as it leaves the ground is controlled by the springing ligament. This flip gives a final increase in speed and thrust to the stride and increases the ungulate's speed afoot without the use of muscular effort.

tarsals. The calcaneum remains important as a point of insertion for extensors of the foot, but it no longer is a major weight-bearing bone of the tarsus. A similar bypassing of the calcaneum occurs in the cursorial peramelid marsupials (Fig. 5–22).

Two distinctive ungulate specializations involve connective tissue. The nuchal ligament is a heavy band of elastin (an elastic protein found in vertebrates) that is anchored posteriorly to the tops of the neural spines of some of the anteriormost thoracic vertebrae and attaches anteriorly high on the occipital part of the skull (Fig. 12–4). This ligament, especially robust in large, heavy-headed ungulates such as the horse and moose, helps support the head so that the burden on the muscles that lift the head is greatly lightened. The elasticity of the ligament allows the animal to lower its head when it is eating or drinking. A second specialized ligament, the springing ligament, occurs in the front and hind feet of ungulates and evolved from muscles that flexed the digits (Camp and Smith, 1942). In the hind foot of the pronghorn antelope *(Antilocapra),* for example, the springing ligament arises from the proximal third of the back of the cannon bone and inserts distally on the sides of the first phalanges of digits three and four (Fig. 13–1). When the foot supports the weight of the body, the phalanges are extended, thereby stretching the springing ligament. As the foot begins to be relieved of the weight of the body toward the end of the propulsion stroke of the stride, however, the elastic

Feeding Specialization

The herbivorous diet characteristic of most ungulates has favored the development of cheek teeth with large and complex occlusal surfaces that finely section plant material as an aid to digestion. Premolars tend to become molariform and thus to increase the extent of the grinding battery, and the anterior dentition becomes variously modified. In advanced types, there is a diastema between the anterior dentition and the cheek teeth.

Many ungulates have become large; indeed, the largest members of most mammalian faunas are ungulates. Large size reduces the number of predators to which an animal is vulnerable and is advantageous in terms of temperature regulation and energy requirements. Probably these and other factors have influenced the size of ungulates.

Diet puts unusual demands on the digestive systems of ungulates. Vegetation is far less concentrated food than is meat, is more difficult to digest, and is often protected by defensive secondary compounds. In addition, plant material is frequently low in protein. A herbivore must break down the cell wall, a fairly rigid structure formed largely of cellulose, not so much for the energy it yields but to gain access to the proteins within the cell. This breakdown is difficult, however, for mammals lack enzymes that digest cellulose. All herbivores must therefore have an alimentary canal that is specialized to cope with cellulose by means other than direct enzymatic action. Both perissodactyls (horses,

rhinoceroses, and tapirs) and ruminant artiodactyls (camels, deer, antelope, sheep, goats,
and cattle) utilize a fermentation process that
breaks down cellulose with the cellulolytic enzymes of microorganisms that live in the alimentary canal.

In perissodactyls, if we can assume that
their digestive systems are all similar to the system of the domestic horse, micororganism-
aided fermentation takes place in the large intestine and enlarged colon. Protein is digested
and absorbed in the relatively small and simple
stomach. In ruminants, in contrast, the enlarged
and several-chambered stomach, called a
rumen, harbors microorganisms that break
down cellulose. Large food particles float on top
of the fluid in the rumen, are passed to the diverticulum (Fig. 2–14), and are then regurgi-
tated and remasticated. In the vernacular, the
animals chew their cud. This cycle is repeated
until the chemical and mechanical breakdown
of the food has reached a point at which the particles sink in the fluid and pass on to the
intestine.

As a further refinement, a complex system
of recycling and reconstitution of food constituents in the stomach enhances the nutritional
yield of poor-quality food. When the ruminant's
food is high in cellulose and lignin (a substance
similar to cellulose that contributes the hard,
woody characteristics to plant stalks and roots),
it is digested slowly and its rate of passage
through the gut is low. The horse's system, however, cannot regulate the rate at which food is
processed. Food takes 30 to 45 hours to pass
through the gut of a horse, as compared with 70
to 100 hours to make the corresponding journey
in a cow. Some ruminants can subsist on remarkably low-quality food. As an example,
black Bedouin goats can survive on a diet of
wheat straw, material high in cellulose and extremely low in protein, by recycling urea (Sil-
anikove, 1980).

Perissodactyls and ruminant artiodactyls
have thus evolved contrasting nutritional strategies. Ruminants can satisfy their nutritional

needs with relatively unnutritious food; in areas
with marked seasonality, the only areas where
ruminants are abundant, they can remain widespread through times when forage is nutrition-
ally poor. During comparable times, perissodactyls must seek sites that support the greatest
quantity of vegetation and the most nutritious
vegetation, and their range is correspondingly
restricted. When ruminants compete with perissodactyls for nutritious food, however, ruminants are the losers. Ruminants are less efficient
in utilizing nutritious food because proteins are
used inefficiently by the ruminant system of microbial fermentation (Reid, 1970; R. H. Smith,
1975). By lengthening their food chain and by
"domesticating microbes" and using microbial
digestion, ruminants have reduced their efficiency in transforming forage into animal biomass while enhancing their ability to survive on
low-quality food (Kinnear, Cockson, et al.,
1979).

An extremely interesting African grazing
succession, strongly influenced by differences
between digestive efficiencies and food requirements, has been described by Bell (1971). He
studied primarily the most abundant ungulates:
the zebra *(Equus burchelli),* a nonruminant,
and the wildebeest *(Connochaetes taurinus)*
and Thompson's gazelle *(Gazella thompsonii),*
both ruminants. He found that, in the Serengeti
Plains, the zebra was the first of these ungulates
to be forced by food shortages to move from the
preferred short-grass area down into the longer,
coarser grasses of the lowlands. After the zebras'
feeding and trampling in the lowlands had removed the coarse upper parts of the grass and
had made the lower, more nutritious parts more
readily available, the wildebeest, a more selective feeder, moved in. By this time, the zebras
were becoming less able to get sufficient quantities of forage and were moving to new tall-
grass pastures. A similar replacement of wildebeest by Thompson's gazelles occurred after the
wildebeest had removed still more grass and
had made available to the small, highly selective gazelles the fruits and leaves of low-grow-

ing forbs. Not only was competition between these abundant ungulates minimized by this grazing pattern, but the activities of the early members of the grazing succession were highly advantageous to the later, more selective members. Bell's study clearly illustrates that differences in the digestive systems of ungulates have pronounced effects on food preferences, migratory patterns, and, in fact, many basic interactions of a grazing ecosystem.

Perissodactyl Evolution

Perissodactyls evolved from herbivorous condylarths of the family Phenacodontidae (Fig. 12–5). The order Condylarthra includes a diverse group of ancient ungulates that occurred from the Cretaceous to the Oligocene, and was probably the basal stock for 18 mammalian orders (Szalay, 1969), including the Recent orders Artiodactyla, Perissodactyla, Hyracoidea, Proboscidea, and Sirenia. Perissodactyls appeared in the late Paleocene in North America and underwent rapid diversification. Eleven of the 12 families appeared in the Eocene, but in addition to the living families Tapiridae, Rhinocerotidae, and Equidae, only the anomalous extinct family Chalicotheriidae survived into the Pleistocene.

The features of several important perissodactylan families illustrate the considerable structural and functional diversity within the group. The dentition and cranial morphology developed in response to herbivorous feeding habits. Living perissodactyls have elongate skulls, owing to an enlargement of the facial region to accommodate a full series of large cheek teeth (often hyposodont), and some have a complete complement of 44 teeth. The teeth are usually lophodont and are either hypsodont in grazing types (all Equidae, and *Ceratotherium* of the Rhinocerotidae) or brachyodont in browsers (all Tapiridae, and *Rhinoceros* and *Didermocerus* of the Rhinocerotidae). Many postcranial specializations further cursorial

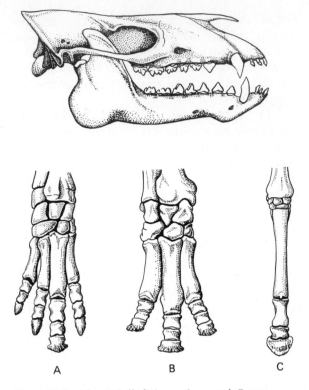

Figure 12-5. (Above) Skull of *Phenacodus,* an early Eocene primitive ungulate (order Condylarthra); length of skull 125 mm. (After Romer, 1966) (Below) Front feet of some perissodactyls: (A) tapir *(Tapirus);* (B) rhinoceros *(Rhinoceros);* (C) horse *(Equus).* (After Howell, 1944)

ability. The clavicle is absent, and usually the manus has three or four digits and the pes, three digits. In the equids, however, only one functional digit is retained on each foot (Fig. 12–5C). The feet are mesaxonic; that is, the plane of symmetry of the foot passes through the third digit, whereas this plane passes between digits three and four in the paraxonic foot of artiodactyls.

Family Equidae. Horses, the most highly cursorial and graceful perissodactyls, now occur wild only in Africa, Arabia, and parts of western and central Asia. In addition, feral populations of domestic horses and burros live in various places. There is but one genus with nine living species.

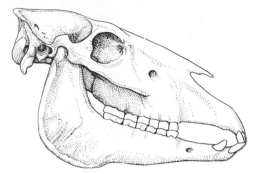

Figure 12-7. The skull of the domestic horse *(Equus caballus);* length of skull 530 mm.

Figure 12-6. Two species of equids: (top) Burchell's zebra *(Equus burchelli);* (bottom) Grevy's zebra *(Equus grevyi).* Note the differences in striping pattern in the two species. (Taken in Kenya, East Africa, by J. E. Varnum)

Wild horses are in general not as large as domestic breeds. The average weight of a female zebra *(Equus burchelli;* Fig. 12–6A) is given by Bell (1971) as 219 kilograms, but some domestic breeds weigh over 1000 kilograms. The skull has a fairly level profile, and the rostrum is long and deep (Fig. 12–7); the dental formula is 3/3, 0-1/0-1, 3-4/3, 3/3 = 36-42. The cheek teeth are hypsodont and have complex patterns on the occlusal surfaces (Fig. 12–8). The limbs are of a highly cursorial type: only the third digit is functional, all but the proximal joints largely restrict movement to one plane, and the foot is greatly elongate. In

Figure 12-8. Right upper molars of equids. (A) *Hyracotherium;* (B) *Mesohippus;* (C) *Parahippus;* (D) *Pliohippus;* (E) *Equus.* These teeth illustrate stages in the evolution of the equid molars. Abbreviations: *hy,* hypocone; *me,* metacone; *mes,* mesostyle; *mts,* metastyle; *pa,* paracone; *pas,* parastyle; *pr,* protocone. (After Romer, 1966)

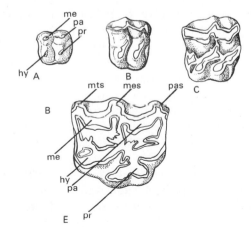

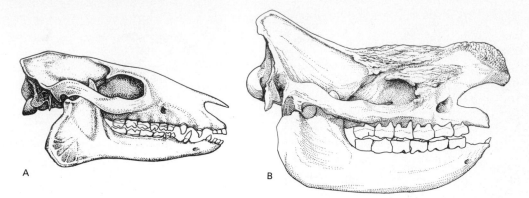

Figure 12-9.   (A) Skull of *Hyracotherium*, first known equid; length of skull 134 mm. (B) Skull of black rhinoceros *(Diceros bicornis)*; length of skull 692 mm. (A after Romer, 1966)

the tarsus, the main weight-bearing bones are the ectocuneiform, navicular, and astragalus; the calcaneum is mostly posterior to the astragalus (Fig. 12–3).

The evolution of horses is well documented by an excellent and largely New World fossil record and is discussed by Simpson (1951). Equids are first represented by *Hyracotherium* from the late Paleocene record of Wyoming (Jepsen and Woodburne, 1969). This primitive type had a generalized skull with 44 teeth (Fig. 12–9A). The upper and lower molars were brachyodont and basically four-cusped. The upper molars bore a protoconule and a metaconule, and the paraconid of the lower molars was reduced (Fig. 12–8A). The premolars were not molariform. The limb structure reflected considerable running ability: the limbs were fairly long and slender; the front foot had four toes and the hind foot had three toes, but the animal was functionally tridactyl. Each digit terminated in a small hoof, and the foot posture was unguligrade. *Hyracotherium* was the size of a small dog and presumably browsed on low-growing vegetation in forested or semiforested areas.

Side branches from the main stem of equid evolution developed at various times, and during part of the late Miocene, North American savannas supported 12 species of horses. The main evolutionary line can be traced through such intermediate genera as *Merychippus* and *Pliohippus* to the Pleistocene and Recent *Equus. Merychippus,* a pony-size Miocene type, was functionally tridactyl but retained short lateral digits. The dentary bone was deep, the face was long, and the orbit was fully enclosed. The cheek teeth were high-crowned and covered with cement and had an occlusal pattern similar to that of *Equus* (Fig. 12–8E). *Pliohippus* occurred in the Pliocene; it had the skull features of its progenitor, *Merychippus,* but was more progressive in having higher-crowned teeth and lateral digits reduced to splintlike vestiges. *Equus,* as well as an extinct evolutionary side branch of short-legged South American horses (typified by *Hippidium*), evolved from *Pliohippus. Equus,* the genus to which all living horses belong, differs from *Pliohippus* in greater size and in a more complex crown pattern of the cheek teeth. Major evolutionary trends of the Equidae listed by Colbert (1980) include increase in size, lengthening of legs and feet, reduction in the size of the lateral toes and emphasis on the middle toe, molarification of premolars, increase in height of the cheek teeth crowns, lengthening of the facial part of the skull to accommodate the large cheek teeth, and deepening of the maxillary and dentary bones to accommodate the high-crowned teeth. In addition, the profile of the angular border of the dentary bone swept progressively farther forward and the origin of the masseter muscles migrated forward. These adaptations increased the force the masseter muscles could exert on the dentary bone.

Cenozoic changes in climate and in the flora of North America may have had a critical influence on the evolution of horses. Especially important was the Miocene development of grasslands and savannas over much rolling or nearly level land in the present Great Plains, the Great Basin, and the southwestern deserts of the United States. Many of the most progressive equid skull and dental features probably arose in response to the shift to a grazing habit. Grass, at least at certain times of the year, has low nutritional value and must be eaten in large quantities to sustain life. High-crowned, persistently growing teeth were necessary to cope with large amounts of grasses made highly abrasive by silica in the leaves and by particles of soil deposited on leaves by wind and the splash effect of rain. Also of great adaptive value were the highly cursorial limbs with single-toed feet, which facilitated rapid and efficient locomotion on the firm, level footing of the grasslands. Cursorial ability was perhaps as advantageous for traveling between widely scattered concentrations of food and distant water holes in semiarid regions as for escaping from predators.

For some unknown reason, horses disappeared from the New World, their place of origin and the primary center of their evolution, before historic times. Their decline began toward the end of the Miocene Epoch, when savannas were being replaced by cooler and drier steppe conditions. Although wild horses now occupy only Africa and parts of Asia, within historic times they occurred throughout much of Eurasia. Wild equids inhabit grasslands in areas ranging from tropical to subarctic in climate. Feral horses and burros are highly destructive to native vegetation in some parts of the western United States.

Family Tapiridae. Tapirs occupy tropical parts of the New World and the Malayan area. The family includes one living genus and four species. Structurally, tapirs are notably primi-

tive and, according to Romer (1966:269), are still very close in many respects to the common ancestors of all perissodactyls. "True" tapirs are known first from the Oligocene, but possible ancestral types occurred in the Eocene of North America.

Tapirs have a stocky build and weigh up to about 300 kilograms. The limbs are short, and both the ulna and fibula are large and separate from the radius and tibia, respectively; the front feet have four toes (Fig. 12–5A) and vestiges of the fifth (the pollex), and the hind feet have three toes. Tapirs retain a full placental complement of 44 teeth. Three premolars are molariform, and the brachyodont cheek teeth retain a simple pattern of cross lophs. The short proboscis (Fig. 12–10) and reduced nasals are among the few specializations of tapirs.

These animals have probably always occupied moist forests, where their primitive feet serve well on the soft soil and their teeth are adequate for masticating plant material that is not highly abrasive. Tapirs today inhabit mostly tropical areas and usually are found near water. They are rapid swimmers and often take refuge from predators in the water. Tapirs are solitary and nocturnal, and their presence is frequently made known chiefly by their systems of well-worn trails between feeding areas, resting places, and water. Their food is largely succulent plant material, including fruit.

Family Rhinocerotidae. This family is represented today by three genera and five species and is restricted to parts of the tropical and subtropical sections of Africa and southeastern Asia. These ponderous creatures—the armored tanks of the mammal world—are surviving members of the spectacular late Tertiary and Pleistocene ungulate fauna. Although apparently a declining group, rhinoceroses have an illustrious past.

The fossil record of the rhinoceroses and their relatives (superfamily Rhinocerotoidea) is remarkably complex and parallels that of the horses in documenting the Early and Middle Tertiary success and the Late Tertiary decline of the group. Two genera that illustrate well the diversity of Early Tertiary rhinocerotoids are

Figure 12–10.   South American tapir *(Tapirus terrestris).* (By R. Garrison; San Diego Zoo)

Hyracodon and *Baluchitherium. Hyracodon* (Hyracodontidae), a small North American Oligocene "running rhinoceros," had slender legs and tridactyl feet and was probably similar in cursorial ability to Oligocene horses. Perhaps as a result of competition with horses, hyracodonts became extinct in late Oligocene. A con-temporary of *Hyracodon* in the Oligocene was the Asian form *Baluchitherium* (Rhinocerotidae), the largest known land mammal. This giant was 6 meters high at the shoulder, and the skull (small in proportion to the great size of the rest of the animal) was 1.3 meters long. The neck was long, and *Baluchitherium* perhaps

Figure 12–11.   The black rhinoceros *(Diceros bicornis)* of East Africa. (Taken in Amboseli National Park, Kenya)

browsed on high vegetation in giraffe-like fashion. The limbs were long and graviportal, but the tridactyl feet were unique in that the central digit was greatly enlarged and terminated in a broad hoof, whereas the lateral digits were much smaller than in any other rhinocerotoid. Rhinoceroses died out in the New World in the Pliocene, but remained common and diverse in Eurasia through the Pleistocene. The Pleistocene woolly rhinoceros *(Coelodonta)* was apparently adapted to cold climates. Entire preserved specimens of this animal have been found in an oil seep in Poland.

All Recent rhinoceroses are large, stout-bodied herbivores with fairly short, graviportal limbs (Fig. 12–11). Weights range up to about 2800 kilograms. The front foot has three or four toes (Fig. 12–5B), and the hind foot is tridactyl. The nasal bones are thickened and enlarged, often extend beyond the premaxillary bones, and support a horn. Where there are two horns, the posterior one is on the frontals. The horns are of dermal origin and lack a bony core. The occipital part of the skull is unusually high and yields good mechanical advantage for neck muscles that insert on the lambdoidal crest and raise the heavy head. The incisors and canines are absent in some rhinoceroses and are reduced in number in others; the dental formula is 0-1/0-2, 0/0-1, 3-4/3-4, 3/3 = 24-36. The cheek teeth have a primitive pattern of cross lophs far simpler than that of equids (Fig. 12–12).

Rhinoceroses inhabit grasslands, savannas, brushlands, forests, and marshes in tropical and subtropical areas. Some species are usually solitary *(Diceros),* whereas others occur in family

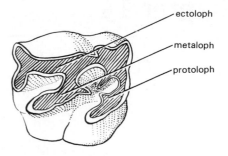

Figure 12–12. The right upper molar of a rhinoceros *(Rhinoceros).*

groups *(Ceratotherium)* or even in assemblages including up to 24 animals (Heppes, 1958). Rhinoceroses practice scent marking by establishing dunghills along well-worn trails. A variety of plant material is eaten; some species are browsers and some are grazers. Adults are nearly invulnerable to predation, except by humans, but young rhinoceroses may be attacked by lions or spotted hyenas. The Asian rhinoceroses *(Rhinoceros* and *Didermocerus)* and the African black rhinoceros *(Diceros)* are probably facing extinction. Because of the supposed medicinal properties of the horn and other parts, Asian rhinoceroses have been hunted persistently for at least 1000 years. Remnants of the formerly more widespread populations of the one-horned rhinoceros *(Rhinoceros)* are restricted to India, Nepal, and Java. The African black rhinoceros continues to decline over broad areas despite efforts to protect it from poaching. Regrettably, the future of rhinoceroses seems extremely dim.

Chapter 13

Order Artiodactyla

Today artiodactyls (pigs, camels, deer, antelope, cattle, and their kin) far overshadow perissodactyls in diversity and abundance. Whereas perissodactyls were abundant and reached their greatest diversity in the late Eocene, artiodactyls underwent their most important evolution much later, in the Miocene. Since this epoch, the perissodactyls have steadily declined but the artiodactyls have remained diverse and successful. One is tempted to relate the decline of the perissodactyls to the rise of the artiodactyls and to regard the latter as the more effective competitors. Although many structural differences between perissodactyls and artiodactyls are apparent, however, the functional advantages conferred by many of the features of the latter are not easily recognized.

In the order Artiodactyla, the structure of the foot is especially diagnostic. The foot is paraxonic; that is, the plane of symmetry passes between digits three and four. The weight of the animal is borne primarily by these digits: the first digit is always absent in living members, and the lateral digits (two and five) are always more or less reduced in size. Four complete and functional digits occur in the families Suidae, Hippopotamidae, and Tragulidae and in the forelimb of the Tayassuidae (the hind limb has the medial digit suppressed). Two complete toes, with the lateral digits absent (Camelidae, some Bovidae, Antilocapridae, and Giraffidae) or with incomplete remnants of the lateral digits (Cervidae and some Bovidae) occur in the more cursorial families. The cannon bone is present in the families Camelidae, Cervidae, Giraffidae, Antilocapridae, and Bovidae. Typically, the terminal phalanges are encased in pointed hoofs. The limbs have springing ligaments (Figs. 13–1 and 13–2), and the astragalus has a "double pulley" arrangement of articular surfaces (Fig. 13–3B) that completely restricts

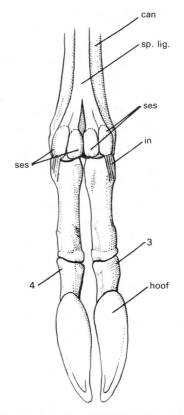

Figure 13-1. Posterior view of the left hind foot of an artiodactyl *(Antilocapra americana),* showing the position of the springing ligament. Abbreviations: *can,* cannon bone; *in,* insertion of the springing ligament; *ses,* sesamoid bone; *sp lig,* springing ligament; 3, third digit (second phalanx); 4, fourth digit (second phalanx).

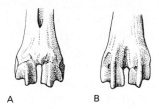

A   B

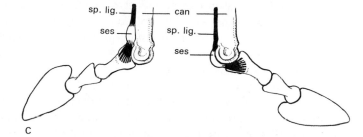

sp. lig. ——   —— can
ses ——   sp. lig.
                       ses

Figure 13-2. Left hind foot of the pronghorn antelope *(Antilocapra americana)*. (A) Anterior view of the distal end of the cannon bone; (B) posterior view of this bone. (C) Position of the phalanges: when the foot is supporting the weight of the body and the springing ligament (shown in black) is stretched and when the foot leaves the ground and the springing ligament flexes the phalanges. Abbreviations as in Figure 13-1.

C

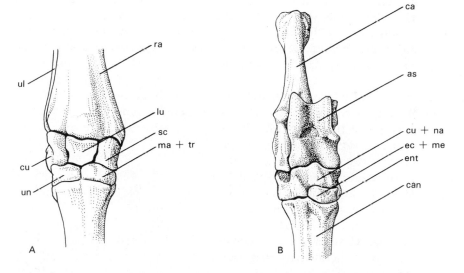

Figure 13-3. Ankle and wrist joints of artiodactyls: (A) right carpus of the mule deer *(Odocoileus hemionus)*; (B) right tarsus of the pronghorn antelope *(Antilocapra americana.)*. Abbreviations: *as,* astragalus; *ca,* calcaneum; *can,* cannon bone (fused third and fourth metatarsals); *cu,* cuneiform of the carpus; *cu + na,* fused cuboid and navicular; *ec + me,* fused ectocuneiform and middle cuneiform; *ent,* internal cuneiform; *lu,* lunar; *ma + tr,* fused magnum and trapezoid; *ra,* radius; *sc,* scaphoid; *ul,* ulna; *un,* unciform.

lateral movement. The proximal articulation (with the tibia) and the distal articulation (with the navicular and cuboid, which are fused in many advanced types) of the astragalus are critical in allowing great latitude of foot and digit flexion and extension, as is the extension of the articular surfaces and keels on the distal ends of the cannon bones to the anterior surfaces (Fig. 13–2A,B). The distinctive artiodactyl astragalus is regarded by some as a key to the success of the group. Perhaps more important, however, is the remarkable efficiency of the artiodactyl (ruminant) digestive system.

The limbs of artiodactyls, especially the distal segments, are usually elongate and fairly slim. The femur lacks a third trochanter. Whereas this prominence serves as a point of insertion of gluteal muscles in perissodactyls, in artiodactyls these muscles insert more distally, on the tibia. The distal parts of the ulna and the fibula are usually reduced and may fuse with the radius and tibia, respectively; this fusion is associated with the restriction of limb movement to one plane. The clavicle is seldom present. The intrinsic muscles of the feet (those that both originate and insert on the feet) are usually absent, being replaced by specialized tendons and ligaments.

The skull usually has a long preorbital section, and a postorbital bar or process is always present. Horns, always of bone or with a bony core, are most often borne on the frontals, which are enlarged at the expense of the parietals. The teeth are brachyodont or hypsodont and vary from 30 to 44 in number. The crown pattern is bunodont or, more often, selenodont. The premolars are not fully molariform, in contrast to the perissodactyl situation, and considerable specialization of the anterior dentition occurs in advanced types.

The system of classification used here is that of Romer (1966) and recognizes the extinct suborder Palaeodonta and the living suborders Suina and Ruminantia, all of which are first known from the Eocene. The Palaeodonta contains the most primitive known artiodactyls from the early Tertiary.

Suborder Suina

In members of the suborder Suina, which includes pigs, peccaries, and hippopotami, the molars are bunodont, the canines (and in the case of the hippopotami, the incisors also) are tusklike (Figs. 13–4 and 13–5), and the feet usually retain four toes with complete and separate digits. The skull contrasts with those of other artiodactyls in having a posterior extension of the squamosal bone that meets the exoccipital and conceals the mastoid bone. The stomach is of a nonruminant type (the animals do not chew their cud) but may have as many as three chambers.

Family Suidae. Swine are omnivorous and lack many structural modifications typical of more specialized artiodactyls. The Suidae is an Old World Family; the present distribution includes much of Eurasia and the Orient, and Africa south of the Sahara. There are nine Recent species of five genera. Suids appeared in the Oligocene. The entelodonts (Entelodontidae), an early branch on the swine evolutionary line, were huge piglike creatures with skulls up to 1 meter in length.

Most suids resemble the domestic swine

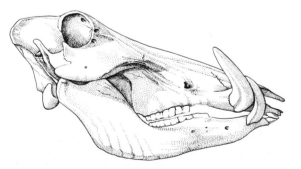

Figure 13–4. Skull of the warthog (*Phacochoerus aethiopicus*) of Africa; length of skull 376 mm.

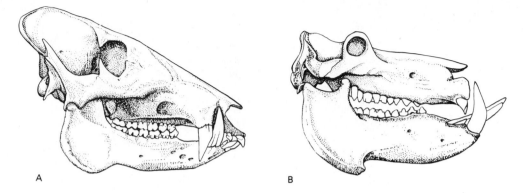

Figure 13-5. Skulls of artiodactyls of the suborder Suina: (A) pecary *(Tayassu tajacu)*; length of skull 225 mm; (B) hippopotamus *(Hippopotamus)*; length of skull 600 mm. (B after Romer, 1966)

(Sus). Adults may weigh as much as 275 kilograms and typically have thick, sparsely haired skin. The skull is long and low and usually has a high occipital area and a concave or flat profile (Fig. 13–4). The large canines are ever-growing, and the upper canines form conspicuous tusks that protrude from the lips and curve upward. In *Babyrousa,* an Indonesian suid, the upper canines protrude from the top of the snout. Suid molars are bunodont, and the last molars are often elongate, with many cusps and a complexly wrinkled crown surface (Fig. 13–6A). The dental formula is variable even within a species, with the total number of teeth ranging from 34 to 44. The limbs are usually fairly short, and the four-toed feet never have cannon bones (Fig. 13–7B).

Swine inhabit chiefly forested or brushy areas, but the warthog *(Phacochoerus)* favors savanna or open grassland and is entirely herbivorous. Most suids are gregarious, and some assemble in groups of up to 50 individuals. Most species eat a broad array of plant food and carrion and, given the opportunity, kill and eat such animals as small rodents and snakes. By comparison with ruminant artiodactyls, cursorial ability in suids is modest. Warthogs (Fig. 13–8) are fairly swift, however, and escape their predators by speed and by taking refuge in burrows.

Family Tayassuidae.   These animals, usually called javelinas or peccaries (Fig. 13–9), are restricted to the New World, where they occur from the southwestern United States to central Argentina. There are but three Recent species of two genera *(Tayassu* and *Catagonus)*. *Catagonus* was first known from Pleistocene fossils and was for many years thought to be extinct, but Wetzel, Dubos, et al. (1975) reported a surviving population in the biologically poorly known Gran Chaco area of Paraguay. The fossil record of tayassuids begins in the Oligocene.

Figure 13-6.   (A) Second and third right upper molars of the swine *(Sus scrofa)*; (B) Comparable teeth of the elk *(Cervus elaphus),* with the enamel ridges unshaded, the enamel-lined depressions stippled, and the dentine cross-hatched. The molars of *Sus* are bunodont; those of *Cervus* are selenodont.

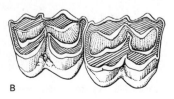

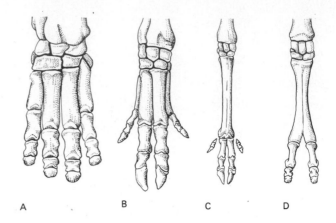

Figure 13-7. Right front feet of some artiodactyls: (A) hippopotamus *(Hippopotamus)*; (B) swine *(Sus)*; (C) elk *(Cervus)*; (D) camel *(Camelus)*. (After A. B. Howell, 1944)

A   B   C   D

Presumably peccaries evolved from primitive Old World pigs, but they are not known from the Old World after the Miocene. They are more progressive in limb structure than are suids and are less carnivorous.

Peccaries are much smaller than suids; weights of the former range up to only about 30 kilograms. The skull has a nearly straight dorsal profile, and the zygomatic arches are unusually robust (Fig. 13–5A). The canines are long and are directed slightly outward; they never turn upward, however, and have sharp medial and lateral edges. These teeth slide against one another, and the anterior surface of the upper and posterior surface of the lower are planed flat by this contact. These interlocking canines form "occlusal guides" that, together with the hinge-like jaw joint, stabilize the jaw joint against forces generated when hard nuts or seeds are cracked by the rear teeth (Kiltie, 1981). The molars are roughly square and have four cusps; they have thick enamel and lack the complex wrinkled and multicusped pattern typical of suids. The dental formula is 2/3, 1/1, 3/3, 3/3 = 38. The feet are slender and appear delicate, and the side toes are small relative to those of

Figure 13-8. A warthog *(Phacochoerus aethiopicus)*. (Taken in Buffalo Springs Game Reserve, Kenya, East Africa, by K. P. Dial)

Figure 13-9. Peccaries *(Tayassu tajacu;* Tayassuidae) in the Verde Valley of Arizona. (A. Richmond)

suids and usually do not reach the ground. Of the two living genera of peccaries, *Catagonus* is the more cursorial. Both genera have four toes on the front foot, but whereas in *Catagonus* digits two and five of the hind foot are vestigial (they lack phalanges and hoofs), digit two is complete and digit five is vestigial in *Tayassu.* In both genera, the medial metatarsals are partly fused. An additional difference is the more elongate distal segments of the limbs in *Catagonus* (Wetzel, 1977). Modern peccaries are not as cursorial as was *Mylohyus,* an extinct Pleistocene species of North America in which the side toes of the forefoot were very strongly reduced and the didactyl hind foot had a fully developed cannon bone.

Peccaries occupy diverse habitats, from deserts and oak-covered foothills in Arizona to dense tropical forests and thorn scrub in southern Mexico, Central America, and South America. They are highly social, and each social group contains up to 12 individuals. Peccaries are omnivorous, but seem to rely more heavily on plant material than do suids. The presence of peccaries is often indicated by shallow excavations where roots have been exposed beneath bushes or patches of prickly-pear cactus. Despite their chunky build, peccaries are rapid and extremely agile runners.

Family Hippopotamidae. This family is represented today by the genera *Hippopotamus* and *Choeropsis* (the pigmy hippopotamus), each with one species. The group first appeared in the late Pliocene in Africa and Asia and occurred widely in the southern parts of the Old World in the Pleistocene. Hippopotami now occur only in Africa; in North Africa they are restricted to the Nile River drainage, but they occur widely in the southern two thirds of the continent.

Hippopotami are bulky, ungraceful creatures with huge heads and short limbs (Fig. 13–10). They are large, with *Hippopotamus* weighing up to 3600 kilograms and *Choeropsis* about 180 kilograms. Some of the distinctive features of these animals probably evolved in association

Figure 13–10. Two male hippopotami *(Hippopotamus amphibius)* fighting. (Taken at Lake Baringo, Kenya)

with their amphibious mode of life. Specialized skin glands secrete a pink, oily substance that protects the sparsely haired body. The highly specialized skull has elevated orbits and enlarged and tusklike canines and incisors (Fig. 13–5B). The bunodont molars are basically four cusped; the dental formula is 2-3/1-3, 1/1, 4/4, 3/3 = 38-44. The limbs are robust, and the feet are four-toed (Fig. 13–7A). The foot posture is semidigitigrade; only the distal phalanx of each toe touches the ground. The broad foot is braced by a sturdy "heel" pad of connective tissue, and the central digits are nearly horizontal.

Hippopotami are restricted to the vicinity of water. *Hippopotamus* is gregarious, and groups spend much of the day in the water. When bodies of water are at a premium during the dry season, hippopotami often concentrate in stagnant ponds which they churn into muddy morasses. They are good swimmers and divers and when submerged are able to walk on the bottom of rivers or lakes by using a slow-motion gait. At night, hippopotami may move far inland to feed on vegetation. *Choeropsis* is solitary or occurs in pairs and inhabits forested areas. Instead of seeking shelter in the water when disturbed, as is characteristic of *Hippopotamus, Choeropsis* seeks refuge in dense vegetation.

Suborder Ruminantia

This suborder includes camels, giraffes, deer, antelope, sheep, goats, and cattle. Members of this most advanced artiodactylan suborder have been in the past, and remain, the dominant artiodactyls. In general, these animals are committed strictly to a herbivorous diet and to highly cursorial locomotion. Ruminants chew their cud; the stomach has three or four chambers (Fig. 2–14) and supports microorganisms that have cellulolytic enzymes. Ruminants have selenodont molars (Fig. 13–6B), and the anterior dentition is variously specialized by loss or reduction of the upper incisors, by the development of incisiform lower canines, and commonly by the loss of upper canines. The skull differs from those of members of the suborder Suina in the exposure of the mastoid bone between the squamosal and exoccipital bones. Antlers or horns, often large and complex structures, are present in the most progressive families. In the limbs, there is a pronounced trend toward elongation of the distal segments, fusion of the carpals and tarsals, and perfection of the two-toed foot.

The ruminants may conveniently be separated into two divisions (infraorders): Tylopoda, the camels and llamas and their extinct relatives, and Pecora, which includes all of the remaining, more progressive ruminants. A diagnostic feature of the pecorans is the fusion of the navicular and cuboid bones, over which the astragalus is nearly centered (Fig. 13–3B).

An early tylopod family (now extinct), the Merycoidodontidae, usually called oreodonts, deserves mention because these are by far the most abundant mammals in some Oligocene and early Miocene strata in North America. Oreodonts were typically piglike in general build, with short limbs, digitigrade and four-toed feet, and a continuous tooth row with no loss of teeth (Fig. 13–11). In contrast to the Suina, however, the cheek teeth were selenodont and became fairly high-crowned in advanced types, suggesting the acquisition of grazing habits. Probably as a result of competition with more advanced artiodactyls, oreodonts declined in the late Miocene, and disappeared in the Pliocene.

Family Camelidae. These primitive ruminants are restricted to arid and semiarid regions. *Camelus,* with two species, occupies the Old World, and wild populations persist in the Gobi Desert of Asia; *Lama* (Fig. 13–12), with two species, occurs in South America from Peru through Bolivia, Chile, Argentina, and Tierra del Fuego. Camels probably arose in the Old World and migrated in the late Eocene to North America, where their fossil record begins. Of special interest, as an example of a reversal of a well-established evolutionary trend, is the development of the camelid foot. By the Oligocene, the

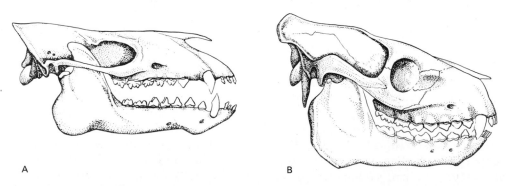

A

B

Figure 13–11.   Skulls of extinct ungulates. A, *Phenacodus,* a Lower Eocene primitive ungulate (Condylarthra); length of skull roughly 230 mm. B, *Oreodon,* a primitive Oligocene ruminant artiodactyl; length of skull roughly 125 mm. (After Romer, 1966.)

Figure 13-12. The vicuña *(Lama vicugna),* a camelid that inhabits the central Andes of South America. (C. B. Koford)

camel's foot was already highly specialized. It was nearly unguligrade in posture and didactyl, and the distalmost phalanges probably bore hoofs. The distinctive distal divergence of the metapodials (Fig. 13–7D), however, was already recognizable. In the Miocene the central metapodials fused to form a cannon bone, but at this same time a retrograde trend toward the secondary development of a digitigrade foot posture began, and from Pliocene times onward camels were digitigrade. They are the only living fully digitigrade ungulates. Because semiarid conditions developed and became widespread in the Miocene, one is tempted to relate the changes in the camelid foot posture to changing soil conditions. In any case, the camelid foot clearly provides effective support on soft, sandy soil, into which the feet of "conventional" unguligrade artiodactyls sink deeply. Taking advantage in the Pleistocene of land bridges between North America and Asia, and between North and South America, camels spread to the Old World and to South America.

Although highly specialized in foot structure, camelids are the most primitive living ruminants. They are fairly large mammals, ranging in weight from about 60 to 650 kilograms, and have long necks and long limbs. The dentition has advanced less toward herbivorous specialization than has that of the Pecora. In camelids, only the lateral upper incisor is present, but it is caniniform; the lower canines are retained and are little modified (Fig. 13–13). The lower incisors are inclined forward and occlude with a hardened section of the gums on the premaxillary bones. A broad diastema is present,

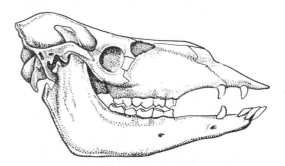

Figure 13-13. Skull of an extinct Pleistocene New World camel *(Camelops);* length of skull 565 mm. (After Romer, 1966)

and the premolars are reduced in number (to 3/ 1-2 in *Camelus* and 2/1 in *Lama*). As in other ruminants, the limbs are long and the ulna and fibula are reduced; the trapezium is absent in the carpus, and the mesocuneiform and ectocuneiform are fused in the tarsus. The digitigrade feet are didactyl, but the cannon bone is distinctive in that the distal ends of the metapodials remain separate and flare outward (Fig. 13–7D). The toes are separate, and each is supported by a broad cutaneous pad, which largely encases the second phalanx and increases the surface area of the foot greatly. The short ungual phalanges do not bear hoofs but have nails on the dorsal surfaces.

Camels are remarkably well adapted to arid areas, and their ability to go without water for long periods in hot weather is remarkable (p. 482). They are grazers and can survive in regions with only sparse vegetation. The guanaco *(Lama glama)* and vicuña *(Lama vicugna)* of South America are gregarious and live in small social groups dominated by an adult male. Guanacos are fairly speedy runners but are especially skillful at moving rapidly over extremely rough terrain. Camelids are highly vocal. The dromedary (Arabian) camel *(Camelus dromedarius)* makes an assortment of snarling sounds, and guanacos give a yammering call, about which Simpson (1965a) comments that "there is something distinctly indecent about the noise as it issues from the beast's protrusile and derisive lips."

Family Tragulidae. This family, which contains the chevrotains (mouse deer), has only two living genera with four species but is of interest because these animals probably resemble in many ways the ancestors of the more advanced pecorans. Chevrotains are small, delicate creatures, weighing from 2.3 to 4.6 kilograms, that occur in tropical forests in central Africa *(Hyemoschus)* and in parts of southeastern Asia *(Tragulus;* Fig. 13–14). The tragulid fossil record begins in the Miocene Epoch.

Although apparently related to higher pecorans, chevrotains combine a unique complex of features. The tragulid skull never bears antlers, but, seemingly in compensation, the upper canines are tusklike and are used by males in intraspecific combat. Otherwise, the dentition resembles that of higher pecorans: the upper incisors are lost, the lower canine is incisiform, and the cheek teeth are selenodont. Unique to tragulids is an ossified plate, derived from an aponeurosis (fusion) of muscles, to which the sacral vertebrae attach. The limb structure is a mosaic of primitive and advanced features. Although the limbs are long and slender and the carpus is highly specialized in having the navicular, cuboid, and ectocuneiform bones fused, the lateral digits are complete, a condition never present in higher pecorans. In addition, although a cannon bone occurs in the hind limb, the metacarpals of the central digits are separate in the African tragulid and are partly fused in the Asian form, whereas the cannon bone is represented by fully fused metapodials in all other pecorans.

Tragulids are secretive, chiefly solitary, nocturnal creatures that inhabit forests and underbrush and thick growth along water courses. They escape predators by darting along diminutive trails into dense vegetation. Their food consists of grass, the leaves of shrubs and forbs, and some fruit.

Advanced Pecorans

The remaining pecorans (Cervidae, Giraffidae, Antilocapridae, and Bovidae) are advanced artiodactyls and share a series of progressive features. The upper incisors are absent, the upper canines are usually absent, the lower canines are incisiform, and the cheek teeth are selenodont. The dental formula is typically 0/3, 0/1, 3/3, 3/3 = 32. The cannon bone is present in fore and hind limbs, its distal articular surfaces are extensive, and the lateral digits are always incomplete (Fig. 13–7C) and are often lacking. Movement of the foot is strongly limited to a single plane by the tongue-in-groove contacts be-

Figure 13-14. A chevrotain (*Tragulus napu*, Tragulidae). Note the large upper canine and the lack of antlers or horns. (By R. Garrison; San Diego Zoo)

tween the astragalus and the bones with which it articulates, and by the specialized articular surfaces of the joint between the cannon bone and the first phalanges (Fig. 13–2). Some fusion of carpal elements always occurs and further restricts movement to one plane. The navicular and cuboid bones are always fused (Fig. 13–3B), and a variety of fusion patterns occur in the other elements. The four-chambered stomach is of a ruminant type. Although all pecorans but the tragulids share this basic structural plan, each family has distinctive features usually related to diet and degree of cursorial ability.

Family Cervidae. Members of this family, which includes the deer, elk, caribou, and moose, occur throughout most of the New World and in the Old World in Europe, Asia, and northwestern Africa; they have been introduced widely elsewhere. Living members include 17

genera and 36 species. Cervids appeared in the early Oligocene in Asia and reached North America in the early Miocene.

Antlers are the most widely recognized characteristic of members of the family Cervidae (Figs. 13–15 and 13–16). Antlers attain spectacularly large size in some species and vary widely from one species to another. People have long been fascinated by their complexity, variety, and symmetry. All but two of the 37 species of cervids have antlers, and they occur only in males, except in caribou *(Rangifer)*. In some antlered cervids, the upper canines are retained but reduced (as in the elk, *Cervus elaphus*). Two cervids with short antlers have enlarged canines *(Cervulus* and *Elaphodus)*, and in two deer that have no antlers *(Hydropotes* and *Moschus)*, the canines are large sabers. Antlers usually arise from a short base on the frontals (the

Figure 13-15.   Two young male caribou *(Rangifer tarandus)* sparring. (Taken in Mt. McKinley National Park, Alaska, by G. C. Bateman)

pedicel, Fig. 13–16) and are entirely bony. Of particular interest is the annual cycle of rapid growth of the antlers, their use during the breeding season in ritualized social interactions, and their subsequent loss.

This annual cycle has been carefully studied in the white-tail deer *(Odocoileus virginianus)* of North America (Wislocki, 1942, 1943; Wislocki, Aub, and Waldo, 1947). The cycle is primarily under the control of the testicular and pituitary hormones. Secretions from the pituitary, activated by increasing daylength in the spring, initiate antler growth in April or May, and some time later pituitary gonadotropin stimulates growth of the testes. The growing antlers are covered by "velvet," a fur-covered skin that carries blood vessels and nerves. In the fall, androgen from the enlarging testes inhibits the action of the pituitary antler-growth hormone, leading to the drying up and loss of the velvet.

At this time, the animals rub and thrash their antlers against vegetation, and as the velvet is removed the antlers are stained by resins and take on a brown, polished look. In the fall and early winter, androgen maintains the connection between the dead bone of the antlers and the live frontal bones, and during the fall breeding season the antlers are used in clashes between males competing for females. In winter, pituitary stimulation of the testes declines as daylength is reduced and androgen secretion diminishes. This results in decalcification in the pedicel, weakness at the point of connection between the antler and the pedicel, and shedding of the antlers. For several months in late winter, before reinitiation of antler growth, the males are antlerless.

During the spring and summer, the rate of growth of the developing antlers is remarkable. In such large cervids as elk *(Cervus elaphus)*

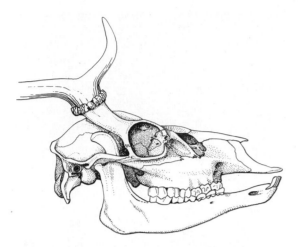

Figure 13–16. Skull of a male fallow deer (*Dama dama,* Cervidae); length of skull 265 mm. The bony antlers are shed yearly.

and caribou (Fig. 13–15), the antlers attain a length of 150 centimeters in 90 days or less. The growth rate of the antler and its attending nervous and vascular tissue is therefore about 1.7 centimeters per day. Touch and pain receptors in the velvet protect the growing antler from injury, and damage to nerves supplying a growing antler may be followed by enough injury to the antler to cause abnormal development.

The cheek teeth of cervids are brachyodont, reflecting a browsing habit. These animals have a wide size range: the musk deer *(Moschus)* weighs but 10 kilograms, whereas the moose *(Alces)* weighs up to roughly 800 kilograms. The feet are always four-toed, but the lateral toes are often greatly reduced. Distal to the astragalus and calcaneum, the tarsus is usually composed of three bones: the fused navicular and cuboid, the fused ectocuneiform and mesocuneiform, and the internal cuneiform (Fig. 13–3).

Cervids occur from the Arctic to the tropics. Many are well adapted to boreal regions and occupy mountainous or subarctic areas with severely cold winters. Effective insulation is provided in many cervids by the long, hollow hairs

of the pelage. Some species are gregarious for much of the year and may assemble in large herds during the winter and during migratory movements. In the western United States, mule deer *(Odocoileus hemionus)* and in the eastern United States, white-tail deer *(O. virginianus)* are common over wide areas and are heavily hunted in many states. Some states employ many biologists to study and manage deer; the expense of this work is generally covered by revenue from the sale of hunting licenses.

Family Giraffidae.   This group is represented today by but two genera, *Giraffa* and *Okapia;* each has a single species. The family occurs in much of Africa south of the Sahara.

The robust cheek teeth are brachyodont and are marked with rugosities. Short, bony horns, covered with furred skin, are borne on the front part of the parietals, and a medial thickening of the bone in the area where the nasals and the frontals join is conspicuous (Fig. 13–17). In some populations from north of the equator in East Africa, this thickening produces a median horn. Horns occur in both sexes and are never shed. The lateral digits of the elongate limbs are entirely gone, and the tarsus is highly specialized. Distal to the astragalus and calca-

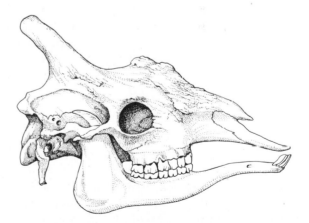

Figure 13–17. Skull of a male giraffe *(Giraffa camelopardalis);* length of skull 708 mm. The heavy deposits of bone on the frontal and nasal bones seemingly protect the skull when the head is used as a weapon in fights between males.

neum, only two tarsal bones are present. One is formed by the fusion of the navicular and cuboid bones, and the other by the fusion of the three cuneiform bones. The okapi lacks the extreme elongation of the neck and legs that is typical of giraffes but has an even more specialized tarsus in which all bones distal to the astragalus and calcaneum are fused. The fossil record of the giraffids begins in the Miocene Epoch. The okapi, not known to zoologists until 1900, is remarkable in its close resemblance to primitive late Miocene and early Pliocene giraffids long known to paleontologists.

Giraffes occurs in savannas and lightly wooded areas, where their exceptional height (Fig. 13–18) enables them to browse on branches of leguminous trees up to 6 meters above the ground. Some of the species of *Acacia* that these animals feed on bear long thorns, but giraffes adroitly use their long tongue and prehensile upper lip to gather leaves from even the most thorny acacias. Despite their considerable weight (to nearly 1820 kilograms in males), giraffes can gallop for short distances at over 55 kilometers per hour (White, 1948). Relative to those of lighter, shorter-limbed artiodactyls, the limbs of giraffes are flexed little during each stride, producing a stiff-legged gait. When the animal is walking or galloping, its center of gravity is partly controlled by fore-and-aft movements of the head and neck (Dagg, 1962). Ritualized fighting by males involves powerful blows by the head against the opponent's head, neck, and body. The okapi lives in dense tropical forests and eats leaves and fruit.

Family Antilocapridae. This family is represented today by one species, the pronghorn antelope (*Antilocapra americana;* Fig. 13–19), which occurs from central Canada to north central Mexico. The fossil record of these animals is entirely North American and begins with *Merycodus* in the middle Miocene. The horns of this antilocaprid were cervid-like in form, that is, they were forked and had a basal burr. In contrast to cervids, however, the cores were never

Figure 13-18.   A giraffe among umbrella trees *(Acacia tortilis)* in Tsavo West National Park, Kenya. The markings on the body enable this huge animal to blend inconspicuously into savanna vegetation.

shed, and both sexes had horns. In addition, the unusually prominent orbits were situated high and far back in the skull, as in *Antilocapra.* The horns of fossil pronghorn antelope were generally more complex than those of *Antilocapra,* but the limbs and teeth of fossil forms were similar to those of the present species.

The fossil record of the pronghorn antelopes is entirely North American, and today this animal occupies open country from central Canada to north central Mexico. The pronghorn fauna was at one time more di-

Figure 13–19. The pronghorn antelope *(Antilocapra americana),* one of the fastest cursorial mammals. The long, slender limbs are typical of the more cursorial ungulates. (By F. D. Schmidt; San Diego Zoo)

verse than it is now. From the middle Pleistocene Tacubaya Formation in central Mexico, which contains numerous fossils of mammals that lived in a small area at one time, Mooser and Dalquest (1975) list four species of extinct pronghorn antelope, ranging in size from a tiny species to one at least as large as the living species.

The pronghorn antelope is unique in being the only mammal that sheds its horn sheaths annually. The sheaths are formed of a specialized growth of skin. The old sheath, beneath which a new sheath is beginning to develop, is shed annually in early winter, and the new sheath is fully grown by July (O'Gara, May, and Bear, 1971). Whereas the mature sheath is forked, the bony core is a single, laterally compressed blade. Both sexes have horns, but those of the females are small and inconspicuous. The dental formula is that typical of pecorans, and al-

though the animals are largely browsers on low shrubs and forbs, the cheek teeth are high-crowned. The abrasive soil particles adhering to the low-growing vetetation they eat probably make high-crowned teeth advantageous. Perhaps as an adaptation allowing the antelope to watch for danger while its head is close to the ground, the orbits are unusually far back in the skull (Fig. 13–20). The legs are long and slender, and all vestiges of lateral digits are gone. The tarsus distal to the astragalus and calcaneum consists of but three bones (Fig. 13–3B).

Pronghorn antelope inhabit open prairies and deserts that support at least fair densities of low grasses, shrubs, and forbs. The numbers of antelope were seriously reduced during the pioneering period of the western United States, but according to Einarsen (1948:7), "at no time has the antelope been driven from its original range, being continuously represented by a few

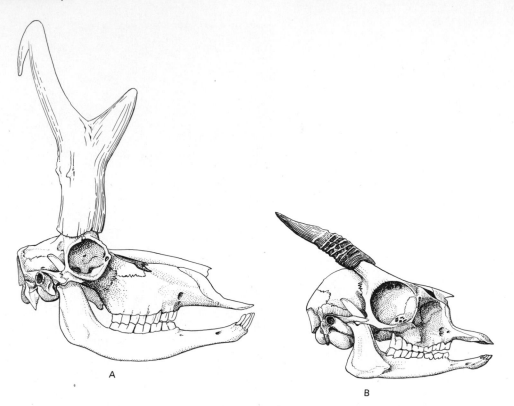

Figure 13-20. Skulls of bovids: (A) pronghorn antelope *(Antilocapra americana);* length of skull 292 mm; (B) Kirk's dik-dik *(Madoqua kirki);* length of skull 108 mm. The receding nasal bones of the dik-dik are an adaptation allowing mobility of the short proboscis.

individuals in widely scattered sections of the plains country." Today pronghorn antelope are common in a number of the western states, and even the most unobservant tourist driving across Wyoming cannot fail to notice these animals. They are among the fastest of cursorial mammals. At full speed on level footing, they can attain a speed of about 95 kilometers per hour for short distances (Einarsen, 1948; McLean, 1944). Their endurance is remarkable. On the short-grass prairie of north central Colorado, I observed antelope that had run more than 3.5 kilometers put on a burst of speed and run away from a closely pursuing light plane that was traveling at 72 kilometers per hour.

Family Bovidae.   This family, which includes African and Asian antelope, bison, sheep, goats, and cattle, is the most important and most diverse living group of ungulates. The family includes 45 genera and 124 species, and wild species occur throughout Africa, in much of Europe and Asia, and in most of North America. (A classification of living bovids is given in Table 13–1). The domestication of bovids began in Asia roughly 8000 years ago (Darlington, 1957:405), and domestic bovids are nearly as cosmopolitan in distribution as are humans.

Bovids derived from traguloid ancestry in the Old World and first appeared in the early Miocene of Europe. Judging from the many kinds of bovids known from the Pliocene, this group underwent a rapid radiation. More Pleistocene than Recent genera of bovids are known (100 versus 45). Toward the end of the Pleis-

Table 13-1: Classification of Bovidae and Distribution of Recent Genera

Subfamily Bovinae
  Tribe Bovini
    Bison   Bison (Europe, North America)
    Bos   Cattle (worldwide)
    Bubalus   Asiatic buffalo (Asia)
    Synceros   African buffalo (Africa)
  Tribe Boselaphini
    Boselaphus   Nilgai (Asia)
    Tetracerus   Four-horned antelope (Asia)
  Tribe Tragelaphini
    Tragelaphus   Bushbuck, nyala, kudu, bongo, etc. (Africa)
    Taurotragus   Eland (Africa)
Subfamily Cephalophinae
  Tribe Cephalophini
    Cephalophus   Duiker (Africa)
    Sylvicapra   Bush duiker (Africa)
Subfamily Reduncinae
  Tribe Reduncini
    Redunca   Reedbuck (Africa)
    Kobus   Waterbuck, kob, lechwe (Africa)
Subfamily Hippotraginae
  Tribe Hippotragini
    Hippotragus   Roan, sable (Africa)
    Oryx   Oryx (Africa)
    Addax   Addax (Africa)
Subfamily Alcelaphinae
  Tribe Connochaetini
    Connochaetes Wildebeest (Africa)
  Tribe Alcelaphini
    Alcelaphus   Hartebeest (Africa)
    Damaliscus   Hartebeest, topi, blesbok (Africa)
Subfamily Aepycerotinae
  Tribe Aepycerotini
    Aepyceros   Impala (Africa)
Subfamily Antiopinae
  Tribe Antilopini
    Antilope   Blackbuck (Asia)
    Gazella   Gazelle (Africa)
    Antidorcas   Springbok (Africa)
    Litocranius   Gerenuk (Africa)
    Procapra   Black-tailed gazelle (Asia)
  Tribe Ammodorcadini
    Ammodorcas   Dibatag (Africa)
  Tribe Neotragini
    Oreotragus   Klipspringer (Africa)
    Madoqua   Dik-dik (Africa)
    Dorcatragus   Beira antelope (Africa)
    Ourebia   Oribi (Africa)
    Rhaphicerus   Steinbuck, grysbuck (Africa)
    Neotragus   Royal, pigmy, and suni antelope (Africa)
Subfamily Peleinae
  Tribe Peleini
    Pelea   Rhebuck (Africa)

Table 13-1:   Classification of Bovidae and Distribution of Recent Genera *(Continued)*

Subfamily Caprinae
  Tribe Saigini

Pantholops	Chiru (Asia)
Saiga	Saiga (Asia, Europe)

  Tribe Rupicaprini

Naemorhedus	Goral (Asia)
Capricornis	Serow (Asia)
Oreamnos	Rocky mountain "goat" (North America)
Rupicapra	Chamois (Southwest Asia)

  Tribe Ovibovini

Budorcas	Takin (Asia)
Ovibos	Musk ox (North America, Greenland)

  Tribe Caprini

Hemitragus	Tahr (Asia)
Capra	Ibex, goat (Asia, Europe, North America)
Pseudois	Nahur (Asia)
Ammotragus	Barbary sheep (North Africa)
Ovis	Mouflon, argali, bighorn sheep, sheep (Asia, Europe, North America, North Africa)

Partly from Ansell, 1971, and Simpson, 1945.

tocene most bovids were driven from Europe by the southward advance of cold climate. A few reached the New World in the Pleistocene via the Bering Strait land bridge. Because this boreal avenue of dispersal was under the influence of cold climates, it functioned as a "filter bridge" (Simpson, 1965b:88) and only animals adapted to the cold dispersed across it.

As a consequence, the New World received from Asia such bovids as the bighorn sheep *(Ovis),* the mountain "goat" *(Oreamnos),* the musk ox *(Ovibos),* and the bison *(Bison),* all animals able to withstand cold. Bovids less able to withstand boreal conditions—the Old World antelopes and the gazelles are prime examples—were forced from the northern parts of Europe and Asia in the Pleistocene to their present strongholds in Africa and Asia and hence did not disperse across the Bering bridge to North America. An exception is the saiga antelope *(Saiga),* which now inhabits arid parts of Asia but which occurred in Alaska in the Pleistocene. Bison were extremely abundant members

of grassland faunas in the Pleistocene and Recent in North and Central America, where they occurred as far south as El Salvador. Some structural divergence occurred in Pleistocene bison, and in some areas several species may have occupied common ground. Some Pleistocene bison were considerably larger than the present *Bison bison.* Specimens of the Pleistocene species *Bison antiquus* from California indicate that this animal was over 2 meters high at the shoulder and had horns that in larger individuals spanned more than 2 meters.

Bovids characteristically inhabit grasslands, and their advanced dentition and limbs probably developed in association with grazing habits. The cheek teeth are high-crowned and the upper canines are reduced or absent. Preorbital vacuities in the skull are present in some bovids and absent in others. The lateral digits are reduced or totally absent, the ulna is reduced distally and is fused with the radius, and only a distal nodule remains as a vestige of the fibula. Horns, formed of a bony core and a keratinized

Figure 13–21. Male desert bighorn sheep *(Ovis canadensis).* (Taken in the Plumosa Mountains of Arizona by J. Witham)

sheath, are present in males of all wild species, and females often also bear horns. The entire horns (including both sheath and core) are never shed and in some species grow throughout the life of the animal. Bovid horns are never branched but are often large and occur in a variety of forms (Figs. 13–21, 13–22, and 13–23). Males of the Indian four-horned antelope *(Tetracerus quadricornis)* are unique in having four short, dagger-like horns. The horns are frequently used in fights between males during the breeding season, but in many species the horns are used in ritualized tests of body strength in such a way as to minimize injury (p. 416). Some bovids, such as oryx and gemsbok *(Oryx),* can use their horns as awesome defensive weapons, respected even by lions.

Horns occur in both sexes of many antelope, but the females of some species are hornless; horns are usually present in both sexes of the larger species but absent in the females of the smaller species. Females have horns in 75 percent of the genera in which the average female weight is more than 40 kilograms, whereas females nearly always lack horns in genera in which females weigh less than 25 kilograms. Further, when both sexes have horns, those of

the males are thicker at the base, more complex in shape, and adapted to withstanding the forces encountered during intraspecific combat. The horns of the female are straighter and thinner, better adapted to stabbing, and thus more effective defensive weapons. The relationship in females between body weight and the possession of horns is probably correlated with antipredator behavior: larger species depend on direct defense but small species use concealment or flight. Large antelope offer more effective defense than do small antelope because the former are larger than most of their predators and are greatly larger than the predators of their young.

The last great strongholds of bovids are the grasslands and savannas of East Africa. Here a diverse bovid fauna occurs (Fig. 13–23), and seemingly every conceivable bovid niche has

Figure 13-22. An ibex *(Capra ibex)* from the Greek island of Crete.

Figure 13–23.   Some members of the diverse bovid fauna of East Africa: (A) klipspringer *(Oreotragus oreotragus)*; (B) wildebeest *(Connochaetes taurinus)*; (C) topi *(Damaliscus lunatus)*; (D) waterbuck *(Kobus ellipsiprymnus)*; (E) gerenuk *(Litocranius walleri)*. (E by K. P. Dial; all taken in Kenya, East Africa.)

been occupied. Some antelope, such as the Bohor reedbuck *(Redunca redunca)* and the lechwe *(Kobus leche),* inhabit river borders and swampy ground, while at the other extreme, the oryx *(Oryx)* and addax *(Addax)* live in arid plains and desert wastes, where they may seldom have access to drinking water. Hopefully, the protection afforded game in some parts of Africa will allow the survival of many species of this group.

Chapter 14

Xenarthrans; Pangolins; the Aardvark

The three orders of mammals considered in this chapter—Xenarthra, Pholidota, and Tubulidentata—share a major structural trend, the loss or simplification of the dentition, but the phylogenetic relationships of these orders remain unclear. Xenarthrans (armadillos, sloths, and anteaters) underwent an impressive Tertiary radiation in South America, whereas pangolins (pholidotes) and aardvarks (tubulidentates) are Old World groups each of which has conservatively maintained a single structural plan. Living xenarthrans exhibit a number of feeding habits, but insect eating is most popular, and the South American anteaters are highly specialized termite eaters. Pangolins and the aardvark are also termite eaters, and their unlikely appearance seems to follow plans drawn by a mischievous designer with tongue in cheek.

lindrical. The coracoid process is unusually well developed, and the clavicle is present. The ischium is variously expanded and specialized (Fig. 14–2) and usually forms an ischiocaudal, as well as an ischiosacral, symphysis. The hind foot is typically five-toed, and the forefoot has two or three predominant toes with large claws. Major xenarthran structural trends are toward reduction and simplification of the dentition, specialization of the limbs for such functions as digging and climbing, and rigidity of the axial skeleton.

Paleontology

Xenarthrans, strictly a New World group, are first known from the Paleocene record of South America. The earliest fossils from this region are

Order Xenarthra

Although the xenarthrans are not of great importance today, including but 14 genera and 31 species, they are remarkably interesting animals because of their unique structure and unusual ecological roles, their large fossil types, and their remarkable Tertiary radiation in South America.

The living xenarthrans share a series of distinctive morphological features. Extra zygapophysis-like (xenarthrous) articulations (Fig. 14–1) brace the lumbar vertebrae. The incisors and canines are absent; the cheek teeth, when present, lack enamel, and each has a single root. The tympanic bone is annular; the brain is small, and the braincase is usually long and cy-

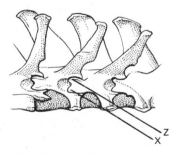

Figure 14-1. Three lumbar vertebrae (viewed from the left; anterior is to the left) of the nine-banded armadillo *(Dasypus novemcinctus)*, showing the xenarthrous articulations (*x*) supplementing the normal articulations between zygapophyses (*z*).

215

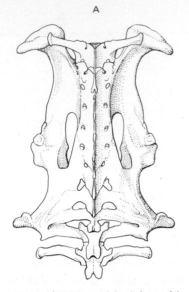
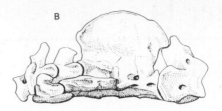

Figure 14–2. Parts of the skeleton of the nine-banded armadillo *(Dasypus novemcinctus)*. (A) Dorsal view of the pelvic girdle, showing the great degree of fusion of vertebrae with the ilium and ischium; (B) lateral view of the cervical vertebrae (anterior is to the right). The axis and cervical three, four, and five are fused.

from the late Paleocene and consist of scutes from the armor of armadillos (family Dasypodidae).

A diverse array of forms resulted from a mid-Tertiary radiation of South American armadillos: a large Pliocene type *(Macroeuphractus)* had enlarged, canine-like teeth and was probably a scavenger; the Miocene *Stegotherium* was most likely a termite eater; another Miocene form *(Peltephilus)* and its relatives had specialized scutes that formed pointed horns on the front of the snout; and rhinoceros-size armadillos of the subfamily Pampatheriinae lived in both North and South America in the Pleistocene.

The structural diversity of extinct South American armadillos indicates that they exploited a variety of foods and together formed an important part of the South American Tertiary biotas. The early dasypodids were armored with ossified dermal scutes, as are all modern species, and are perhaps the basal group from which other xenarthrans evolved.

Dispersal of mammals back and forth between North and South America was restricted from Paleocene until late Pliocene times (p. 360); during this interval not only dasypodids but also other xenarthrans underwent a radiation in South America. Several now extinct evolutionary lines arose. The Glyptodontidae appeared in the late Eocene Epoch and represent one line that probably evolved from armadillo stock. These ponderous creatures, some of which were 3 meters long, had unusually deep skulls (Fig. 14–3A). Many of the unique structural features of the glyptodonts are associated with their development of a nearly impregnable, turtle-like carapace composed of many fused polygonal scales. These are the most completely armored vertebrates known. The limbs are distinctive and highly specialized, and the last two thoracic vertebrae and the lumbar and sacral vertebrae are fused into a massive arch that, together with the ilium, supports the carapace. Patterson and Pascual (1972:267) suggest that the post-Miocene diversification of the

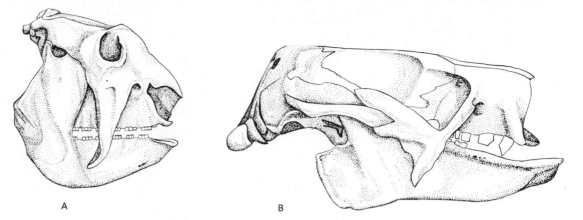

Figure 14–3. Skulls of extinct xenarthrans: (A) *Glyptodon,* length of skull 560 mm; (B) *Paramylodon,* length of skull 510 mm. (After Romer, 1966)

glyptodonts was favored by the spread of pampas (grassland) in South America.

Additional evolutionary lines are represented by ground sloths of several extinct families. Ground sloths first appear in South American Oligocene deposits. These animals were herbivores; their teeth lacked enamel and were ever-growing. The family Megatheriidae appears in the Oligocene record of South America and includes *Megatherium,* a massive ground sloth larger than an elephant, as well as smaller types 1 to 3 meters in length. These animals were covered with hair, lacked upper canine-like teeth, and walked on the outer edges of the unusually specialized hind feet. A second family of ground sloths, the Megalonychidae, is closely related to the megatheriids but differs from that group in having the anteriormost cheek teeth modified into "canines." *Megalonyx,* a Pleistocene genus that reached the size of a cow, was widely distributed in North America. The remains of smaller species of megalonychids have been found in the West Indies in association with human artifacts. These sloths seemingly survived into the Recent. A third family of ground sloths, the Mylodontidae, appeared in the Oligocene, and is characterized in part by the development of upper "canines" (Fig. 14–3B) and remarkably robust limbs. Protection was af-

forded some members of this group by round dermal ossicles embedded in the presumably thick skin. A trend toward large size is apparent in the mylodonts, as in other ground sloths.

The glyptodonts, megatheriids, megalonychids, and mylodonts underwent much of their evolution in the Tertiary isolated from the progressive North American mammalian fauna. Nevertheless, when the land bridge between the Americas was reestablished in the Pliocene, the glyptodonts and ground sloths were remarkably successful in invading North America. By some means of chance dispersal, ground sloths reached North America before it was joined with South America, and one genus *(Megalonyx)* seemingly evolved in North America.

The plains-dwelling mylodont *Paramylodon* was widespread in North America and is the most common xenarthran in the Pleistocene deposits of Rancho La Brea in Los Angeles. This ground sloth had large claws on digits two and three (Fig. 14–4B), an arrangement similar to that in the living armadillo (Fig. 14–5B).

The common megatheriid *Nothrotheriops* (Fig. 14-6) occurred in North and South America in the Pleistocene, and its remains from Gypsum Cave in Nevada include bones, skin, and hair. *Nothrotheriops* probably walked on the sides of its highly modified hind feet (Fig. 14–

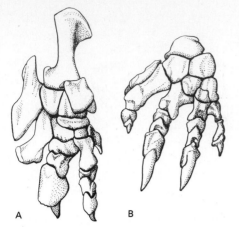

Figure 14-4. (A) The right pes of *Nothrotheriops;* (B) the right manus of *Paramylodon.* (After Romer, 1966)

Figure 14-6. How *Nothrotheriops* might have looked. This extinct megatheriid ground sloth survived until 11,000 years ago in the southwestern United States.

4A), as did other large ground sloths. The powerful claws of the forelimbs were probably used to grasp and tear down vegetation in preparation for ingestion. Dung of this animal is well preserved in dry caves in Nevada and Arizona.

Dung from Rampart Cave in Arizona contains such plants as Mormon tea *(Ephedra)* and globe mallow *(Sphaeralcea)* (Hansen, 1978); these plants remain common today in dry parts of the southwestern United States. This ground sloth persisted into the Recent in this area and probably did not disappear until 11,000 years ago. Thus ended a most fascinating and spectacular cycle of edentate evolution.

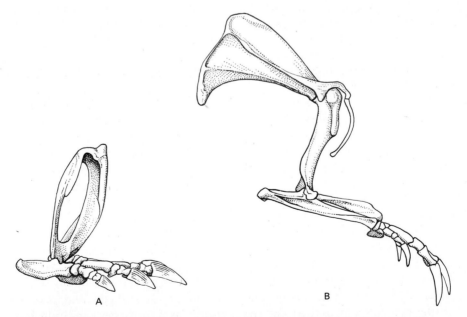

Figure 14-5. The right limbs of the nine-banded armadillo. (A) Part of hind limb; (B) forelimb. The flattening of the bones of the forearm and shank increases the surface area for attachment of muscles, and the elongation of the olecranon and the calcaneum give added mechanical advantage to the muscles that insert on them.

Figure 14-7. A southern tamandua (*Tamandua tetradactyla;* Myrmecophagidae) feeding on termites. (Taken by R. M. Warner in southeastern Peru)

Highly developed protective devices (as in Dasypodidae) and narrow, specialized feeding habits (Bradypodidae and Myrmecophagidae) have perhaps been important in allowing Recent xenarthrans to survive under competitive pressure from more "advanced" eutherians.

Family Myrmecophagidae.   Members of this family, the anteaters, are highly specialized ant and termite eaters (Fig. 14-7). They occur in tropical forests of Central and South America and in South American savanna. Although unimportant in terms of numbers of species (there are four Recent species of three genera), anteaters are locally common and "own" their narrow Neotropical feeding niche.

The most obvious structural features of anteaters are associated with their ability to capture insects, to dig into or tear apart insect nests, and to climb. The skull is long and roughly cylindrical (Fig. 14-8A), the zygoma are incomplete, and the long rostrum contains complex, double-rolled turbinal bones. Teeth are absent, and the dentary bone is long and delicate. The long, vermiform tongue is protrusile and covered with sticky saliva secreted by the enlarged and fused submaxillary and parotid salivary glands. The forelimbs are powerfully built; the third digit is enlarged and bears a stout, recurved claw, and the remaining digits

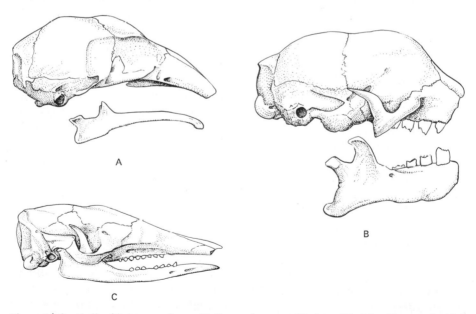

Figure 14-8. Skulls of living xenarthrans: (A) Two-toed anteater (*Cyclopes didactylus,* Myrmecophagidae); length of skull 46 mm; (B) three-toed sloth (*Bradypus griseus,* Bradypodidae); length of skull 76 mm; (C) nine-banded armadillo (*Dasypus novemcinctus,* Dasypodidae); length of skull 95 mmm. (A and B after E. R. Hall and K. R. Kelson, 1959)

are reduced. The giant anteater *(Myrmecophaga)* walks on its knuckles with its toes partly flexed as did some Pleistocene ground sloths, whereas the other anteaters *(Cyclopes* and *Tamandua)*, which are fully or partly arboreal, walk on the side of the hand with the toes inward. The plantigrade foot has four or five clawed digits. In size, anteaters range from that of a squirrel (in *Cyclopes,* 350 grams) to that of a large dog (in *Myrmecophaga,* 25 kilograms). *Myrmecophaga* is covered with long, coarse fur, and its nonprehensile tail has long hairs that hang downward, but in the other anteaters, the fur on the body and tail is shorter and the tail is prehensile.

Anteaters use the powerful forelimbs to expose ants and termites by tearing apart their nests; the insects are captured by the long tongue, swallowed whole, and ground up by the thickened pyloric portion of the stomach. Terrestrial locomotion in all anteaters is slow and fairly awkward. In the two genera that climb trees, the claws of the manus are used as grappling hooks or for grasping as the animal travels along branches hand over hand. *Cyclopes* is nocturnal and entirely arboreal and forages for insects high in the trees. The two species of *Tamandua* are largely arboreal but are terrestrial to some extent and mostly nocturnal. *Myrmecophaga* is entirely terrestrial and seems largely diurnal. The defensive behavior of these animals involves standing on the hind limbs, bracing the body with the tail and hind limbs, and slashing at the enemy with the forelimb claws.

Family Bradypodidae. The members of this family, called tree sloths (Fig. 14–9), are so highly modified for a specialized form of arboreal locomotion that they have nearly lost the ability to move on the ground. The five Recent species of tree sloths belong to two genera and range from Central America (Honduras) through the northern half of South America to northern Argentina. These animals primarily inhabit tropical rain forests.

The adaptive zone of tree sloths is quite different from that of the anteaters and involves ar-

Figure 14–9. A two-toed tree sloth *(Choloepus hoffmanni).* (San Diego Zoo)

boreal herbivory. The bradypodids differ strongly from the myrmecophagids, especially in skull characteristics. The tree sloth skull is short and fairly high, with a strongly reduced rostrum. The zygomatic arch is robust but incomplete, and its jugal portion bears a ventrally projecting jugal process similar to that present in many extinct xenarthrans (Fig. 14–8B). The premaxillary bones are greatly reduced, and the turbinal bones are complexly rolled, as in myrmecophagids. Five maxillary and four or five mandibular teeth are present; the anteriormost teeth in *Choloepus* are caniniform and are kept sharp by abrasion between the posterior surfaces of the upper teeth and the anterior surfaces of the lower teeth. (The two species of *Choloepus* are separated from the family Bradypodidae by some experts and regarded as the only living members of the family Megalonychidae.) The persistently growing teeth are roughly cylindrical and have a central core of soft dentine surrounded successively by hard dentine and cement.

A departure from the usual mammalian pattern of seven cervical vertebrae occurs in the bradypodids, in which from six to nine occur, the number differing from species to species and in some cases even from one individual to another of the same species. Xenarthrism (Fig. 14–1) is strongly developed in the thoracic and lumbar vertebrae, and as in some extinct ground sloths, the coracoid and acromion processes of the scapula are united. The externally visible digits do not exceed three in number and, except for the long and laterally compressed claws, are syndactylous (bound together). Tree sloths are of moderate size, weighing from 4 to 7 kilograms, and are covered with long, coarse hair. This fur provides a habitat for algae, which grow in the flutings on the surface of the hairs during the rainy season and tint the fur green. In addition, the adults of two genera of moths (*Bradypodicola* and *Cryptoses;* Pyralididae, Microlepidoptera) hide in large numbers in the dense pelage. The tail is short.

These remarkably specialized animals eat leaves and descend to the ground only to defecate. The stomach is chambered, and digestion is enhanced by fermentation aided by a symbiotic microbiota. Climbing is done in an upright position by embracing a branch or by hanging upside down and moving along hand over hand. In both genera, the forelimbs are considerably longer than the hind limbs (Fig. 14–9). The animals are unable to support their weight on their limbs and progress on the ground by slowly dragging their bodies forward with the forelimbs.

Although tree sloths occupy tropical environments, in which temperature extremes virtually never occur, they possess some adaptations typical of animals that are subjected to cold stress in boreal areas. The limbs have retia mirabilia (p. 143), which by performing countercurrent heat exchange, allow the limbs to be cooler than the core of the body and thereby reduce thermoregulatory energy losses. In addition, tree sloths are insulated by long guard hairs and short, fine-textured underfur, a type of pelage characteristic of inhabitants of cold climates.

What is the adaptive importance to a tropical animal of such specializations for the retention of body heat? Leaves are not a concentrated source of energy, and many leaves contain defensive secondary compounds that inhibit their digestion by animals. Tree sloths have probably been under chronic selective pressure to reduce their energy needs. The low and variable body temperature and very low metabolic rate may thus be vital in conserving energy, and the retia mirabilia and long pelage are closely associated adaptations that minimize heat losses during cold stress.

Family Dasypodidae. This family includes the armadillos, which differ from the South American anteaters and tree sloths in having protective bony armor. Dasypodidae is the most diverse and widespread xenarthran family, with 20 Recent species of eight genera and a distribution from Kansas and much of the southeastern United States through Mexico and Central America into South America to near the southern end of Argentina. Armadillos occupy many ecological settings, from temperate and tropical forests to deserts.

The most obvious and unique structural feature of armadillos is the jointed armor. This consists of plates (bony scutes covered by horny epidermis; Fig. 14–10) that occur in a variety of patterns but always include a head shield and protection for the neck and body. Sparse hair usually occurs on the flexible skin between the plates and on the limbs and the ventral surface of the body. Individuals of some species can curl up into a ball so that their limbs and vulnerable ventral surfaces are largely protected by the armor. The largest species, the giant armadillo *(Priodontes giganteus),* weights up to 60 kilograms; the smallest, the pigmy armadillo *(Chlamyphorus truncatus),* is roughly the size of a small rat (120 grams).

The skull is often elongate and is dorsoventrally flattened; the zygomatic arch is complete, and the mandible is slim and elongate

Figure 14-10. A naked-tail armadillo *(Cabassous centralis).* (L. G. Ingles)

(Fig. 14–8C). The teeth are borne only on the maxillary bone (except in one species), are nearly cylindrical, and vary from 7/7 to 18/19. The teeth are frequently partially lost with advancing age. The axial skeleton is fairly rigid and is partially braced against the carapace; the second and third cervical vertebrae, and in four species other cervical vertebrae as well, are fused (Fig. 14–2B); eight to 13 sacral and caudal vertebrae form an extremely powerfully braced anchor for the pelvis (Fig. 14–2A); xenarthral articulations between thoracic and lumbar vertebrae (Fig. 14–1) produce a rigid vertebral column. The skeleton braces but does not come into contact with the carapace. For example, in the familiar nine-banded armadillo *(Dasypus novemcinctus),* the species that enters the United States, the carapace is partially supported by prominent dorsolateral processes from the ilium and ischium (Fig. 14–2A), and by the modified tips of the neural spines of all the thoracic and lumbar vertebrae. In this species, the powerfully developed panniculus carnosus muscles and a broad ligament from the expanded dorsolateral flange of the ilium attach to the inner surface of the carapace and bind it to the body, but no bony contact is made between the carapace and the skeleton. The rib cage is made rigid by a broadening of the ribs, by heavy intercostal muscles, and by ossification of the greatly enlarged parts of the ribs that in most mammals are the costal cartilages.

In all armadillos, the limbs are powerfully built and the fore and hind feet bear large, heavy claws (Fig. 14–10). The feet are five-toed in all but one genus, and the foot posture is usually plantigrade. The tibia and fibula are fused proximally and distally and are highly modified for giving origin to powerful shank muscles (Fig. 14–5A). Retia mirabilia occur in the limbs.

Armadillos are more generalized in their feeding habits and locomotion than are other xenarthrans. Most species feed primarily on insects, but a variety of invertebrates, small vertebrates, and vegetable material are also taken. All armadillos are at least partly adapted for digging, and some species are highly fossorial. Such a fossorial creature is the pygmy armadillo *(Chlamyphorus truncatus),* which utilizes a style of digging seemingly unique among mammals. The soil is dug away and pushed beneath the animal by the long claws of the forepaws, and the hind feet rake the soil behind the animal. The pelvic scute is then used to pack the soil behind the body. During the packing, the front limbs push the animal backward and the hind quarters vibrate rapidly from side to side (Rood, 1970). No permanent burrow is formed.

Running is limited in armadillos by a structural pattern that achieves power but no speed; some species can be run down by a human. The "pichi" *(Zaedyus pichiy)* was observed in southern Argentina by Simpson (1965a:201), who found this armadillo to depend heavily on its sense of smell. A captive individual "seemed to pay practically no attention to anything she saw or heard but lived by and for her nose." The pichi cannot curl up into a ball and escapes from enemies either by burrowing beneath a thorn bush or by clutching a solid surface with the claws and pulling the carapace down against the ground.

As indicated by their wide range, armadillos are more resistant to cold than are tree sloths, but the possession of retia mirabilia in the limbs suggests that armadillos share the general xenarthran sensitivity to cold. Indeed, in the northern parts of their range, armadillos may suffer 80 percent mortality during prolonged cold spells (Fitch, Goodrum, and Newman, 1952). Nevertheless, armadillos have extended their range northward in the last 100 years (Fig. 19–14).

Order Pholidota

The pangolins, or scaly anteaters, members of the family Manidae, are represented today by a single genus *(Manis)* with eight species. Pangolins occur in tropical and subtropical parts of the southern half of Africa and in much of southeastern Asia. Their fossil record is poor but documents the probable Oligocene occurrence of these animals in Europe.

At a glance, pangolins seem more reptilian than mammalian (Fig. 14–11). They are of moderate size, weighing from about 5 to 35 kilograms. The skull is conical and lacks teeth; the dentary bones are slender and lack angular and coronoid processes. The tongue is extremely long and vermiform and originates on the last pair of cartilaginous ribs, which have lost their vertebral connections and have formed an enormously elongate posterior extension of the sternum. This extension passes into the posterior part of the abdominal cavity, curves upward, and ends near the kidneys. The fused and spat-

Figure 14–11. A pangolin (*Manis gigantea,* Pholidota). (American Museum of Natural History)

ulate end of these ribs gives origin to the accessory musculature of the tongue. The tongue and its accessory musculature are therefore longer than the head and body together, allowing the tongue to be extremely protrusile. The scales are the most distinctive feature of pangolins; they cover the dorsal surface of the body and the tail and are composed of agglutinated hair. The skin and scales account for a large share of the weight of these animals: Kingdon (1971:356) reports that these parts constitute one third to one half of the weight of the ground pangolin *(Manis temmincki)*. The manus and pes have long, recurved claws; the pes has five toes, and the manus is functionally tridactyl. The walls of the pyloric part of the stomach are thickened. This part of the stomach usually contains small pebbles and seems to grind food as does the gizzard of a bird.

The food of pangolins is mostly termites, but ants and other insects are also eaten. The insects are located by smell, and pangolins seem highly selective in their choice of food. Sweeney (1956) found that only rarely would a pangolin dig for the "wrong" species of ant or termite. Some species are strictly terrestrial, some are semiarboreal, and two species (one in Java and one in Africa) are quite arboreal and have semiprehensile tails. Pangolins roll up into a ball when disturbed, erect the scales, flail the tail, and move the sharp scales in a cutting motion. Some species spray foul-smelling fluid from the anal glands.

Order Tubulidentata

This order, the aardvarks, includes but one family (Orycteropodidae). The one Recent species *(Orycteropus afer;* Fig. 14–12) inhabits Africa south of the Sahara. The earliest record of tubulidentates is from the Miocene of Africa, and in the Pliocene an extinct member of *Orycteropus* occupied parts of Europe and Asia. Skeletal similarities between tubulidentates and condylarths (primitive ungulates; order Condylarthra) indicate that these groups may be related.

Aardvarks are powerful diggers and feed on ants and termites. They weigh up to 82 kilograms, and the thick, sparsely haired skin provides protection from insect bites. The skull is elongate, and the dentary bone is long and slender (Fig. 14–13). In the adult dentition, inci-

Figure 14–12. A juvenile aardvark *(Orycteropus afer).* (San Diego Zoo)

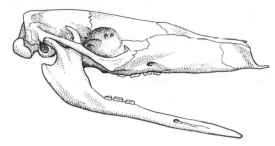

Figure 14–13. The skull of the aardvark *(Orycteropus afer).* (After Hatt, 1934)

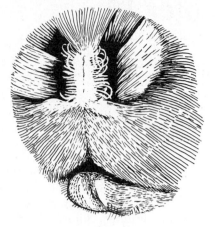

Figure 14-14. The complex nose of an aardvark *(Orycteropus afer).* Note the fleshy tentacles on the nasal septum and the dense tracts of hair that can seal the nostrils. (After Kingdon, 1971)

sors and canines are lacking; the cheek teeth are 2/2 premolars and 3/3 molars. Each tooth is rootless and consists of many (up to nearly 1500) hexagonal prisms of dentine, each surrounding a slender, tubular pulp cavity. The columnar teeth lack enamel but are surrounded by cement. The anteriormost teeth erupt first and are often lost before the posterior molars are fully erupted. The slender tongue is protrusile.

Olfaction is used in finding insects; the olfactory centers of the brain are unusually well developed, and the turbinal bones are remarkably large and complex. The nostrils are highly specialized in a fashion unmatched by any other mammal. Fleshy tentacles, which presumably have an olfactory function, occur on the nasal septum (Fig. 14–14), and dense hair surrounds the nostrils and can seal them when the aardvark digs. The pollex is absent, and the hind feet are five-toed; the robust claws are flattened and blunt.

The powerful forelimbs are used in burrowing and in dismantling termite and ant nests, and the hind limbs thrust accumulated soil from the burrow. Although the foot posture is digitigrade, aardvarks are slow runners and can be run down by a human. Burrows dug by aardvarks are numerous in some areas and are used as retreats by a variety of mammals, including the warthog *(Phacochoerus africanus).*

Chapter 15

Cetaceans: Whales, Porpoises, Dolphins

Cetaceans (order Cetacea) are notable for being the mammals most fully adapted to aquatic life. Baleen whales (suborder Mysticeti), the largest living or fossil animals known, are mainly plankton feeders (although some eat fish) and thus draw from the tremendous productivity near the bottom of the marine food chain. Some of the large odontocetes (suborder Odontoceti), in contrast, exploit the top of the marine food chain. Sperm whales, for example, eat giant squid, large sharks, and bony fish; the food of killer whales includes fish, seals, porpoises, and baleen whales. Remarkable swimming and diving ability, the capability of many (perhaps all odontocetes) to echolocate, considerable intelligence, and complex social behavior have all contributed to the great success of the cetaceans.

Cetaceans have long intrigued and inspired humans. Leaping dolphins have been the embodiment of beauty and exuberance to seafarers for centuries: 4000 years ago Minoan artists included graceful drawings of dolphins in their frescoes at the Palace of Knossos on Crete (Fig. 15–1). It has remained for modern humans, with our use of technical skills for devastation, to deplete the cetacean populations that once seemed limitless.

Morphology

All cetaceans are completely aquatic, and their structure reflects this mode of life. The body is fusiform (cigar-shaped), nearly hairless, insulated by thick blubber, and lacks sebaceous glands. Most vertebrae have high neural spines (Fig. 15–2), and the cervical vertebrae are highly compressed. The clavicle is absent, the forelimbs (flippers) are paddle-shaped, and no external digits or claws are present. Little movement is possible between the joints distal to the shoulder. The proximal segments of the forelimb are short, whereas the digits are frequently unusually long because of the development of more phalanges per digit than the basic eutherian number (Fig. 15–3). The hind limbs are vestigial, do not attach to the axial skeleton, and are not visible externally. The flukes (tail fins) are horizontally oriented. The skull is typically highly modified as a result of the posterior migration of the external nares (Fig. 15–4). The

Figure 15-1. Dolphins forming part of a fresco dating from 1600 B.C. at the Palace of Knossos, Crete.

Figure 15-2. The skeleton of the Tasmanian beaked whale (*Tasmacetus shepherdi*, Ziphiidae).

premaxillary and maxillary bones form most of the roof of the skull, and the occipitals form the back. The nasals and parietals are telescoped between these bones and form only a minor part of the skull roof (Figs. 15-4 and 15-5), and large frontal bones are mostly covered by the maxillaries and premaxillaries. The tympano-periotic bone (the bone that houses the middle and inner ear) is not braced against adjacent bones of the skull in most cetaceans and is partly insulated from the rest of the cranium by surrounding air sinuses (Fig. 15-6). Members of the families Ziphiidae and Physeteridae have a pneumatic, bony strut that braces the tympanoperiotic bone against the skull.

Experimental work on the bottlenose dolphin *(Tursiops truncatus)* by L. M. Herman, Peacock, et al. (1975) showed that visual acuity is similar above and below the water. The above-water acuity may be due to the "pinhole camera" effect of the cetacean's pupil. Mechanisms for fine adjustment of lens shape or displacement are seemingly absent in cetaceans, but in bright light the central part of the pupil

closes completely, leaving two tiny apertures which yield great depth of field, as does a pinhole camera. This allows the dolphin to receive a sharp image from distant objects when its head is above water.

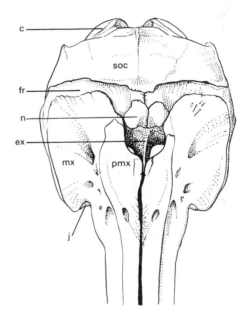

Figure 15-4. Dorsal view of part of the skull of a bottlenose dolphin (*Tursiops truncatus*, Delphinidae). Note the asymmetry of the bones surrounding the external nares *(ex)*. The function of this remarkable asymmetry is not known.
Abbreviations: *c*, occipital condyle; *fr*, frontal; *j*, jugal; *mx*, maxillary; *n*, nasal; *pmx*, premaxillary; *soc*, supraoccipital.

Figure 15-3. Dorsal view of the right forelimb of the bottlenose dolphin (*Tursiops truncatus*, Delphinidae).

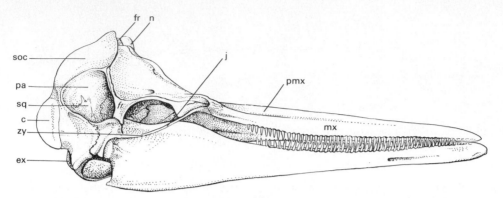

Figure 15-5. Skull of a dolphin (*Delphinus delphis*, Delphinidae); length of skull 475 mm. Note the highly telescoped skull with the maxillary and frontal bones roofing the small temporal fossa. The frontal bone is barely exposed on the skull roof. Abbreviations: *c,* occipital condyle; *ex,* exoccipital; *fr,* frontal; *j,* jugal; *mx,* maxillary; *n,* nasal; *pa,* parietal; *pmx,* premaxillary; *sq,* squamosal; *soc,* supraoccipital; *zy,* zygomatic arch.

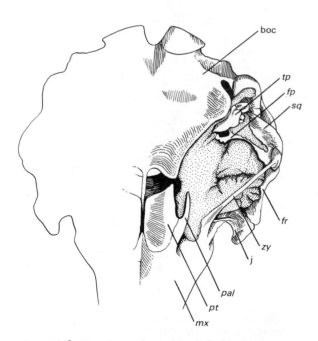

Figure 15-6. Ventral view of part of the skull of the grampus (*Grampus griseus*, Delphinidae), showing the large air sinuses (stippled) anterior to the tympanoperiotic bone *(tp)* and partly surrounding it. Abbreviations: *boc,* basioccipital; *fp,* falciform process of the squamosal; *fr,* frontal; *j,* jugal; *mx,* maxillary; *pal,* palatine; *pt,* pterygoid; *sq,* squamosal; *zy,* zygomatic arch. (After Purves, 1966)

Paleontology

The earliest known cetacean, *Pakicetus inachus,* from the early Eocene record of Pakistan, was found in association with various land mammals in river sediments deposited at the border of a shallow sea. This fossil provides strong evidence that whales evolved from terrestrial carnivorous mammals and that the progenitors of whales (probably mesonychid Condylarthra) gradually spent progressively more time in shallow seas feeding on plankton-feeding fish (Gingerich, Wells, et al., 1983). *Pakicetus* was not fully adapted to aquatic life, for the ear region lacked the characteristic specializations that facilitate directional underwater hearing in advanced cetaceans. Some archaeocetes attained large size. The skull of *Basilosaurus,* an elongate Eocene type, is 1.5 meters long, and the slim body is roughly 17 meters long. In primitive cetaceans (Fig. 15–7), the external nares had not migrated so far toward the back of the skull as they have in modern cetaceans, but the hind limbs were vestigial.

The earliest records of the suborder Mysticeti, the baleen whales, are from the early Oligocene record of New Zealand. A relict, late

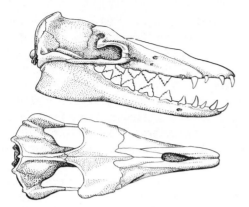

Figure 15-7. Skull of a primitive cetacean *(Prozeuglodon),* a fossil archaeocete from the Eocene. Length of skull approximately 600 mm. Note the lack of telescoping of the skull and the heterodont dentition. (After Romer, 1966)

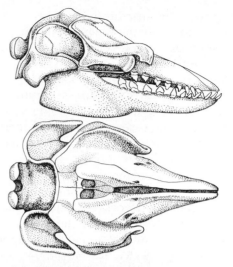

Figure 15-8. Skull of *Prosqualodon,* a fossil porpoise from the Miocene. Length of skull approximately 450 mm. The skull is highly telescoped, and the maxillary and frontal bones form a roof over the temporal fossa (as in the modern delphinid shown in Figure 15-5). (After Romer, 1966)

Oligocene species *(Aetiocetus cotylalveus)* from the coast of Oregon is roughly intermediate between primitive toothed forms and true baleen whales. The suborder Odontoceti, the toothed whales, appears in the early Oligocene record of North America. By the Miocene, cetaceans had undergone considerable radiation; over half of the known cetacean genera appeared in this epoch. Advanced odontocetes with highly telescoped skulls, homodont dentition, and many more teeth than the primitive eutherian complement are known from the Miocene (Fig. 15-8).

The two Recent suborders of cetaceans can be distinguished on the basis of a number of morphological characters. Members of the suborder Mysticeti lack teeth as adults but have baleen plates of horny material that grow from the palate and serve as sieves (Fig. 15-9). The ascending processes of the maxillae are narrow, interlock with the frontals, and do not spread

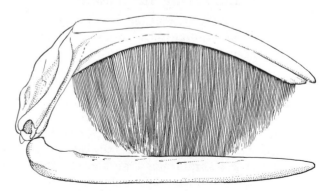

Figure 15-9. Skull of the Atlantic right whale (*Eubalaena glacialis,* Balaenidae); length of skull roughly 4 m. Note the baleen plates attached to the maxilla.

laterally over the supraorbital processes. The respiratory and alimentary canals are not permanently separated. Mysticetes are not known to use echolocation.

Members of the suborder Odontoceti have teeth frequently far exceeding the typical eutherian maximum number or secondarily reduced in extreme cases to 0/1. These teeth are usually simple and conical, but the rear teeth of the Amazon dolphin *(Inia)* have crushing cusps, seemingly adapted to dealing with armored catfish, which this dolphin is known to eat. The ascending processes of the maxillae are large, do not interlock with the frontals, and spread laterally over the supraorbital processes (Figs. 15–4 and 15–5). The respiratory and alimentary tracts are permanently separated by specializations of the glottis and the laryngeal apparatus. In odontocetes, but not in mysticetes, the bones surrounding the blowhole depart from the usual mammalian pattern of bilateral symmetry (Fig. 15–4). Probably all odontocetes use echolocation.

Magnetic material is associated with nerve fibers in a membrane covering the brain in the Pacific dolphin *(Delphinus delphis)*. This suggests that these cetaceans (and perhaps others) may have the ability to navigate by detecting the earth's magnetic field (Zoeger, Dunn, and Fuller, 1981).

Swimming and Diving Adaptations

The Achilles' heel of cetaceans is their need to breathe air, but, unlike most other mammals, many cetaceans are able to alternate between periods of eupnea (normal breathing) and long periods of apnea (cessation of breathing). Some whales remain submerged for over 1 hour; a sperm whale *(Physeter catadon)* has been timed at 70 minutes. Most delphinids can hold their breath 4 to 5 minutes, but they often surface to breathe several times a minute. The ability of cetaceans to remain active during periods of apnea probably depends on many adapta-

tions. Rapid gaseous exchange is enhanced by two layers of capillaries in the interalveolar septa. During expiration, most of the air can be exhausted from the lungs, and up to 12 percent of the oxygen from inhaled air is utilized (the corresponding figure for terrestrial mammals is only 4 percent). Cetaceans have up to at least twice as many erythrocytes per volume of blood as terrestrial mammals do, and about two to nine times as much myoglobin (a molecule able to store oxygen and release it to tissue) in the muscles. During deep dives, the heart rate drops to roughly half the surface rate and vascular specializations allow blood to bypass certain muscle masses. Important physiological adaptations to prolonged submersion include tolerance to high levels of lactic acid and a relative insensitivity to carbon dioxide. Good discussions of cetacean adaptations to deep diving are given by Elsner (1969), Irving (1966), Kooyman and Andersen (1969), and Lenfant (1969).

Most small odontocetes are seemingly shallow divers, but some large odontocetes and some mysticetes can perform deep dives (Table 15–1). During deep dives, cetaceans are subjected to tremendous pressures, for with every 10-meter increase in depth an additional 1 atmosphere of pressure is exerted on the body. A cetacean swimming at a depth of only 200 meters, probably a common depth for many species, is subjected to 20 atmospheres of pressure, or 294 pounds per square inch. Any gases that remain within body cavities are therefore subjected to great pressure, resulting in a decrease in their volume and an increase in the amount of gases that dissolve in the body solvents, such as blood. One serious result in humans is a condition known as "the bends," or decompression sickness. When humans use equipment that allows them to breathe under water and undergo prolonged exposure to high pressures during diving, greater than normal amounts of gases in the lungs are dissolved in the tissues and the blood. If decompression is too rapid, these gases cannot be carried to the lungs rapidly enough to be removed from the body; instead,

Table 15-1: Depths at Which Cetaceans Have Been Recorded

Species	Depth (m)	Method of Observation
Balaenopteridae		
Fin whale, *Balaenoptera physalus*	500	Harpooned and collided with bottom
	355	Depth manometer on harpoon
Physeteridae		
Sperm whale, *Physeter catodon*	900	Entangled in deep-sea cable
	1134	Entangled in deep-sea cable
	520	Echo-sounder
Stenidae		
Rough-toothed dolphin, *Steno bredanensis*	30	Attached to depth recorder
Delphinidae		
North Pacific pilot whale, *Globicephala macrorhyncha*	366	Inferred from feeding behavior
Bottlenose dolphin, *Tursiops truncatus*	92	Visual observations from underwater craft
	185	Vocalizations near underwater craft
	170	Trained to activate buzzer

Data from Kooyman and Andersen (1969:72), who cite the sources of the observations.

they quickly leave solution and appear as bubbles in the tissues. Intravascular bubbles may occlude capillaries and result in injury to tissues or even death.

Cetaceans are not known to have these problems, probably because their lungs collapse in the first 100 meters of a dive. The following anatomical specializations are seemingly adaptations to deep diving: (1) a large proportion of the ribs lack either sternal attachment or attachment to the sternum or to other ribs and allow rapid lung collapse; (2) the lungs are dorsally situated above the oblique diaphragm; (3) in deep divers the lungs are small and the volume of the air passages is relatively large; (4) the trachea is short and often of large diameter, and the cartilaginous rings bracing the trachea are nearly complete, have small intermittent breaks, or are fused (Slijper, 1962); (5) the bronchioles are reduced in length, and the entire system of bronchioles, to the very origin of the alveolar ducts, is braced by muscles and cartilaginous rings; (6) the lungs, especially the walls of the alveolar ducts and the septa, contain unusually high concentrations of elastic fibers; (7) in some of the small odontocetes, a series of myoelastic sphincters occur in the terminal sections of the bronchioles.

The adaptive importance of all these features is not yet completely understood. Clearly, the specializations of the ribs and the placement of the lungs relative to the diaphragm permit the lungs to collapse and air to move from space where gas exchange occurs to space where it does not occur. Bracing of the respiratory passages may facilitate alveolar collapse and rapid inhalation and exhalation at the surface. The alveoli fully collapse during deep dives, forcing air into respiratory passages. Experimental work with trained dolphins by Ridgway and Holland (1979) indicated that lung collapse occurs at a depth of 70 meters; little or no pulmonary respiration occurs at greater depths. These researchers conclude, however, that bottlenose dolphins, and probably other cetaceans, are seemingly not protected from decompression sickness during dives shallower than 70 meters and of durations that would cause this problem in humans. Other workers disagree. R. S. Mackay (1982) postulates that nitrogen bubbles are generated in cetacean tissues but are crushed by pressure at depths, thus avoiding decompression problems. Lettvin, Gruber, et al. (1982), on the other hand, believe that cetaceans' smooth contours and "quiet" internal function (such as heart beat and heart valve slap) avoid minor shock waves in the system. The avoidance of such shocks reduces the tendency to form bubbles (compare the gently opened bottle of beer with one shaken before opening) and is regarded by Lettvin and his coworkers as the key to immunity to the bends. Ridgway and Howard (1982) refute these ideas by indicating that no experimental evidence supports either theory and conclude that how cetaceans avoid decompression problems during repeated shallow dives remains poorly understood.

Cetaceans are fast swimmers. Powerful dorsoventral movements of the tail provide propulsion, and the flippers are used for steering. T. G. Lang (1966) cites the following top speeds for cetaceans: up to 36 kilometers per hour for dolphins and 55 and 27 kilometers per hour for a killer whale and a pilot whale, respectively. Gawn (1948) reports the huge blue whale's speed as 37 kilometers per hour for 10 minutes.

This remarkable swimming performance of cetaceans has proved difficult to explain. Recent studies have demonstrated that their speed is due not to muscles that are vastly more powerful than those of other mammals but to specializations that greatly reduce resistance (drag) as the animals swim. The amount of resistance depends on the type of water flow over the body surface. If the flow is smooth—parallel to the surface—it is said to be laminar. When such smooth flow is interrupted by water movements that are not consistently parallel to the body surface, however, turbulent flow occurs. All other conditions being equal, laminar flow creates much less resistance than does turbulent flow. If the bodies of small dolphins were subjected to turbulent flow, swimming at 38 kilometers per hour would require their muscles to

be five times as powerful as those of humans (T. G. Lang, 1966:426). Assuming flow to be nearly laminar, however, this speed is approximately that expected if their power output were that of a well-trained human athlete. Scientists for many years have attempted without success to design bodies shaped so that air or water flow over the surface is laminar or nearly so. What is the cetacean solution to this problem?

Several factors seemingly contribute importantly to reducing resistance as a porpoise swims rapidly (these are discussed by Hertel, 1969:33). The body is hairless, and no obstructions except the streamlined appendages break the extremely smooth surface. In addition, the body form of the dolphin is approximately parabolic; this form creates even less resistance than the rounded (elliptical) head end and tapered body of such a rapid swimmer as a trout. According to Hertel, further reduction of resistance may be associated with movements of the flukes and body during swimming.

Acoustic Debilitation of Prey by Odontocetes

"The hypothesis is that some odontocete cetaceans may emit sounds so intense that their prey is debilitated and capture made easier." This statement, by Norris and Møhl (1983), was made after these scientists carefully reviewed diverse lines of highly suggestive evidence. These authors further speculate that such an ability has strongly affected the evolution of the anatomy and behavior of odontocete cetaceans.

That odontocetes acoustically stun their prey has been mentioned by several researchers. This ability was first considered by Bel'kovich and Yablokov (1963), who suggested that an intense high-frequency sound of the sort that could be made by even a small dolphin could produce a shock sufficient to stun prey. Berzin (1971), who carefully studied sperm whales, was impressed by the fact that large squid and sharks found in the stomachs of sperm whales bore no teeth marks and that even sperm whales with deformed or injured lower jaws, incapable of grasping prey, had normally full stomachs. Berzin proposed that sperm whales acoustically stun their prey and that this reduced the importance of jaws. More recently, Hult (1982) observed disorientation among schooling fish and hypothesized that this resulted from the high-intensity sounds made by approaching bottlenose dolphins.

The evolution of odontocete feeding structures and functions in relation to acoustic prey debilitation was discussed by Norris and Møhl (1983). Early cetaceans, from the early Eocene, generally had laterally compressed, serrated teeth (Fig. 15–7) that probably functioned, as do the very similar teeth of the crab-eater seal (Fig. 10–30) today, to allow water to escape when the jaws are closed and to trap small prey under water. As odontocetes radiated in the Miocene, these serrated teeth were progressively replaced by many simple conical teeth borne on slim, elongate jaws. Such pincer-type jaws probably enabled Miocene odontocetes to capture small underwater prey by quick thrusts and snaps of the jaws. Many living odontocetes retain such teeth and beaklike jaws. In living odontocetes, engulfment of prey is caused by a rapid retraction of the piston-like tongue and the resultant rush of water and prey into the mouth. In many living odontocetes, the beak pincers are lost or reduced, perhaps in association with increasing perfection of the ability to stun prey acoustically.

Two additional important and obvious evolutionary trends in odontocetes may also be related to this ability. The first is a trend among advanced odontocetes (such as the bottlenose dolphin) toward the focusing of sound energy into a concentrated beam. Such a specialization furthers long-range echolocation ability, and prey debilitation could be its byproduct. A second suggestive trend involves the loss of teeth, a reduction in their size, or the development of specialized teeth unsuitable for capturing prey (such as the long narwhal tusk). A specialized means of debilitating prey would open the door to such dental changes.

The sperm whale offers a special and suggestive case. The ability of this slow-moving whale to capture large squid, among the swiftest of marine creatures and capable of bursts of speed up to 55 kilometers per hour, and the fact that squid in the stomachs of sperm whales typically bear no teeth marks and are occasionally still alive, point to some specialized prey-immobilization ability.

Experimental trials with trained bottlenose dolphins showed that they can indeed produce sounds of intensities equal to the lethal thresholds of some marine fish and close to those which have been observed to kill moderately large squid. These intense sounds were not harmful to the dolphins but were some five orders of magnitude above levels usually reported for wild or captive dolphins.

Although acoustic debilitation of prey by odontocetes has not been proved, it seems probable and could have evolved by a series of entirely plausible steps. Highly specialized sound-producing structures evolved in response to sociality and the importance of communication. Sound production later became used for echolocating obstacles and prey. With an increasing ability to produce intense and focused sounds in the interest of long-range echolocation, a remarkable new ability appeared: prey subjected to these sounds could be debilitated. This innovation would free the jaws and dentition from many selective pressures associated with killing or grappling with prey and could allow the shortening of beaks, reduction of dentition, or development of teeth primarily adapted to ritualized combat, as in the case of the male narwhal.

Suborder Mysticeti

In this suborder are the huge baleen whales, which inhabit all oceans. There are ten Recent

Right whale

Rorqual

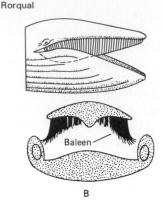

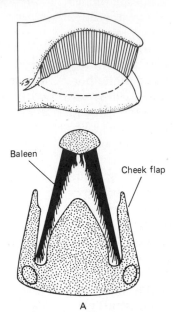

Baleen

Baleen

Cheek flap

B

Figure 15-10. Cross sections of the heads of two kinds of whales, showing contrasting arrangements of baleen: (A) right whale; (B) rorqual. (Reprinted from Pivorunas, 1979, vol. 67, by permission of American Scientist and Sigma Xi, The Scientific Research Society)

A

species, grouped in six genera and three families. Before intensive whaling decimated their populations, the mysticetes were perhaps more important than the odontocetes in terms of biomass, although there are fewer species of mysticetes.

Although all baleen whales are filter feeders, they utilize three rather distinct feeding styles (Pivorunas, 1979). The first is typical of

the right whales (Balaenidae). These are huge-headed animals with long baleen plates and conspicuous lips (cheek flaps; Fig. 15–10). Right whales generally graze on small plankton, usually copepods less than 1 centimeter in length, that concentrate in layers at or near the water surface. The whales swim slowly through these concentrations (Fig. 15–11); the water and plankton flow in the front of the mouth, and

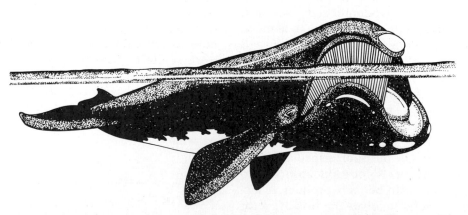

Figure 15-11.   Right whale feeding on plankton at the surface of the sea. (Reprinted from Pivorunas, 1979, vol. 67, by permission of American Scientist and Sigma Xi, The Scientific Research Society)

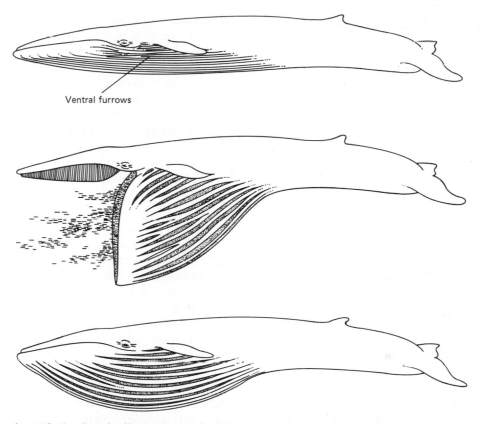

Ventral furrows

Figure 15–12. The style of feeding of rorquals and the use of the furrowed, expandable pouch. (Reprinted from Pivorunas, 1979, vol. 67, by permission of American Scientist and Sigma Xi, The Scientific Research Society)

the water passes along the side of the tongue and out through the baleen, which traps the plankton. This sifting of the water can involve skimming a surface layer of plankton or moving through plankton swarms at greater depths.

The second method is that of the rorquals (Balaenopteridae). These whales have huge mouths and heads, relatively short baleen (Fig. 15–10), and a furrowed pouch that is highly distensible (Fig. 15–12). Rorquals engulf food that occurs in dense swarms, usually krill (shrimplike crustaceans) or fish. One of the rorquals, the blue whale *(Balaenoptera musculus),* has been observed to swim up to its prey and engulf it, along with huge amounts of water (up to 6400 kilograms). The food and water are contained briefly in the capacious pouch. The pouch is then contracted, the water passes through the baleen, and the food trapped by the baleen is swallowed. A number of variations on the basic feeding pattern are used. On occasion, for example, the humpback whale *(Megaptera novaeangliae)* blows a ring of bubbles, a bubble "net," near the surface and, by rising upward within the net, engulfs animals that hesitate to pass through the net and are trapped at the surface.

The third feeding style is that of the gray whale *(Eschrichtius robustus),* which is thought to feed by scooping material from the bottom and filtering out bottom-dwelling organisms.

Family Balaenidae. These, the right whales (Figs. 15–10 and 15–11), were killed in such great numbers during the height of whaling activities that they are rare today and protected by international treaty. The family includes three Recent species of two genera and inhabits most marine waters except tropical and south polar seas. Balaenids are known as fossils from the Miocene to the Recent.

These are large, robust whales that reach about 18 meters in length and over 67,000 kilograms in weight. The head and tongue are huge—the head amounts to nearly one third of the total length. The flippers are short and rounded, and the dorsal fin is usually absent. There are more than 350 long baleen plates on each side of the upper jaw; these plates fold on the floor of the mouth when the jaws are closed. No furrows are present in the skin of the throat or chest. The cervical vertebrae are fused, and the skull is telescoped to the extent that the nasal bones are small and the frontal bones are barely exposed on the top of the skull.

Right whales feed largely on copepods and are most common near coastlines or near pack ice. The bowhead whale *(Balaena mysticetus)* never ventures far south of the Arctic Circle, and Eskimos claim it can break through ice nearly 1 meter thick to reach air (Rice, 1967:299). The southern right whale *(Eubalaena australis)* of the southern oceans makes long annual migrations from temperate or tropical waters to spend the austral summer in Antarctic waters. The northern right whale *(E. glacialis)* remains in North Atlantic or North Pacific waters.

Family Eschrichtiidae. This family is represented today only by the gray whale *(Eschrichtius robustus),* which occupies parts of the North Pacific. There is no fossil record of this family, but there are subfossil records from the North Atlantic (Rice and Wolman, 1971:20).

The gray whale is fairly large, weighing up to about 31,500 kilograms and measuring 15 meters in length, and has a slender body with no dorsal fin. The baleen plates are short, and the telescoping of the skull is not extreme; the nasal bones are large and the frontals are broadly visible on the roof of the skull. The throat usually has two longitudinal furrows in the skin.

Gray whales migrate extremely long distances, but probably no farther than blue whales. The round trip distance of the migration is from 10,000 to 22,000 kilometers. Gray whales occupy parts of the North Pacific (the Bering, Chukchi, and Okhotsk seas) in the summer. Here they feed largely on bottom-dwelling crustaceans (amphipods), which they take by stirring up sediment with their snouts. In late autumn, they migrate southward along the coastlines. The western Pacific population winters along the coast of Korea, and the eastern Pacific gray whales winter along the coast of Baja California and Sonora. Young are born in shallow coastal lagoons in the wintering areas. Many people each year watch migrating gray whales from a vantage point at Cabrillo National Monument, near San Diego, and from many whale-watching vessels.

The future of the gray whale today seems reasonably bright. Driven nearly to extinction by whaling activities between 1850 and 1925, they are now protected by the International Convention for the Regulation of Whaling, and their numbers have increased greatly in recent years.

Family Balaenopteridae. This group of whales, usually known as rorquals, includes six Recent species of two genera. The distribution includes all oceans.

These whales vary in size from fairly small (for whales) to extremely large. The huge blue whale *(Balaenoptera musculus),* a giant even among whales, reaches a length of about 31 meters and a weight estimated at 160,000 kilograms. In some species of rorquals the body is slender and streamlined, but it is chunky in others. The baleen plates are short and broad, and the skin of the throat and chest is marked by numerous longitudinal furrows (Figs. 15–10 and 15–12). The nasals are small, and the frontals are either not exposed or only barely exposed on the skull roof.

Some of these whales feed in cold waters,

near the edges of the ice where up-welling, nutrient-rich water results in great growths of plankton in summer. Planktonic crustaceans and small schooling fish are eaten. During the northern winter, the Northern Hemisphere populations move southward toward equatorial areas, and during the southern winter, southern populations move northward. Wintering adults do not feed but live off stored blubber. Breeding occurs in the wintering areas, but because the southern and northern winters are six months out of phase, no interbreeding between populations occurs. The humpback whale, an animal given to spectacular leaps, makes remarkably melodious and varied underwater sounds that have been beautifully recorded by Payne (1970) and discussed by Payne and McVay (1971).

Excessive commercial exploitation has resulted in a tremendous decline in the populations of such rorquals as fin whales *(Balaenoptera physalus)*, humpbacks, and blue whales. Over broad areas, blue and humpback whales are so scarce as to be "commercially extinct" (Rice, 1967:304), but intense hunting of the smaller kinds of rorquals continues over broad areas and some populations continue to decline at alarming rates.

Suborder Odontoceti

The odontocetes—the toothed whales, porpoises, and dolphins—constitute the most important suborder of cetaceans in terms of abundance, species diversity, and distribution. There are 67 Recent species within 43 genera and six families, and they occur in all oceans and seas connected to oceans. Some members of three families inhabit some rivers and lakes in North America, South America, Asia, and Africa. Odontocetes are readily observed; they frequently forage close to shore, often make spectacular leaps, and roll repeatedly out of the water. Some ride the bow waves of ships much as humans ride shore waves.

The classification used here is that given in the list of Recent "Marine Mammal Names" prepared by the U.S. Marine Mammal Commission. **Family Physeteridae.** The sperm whales occur in all but arctic oceans, and the giant sperm whale *(Physeter catodon)*, of Moby Dick fame, has long been an important species to the whaling industry. There are two genera and three species of living sperm whales, and fossil forms are known from the early Miocene.

Physeter is large, attaining a length of over 18 meters and a weight in excess of 53,000 kilograms. The pigmy sperm whales *(Kogia)* are small, reaching about 4 meters in length and 320 kilograms in weight. The head is huge in *Physeter,* accounting for about one third of the total length. In both genera, the rostrum is truncate, broad, and flat. The facial depression of *Physeter* contains a huge spermaceti organ, which contains great quantities of oil. This organ probably functions as a reverberation chamber for the production of bursts of sound used for long-range echolocation (Norris and Harvey, 1972). The blowhole is toward the end of the left side of the snout, and the nasal passages are highly specialized. The upper jaw (except in occasional individuals) lacks functional teeth; the lower jaw has 25 functional teeth on each side in *Physeter,* and from 8 to 16 in *Kogia*. All of the cervicals are fused in *Kogia,* and all but the atlas are fused in *Physeter.*

The habits of *Kogia* are not well known, but those of *Physeter* are better understood, probably because humans have persistently hunted this animal for many years. *Physeter* is social and assembles in groups with occasionally as many as 1000 individuals. Schools of females with their calves, together with male and female subadults, are overseen by one or more large adult males, whereas younger males congregate in "bachelor schools." Some adult males are solitary. Sperm whales generally forage in the open sea at depths where little or no light penetrates (the use of echolocation by *Physeter* is discussed on p. 508). Dives to depths of 1000 meters are probably usual, and dives of 1130 meters have been recorded (Heezen, 1957). *Physeter* feeds largely on deep-water squid, including giant squid, such bony fish as tuna and barracuda, and sharks and skates. Males commonly migrate north in summer and are occasional in the Bering Sea, but females remain in temperate and tropical waters. *Kogia* occurs in small schools and feeds largely on cephalopods such as squid and cuttlefish.

Family Monodontidae.   This family contains three species: the narwhal *(Monodon monoceros),* remarkable for its long, straight, forward-directed tusk; the beluga *(Delphinapterus leucas),* also called the white whale; and the Irrawaddy River dolphin (*Orcaella brevirostris;* Kasayu, 1973). The first two species occur in the Arctic Ocean and the Bering and Okhotsk seas, in Hudson Bay, in the St. Lawrence River in Canada, and in some large rivers in Siberia and Alaska. The Irrawaddy River dolphin occupies the warm coastal waters of southeastern Asia and the Irrawaddy River in Burma as far as 1440 kilometers from the sea. The earliest monodontids are from the Miocene of California and Baja California.

These are small to medium-size cetaceans; belugas reach about 6 meters in length and 2000 kilograms in weight, and narwhals, without the tusk, are of similar length. The Irrawaddy River dolphin is about 2 meters in length. In all species, the facial depression in the skull is large, and the maxillary and frontal bones roof over the reduced temporal fossa (Figs. 15–4 and 15–5); the zygomatic process

of the squamosal bone is strongly reduced; and the cervical vertebrae are not fused. The beluga has 11/11 teeth, the narwhal has 1/0, and the Irrawaddy River dolphin has 12-19/12-19. One upper tooth of the male narwhal (usually the left) forms a straight, spirally grooved tusk up to 2.7 meters long; the corresponding tooth in the other upper jaw is rudimentary.

The gregarious belugas and narwhals are characteristic of northern seas, where in winter they assemble in open-water areas. In summer, belugas move far up large rivers. They feed largely on fish, both benthic (bottom-dwelling) kinds and those that live at intermediate depths, and squid. Narwhals are seemingly largely pelagic (open-sea dwellers). Male narwhals fence with their long tusks, and a tusk occasionally becomes imbedded and broken off in the head of one of the combatants (Silverman and Dunbar, 1980). Both belugas and narwhals are quite vocal, and the trilling sounds made by belugas account for their common name of sea canary.

Family Ziphiidae.   The beaked whales are widely distributed—they occupy all oceans—but are rather poorly known; some species have never been seen alive. Eighteen Recent species of six genera are recognized. The earliest fossil record of ziphiids is from the early Miocene.

These are medium-size cetaceans with fairly slender bodies. The length varies from 4 to over 12 meters, and the weight reaches 11,500 kilograms. The snout is usually long and

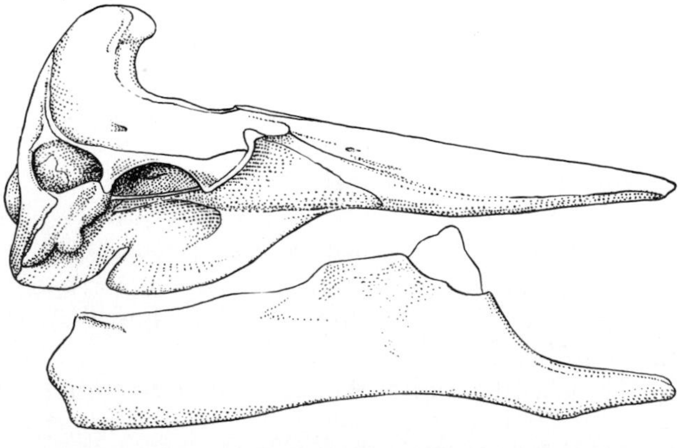

Figure 15–13.   Skull of a beaked whale (*Mesoplodon* sp., Ziphiidae); length of skull about 590 mm. Note the single large tooth in the dentary bone.

narrow, and in some species the forehead bulges prominently. One species *(Tasmacetus shepherdi)* has a large number of teeth; in the others the dentition is strongly reduced. Only two lower teeth on each side occur in the two species of *Berardius;* in all remaining ziphiids, there is only a single functional tooth, a lower one, on each side (Fig. 15–13). In some species, the lower jaw is "undershot" and the teeth are outside the mouth. Two to seven cervical vertebrae are fused. The stomach is divided into from four to 14 chambers.

Beaked whales are deep divers able to remain submerged for long periods. The North Atlantic bottlenose whale *(Hyperoodon ampullatus)* can dive for periods well over 1 hour. Some species forage in the open ocean far from land. Most beaked whales are highly social and travel in schools in which all members surface and dive in synchrony. The teeth in those species with reduced dentition may be used primarily during intraspecific social interactions and may be of little use during feeding. The primary food is squid, but deep-sea fish are also taken. The North Atlantic bottlenose whale is known to make annual migrations, and other species are probably also migratory.

Family Delphinidae.   This is by far the largest and most diverse group of cetaceans (Figs. 15–14 and 15–15). Because some species come close to shore and roll and jump conspicuously, they are the most frequently observed cetaceans. About 34 Recent species representing 17 genera are known. Delphinids inhabit all oceans

Figure 15–14.   Dorsal view of the head of the Pacific pilot whale (*Globicephala scammoni,* Delphinidae), showing the blowhole (the opening of the external nares). (Marineland of the Pacific)

Figure 15–15.   Bottlenose dolphin (*Tursiops truncatus,* Delphinidae) giving birth. (Marineland of the Pacific)

and some large rivers and estuaries in southern Asia, Africa, and South America. Fossil delphinids appear first in the late Miocene record.

Small delphinids are roughly 1.5 meters in length and 50 kilograms in weight, but the killer whale *(Orcinus orca)* reaches 9.5 meters and at least 7000 kilograms. The facial depression of the skull is large, and the frontal and maxillary bones roof over the reduced temporal fossa (Fig. 15–5).

The "melon," a lens-shaped fatty deposit that lies in the facial depression, is well developed and gives many delphinids a forehead that bulges prominently behind a beaklike snout. Some delphinids, such as the killer whale, lack a beak and have a rounded profile. The number of teeth varies from 65/58 to 0/2. From two to six cervical vertebrae are fused (Fig. 15–16). Males are typically larger than females, and in

some species there is considerable sexual dimorphism in the shape of the flippers and dorsal fin. Coloration is varied: some species are uniformly black or gray, some have striking patterns of black and white, and still others have yellowish or tan stripes or spots.

Delphinids characteristically feed by making shallow dives and surfacing several times a minute. They are rapid swimmers, and some species regularly leap from the water during feeding and traveling. In the Gulf of California, I observed a bottlenose dolphin *(Tursiops truncatus)* leaping completely out of the water and catching mullet in midair, much as a trout catches a fly. Pacific striped dolphins *(Lagenorhynchus obliquidens)* have been trained at Marineland of the Pacific to leap over a wire 4.8 meters above the water. Most small delphinids eat fish and squid, but the killer whale is known

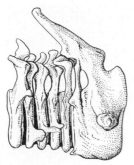

Figure 15–16. Cervical vertebrae of a dolphin (*Delphinus delphis*, Delphinidae). Only the axis and atlas are fused, whereas most of the series are fused in some cetaceans.

to take a great variety of items, including large bony fish, sharks, sea birds, sea otters, seals and sea lions, porpoises, dolphins, and whales.

Delphinids are typically highly gregarious, and assemblages of approximately 100,000 individuals have been observed. Some groups of delphinids kept in large tanks establish a dominance hierarchy, with an adult male having the highest position (Bateson and Gilbert, 1966). Most dolphins spend their entire lives in schools (Norris and Dohl, 1980); a solitary dolphin is often frightened or ill at ease. Schooling behavior enhances the effectiveness of food searching, prey capture, and predator avoidance and may increase reproductive synchrony and efficiency. Dolphins have highly evolved systems of communication and social behavior. They display altruistic behavior and, apart from humans, are the only animals that assist a member of another species in distress. Repeated observations have been made of one species of delphinid supporting at the surface a distressed individual of another species. Many recent studies have indicated that cetaceans are remarkably intelligent, inventive, and capable of "higher-order" learning. L. H. Herman (1980) compares the learning ability of odontocetes with that of primates: "Both taxa have advanced capabilities for classifying, remembering, and discovering

relationships among events. . . . " In cetaceans, these capabilities are based on auditory perception, whereas primates use mostly visual perception. The behavioral adaptability of dolphins was observed by Hoese (1971), who watched two bottlenose dolphins cooperatively pushing waves onto a muddy shore and stranding small fish. The dolphins rushed up the bank, snatched the fish from the mud, and then slid back into the water.

Dolphins seem to make friends with humans easily. "Opo," of Opononi Beach near Auckland, New Zealand, was a wild dolphin that allowed children to ride on its back and would play catch with a ball (Slijper, 1979).

Family Phocoenidae. The members of this family are generally called porpoises. Six Recent species of three genera are recognized. They occur widely in coastal waters of all oceans and connected seas of the Northern Hemisphere, as well as in some coastal waters of South America and some rivers in southeastern Asia. The earliest fossil record of phocoenids is from the middle Miocene.

Phocoenids are small, from about 1.5 to 2.1 meters in length and 90 to 118 kilometers in weight, and have fairly short jaws and no beak. The dorsal fin is either low or absent. The skull resembles that of the Delphinidae but has conspicuous prominences anterior to the nares. The teeth of most phocoenids are distinctive in being laterally compressed and spadelike; the crowns have two or three weakly developed cusps. *Phocoenoides,* however, has conical teeth. The number of teeth varies from 15/15 to 30/30. From three to seven cervical vertebrae are fused.

Some phocoenids (*Phocoena* and *Neophocaena*) inhabit in-shore waters, such as bays and estuaries, whereas the swift white-flank porpoises (*Phocoenoides*) generally inhabit deeper water. Schools of at least 100 phocoenids may assemble, and crescentic formations associated with feeding have been noted (Fink, 1959). A variety of food is taken, including cuttlefish and squid, crustaceans, and fish.

Figure 15-17.   The Ganges dolphin (*Platanista gangetica,* Platanistidae), which normally swims on its side. (From Herald, Brownell, et al., 1969)

Family Platanistidae.   This group, the long-snouted river dolphins, is remarkable because its members live largely in rivers. The distribution includes some large river systems in India and Pakistan *(Platanista gangetica),* the Amazon and Orinoco river systems of South America *(Inia geoffrensis),* coastal waters along the eastern coast of South America *(Pontoporia blainvillei),* and some large rivers in China *(Lipotes vexillifer).* Some experts place each of these species in a separate family as follows: *P. gangetica,* Platanistidae; *I. geoffrensis,* Iniidae; *P. blainvillei,* Pontoporiidae; *L. vexillifer,* Lipotidae. There are four Recent genera and five species, and fossil members are recorded back to early Miocene times.

These are small cetaceans, from 1.5 to 2.9 meters in length and 40 to 125 kilograms in weight. The jaws are unusually long and narrow and bear numerous teeth (from about 26/26 to 61/61); the forehead rises abruptly and is rounded, giving the head an almost birdlike aspect. In *Platanista,* there are large crests that are probably of sesamoid origin (p. 508). The large temporal fossa is not roofed by the maxillary and frontal bones. None of the cervical vertebrae are fused. The eyes of all members of the family are reduced, and presumably food and obstacles are detected largely by echolocation. *Platanista,* which usually swims on its side (Fig. 15–17), lacks eye lenses and can perhaps detect only light and dark. The eyes of the white-flag dolphin *(Lipotes)* are greatly reduced, and vision is presumably poor. The eyes of the other river dolphins are small but presumably functional.

These cetaceans often inhabit rivers that are made nearly opaque by suspended sediment, and under these conditions echolocation may completely supplant vision. A variety of fish and crustaceans are eaten, some of which are captured by probing muddy river bottoms. The Amazon dolphins *(Inia)* feed largely on fish and during the rainy season may move deep into flooded tropical forests (Humboldt and Bonpland, 1852). River dolphins are seemingly not as social as many other cetaceans. Only 14 percent of Layne's (1958:18) observations of Amazon dolphins were of groups containing more than four individuals. Layne made observations that suggest fairly acute vision above water. Individuals approaching a narrow channel used their eyes above water, presumably to scan the banks for danger.

Chapter 16

Order Rodentia

Because plants are the most abundant food source for terrestrial mammals, it is not surprising that most members of the largest mammalian order, Rodentia, are herbivorous. There are some 30 living families, 418 genera, and roughly 1750 species. Rodents have been, and remain today, spectacularly successful: they are nearly cosmopolitan in distribution, exploit a broad spectrum of foods, are important members of most terrestrial faunas, and often reach extremely high population densities. Repeated rodent radiations have occurred at many times in many places. As a result of convergent evolution, many rodents have styles of life and morphological features similar to those of members of other orders. Among rodents that radiated in South America, for example, are species that resemble rabbits or small antelope, and one species resembles a miniature hippopotamus. Rodents are an unusually complex group with respect to morphological diversity, lines of descent, and parallel evolution of similar features in different evolutionary lines. Parallelism is regarded by Wood (1935) as "the evolutionary motto of the rodents in general." Because of these complexities, disagreement among zoologists as to relationships among taxa and the systematics of rodents has been the rule.

The terms "sciuromorph," "myomorph," and "hystricomorph" have been used repeatedly to designate major divisions within the order Rodentia and refer to basic patterns in the arrangement of the masseter muscles, the skull, and the zygomatic arch. Regarding the use of these terms in the taxonomic scheme of rodents, however, there has been little agreement. The classification used here (Table 4–1) is that of Carleton (1984) and recognizes two main divisions, the suborders Sciurognathi and Hystricognathi.

Rodents are fascinating in part because of the very features that make them difficult to classify. Their complex patterns of evolution, different morphological solutions to similar basic functional problems, intricate systems of resource allocation, and finely tuned adaptations to such extreme environments as those in arctic and desert areas make rodents a rewarding group to study.

Morphology

Recent members of the order Rodentia share a series of distinctive cranial features. The upper and lower jaws each bear a single pair of persistently growing incisors, a feature developed early in the evolution of rodents and one that committed them to a basically herbivorous mode of feeding but permitted the exploitation of such often abundant foods as insects. Because only the anterior surfaces are covered with enamel, the incisors assume a characteristic beveled tip as a result of wear. The occlusal surfaces of the cheek teeth are often complex and allow for effective sectioning and grinding of plant material. In some rodents, the cheek teeth are ever-growing. The dental formula seldom exceeds 1/1, 0/0, 2/1, 3/3 = 22, and a diastema is always present between the incisors and the premolars. The incisors and canines are always 1/1, 0/0.

The glenoid fossa of the squamosal bone is elongate and allows anteroposterior and trans-

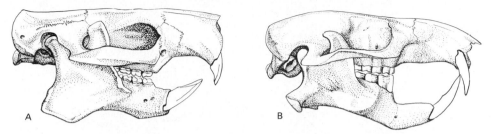

Figure 16–1. Skulls of primitive rodents. (A) *Paramys* (Paramyidae), a primitive late Paleocene and early Eocene sciuromorph; length of skull 89 mm. (B) Mountain beaver (*Aplodontia rufa,* Aplodontidae), the most primitive living rodent; length of skull 68 mm. (A after Romer, 1966)

verse jaw action. The mandibular symphysis has sufficient give in many species to enable the transverse mandibular muscles to pull the ventral borders of the rami together and spread the tips of the incisors. The masseter muscles are large and complexly subdivided, provide most of the power for mastication and gnawing, and in all but one species have at least one division that originates on the rostrum. The temporal muscles are usually smaller than the masseters, and their point of insertion, the coronoid process, is usually reduced. In general, rodents have undergone little postcranial specialization. There are notable exceptions, however, among saltatorial, fossorial, and gliding rodents.

Paleontology

Rodents are an old group, dating back to the late Paleocene of North America. The earliest fossils are but scattered teeth representing the primitive family Ischyromyidae. Early Eocene ctenodactyloid rodents from Asia seem equally primitive in zygomasseteric structure (the structure of the masseter muscles and the parts of the skull from which they originate) but are probably more primitive dentally (M. R. Dawson, Li, and Qi, 1985). The structure of the skull of primitive rodents (Fig. 16–1A) indicates that the temporalis muscle was large and the masseter muscles were not highly specialized and originated entirely from the zygomatic arch

(Fig. 16–2A). The primitive dental formula is 1/1, 0/0, 2/1, 3/3 = 22, and the cheek teeth are brachyodont.

Rodents may have diverged early into two major groups. This hypothetical dichotomy is recognized systematically by the suborders Sciurognathi and Hystricognathi, names that refer to contrasting types of mandibles. In the *sciurognathous type* (Fig. 16–3), the angular process of the dentary bone originates in the vertical plane that passes through the alveolus of the incisor and is ventral to the alveolus. In the *hystricognathous type* (Figs. 16–2D and 16–4), the angular process originates lateral to the vertical plane of the alveolus. In contrast to the sciurognathous dentary bone, the hystricognathous dentary bone tends to have a more strongly reduced coronoid process and its lower border is generally marked by a more prominent projection at the base of the incisor root. These characters, however, vary widely within each suborder.

The early success of rodents is indicated by their Eocene abundance in Eurasia and North America and by their rapid radiation. By the Eocene, when they were seemingly replacing the ancient and formerly highly successful multituberculates, rodents were abandoning the primitive zygomasseteric arrangement. The time between the late Eocene and middle Oligocene was one of accelerated rodent evolution (R. W. Wilson, 1972), and it was then that the major patterns of jaw muscle specialization typ-

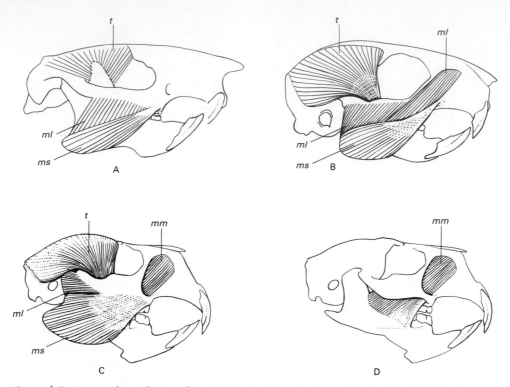

Figure 16-2. Patterns of specialization of the skull, jaws, and jaw musculature of some rodents. (A) *Ischryotomus* (Paramyidae), a primitive Eocene rodent; the jaw muscles are restored. Note that the masseter muscles originate entirely on the zygomatic arch. (B) Abert's squirrel (*Sciurus aberti,* Sciuridae); the anterior part of the lateral masseter originates on the rostrum and the zygomatic plate. (C and D) Porcupine (*Erethizon dorsatum,* Erethizontidae); the anterior part of the medial masseter originates largely on the rostrum and passes through the greatly enlarged infraorbital foramen. (The temporalis muscle, typically reduced in size in hystricomorphous rodents, is unusually large in porcupines.) Abbreviations: *ml,* lateral masseter; *mm,* medial masseter; *ms,* superficial masseter; *t,* temporalis. (A after A. E. Wood, 1965)

ical of modern rodents (Figs. 16–2 and 16–3) were established. Over half of the living families of rodents appeared by the end of the Oligocene. The Miocene spread of grasslands or savannas in both the Old World and the New World provided new adaptive zones for rodents, and the evolution of the jerboas (Dipodidae) in the Old World and the kangaroo rats and pocket mice (Heteromyidae) in the New World was probably decisively affected by increasing aridity in the late Miocene and the appearance of deserts in the Pliocene. M. R. Dawson and Krishtalka (1984) summarize broad patterns of relationships among rodents as follows: (1) ctenodactyloids (the living family Ctenodactylidae and the extinct family Chapattimyidae) are the most primitive rodents morphologically and represent a very early line of rodent evolution; (2) early members of the primitive family Ischyromyidae gave rise to several groups of rodents, to which the living families Aplodontidae, Sciuridae, and Gliridae are most closely related; (3) similarities between early geomyoids (represented today by the Heteromyidae and Geomyidae), muroids (Muridae), and dipodoids (Dipodidae) indicate that these groups share a common ancestral stock; (4) the Hystricognathi may have originated in Asia and in-

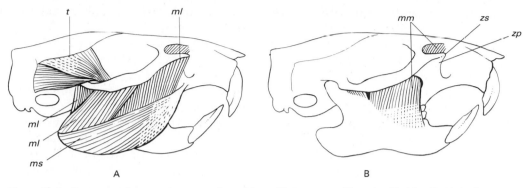

Figure 16-3. Zygomasseteric pattern in myomorphous rodents. (A) A cotton rat (*Sigmodon hispidus,* Muridae); the superficial masseter originates on the rostrum, and the anterior part of the lateral masseter originates on the anterior extension of the zygomatic arch. (B) The superficial muscles have been removed; the medial masseter originates partly on the rostrum and passes through the slightly enlarged infraorbital foramen. Abbreviations are the same as in Figure 16-2. (After Rinker, 1954)

cludes a number of relatively closely related groups.

Some of the accounts of families that follow include brief additional comments on the rodent fossil record.

Jaw Muscle and Skull Specializations

Despite their diversity and success in adapting to contrasting environments and styles of life, rodents have rather consistently followed certain basic trends in the evolution of the jaw muscles, the bones from which these muscles take origin or insertion, and the teeth. Even in

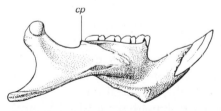

Figure 16-4. The hystricognathous dentary bone of the nutria (*Myocastor coypu,* Myocastoridae). Note the strongly reduced coronoid process *(cp).* Compare this dentary to that of *Aplodontia,* shown in Fig. 16-1.

the early stages of their evolution, selective pressures apparently favored forward migration of the jaw muscles. As pointed out by A. E. Wood (1959:171), however, there were seemingly a very limited number of ways in which this could occur, given the structural constraints imposed by the skull design and jaw musculature of rodents. Accordingly, the trend in the evolution of the jaw musculature seems to have been toward "myriad detailed variations on a limited number of basic patents" (Wood, 1974:50). This variation, involving repeated and complex parallelism and intricate variation on the basic theme, has obscured both the relationships among rodents and their evolutionary patterns.

Since their appearance in late Paleocene times, rodents have had a dentition featuring a division of labor between incisors and cheek teeth. This dentition serves chiefly two functions. The incisors serve as chisels with which food is gnawed, vegetation is clipped, or, in some fossorial forms, soil and rocks are dug away. These teeth are subject to heavy wear and became ever-growing early in the evolution of rodents. The cheek teeth, separated from the incisors by a broad diastema, perform a different function, that of mastication of food. A complicated jaw action allows the lower cheek teeth to move transversely or anteroposteriorly against the upper teeth, producing a crushing and grinding action. Not only are the gnawing and grinding functions performed by different teeth, but they must be performed separately. When the cheek teeth are in position for grinding, the tips of the incisors do not meet; the lower jaw must therefore be moved forward for the incisors to be in position for gnawing.

This division of labor between incisors and cheek teeth clearly "guided" the evolution of the rodent jaw musculature. During gnawing, muscles that attach far forward on the jaw and skull are advantageous because they confer great power on the jaw action through increased mechanical advantage. Furthermore, because forward movement of the lower jaw is a prerequisite for gnawing, selection probably favored the attachment of some of the jaw muscles far forward on the rostrum; contraction of these muscles caused a forward shift of the jaw. During the grinding of food by the cheek teeth, jaw muscles with mechanical advantage for power are also important, and the complex jaw action associated with grinding demands jaw musculature with precise control over anterior, posterior, and transverse jaw movement. The jaw musculature and skull specializations to be considered can best be put in functional perspective if the importance of complex grinding movements and powerful forward movements of the rodent jaw are kept in mind.

Not every rodent can readily be classified as sciuromorph, hystricomorph, or myomorph, and, as mentioned above, experts do not always agree on how the types evolved. The myomorph pattern, for example, appears to have evolved through a hystricomorphous ancestry (Klingener, 1964). The broad trend in rodents was away from the primitive condition, in which the masseter muscles originated entirely on the zygomatic arch, toward the placement of the origin of at least one division of the masseter on the rostrum. The primitive condition, termed *protrogomorphous* by A. E. Wood (1965), is

retained today by only one rodent (*Aplodontia rufa,* Aplodontidae).

Presumably as a result of competition among the rapidly diversifying rodents, in late Eocene the skull and jaw musculature were altered in some phylogenetic lines in a way that increased the effectiveness of gnawing and grinding. These specializations involved primarily the lateral and medial masseter muscles and their areas of attachment. In some rodents, the insertion of the anterior part of the lateral masseter shifted onto the anterior surface of the zygomatic arch and the adjacent part of the rostrum (Fig. 16–2B). This pattern is termed *sciuromorphous* and, as defined here, occurs in the Sciuridae and Castoridae. In these families, the temporalis muscle is relatively large and the coronoid process is moderately well developed.

A second pattern of zygomasseteric specialization involves the shift of the origin of the medial masseter from the zygomatic arch to an extensive area on the side of the rostrum. This muscle passes through the often greatly enlarged infraorbital foramen (Fig. 16–2C,D), and the arrangement is termed *hystricomorphous.* It occurs in the suborder Sciurognathi in the families Dipodidae, Ctenodactylidae, and Pedetidae and in most families in the suborder Hystricognathi. Although included in the Hystricognathi, members of the African family Bathyergidae are not hystricomorphous: in this family the infraorbital foramen is reduced and transmits no part of the medial masseter. Lavocat (1973) believes that ancestral bathyergids were hystricomorphous but that they secondarily reduced the infraorbital foramen. In living bathyergids, the large anterior part of the medial masseter originates in the orbit, a condition perhaps permitted by the reduction of the eye in these highly fossorial rodents.

Some rodents have utilized a third type of zygomasseteric specialization, termed *myomorphous.* In such rodents, the anterior part of the lateral masseter originates on the highly

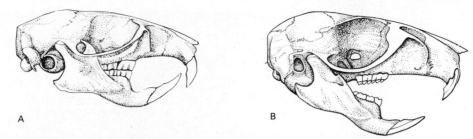

Figure 16–5.   Myomorphous rodent skulls: (A) Stephen's woodrat (*Neotoma stephensi*, Muridae, Sigmodontinae); length of skull 42 mm; (B) gerbil (*Tatera humpatensis*, Muridae, Gerbillinae); length of skull 38 mm. (B after J. E. Hill and Carter, 1941)

modified anterior extension of the zygomatic arch (the zygomatic plate and zygomatic spine; Fig. 16–3) and the anterior part of the medial masseter originates on the rostrum and passes through the somewhat enlarged infraorbital foramen. The temporalis is typically reduced, and the coronid process ranges from well developed to vestigial (Fig. 16–5). Many rodents belonging to the suborder Sciurognathi, including all members of the huge family Muridae (1135 species), are myomorphous.

Suborder Sciurognathi

Family Aplodontidae.   This family is of interest primarily because of the unique primitive morphological features that characterize its one living member. *Aplodontia rufa,* the mountain "beaver," is restricted to parts of the Pacific Northwest. This animal is roughly the size of a small rabbit and has a robust, short-legged form.

Aplodontia is generally regarded as the most primitive living rodent and is the only living member of the primitive infraorder Protrogomorpha, all members of which are characterized by masseters with an entirely zygomatic origin. The skull is flat, and the coronoid process of the dentary bone is large (Fig. 16–1B). The cheek teeth are ever-growing (a specialized feature) and have a unique crown pattern (Fig. 16–6A). The dental formula is 1/1, 0/0, 2/1, 3/3 = 22.

The earliest records of aplodontids are from the Oligocene of western North America. Derived from aplodontid ancestry, the highly specialized family Mylagaulidae appeared in late Miocene. These extinct woodchuck-size rodents were fossorial and are notable for their prominent nasal "horns." The aplodontids spread later to Europe and Asia, but since the middle Pliocene have lived only in the moist, forested parts of the Pacific slope of North America, where they occur today from central California

Figure 16–6.   Crowns of first two right upper molars of two rodents. (A) Mountain beaver (*Aplodontia rufa,* Aplodontidae); note the simplified and unique crown pattern. (B) Merriam's kangaroo rat (*Dipodomys merriami,* Heteromyidae); note the highly simplified crown pattern. The outer border of the tooth is above; anterior is to the right. The unshaded part is enamel, and the stippled part is dentine.

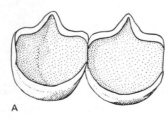

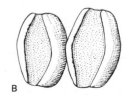

Figure 16-7. Yellow-bellied marmot (*Marmota flaviventris,* Sciuridae). (O. D. Markham)

to southern British Columbia. Widespread Late Tertiary aridity may have restricted the aplodontids to their present range.

Aplodontia occurs in small colonies, favors moist areas supporting lush growths of forbs, and often builds its burrows next to streams. The diet includes a variety of forbs and the buds, twigs, and bark of such riparian plants as willow *(Salix)* and dogwood *(Cornus)*. On occasion, *Aplodontia* builds "hay piles" of cut sections of forbs (J. Grinnell and Storer, 1924:157). Although usually terrestrial, this animal is known to climb to some extent in search of food.

Family Sciuridae. This successful and widespread family includes 262 Recent species representing 49 genera. Squirrels, chipmunks, marmots (Fig. 16–7), and prairie dogs (Fig. 16–8) belong to this family. Sciurids appeared first in the middle Oligocene of North America, and ground and tree squirrels can be distinguished by the end of the Oligocene. Sciurids remain widespread today, absent only from the Australian region, Madagascar, the polar regions, southern South America, and certain Old World desert areas.

Sciurids are fairly distinctive structurally. The skull is usually arched in profile, and the front of the zygomatic arch is flattened where the anterior part of the lateral masseter rests against it. The dental formula is 1/1, 0/0, 1-2/1, 3/3 = 20-22. The cheek teeth are rooted and usually have a crown pattern that features trans-

Figure 16-8. Blacktail prairie dogs (*Cynomys ludovicianus,* Sciuridae). (O. J. Reichman)

verse ridges. Sciurids have relatively unspecialized bodies: a long tail is usually retained, and the limbs seldom have a loss of digits or reduction of freedom of movement at the elbow, wrist, and ankle joints. Several semifossorial types, including ground squirrels *(Spermophilus),* prairie dogs *(Cynomys),* and marmots *(Marmota),* have variously departed from this plan in the direction of greater power in the forelimbs and, in some cases, reduction of the tail.

Sciurids are basically diurnal herbivores, but a great variety of food is utilized. Tree squirrels occasionally eat young birds and eggs; chipmunks *(Eutamias)* and the antelope ground squirrels *(Ammospermophilus)* are seasonally partly insectivorous in some areas. Sciurids are tolerant of a great range of environmental conditions. Some, such as marmots, prairie dogs, chipmunks, and some ground squirrels, hibernate during cold parts of the year. Red squirrels *(Tamiasciurus)* in boreal coniferous forests of North America remain active throughout the winter, when temperatures frequently stay well below 0°F for days at a time. The antelope ground squirrel is adapted to living under demanding desert conditions (p. 464).

Styles of locomotion vary among sciurids; the most specialized style occurs in the flying squirrels. This group of 13 genera, constituting the subfamily Petauristinae, is characterized by gliding surfaces formed by broad folds of skin between the forefoot and hind foot. These animals are able to glide fairly long distances between trees. The giant flying squirrel *(Petaurista)* of southeastern Asia, for instance, can glide up to 450 meters and can turn in midair (Walker, 1968:716).

Family Castoridae. To this family belong the beavers. The family is represented today by but two species, *Castor canadensis* of the United States, Canada, and Alaska and *Castor fiber* of northern Europe and northern Asia. Beavers had a pronounced effect on the history of the United States, for much of the early exploration of some of the major river systems in the western

United States was done by trappers in quest of valuable beaver pelts.

The fossil record of beavers begins in the Oligocene of North America, and beavers reached Europe in the Miocene. Several lines of descent developed in the Tertiary. One line developed fossorial adaptations; the fossil remains of the North American Miocene beaver *Paleocastor* have been found in the spectacular corkscrew-shape burrows apparently dug by these animals. Another evolutionary line led to the bear-size giant beaver *(Castoroides)* of the North American Pleistocene. Throughout their history, castorids have been restricted to the Northern Hemisphere.

Living beavers are semiaquatic, and some of their distinctive structural features are adaptations to this mode of life. The animals are large, reaching over 30 kilograms in weight. Their large size is associated with a mass-to-surface ratio that is more advantageous in terms of heat conservation than that of smaller rodents. In addition, the body is insulated by fine underfur protected by long guard hairs. These are important adaptations in animals that frequently swim and dive for long periods in icy water. The large hind feet are webbed, the small eyes have nictating membranes (membranes that arise at the inner angle of the eye and can be drawn across the eyeball), and the nostrils and ear openings are valvular and can be closed during submersion.

Because of two structural specializations, beavers can open their mouth when gnawing under water; while they are swimming, they can carry branches in the submerged open mouth without danger of taking water into the lungs (Cole, 1970). The epiglottis is internarial (it lies above the soft palate); this arrangement allows efficient transfer of air from the nasal passages to the trachea but does not allow mouth breathing or panting. Also, the mid-dorsal surface of the back part of the tongue is elevated and fits tightly against the palate and, except when the animal is swallowing, blocks the passage to the pharynx (Cole, 1970).

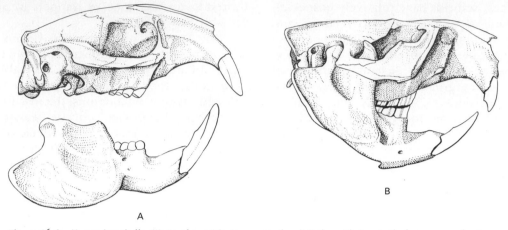

Figure 16-9. Two rodent skulls. (A) A mole rat (*Cryptomys mechowii,* Bathyergidae); note the large, procumbent incisors, used for digging; length of skull 57 mm. (B) A beaver (*Castor canadensis,* Castoridae); note the depression in the side of the rostrum from which the anterior part of the lateral masseter originates; length of skull 139 mm. (A after J. E. Hill and Carter, 1941)

The tail is broad, flat, and largely hairless. The skull is robust. The zygomasteric structure is specialized in that the rostrum is marked by a conspicuous lateral depression (Fig. 16–9B) from which a large part of the lateral masseter muscle originates. The jugal is broad dorsoventrally, and the external auditory meatus is long and surrounded by a tubular extension of the auditory bulla. The dental formula is 1/1, 0/0, 1/1, 3/3 = 20. The premolars are molariform, and the complex crown pattern features transverse enamel folds (Fig. 16–10E).

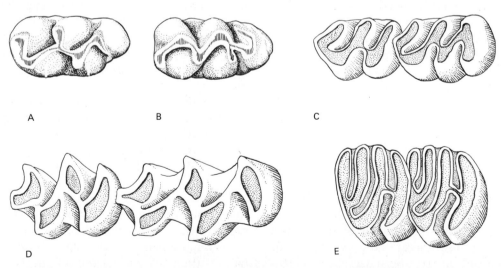

Figure 16-10. Crown patterns of rodent molars. (A) First right upper and (B) first left lower molar of a harvest mouse (*Reithrodontomys megalotis,* Muridae). First two right upper molars of (C) a cotton rat (*Sigmodon alleni,* Muridae), (D) a mexican vole (*Microtus mexicanus,* Muridae), and (E) a beaver (*Castor canadensis,* Castoridae). Unshaded areas on occlusal surfaces are enamel; stippled areas within the enamel folds are dentine.

Beavers are always found along waterways. Although they are most typical of regions supporting coniferous or deciduous forests, they also live in some hot desert regions, as, for example, along the lower Colorado River of Arizona and California. In the southwestern and middle-western United States, beavers dig burrows in the river banks, but in northern and mountainous regions they build lodges of sticks and mud in ponds formed behind their dams. They remain active beneath the ice throughout the winter, feeding on the cambium of aspen and willow branches that they have stuck in the mud bottoms of the ponds.

Beavers are remarkable in their ability to modify their environment by felling trees and building dams. Many high valleys in the Rocky Mountains have been transformed by beaver dams from a series of meadows through which a narrow, willow-lined stream meandered to a terraced series of broad ponds bordered by extensive willow thickets and soil saturated with water. A valley suitable for grazing by cattle before occupancy by beavers may be more suitable afterward for trout and waterfowl.

Family Geomyidae. Members of this family, the pocket gophers, are the most highly fossorial North American rodents. They are distributed from Saskatchewan to northern Columbia. The family includes roughly 35 Recent species in five genera. Pocket gophers appeared first in the early Miocene of North America. Although they are not restricted to semiarid habitats today, many of their most characteristic specializations probably evolved in response to the soil conditions and floral assemblages of the semiarid and plains environments that developed in the Miocene.

The most obvious structural characteristics of pocket gophers were developed in response to fossorial life. These animals are moderately small, weighing from 100 to 900 grams. They have small pinnae, small eyes, and a short tail. The head is large and broad, and the body is stout. External fur-lined cheek pouches are used for carrying food. The dorsal profile of the geomyid skull is usually nearly straight, the zyomatic arches flare widely, and in the larger species the skull is angular and features prominent ridges for muscle attachment. The rostrum is broad, robust, and marked laterally by depressions from which the lateral masseter muscles take origin. The large incisors often protrude forward, in some species beyond the anteriormost parts of the nasals and premaxillae; the lips close behind the incisors, which are therefore outside the mouth. The dental formula is 1/1, 0/0, 1/1, 3/3 = 20. The cheek teeth are ever-growing and have a highly simplified crown pattern. There is no loss of digits. The forelimbs are powerfully built and bear large, curved claws; the toes of the forepaw have fringes of hairs that presumably increase the effectiveness of this foot during digging (Fig. 16–11A).

Pocket gophers occupy friable soils in environments ranging from tropical to boreal, eat a variety of above- and underground parts of forbs, grasses, shrubs, and trees, and strongly affect their environment. Pocket gopher burrows provide channels that allow deep penetration of water and partly avoid surface erosion during periods of snow melt in mountainous areas. The disturbance of the soil and the mounds of soil

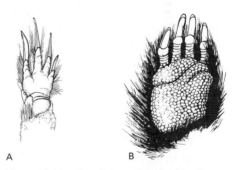
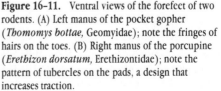

Figure 16-11. Ventral views of the forefect of two rodents. (A) Left manus of the pocket gopher (*Thomomys bottae,* Geomyidae); note the fringes of hairs on the toes. (B) Right manus of the porcupine (*Erethizon dorsatum,* Erethizontidae); note the pattern of tubercles on the pads, a design that increases traction.

thrown up by pocket gophers strongly influence vegetation by favoring pioneer plants. In some mountain meadows, roughly 20 percent of the ground surface is covered with mounds, and in such meadows in Utah pocket gophers may eat more than 30 percent of the annual below-ground productivity of forbs (D. C. Andersen and MacMahon, 1981). Because their preference for alfalfa and other cultivated plants results in great crop damage, large amounts of money have been spent by farmers and federal agencies to control pocket gophers on cultivated land in the western United States.

Family Heteromyidae. Members of this family are the North American rodents most fully adapted to desert life. Heteromyids are restricted to the New World, where they range from southern Canada through the western United States to Ecuador, Colombia, and Venezuela. Although they occupy areas ranging from temperate to tropical, they reach their greatest diversity and density in arid and semiarid regions. This family contains 63 Recent species of five genera.

Heteromyids first appeared in the Oligocene of North America. The kangaroo rats (Fig. 16–12) are known from the Pliocene, when the deserts and semiarid brushlands that heteromyids now frequently inhabit became widespread in western North America. Certain diagnostic characteristics of kangaroo rats, such as the greatly enlarged auditory bullae (Fig. 16–13) and features of the hind limbs that favor saltation, probably evolved under constraints associated with desert or semidesert conditions.

Heteromyids are specialized for jumping. Such adaptations are most strongly developed in the kangaroo rats and kangaroo mice *(Microdipodops)*. In these heteromyids, the forelimbs are small, the neck is short, and the tail is long and serves as a balancing organ. The hind limbs are elongate, and the thigh musculature is powerful. The hind foot is elongate, but except for the almost complete loss of the first digit in some kangaroo rats (Fig. 16–14E), there is no loss of digits. The cervical vertebrae are largely

Figure 16–12. A chisel-toothed kangaroo rat in a saltbush (*Atriplex confertifolia,* Heteromyidae). This kangaroo rat is unusual both in its ability to climb and in its use of leaves for food. It uses its chisel-like lower incisors to scrape off the hypersaline peripheral tissues of the saltbush leaves and eats the starch-rich inner parts (Kenagy, 1972). (G. J. Kenagy)

fused in *Microdipodops,* and in *Dipodomys* they are strongly compressed and partly fused (Fig. 16–14C), producing a short, rigid neck. These species are mostly bipedal when moving rapidly, and when frightened they move by a series of erratic hops.

As in the pocket gophers, fur-lined cheek pouches are present in heteromyids. The skull is delicately built, with thin, semitransparent bones; the zygomatic arch is slender. The auditory bullae are usually large and in some genera are enormous, formed largely by the mastoid and tympanic bones (Fig. 16–13). The enlargement of the bullae in heteromyids (and in jerboas, Dipodidae) greatly increases auditory sensitivity (p. 511). The nasals are slender and usually extend well forward of the slender upper incisors. The dental formula is 1/1, 0/0, 1/1, 3/3 = 20. The ever-growing cheek teeth have a strongly simplified crown pattern (Fig. 16–6B) resembling that of pocket gophers, to which heteromyids are seemingly closely related.

Most heteromyids live in areas with strongly seasonal patterns of precipitation, and many must cope with the annual cycles of

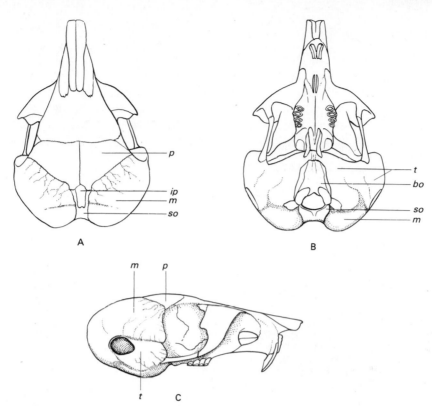

Figure 16–13.   (A) Dorsal, (B) ventral, and (C) lateral views of the skull of Merriam's kangaroo rat (*Dipodomys merriami*, Heteromyidae). Note the great enlargement of the auditory bulla, the chamber surrounding the middle ear; length of skull 45 mm. Abbreviations: *bo*, basioccipital; *ip*, interparietal; *m*, mastoid part of the bulla; *p*, parietal; *so*, supraoccipital; *t*, tympanic part of the bulla (petrosal and tympanic bones). (After J. Grinnel, 1922)

weather and plant growth that occur in deserts. In such areas, brief periods of precipitation and long dry periods are typical. Small annual plants in deserts are able to make the most of irregular moisture by germinating, growing, and flowering rapidly and by producing abundant seeds that remain dormant for various periods. This enormously abundant seed crop is the major food source of heteromyids. Perhaps the most remarkable heteromyid adaptation is the ability to survive for long periods on a diet of dry seeds with no free water. This capability probably does not occur in all heteromyids, nor is it developed to the same degree in all species adapted to dry climates. In the species of kan-garoo rats and pocket mice *(Perognathus)* that occupy the desert, however, this ability is well developed (p. 485).

Family Dipodidae.   This family includes the jerboas (strongly saltatorial), the jumping mice (moderately saltatorial), and the birch mice (nonsaltatorial). Jerboas occur in arid and semi-arid areas in northern Africa, Arabia, and Asia Minor and in southern Russia eastward to Mongolia and northeastern China. The Nearctic jumping mice occupy Alaska and much of Canada and in the United States live (mostly in boreal habitats) as far south as New Mexico and Georgia. Palearctic jumping mice and birch mice are found from Scandinavia into central

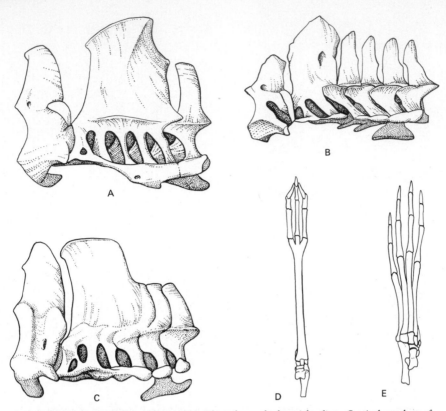

Figure 16-14. The cervical vertebrae and hind feet of several saltatorial rodents. Cervical vertebrae of (A) a jerboa (*Jaculus* sp., Dipodidae), (B) the springhaas (*Pedetes capensis,* Pedetidae), (C) Heermann's kangaroo rat (*Dipodomys heermanni,* Heteromyidae). (D) Dorsal view of hind foot of a jerboa (*Scirtopoda* sp., Dipodidae); note the reduction of digits and the cannon bone formed by metatarsals two, three, and four. (E) Left hind foot of the desert kangaroo rat (*Dipodomys deserti,* Heteromyidae); note the near loss of the first digit and the elongation of the foot. (A, B, C after Hatt, 1932; D after A. B. Howell, 1944; E after J. Grinnell, 1922)

Europe and in Russia, Mongolia, and China. The family first appeared in the Oligocene of Europe, Asia, and North America and is represented today by 14 genera and 44 species.

Jerboas have a compact body, large head, reduced forelimbs, and elongate hind limbs—features associated with saltatorial locomotion. The tail is long and usually tufted, and, as in the New World kangaroo rats (Heteromyidae), the tuft is frequently conspicuously black and white. The posterior part of the skull is broad (Fig. 16–15A), owing mostly to the enlargement of the auditory bullae, which are huge in some species. The rostrum is usually short, the orbits are large, and through the enlarged infraorbital canal passes most of the anterior part of the medial masseter, which originates largely on the side of the rostrum. The zygomatic plate is narrow and below the infraorbital canal. The dental formula is 1/1, 0/0, 0-1/0, 3/3 = 16-18, the cheek teeth are hypsodont, and the crown pattern usually involves re-entrant enamel folds.

The hind limbs are elongate in all genera, but varying stages of specialization for saltation are represented. In members of the subfamily

Cardiocraninae, the toes vary in number from three to five and the metatarsals are not fused. At the other extreme are such genera as *Dipus* and *Jaculus* (subfamily Dipodinae), which represent the greatest degree of hind limb specialization for saltation in rodents. In these genera, only three toes (digits two, three, and four) remain and the elongate metatarsals are fused into a cannon bone (Fig. 16–14D). An additional specialization that occurs in some species is a brush of stiff hairs on the ventral surface of the phalanges. The ears of jerboas vary from short and rounded to long and rabbit-like.

Jumping mice and birch mice are small and graceful, from roughly 10 to 25 grams in weight, with a long tail and, in all genera but *Sicista,* elongate hind limbs. The coloration in most species is striking: the belly is white and the dorsum is bright yellowish or reddish brown. Much of the anterior part of the medial masseter muscle originates on the side of the rostrum and passes through the enlarged infraorbital foramen (Klingener, 1964:11). Although all living dipodids have an hystricomorphous zygomasseteric arrangement, they are regarded as related to the myomorphous Muridae. Both R. W. Wilson (1949) and Klingener (1964) concluded that the murid myomorphous masseter was derived from a dipodoidlike ancestor. The dental formula is 1/1, 0/0, 0-1/0, 3/3 = 16-18; the cheek teeth are brachyodont or semihypsodont and have quadrituber-cular crown patterns with re-entrant enamel folds. The hind limbs in jumping mice are elongate and somewhat adapted for hopping, but unlike the case with more specialized saltatorial rodents, all digits are retained. As an additional contrast with jerboas and other specialized saltators, the cervical vertebrae of jumping mice are unfused.

Jerboas occupy arid areas and lead lives that resemble in some ways those of members of the New World family Heteromyidae. They live in burrows that are frequently kept plugged during the day, a habit that favors water conservation by keeping the humidity in the burrow as high as possible. They are nocturnal, and many species sift seeds from sand or loose soil with the forefeet, although some species depend largely on insects for food. Unlike kangaroo rats, jerboas hibernate during the winter in fairly deep burrows. Locomotion in jerboas is chiefly bipedal, but when they are moving slowly, the forefeet may be used to some extent. When frightened, jerboas move rapidly in a series of long leaps, each of which may cover 3 meters. Such a rapid and, more important, erratic mode of escape from predation is especially effective in the barren terrain jerboas occupy.

Jumping mice and birch mice usually inhabit boreal forests. Some species occur typically in coniferous forests, while others appear in birch stands or in mixed deciduous forests.

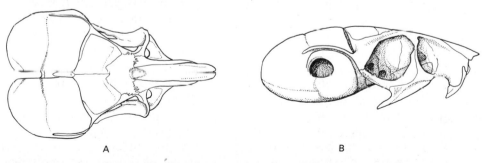

A   B

Figure 16–15. (A) Dorsal and (B) lateral views of the skull of a jerboa (*Salpingotus kozlovi,* Dipodidae); length of skull 27 mm. Note the greatly enlarged auditory bullae and the general resemblance between this skull and that of the kangaroo rat (Fig. 16–13). (After G. M. Allen, 1940)

Table 16-1: The Subfamilies of the Family Muridae

Subfamily (no. of species)	Common Name	Distribution
Sigmodontinae (369)	New World rats and mice	North and South America
Cricetinae (22)	Hamsters	Palearctic
Arvicolinae (125)	Voles, lemmings, muskrat	Holarctic
Gerbillinae (83)	Gerbils, jirds, sand rats	Africa, southern Asia
Cricetomyinae (6)	Pouched rats and mice	Africa south of Sahara
Petromyscinae (3)	Rock mice, swamp mouse	Parts of Africa
Dendromurinae (20)	Climbing, gerbil, rat, and forest mice	Africa south of Sahara
Lophiomyinae (1)	Maned rat	Eastern Africa
Nesomyinae (11)	Malagasy rats and mice	Madagascar
Murinae (460)	Old World rats and mice	Nearly worldwide
Otomyinae (12)	Vlei rats, karoo rats	Parts of Africa
Platacanthomyinae (2)	Spiny mouse, blind tree mouse	India, southeastern Asia
Myospalacinae (4)	Zokors	Siberia, northern China
Spalacinae (5)	Blind mole rats	Mediterranean region
Rhizomyinae (6)	Mole rats, bamboo rats	East Africa, southern Asia

Usually, jumping mice favor moist situations, and *Zapus princeps* of the western United States is most abundant in dense cover adjacent to streams or in wet meadows. These mice hibernate in the winter and emerge during or after snow melt. Food consists of a variety of seeds, fungus *(Endogone),* and other plant material, but insect larvae and other animal material made up approximately half of the food of *Z. hudsonius* in New York (Whitaker, 1963) and roughly one third of the diet of *Z. princeps* in Colorado (Vaughan and Weil, 1980).

Family Muridae.   This huge family includes 65 percent of the living species of rodents (roughly 1130 species and 261 genera) and is nearly worldwide in distribution; its members occupy environments ranging from high arctic tundra to tropical forests to desert sand dunes. Table 16–1 lists the subfamilies of murids, their common names, and (roughly) their distributions. Excellent coverage of murid rodents is given by Carleton and Musser (1984).

Most murids retain a "standard" mouselike form, with a long tail, generalized limb structure, and no loss of digits. Murids range in size from about 10 grams in weight and 100 milli-meters in total length, as in the pygmy mouse *(Baiomys),* to approximately 1500 grams and 600 millimeters, as in the muskrat *(Ondatra).* The skull varies widely in shape (Figs. 16–5, 16–16, 16–21) but the infraorbital foramen is always above the zygomatic plate and is enlarged dorsally for the transmission of part of the medial masseter, which originates on the side of the rostrum. Through the narrowed ventral part of this foramen pass blood vessels and a branch of the trigeminal nerve. The maxillary root of the zygomatic arch is platelike and provides surface for the origin of part of the lateral masseter. This myomorphous zygomasseteric arrangement was perhaps derived through a hystricomorphous ancestry. The dental formula is generally 1/1, 0/0, 0/0, 3/3 = 16; in some species the molars are reduced in number. Molars range from brachyodont to hypsodont and ever-growing. The basic cusp pattern involves transverse crests (Fig. 16–17), but crests are absent in some species and molar crown patterns vary widely (Fig. 16–10).

Within the array of murid species, a variety of modes of life and morphological and behavioral specializations are represented. The fol-

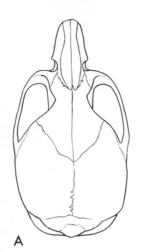 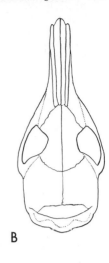

Figure 16–16. Extremes in skull shape in rodents of the family Muridae. (A) Rock mouse *(Delanymys brooksi)*, a rock-dwelling omnivore; (B) shrew-like rat *(Rhynchomys soricoides)*, a rare species that apparently feeds on invertebrates. (After Walker, 1968)

A   B

lowing paragraphs deal with several subfamilies that display some of this broad structural and functional variety.

The Sigmodontinae is the second largest murid subfamily (369 species) and occupies South America and most of North America. *Peromyscus*-like rodents *(Copemys)* entered North America from Eurasia in early Miocene but underwent little radiation until late Miocene and Pliocene, when such sigmodontine genera as *Peromyscus, Neotoma, Onychomys,* and *Reithrodontomys* appeared. Neotropical sigmodontines may have been derived from a separate invasion from the Old World in late

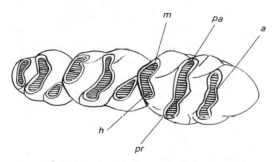

Figure 16–17. Occlusal surfaces of the right upper molars of a murid rodent *(Rattus)*. With wear, the cross lophs become lakes of dentine (cross-hatched areas) rimmed with enamel. Abbreviations: *a,* anterior loph; *h,* hypocone; *m,* metacone; *pa,* paracone; *pr,* protocone.

Miocene *(Calomys)*. They also underwent rapid radiation in the late Miocene and Pliocene, and by early Pliocene the living genus *Sigmodon* had appeared. Sigmodontines probably entered South America when the Panamanian land bridge formed in late Pliocene (the majority opinion) but could have entered earlier by waif dispersal across the seaway separating the Americas. Experts disagree as to whether the great number of living South American sigmodontine species resulted from a tropical North American radiation before access to South America was gained, or whether this radiation occurred in South America after the land bridge formed. In any case, of the 73 living sigmodontine genera, 86 percent live in South America or no farther north than tropical North America. Today sigmodontines occupy habitats ranging from subarctic to tropical and are variously terrestrial, amphibious, fossorial, or arboreal. In many parts of the western United States, the large stick-and-debris houses of woodrats *(Neotoma)* are conspicuous features, and white-footed mice *(Peromyscus)* seem to occupy every conceivable North American habitat.

The Murinae (Fig. 16–18) is the largest murid subfamily (460 species), occurs nearly worldwide, and includes a wide diversity of types adapted to terrestrial, fossorial, largely

Figure 16-18. The Australian bush rat (*Rattus fuscipes,* Muridae, Murinae). (J. Hudson)

aquatic, or arboreal life. Some murines live in close association with humans in situations ranging from isolated farms to the world's largest cities. As a result of introductions by humans, these animals have become nearly cosmopolitan in distribution and are probably the rodents most familiar to us. Murines that are not commensal with humans occur in much of southeastern Asia, Eurasia, Australia, Tasmania and Micronesia, and Africa. Tropical and subtropical areas are centers of murine abundance, but these animals have occupied a wide variety of habitats, and some genera are highly adapted to specialized modes of life. Murines range in size from that of a small mouse to that of *Mallomys,* a New Guinean rat that weighs about 2 kilograms. Although the tail is usually more or less naked and scaly, it is occasionally heavily furred and bushy. The molars are rooted or evergrowing and usually have crowns with crests or chevrons (Fig. 16–17); great simplification of the crown pattern occasionally occurs. The dental formula is usually 1/1, 0/0, 0/0, 3/3 = 16. In some murines, the reduction of the cheek teeth has become extreme. The greatest reduction occurs in *Mayermys,* a rare mouse from New Guinea, in which only one molar is retained on each side of each jaw.

The feet retain all of the digits, but the pollex is rudimentary. Murines appear fairly late in the fossil record (middle Miocene), but the subfamily has been remarkably plastic from an evolutionary point of view. In both Africa and the Australian faunal region, murines have undergone impressive radiations and are variously amphibious, terrestrial, semifossorial, arboreal, and saltatorial. The water rats *(Crossomys)* have greatly reduced ears, large, webbed hind feet, and nearly waterproof fur. These animals live along waterways in New Guinea. At the other extreme is the hopping mouse *(Notomys alexis),* a saltatorial inhabitant of extremely arid Australian deserts. This rodent needs no drinking water and has the greatest ability of any animal in which water metabolism has been studied to concentrate urine as a means of conserving water (MacMillen and Lee, 1969). Murines feed on a variety of plant material and on invertebrate and vertebrate animals. In association with the great diversity of feeding habits of murines, the skull form varies widely within the subfamily (Fig. 16–16), with a shrewlike elongation of the rostrum occurring in some genera. Extremely high population densities have been recorded for feral populations of some murines, often commensal with humans. A 35-acre area near Berkeley, California, which had only an occasional house mouse for a number of years after population studies began in 1948, supported 7000 *Mus* in June 1961 (Pearson, 1964). Among other factors, the high reproductive rate of *Mus* contributes to its ability to reach high densities quickly. Murines that live with humans are of great economic importance. Not only do they spread such serious diseases as bubonic plague and typhus, but also the damage they do to stored grains and other foods is so severe that in many countries *Rattus* and *Mus* compete effectively and devastatingly with humans for food.

The subfamily Arvicolinae includes the voles, lemmings (Fig. 16–19), and muskrat, a group of 17 genera and 125 species of rodents distributed throughout the Northern Hemisphere. These rodents frequently have short tails, ear openings that are partially guarded by fur, and a chunky, short-legged appearance. The cheek teeth often feature complex crown patterns (Fig. 16–10D) adapted to masticating forbs and grasses. The voles and lemmings have high reproductive rates and undergo remarkable population fluctuations in some areas (p. 338). Because many northern carnivores rely on arvicolines for food, the densities and even the distribution of these predators are partially controlled by cycles of arvicoline abundance (Table 18–14).

The subfamily Gerbillinae, the gerbils, includes 16 genera and 23 species and is a group of rodents that resembles jerboas (Dipodidae) and kangaroo rats (Heteromyidae) in being semifossorial and more or less saltatorial and in inhabiting mainly desert regions. Gerbils now occur in arid parts of Asia, in the Middle East, and in Africa. The hind limbs are large, the central three digits are larger than the lateral ones, and the tail is often long and functions as a balancing organ. The skull does not depart strongly from the general murid plan (Fig. 16–5B). In their ability to hop and in their choice of habi-

Figure 16-19.   Norwegian lemmings (*Lemmus lemmus,* Muridae, Arvicolinae). The animal on the right is in a threatening posture; the one on the left is in a submissive posture. (G. C. Clough)

tats, gerbils resemble heteromyid rodents. Gerbils maintain water balance in hot, arid conditions partly by eating food with a high water content and partly by concentrating urine, as do heteromyids.

The subfamily Spalacinae contains animals usually called mole rats. There are five Recent species of one genus *(Spalax)*. These rodents occur in the eastern Mediterranean region and southeastern Europe and are fossorial. The eyes are small and covered with skin, and the eye muscles and optic nerve are degenerate. The ears are absent, and the tail is vestigial. Unlike many fossorial rodents, *Spalax* digs primarily with its large incisors and bulldozes soil with its blunt head. As adaptations to this style of digging, the neck and jaw muscles are powerful, the incisors are robust, and the nose is protected by a broad, horny pad. The feet, surprisingly, are not unusually large; the claws have been described as blunt, round nubbins. These nocturnal rodents burrow in both alluvial and stony soils and eat both below- and above-ground parts of plants. *Spalax* often lives in burrows in water-saturated or snow-covered soil; these burrows have extremely high levels of carbon dioxide and extremely low levels of oxygen. Whereas in most mammals these conditions interfere seriously with heart action, *Spalax* can raise its heart rate and maintain a stable rate at low oxygen levels (Arieli and Ar, 1981a). As an adaptation to improve oxygen delivery to tissues under these conditions, the capillary density in the heart and skeletal muscles of *Spalax* is nearly twice as high as in the laboratory rat (Arieli and Ar, 1981b).

Members of the subfamily Rhizomyinae, called root rats (Fig. 16–20), occur in southeastern Asia and East Africa and are first known from the middle Miocene of southern Asia. The subfamily includes three genera represented by six species. These rodents range from a total length of 200 to 500 millimeters and have short, robust limbs and a compact body. In two genera, the incisors are procumbent (Fig. 16–21). Root rats live primarily in areas with at least 500 millimeters of annual precipitation and occupy habitats ranging from dense bamboo thickets in Asia to subalpine slopes at 4000 meters on Mount Kenya in East Africa. The dig-

Figure 16–20. A root rat (*Tachyoryctes splendens,* Muridae, Rhizomyinae) from Kenya. (J. U. M. Jarvis)

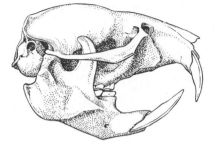

Figure 16–21. Skull of an African root rat (*Tachyoryctes splendens,* Muridae, Rhizomyinae). Note the procumbent incisors, which are used for digging; length of skull 41 mm.

ging behavior of the African root rat, *Tachyoryctes,* has been described by Jarvis and Sale (1971). This animal burrows by slicing away the soil with powerful upward sweeps of the protruding lower incisors. The dislodged soil is moved behind the animal by synchronous thrusts with the hind limbs. When the burrow becomes blocked with freshly dug soil, *Tachyoryctes* turns and pushes the load to the surface with the side of its head and one forefoot. The conspicuous mounds (up to 6 meters in diameter) associated with the activities of *Tachyoryctes* on Mount Kenya resemble the Mima mounds that occur in some parts of the western United States and are thought to be formed by pocket gophers (Fig. 18–36). *Tachyoryctes* is solitary and aggressive and eats a variety of below- and above-ground parts of plants.

Family Anomaluridae. This family, composed of seven Recent species of three genera, includes the scaly-tailed squirrels. These animals occupy forested, tropical parts of western and central Africa, where they are locally common. They resemble the flying squirrel (Sciuridae) in some structural features and in gliding ability.

Increased surface area for gliding is provided in anomalurids by a fold of skin that extends between the wrist and the hind foot and is supported and extended during gliding by a long cartilaginous rod, roughly the length of the forearm, that originates on the posterior part of the elbow. (In "flying" members of the Sciuridae, in contrast, a short cartilaginous brace arises from the wrist.) In anomalurids, folds of skin similar to the uropatagium of bats occur between the ankles and the tail a short distance distal to its base. The fur over most of the membrane is fine and soft, but a tract of stiff hairs occurs along the outer edge of the membrane behind the cartilaginous elbow strut. This tract may possibly improve the efficiency of gliding by controlling the flow of the boundary layer of air sweeping over the membrane. Relative to those of most rodents, the limbs of anomalurids are unusually long and lightly built; they provide for a wide spreading of the flight membranes. One genus *(Zenkerella)* does not have a gliding membrane. The anomalurid tail is usually tufted and has a bare ventral area near its base that has two rows of keeled scales. These scales seemingly keep the animals from losing traction when they cling to the trunks of trees. The feet are strong and bear sharp, recurved claws.

In the anomalurid skull, the infraorbital canal is enlarged and transmits part of the medial masseter muscle. The dental formula is 1/1, 0/0, 1/1, 3/3 = 20, and the cheek teeth are rooted.

Anomalurids are handsome animals beautifully adapted to an entirely arboreal life. They are apparently largely vegetarians. Derby's anomalure *(Anomalurus derbianus)* feeds primarily on fruit and bark of numerous kinds (Rahm, 1970) but is known to also eat flowers and some insects. This species is a graceful and highly maneuverable glider, capable of glides of over 100 meters and of mid-air turns. The dwarf anomalure *(Idiurus zenkeri)* has been described by Durrell (1954) as gliding "with all the assurance and skill of hawking swallows." Kingdon (1974b:454) has observed that animals launching themselves do not immediately spread their membranes but gain speed by leaping powerfully out from the trunk of a tree and dropping at least 1 meter before beginning to glide.

Groups of anomalurids often take shelter during the day in cavities in trees. Group size varies from six or eight animals to colonies of over 100 anomalurids of several species. Rosevear (1969) reports four species occupying the same hole in a tree. On occasion, these rodents share a hollow tree with dormice *(Graphiurus)* and with several species of bats. The propensity of anomalurids to sleep during the day in densely packed clusters and their use of sunbathing suggest that they may at times utilize adaptive hypothermia to conserve energy.

Family Pedetidae.   This family is represented by one distinctive species, *Pedetes capensis,* the spring hare. Its distribution includes East Africa and southern Africa, where it inhabits sandy soils in semiarid regions. Sparsely vegetated areas or places where the vegetation has been heavily grazed by ungulates are preferred.

The spring hare is saltatorial and roughly the size of a large rabbit, weighing up to about 4 kilograms. The eyes are extremely large, suggesting perhaps a reliance on vision for detecting predators. The forelimbs are short but robust and bear long claws that are used in digging. The hind limbs are long and powerfully built, the fibula is reduced and fused distally to the tibia, and the feet have only four toes. The long tail is heavily furred throughout its length. Through the enormous infraorbital foramen (Fig. 16–22) passes the large anterior division of the medial masseter. As in a number of saltatorial rodents, the cervical vertebrae are partly fused (Fig. 16–14B). The dental formula is 1/1, 0/0, 1/1, 3/3 = 20, and the cheek teeth are ever-growing with a simplified crown pattern. A tragus fits against the ear opening and keeps out sand and debris when the animal digs.

Spring hares dig fairly elaborate burrows. Because of their restriction to friable soils, these animals are not evenly distributed and appear to occur in colonies. When frightened, spring hares can make tremendous bipedal leaps of over 6 meters, but when foraging and moving

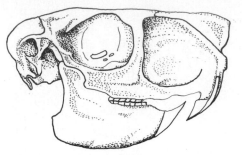

Figure 16–22.   The skull of the spring hare (*Pedetes capensis,* Pedetidae). Note the enormously enlarged infraorbital foramen. (R. P. Vaughan)

slowly, they are quadrupedal. A variety of plant material is eaten, including bulbs, seeds, and leaves. Water balance may be maintained during some seasons by eating succulent vegetation or insects. The spring hare has an unusually low reproductive rate for a rodent. There are but two pectoral mammae, and typically the female bears only one young. Newborn young are large—roughly one third the size of the adult (Coe, 1967; Hediger, 1950)—and well developed and remain in the maternal burrows until they weigh at least half as much as adults.

Family Ctenodactylidae.   Members of this family, commonly called gundis, inhabit arid parts of northern Africa from Senegal, Chad, Niger, and Mali on the west to Somaliland on the east. There are four Recent genera and five Recent species. The earliest known and most primitive rodents are from the early Eocene of China; they are ctenodactylid-like rodents. The fossil record of the Ctenodactylidae begins in the Oligocene of Asia.

These are small, compact, short-tailed rodents with long, soft pelages. The ears are round and short and are protected from wind-blown debris in some species by a fringe of hair around the inner margin of the pinnae. The infraorbital canal is enlarged, and through it passes part of the medial masseter muscle. The skull is flattened, and the auditory bullae and external auditory meatus are enlarged. The cheek teeth are ever-growing, the crown pattern is simple, and the premolars are nonmolariform. The dental formula is 1/1, 0/0, 1-2/1-2, 3/3 = 20-24. The limbs are short; the manus and pes each have four digits.

These herbivorous rodents occur in arid and semiarid areas, where they are restricted to rocky situations. They are diurnal and crepuscular and scurry into jumbles of rock or fissures when threatened.

Family Gliridae. This family includes the dormice, a group (seven genera with 13 Recent species) of squirrel-like Old World rodents known first from the Eocene of France. Glirids are an old and distinctive group that was one of the first to branch off from the primitive and extinct family Ischyromyidae. Dormice are entirely Old World in distribution, occurring in much of Africa south of the Sahara, England, Europe from southern Scandinavia southward, Asia Minor, southwestern Russia, southern India, southern China, and Japan.

Dormice are small (up to 325 millimeters in length), and most genera have bushy or well-furred tails. The skull has a smooth, rounded braincase, a short rostrum, and large orbits. The dental formula is 1/1, 0/0, 1/1, 3/3 = 20. The crowns of the brachyodont molars have parallel cross ridges of enamel, or in some cases the ridges are reduced and the crowns have basins. The infraorbital foramen is somewhat enlarged and transmits part of the medial masseter muscle. The limbs and digits are fairly short, and the sharp claws are used in climbing. The manus has four toes, and the pes has five.

Glirids are typically climbers that occupy trees and shrubs, rock piles, or rock outcrops. These rodents are omnivorous but are able little predators capable of killing small birds and large insects. In temperate areas, the animals are active and breed in spring and summer but hibernate in winter. A unique feature of glirids is their ability to lose and regenerate the tail (Mohr, 1941:63).

Family Seleviniidae. This family is represented by one species, the desert dormouse or dzhalman *(Selvinia betpakdalaensis),* and is restricted to the Betpak-dala Desert of Russia. This small rodent eats primarily insects and spiders and, as is typical of many desert mammals, has greatly enlarged auditory bullae. The dental formula is 1/1, 0/0, 0/0, 3/3 = 16; the cheek teeth are small and short-crowned, and the much-simplified crown pattern features smooth, concave surfaces. This unusual desert rodent may have evolved from dormouse-like (gliroid) ancestors.

Suborder Hystricognathi

Although the African hystricognaths (Bathyergidae, Hystricidae, Petromuridae, Thryonomyidae) and those of South America (termed caviomorphs) share many morphological characters and are regarded by most students of rodents to be closely related, their origins and relationships remain a source of controversy and speculation. The controversy centers in part on the geographic origins of these groups. In the view of A. E. Wood (1980), the Franimorpha of North America are near the base of hystricognath evolution. He holds that early in the Tertiary members of this basal stock immigrated to Eurasia and also moved south to South America. Lavocat (1976, 1980), on the other hand, believes that both African and South American hystricognaths were ultimately derived from a European ancestral stock (Theridomorpha) and that South American hystricognaths had a southern origin by way of rafting from Africa. Because rafting is a chancy means of transport and cannot be documented by fossils, a northern origin for the South American hystricognaths seems more probable.

Regarding the South American hystricognaths, except for the New World porcupines (Erethizontidae), all almost certainly have a common ancestry within an immigrant stock that entered South America in the early Tertiary. Competition from other orders of mammals was apparently not intense, for the caviomorphs rapidly radiated; among fossils from the Early Oligocene (Deseadan), when rodents first appear in South America, eight of the 14 living caviomorph families listed in Table 4–1 can be distinguished. The Miocene climatic changes in South America that were accompanied by an expansion of grasslands strongly affected the fortunes of the caviomorphs. The Octodontidae increased its range in the Pliocene, whereas the range of the Echimyidae shrank. The Dasyproctidae became less common in the Pliocene, and the Chinchillidae became less diverse. From South America, the hutias (Capromyidae) probably reached the Lesser Antilles in the Oligocene by rafting and had varying degrees of success there. The New World porcupines have been extremely successful in moving northward from their ancestral home in South America, are now widespread in the North American tropics, and one species is widely distributed in North American coniferous forests.

Family Bathyergidae. This family contains the African mole rats, a group of highly specialized fossorial rodents. The family includes five genera with approximately nine Recent species. Mole rats occupy much of Africa from Ghana, Sudan, Ethiopia, and Somaliland southward. The earliest fossil records of bathyergids are from the late Oligocene of Asia.

Bathyergids are from 120 to 330 millimeters in total length, and they possess a number of unique structural features associated with their fossorial life. The eyes are small in all species, and vision is apparently poorly developed, but the unusually thick cornea is sensitive to air currents and seemingly allows the animals to detect an opening in the burrow system (G. Eloff, 1958). The ears lack or nearly lack pinnae. The skull is robust, the powerful incisors are procumbent in all species (Fig. 16–9A), and the roots of the upper incisors usually extend behind the molars. The lips close tightly behind the incisors such that dirt does not enter the mouth when the animal is burrowing. The cheek teeth are hypsodont but rooted and typically have a simplified crown pattern. The dental formula is variable (1/1, 0/0, 2-3/2-3, 0-3/

0-3 = 12-28). In *Heliophobius* there are six cheek teeth, but not all are functional simultaneously. The zygomasseteric structure is distinctive. The infraorbital foramen transmits little or no muscle. The masseter muscles, however, are highly specialized: the large anterior part of the medial masseter originates from the upper part of the medial wall of the orbit, and the superficial part of the lateral masseter originates partly on the anterior face of the zygoma. The mandibular fossa and angular part of the dentary bone are greatly enlarged (Fig. 16–9A) and provide an extensive area for the insertion of the masseter muscles.

The limbs are fairly robust in all species but are apparently used for digging only in *Bathyergus.* The hind feet are broad, and the animals back up against a load of soil and push it from the burrow with the hind feet. The tail is short and is used as a tactile organ. The pelage is normal in most species, but in *Heterocephalus glaber,* the skin is nearly naked with only a sparse sprinkling of long hairs (Fig. 16–23).

The African mole rats are basically herbivorous and eat largely bulbs, roots, and rhizomes reached by burrowing. They seldom appear above ground and typically occupy soft loamy or sandy soils in desert and savanna areas. The huge incisors are the major digging tools in all species except *Bathyergus suillus,* which seems to use its feet also. *Heterocephalus* and *Cryptomys* are colonial, and their burrow systems are far more extensive than those of solitary types. The remarkable eusocial behavior of *Heterocephalus* is discussed on p. 416.

Family Hystricidae. To this family belong the Old World porcupines, a widely distributed group of rodents (three genera with 11 Recent species) that resemble the New World porcupines (Erethizontidae) in having quills for protection. Hystricids occur throughout most of Africa, in southern and central Italy, in southern Asia and South China, and in Borneo, southern Celebes, Flores, and the Philippines. They appear first in the Miocene of India.

These large rodents weigh up to 27 kilo-

Figure 16–23. The naked mole rat (*Heterocephalus glaber,* Bathyergidae) of East Africa. The lips close behind the incisors, and a fold of skin guards the nostrils; these adaptations seemingly keep soil from being inhaled and ingested when the incisors are involved in digging. (J. U. M., Jarvis)

grams and have a stocky build. The occipital region of the skull is unusually strongly built and provides attachment for powerful neck muscles. The zygomasseteric arrangement is hystricomorphous, with the large anterior part of the medial masseter originating on the deep rostrum. The nasoturbinal, lacrimal, and frontal bones are highly pneumatic in some species (Fig. 16–24). The dental formula is 1/1, 0/0, 1/1, 3/3 = 20. The hypsodont cheek teeth have re-entrant enamel folds that, with wear, become islands on the occlusal surfaces. Some of the hairs are stiff, sharp spines that reach at least 40 centimeters in length in some species; in some species, open-ended, hollow spines make a noise when rattled that appears to have a warning function. One genus (*Trichys,* of Borneo, the Malay Peninsula, and Sumatra) lacks stiff spines. The large, plantigrade feet of hystricids are five-toed, and the soles are smooth.

Hystricids eat a wide variety of plant material and locally may damage crops. In contrast to New World porcupines, hystricids are terrestrial rather than partly arboreal and dig fairly extensive burrows that are used as dens. The quills are often conspicuously marked with black and white bands, and this visual signal, together with the rattling of the quills, deters some predators. When threatened, hystricids erect their quills and may rush an attacker. Aside from humans, the larger cats are the main predators of hystricids.

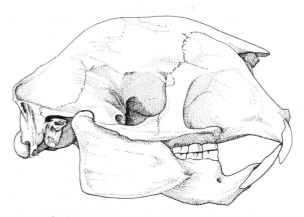

Figure 16–24. Skull of the African crested porcupine (*Hystrix cristata,* Hystricidae). Note the greatly inflated rostrum and frontal part of the skull.

Family Petromuridae. This relict family includes but a single species, *Petromus typicus,* the dassie rat. This animal is restricted to parts of southwestern Africa. The family first appears in the Oligocene of Africa.

The dassie rat is a small rodent with a squirrel-like appearance. The infraorbital foramen is enlarged and transmits part of the medial masseter. The rooted, hypsodont cheek teeth have a simplified crown pattern; the dental formula is 1/1, 0/0, 1/1, 3/3 = 20. Structurally, these animals are most remarkable for specializations enabling them to seek shelter in narrow crevices. Such specializations include a strongly flattened skull, flexible ribs that allow the body to be dorsoventrally flattened without injury, and mammae situated laterally, at the level of the scapulae, where the young can suckle while the female is wedged in a rock crevice.

Dassie rats are diurnal and feed largely on leaves. They are restricted to rocky sections of foothills and mountains, where shelter in the form of rock crevices is available.

Family Thryonomyidae. One genus with but two species makes up this small family, the members of which are known as cane rats. These animals are broadly distributed in Africa south of the Sahara. The earliest records of cane rats are from the Oligocene of Africa. Fossil cane rats, from the Pliocene of Asia and Europe, and the occurrence of an extinct species of the Recent genus *Thryonomys,* from the central Sahara, indicate that the range of cane rats was once far greater than it is now.

Cane rats are large rodents, from 4 to 6 kilograms in weight, with a coarse, grizzled pelage. The snout is blunt, and the ears and tail are short. The robust skull has prominent ridges, a heavily built occipital region, and a large infraorbital foramen. The cheek teeth are hypsodont, and the large upper incisors are marked by three longitudinal grooves. The dental formula is 1/1, 0/0, 1/1, 3/3 = 20. The fifth digit of the forepaw is small, and the claws are strong and adapted to digging.

Cane rats are capable swimmers and divers and are largely restricted to the vicinity of water, where they take shelter in matted vegetation or in burrows. Males indulge in ritualized snout-to-snout pushing contests, and the blunt shape of the snout seems to enable the animals to avoid damage during these bouts. They are herbivorous and do considerable local damage to crops, particularly sugar cane. Cane rats are prized for food in many parts of Africa. The animals are often taken during organized drives using dogs or are driven from their hiding places and captured when hunters set fire to reeds.

Family Erethizontidae. To this small family belong the New World porcupines, a group including four Recent genera with ten living species. These animals are widely distributed in forested areas and occur from the Arctic Ocean south through much of the forested part of the United States into Sonora, Mexico, in the case of *Erethizon,* and from southern Mexico through much of the northern half of South America in the case of the other genera. Porcupines are of interest to most people because of their remarkable coat of quills and because the animals often have little fear of humans and can be observed easily.

New World porcupines are large, heavily built rodents, weighing up to 16 kilograms; all species have quills on at least part of the body. The stiff quills are usually conspicuously marked by dark- and light-colored bands, and the sharp tips have small, proximally directed barbs. These barbs make the quills difficult to remove from flesh and aid in their penetration, which may be at the rate of 1 millimeter or more per hour. The skull is robust, the rostrum is deep, and the greatly enlarged infraorbital foramen is nearly circular in some species (Fig. 16–25B) and accommodates the highly developed medial masseter. The dental formula is 1/1, 0/0, 1/1, 3/3 = 20; the rooted cheek teeth have occlusal patterns dominated by re-entrant enamel folds (Fig. 16–26A). New World porcupines have some arboreal adaptations that are lacking in their more terrestrial Old World counterparts. The feet of erethizontids have broad soles marked by a pattern of tubercles

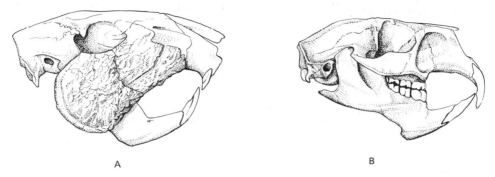

Figure 16–25. Skulls of two hystricomorph rodents. (A) *Agouti paca,* Agoutidae; length of skull 150 mm. Note the great enlargement of the zygomatic arch. (B) Porcupine (*Erethizon dorsatum,* Erethizontidae); length of skull 115 mm. (A after E. R. Hall and Kelson, 1959)

that increase traction (Fig. 16–11B); in some species, the hallux is replaced by a large, movable pad. The toes bear long, curved claws, and the limbs are functionally four-toed. In *Coendou,* the long tail is prehensile and curls dorsally to grasp a branch.

Beginning in the Oligocene, this family underwent its early evolution in South America, becoming established in North America only after the emergence of the previously inundated Isthmus of Panama in the Pliocene.

New World porcupines eat a variety of plant material. *Erethizon* feeds extensively on the cambium layer of conifers, and in many timberline areas, trees missing large sections of bark and cambium give evidence of long-term occupancy by porcupines. Cambium is a staple winter food in the Rocky Mountains, but in summer a variety of plants are eaten. Porcupines often "graze" at this time on sedges *(Carex)* along the borders of meadows. Most species are able climbers, and *Coendou* spends most of its life in trees. *Erethizon* is inoffensive and at times almost oblivious to humans, but when in danger the animal directs its long dorsal hairs forward, exposing the quills; it erects the quills and humps its back. The tail is flailed against an attacker as a last resort. Surprisingly, *Erethizon* is killed by a variety of carnivores. Some mountain lions learn to flip porcupines on their backs and kill them by attacking the unprotected belly. Occasionally, however, dead or dying carnivores are found with masses of quills penetrating the mouth and face, indicating that learning to prey on porcupines may be a fatal undertaking. Erethizontids characteristically take shelter in rock piles, beneath overhanging rocks, or in hollow logs but do not dig burrows as do Old World porcupines.

Family Chinchillidae. One member of this family, the chinchilla *(Chinchilla),* is somewhat familiar to many because of the publicity given to chinchilla fur farming. The family also includes the viscachas (*Lagidium* and *Lagostomus*). Three genera with six Recent species represent the family, which occurs in roughly

Figure 16-26. Crowns of hystricomorph molars. (A) Upper right molars one and two of the porcupine (*Erethizon dorsatum,* Erethizontidae); (B) third lower left molar of the capybara (*Hydrochoerus hydrochoeris,* Hydrochoeridae). The cross-hatched areas on the porcupine teeth are dentine; the stippled areas on the capybara tooth are cement. (B after Ellerman, 1940)

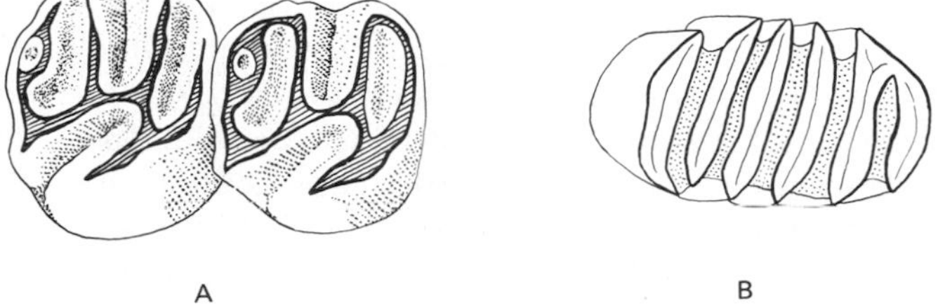

A   B

the southern half of South America in the high country of Peru and Bolivia and throughout much of Argentina to near its southern tip. The fossil record of this group is entirely South American and extends from the Oligocene to the Recent.

Chinchillids are densely furred and of moderately large size (1 to 9 kilograms), with a long, well-furred tail. Mountain viscachas (*Lagidium*) and chinchillas have fairly large ears and a somewhat rabbit-like appearance, whereas the plains viscacha (*Lagostomus*) has short ears. The cheek teeth are ever-growing and the occlusal surfaces are formed by transverse enamel laminae with intervening cement. The dental formula is 1/1, 0/0, 1/1, 3/3. There are some cursorial adaptations, but the clavicle is retained. The forelimbs are fairly short and tetradactyl; the hind limbs are long, however, and the elongate feet have four (*Chinchilla* and *Lagidium*) or three (*Lagostomus*) toes.

Chinchillids are herbivorous and occupy a variety of situations, including open plains (pampas), brushlands, and barren, rocky slopes at elevations ranging from 800 to 6000 meters. The mountain viscachas and chinchillas are diurnal and seek shelter in burrows or rock crevices. Although adept at moving rapidly over rocks and broken terrain, they seem not to depend on speed in the open to escape enemies. The plains viscacha (*Lagostomus*), in contrast, occurs in open pampas areas with little cover, where colonies live in extensive burrow systems marked by low mounds of earth and accumulations of such debris as bones, livestock droppings, and plant fragments. The habit of collecting items is displayed even by captive animals. Colonies may occupy large areas; one such area measured 20 by 300 meters, and this colony was known to have been in existence for at least 70 years (Weir, 1974). Cursorial ability is highly developed: these animals are able to make long leaps and to evade a pursuer by abrupt turns. They have considerable endurance and can run at speeds up to at least 40 kilometers per hour. Plains viscachas occupy a rabbit-

less area and in locomotor style resemble jackrabbits.

Family Dinomyidae. This family includes a single South American species, *Dinomys branickii,* the pacarana. This seemingly rare animal inhabits the foothills of the Andes and adjacent remote valleys in Peru, Colombia, Ecuador, and Bolivia. An extinct member of this family reached spectacular size: the Pliocene dinomyid *Telicomys* was the size of a rhinoceros. Although dinomyids appear to be near extinction today, they may have been more successful and diverse in the past. Fields (1957:359) assigned eight fossil genera to the family Dinomyidae, the oldest of which appears in the Oligocene of South America.

The pacarana weighs up to 15 kilograms; the dark brown pelage is marked by longitudinal white stripes and spots. Pacaranas lack the cursorial adaptations of the Caviidae and Hydrochoeridae. Instead, the broad tetradactyl feet of pacaranas have long, stout claws seemingly adapted to digging, and the foot posture is plantigrade. The clavicle is complete, another departure from the conventional cursorial plan. The unusually hypsodont cheek teeth consist of a series of transverse plates. The dental formula is 1/1, 0/0, 1/1, 3/3.

These unusual rodents feed on a variety of plant material and are slow-moving and docile in captivity.

Family Caviidae. This fairly small family, including just five genera and 14 species, contains the familiar guinea pig (*Cavia*), as well as several similar types, and the Patagonian "hare" (*Dolichotis*), an animal remarkable in having many cursorial adaptations. Caviids occur nearly throughout South America, except in Chile and parts of eastern Brazil, and first appeared in the Miocene of South America.

The guinea pig-like caviids (subfamily Caviinae) are chunky and moderately short-limbed, and weigh from 400 to 700 grams. *Dolichotis* (Dolichotinae), in contrast, has an antelope-like form, with long, slender legs and feet, and weighs up to approximately 16 kilograms. All caviids have ever-growing cheek

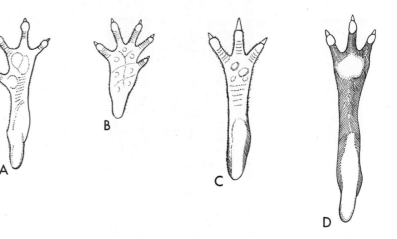

Figure 16-27. Ventral views of the left hind foot of some South American rodents. (A) Chinchilla (*Chinchilla* sp., Chincillidae); (B) degu (*Octodon* sp., Octodontidae); (C) *Dasyprocta* sp., (Dasyproctidae); (D) Patagonian cavy (*Dolichotis* sp., Caviidae). (After A. B. Howell, 1944)

teeth with occlusal patterns consisting basically of two prisms. The dental formula is 1/1, 0/0, 1/1, 3/3. The dentary bone has a conspicuous lateral groove into which insert the temporal muscle and the anterior part of the medial masseter. Although only *Dolichotis* is strongly cursorial, all caviids have certain features typical of cursorial mammals: the clavicle is vestigial, the tibia and fibula are partly fused, and the digits are reduced to four on the manus and three on the pes. Members of the subfamily Caviinae, despite these cursorial adaptations, have a plantigrade foot posture and scuttle about in mouse-like fashion. Locomotion in *Dolichotis,* however, is rapid and involves long bounds. The foot posture of *Dolichotis* during running is digitigrade, and specialized pads beneath the digits (Fig. 16-27D) cushion the impact when the feet strike the ground. *Dolichotis* resembles a rabbit because of its deep, somewhat laterally compressed skull and large ears.

Caviids are herbivorous, and some have complex systems of social behavior and vocal communication (Lacher, 1979; Rood, 1970, 1972). They occupy habitats ranging from grassland and open pampas to brushy and rocky areas and forest edges. Most are nocturnal or crepuscular and often live in large colonies. *Dolichotis,* an inhabitant of open, arid regions, is diurnal, and large groups have been observed on occasion. In some ways, *Dolichotis* is remarkably close in form and escape behavior to

some of the small African antelope (Smythe, 1970).

Family Hydrochoeridae. This family contains the largest living rodent, the capybara *(Hydrochoerus).* The single species occupies Panama and roughly the northern half of South America east of the Andes. The capybara is known from the Pliocene and Pleistocene in North America and the Pliocene to Recent in South America. An extinct Pliocene type *(Protohydrochoerus)* probably weighed over 200 kilograms.

Capybaras are large (up to 50 kilograms in weight), robust, rather short-limbed rodents with a coarse pelage. The head is large and has a deep rostrum and truncate snout. The skull and dentary bone are similar to those of members of the Caviidae, but the paroccipital processes are unusually long. The teeth are ever-growing. Both upper and lower third molars are much larger than any other cheek tooth in their respective rows, and are formed by transverse lamellae united by cement (Fig. 16-26B). The dental formula is 1/1, 0/0, 1/1, 3/3. The tail is vestigial, and the same cursorial features listed for the caviids occur. The digits are partly webbed and unusually strongly built, an adaptation that probably allows the support of the considerable body weight.

Capybaras occur along the borders of marshes or the banks of streams and forage on succulent herbage. They are largely crepuscu-

Figure 16-28. *Dasyprocta* sp., Dasyproctidae. (L. G. Ingles)

lar, and although they can run fairly rapidly, usually seek shelter in the water. They swim and dive well and can remain submerged beneath water plants with only the nostrils above water. In both form and function, a capybara resembles a miniature hippopotamus.

Family Dasyproctidae.   Members of this family, the agoutis (Fig. 16–28) and their relatives, occur in the Neotropics from southern Mexico south to Ecuador, Bolivia, Paraguay, and northeastern Argentina. Two genera and 15 species are included in this family. The earliest records are from the Oligocene of South America.

These rodents are fairly large, up to 2 kilograms in weight. The tail ranges from short to long. The skull is robust, the incisors are fairly thin, and the crowns of the hypsodont cheek teeth are flat and bear five crests. The dental formula is 1/1, 0/0, 1/1, 3/3 = 20. Although these rodents are compactly built, the limbs are slim and have many cursorial adaptations. The forefeet are tetradactyl, and the plane of symmetry passes between digits three and four (as in the Artiodactyla). The hind feet have three toes, and the plane of symmetry passes through digit three (Fig. 16–27C), as in the Perissodactyla. The clavicle is vestigial, and the claws are sharp and hooflike.

These rodents are herbivorous and typically inhabit tropical forests, where they are largely diurnal. Some species take refuge in burrows that they dig in the banks of arroyos, beneath roots, or among boulders. They are rapid and agile runners and usually travel along well-worn trails. *Dasyprocta punctata* is territorial and scatter-hoards fruit and nuts in times of plenty to be used during lean times. Agoutis have a habit of remaining still when approached by a predator and then bursting from cover and running away after the fashion of a small antelope.
Family Agoutidae.   This family includes two species of a single genus *(Agouti)*. These rodents live in tropical forests from central Mexico to southern Brazil and are known only from the Recent.

These rodents, often called pacas, are large, weighing up to about 12 kilograms, and nearly tail-less and have a conspicuous pattern of white spots and stripes on the body. They have an exceptionally ungraceful form, with short legs and a blunt head (Fig. 16–29). There are four digits on the forefeet and five digits on the hind feet. The cheek teeth are high-crowned, and the dental formula is 1/1, 0/0, 1/1, 3/3 = 20. Resonating chambers are formed by concavities in the maxillaries and by greatly broadened zygomatic arches (Fig. 16–25A); air is forced through associated pouches, producing a resonant, rumbling sound. The massively enlarged zygoma are unique to these animals.

These terrestrial rodents live in tropical forests along streams and rivers, where they dig burrows in banks. The diet consists of a variety of plant material, including fallen fruit. They are nocturnal and not particularly swift on land; they are good swimmers, however, and often escape enemies in the water. Because pacas are highly prized as food by humans, they are rare in many areas. These rodents seldom bear more than one young.
Family Ctenomyidae.   Members of this family, called tuco-tucos, are fossorial and resemble pocket gophers (Geomyidae). They occupy much of the southern two thirds of South America, from Peru to Tierra del Fuego. They inhabit

Figure 16–29. *Agouti paca*, Agoutidae. (L. G. Ingles)

the Andes Mountains to elevations of 4000 meters. There is a single genus *(Ctenomys)* with 32 species. Ctenomyids first appear in the middle Pliocene of South America.

These rodents range in size from 100 to 700 grams and are unusual in having simplified cheek teeth that are roughly kidney-shaped; the dental formula is 1/1, 0/0, 1/1, 3/3 = 20, and the third molar is vestigial. The skull is broad and dorsoventrally flattened, and the robust incisors are pigmented (orange). Among South American hystricognathous rodents, only in the Ctenomyidae and in one species of the Octodontidae *(Spalocopus cyanus)* are fossorial adaptations strongly developed. The head of the tuco-tuco is large and broad, and the stout incisors protrude permanently from the lips. The eyes and ears are small, the neck is short and powerfully built, the forelimbs are powerful, the manus has long claws, and the tail is short and stout. In contrast to pocket gophers, tuco-tucos have greatly enlarged hind feet with powerful claws, and they lack external cheek pouches. Fringes of hair on the toes of the fore and hind feet in tuco-tucos are presumably an aid to the animals when they are moving soil.

Tuco-tucos are herbivorous and eat such underground parts of plants as roots, tubers, and rhizomes. They dig extensive burrow systems in open, often barren areas and live in colonies composed of many solitary individuals, each with its burrow systems spaced widely apart from those of its neighbors. An animal typically occupies a given burrow system permanently but periodically seeks adjacent foraging areas by digging new burrows. Tuco-tucos occasionally make short forays from their burrows to gather leaves and stems (Pearson, 1959). In contrast to pocket gophers, tuco-tucos are quite vocal and give distinctive cries from burrow entrances.

Family Octodontidae. These small rodents, although mostly burrow dwellers, are ratlike in general appearance, and most species lack the fossorial specializations typical of the Ctenomyidae. Octodontids, variously called degus, cururos, or rock rats, have a restricted range near the western coast of South America from southwestern Peru south to northern Argentina and northern Chile. There are five genera and eight species. The earliest fossil records are from the Oligocene of South America.

These small rodents (200 to 300 grams) derive their family name from the "eight-shaped" crown pattern of the cheek teeth; the dental formula is 1/1, 0/0, 1/1, 3/3. Most species have large ears, large eyes, long vibrissae, and the familiar form of a small rat. The forefeet have four digits, and the hind feet have five; the tail varies from long to rather short.

Octodontids occupy a variety of habitats, from grassy areas to high Andean forests to dry cactus and acacia slopes. *Octodon* is an able climber and takes shelter in rocks or the burrows of other animals, and *Octodontomys* lives in burrows and in rock crevices or caves and feeds on acacia pods and cactus. Neither of these genera is specialized for fossorial life, nor is *Octomys*. Of the remaining two genera,

Aconaemys is somewhat modified for fossorial life and *Spalacopus* is strongly so. In Chili, *Spalacopus cyanus* occupies sandy coastal areas where it occurs in colonies, all members of which occupy a common burrow system. The animals feed entirely below ground, and the tubers and underground stems of huilli, a species of lily *(Leucoryne ixiodes)*, form the bulk of the diet. *Spalacopus* is nomadic, an exceptional mode of life for a rodent. When a colony exhausts the supply of huilli roots at one place, the animals abandon this foraging site and move to a nearby undisturbed area (Reig, 1970). *Spalacopus* is unusually vocal for a rodent and gives distinctive calls at burrow openings. It uses its forelimbs and teeth to loosen soil and its large hind feet to throw dirt from the mouth of the burrow. The ranges of the tuco-tucos (Ctenomyidae) and the similarly adapted *Spalacopus* do not overlap.

Family Abrocomidae.   Members of this family, the chinchilla rats, occur in parts of west central South America. Their range includes southern Peru, Bolivia, and northwestern Argentina and Chile. The family is represented today by one genus with two species. This family appears first in the South American Miocene.

Chinchilla rats look roughly like large woodrats *(Neotoma)*, reach over 400 millimeters in total length, and resemble octodontids in many ways. The pelage is long and dense. The skull has a long, narrow rostrum, and the bullae are enlarged. The cheek teeth are ever-growing; the upper teeth have an internal and an external enamel fold, while the lowers have two internal folds. The dental formula is 1/1, 0/0, 1/1, 3/3. The limbs are short and have short, weak nails. The pollex is absent.

These herbivorous rodents are poorly known. They are seemingly colonial, climb well, and usually seek shelter beneath or among rocks. They live in cold, bleak, rocky areas in the Andes at elevations between 3500 and 5000 meters.

Family Echimyidae.   Members of this important Neotropical family, which includes a variety of roughly rat-size rodents, are called spiny rats. Most of the living species have flattened, spinelike hairs with sharp points and slender basal portions. Approximately 54 living species of 19 genera are recognized. Spiny rats are widely distributed in the Neotropics, occurring from Nicaragua southward through the northern half of South America to Paraguay and southeastern Brazil.

Echimyids are normally proportioned rodents with prominent eyes and ears. The tail, which in some genera is longer than the head and body, is lost readily, a feature perhaps of value in aiding escape from predators. The point of weakness is at the centrum of the fifth caudal vertebra. Among 637 *Proechimys* taken in Panama, 18 percent were tail-less (Fleming, 1970:486). The cheek teeth are rooted, and the occlusal surfaces in most species are marked by transverse re-entrant folds. The dental formula is 1/1, 0/0, 1/1, 3/3. The feet are not highly specialized in most genera. In several arboreal genera, however, the digits are elongate and partially syndactylous. When an animal is climbing, the first two digits grasp one side of a branch in opposition to the remaining digits, which grasp the other side.

Echimyids are an old group, appearing first in the early Oligocene of South America. Two extinct genera are known from skeletal material found in Indian kitchen middens in Cuba and Haiti. These genera seemingly became extinct fairly recently. In the case of the genus from Haiti *(Brotomys)*, extinction may have resulted from the introduction of predators by Europeans.

As far as is known, spiny rats are completely herbivorous. In Panama, fruit was the primary food found in the stomachs of many *Proechimys semispinosus* (Fleming, 1970:486). Many species are at least partly arboreal. The bamboo rat *(Dactylomys dactylinus)* of South America is completely nocturnal and arboreal, eats leaves and buds, and gives explosive calls that presumably play a role in territoriality (Emmons, 1981).

Figure 16–30.   Bahamian hutias (*Geocapromys ingrahami*, Capromyidae). (G. C. Clough)

Family Capromyidae.   Members of this family are known locally as hutias, zagouties, cavies, or coneys and are restricted to the West Indies, where the living species (11) occupy the Bahama Islands, Cuba, Isle of Pines, Hispaniola, Puerto Rico, and Jamaica. The only fossils are those of recently extinct species. These mostly herbivorous rodents weigh up to about 7 kilograms and look like unusually large rats (Fig. 16–30). They closely resemble the nutria (Myocastoridae) structurally and are often included in the same family.

Capromyids are of little importance today except as interesting and, to some people, alarming examples of a group on its way toward extinction at the hands of humans. These rodents, adapted to the insular conditions of the West Indies before the coming of Europeans, were unable to cope with predation by the introduced mongoose *(Herpestes)* or by humans and their dogs. Of the 35 Recent species of capromyids, 24 are extinct and the remaining ones are restricted to steep or inaccessible areas. One living species *(Capromys nana)* was first described from bones found in a cave but was later found alive.

Family Myocastoridae.   The nutria *(Myocastor coypu),* the only living member of this family, is familiar to many people in North America, Europe, and Asia because this South American rodent has been introduced widely and has thrived in certain areas. In some places it has become a serious pest because of its destruction of aquatic vegetation and crops and its disruption of irrigation systems, and in others it has caused a deterioration of waterfowl habitat. Costly federal study and local control of the nutria has become necessary in some parts of the United States. This animal is native to southern South America, from Paraguay and southern Brazil southward, but now also occurs in some 15 states in the United States as well as many countries in Europe. The family is also represented by nine extinct genera ranging from the Oligocene to the Recent in South America.

The nutria is large, up to roughly 8 kilograms, and looks like a rat-tail beaver *(Castor).* The skull is heavily ridged and has a deep rostrum. The zygomasseteric structure is hystricomorphous; in association with the reduction of the temporal muscles, the coronoid process of the dentary bone has nearly disappeared and is

Figure 16-31. First and second upper right molars of the nutria (*Myocastor coypu,* Myocastoridae). Note the tremendous changes in the crown pattern due to wear: (A) lightly worn molars; (B) heavily worn molars. Stippled areas on the occlusal surfaces surrounded by enamel (unshaded) are dentine.

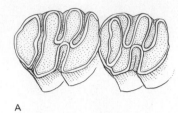

A

B

represented by a small knob (Fig. 16–4). The hypsodont cheek teeth well illustrate changes in crown pattern that occur with increasing age and wear (Fig. 16–31). The dental formula is 1/1, 0/0, 1/1, 3/3. The feet have heavy claws, and a web joins all but the fifth toe of the pes.

Nutrias resemble beavers in some of their habits. They dig burrows in banks, use cleared trails through vegetation, are extremely destructive to plants near their dens, and are skillful swimmers and divers. They have dense, fine underfur, have been raised in some fur farms in the United States, and are trapped for their fur in some states. In the trapping season of 1975–76, the nutria sold to the fur trade in Louisiana yielded $8 million. Most biologists strongly oppose the indiscriminate introductions of such non-native animals as the nutria. The activities of these species occasionally result in the alteration of the vegetation, with the resultant disappearance of native species and the destruction, perhaps irretrievably, of the original biotic community.

Family Heptaxodontidae. This family includes several kinds of West Indian mammals (seven genera and eight species) that became extinct in Recent or sub-Recent times and are known from fragmentary skeletal material from caves or kitchen middens. Heptaxodontids are recorded from Puerto Rico, Jamaica, Hispaniola, Anguilla, and the St. Martins Islands. These were large rodents; the length of the skull of the largest genus *(Amblyrhiza)* suggests an animal approaching the size of an American black bear. The skull is robust, with strongly developed ridges for muscle attachments, and the cheek teeth have four to seven laminae oblique to the long axis of the anteriorly converging tooth row. These rodents were probably terrestrial and herbivorous, and were eaten by humans. Nothing else is known of their biology.

Chapter 17

Order Lagomorpha

Although lagomorphs—the rabbits (Leporidae) and pikas (Ochotonidae)—are not a diverse group, including but ten genera with 63 Recent species, they are important members of many terrestrial communities and are nearly worldwide in distribution. Considering large land masses, lagomorphs were absent only from Antarctica, the Australian region, and southern South America before recent introduction by humans. Lagomorphs occupy diverse terrestrial habitats from the Arctic to the tropics, and in some temperate and boreal regions, rabbits are subject to striking population cycles marked by periods of great abundance alternating with times of extreme scarcity. In such regions, the population cycles of many carnivores are influenced strongly by changes in rabbit population densities.

Many important diagnostic features of Recent lagomorphs are related to their herbivorous habits and, in the case of leporids, to their cursorial locomotion. Lagomorphs have a fenestrated skull, a feature highly developed in some leporids (Fig. 17–1A). The anterior dentition resembles that of a rodent, but whereas rodents have 1/1 incisors, rabbits have 2/1 incisors; the second incisor is small and peglike, and lies immediately posterior to the first (Fig. 17–1A). As in rodents, the lagomorph incisors are evergrowing. A long postincisor diastema is present in lagomorphs, and the canines are absent. The cheek teeth are hypsodont and rootless, and the crown pattern features transverse ridges and basins (Figs. 17–2A and 17–3B). The distance between the upper tooth rows is greater than that between the lower rows, accentuating occlusion of upper and lower cheek teeth only on one side at a time and requiring a lateral or oblique jaw action. The masseter muscle is large, and the pterygoideus muscles are well de-

A　　　　　　　　　　　　B

Figure 17-1.　(A) Skull of the antelope jackrabbit *(Lepus alleni);* note the highly fenestrated maxillary and occipital bones. (B) Anterior part of the skull of the arctic hare *(L. arcticus);* note the procumbent incisors and the receding nasals, specializations associated with this animal's habit of using the incisors to scrape away ice and snow to reach food. (B after E. R. Hall and Kelson, 1959)

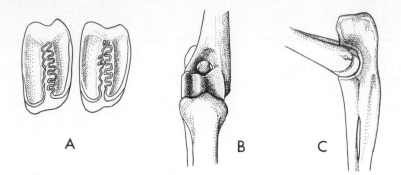

A   B   C

Figure 17-2. Leporid structural features. (The drawings are of the antelope jackrabbit, *Lepus alleni*.) (A) Occulusal view of upper right premolars three and four. (B) Anterior view of right elbow joint; movement is limited to a single (anteroposterior) plane by this "tongue and groove" articulation. When the forearm is fully extended, a process on the olecranon of the ulna locks into the conspicuous hole in the humerus and braces the joint. (C) Medial view of right elbow, showing the tight fit between the articular surface of the humerus and those of the radius and ulna. The radius and ulna are partially fused.

veloped and help control transverse jaw movements; the temporalis is small, and the coronoid process, its point of insertion, is rudimentary.

The skull of leporids is unique among mammals in having a clearly defined joint at which slight movement occurs. This joint fully encircles the skull just anterior to the occipital and otic bones. This unusual cranial specialization appears first in Miocene leporids and is a mechanism that absorbs shock to the skull while the animal is bounding at high speeds (Bramble, 1982). The clavicle is either well developed (Ochotonidae) or rudimentary (Leporidae), and the elbow joint limits movement to a single anteroposterior plane (Fig. 17–2B, C). The tibia and fibula are fused distally; the front foot has five digits, and the hind foot has four or five digits. The soles of the feet, except for the distalmost toe pads in *Ochotona,* are covered with hair. The foot posture is digitigrade during running but plantigrade during slow movement. The tail is short and in *Ochotona* is not externally evident.

The first fossil record of mammals with lagomorph-like characters is from the Paleocene of China. Although rodents and rabbits were long regarded as unrelated, recent evidence indicates that these groups share a common ancestry within the Paleocene order Anagalida (Li, 1977). The family Leporidae probably originated in Asia but underwent most of its early (Oligocene and Miocene) evolution in North America. Leporids became well established in the Old World in the Pliocene, and the advanced subfamily Leporinae arose there. The pikas appeared first in the Oligocene of Eurasia and spread in the Pliocene to Europe and North America. The Recent genus *Ochotona* is known from the late Miocene. In contrast to the leporids, which have remained widespread since the Pliocene, the ochotonids reached their greatest diversity and widest distribution in the Miocene, when they occupied Europe, Asia, Africa, and North America (M. R. Dawson, 1967b:305), and have declined since. In North America, pikas are now of local occurrence in high mountains north of Mexico. They occur more widely and are more diverse in the Old World, where they inhabit eastern Europe and much of northern and central Asia.

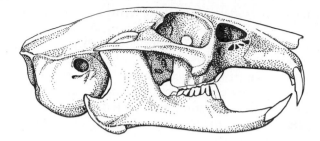

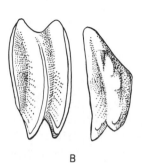

Figure 17-3. (A) Skull and (B) occlusal view of upper right premolars three and four of the pika *(Ochotona princeps).*

The environmental tolerances, or perhaps the competitive success, of ochotonids has seemingly changed in the New World since the Miocene. For example, in the Miocene an ochotonid occupied riparian communities in the Great Plains, Rocky Mountains, and Great Basin regions of North America (R. W. Wilson, 1960:7); riparian situations in these areas no longer support pikas. The factors influencing the striking post-Miocene decline of the ochotonids are unknown.

Why, although they are an old and thriving group, have lagomorphs not undergone a greater adaptive radiation? Perhaps their conservatism is related to the limitations of their functional position as "miniature ungulates." Competition with members of the larger and more diverse order Artiodactyla, a group highly adapted to a herbivorous diet and cursorial locomotion, may have limited lagomorphs to the exploitation of but a single limited adaptive

zone, although this zone was occupied with great success over broad areas. Of interest in this regard are the scarcity and local occurrence of lagomorphs in many parts of East Africa, where there is an extremely rich ungulate fauna.

Family Ochotonidae. The pikas are represented today by one genus with roughly 14 species. Pikas are less progressive with regard to cursorial adaptations than are rabbits and usually venture only short distances from shelter. Pikas occur in the mountains of the western United States and south central Alaska, and over a wide area in the Old World, including eastern Europe and much of Asia south to northern Iran, Pakistan, India, and Burma.

Pikas are smaller than rabbits, weighing about 100 to 150 grams. They have short, rounded ears, short limbs, and no externally visible tail (Fig. 17–4). The ear opening is guarded by large valvular flaps of skin that may protect it during severe weather. The skull is strongly constricted between the orbits and

Figure 17-4. A pika *(Ochotona princeps)* at a lookout point on a rock, with a mouthful of plant material for its hay pile. (O. D. Markham)

lacks a supraorbital process; the rostrum is short and narrow. The skull is less strongly arched in ochotonids than in leporids (Fig. 17–3A), and the angle between the basicranial and palatal axes is lower. The maxilla has a large fenestra. The dental formula is 2/1, 0/0, 3/2, 2/3 = 26. The third lower premolar has more than one re-entrant angle, and the re-entrant enamel ridges of the upper cheek teeth are straight (Fig. 17–3B). The anal and genital openings are enclosed by a common sphincter, and males have no scrotum.

In North America, pikas usually inhabit talus slopes in boreal or alpine situations and occur from near sea level in Alaska to the treeless tops of some of the highest peaks in the Rocky Mountains and Sierra Nevada–Cascade chain. When frightened, pikas seek shelter in the labyrinth of spaces and crevices between rocks and seldom forage far from such shelter. Large "hay piles" are built each summer in the shelter of large, usually flat-bottomed boulders and provide food when snow covers the ground. In Eurasia, pikas occupy an extensive geographic range and a wide variety of habitats, including talus, forests, rock-strewn terrain, and open plains and desert-steppe areas. Unusually large hay piles, weighing up to 20 kilograms, are made by pikas inhabiting dry areas in southern Russia (Formozov, 1966).

Family Leporidae.   The rabbits are a remarkably successful group in terms of ability to occupy a variety of environments over broad areas and are now nearly cosmopolitan. Their distribution before introduction by humans included most of the New and Old Worlds, and rabbits have been introduced into New Zealand, Australia, parts of southern South America, and various oceanic islands in both the Atlantic and the Pacific Ocean. Nine Recent genera represented by 49 Recent species are known.

Several major leporid evolutionary trends in structure are recognized by M. R. Dawson (1958:6). The cheek teeth have become hypsodont, some of the premolars have become mo-lariform, and the primitive crown pattern has been modified into a simple arrangement in which most traces of the primitive cusp pattern have been lost. These changes resemble those in some groups of strictly herbivorous rodents. The skull has become arched, and the angle between the basicranial and palatal axes has increased. The changes are associated with a posture involving a greater angle between the long axis of the skull and the cervical vertebrae than that typical of primitive leporids. Trends in limb structure leading to increased cursorial ability include elongation of the limbs and articulation specializations that limit movement to one plane.

The leporid skull (Fig. 17–1A) is more or less arched in profile, and the rostral portion is fairly broad. The maxillae, and often the squamosal, occipital, and parietal bones are highly fenestrated, and a prominent supraorbital process is always present. The auditory bullae are globular, and the external auditory meatus is tubular. The dental formula is usually 2/1, 0/0, 3/2, 3/3 = 28; the re-entrant enamel ridges of the upper cheek teeth are usually crenulated (Fig. 17–2A). The clavicle is rudimentary and does not serve as a brace between the scapula and the sternum. The limbs, especially the hind limbs, are more or less elongate; movement at the elbow joint is limited to the anteroposterior plane (Fig. 17–2B, C). The tail is short. The ears have a characteristic shape: the proximal part is tubular and the lower part of the opening is well above the skull (Fig. 17–5). The testes become scrotal during the mating season. In some species that inhabit regions with snowy winters, the animals molt into a white winter pelage in the fall (Fig. 17–5) and into a brown summer pelage in the spring. Wild leporids weigh from 0.3 to 5 kilograms.

Leporids inhabit a tremendous array of habitats, from arctic tundra and treeless and barren situations on high mountain peaks to coniferous, deciduous, and tropical forests, open grassland, savanna, and deserts. Some species, such

Figure 17–5. A white-tail jackrabbit *(Lepus townsendii)* in its partially white winter pelage. This particular animal is from Colorado, but in more northerly parts of its range, *L. townsendii* is almost entirely white in winter. (G. D. Bear)

as *Sylvilagus palustris* and *S. aquaticus* of the southeastern United States, are excellent swimmers and lead semiaquatic lives. Leporids are entirely herbivorous and eat a wide variety of grasses, forbs, and shrubs. Several species are known to reingest fecal pellets and are thought to obtain essential nutrients (proteins and some vitamins) from material as it passes through the alimentary canal a second time.

Habitat preference and cursorial ability differ markedly from species to species and are strongly interrelated. Broadly speaking, species with relatively poor cursorial ability, such as *Brachylagus idahoensis* and *S. bachmani* of the western United States, scamper short distances to the safety of burrows or dense vegetation when disturbed and typically occur in stands of big sagebrush *(Artemesia tridentata)* or dense chaparral, respectively. Cottontails,

such as *S. floridanus* of the eastern and *S. audubonii* of the western United States, are intermediate in cursorial ability and typically inhabit areas with scattered brush, rocks, or other cover and do not run long distances to reach a hiding place. Representing the extreme in cursorial specialization among lagomorphs are some members of the genus *Lepus,* such as the New World jackrabbits *(L. californicus, L. townsendii,* and *L. alleni* and their relatives) and some hares of the Old World (such as *L. capensis* of Africa). These animals, which have greatly elongate hind limbs, have adopted a bounding gait and occupy areas with limited shelter, such as deserts, grasslands, or meadows, where they take shelter in "forms" (Fig. 17–6). Instead of taking cover at the approach of danger, they depend for escape on their running ability. (Adaptations that contribute to running ability are

Figure 17-6. A white-tail jackrabbit *(Lepus townsendii)* in its "form," a hollowed-out hiding place beneath a bush or other (often scanty) shelter. This animal is in its brown summer pelage. (G. B. Bear)

discussed on p. 184). Jackrabbits and other similarly adapted members of the genus *Lepus* are extremely rapid runners for their size; some attain speeds of up to 70 kilometers per hour. This speed allows them to occupy open areas with little cover, where safety depends upon outrunning predators. The arctic hare *(L. arcticus)* of the North American Arctic often uses bipedal locomotion and can stand and jump using only its hind legs.

Although rabbits are seemingly peaceful and nonaggressive animals, they are strong competitors and remarkably adaptable. In some parts of Australia, the extinction or near extinction of certain marsupials is perhaps due primarily to competition with introduced European rabbits *(Oryctolagus)*. In addition, these prolific rabbits have caused great damage to crops and rangeland and at various times have been a primary agricultural pest in many parts of Australia as well as in New Zealand, where they were also introduced. The range of environmental conditions to which leporids have adapted is tremendous. Populations of *Lepus arcticus* along the arctic coasts of Greenland use their protruding incisors (Fig. 17–1B) to scrape through snow and ice to reach plants during the long arctic winters, whereas far to the south, in the deserts of northern Mexico, jackrabbits *(L. alleni)* maintain their water balance through hot, dry periods by eating cactus and yucca.

Chapter 18

Ecology

One of the most remarkable attributes of humans is our ability to recognize relationships between disparate phenomena or events—to discover order, symmetry, predictability, and beauty in an apparently disorderly world. The study of ecology demands such ability, for of central interest to ecologists is an understanding of the often complex relationships between living things. Kendeigh (1961:1) describes ecology as "a study of animals and plants in their relations to each other and to their environment." Early ecological work featured descriptive field studies, and later investigations involved controlled field or laboratory experiments. These provide a base of knowledge on which the modern theoretical ecologists depend.

Our interest in ecology is not new. Early peoples understood many relationships between the major game animals and their environments. These peoples could predict where and when certain species could be found and used this knowledge to increase hunting success and hence survival. Many thousands of years later, we are belatedly understanding that our very survival may depend on an appreciation of basic ecological principles.

It is difficult to overemphasize the value of an ecological approach to the study of mammals. Knowledge of the biology of any mammalian species is clearly incomplete if the relations of the animal to its environment are unknown. An understanding of mammalian ecology has been long in emerging, however, not because mammalogists lack interest in ecology but because a study of the ecology of even a single species involves detailed knowledge of many aspects of that species' biology, of its physical environment, and of the biology of species with which it is associated. The problem is more acute when one considers interactions among many species in a natural community. Because mammals are far more difficult to observe than are birds, avian ecology has led the way.

The scope of ecology is extremely broad, and an incomplete and selective coverage is given here. This chapter concentrates on ecological relationships and principles basic to an understanding of mammalian biology.

The environments of animals can be characterized in terms of physical and biotic factors. Physical factors include temperature, humidity, climatic patterns, precipitation, and soil types; biotic factors are those associated with interactions between organisms.

Physical Factors of the Environment and the Distribution of Mammals

Temperature and Climate

Solar radiation provides the energy, in the form of heat and light, on which living organisms depend. The intensity of solar radiation at the earth's surface is influenced largely by the directness with which the sun's rays strike the earth. The angle of these rays decreases, and the climate becomes progressively cooler the farther north or south of the equator areas are situated.

Warm air holds more moisture than does cool air, and equatorial areas, especially areas

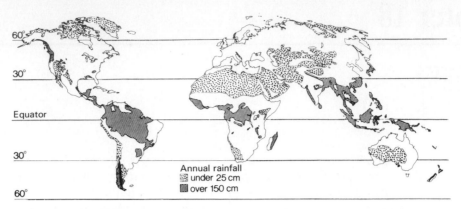

Figure 18-1. The important desert regions (those with less than 25 centimeters of precipitation annually) and the major wet regions (those with more than 125 centimeters of precipitation annually). (After Espenshade, 1971)

near 25° N and 25° S latitude, receive relatively heavy precipitation (Fig. 18–1). In addition, major global patterns of air circulation are set up as the warm air that rises from equatorial regions moves northward and southward. In a belt centered 30 degrees north and south of the equator, the equatorial air masses reach a stage of cooling at which they tend to sink; as they sink, they are warmed and their ability to carry moisture is increased. Some of the major deserts of the world, such as those in the southwestern United States and northern Mexico, are roughly 30 degrees north of the equator and are under the influence of this system of descending air.

Even in tropical areas, rainfall is seasonal (Fig. 18–2), and animals and plants must adapt to times of relative scarcity of food. Migrations of some tropical bats coincide with seasonal

shifts in the abundance of insects, fruits, or flowers, and the dramatic migrations of wildebeest in East Africa are in response to seasonal changes in the availability of nutritious forage.

Superimposed on global or regional weather patterns are myriad local variations and complexities. In winter in the western United States, for example, moist air masses that sweep inland from the west or northwest are forced upward as they pass over high, north-south–oriented mountain ranges. As the air flows up the western slopes of the mountains, it is cooled and loses moisture; as it descends the eastern faces of the mountains, it becomes warmer and its ability to hold moisture increases. Typically, then, the western slopes receive high precipitation, the eastern slopes receive lower precipitation, and the basins at the eastern bases of the

Figure 18-2. Mean monthly rainfall in a tropical area (Bushwhackers Safari Camp, Kenya). Note the sharply bimodal annual pattern of precipitation. (From Vaughan, 1976)

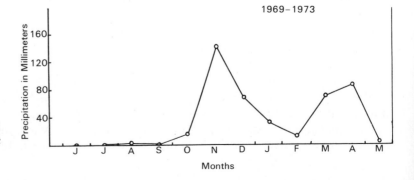

mountains are often deserts. This "rain shadow" effect strongly influences the distribution of both plants and animals.

Even minor topographic features may have pronounced local effects on precipitation and on plant and animal communities. As an example, the Santa Ana Mountains of southern California are often enshrouded by fog sweeping inland from the Pacific Ocean. The fog funnels through the major passes in the mountains, condenses on vegetation, and drops of water fall to the ground. In the spring and early summer, when fogs are most frequent but there is typically little rain, condensation drip from the needles of the knob-cone pine *(Pinus attenuata)* in these mountains can total a remarkable 10.2 centimeters of precipitation per month. Such precipitation supplies sufficient moisture to allow the pine to extend its growth into this otherwise dry period (Vogl, 1973), and Coulter pine *(Pinus coulteri)* is restricted to these passes (Pequegnat, 1951:9). The influence of these pockets of relatively heavy precipitation on the distributions of small mammals remains to be studied but may well be important.

Local conditions may also strongly affect the amount of heat the surface of the earth receives. In many mountainous sections of the western United States, steep slopes and precipitous canyon walls are common. Because the main axes of most mountain ranges lie north and south, the drainage systems are oriented more or less east and west and the canyon walls face roughly north or south. In northern latitudes, no matter what the time of year, the sun's rays strike a south-facing slope more directly than a north-facing slope. As a result, south-facing slopes are considerably drier and warmer than are nearby north-facing slopes. The effects of slope (steepness of incline) and exposure (direction the slope faces) are strongly reflected by the flora, and the compositions of small-mammal faunas are frequently as conspicuously different on adjacent opposing slopes as are the assemblages of plants. In the precipitous chaparral-covered mountains of southern California, for example, contrasting biotas occupy adjacent north- and south-facing slopes (Vaughan, 1954). Some species of mammals that occur on one slope are absent from the opposing one (Fig. 18-3).

Because the cover provided by such fea-

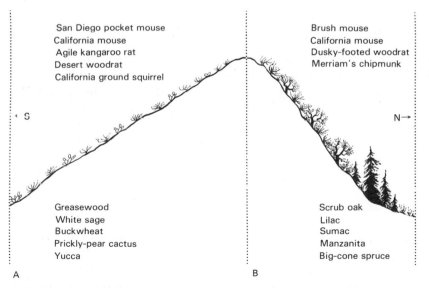

Figure 18-3. Assemblages of plants and mammals inhabiting (A) a south-facing slope and (B) a north-facing slope in lower San Antonio Canyon, San Gabriel Mountains, Los Angeles County, California. (Data from Vaughan, 1954)

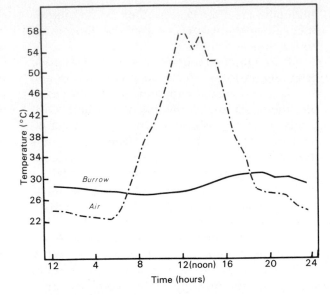

Figure 18-4. Temperature fluctuations of the air in a pocket gopher burrow (solid line) and just above the surface of the ground (dashed line). Temperatures were recorded on 24 June 1961, in McLennan County, Texas. (After T. E. Kennerly, 1964)

tures as vegetation and rock alters it locally, the environment at a terrestrial locality is not uniform but consists of a complex mosaic of microenvironments. As a general rule, few terrestrial mammals can withstand the most extreme temperatures (or other conditions) that occur in the habitats they occupy, but all are able to select microenvironments in which temperature extremes are moderated or eliminated. A notable example of such a microenvironment is shown in Fig. 18-4. Although occupying a temperate region, a pocket gopher may live for much of the year in a "tropical" microenvironment that features even, fairly high temperatures and high humidities. A group of beavers occupying a beaver lodge in the winter is by no means subjected to the extreme air temperatures outside the lodge (Fig. 18-5). Similarly, shrews forage beneath litter, under logs or rocks, or beneath dense foliage; not only is their food abundant in such places, but temperature and humidity are moderated by such cover. These animals cannot tolerate the general climatic conditions of the regions they occupy but are instead adapted to a limited set of conditions which occur in a chosen microenvironment.

Vertical temperature gradients are encountered as one ascends a mountain. The lowering of the temperature with increased elevation is of the magnitude of approximately 1 C degree for every 150 meters. This effect, coupled with increased precipitation at higher elevations, shorter growing seasons for plants, and drastic diurnal-nocturnal fluctuations in temperature, is associated with a distinct separation of climatic zones in high mountains throughout the world. The distributions of some mammals clearly reflect this zonation. In northern Arizona, Abert's squirrel *(Sciurus aberti)* occupies the ponderosa pine belt but is abruptly replaced by the red squirrel *(Tamiasciurus hudsonicus)* where spruce and fir forests appear at higher elevations. In some areas of the western United States, an assemblage of "desert" mammals resembling those typical of arid lands as far south as central Mexico may occupy the arid or semiarid land at the foot of a mountain range, while the crests of the mountains, a few airline miles away, may support boreal genera that occur as far north as northern Canada or Alaska (Fig. 18-6).

Temperature even imposes rigid constraints on the structure and behavior of whales, the largest mammals. The enormous body size of many whales allows them to conserve heat more effectively than can small mammals, but because heat is dissipated from a warm body many times faster in water than in air, temperature has exerted strong selective pressures on marine mammals.

Brodie (1975) discussed some of the thermoregulatory and energetic problems faced by the fin whale *(Balaenoptera physalis)*. This whale is a plankton feeder of great size: its length is about 20 meters, and its weight is up to 48 metric tons. Fin whales feed in summer in the cold (near 0 degrees C), plankton-rich waters of Antarctica and the far north, spending an average of 120 days and 183 days respectively in these regions. The whales migrate slowly to subtropical waters in the winter, where they eat little or nothing, deriving energy from the layer of blubber deposited in the sum-

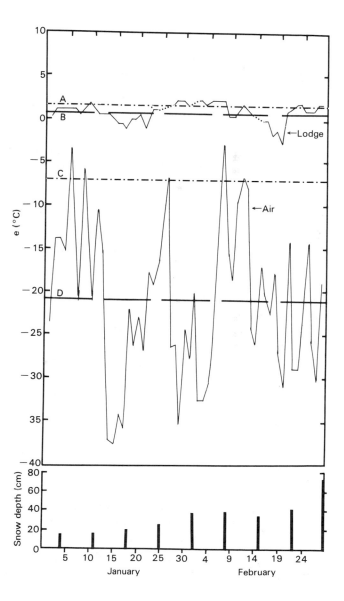

Figure 18-5. Daily minimum temperatures inside and outside a beaver lodge in Algonquin Park, Ontario, Canada. (After Stephenson, 1969)

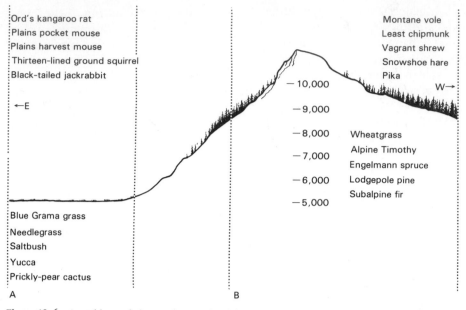

Figure 18-6. Assemblages of plants and mammals inhabiting (A) short-grass prairie and (B) subalpine habitats in northern Colorado (Larimer County).

mer. In the case of the Antarctic fin whales, the period of fasting lasts 245 days. This migration allows the animals to spend their fasting period in warm waters (25 to 30 degrees C), where energy can be conserved because of the greatly reduced cost of thermoregulation. Brodie argues that the demands of the thermal environment of this whale have influenced the evolution of body size. The optimum body size and surface area in these whales are those allowing sufficient subcutaneous fat to be deposited to maintain the whales through the long fasting period in warm waters poor in food.

Water

In mammals, as in other animals, water forms an essential part of protoplasm and body fluids, and the maintenance of water balance is basic to life. The availability of free water and the amount of water in the air affect the habitat selection of mammals.

Fossorial mammals occupy microenvironments generally characterized by high humidities. Under such conditions, pulmocutaneous water loss (water loss from breathing and through the skin) is minimal, and fossorial mammals that eat moist food can maintain water balance without drinking water. Careful studies of the microenvironmental conditions of burrows of the plains pocket gopher *(Geomys bursarius)* in Texas by T. E. Kennerly (1964) showed that the relative humidity in burrows was usually between 86 and 95 percent. Relative humidity levels in the sealed burrows may be high despite low soil moisture: Kennerly recorded a relative humidity of 95 percent in a burrow in July when soil from the floor of the burrow contained but 1 percent water. Although the microenvironment in burrows is such that stresses resulting from high temperatures and low humidities are avoided, other stresses may occur. Kennerly recorded burrow carbon dioxide concentrations from ten to 60 times that of atmospheric air!

Flooding or high soil moisture may cause seasonal changes in mammal distributions. The mole rat *(Cryptomys hottentotus)* in Zimbabwe centers its activity during the rainy season in the bases of large termite mounds that rise a meter or so above the surrounding grassland and provide relatively dry islands in a sea of water-logged terrain (Genelly, 1965).

Snow is an environmental feature of great importance in some boreal areas. The continuous snow cover that persists through the winter in many areas is a severe hardship for some large mammals. To most North America artiodactyls, including deer, elk, bighorn sheep, and moose, even moderately deep snow imposes a burden. In mountainous areas, deer and elk avoid deep snow by abandoning summer ranges and moving to lower elevations, whereas in other areas they restrict their winter activity to "yards." Prolonged winters with deep snow commonly cause high mortality among deer and elk. On Vancouver Island, British Columbia, in the winter of 1948–1949, snow covered the ground even at low elevations to depths of about 1 meter; deer were denied access to their accustomed winter foods and mortality from starvation was high. Forestry crews in the summer of 1949 counted 18 deer carcasses per square mile in some areas (Cowan, 1956:560).

Rather than being a source of winter hardship for small mammals, snow is actually a blessing. It forms an insulating mantle that provides a microenvironment at the surface of the ground where activity, including breeding in some species, continues through the winter. To these small mammals, such as shrews *(Sorex),* pocket gophers *(Thomomys),* voles *(Microtus, Clethrionomys, Phenacomys),* and lemmings *(Lemmus, Dicrostonyx),* the most stressful periods are in the fall, when intense cold descends but snow has not yet moderated temperatures at the surface of the ground (Formozov, 1946), and in the spring, when rapid melting of deep snow often results in local flooding (Ingles, 1949; H. O. Jenkins, 1948; Vaughan, 1969).

Even in the summer, snow may be impor-

tant to some mammals. Caribou *(Rangifer tarandus)* and bighorn sheep *(Ovis* spp.) often congregate in summer on persistent snow patches to seek relief from warble flies.

Substrates

Many small mammals seek diurnal refuge in burrows, and many terrestrial mammals of a variety of sizes have specialized modes of locomotion that are effective on reasonably smooth surfaces. To these mammals, the type or texture of the soil or substrate is a critical environmental feature. Burrowing species may be narrowly restricted to a particular type of soil; for example, some heteromyid rodents that are weak diggers occur only where the soil is sandy. Some scansorial (climbing) species occur only where there are large rocks or cliffs, and many kinds of bats depend on such places for daytime roosting.

Biotic Environmental Factors

Vegetation

Not only are plants important as food for many mammals, but the cover, escape routes, and retreats they provide, as well as the degree to which they facilitate or obstruct rapid locomotion, are important aspects of the environment of many terrestrial mammals. Therefore, plants that are never used as food by a mammal are often as essential a part of its environment as are staple food plants.

A species of mammal is seldom evenly distributed, even within an area of seemingly homogeneous vegetation. On the contrary, the distributional patterns of most mammals are discontinuous, indicating that all requirements for the species are not met over broad areas. Close observation of even a limited area usually indicates local changes in the relative density and spacing of plants; further, a given species of mammal is usually restricted to a habitat characterized by plants of a certain life form. The size, shape, foliage density, and pattern of branching of a plant determine its life form. Analyses of the environmental requirements of a mammal, therefore, must include considerations of not only the species of plants with which the animal is associated, but also (and frequently equally important) the life forms of these plants and the "aspect" they give to the habitat.

Consider, for example, two species of East African ungulate. Kirk's dik-dik *(Madoqua kirki)* is a small, delicately built bovid (Fig. 20–9). When disturbed, it dashes into the nearest patch of brush, where it "freezes" and stares at the source of danger. This animal is restricted to brushy areas, and although when in cover it may be approached fairly closely, one's view of the dik-dik is often through a formidable screen of thorny branches. At the other extreme is Grant's gazelle (*Gazella granti;* Fig. 20–11). This extremely swift animal seeks to escape its predators by outrunning them, at times over long distances. Although this gazelle occupies a variety of settings, from semideserts to grasslands, it is restricted to areas where escape can be sought in the open, where shrubs are scattered or absent, and where the grasses are not tall enough to limit high-speed running.

Food

Food is necessary to animals as a source of energy and for building and maintaining protoplasm and is therefore one of the most important biotic factors in the environment of a mammal. Mammalian adaptive radiation has involved a progressively broader exploitation of food sources, but although mammals as a group utilize many types of food, a single species usually eats a fairly limited array of foods that it is structurally, physiologically, and behaviorally capable of utilizing efficiently. Much of mammalian evolution has been "guided" by the ne-cessity of achieving the most favorable balance between the energy and time expended in securing and metabolizing food on the one hand and the energy gained from the food on the other. The relationships of mammals to their environment must ideally be considered against a background of knowledge of feeding biology.

The most abundant and omnipresent foods for terrestrial mammals are plants and insects. It is not surprising that the two most important mammalian orders in terms of numbers of species—Rodentia and Chiroptera—depend primarily on these major food sources. In addition, members of the orders Edentata, Pholidota, and Tubulidentata are primarily insect eaters, whereas some members of the Marsupialia, Primates, and Rodentia, as well as the Lagomorpha, Proboscidea, Hyracoidea, Sirenia, Perissodactyla, and most artiodactyls are herbivores.

Although many herbivores are selective in their feeding, a fairly wide variety of food is generally utilized, and seasonal variations in feeding habits are typical of temperate-zone species. Western wheat grass (*Agropyron smithii*) forms about 35 percent of the diet of the plains pocket gopher *(Geomys bursarius)* in July but is not eaten in December (Fig. 18–7). A number of studies have shown that, given a wide variety of plants to choose from, most herbivorous mammals are selective foragers (see, for example, Ward and Keith, 1962; Yoakum, 1958; Zimmerman, 1965); accordingly a herbivore may show a great preference for one of the least abundant plants in its habitat (Fig. 18–8).

The nutrient content of plants may determine in part their choice as food by herbivores. Lindlöf, Lindström, and Pehrson (1974) found that, in February, mountain hares *(Lepus timidus)* in Sweden preferred plants high in crude protein and phosphorus; in addition, in habitats preferred by hares the concentrations of these nutrients in the major food plants were unusually high. The nutrient content of various deciduous browse plants utilized by mule deer *(Odocoileus hemionus)* was shown by Short,

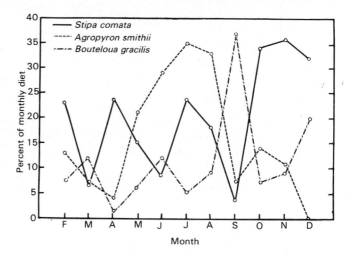

Figure 18-7. Seasonal differences in the diet of the plains pocket gopher *(Geomys bursarius)* in eastern Colorado. (After G. T. Myers and Vaughan, 1964)

Dietz, and Remmenga (1966) to vary seasonally, whereas nutrient levels of evergreen plants were less variable. The deer in many areas eat a great variety of plants but may select those from which nutrients can be extracted most easily by digestion, rather than those in which levels of certain critical nutrients are highest. In rodents, food preferences of different sympatric species (species that live in the same area) may be highly specific and may differ to the extent that

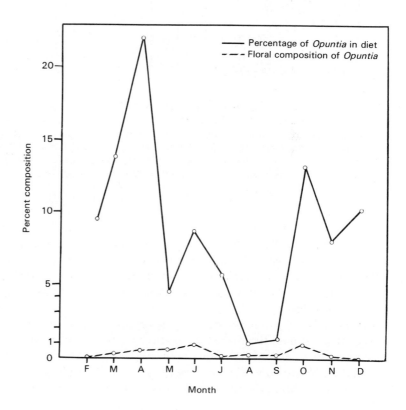

Figure 18-8. Seasonal changes in the utilization of prickly-pear cactus *(Opuntia humifusa)* by the plains pocket gopher *(Geomys bursarius)*. The percent composition of prickly-pear cactus in the diet varies markedly, whereas the floral composition of prickly-pear (an expression of the percentage of the total coverage of vegetation contributed by *Opuntia*) is nearly constant. (After G. T. Myers and Vaughan, 1964)

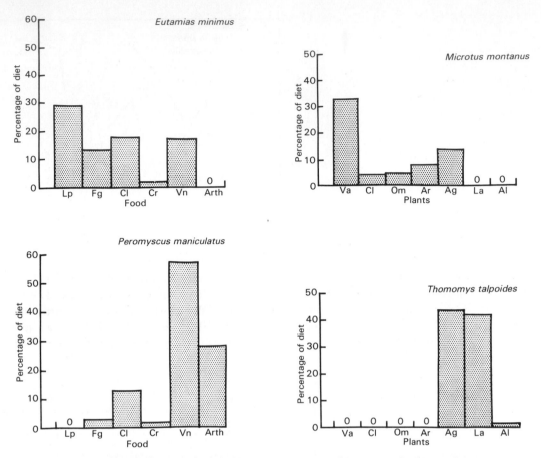

Figure 18-9. Diets of four partly or completely herbivorous rodents in part of the summer of 1965 in a subalpine area in Routt County, Colorado. The rodents are the least chipmunk *(Eutamias minimus)*, the montane vole *(Microtus montanus)*, the deer mouse *(Peromyscus maniculatus)*, and the northern pocket gopher *(Thomomys talpoides.)*. Note the lack of competition among the species for food. Abbreviations: Ag, *Agoseris glauca;* Al, *Achillea lanulosa;* Ar, *Arnica cordifola;* Arth, *arthropods;* Cl, *Collomia linearis;* Cr, *Carex* sp.; Fg, fungus; La, *Lupinus argenteus;* Lp, *Lewisia pygmaea;* Om, *Oenothera micrantha;* Va, *Vicia americana;* Vn, *Viola nuttali.* (Based on unpublished data)

there is remarkably little interspecific dietary overlap (Fig. 18–9). The choice of specific plants or parts of plants by herbivorous animals is also strongly influenced by the occurrence of defensive chemicals, which are discussed in the following section.

Insects have probably been a major food source of mammals for over 180 million years and have had a marked influence on patterns of mammalian evolution. The dentitions of some Late Triassic mammals were well adapted to masticating insects, and the radiation of the Lepidoptera (moths and butterflies) and Isoptera (termites) in the Cretaceous broadened the food base for insectivorous mammals. The Early Tertiary adaptive radiation of the highly successful microchiropteran bats probably occurred primarily because of the great nocturnal abundance of moths. In tropical areas, termites are of tremendous importance to mammals today. At least some members of ten of the 19 orders of mammals commonly eat termites, and the spiny anteaters (Tachyglossidae), South American anteaters (Myrmecophagidae), pangolins (Manidae), and aardvark (Orycteropidae) specialize on termites, as do some members of the order Insectivora. Members of a community of insectivorous bats choose prey not only on the basis of size but on the basis of type: some are "moth strategists" and some are "beetle strategists" (Table 18–1).

Protective devices such as thick, chitinous exoskeletons, toxic sprays, and potent stings protect some insects from predation, but some mammals have evolved countermeasures. In the Serengeti Plains of Africa, the primary food of the aardwolf is the termite *Trinervitermes bettonianus*. Aardwolves feed on foraging parties of this termite for periods averaging only 22 seconds, probably in an effort to avoid the soldier caste of the termite colony. As a protective device, the soldiers spray distasteful terpenoids from a pointed projection of the head capsule. When the column of workers is disturbed, the soldiers stream from the nest and mix in large numbers with the retreating workers. Apparently, the aardwolf adjusts its foraging time accordingly and moves on to undisturbed termites before this occurs (Kruuk and Sands, 1972).

The striped skunk *(Mephitis mephitis)* handles "stink bugs" in a unique way. Some large beetles of the family Tenebrionidae discourage predators by spraying toxic quinones from abdominal glands. The striped skunk takes such a beetle between its front feet and rolls it roughly in the soil; when the beetle has exhausted its spray and the spray has been absorbed by the soil, the skunk eats the beetle (Slobodchikoff, 1977).

Many mammals prey on higher vertebrates, including reptiles, birds, and other mammals. Such carnivores occur in the orders Marsupialia, Insectivora, Chiroptera, Cetacea, and Carnivora. The size of a predator obviously determines the size range of its prey. Many mammalian faunas consist of a wide variety of prey species fed upon by various carnivores; in such faunas, the tendency is for each species of carnivore to differ from the others in size and to take prey of a different size class. Many carnivores are highly specialized structurally for killing and eating their prey, and most have behavioral specializations that further their predatory ability. The learning of efficient hunting and killing methods is critical to the survival of carnivores. Prey must be captured without an excessive outlay of

Table 18–1: Density of Moth Scales and Percent Frequency of Beetles and Moths in Bat Feces in a North Temperate Bat Community

Type of Prey	Bat Species	Sample Size	Mean Scales/Gram	Moth Frequency (%)	Beetle Frequency (%)
Moth	*Lasionycteris noctivagans*	19	145,591	100	0
Moth	*Lasiurus cinereus*	39	10,943	100	5
Moth	*Pipistrellus hesperus*	7	5016	100	0
Moth	*Idionycteris phyllotis*	3	2397	100	0
Beetle	*Eptesicus fuscus*	165	679	61	84
Beetle	*Antrozous pallidus*	12	32	17	92
?	*Myotis californicus-leibii*	16	19,929	94	69
Moth	*M. volans*	29	4031	96	17
Moth	*M. auriculus*	10	1868	90	20
?	*M. yumanensis*	16	913	53	24
Beetle	*M. evotis*	13	98	62	92
Beetle	*M. thysanodes*	11	48	36	73

From Black, 1974.

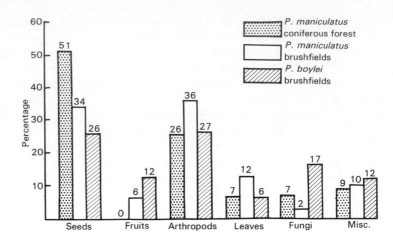

Figure 18–10. Foods of deer mice *(Peromyscus)* in different habitats in the northern Sierra Nevada of California. (After Jameson, 1952)

energy. Of equal importance, particularly in the larger carnivores that kill prey as large as or larger than themselves, hunting and killing behavior must be geared to avoiding serious injury to the predator. The high incidence of skeletal damage in the fossil remains of saber-tooth cats *(Smilodon)* suggests that in the Pleistocene, as it is today, preying on large game was a dangerous undertaking.

Still other mammals are omnivorous and are opportunistic feeders; such mammals eat a wide variety of plant and animal material. Omnivores occur among the orders Marsupialia, Insectivora, Chiroptera, Primates, Rodentia, Carnivora, and Artiodactyla. Omnivores are typically less specialized in structure than are mammals adapted to narrower diets, and in some orders the omnivorous mode of life has been highly successful. Among rodents, for example, the nearly ubiquitous North American genus *Peromyscus* includes species seemingly adapted to a wide variety of plant and animal foods (Fig. 18–10). Some terrestrial sciurids are also omnivorous, and in the order Carnivora, such widespread and successful families as Canidae, Ursidae, and Procyonidae have many omnivorous members.

A variety of additional foods are eaten by mammals. Planktonic organisms are the major food of the filter-feeding baleen whales and of the enormously abundant crab-eater seal *(Lobodon carcinophagus)* of the Antarctic. Most odontocete cetaceans eat fish, squid, and, less commonly, a variety of other invertebrates, but the killer whale *(Orcinus)* preys on porpoises, whales, diving birds, and a variety of large sharks and bony fish. Fish, squid, and mollusks are taken by pinnipeds; fish are also eaten by some members of Carnivora and even by some bats. The leopard seal *(Hydrurga)* of Antarctica preys on a variety of marine vertebrates and is an important enemy of penguins. A number of marine invertebrates, including sea urchins *(Strongylocentrotus),* provide food for sea otters *(Enhydra).* A few mammals do considerable scavenging; hyaenas *(Crocuta* and *Hyaena)* and jackals *(Canis* spp.) frequently feed on the leftovers of kills by larger carnivores. Nectar and pollen feeding are common among pteropodid and phyllostomid bats, and one phalangerid marsupial *(Tarsipes)* is highly specialized for a diet that consists partly of nectar. Vampire bats, which have one of the most specialized mammalian feeding techniques, feed entirely on blood. Some carnivores—the coyote is a good example—are highly opportunistic and, under the pressure of hunger, may take almost any vulnerable animal, vertebrate or invertebrate, as well as plant material.

An unusual example of opportunism is described by Huey (1969). Part of the range of the Peruvian desert fox *(Dusicyon sechurae)* in-cludes the extremely barren Sechura Desert of northwestern Peru, where the only other mam-mal is the rodent *Phyllotis gerbillinis.* On an extremely sterile area away from the Pacific coast, this fox subsists almost exclusively on the seeds of shrubs. All of the fecal material from this area contained seeds, and 77 percent con-tained seeds alone; the only other food of any importance was tenebrionid beetles, which oc-cured in small amounts in 23 percent of the stomachs. Near the coast, this fox had access to animals washed up on the beach; here, although seeds still occurred in most feces, invertebrates such as crabs and beetles and vertebrates such as gulls, mice, and lizards were important foods.

Plant Secondary Compounds: The Wages of Herbivory

Plants being the most abundant source of food, herbivory has been popular among mammals for at least 65 million years, but herbivores face several problems. Many plant tissues are diffi-cult to digest, they often contain low levels of protein, and they are protected by defensive chemicals. Such compounds, usually called sec-ondary compounds, affect the diets and feeding strategies of mammals and in some cases have influenced other aspects of their life histories.

Plant defensive chemicals are of two major types. Ephemeral plant parts (those available briefly), such as flowers, fruits, or new leaves, are typically protected by toxins that have prob-ably evolved in response to pressure from di-etary generalists. Toxins exhibit little individ-ual variation from plant to plant within a species. Mature leaves, on the other hand, are a predictable and abundant food that is available over a relatively long period of time. Probably in response to dietary specialists, mature leaves have evolved secondary compounds with a high degree of individual variation (Rhoades, 1979). These chemicals reduce digestibility (Cates and

Rhoades, 1977; Feeny, 1975). Tannins occur in the leaves of a wide array of plants and combine chemically with and denature many mammalian digestive and nondigestive enzymes (Pridham, 1965). The volatile oils of conifers contain ter-penoids. Some juniper terpenoids are known to have an antimicrobial action in the rumen of deer (Schwartz, Nagy, and Regelin, 1980a). Be-cause microbial fermentation in the rumen yields the largest share (50 to 70 percent) of the energy required by ruminants (Annison and Lewis, 1959), antimicrobial secondary com-pounds markedly inhibit ruminant digestion (Nagy, Steinhoff, and Ward, 1964) and may also inhibit digestion in the caecum of rodents (Vaughan, 1982).

Plant tissues with high levels of defensive chemicals are not popular foods among mam-mals. Conifer foliage, for example, is generally either an emergency food or one taken in small amounts with an array of other plants. In North America, only two species of mammals, the red tree vole *(Phenacomys longicaudus;* W. J. Hamilton III, 1962) and Stephens' woodrat *(Neotoma stephensi;* Vaughan, 1982) are coni-fer-leaf specialists. Plants other than conifers can also produce such high levels of defensive chemicals that herbivores are deterred. After in-tense browsing by snowshoe hares *(Lepus americanus),* four nonconiferous species of Alaskan trees produced adventitious shoots (shoots initiated from more or less mature tis-sue without connection with apical meristem) that had exceptionally high levels of terpene and phenolic resins. Because these resins are re-pellent to snowshoe hares, the animals avoid the adventitious shoots (Bryant, 1981). These browsing-induced chemical defenses drastically reduce the forage available to the hares at the very time when high populations require large amounts of food. This may trigger the popula-tion crash. These plants do not produce usable forage until three years after intense browsing, and only after this time can hare populations begin to recover. Bryant speculates that the chemical defenses of the food plants are thus strongly influencing the ten-year population cycle of the snowshoe hare.

Because defensive chemicals in plants are of broad occurrence, they have forced herbivores to evolve countermeasures. One such measure is microbial breakdown. Oxalates occur in many plants important in the diets of herbivores, and to nonadapted mammals oxalates can be lethal. Some mammals, including rabbits, rodents, pigs, horses, some ruminant artiodactyls, and humans, are known to degrade oxalates by microbial action in the large intestine or rumen. Oxalates in the diet favor intestinal or rumen bacteria that utilize these compounds for growth; higher populations of such bacteria result in increased rates of oxalate breakdown (Allison and Cook, 1981). Herbivores adapted in this way can tolerate levels of dietary oxalates that would kill nonadapted individuals. The ability of a number of wild rodents to eat large quantities of plants containing oxalates is probably due to adaptation of their intestinal or caecal bacteria.

A second countermeasure is selective foraging, the ability to discriminate among individual plants and eat only those with relatively low levels of defensive chemicals. Such finely tuned selective foraging depends on intraspecific variation in secondary compounds among plants. This type of variation has been well documented (Hanover, 1966, 1971; Schwartz, Nagy, and Regelin, 1980), as has selective foraging by mammals. Glander (1977) found that, in Costa Rica, leaf-eating howler monkeys *(Alouatta palliata)* were forced to be selective by defensive chemicals in the leaves of trees. The leaves of only certain individuals of some species of trees were eaten, and the monkeys generally ate the petioles (leaf stalks), which have lower concentrations of defensive chemicals than do the leaves. The price of carelessness is high: three of six dead howler monkeys that Glander examined had been eating the leaves of either of two trees with toxic leaves, as had a female that went into convulsions and fell from a tree. There are also less dramatic examples. Deer are

able to choose experimental foods with low levels of the volatile oils of juniper (Schwartz, Regelin, and Nagy, 1980), and Stephens' woodrats feed repeatedly on specific juniper trees but never on others (Vaughan, 1980). The chemical governing the selectivity of Abert's squirrels *(Sciurus aberti)* has been isolated by careful experimental work. This squirrel depends in winter on the cortical tissue of ponderosa pine twigs. Twigs of trees fed upon by these squirrels had smaller amounts of monoterpenes (compounds that have a deterrent or toxic effect on some animals) than did nearby trees not fed upon, and feeding trials in the laboratory showed that Abert's squirrels chose twigs with low levels of the monoterpene α-pinene (Farentinos, Capretta, et al., 1981).

A third way in which mammals avoid the effects of defensive chemicals is by having a diverse diet. By eating a great variety of plants, an herbivore can keep the levels of defensive chemicals ingested low enough to be tolerated.

The life histories of some folivorous (leaf-eating) mammals have probably evolved under selective pressures associated with chemicals that reduce digestibility and a low-energy diet. The low metabolic rate of Neotropical tree sloths and howler monkeys is perhaps an adaptation to their folivorous diet. The two North American conifer-leaf specialists offer interesting examples. Their life histories depart markedly from the norm for rodents; these departures may be the price paid for eating food that is abundant and predictable but protected by digestibility-reducing chemicals. Red tree voles, which feed on the leaves of Douglas fir, never reach high population densities and have small litters (usually two young) and the young grow slowly, whereas most other voles attain high densities and have large litters, and their young grow rapidly. Similarly, Stephens' woodrats, which eat juniper leaves, are never abundant and usually have but a single young per litter, and the young grow and reach sexual maturity slowly (Vaughan, 1985).

Grasses use silica as a defense against herbivores, and McNaughton et al. (1985) found that grasses in the Serengeti Plains of Tanzania, which are subjected to the heaviest sustained grazing pressure of any terrestrial habitat (McNaughton, 1979a), have remarkably high levels of silica in their tissues. Silica contents were higher in grasses of more heavily grazed areas and were highest in the least expendable parts of the plants (stems and roots). Adverse effects of dietary silica on grazing animals include accelerated tooth wear, esophageal cancer, and deposits of silica in the urinary tract. Such deposits can be fatal to cattle (Iler, 1979). The high-crowned teeth of grazing ungulates and of such grass-eaters as many arvicoline rodents are probably a response to rapid abrasion by silica, but little is known of other ways in which these mammals have adapted to dietary silica.

In summary, mammalian herbivores have been forced to adapt to the nearly ubiquitous defenses of plants. Adaptation of intestinal, caecal, or rumen microbes increases the rate of degradation of some defensive chemicals, and highly selective foraging enables some mammals to feed on those individual plants within a species with relatively low levels of these chemicals. Some herbivorous mammals avoid ingesting large amounts of any defensive chemical by eating a broad array of plants.

Environmental Impact of Mammals

No animals have had, nor now have, a greater impact on terrestrial environments than mammals. The ability of humans to modify, or in some cases devastate, their environment is of ever-increasing importance as human populations continue to expand. Other mammals also strongly effect their environment, however, and in "natural" areas, where the human hand has lain lightly on the land, the activities of wild mammals often drastically alter the character of the vegetation, the availability of water, the patterns of erosion, and the diversity and nature of the vertebrate and invertebrate fauna. If the im-

portance of a group of animals is equated with its effect on the environment, mammals are clearly the most important terrestrial animals.

In part as a result of endothermy, with its attending high energy requirements, mammals have a great impact on their environment. This impact results from a variety of activities, including feeding, patterns of migration or daily movement, the quest for water, and the construction of shelters or refuges. These activities are typically interrelated, and an understanding of the impact of mammals can best be approached by a consideration of daily or annual cycles of activity. Even the capture of a moth larva by a shrew or the hoarding of a seed by a white-foot mouse has an effect on the environment, but as might be expected, large mammals can modify their environment most drastically and conspicuously.

In some of the national parks and game preserves of Africa, there are high elephant populations, and because of the encroachment of agriculture, the elephants are no longer free to range widely when pressed by local or seasonal shortages of food or water. Studies on elephants in various parts of East Africa document the impact elephants can have on the landscape (Laws, 1970) under these conditions. The diet of elephants seems ideally to consist of a mixture of grass and browse from trees and shrubs (Laws and Parker, 1968), and the preferred habitat is thus forest edge, the woodland, or bush-grass mosaic. During the dry seasons in the drier areas, such as Tsavo National Parks in Kenya, sources of water are not generally distributed; because elephants at these times need water daily, they concentrate within a radius of 20 to 30 kilometers of water. A very large group of elephants may have a daily food requirement of 50,000 kilograms or more, and when high densities of elephants occupy restricted areas near water, rapid destruction of trees and shrubs results.

In Tsavo National Parks, elephants are transforming the original bushland to grassland. Aerial photographic transects studied by R. M.

Figure 18-11. Elephants destroying a baobab tree in Tsavo West National Park, Kenya.

Watson (1968) indicated that, in a period of five years, elephants killed from 26 to 28 percent of the trees above 65 centimeters in crown diameter. To the south, in Lake Manyara National Park of Tanzania, Douglas-Hamilton (1973) observed similar destruction: in one area of especially acute damage, elephants killed 8 percent of the umbrella trees *(Acacia tortilis)* in one year. Particularly notable in Tsavo is the destruction of baobab trees (Fig. 18-11). These huge, picturesque trees live to be several hundred years old, and their reproductive rate is low. At the present rate of destruction by elephants, the baobabs will be eliminated from Tsavo National Parks within 15 years.

In Tsavo, as elsewhere, elephants have not affected the vegetation alone. There have also been striking changes in the fauna. Ungulates such as zebras *(Equus burchelli),* Grant's gazelles *(Gazella granti),* and oryx *(Oryx gazella)* have been favored by the shift toward grassland, and as pointed out by Sheldrick (1972), the visitor to Tsavo is now able to observe a much greater diversity of game animals than could be seen when the area was bushland. Other effects cited by Sheldrick include enlargement of wallows to form water holes in the wet season and local compaction of riverbeds

resulting in the surfacing of previously subsurface water. These processes are clearly dynamic, and the rate and direction of change are controlled by a great many factors, including fluctuating patterns of rainfall and changes in elephant population densities. Seemingly, the grassland cannot continue for long to support high densities of elephants, and Sheldrick (1972, 1973) suggests that we may be observing but one phase of a cycle that has recurred over and over again.

Feeding activities of a variety of mammals have pronounced effects on vegetation. Randolph S. Peterson (1955:162) noted that moose *(Alces alces)* were responsible for the local decimation of ground hemlock, quaking aspen, and balsam on Isle Royale, Michigan, and these animals have caused considerable damage to vegetation in Finland (Kangas, 1949). In Rocky Mountain National Park in Colorado, heavy browsing by deer and elk in some areas caused the death of 85 percent of the sagebrush and 35 percent of the bitterbrush plants in a five-year period (H. M. Ratcliff, 1941). In California, a tract of bitterbrush was killed by overbrowsing by deer within five years (Fischer, Davis, et al., 1944). Small mammals may also have a marked effect on vegetation. Batzli and Pitelka (1971) found that, during cyclic changes in density,

California voles *(Microtus californicus)* had a significant effect on preferred food plants. During high mouse densities (160 per acre), the mouse's major food plants contributed 85 percent less volume to the vegetation outside experimental plots from which the mice were excluded than to the vegetation inside the plots. In addition, the fall of seeds of preferred grasses was reduced by 70 percent on grazed areas.

Patterns of Ecological Distribution

Competition and the Ecological Niche

Just as no two species of animals are structurally identical, no two are functionally identical or have exactly the same environmental requirements. The very morphological, physiological, and behavioral characters that determine the distinctness of a species also determine the distinctness of its habitat requirements. Each species requires a specific environment—a particular combination of physical and biotic factors—and each is functionally unique, pursuing a particular mode of life within its environment. The specific environmental setting a species occupies and the functional role it plays in this habitat constitute the animal's ecological niche.

The fundamental niche of Hutchinson (1957) is an abstract formalization of the usual concept of ecological niche. The fundamental niche is an "*n*-dimensional hypervolume" defined by all the values limiting the survival of a species and within which every point "corresponds to a state of the environment which would permit the species S to exist indefinitely."

Niche segregation avoids competition or reduces it to limits that sympatric species can bear. Competition occurs when two or more individuals occupying the same habitat at the same time are utilizing some environmental resource in short supply. Competition can be between members of the same species or between members of different species. Competition may also be direct or indirect (G. J. Miller, 1969). Individuals competing indirectly may never come in contact—a chipmunk may eat so many cutworms during the day that it becomes unprofitable for a white-foot mouse to search for this preferred food at night. This sort of mutual use of the same resource is an important type of competition. Individuals competing directly, on the other hand, are in direct confrontation, as when a pair of lions take over a freshly killed wildebeest from a group of spotted hyenas. Direct competition often involves the defense of space. Typically, competition has one of the following outcomes: one species becomes extinct and is replaced by the other (a process repeated countless times during the three or four billion years of the history of life on earth); one species emigrates to another area; or one species changes with regard to its use of the disputed resource (Fig. 18–12).

Competition, because it has long been recognized as a selective force leading to ecological, structural, and physiological changes in species, has been the focal point of much research. The ecological consequences of competition between closely related species, between sexes of the same species, and between distantly related taxa have been studied, in some cases experimentally.

Competition between closely related species may lead to several different types of segregation. The niches of Merriam's kangaroo rat *(Dipodomys merriami)* and the Arizona pocket mouse *(Perognathus amplus),* which live on common ground in the Sonoran Desert in Arizona, are separated in several ways (Reichman, 1975). Merriam's kangaroo rat eats large numbers of insects but few creosote bush seeds, reduces summer activity but is active all winter, and forages mostly in the open. The Arizona pocket mouse eats few insects but many creosote bush seeds, is active all summer but below ground in winter, and forages near shrub cover. A group of sympatric ungulates in Zimbabwe practice asynchronous breeding (Dasmann and

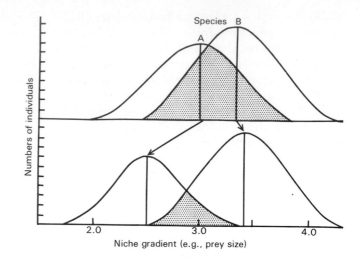

Figure 18–12. (Above) When two species overlap broadly in some niche characteristic, such as choice of prey size, there is reduced survival of the most strongly competing members of each species. (Below) An evolutionary trend toward niche divergence typically results. (After Whittaker, 1970)

Mossman, 1962) and thus offset the times at which females of the different species have increased food requirements (Table 18–2). In such demanding environments as subalpine meadows, where the growing season is short and the winter snow deep, dietary segregation may be strongly developed. Although the foraging areas of four species of subalpine rodents in Colorado overlap broadly, the diets of no two species are the same (Fig. 18–9). The altitudinal separation of three species of chipmunks in the Sierra Nevada of California is the result of different physiological tolerances and levels of dominance (Chappell, 1978). The lodgepole chipmunk *(Eutamias speciosus),* which is sensitive to heat, is dominant over the yellow pine

Table 18–2: Reproductive Seasons of Some Ungulates in Zimbabwe

Species	Wet Season Jan	Feb	Mar	Dry Season Apr	May	Jun	Jul	Aug	Sep	Oct	Wet Season Nov	Dec
Burchell zebra	Y	Y	Y	Y	Y	Y	XP	L	L	Y	Y	YP
Warthog	X						P	P				B
Giraffe					B	Y	Y	Y	X			
African buffalo		YP			Y				LY			
Blue wildebeest			X							P	P	
Waterbuck		YB	YB	X	L							
Eland								Y				
Kudu		P	BY	X				LP	L		LY??	
Bushbuck							Y					
Impala	X	L	L	L	R	R	LP	P	P	P	P	B
Common duiker			X		X	X		YP			PY	
Steenbok		Y	Y		Y	Y	X	YP			YP	
Klipspringer								Y				

Key: Y = Young, estimated under one month of age, observed
  X = maximum number of young observed
  B = most births occur
  L = lactating females observed or collected
  R = rutting-season behavior observed
  P = pregnant females collected

After Dasmann and Mossman, 1962.

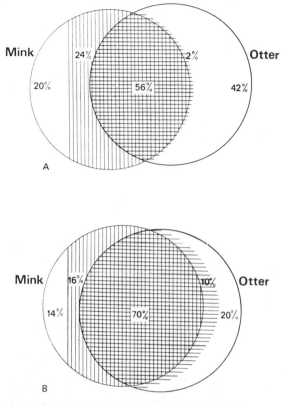

Figure 18-13. Dietary overlap between the European otter *(Lutra lutra)* and the mink *(Mustela vison)* in Sweden, (A) in summer and (B) in winter. Based on data obtained from 709 otter scats and 743 mink scats. (From Erlinge, 1972)

chipmunk *(E. amoenus)* and the least chipmunk *(E. minimus)* and excludes these two species from its forest habitat. The yellow pine chipmunk tolerates dry, hot woodlands where the lodgepole chipmunk cannot live, and by being dominant over the least chipmunk forces this heat-adapted species to live in sagebrush habitats.

Erlinge (1972) described an interesting competitive situation resulting from the introduction in the late 1920s of mink *(Mustela vison)* from the United States into the range of the European otter *(Lutra lutra)* in Sweden. The establishment of mink seemed to cause a decline in otter populations, but the two carnivores continue to coexist despite their heavy

mutual use in winter of such foods as fish and crayfish. Erlinge concluded that the mink is able to exploit a wide array of foods, whereas the otter specializes on aquatic vertebrates. Also, otters may directly interfere with mink and are even known to kill them occasionally (Egorov, 1966); this may reduce the extent to which the mink use the habitat and preferred foods of the otters. The specialized otter thus seems to occupy a relatively narrow niche within the broader niche of the less specialized mink (Fig. 18–13).

Competition between the sexes may be reduced in several ways. In heterogeneous habitats in Utah, female deer mice *(Peromyscus maniculatus)* are up to 15 percent larger than males, are dominant over males, and occupy the moister microhabitats; males occupy the drier, less productive sites (Bowers and Smith, 1979). This displacement presumably reduces competition between the sexes and provides females with more energy for survival and reproduction. Sexual size differences in some predators may enable males and females to exploit different sizes or kinds of prey (Fig. 18–14). The males of two species of weasel *(Mustela erminea* and *M. frenata)* are roughly twice as heavy as the females, and the skulls of the males are more robust (Hall, 1951:26). This situation seems to parallel that described for some birds (Selander, 1966; R. W. Storer, 1955), in which pronounced sexual dimorphism allows the sexes to utilize different feeding niches. Where *M. erminea* and *M. frenata* occur together, they differ considerably in size. Competition for food between the sexes of the greater galago in South Africa is reduced because groups of female kin concentrate their foraging on prolific food patches, whereas males seldom use these patches and forage over wide areas (A. B. Clark, 1978). Competition for food among related female galagos may heighten female mortality, for it is associated with a male-biased sex ratio.

Competition between distantly related taxa is probably a major factor shaping community organization and may markedly influence pop-

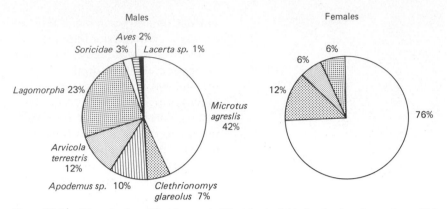

Figure 18–14.   Diets of male and female weasels *(Mustela nivalis)* in Sweden, based on analyses of 94 male scats and 15 female scats. (From Erlinge, 1975)

ulation levels of competing organisms. Experimental studies suggest that rodents and ants compete for food and that their populations are food-limited (p. 308; Fig. 18–15).

Biotic Communities

It has long been recognized that animals and plants with similar environmental requirements form identifiable communities. A perceptive description of a community was given by Mobius (1877), who considered an oyster bed as "a community of living beings, a collection of species and a massing of individuals, which find here everything necessary for their growth and continuance." A community is characterized not only by its unique plant and animal assemblage, however, but also by complex interactions between organisms and by the effects the physical environment has on the biota. The term "community" has been used to designate plant-animal assemblages of differing size and importance (Odum, 1971:140). This term can be appropriately used in reference to the biota of a woodrat nest or to that of the extensive deciduous forests of the eastern United States. As described by Ricklefs (1973:603), "community structures and dynamics are the sum of the fates of the organisms in the community." The struc-

ture of a community and its dynamics are determined by a web of interactions between physical and biotic factors.

The highest numbers of species are encountered in tropical communities, with decreasing diversity in areas progressively closer to the poles. As an example of this pattern in mammals, an area of some 40 square kilometers near Point Barrow, Alaska, supports only about 16 species of land mammals, whereas comparable figures for eastern Kansas and an area near Panama City, Panama, are 55 and 140, respectively (E. R. Hall and Kelson, 1959:xxiv). A number of factors seem to influence this trend. Mammals near Point Barrow can live only on the surface of the ground or in a shallow stratum of soil above the permafrost, which extends locally to depths of about 300 meters. Large bushes and trees are absent, and bodies of water are frozen throughout the winter. There are therefore neither arboreal mammals nor those that depend on open water. Insects are available for only two months of the year; understandably, there are no bats. Primary productivity is contributed entirely by low-growing layers of vegetation.

The contrast between this community and that of a tropical rain forest is extreme. In the rain forest, an abundance of evergreen vegetation extends from ground level to over 30 meters above the ground, and temperatures are

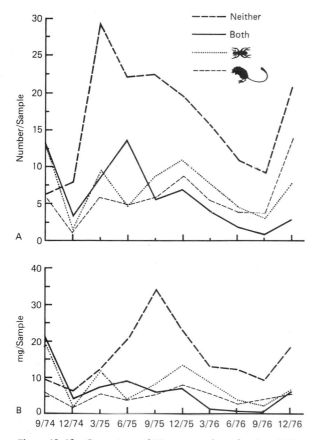

Figure 18–15. Comparisons of (A) mean numbers of seeds and (B) their biomass in soil samples from plots where neither ants nor rodents were excluded, where both were excluded, where ants only were excluded, and where rodents only were excluded. Note that seed biomass increased markedly when granivores (seed-eaters) were excluded. (From J.H. Brown, Davidson, and Reichman, 1979)

moderate year long. Throughout the year, there is available a variety of food that never occurs even briefly in the Arctic. Large nectar-bearing flowers and fruit are present at all seasons and support nectar-feeding and fruit-eating bats, rodents, and primates; green leaves are perennially available to primates, rodents, ungulates, and (in the Neotropics) tree sloths. Of tremendous importance are kinds of insects absent or rare in the far north. Immense numbers of moths and beetles support a great variety of insectivorous bats. Termites support anteaters (in

some Neotropical forests) and elephant shrews (in some African forests) and contribute to the diets of many small mammals. Dung beetles are eaten by everything from bats to small carnivores. In addition, many kinds of shelter, from beneath the ground to high in the trees, are provided. For mammals, and for animals in general, the Arctic offers a restricted assortment of modes of life, whereas the tropics offer an extraordinarily rich variety of possibilities.

The stability of a community seems to depend in part on its complexity. The greater the species diversity among both plants and animals and the greater the number of energy pathways, the more resistant the community is to such changes as strong shifts in the densities of common species. MacArthur (1955) suggested that "stability increases as the number of links increases," and more recently Goodman (1975) has discussed the positive relationship between species diversity and community stability.

At the interface where one community adjoins another, as where a grassy meadow meets a coniferous forest, an "edge effect" is produced. At this edge where the communities adjoin, a greater diversity and density of animals may occur than within either adjacent community, owing to the increased diversity of vegetation and types of shelter. In chamise *(Adenostoma)* chaparral in California, for example, rodents are concentrated near the chaparral edge where herbaceous seeds and seedlings are most abundant (Bradford, 1976). Carnivores frequently concentrate their efforts on these edge situations, and habitat manipulation by game managers often includes making maximum use of the edge effect.

Often there are no sharp dividing lines between adjacent communities. A subalpine forest, for example, may become progressively more interrupted by open "parks" or alpine meadows until the treeless alpine tundra predominates. The zone of gradation between communities, whether broad or narrow, is called an ectone. Ecotonal belts between communities are broad in some regions, as in western Mex-

Table 18-3: Some Major Plant Communities and Their Distributions

Community	Distribution
Tropical rain forest (Fig. 18–16)	South and Central America, Africa, S.E. Asia, East Indies, N.E. Australia
Tropical deciduous forest (Fig. 18–17)	Mexico; Central and South America; Africa; S.E. Asia
Temperate rain forest (Fig. 18–18)	Pacific Coast from northern California to northern Washington; parts of Australia, New Zealand, and Chile
Temperate deciduous forest (Fig. 18–19)	Eastern United States, parts of Europe and eastern Asia
Subarctic-subalpine coniferous forest (Fig. 18–20)	Northern North America, Eurasia; high mountains of Europe and North America
Thorn scrub forest (Fig. 18–21)	Parts of Mexico; Central and South America; Africa; S. E. Asia
Temperate woodlands (Fig 18-22)	Parts of western and southwestern United States and Mexico; Mediterranean area; parts of Southern Hemisphere
Temperate shrublands (Fig. 18–23)	California; Mediterranean area; South Africa; parts of Chile; western and southern Australia
Savanna (tropical grasslands) (Figs. 18–24 and 18–25)	Parts of Africa, Australia, southern Asia, South America
Temperate grasslands (Fig. 18–26)	Plains of North America; steppes of Eurasia; parts of Africa and South America
Arctic and alpine (Fig. 18–27)	Tundras north of treeline in North America and Eurasia; some areas in Southern Hemisphere
Deserts (Figs. 18–28 and 18–29)	On all continents; widespread in North America, North Africa, and Australia

ico, where one may travel for many kilometers through the transitional region between the Sonoran desert community and the tropical thorn forest community. Under such conditions, there is a continuous gradient from one major community to another. Despite the difficulty in assigning geographic limits to some communities, the major terrestrial communities are usually readily recognized. Some terrestrial communities and their distributions are shown in Table 18-3, which is based on the community and plant-formation types of Whittaker (1970:52). Some of these communities are illustrated in Figs. 18–16 to 18–30.

Over broad areas, this pattern of gradual clinal changes in the biota in response to gradual climatic and edaphic changes is more typical than is a sudden shift from one community to another. Because the classification of units (areas) within such a continuum involves arbitrary choices as to limits and takes little account of variation within units, systems of classifica-tion of broad environmental units have not been completely satisfactory. Nevertheless, some of them have been widely used. Such a classification is one that recognizes terrestrial biomes. According to Odum (1971:378), "The biome is the largest land community unit which it is convenient to recognize." The biome is defined in terms of climate, biota, and substrate. The following biomes are recognized by Odum (1971:379): tundra, northern coniferous forest, temperate deciduous and rain forest, temperate grassland, chaparral, desert, tropical rain forest, tropical deciduous forest, tropical scrub forest, tropical grassland and savanna, and mountains, with complex zonation of plants and animals.

In North America, the "life zone" has in the past been used widely in descriptions of vertebrate distributions. Life zones were originally described as temperature zones by Merriam (1894) but later were used as community zones characterized by assemblages of plants and animals.

Figure 18–16. (A) Tropical rain forest near Catemaco, southern Veracruz, Mexico. Common mammals in this area include the Mexican mouse-opossum *(Marmosa mexicana)*, opossums (*Didelphis marsupialis* and *Philander opossum*), many kinds of leaf-nosed bats (Phyllostomidae), the howler monkey *(Alouatta villosa)*, agouti *(Dasyprocta mexicana)*, and coati *(Nasua narica)*. (B) Tropical rain forest in the Chyulu Range, Tsavo West National Park, Kenya. In addition to a diverse group of smaller mammals, this forest supports African elephants *(Loxodonta africana)* and African buffalo *(Syncerus caffer)*.

Figure 18–17. Tropical deciduous forest near Kibwezi, southern Kenya. This forest, shown in the dry season, is dominated by several species of *Commiphora*. Common mammals in the area when the photograph was taken were elephant shrews *(Elephantulus rufescens)*; vervet monkeys *(Cercopithecus aethiops)*; baboons *(Papio anubis)*; bushbabies *(Galago senegalensis)*; genets *(Genetta tigrina)*; African civets *(Viverra civetta)*; a variety of bovid ungulates, including dik-diks *(Madoqua kirki)* and bushbuck *(Tragelaphus scriptus)*; and bats of the families Pteropodidae, Emballonuridae, Nycteridae, Megadermatidae, Rhinolophidae, Vespertilionidae, and Molossidae.

Figure 18-18. Temperate evergreen rain forest in Olympic National Park, Washington. Mammals typical of this area are the shrew mole *(Neurotrichus gibbsii)*, Townsend's mole *(Scapanus townsendii)*, mountain beaver *(Aplodontia rufa)*, western red-backed mouse *(Clethrionomys occidentalis)*, marten *(Martes americana)*, black bear *(Ursus americana)*, and elk *(Cervus elaphus)*. (R. Atkeson)

Merriam's thinking was influenced by his recognition of the sharp elevational zonation of biotas in some mountains of the western United States (Fig. 18–30). Although the life zone system is useful descriptively, it is not a precise scheme for delineating plant-animal communities and is therefore not in general use today.

Biotic Interactions

Considerations of interactions between one species of animal and another and between plants and animals are essential to an understanding of mammalian ecology. A biotic community is a tremendously complex functional unit within which animals live, feed, reproduce, and die; it has an evolutionary history and through time maintains a state of dynamic equilibrium. The role of an organism in a community depends on that organism's interactions with other members of the community and with the physical environment. The fabric of the entire community depends on the combined effects of the interlacing threads of interaction.

Mammals and Community Structure

Many ecologists regard competition and predation as important forces structuring communities. Both of these forces set limits on the pat-

Figure 18–19. Temperate deciduous forest in Indiana in summer. Mammals typical of this type of community are the short-tail shrew *(Blarina brevicauda),* eastern chipmunk *(Tamias striatus),* gray squirrel *(Sciurus carolinensis),* flying squirrel *(Glaucomys volans),* white-foot mouse *(Peromyscus leucopus),* gray fox *(Urocyon cinereoargenteus),* and white-tail deer *(Odocoileus virginianus).* (U.S. Forest Service)

Figure 18–20. Subalpine coniferous forest near Rabbit Ears Pass, Routt County, Colorado. The following mammals are common in this community: shrews *(Sorex vagrans* and *S. cinereus),* the red squirrel *(Tamiasciurus hudsonicus),* least chipmunk *(Eutamias minimus),* pocket gopher *(Thomomys talpoides),* montane vole *(Microtus montanus),* red-backed vole *(Clethrionomys gapperi),* beaver *(Castor canadensis),* porcupine *(Erethizon dorsatum),* red fox *(Vulpes vulpes),* mule deer *(Odocoileus hemionus),* and elk *(Cervus elaphus).*

Figure 18–21. Thorn scrub near San Carlos Bay, Sonora, Mexico. Typical mammals are bats (including several members of the Neotropical families Phyllostomidae and Mormoopidae), the antelope jackrabbit *(Lepus alleni),* Merriam's kangaroo rat *(Dipodomys merriami),* hispid cotton rat *(Sigmodon hispidus),* white-throated woodrat *(Neotoma albigula),* and javelina *(Tayassu tajacu).*

terns of resource use by the members of a community. Some of the best studies of mammals and community structure have included experimental habitat manipulation.

The studies of J. H. Brown and his colleagues (Brown, Davidson, and Reichman, 1979; Davidson, Brown, and Inouye, 1980) have suggested the importance of competition between distantly related granivores (seed-eaters) in structuring desert communities. Their study area in southern Arizona supported granivorous heteromyid rodents, which preferred large seeds, and harvester ants, which preferred small seeds. A series of experimental plots were manipulated by removing ants and leaving rodents, by removing rodents and leaving ants, or by removing both. By comparison with undisturbed control plots, numbers of ant colonies increased sharply (71 percent) in the plots with

no rodents, and rodent biomass increased 29 percent in the ant-free plots. Whereas seed densities were roughly equal in the ant-free, rodent-free, and undisturbed control plots, they were four times higher in plots with neither ants nor rodents (Fig. 18–15). Plant density only doubled on the latter plots, indicating competition between seedlings. These results indicate that rodents and ants compete for seeds and affect seed and plant densities. They also show that plant densities are partly controlled by competition between plants and that the composition of the plant community may be influenced by the seed preferences of rodents and ants.

Further, rodents and ants interact indirectly. By concentrating on large-seed plants, rodents reduce the abundance of these plants and allow a reciprocal increase in the abundance of small-seed species. This shift benefits

Figure 18–22. Temperate woodlands. (A) Piñon-juniper woodland near Flagstaff, Arizona. Most of the trees are piñons; the tree at the extreme left is a juniper. Common mammals include several bats of the genus *Myotis*, the desert cottontail *(Sylvilagus audubonii)*, piñon mouse *(Peromyscus truei)*, northern grasshopper mouse *(Onychomys leucogaster)*, Stephen's woodrat *(Neotoma stephensi)*, gray fox *(Urocyon cinereoargenteus)*, bobcat *(Lynx rufus)*, and mule deer *(Odocoileus hemionus)*. (B) Oak woodland near Claremont, California. A number of vespertilionid bats, the gray squirrel *(Sciurus griseus)*, brush mouse *(Peromyscus boylii)*, dusky-foot woodrat *(Neotoma fuscipes)*, and raccoon *(Procyon lotor)* are common in this habitat.

Figure 18–23. Temperate shrubland (chaparral) near San Antonio Canyon, Los Angeles County, California. (A) Steep slopes covered by dense brush. (B) Interlacing stems and branches beneath the chaparral in the right foreground of A. Typical mammals in this area are the western pipistrelle *(Pipistrellus hesperus)*, Merriam's chipmunk *(Eutamias merriami)*, brush mouse *(Peromyscus boylii)*, California mouse *(P. californicus)*, dusky-foot woodrat *(Neotoma fuscipes)*, gray fox *(Urocyon cinereoargenteus)*, bobcat *(Lynx rufus)*, and mule deer *(Odocoileus hemionus)*.

Figure 18–24. (A) Savanna in the Serengeti Plains, Tanzania, Africa, This area supports large numbers of ungulates. Some of the typical kinds are zebra *(Equus burchelli),* buffalo *Syncerus caffer),* wildebeest *(Connochaetes taurinus),* and Thompson's gazelle *(Gazella thompsonii).* Lions *(Panthera leo),* hyenas *(Crocuta crocuta),* and African hunting dogs *(Lycaon pictus)* prey on these ungulates. The African savanna supports a richer ungulate fauna than any other area in the world. (B) Twig of *Acacia elatior,* an East African riverine tree that has huge thorns (some 6 cm long) as defense against browsing ungulates. This defense is typical also of many savanna-dwelling acacias.

Figure 18–25. Savanna (mallee scrub) in the southern part of New South Wales, Australia. The trees are eucalyptus. In this area occur a number of marsupials, including marsupial "mice" *(Sminthopsis crassicaudata* and *Antechinus flavipes),* the vulpine phalanger *(Trichosurus vulpecula),* the "mallee gray" kangaroo *(Macropus giganteus),* and the red kangaroo *(Megaleia rufa).* (D. Harrison)

311

Figure 18-26. Temperate grassland: short-grass prairie near Nunn, Colorado. Among the common mammals are the white-tail jackrabbit *(Lepus townsendii)* and black-tail jackrabbit *(L. californicus)*, the thirteen-line ground squirrel *(Spermophilus tridecemlineatus)*, prairie dog *(Cynomys ludovicianus)*, Ord's kangaroo rat *(Dipodomys ordii)*, northern grasshopper mouse *(Onychomys leucogaster)*, coyote *(Canis latrans)*, badger *(Taxidea taxus)*, and pronghorn antelope *(Antilocapra americana)*. (R. E. Bement, Agricultural Research Service)

ants. Clearly this system is dynamic. Even small year-to-year differences in rainfall, which are known to affect the germination and growth of desert annuals, would alter the food supply of the granivores and might favor either small-seed or large-seed plants. Such changes in food supply would be reflected by shifts in the populations of the granivores and would thus alter the intensity of seed harvesting and the crop of seeds that survive to germinate the following year.

The role of microhabitat in structuring rodent communities has been studied in experimental enclosures. Each of four species of heteromyid rodents occupied a different microhabitat, and each became more abundant when its particular microhabitat was increased. Each species shifted its microhabitat use when its competitors were manipulated experimentally. When its competitors were removed, a species used a greater breadth of habitat, whereas with competitors present it concentrated its activity

Figure 18-27. Alpine moorlands (tundra) on Mt. Kenya in Kenya. This site lies within one degree of the equator and is at an elevation of 4600 meters. Typical mammals include a mole rat *(Tachyoryctes splendens)*, rock hyrax *(Procavia* sp.), the black-fronted duiker *(Cephalophus nigrifrons)*, and the African buffalo *(Synceros caffer)*.

Figure 18-28.   Sonoran Desert scrub near the Kofa Mountains, Arizona. Common mammals of the desert floor are pocket mice (*Perognathus amplus* and *P. penicillatus*), kangaroo rats (*Dipodomys merriami*), woodrats (*Neotoma albigula*) and antelope ground squirrels (*Ammospermophilus harrisi*). Typical predators are coyotes (*Canis latrans*) and kit foxes (*Vulpes macrotis*).

in its preferred microhabitat (Price, 1978). In a small-mammal community in the Kalahari Desert of Africa, marked changes occurred in association with shifts in rainfall during the six-year period of a study (Fig. 18–31; Nel, 1978). Some changes in niche dimensions and niche overlap between the rodents were similar to those Price observed experimentally.

Some mammals are known to have an important influence on a specific aspect of community structure. Following are several examples. Frugivorous bats in Neotropical forests carry fruit to feeding trees or roosts, and seeds are discarded at these sites. By producing local concentrations of seeds of fruit-bearing trees, these bats favor the occurrence of mixed-species clumps of fruit trees (Fleming and Heithaus, 1981). For Abert's squirrel (*Sciurus aberti*), an inhabitant of ponderosa pine (*Pinus ponderosa*) forests, hypogeous fungi are the

Figure 18-29.   Sandy desert near Lake Victoria, southwestern New South Wales, Australia. Mammals of this community include a marsupial "mouse" (*Sminthopsis crassicaudata*), a kangaroo (*Macropus giganteus*), and a rodent, the Australian kangaroo mouse (*Notomys mitchelli*). (D. Harrison)

Figure 18-30. Pronounced zonation of vegetation on the San Francisco Peaks in northern Arizona. Observations he made in this area influenced Merriam's thinking on life zones. In the foreground, at about 1900 meters elevation, is short-grass prairie; in the near distance are flats supporting piñon and juniper trees. The mesas and slopes in the middle distance are covered with ponderosa pine, the higher slopes in the distance have spruce and fir forests, and on the treeless and barren tops of the peaks, between 3350 and 3660 meters elevation, grow small alpine forbs and grasses.

primary food in spring, summer, and fall. As the most important dispenser of these fungi, this squirrel favors the symbiotic relationship between pine and fungi (States, 1980). The fungi gains optimal habitat (shaded ground with a thick layer of ponderosa pine needles) from the pines, and the pines get essential nutrients from the metabolic byproducts of the fungi. The herbivorous cotton rat *(Sigmodon hispidus),* perhaps through its style of clipping shoots, has an effect on the productivity of rye grass *(Lolium perenne).* Rye grass clipped by this rodent has a higher regrowth rate than mechanically cut grass (Howe, Grant, and Folse, 1982).

Many ecologists believe that predation plays a critical role in structuring communities. Predators are responsive to the distribution and abundance of their prey: individuals of a prey species that are not discriminate in their use of space, time, and food resources are unusually vulnerable to predation. As an extreme example, the vole that appears on the surface of snow in daylight is easily captured by an array of birds of prey and by coyotes, foxes, and weasels, whereas weasels are the only predators consistently able to kill voles beneath a snowpack. Predators thus exert chronic selection on each prey species for optimal and discriminate use of the habitat. Each prey species tends to occupy the microhabitat where it is least vulnerable to predators, to forage when it is least conspicuous, and to maintain population levels below limits set by environmental resources. Because more species with discriminate patterns of resource utilization can exist in a community than can species with indiscriminate patterns, predation fosters species diversity. Predation shapes and maintains community structure,

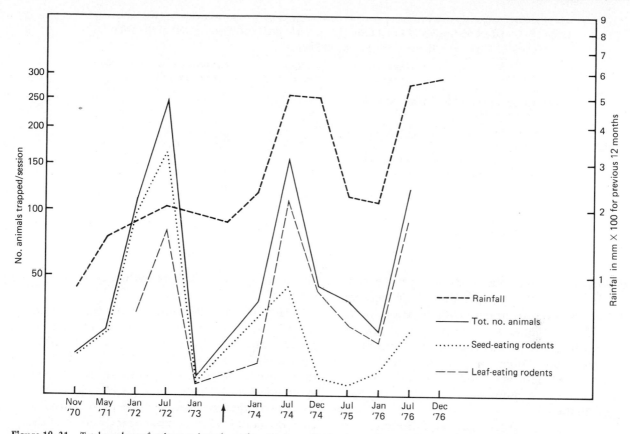

Figure 18–31.   Total numbers of rodents and numbers of seed-eating and leaf-eating rodents trapped over a several-year period in the Kalahari Gemsbok National Park, South Africa. (From Nel, 1978)

then, by limiting the ways in which a prey species can utilize resources.

Keystone Species

Some species of mammals affect their communities so strongly that can be viewed as "keystone species." Such species "are the keystone of the community's structure, and the integrity of the community and its unaltered persistence through time, that is, stability, are determined by their activities and abundance" (Paine, 1969).

The wildebeest *(Connochaetes taurinus)* is a keystone species. The annual migration of

wildebeest from the Serengeti Plains in Tanzania, northward and westward toward the bush country of northern Tanzania and the Maasai-Mara Game Reserve of Kenya, is a justly famous spectacle. The movement begins at the end of the rainy season, in May or June, and the animals return to the short grass in November. In May 1974, some half million wildebeest were concentrated in the short-grass country along the western border of the Serengeti Plains. On May 22, the great herd began to move northward and, during the four-day period of the passage of the main mass, the animals had a tremendous impact on the dynamics of this short-grass community (McNaughton, 1976). In this brief time, grazing reduced the green biomass of this *The-*

Table 18–4: Effect on Grassland Vegetation (Largely *Themeda* and *Pennisetum*) of the Four-Day Passage of Wildebeest Herds in the Serengeti Plains

	Biomass (g/m²)	Height (cm)	Biomass Concentration (mg/10 cm²)
Fenced vegetation, wildebeest excluded			
Before passage	501.9	64	7.9
After passage	449.2	63	7.1
	N.S.	N.S.	N.S.
Vegetation subject to wildebeest grazing			
Before passage	457.2	66	6.9
After passage	69.0	29	2.4
	P = .005	P = .005	P = .05

P = level of significance; N.S. = not significant.
After McNaughton, 1976.

meda-Pennisetum grassland by 84.9 percent and the height of the vegetation by 56 percent (Table 18–4). This apparent devastation markedly affected the subsequent growth of grasses and, indirectly, the dry-season distribution of the abundant Thompson's gazelle *(Gazella thompsonii)*. This antelope, next to the wildebeest, is the most abundant ungulate in the Serengeti Plains.

During the 28-day period following the main migration, areas grazed by wildebeest had a net productivity of green vegetation of 2.6 grams per square meter per day, whereas in experimental plots protected from grazing, green biomass declined at a rate of 4.9 grams per square meter per day. A dense mat of new and nutritious vegetation was produced in the grazed area by abundant tillering (growth of new shoots from nodes of stems or of rhizomes) of the grasses, while in the ungrazed plots the bulk of the biomass was tall, relatively non-nutritious grass stems. One month after the exodus of the wildebeest, the area was occupied by Thompson's gazelles, which selectively grazed the areas of vigorous regrowth, that is, the areas previously grazed heavily by wildebeest. As an indication of the high selectivity of the gazelles, consumption of vegetation in these areas averaged 1.05 grams per square meter per day, whereas virtually no grazing by gazelles occurred in the plots where wildebeest had not grazed.

McNaughton concluded that, by their intense grazing, the wildebeest transformed a senescent grassland into a productive community on which Thompson's gazelle depended in the dry season. Further, the coexistence of the wildebeest and Thompson's gazelle is favored by coevolution resulting in the partitioning of the grassland in the dry season, the period of the year when competition between the two most abundant ungulates would be mutually disadvantageous. On the basis of his observations of the grass-ungulate ecosystem of the Serengeti, McNaughton (1979) concluded that grazing there is an optimization process. Moderate grazing stimulates grass productivity to levels twice those in ungrazed control plots, and even under intense grazing pressure, productivity remains high. Heavy grazing by ungulates for many thousands of years has selected for rapid compensatory growth by grasses after defoliation. The conventional wisdom of range managers as to the harmful effects of "overgrazing" seems not to apply to the Serengeti ecosystem.

The activities of a single mammal, the sea otter *(Enhydra lutris),* seem to play a critical role in structuring nearshore marine communities along the Aleutian Islands of Alaska (J. A. Estes and Palmisano, 1974). The Rat Islands

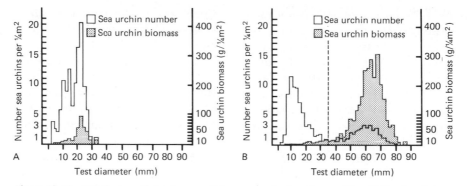

Figure 18-32. Distributions of size classes and biomass contributions of sea urchins. (A) Based on data from Amchitka Island, which supports high densities of sea otters; (B) based on data from Shemya Island, where sea otters are absent. The dotted line indicates the largest class of sea urchins present at Amchitka Island. (From J. A. Estes and Palmisano, 1974)

support high populations of sea otters, and the nearshore community is characterized by abundant beds of macrophytes, consisting mostly of brown algae (kelp) and red algae; filter feeders, such as barnacles and mussels, are scarce, as are motile herbivores, such as sea urchins and chitons. In the Near Islands, 400 kilometers to the northwest, sea otters are absent; here macrophytes are scarce below the lower intertidal zone, and barnacles, mussels, sea urchins, and chitons are many times denser than along the Rat Islands (Fig. 18–32). These differences seem related to the activities of otters. These animals prey heavily on sea urchins (*Strongylocentrotus* spp.), which graze on macrophytes. When high populations of otters drastically reduce the abundance of urchins, kelp beds flourish and the complexion of the nearshore community is conspicuously altered. The abundance of vertebrates is also affected. Probably because of the relative abundance of fish around kelp beds, harbor seals *(Phoca vitulina)* and bald eagles *(Haliaeetus leucocephalus)* are common in the Rat Islands but rare or absent in the Near Islands. It seems, then, that the character of the nearshore communities depends on the presence or absence of otters, and it may well be that productive kelp beds and a stable nearshore community on the Pacific coast of the United States are maintained only in the pres-

ence of the sea otter. The otter thus seems to be the keystone of the nearshore community.

Plant-Mammal Coevolution

Through many millions of years of plant-herbivore interactions, plants have become adapted to animals, often by taking advantage of them as agents of pollination or seed dispersal, and animals have developed means of coping with plant defenses and of selecting plant material with high energetic-nutritional yields per unit of foraging time. Such coevolution has resulted in many remarkable patterns of plant-animal interdependence and probably in some far-reaching changes in land floras. The Cretaceous ecological revolution that resulted in the domination of terrestrial floras by angiosperms (flowering plants) was perhaps facilitated by the diversification of birds and mammals and their effectiveness as seed dispersers (Regal, 1977).

For a particularly interesting example of mutually beneficial interactions between mammals and a plant, we return to Serengeti National Park. Here, and over wide areas of East Africa, the umbrella tree *(Acacia tortilis)* is a conspicuous, picturesque, and important savanna tree. The green, leathery pods of the plant are eaten avidly by a variety of herbivores, ranging from the tiny dik-dik to the elephant, and elephants often feed heavily on the foliage. Lamprey, Halevy, and Makacha (1974) found that the germination rates of seeds that had been ingested and eliminated by herbivores were strikingly higher than the germination rates of uningested seeds. In some vertebrate-adapted seeds, the digestive process is known to erode the seed coat and hasten germination, but the high germination rate in ingested acacia seeds is thought by Lamprey and his coworkers not to depend primarily on this effect. Beetles of the family Bruchidae lay their eggs on acacia seed pods, and the larvae feed and grow within the seeds. If the larvae damage the embryo or destroy a large amount of the cotyledon material, the seed will not germinate, but if the seed pods are eaten by herbivores soon after they fall to the ground, as is typically the case in the Serengeti, the digestive process kills the larval bruchids at an early stage of development, before they have killed the seeds. Some 500 seed samples were collected from the ground and stored for one year; over 95 percent of these seeds had bruchid damage, and the germination rate was only 3.0 percent. In seeds eaten by impalas *(Aepyceros melampus)*, however, the damage rate was 26 percent and the germination rate was 28 percent; in seeds eaten by dik-diks, these figures were 45 percent damage rate and 11 percent germination rate. The interactions seem clearly to be mutually advantageous: from the acacia, the herbivores get a seasonally important food; from the mammals, the acacia gains an effective means of escape from a seed predator and wide dissemination.

Some mutually advantageous plant-animal associations are long-standing, and in some cases, remarkable mutual adaptations have evolved. Especially noteworthy examples of such coevolution are offered by nectar-feeding bats and the plants on which they feed. In the Old World, a number of bats of the family Pteropodidae feed on nectar; in the Neotropics, nectar feeding occurs in the family Phyllostomidae.

Chiropterophily (the dependence of a plant largely on bats for pollination) has been discussed by Alcorn, McGregor, et al. (1959), H. G. Baker (1961, 1973), and Faegri and Van Der Pijl (1966:111), and laboratory studies by D. J. Howell (1974b) have demonstrated several physiological adaptations of a bat *(Leptonycteris sanborni)* that feeds from flowers. Pollination of plants by mammals may originally have involved marsupials and lemurs, but nectar-feeding bats probably displaced these animals early in the Tertiary (Sussman and Raven, 1978). Only in areas where nectar-feeding bats are uncommon or absent, as in Madagascar and parts of Australia, does this archaic type of pollination by lemurs or marsupials persist.

Rarely is pollination in a plant that depends on animal pollination accomplished by a single species. A common situation is that involving pollination by many agents, including perhaps a number of insects and birds. Plants of this sort are termed polyphilous. Some plants, on the other hand, are pollinated by relatively few agents, and the flowers are open only during the day or night. In these plants, characteristics of the flower that favor one or several pollinators would be expected to be under strong positive selective pressure; the plants would be expected to evolve features making their flowers especially attractive to a small group of pollinators. This seems to have occurred in a number of bat-pollinated plants, but in no case is the bat the sole pollinating agent.

Most chiropterophilous plants occur in the tropics and subtropics. H. G. Baker (1961) ascribes this to the fact that the form of tropical trees and their inflorescences encourage nectar-feeding bats, whereas trees of temperate regions are less hospitable.

Bat-pollinated plants typically have the following features: the flowers occur on spreading, often leafless branches and are unobstructed by vegetation; they are pale in color and offer visually conspicuous targets at night; they have a strong smell, often resembling that of bats; copious amounts of nectar are produced in early

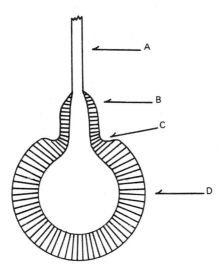

Figure 18-33. Schematic cross section of an inflorescence of *Parkia clappertoniana:* (A) stout peduncle; (B) sterile, nectar-producing flowers; (C) ring in which nectar collects and where bats feed; (D) potentially fertile flowers. (After H. G. Baker and Harris, 1957)

evening when the bats forage; the plants produce flowers over long periods, but only a few buds open each night (this favors traplining, a nightly patrolling of the same series of plants by the bats).

Two general types of flowers are chiropterophilous. The first type, an example of which is the large flower of the African baobab tree, accommodates large bats. The feeding bat clutches the ball of stamens of this flower while lapping nectar from the pillar-like stamen column. An African species of *Parkia* studied by H. G. Baker and Harris (1957) displays a division of labor between nectar-producing and reproductive stamens (Fig. 18-33). Bats of the genera *Epomophorus* and *Nanonycteris* were observed lapping nectar from the circular depression in the top of the ball of the flowers, which was grasped by the bat's feet (Fig. 18-34). Most of the flowers of a tree are in a staminate condition one night and a pistillate condition the next, thus ensuring cross-fertilization. Evidence of the importance of bats as

agents of pollination in *P. clappertoniana* is provided by the position of the fruit, which tends to form only on the side of the inflorescence that faces the periphery of the tree, the very side from which bats characteristically approach.

The second group of flowers includes many Neotropical types adapted to smaller bats; these flowers often have corolla tubes into which the bat pushes its head. During approaches to these flowers, the bat's back is liberally dusted with pollen.

Sanborn's long-nosed bat *(Leptonycteris sanborni)* is a New World bat of the family Phyllostomidae that feeds on nectar (Fig. 8-9). Nectar is rich in carbohydrates, but has no more than trace amounts of protein. Vertebrate animals are known to need protein in their diet. Has *Leptonycteris* developed the remarkable and unique ability to subsist on a pure carbohydrate diet, or does this bat take from flowers other food in addition to nectar? D. J. Howell (1974b) showed that *Leptonycteris* must indeed have protein in its diet: experimental bats fed on artificial or real nectar fortified with vitamins and sodium chloride but containing no

Figure 18-34. An African pteropodid bat *(Epomophorus gambianus)* feeding at an inflorescence of *Parkia clappertoniana.* (H. G. Baker and B. J. Harris)

Table 18–5:   Protein in Pollen of Plants Pollinated by Bats and in Pollens Dispersed by Other Means

Plant Species	Protein Content of Pollen (%)
"Bat-adapted" pollen	
Agave palmeri	22.9
Carnegeia gigantea	43.7
"Nonbat" pollen	
Opuntia versicolor	8.93
Ferocactus wislizeni	10.17
Agave schottii	8.30
Agave parviflora	15.90

Data from D. J. Howell, 1974b.

proteins lost weight rapidly, and most died within eight days, whereas bats given nectar and a protein source remained healthy.

Under natural conditions, the source of protein for these bats (and presumably for most nectar-feeding bats) is pollen, which these bats ingest in large quantities. Some pollen may be taken directly from flowers, but probably the bats ingest pollen largely when grooming their fur. Nectar-feeding bats have specialized hairs to which pollen adheres extremely readily (Fig. 8–19). Of particular importance is the fact that the pollen of the plants visited by Leptonycteris (saguaro, Carnegeia gigantea, and agave, Agave palmeri) is much higher in protein than the pollen of closely related plants that are pollinated by a variety of agents (Table 18–5). Of additional interest, the "bat pollens" contain at least 18 amino acids and are unusually high in the amino acid proline, which makes up over 80 percent of the protein collagen. Because it composes much of the extensive network of connective tissue that braces the wings and tail membrane, collagen is especially important in bats.

But Leptonycteris faces another problem, for pollen grains are notoriously resistant to being broken down mechanically or biologically. In a sugar solution, however, such as that in the stomach of a bat that has fed on nectar, the cellular contents of pollen are known to ex-

trude through the pores. Also, hydrochloric acid solutions extract proteins from some kinds of pollen, and unusually high concentrations of HCl-secreting glands are present in the gastrointestinal submucosa of Leptonycteris. Finally, the habit that this bat has of ingesting its urine on occasion could possibly be of importance in connection with digestion (D. J. Howell, 1974b:270) since urea is known to degrade pollen protein.

In the particular case of Leptonycteris, then, the "bat flowers" have seemingly evolved a type of pollen that supplements the nectar and provides the bat with a balanced diet. The bat, in turn, has specialized hairs that facilitate rapid accumulation of this pollen and a digestive system able to utilize it. The plants are effectively pollinated, and the bat is well fed.

Assorted Interactions

Mammals interact with many other kinds of animals and benefit from some of these interactions. Murie (1940) found that coyotes were often attracted to carrion by concentrations of ravens in Yellowstone National Park, and in Africa, jackals (Canis spp.) and spotted hyenas (Crocuta crocuta) are avid vulture watchers. The high leaps made by dolphins allow them to see assemblages of sea birds and thereby locate large schools of fish. Coyotes have been observed following a foraging, low-flying golden eagle, presumably to take advantage of prey flushed by the raptor.

One of the most remarkable instances of mutually beneficial interaction concerns a bird, the African honey guide (Indicator indicator), and the African honey badger (Mellivora capensis). The bird attracts the badger's attention by raucous chattering and then leads the way to a bee's nest. After the nest is torn apart by the mammal (African tribesmen sometimes perform this service), the honey guide eats fragments of the wax, which it can digest. Another example is found in two species of African birds, the oxpeckers *(Buphaga),* which eat ticks that they remove from big game mammals. The mammals are oblivious to the birds' attention but seem to derive benefit from the loss of ticks and from the alarm calls given by the birds when predators approach.

Some species of mammals profit in various ways from the activities of other mammals. Often one species will use shelter provided by another species: certain white-foot mice *(Peromyscus)* and shrews *(Notiosorex)* use woodrat houses for shelter, and warthogs *(Phacochoerus)* seek refuge in the burrows of aardvarks *(Orycteropus).* The burrows of pocket gophers are used by a variety of vertebrates, including amphibians, reptiles, and other mammals (Vaughan, 1961).

Of particular interest are the close and seemingly mutually beneficial associations that occur between two species of mammals. A badger and a coyote have been observed hunting together on many occasions in widely scattered areas (S. P. Young and Jackson, 1951:95). Herds of impala *(Aepyceros)* were observed staying with baboons *(Papio)* through much of the day by DeVore and Hall (1965:48), who judged that the excellent eyesight of the baboons supplemented the acute senses of smell and hearing of the impalas and made the mixed group difficult for a predator to approach undetected. On one occasion, these authors observed a large male baboon discouraging three cheetahs that were approaching a mixed group of impalas and baboons.

Territoriality and Home Range

Because individuals of the same species usually have identical niches and are therefore potentially competing for the same environmental resources at the same place and time, intraspecific competition might be expected to be intense. Indeed, this seems to be the case, with the result that individuals of the same species customarily occupy separate, or nearly separate, home

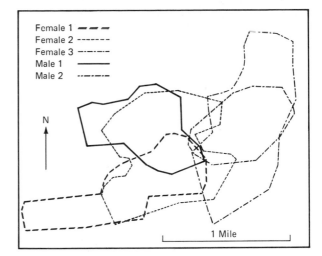

Figure 18–35. Distributions of red fox home ranges at the University of Wisconsin Arboretum. (After Ables, 1969)

ranges. Burt (1943) has described the home range of a mammal as "that area traversed by the individual in its normal activities of food gathering, mating, and caring for the young." Home ranges may have irregular shapes and may partially overlap (Fig. 18–35). Within the home range of some animals is an area that is actively defended against other members of the species. This area, which usually does not include the peripheral parts of the home range, is called the territory, and species that apportion space in this fashion are termed territorial. A home range or a territory may be occupied by one individual, by a pair, by a family group, or by a social group consisting of a number of families.

To solitary animals or to members of a group, the occupancy of a home range has several important advantages. Each home range provides all the necessities of life for an individual or group, permitting self-sufficiency within as small an area as possible; the less extensively the animal must range, the less chance there is for encounters with predators. The home range quickly becomes familiar to the individual, who can then find food and shelter with the least possible expenditure of energy

Table 18-6:   Sizes of Home Ranges of Some Mammals

Species	Home Range (acres)	Source
Common shrew *(Sorex araneus)*	0.7	Buckner, 1969
Varying hare *(Lepus americanus)*	14.5	T. P. O'Farrell, 1965
Mountain beaver *(Aplodontia rufa)*	0.3	P. Martin, 1971
Least chipmunk *(Eutamias mimimus)*	2.1–4.7 (summer only)	Martinsen, 1968
Yellow-pine chipmunk *(E. amoenus)*	3.89 (males); 2.49 (females)	Broadbooks, 1970
White-foot mouse *(Peromyscus)*	0.08–10.66	Redman and Sealander, 1958; Blair, 1951
Red-backed mouse *(Clethrionomys gapperi)*	0.25 (winter only)	Beer, 1961
Prairie vole *(Microtus ochrogaster)*	0.11 (males); 0.02 (females)	Harvey and Barbour, 1965
Timber wolf *(Canis lupus)*	23,040 (pack of 2)	Stenlund, 1955
	345,600 (pack of 8)	Rowan, 1950
Red fox *(Vulpes vulpes)*	1280	Ables, 1969
Grizzly bear *(Ursus arctos)*	50,240 (1 mother + 3 yearlings)	Murie, 1944
Russian brown bear *(U. arctos)*	6400–8320	Bourliere, 1956
Raccoon *(Procyon lotor)*	13.3–83.4	Shirer and Fitch, 1970
Badger *(Taxidea taxus)*	2100	Sargent and Warner, 1972
Mountain lion *(Felis concolor)*	9600–19,200 (males)	Hornocker, 1970b
	3200–16,000 (females)	
Lynx *(Lynx canadensis)*	3840–5120	Saunders, 1963
Black-tail deer *(Odocoileus hemionus)*	90 (winter); 180 (summer)	Leopold, Riney, et al., 1951
Mule deer *(O. hemionus)*	502–2534	Swank, 1958
White-tail deer *(O. virginianus)*	126–282	Ruff, 1938
Pronghorn antelope *(Antilocapra americana)*	160–480	Bromley, 1969

and can escape predators more effectively because escape routes and retreats are familiar and no time or movement is lost in seeking shelter. Some mammals, such as rabbits and meadow voles *(Microtus),* maintain trails that serve as routes to food and as avenues of escape. To some male mammals that guard harems, the territory is an exclusive mating area.

Reproductive success may be increased by an animal's knowledge of areas occupied by animals inhabiting adjoining home ranges (in the case of solitary species) or by familiarity with animals sharing its home range (in the case of social species). During early life, young can develop under parental care largely free from interference by other individuals of their own species. The spacing of home ranges is often such that the individual or the group is assured an exclusive food supply.

Home range size varies tremendously (Table 18–6). Many mammals in the orders In-

sectivora, Primates, Rodentia, Lagomorpha, Carnivora, Perissodactyla, and Artiodactyla are known to be territorial. The recognition of territorial boundaries in some species depends on scent marking and other means of territorial marking, and much remarkable behavior is associated with the maintenance of territories (some of this behavior is discussed in Chapter 20).

Some territorial species are distributed according to a pattern of home ranges that may persist throughout the lives of many generations. Hansen (1962) found such a pattern to be typical of northern pocket gophers *(Thomomys talpoides)* in some areas. Each animal occupies an area of raised ground called a mima mound (Fig. 18–36), which is some 10 meters in diameter. The mima mounds are more productive of food than are the relatively narrow intermound areas, which usually have shallow soil. Except in the winter, the intermound areas are

Figure 18-36. Mima mounds in Mima Prairie, Thurston County, Washington. These mounds, some 10 meters in diameter, are probably formed by the burrowing activities of pocket gophers. (V. B. Scheffer)

used little by pocket gophers, and the chances of survival are slim for an animal that is unable to establish itself in a mima mound. Likewise, woodrat houses may be used over periods of thousands of years (P. V. Wells and Jorgensen, 1964), as indicated by the presence in them of plants that no longer occur in the area but did thousands of years ago.

Feeding Strategies

Biologists stress that the success of organisms can be measured in terms of relative reproductive rates; the organism with the greatest fitness from an evolutionary point of view is the one that leaves the greatest number of progeny. A morphological, physiological, or behavioral character and the genotype determining it are adaptive if the organisms possessing the feature leave more progeny than do those without it.

Evolutionary change thus depends in part on the differential fitness of organisms with different genotypes. The primacy of reproduction in determining the success of a species is virtually axiomatic—but what about energy?

For an individual to survive to reproductive age requires energy, and the reproductive process itself is energetically expensive. Clearly, no aspect of an animal's life more directly affects its survival and reproductive success than its efficiency in gaining energy. The increased reproductive success with age in some species is probably a reflection of the animals' learning to forage more efficiently (Ricklefs, 1973).

Feeding strategies are most appropriately viewed in terms of cost and benefit. The optimal strategy assures the greatest yield of energy from the food in relation to the energy cost of pursuit, handling, and eating, per unit of time spent in these activities. Put more gracefully by MacArthur and Pianka (1966), natural selection will achieve optimal allocation of time and energy expenditures. Schoener (1971) gives four key aspects of feeding strategies: optimal diet, optimal foraging space, optimal foraging period, and optimal foraging-group size. A great variety of strategies occurs among mammals; following are a few examples.

The pronghorn antelope *(Antilocapra americana)*, an herbivore that feeds primarily on forbs and small shrubs, utilizes a strategy common to many ungulates. Most of the animal's time is spent eating or processing food. It alternately feeds and beds down through the day and night, and ruminating (chewing the cud) occupies 60 to 80 percent of the bedding time (Kitchen, 1974). The use of open country and group foraging allows the animals to make the most of their sharp vision and speed in avoiding predation.

Herbivory is popular among mammals and has many advantages: the biomass of plants vastly exceeds that of animals, plants need not be pursued, and they are equally available throughout the 24-hour cycle. Because plants with high levels of secondary compounds must be detected and avoided, however, and because the energy yield of plant tissue per unit of weight is low relative to that of animal material, herbivores must invest a great deal of time in feeding. Among granivores, however, the allocation of time is markedly different. Seeds are the most concentrated source of energy that plants offer; granivores therefore spend relatively little time foraging. Kangaroo rats and pocket mice, for example, forage solitarily at night. The finely tuned tactile ability of pocket mice, which enables them to discriminate between seeds and seed-size nonedible items, is in keeping with their ability to forage for dispersed seeds, whereas the more poorly developed tactile ability of kangaroo rats is adequate because they tend to utilize clumps of seeds (Lawhon and Hafner, 1981). Both these granivores make a series of foraging trips, returning with bulging cheek pouches to deposit seeds in their burrows. This gathering phase requires

considerable energy, and during gathering the rodents are exposed to predation and perhaps to low temperatures, but foraging takes as little as an hour or two per night. Selectively in the choice of foraging microhabitat, tactile ability of the forepaws, and the use of high-energy food allow heteromyids to remain in the safety of their burrows for most of the 24-hour cycle. However, they are at the mercy of the vagaries of desert climates. During years when fall and winter rains fail, there may be no green annuals or seed production by annuals, and at these times heteromyids do not reproduce and their populations decline (Beatley, 1969).

Predatory animals employ a variety of feeding strategies, but two major types represent the extremes. The sit-and-wait predator (Pianka, 1966) remains quietly at a vantage point and surveys its surroundings; when prey is detected, the predator makes a brief attack and typically avoids lengthy pursuit. The giant left-nosed bat *(Hipposideros commersoni)* of Africa exemplifies this predatory style (Vaughan, 1977). This bat hangs from an acacia branch and uses echolocation to scan for insects. It is discriminating in its choice of prey and feeds exclusively on large, straight-flying insects. These are captured during brief and precise interception flights, and the bat returns to its perch to consume the insect. Such predators are typically solitary and territorial and gear their watchfulness to periods when prey is most active and consequently most vulnerable.

The sit-and-wait strategy is profitable when prey is common and motile. Although the predator spends considerable time waiting for appropriate prey, it is at rest and probably uses little more energy than it would if simply remaining alert for predators or competitors. The investment in searching time is great, therefore, but the energy outlay for this activity is low. In contrast, the rate of energy expenditure is high during the capture of prey, but the attack and capture occupy little time. This strategy is used by a few bats and many kinds of birds but is not widely used by mammalian carnivores.

More common among mammalian predators is the search-and-chase system. Wolves, for example, form social groups capable of dealing with large prey and search over wide areas. The pursuit and capture of prey is often a lengthy process that taxes the endurance of the predators and clearly involves the outlay of considerable energy. Although the cost is high, however, the benefit is great. Packs of wolves probably average only one large kill every two or three days, but if the prey is an adult moose weighing 400 kilograms, as it often is in the case of the wolves studied by Mech (1966) on Isle Royale, they can gorge on the carcass for at least two days. Small carnivores typically hunt solitarily, but here again search is lengthy, although a series of chases of small prey is usual.

Associated with differing predatory strategies are differences in locomotor abilities. The sit-and-wait predator must be capable of brief bursts of activity and often employs a highly stereotyped style of attack, but it need not be well adapted to efficient, enduring locomotion. For the search-and-chase predator, however, refinements of locomotor efficiency are highly adaptive, and the animals may be more opportunistic in their methods of capture.

The development of feeding strategies can be considered as a case of evolutionary gamesmanship between resourceful opponents. The success of every evolutionary move of one species is tested against a position or a countermove by the competing opponent species. A species heading for a desirable square on the evolutionary chessboard may be forced to occupy another square if the first is already occupied by a competing species. Or mutual vulnerability to devastating countermoves may make the simultaneous occupancy of adjacent squares by two opponents extremely costly.

In areas with high species diversity, the game is complex, involving many opponents utilizing numerous complementary feeding strategies and diets. Pulliam (1974) defines the optimal diet as "a set of successive prey choices which maximizes the rate of caloric intake or,

alternately, minimizes the time required to find the food ration." In the case of Black's (1974) bat community, for example, each species of bat is making a unique set of prey choices (Table 18–1).

Feeding Interactions

Just as energy transfers from one part of an organism to another are vital to life, complex patterns of energy transfer within a biotic community maintain this "superorganism" of interdependent and interacting species. The organisms involved in the transfer of energy within a community—from photosynthetic plants that utilize solar energy and inorganic materials to produce protoplasm, to animals that eat the plants, and thence to animals that eat animals—constitute the food chain. Typically, the transfer in a food chain goes from photosynthetic plants (primary producers) to herbivores (primary consumers) to primary carnivores (secondary consumers) that eat the herbivores, to secondary or perhaps tertiary carnivores in some extended chains (Fig. 18–37).

The term "food web" has been used to describe the complex pathways of energy transfer that usually occur in nature, often involving predators and primary consumers that figure importantly in more than one food chain. The intricacy of a food web is suggested by Fig. 18–38. Animals that occupy comparable functional positions in the food chain—the position of primary consumers, for example—are at the same trophic level. Green plants occupy the first trophic level and are referred to as autotrophs (self-feeders). Using mammals as examples, the second trophic level is occupied by herbivorous rodents, rabbits, and ungulates. Small carnivores, such as weasels, occupy the third trophic level, and large carnivores are in the third or fourth level.

The food chain is often depicted as a pyramid in an attempt to stress the relationships of biomass (the total weight of organisms of a given type in the community), numbers of or-

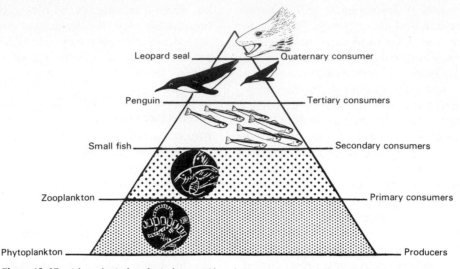

Figure 18-37. A hypothetical ecological pyramid based on an antarctic food chain. The higher the step in the food chain (the higher the trophic level), the larger the individuals and the lower their numbers. Ultimately, gigantic numbers of tiny planktonic plants and animals are necessary to support, through several intermediate steps, one leopard seal.

ganisms, and available energy at the different trophic levels. Food chains typically rest on a broad food base of plant material, but the amount of energy available to animals in each successively higher trophic level becomes progressively reduced because energy is lost through respiration and when organisms die and are utilized by decomposers (organisms that decompose organic material) rather than by animals of the next higher trophic level. Also, energy is lost because of inefficient transfer from one level to another. Consider, for example, the pyramids of numbers, caloric content, and energy utilization shown in Fig. 18–39 and Table 18–7.

The typical relationship of size and abundance of animals in a food chain involves small but numerous primary consumers, larger but much less abundant secondary consumers, and still larger but relatively scarce tertiary consumers. Although the animals at the top of the food chain seem to be in the commanding position of potentially being able to prey on all an-

imals at lower trophic levels without being vulnerable to predation themselves, the top predators occupy precarious positions because they frequently depend on animals from trophic levels with low productivity. Under conditions of food stress, therefore, the fate of the species at the top of the food chain will not be death at the hands of a predator but death by starvation as a result of low availability of food.

Because of the great loss of energy accompanying food transfer between successive trophic levels, the total available energy is largest for consumers at the lower levels. Thus, predators feeding on primary consumers have more energy available to them than do predators feeding on secondary consumers. Considered in this light, the adaptive importance of filter feeding in some marine mammals becomes clear. Plankton feeders, such as the baleen whales and the enormously abundant antarctic crab-eater seal *(Lobodon carcinophagus),* exploit tremendously larger sources of energy in primary-consumer plankton than they could if

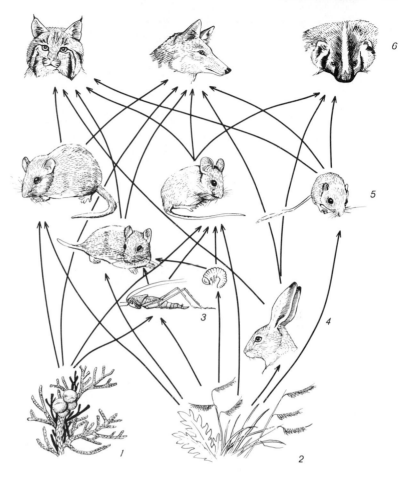

Figure 18-38. A simplified food web involving the mammals of a piñon-juniper community in Coconino County, Arizona. The arrows indicate the foods utilized by the mammals. The plants (1, juniper; 2, grasses and forbs) support arthropods (3), rabbits (4), and rodents (5). The rodents *(Neotoma stephensi, Onychomys leucogaster, Peromyscus truei, Perognathus flavus)* and the rabbits *(Lepus californicus* and *Sylvilagus audubonii)* are preyed upon by bobcats, coyotes, and badgers (6).

they fed on large fish that are secondary or tertiary predators. Only 10 to 20 percent of the energy entering a trophic level can be utilized by the next higher level; this factor limits the length of food chains.

Diagrams of food chains or food webs, although valuable for purposes of illustration, usually of necessity strongly simplify what is really an extremely intricate meshwork of interactions. A broadly adapted carnivore like a coy-

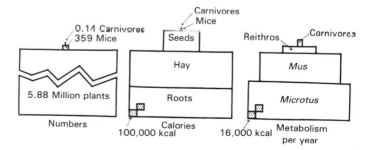

Figure 18-39. Pyramids of numbers, calories, and energy utilization for 1 acre of annual grassland near Berkeley, California. The pyramid showing calories is also approximately to scale for biomass. (After Pearson, 1964)

Table 18–7: Standing Crop of Plants, Prey, and Predators on One Acre of California Grassland and Rate of Use of Vegetation by Rodents and of Prey by Carnivores at Peak Population Levels

	Standing Crop		Annual Rate of Use	
	Kg (dry wt)	Kcal	Kcal	% of Crop
Roots	2131	7,269,000		
Hay	2097	8,141,000		
Seeds	442	1,920,000		
Microtus	1.24	6,402	1,368,750	71
Mus	0.88	4,543	876,000	46
Reithrodontomys	0.084	434	81,650	4
Other prey	0.13	671	27,000	
Carnivores	0.126	650	11,700	97

After Pearson, 1964; figure sources given in Pearson's paper.

ote, or an omnivore such as the possum *(Didelphis),* may function in all trophic levels above that of the primary producer. For a coyote, the fruit of the prickly-pear cactus or juniper berries may form one meal, while a jackrabbit or deer fawn may be the next. More frequently, seasonal differences in the position of an animal in the food chain may occur. Johnson (1961, 1964) found that the deer mouse *(Peromyscus maniculatus)* in Colorado and Idaho became strongly insectivorous in the summer and thus functioned during this season as a secondary consumer, whereas it ate largely plant material during the cooler seasons, functioning as a primary consumer at those times.

Recent ecological research has revealed important differences in small-mammal energetics in various ecosystems and in the proportions of energy utilized by rodents from different trophic levels. An excellent study by N.R. French, Grant, et al. (1976) compared the small-mammal energetics in a series of grassland ecosystems: a tall-grass site in Oklahoma, a mid-grass site in South Dakota, a northern short-grass site in Colorado, a southern short-grass site in Texas, and a desert grassland site in New Mexico. The rodent communities of the ecosystems differed strikingly in composition and in biomass (Fig. 18–40). Of particular interest is the fact that foods from different trophic levels

were utilized to different extents by the rodents at different sites: arvicoline herbivores specializing on green primary production dominated (in terms of biomass) at the tall-grass site; sciurid omnivores utilizing both primary production and primary consumers (insects) dominated in the northern short-grass prairie; and heteromyid seed eaters dominated the desert site. Broadly speaking, small mammals depended chiefly on herbage in the tall-grass and mid-grass site and on herbage, seeds, and invertebrates in the desert site.

Although the energy consumed by rodents was highest in the tall-grass prairie (where 172×10^3 kilocalories per hectare supported 935 grams live weight of small mammals per hectare), the amount of small-mammal biomass relative to consumption by rodents was greatest in the northern short-grass prairie (where 32×10^3 kilocalories per hectare supported 277 grams of small mammals per hectare). This was probably due largely to the high assimilation and digestion rates for granivores and omnivores (Table 18–8). Usually less than 10 percent of the primary production at the various sites was utilized by small mammals, but, in striking contrast, invertebrates were heavily utilized, in extreme cases to close to 100 percent (Table 18–9). This information suggests that arvicoline herbage eaters underutilize their food

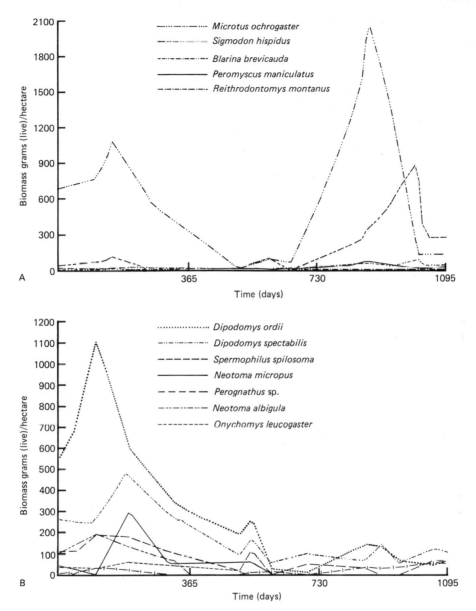

Figure 18-40. A plotting through time of the biomass of small mammals at (A) a tall-grass prairie site and (B) a desert grassland site. (From N. R. French, Grant, et al., 1976)

resources and seed-eating heteromyids over-exploit food reserves, an idea presented by R.H. Baker (1971).

Herbage eaters seem in general to be r-selected types; in r-selected animals, selection has favored high rates of reproduction and growth, and with this pattern goes high mortality. Thus, r-selected species are expected to have a high population turnover and to be able to respond rapidly to optimal conditions by increased pop-

Table 18–8: Digestibility and Assimilation of Natural Foods in Some Small Mammals

Group	Number of Species	Digestibility[a]	Assimilation[a]
Grazing herbivore (arvicolines)	14	67	65
Omnivore (sciurids, sigmodontines)	8	75	73
Granivore (heteromyids)	6	90	88
Insectivore (Blarina)	1	—	80

[a]Expressed as organic matter consumed or percent of energy.
From N. R. French; Grant, et al., 1976.

ulation densities. Seed-eating rodents are seemingly k-selected; selection in k-selected rodents has favored characters furthering survival of the individual, and, typically, these mammals have lower reproductive rates than do herbage eaters. In heteromyids, for example, periodic torpor and specialization favoring predator avoidance tend to preserve the individual and lead to relatively low rates of mortality and population turnover. Under conditions of the tight food supplies typical of arid regions, k-selected rodents would seem to have the adaptive edge, whereas the boom-or-bust productivity of some mesic grasslands perhaps favors r-selected arvicoline species.

N.R. French, Grant, et al. (1976) point out that, where mesic grasslands are receding northward, as they are in some parts of the mid-grass areas in the Great Plains of the United States, herbage-eating arvicolines are being replaced by the specialized seed eaters. Sharp between-habitat differences in the relative importance of herbivorous, omnivorous, granivorous, and insectivorous small mammals can also be ob-

Table 18–9: Fraction of Available Energy Utilized by Small Mammals, According to Diet

Grassland Site	Year	Herbage	Seed	Animal
Tall-grass	1970	0.063	—	0.88
(Oklahoma)	1971	0.010	—	0.38
	1972	0.047	0.080	All?
Mid-grass	1970	0.002	—	0.08
(So. Dakota)	1971	0.001	—	0.08
	1972	0.002	—	0.02
Northern short-grass	1970	—	0.002	All?
(Colorado)	1971	—	—	0.42
	1972	0.005	—	0.83
Southern short-grass	1970	—	—	0.96
(Texas)	1971	0.013	—	0.61
	1972	0.091	—	0.11
Desert	1970	0.032	—	All?
(New Mexico)	1971	0.190	—	All?
	1972	0.004	—	0.68

From N. R. French; Grant, et al., 1976.

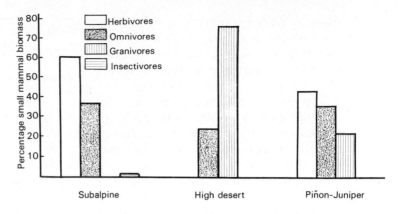

Figure 18-41. Relative biomass of herbivorous, omnivorous, granivorous, and insectivorous small mammals at a subalpine site (Rabbit Ears Pass, Routt County, Colorado), a high desert site, and a piñon-juniper site (both in Wupatki National Monument, Coconino County, Arizona. (Subalpine data from Vaughan, 1974; biomass for other sites based on unpublished data from the summer of 1976, courtesy G. C. Bateman)

served in areas other than grasslands (Fig. 18-41) and reflect differences in the composition of the vegetation.

Factors Influencing Population Densities

The abundance of an animal species at a given time and a given locality depends on (1) the carrying capacity of the habitat and (2) the relationship between the rate at which the animals are added to the population (by reproduction or immigration) and the rate at which they are lost from the population (by death or emigration). Many mammalian populations are in a state of dynamic equilibrium and tend to be stabilized within certain density limits by such interacting processes as competition, reproduction, predation, dispersal, and disease. These factors are usually regarded as density-dependent; that is, they change in relation to changes in population density. Other factors, such as

food requirements and weather, are generally density-independent; that is, they do not change in response to changes in population density.

Natality

The number of individuals added to a population through reproduction depends on the reproductive potential of a species, which refers to the greatest number of individuals that a pair of animals or a population can produce in a given span of time. Reproductive potential is a function of age and sex ratios, age at which a female first bears young, litter size, and frequency of litters. Even species of the same genus occupying the same area can have markedly different reproductive potentials, as indicated by litter size (Table 18-10). Reproductive performance is also responsive to environmental differences, such as temperature or rainfall, for sharp regional shifts in litter size occur in some species (Table 21-6).

Table 18-10: Differences in Reproductive Patterns of Three Species of *Peromyscus* That are Sympatric in Some Areas

Characteristic	*P. maniculatus*	*P. truei*	*P. californicus*
Number of litters per season	4.00	3.40	3.25
Number of young per litter	5.00	3.43	1.91
Number of offspring per breeding female per season	20.00	11.66	6.21

From McCabe and Blanchard, 1950.

The ability of some species to vary reproductive performance in response to environmental conditions or population levels may be of considerable adaptive importance. For example, the reproductive potential of mule deer is lower in poor habitats than in habitats productive of high-quality browse; whereas well-nourished does may breed first at 17 months of age, those that occupy poor ranges may not breed first until as late as 41 months of age (Taber and Dasmann, 1957). The litter size of carnivores is also affected by food supply; Stevenson-Hamilton (1947) reported that the litter size of the African lion dropped when food was scarce. Batzli and Pitelka (1971) found delays in the start of breeding in the California vole (*Microtus californicus*) following times of peak population densities; these delays may have resulted from decreased availability of preferred foods during the period of high vole densities. These authors and other workers (Greenwald, 1957; Hoffmann, 1958) have observed seasonal changes in litter size that are presumably caused by changes in forage quality.

In Finnish Lapland, Kalela (1957) found that reproductive maturity was delayed in young male voles *(Clethrionomys rufocanus)* in several localities in a year of high populations and that, in one area supporting especially high densities, breeding was also delayed in nearly all young females. Such delays usually reduce the numbers of young produced during a season; also, individuals of late litters may have too little time to mature before the critical winter period and may suffer unusually high mortality. Muskrats (*Ondatra zibethica*) in Iowa continued breeding through only part of the usual breeding season in a year of high populations (Errington, 1957), and wild rats *(Rattus norvegicus)* in Baltimore had a markedly low pregnancy rate during population highs (D.E. Davis 1951).

Survival rates of young also strongly affect population levels. Young are clearly the expendable part of a population and show the greatest fluctuations during population

Table 18–11:   Litter Size of Consecutive Litters of the Montane Vole (*Microtus montanus*)

Litter	N	Mean Litter Size	Range
1	12	4.2	2–6
2	12	4.7	3–7
3	10	5.0	3–7
4	9	4.2	2–6
5	6	5.8	3–10
6	6	5.5	3–7
7	5	3.4	1–6

Note that young and old animals have relatively small litters. After Negus and Pinter, 1965.

changes. A 74 percent decline in pocket gopher density in western Colorado in 1958 was associated with an extraordinary drop in the survival rate of young (Hansen and Ward, 1966), and in southern Colorado, Hansen (1962) found that whereas high survival rates of young pocket gophers were characteristic of periods of high densities, low survival rates of young were associated with a declining population. Similarly, C. J. Krebs (1966) reported better survival for expanding than for declining populations of the California vole.

In mammals, the contribution made to a population by reproduction clearly depends on a variety of factors and is seldom constant within a species from year to year. As mentioned, variation in litter size, number of litters, and length of breeding season; age at which young animals breed; and survival rates of young are all important variables. In addition, the litter size (Table 18–11) and the percentage of females that become pregnant change with age distribution of a species; the age composition of a population may therefore have a marked effect on its reproductive performance.

Predation

Rates and age distributions of mortality are important population characteristics. Mortality varies with age and has been carefully studied in some mammals. Specific mortality is the num-

Table 18-12: Life Table for Dall Sheep (*Ovis dalli*) in Mount McKinley National Park, Alaska

Age (years)	Age as Percent Deviation from Mean Life Length	Number Dying in Age Interval per 1000 Born	Number Surviving at Beginning of Age Interval per 1000 Born	Mortality Rate per 1000 Alive at Beginning of Age Interval	Life Expectation, or Mean Lifetime Remaining to Those Attaining Age Interval (years)
0–0.5	−100.0	54	1000	54.0	7.06
0.5–1	− 93.0	145	946	153.0	—
1–2	− 85.9	12	801	15.0	7.7
2–3	− 71.8	13	789	16.5	6.8
3–4	− 57.7	12	776	15.5	5.9
4–5	− 43.5	30	764	39.3	5.0
5–6	− 29.5	46	734	62.6	4.2
6–7	− 15.4	48	688	69.9	3.4
7–8	− 1.1	69	640	108.0	2.6
8–9	+ 13.0	132	571	231.0	1.9
9–10	+ 27.0	187	439	426.0	1.3
10–11	+ 41.0	156	252	619.0	0.9
11–12	+ 55.0	90	96	937.0	0.6
12–13	+ 69.0	3	6	500.0	1.2
13–14	+ 84.0	3	3	1000.0	0.7

After Deevey, 1947.

ber of individuals of a population that have died by the end of a given time span. Specific mortality at given ages can be expressed in a life table. Table 18–12 is based on data assembled by Murie (1944) during his classic study of Alaskan wolves. Ages of some small mammals are difficult to establish, and a life table for females may be based on numbers of litters (Table 18–13).

Of the mortality factors to which mammals are susceptible, predation looms high in importance. There has been much heated debate on the ability of predators to control or influence densities of prey species or to influence population cycles, and the final word has not been heard. The degrees of impact that predators have on prey have been summed up by Pearson (1971:41): "The effectiveness of predation var-

Table 18-13: Life Table for Female African Root Rats (*Tachyoryctes splendens*)

Litter/Age Group	Number of Root Rats	Survivorship (lx)	Death Rate (dx)	Mortality per 1000 Living (qx)
0	412	1000	750	760
1	99	240	19	79
2	91	221	29	131
3	79	192	136	708
4	23	56	41	732
5	6	15	8	553
6	3	7	5	714
7	1	2	2	1000

lx = number surviving at beginning of each age interval out of 1000 born
dx = number dying in each age interval out of 1000 born
qx = mortality rate per thousand alive at beginning of each age interval
From Jarvis, 1973.

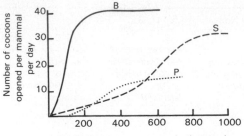

Figure 18–42. Functional responses to density of prey by three predators, the short-tail shrew, *Blarina brevicauda* (B), the masked shrew, *Sorex cinereus* (S), and the deer mouse, *Peromyscus maniculatus* (P). The prey are sawfly larvae, which are removed from their cocoons by the predators. (After Holling, 1959)

ies from the relatively ineffective predation of rats on man, in which rats are able occasionally to kill infants or incapacitated adults, through the mink-muskrat system described by Errington (1967), in which mink take a significant proportion of homeless and stressed muskrats, to the almost total effectiveness of carnivore predation on *Microtus* until almost the last one has been killed." A predator-prey relationship must obviously have some stability; as indicated by Lack (1966:301), "only those predatory species which have not exterminated their prey survive today, hence we observe in nature only those systems which have proved sufficiently stable to persist, and many others were presumably terminated in the past by extinction."

Although the effectiveness of even a single species of predator seems to vary according to situation, some general predator-prey relationships that apply to mammals as well as to other animals can be recognized. The observed responses of a predator to changes in the density of a prey species indicate that predation is influenced by density. The numbers of a preferred prey taken by a carnivore increase as the density of the prey increases (Fig. 18–42) because the greater the number of prey animals per unit area, the greater the opportunity for predators to encounter and capture them. This is a functional response on the part of the predator (Holling, 1959, 1961). Errington (1937) noted such a response in the predators of muskrats and suggested that the intensity of predation is a function of prey population levels. There may also be a numerical response, involving an increase in the predator density with a rise in the prey population. The numerical response may be the result of immigration of predators, as in the case of the striking responses to lemming abundance on the part of the pomarine jaeger (Table 18–14), or it may be due to increased breeding success, as in the case of the masked shrew (Fig. 18–43).

The relative populations of predators and

Table 18-14: Densities of Breeding Pomarine Jaegers (*Stercorarius pomarinus*) and Nesting Success Near Point Barrow, Alaska

Year	Spring Lemmus Density (no./acre)	No. of Pairs of Jaegers	Census Area. (mi²)	Density (pairs/mi²)	Maximum Density (pairs/mi²)	Breeding Success (% of eggs)
1952	15–20	34	9	3.8	5–6	30–35
1953	70–80	128	7	18.3	25–26	20–25
1954	<1	0	—	—	—	—
1955	1–5	2	15±	0.13	—	0
1956	40–50	114	6	19.0	22–23	4
1957	<1	0	—	—	—	—
1958	<1	0	—	—	—	—
1959	1–5	3	15±	0.20	—	0
1960	70–80	118	5.75	20.5	25	55

Note the correlation between high densities of lemmings *(Lemmus trimucronatus)* and high populations of nesting jaegers. After Maher, 1970.

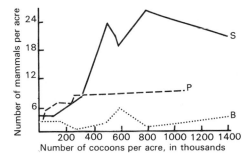

Figure 18-43. Numerical responses to high densities of sawfly larvae by the short-tail shrew (B), the masked shrew (S), and the deer mouse (P). (After Holling, 1959)

their prey have often been regarded as being in dynamic equilibrium. The degree to which a balance between prey population and predator population is reached, and the extent to which the relationship is dynamic, vary widely in situations involving mammals. These factors depend in part on the ratio of predator density to prey density, on the relative sizes of the predator and the prey and the ease with which prey can be captured, and on the degree to which the prey populations are cyclic. Studies by Mech (1966) on Isle Royale in Lake Superior during the period from 1959 to 1961 indicated that the ratio of moose to wolves was roughly 30 to 1, with 20 wolves being supported by approximately 600 moose, the wolves' primary food. The weight differential between an adult moose and an adult wolf was 14 to 1 (450 kilograms to 33 kilograms), and the wolves had difficulty killing moose. A strongly contrasting situation was studied in California by Pearson (1971), who found that the prey-predator ratio varied from 72 to 1 in 1962, during a period of low vole populations, to 5410 to 1 during a peak in vole numbers (Fig. 18–44). In this case the prey was a cyclic vole *(M. californicus)* with an adult weight of roughly 45 grams. The predators—feral cats, raccoons, gray foxes, and skunks—averaged perhaps 2.25 kilograms in weight, yielding a rough estimate of prey-to-predator weight ratio of 0.02 to 1. The predators could catch voles easily.

These examples are based on two very different patterns of predator-prey interaction. The wolf-moose interaction resulted in relatively stable predator and prey densities, whereas the situation involving the California vole was one of great instability. Because of the difficulty wolves have in bringing down moose, the pressure they exert on the moose population is highly selective in that primarily young or old animals are taken; adult moose in the prime of their reproductive life are not killed (Mech, 1966). The predators of the vole, on the other hand, show a high preference for this prey and find it easy to catch; their kill is far more nearly random and they are able to kill almost every vole during times of vole scarcity (Pearson, 1966). As suggested by the above examples, predator-prey interactions are complex; so many variables are involved that few generalizations relating to such interactions can be universally applied.

Some predators tend primarily to remove vulnerable individuals from the prey population. These latter are individuals that, because of inexperience, old age, injury, or sickness, are

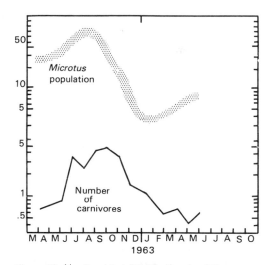

Figure 18-44. Densities of California voles *(Microtus californicus)* compared with densities of predators (feral cats, raccoons, foxes, and skunks) during a population cycle of voles. Multiply vertical scale by 100 to obtain numbers of voles present. (After Pearson, 1971)

readily captured, or they are animals forced by intraspecific competition for space into marginal habitats in which their vulnerability to predators is increased. Work on wolves by Crisler (1956), Mech (1966), and Murie (1944) has shown that vulnerable individuals of the prey species, in these cases Dall sheep, caribou, and moose, were taken more frequently than were healthy, mature individuals. Careful field studies by Hornocker (1970a, 1970b) on the mountain lion in Idaho showed that, of 53 lion-killed elk and 46 lion-killed deer, 75 percent of the elk and 62 percent of the deer were young (less than 1.5 years) or old (more than 8.5 years). These percentages of young and old animals were considerably greater than would be expected if the lions had killed randomly. On Isle Royale, 18 of 51 wolf-killed moose were calves; most of the remainder of the animals killed were from 8 to 15 years old, and 39 percent of these old animals showed evidence of debilitating conditions (Mech, 1966). The wolves had killed no adults from 1 to 6 years old. Errington (1943, 1946, 1963) found that in Iowa, as muskrat populations rose above a "threshold of security," a number of animals were forced into marginal habitats by intraspecific competition for space. This "vulnerable surplus" was preyed upon heavily by mink and red foxes, which made a marked functional response to this available food source. This pattern of predation has been called compensatory predation.

But can predators control densities of mammals, or are they simply killing individuals that would quickly be removed from the population by other means? The answer clearly depends on the particular predator-prey interaction being considered. Murie (1944:230) concluded that in Alaska wolf predation on Dall sheep lambs was the most important factor limiting numbers of sheep, and on Isle Royale the wolves seemingly kept moose densities below the level at which food supply would be the limiting factor (Mech, 1966:167). In Idaho, however, predation by mountain lions had little impact on the populations of deer and elk, which were controlled instead by the winter food supply (Hornocker, 1970b).

Heavy predation on populations of small mammals has been shown to affect population levels. Data reflecting the severity of predation on lemmings demonstrate how important this source of mortality may be locally. On the coastal plain near Point Barrow, Alaska, during times of high lemming densities, the combined impact of several predators deals a staggering blow to lemming populations and is an important factor causing the population crash (Pitelka, 1957a; Pitelka, Tomich, and Treichel, 1955). The combined kill of lemmings by the major predators amounted to at least 49 per acre during a cycle of abundance (Table 18–15). Studies in California by Pearson (1963, 1964, 1966) and by Fitzgerald (1977) on *Microtus* have demonstrated that predators preying upon these cyclic rodents are unable to control a rising prey population but that "carnivore predation during a crash and especially during the early stages of the subsequent population low determines to a large extent the amplitude and timing of the microtine cycle of abundance" (Pearson, 1971:41).

Stability in a predator-prey system seems to be due generally to environmental heterogeneity (Murdoch and Oaten, 1975:117); the predator and prey occupy separate niches and are therefore partially separated. Refuges allow a segment of the prey population to escape predation. Carefully controlled laboratory experiments using a protozoan predator and a bacterium as prey suggest that, even with extremely limited environmental heterogeneity, the predator is unable to exploit the prey to the extent of causing the extirpation of either species (Van den Ende, 1973). The bacteria in the liquid culture could be captured by the protozoan, but because the bacteria adhering to the walls of the culture vessel could not be dislodged by the ciliary action of the protozoan, the two species persisted and their populations approached stability.

Table 18–15: Impact by Predators on High Population of Lemmings (*Lemmus trimucronatus*) Near Point Barrow, Alaska

Predator	Age Class	Density (ind./mi²)	Daily Food Consumption (g/ind.)	Per ind.	Season's Lemming Consumption Per Acre		
					25 May to 15 July	16 July to 31 Aug	Total
Pomarine jaeger	Adult	38	250	338	10	21	31
	Young	38	200	167	—	—	
Snowy owl	Adult	2	250	350	1.3	1.6	3
	Young	7	150	160	—	—	
Least weasel		64	50	100	5	5	10
Glaucous gull		20	250	125	0.7	—	1
Waste					4	—	4
Totals					21	28	49

Data for least weasel from D. Q. Thompson, 1955; for snowy owl from A. Watson, 1958. Table after Maher, 1970.

Disease

Parasitism and disease are known to be significant causes of mortality (Elton, 1942) among mammals and may occasionally cause dramatic population crashes, as in the case of a die-off of prairie dogs *(Cynomys gunnisoni)* in Colorado (Lechleitner, Tileston, and Kartman, 1962) caused by bubonic plague. Talbot and Talbot (1963) estimated that 47 percent of the total mortality suffered by wildebeest was caused by diseases, of which the disease rinderpest seemed most important. The blood parasite *Babesia* is a source of mortality among African lions. Disease in relation to population regulation, however, has been a difficult factor to assess (Chitty, 1954). Disease as the single cause of death may be relatively unimportant, but it may be important in contributing to the vulnerability of an animal to predation or to stressful environmental conditions.

Parasitism has been regarded periodically as an important cause of mortality, but careful obervation indicates that otherwise healthy animals can often tolerate a moderately heavy parasite load. Heavy parasitism has been found to accompany a general health decline in rodents and rabbits during or following times of high density (Batzli and Pitelka, 1971; Erickson,

1944). Work in Colorado by Woodard, Gutierrez, and Rutherford (1974) indicated that high late-summer mortality among young bighorn sheep *(Ovis canadensis)* was caused by heavy lungworm infestations and by bronchopneumonia.

Weather

Flooding and heavy storms are known to result in significant mortality among mammals. A series of beaver colonies was decimated during a flood in Colorado (Rutherford, 1953), and a 70 percent decrease in the combined population of golden mice *(Ochrotomys nuttalli)* and cotton mice *(Peromyscus gossypinus)* was caused by a flood of three weeks' duration in eastern Texas (McCarley, 1959). A spring snowstorm in the Sierra Nevada of California caused heavy mortality among Belding's ground squirrels *(Spermophilus beldingi)* emerging from hibernation. Sixty percent of the females that occupied the study area the previous fall disappeared, the squirrels were highly vulnerable to predators, several were known to have starved, and one case of cannibalism was observed (Morton and Sherman, 1978).

Weather may have strong but indirect effects on mortality. In 1971 in Norway, for example, an estimated 41 percent of the calf crop of red deer *(Cervus elaphus,* called elk in the United States) in the Aure area died, apparently as a result of poor nutrition of the pregnant cows because of unusually heavy late winter and spring precipitation (Wegge, 1975).

Population Cycles

Mammalian population cycles are among the most impressive biological phenomena. Striking changes in density occur primarily in temperate, subarctic, and arctic areas but are not known to occur in tropical or subtropical regions. This difference is probably related to differences in species diversity in the two types of areas. High-latitude areas are characterized by biotic assemblages and food webs that are simpler than those of tropical areas. The typical boreal community has a limited biota and supports few species of vertebrates, but some species may, at least periodically, be remarkably abundant. The simplicity of the northern community is seemingly partly responsible for its instability, for, where so few kinds of organisms exist, any marked fluctuation in the density of one species seems to disrupt the entire community.

In the complex tropical community, by contrast, the heterogeneity of the environment, the complexity of the food web, the diversity of carnivores, the intricate patterns of niche displacement and potential competition, and the relatively small percentage of the energy resources available to any one species provide a buffer system against population outbreaks by any species.

Even in northern Alaska, population cycles are more pronounced in coastal areas, where only two arvicolines occur, than in the foothills, where there are five species. Pitelka (1957a:85) states, "Similarities in their feeding and sheltering activities strongly suggest that where more than one species is important, competition may act to depress their respective populations and hence to depress the likelihood of strong fluctuations."

Characteristics of Population Cycles

In areas where well-marked cycles of arvicolines occur, population peaks have been reported to occur at intervals of three or four years (Fig. 18–45). After a 19-year study of the California vole *(Microtus californicus),* how-

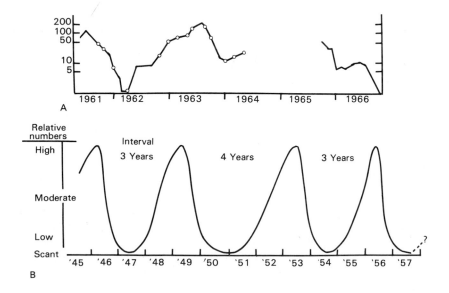

Figure 18-45. Examples of fluctuating populations of arvicoline rodents. (A) Estimates of densities of voles *(Microtus californicus)* in an area near Berkeley, California, during a six-year period. (B) Generalized curves (the amplitudes of successive cycles are actually not the same) showing fluctuations in brown lemming *(Lemmus trimucronatus)* populations near Barrow, Alaska. (A after Pearson, 1971; B after Pitelka, 1957b)

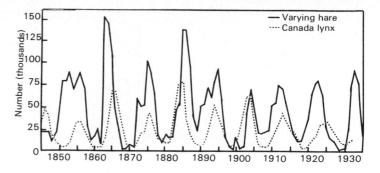

Figure 18-46. Cycles of population density of the varying hare *(Lepus americanus)* and the Canada lynx *(Lynx canadensis),* based on the numbers of pelts purchased by the Hudson Bay Company. (After MacLulich, 1937)

ever, Garsd and Howard (1981) concluded that vole populations do not cycle regularly or periodically; instead they fluctuate randomly or "pseudoperiodically." No one disputes that drastic fluctuations in arvicoline populations do occur, and to these fluctuations, I will apply the familiar term "cycles" in the following discussion. Bear in mind, however, that the question as to whether or not these fluctuations are actually periodic or regularly cyclic remains unsolved.

On the coastal slope of northern Alaska, oscillations in populations of lemmings *(Lemmus trimucronatus),* the chief herbivore, are characterized by a precipitous drop in density in the late summer and winter following a popu-

lation peak, a period of one or two years of extremely low populations and localized distribution, and an upsurge in population in the winters following the low population, with peak numbers occurring early in the third or fourth summer (Pitelka, 1957b:85). The short-term lemming cycles also occur in temperate and boreal parts of the Old World (Elton, 1942; Siivonen, 1954). Longer-term cycles, from eight to 12 years, are known in populations of the varying hare and the Canada lynx (Fig. 18–46). Mammalian population cycles occur in some temperate areas, where perhaps the most striking cycles involve arvicolines, which occasionally reach amazingly high densities (Table 18–16).

Table 18–16: Population Densities of Several Species of Arvicolines

Density (per acre)	Species	Region	Reference
1–20	*M. pennsylvanicus*	N. Minnesota	Beer, Lukens, and Olson, 1954
6–67	*M. pennsylvanicus*	New York	Townsend, 1935
3000	*M. montanus*	N.W. U.S.	D. A. Spencer, 1958a
200–4000	*M. montanus*	Oregon	D. A. Spencer, 1958b
25–81	*M. californicus*	N. California	Greenwald, 1957
425	*M. californicus*	N. California	Lidicker and Anderson, 1962
25–145	*M. ochrogaster*	Kansas	E. P. Martin, 1956
250–300	*M. agrestis*	England	Chitty and Chitty, 1962
1900	*M. arvalis*	France	Spitz, 1963
1004	*M. guentheri*	Israel	Bodenheimer, 1949
2400	*M.* spp.	U.S.S.R.	W. J. Hamilton, 1937
50–100	*L. trimucronatus*	Alaska	Rausch 1950
200–300	*L. lemmus*	Sweden	Curry-Lindahl, 1962

After Aumann, 1965.

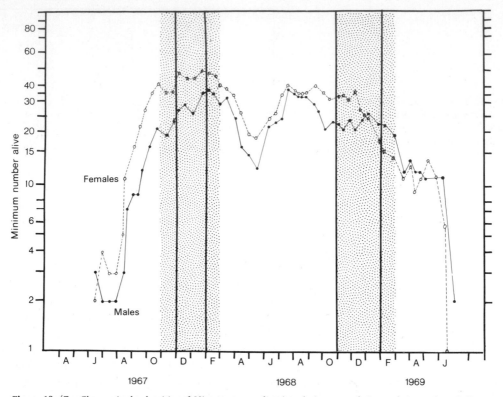

Figure 18-47. Changes in the densities of *Microtus pennsylvanicus* during a population cycle in southern Indiana. The shaded period is winter. (After Gaines and Krebs, 1971)

Arvicoline cycles are of special interest both because they are extremely dramatic and because they have been carefully studied. Population cycles in lemmings and voles can be divided into four phases:(1) increase, (2) peak, (3) decline, and (4) low density, with each phase being to some extent distinct. C. J. Krebs and Myers (1974) present an excellent treatment of population cycles in small mammals. I have used their work repeatedly in preparing the following material on arvicoline population cycles.

The increase phase is a time when densities rise markedly from one spring to the next. This increase may continue over several years and may be be interrupted annually by short-term population declines (W. J. Hamilton, 1937), or, more typically, it may occur within one year,

with extremely sharp increases over a period of three or four months (Fig. 18–47). Arvicolines are flexible with regard to the duration of breeding, and during the increase phase, breeding begins early in the spring and often continues into the winter. At this phase, animals quickly reach sexual maturity. During an increase phase, some Norwegian lemmings *(Lemmus lemmus)* were found by Koshkina and Kholansky (1962) to be pregnant at 20 days of age. Koshkina (1965) found that the rate of sexual maturation was affected by population density, with early maturation being typical of an increasing population. Survival is generally high during the increase phase. There is more dispersal at this time than during the decline phase (Gaines, Vivas, and Baker, 1979). Dispersal during the increase phase is termed presaturation dispersal because it occurs before the habitat is saturated (has reached its carrying capacity) and involves vigorous individuals with a good chance of survival (Lidicker, 1975).

The peak phase is a time of relatively little change in density. The population increase ceases, and the population may remain fairly stable for at least one year or may abruptly swing into a decline. During the peak phase, the breeding season in summer is typically brief and no winter breeding occurs. Animals attain sexual maturity late, and young born at peak times may not mature sufficiently to breed during their first summer. Growth rates are relatively high at this time, and animals are generally 20 to 50 percent larger at the peak phase than at other times (C. J. Krebs and Myers, 1974). Mortality rates are relatively low during this phase, but dispersal is again high. Dispersal at this time, when the habitat is saturated with voles, is termed saturation dispersal. This dispersal involves surplus animals (old, young, or social outcasts) that have little chance of survival (Lidicker, 1975).

The decline phase varies widely, from precipitous drops in density (population crashes) to uneven declines lasting one year or more. As is the peak phase, the decline phase is typified by brief summer breeding and no winter breeding, and animals reach sexual maturity late. Little dispersal occurs during the decline, but mortality rates are high within the relatively sedentary population (Fig. 18–48).

The phase of low densities may last from one to three years, and annual shifts in abundance may occur (C. J. Krebs, 1966). At this time, the breeding season is short, animals seem not to reach sexual maturity early, and mortality is high.

Mention should be made of the famous mass wanderings of Norwegian lemmings that accompany population highs and declines. During these wanderings, the lemmings appear far from their preferred habitats. One such movement occurred in Norway in the summer and fall of 1963 and is well described by Clough

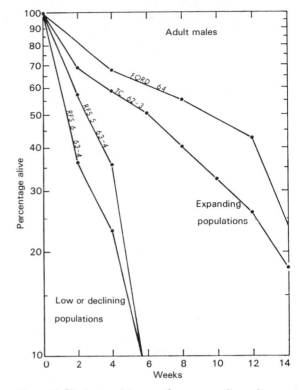

Figure 18–48. Survivorship curves for two expanding and two low or declining populations of *Microtus californicus* from time of first capture in live traps. Data for adult males only. (After C. J. Krebs, 1966)

(1965). The movement began in mid-July, when as many as 40 lemmings per hour passed an observation point; all animals were heading downhill. Lemmings also began to appear on the streets of a nearby town and in pine forests 19 kilometers down the valley. Clough was struck by the complete intolerance of one lemming for another. Each animal moved alone. Every time lemmings met one another, some aggressive action or strong avoidance behavior occurred, and large, mature females were dominant in all conflicts involving them. The wandering lemmings, however, were chiefly middle-size or small and often sexually immature. By October, lemmings were living in virtually all habitats, including those that were highly unsuitable, and the population was on the decline. The following March, Clough did not find a single lemming, nor were any found during his intensive work in May and June in areas where lemmings had been abundant the previous summer. The winter's die-off had seemingly been complete.

Factors Controlling Population Cycles in Small Mammals

Although population cycles in mammals have been recognized and studied for many years, considerable controversy remains as to what factors control them. Some factors seem not to be responsible for population cycles. In populations of lemmings and voles, litter size, pregnancy rates, and sex ratios are often rather constant during the phases of a cycle (C. J. Krebs and Myers, 1974). Lidicker (1973), however, found striking seasonal differences in litter size and sex ratios in *Microtus californicus*. Weather also seems not to be responsible, although Fuller (1967, 1969) and others have considered weather to have important effects on arvicoline populations. Not all arvicolines respond similarly to a given weather pattern, however, and no clear evidence indicates that weather has a major influence on arvicoline cycles.

Food has been considered by some as a dominant factor influencing arvicoline populations. Lack (1954b) regarded overexploitation of food as a major factor triggering changes in arvicoline densities, and Pitelka (1958) suggested that reduced availability of food could cause poor reproductive success and population declines. Because arvicolines are known to prefer certain foods and generally do not feed randomly on the most abundant plants, food might be expected to be limiting, but a number of studies have shown that voles and lemmings seldom consume more than 5 percent of the energy available from the vegetation. A. M. Schultz (1965, 1969) showed that neither the quantity nor the quality of tundra vegetation limited ar-

vicoline populations, and he observed greater vegetative productivity in places with arvicolines than in experimental plots without them. Pieper (1964) hypothesized that the nutrients in the large amounts of urine and feces deposited beneath the snow by large numbers of lemmings rapidly become available to plants when the snow melted in the spring. The plants respond with high nutrient levels, which, at least temporarily, support high densities of lemmings. There is no convincing evidence, however, that plant nutrient levels influence arvicoline densities.

Studies in separate areas on different species of *Microtus* by Hoffmann (1958) and C. J. Krebs and DeLong (1965) involved fertilization of the vegetation and supplementary feeding of the rodents. Neither study demonstrated that the amount of food or its quality markedly affected the population densities. Furthermore, Batzli and Pitelka (1970, 1971) found that, although in one area the favorite food plant of *M. californicus* was ten times more abundant than in another area, the two populations of voles underwent similar declines at the same time.

As discussed in the section on predation, the timing and amplitude of the population cycles of some arvicolines are thought to be influenced by predation. The population crash that terminates the brown lemming cycle in northern Alaska is influenced by the concentrated efforts of a number of predators (Pitelka, 1957a:88). Pearson (1966, 1971) has presented data suggesting that the amplitude and timing of the arvicoline cycle are determined by intense predation during and immediately after the population crash. Lidicker (1973) thought that the "quasi" two-year cycles of *Microtus californicus* on an island occurred because no mammalian predator was present to depress low populations and to retard their recovery. The impact of even a single avian predator (the pomarine jaeger, *Stercorarius pomarinus*) on high lemming populations in summer has been shown to be heavy (Maher, 1970), and intense winter predation by a weasel *(Mustela nivalis)* may be responsible for nearly wiping out lemmings and delaying the recovery of the population until the weasels themselves decline (Maher, 1967; D. Q. Thompson, 1955:173). Fitzgerald (1977) estimated that, during one winter in a study area in the Sierra Nevada, weasels killed 40 percent of the declining population of *Microtus montanus*.

Although predators clearly affect populations, they seem not to be the primary cause of population declines, nor is even intense pedation able to stop population increases. Krebs, Keller, and Tamarin (1969) removed one third of the adults every two weeks from a fenced population of two species of *Microtus* and found that the population continued to increase. Even the stabilization of a population may require severe cropping: a population of *M. californicus* in an outdoor pen was not stabilized until more than half the animals were removed each month (Houlihan, 1963).

Stress is known to have pronounced effects on the mental and physical health of humans (witness the high incidence of mental illness and ulcers among people living for long periods under stressful conditions) and has been shown to have pronounced physiological effects on some other mammals. Such pioneer works as those of Selye and Collip (1936, 1955), Green and his coworkers (1938, 1939), and J. J. Christian (1950) demonstrated increased adrenal activity and other physiological responses in mammals living under conditions of high population densities. Christian proposed that these changes were important in controlling population cycles in mammals. According to his scenario, high populations and frequent interactions between individuals cause increased adrenocortical secretion; this leads to heightened activity, increased intolerance between individuals, reduced reproductive performance, and a population crash. Increased adrenocortical activity, as indicated by increased adrenal weight, was found to be associated with high populations of Norway rats (J. J. Christian and Davis, 1956), voles (Christian and Davis,

1966), short-tail shrews (Christian, 1954), muskrats (Beer and Meyer, 1951), and sika deer (Christian, Flyger, and Davis, 1960). More recent studies, however, show that the physiological responses to crowding described by Christian are not consistently associated with high population levels under natural conditions. Adrenal glands of Canadian lemmings were found not to enlarge in times of high density (C. J. Krebs, 1963), and To and Tamarin (1977) found no increases in adrenal weights with increasing densities of meadow voles *(Microtus pennsylvanicus)*. Clough (1965) observed no difference in adrenal weights or resistance to stress between voles from high populations and those from low populations. It seems, then, that although the stress syndrome occurs under some conditions, it is not important in controlling small-mammal population cycles.

Behavioral changes during population cycles may affect population densities. Paired experimental encounters between voles on neutral ground showed that the more aggressive animals were from peak populations (C. J. Krebs, 1970), and survival of large adults was greatest in such populations (Boonstra and Krebs, 1979). Such populations live under conditions of small home ranges and large individuals, and under these conditions male voles were highly aggressive (Turner, 1971). Both male and female *Microtus longicaudus* in New Mexico were more aggressive at peak population levels than during population declines (Conley, 1971). The tendency of voles to disperse, on the other hand, is reduced at peak densities but is strong during the increase phase (J. H. Myers and Krebs, 1971). Losses from a population are primarily due to emigration during the increase phase and to mortality of sedentary animals during the peak and decline phases. The evidence leaves no doubt about the reality of behavioral changes during population cycles—but what causes these changes?

Well before the studies cited in the above paragraph appeared in print, Chitty (1958, 1960) hypothesized that, in arvicolines, selection of individuals for genetically determined behavioral features changed with changes in density. Thus arvicoline behavior shifted during a cycle and influenced the cycle. The complexity of such a system, however, made Chitty's ideas unattractive to some, and the relatively rapid phenotypic and genotypic changes suggested by the hypothesis were questioned.

Critically important support was forthcoming, however. C. J. Krebs (1964b) found differences in skull size in relation to body size to be associated with different densities in some lemmings, and Chitty and Chitty (1962) recorded animals of both high and low growth potentials in a declining population of voles. Shifts in gene frequencies during changes in population densities were first demonstrated by Semeonoff and Robertson (1968) for declining populations of *Microtus agrestis* in Scotland, and results of careful experimental work on *M. pennsylvanicus* and *M. ochrogaster* by J. H. Myers and Krebs (1971) showed genetic differences between dispersing voles and resident animals. During the increase phase, females that had just reached sexual maturity—individuals with the highest reproduction potential—were dispersing. Electrophoretic analyses of polymorphic plasma proteins (shown by Gaines and Krebs, 1971, to be under the genetic control of a single locus with multiple alleles) demonstrated a genetic component to the dispersal: both male and female dispersers differed from residents not only behaviorally but genetically.

Hillborn (1975) has further shown that, during the increase phase of the populations of four species of *Microtus,* families tend to disperse as units; significant differences in the probability of dispersal occurred between families. Lidicker (1973), who carried out careful long-term studies on island populations of *M. californicus,* suggested that the greater opportunities for emigration in mainland situations are an important factor in extending the mainland vole cycles to three years. Chitty's hypothesis of genetic-behavioral control of arvicoline

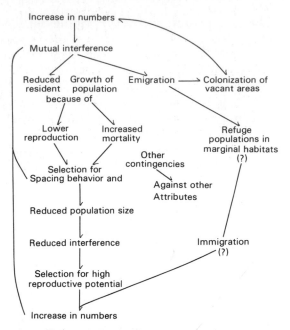

Figure 18-49. The behavioral-genetic hypothesis for control of arvicoline cycles as presented by C. J. Krebs and Myers (1974). These authors explain that "dispersal is viewed as being a more important aspect than originally proposed by Chitty. Central to the hypothesis is selection acting through behavioral interactions and changing the genetic composition of the population with fluctuating densities."

cycles has thus gained respectability through the support of an impressive body of evidence.

How, then, do biologists view the matter of population cycle regulation in small mammals? There is no universal agreement about the mechanisms of regulation, but the weight of considerable evidence is behind the hypothesis that, at least in arvicolines, some self-regulatory mechanism has evolved. C. J. Krebs, Gains, et al. (1973) have presented a modification of Chitty's hypothesis that takes into account recently accumulated evidence (Fig. 18–49). They stress the importance of dispersal on genetic change. During the increase phase of the cycle, animals intolerant of crowding disperse and establish refuge populations. The rising population comes to consist more and more of density-tolerant, aggressive individuals, and these form the peak population. Extreme aggressiveness, most importantly on the part of adult females (Boenstra, 1978), results in both reduced reproduction and poor survival rates of young, and the population declines. When the population reaches a low density, the density-intolerant, less aggressive voles from refuge populations immigrate, reproductive rates increase, and densities increase. The different behavioral patterns of the voles—density intolerance and the tendency to disperse versus aggressiveness and the tendency to be sedentary—are thought to be genetically controlled. Dispersal, therefore, generates a selective force that changes the genetic composition and behavior of the expanding population.

Such complex phenomena as small-mammal population cycles seldom have simple explanations, and the roles of predation and food quality should not be discounted. By acting most strongly on populations during and after a crash, predation probably controls the amplitude and timing of the cycles (Pearson, 1971) but seemingly does not itself cause the cycles. Secondary plant compounds may also play a role. The production by several Alaskan trees and shrubs of unpalatable adventitious shoots in response to severe browsing by snowshoe hares (p. 295) may be a common defensive pattern. Perhaps some perennial plants that are important foods for voles are similarly adapted. When heavily damaged during the peak of an arvicoline cycle, such plants may develop adventitious growth with high levels of repellent defensive chemicals, thus reducing the voles' food supply at a critical time. This plant-herbivore cycle may be of basic importance in many systems, with the cycles of predators roughly tracking prey abundance.

Chapter 19

Zoogeography

One of the most familiar kinds of biological information concerns zoogeography. Children learn that lions and zebras live in Africa and not in North America, and that kangaroos are typical only of Australia. This same type of knowledge of the presence or absence of various kinds of animals in different parts of the world is the substance of zoogeography, the study of animal distribution.

Considerations of zoogeography include two major approaches. The first is descriptive and static, and seeks to delineate the distributions of living species. Such information can be gained by field work and careful observation; it can be dealt with directly by using presently available evidence. The second approach is ecological and historical, and attempts to explain the observed distributions. Such inquiry often involves syntheses based on diverse lines of evidence. The ecologist, for example, may try to explain past or present distributions of animals on the basis of their environmental requirements. However, those scientists studying what Udvardy (1969:6) calls "dynamic zoogeography" ask the most demanding question: How, when, and from where did animals reach the areas they now occupy? Virtually every fauna consists of animals that reached the area at different times, from different regions, and by different means. Our knowledge of the complex history of a fauna depends basically on the completeness of the worldwide fossil record and on our understanding of the geologic history and paleoecology of the major land masses. Regrettably, however, our knowledge in these areas is incomplete, and even some of the major questions can be answered only tentatively.

Mammals occupy all continents, from far beyond the Arctic Circle in the north to the southernmost parts of the continents and large islands in the south. (Antarctica has no land mammals.) In the New World, the northernmost lands—the northern coast of Greenland and of Ellesmere Island—are inhabited by the arctic hare *(Lepus arcticus),* collared lemming (*Dicrostonyx groenlandicus),* wolf *(Canis lupus),* arctic fox *(Alopex lagopus),* polar bear *(Ursus maritimus),* short-tail weasel *(Mustela erminea),* caribou *(Rangifer tarandus),* and musk ox *(Ovibos moschatus).* A similar group of mammals, but lacking the musk ox, lives on the northern coast of the Taymyr Penninsula in the Soviet Union, which is the northernmost coast of Asia (Berg, 1950:19). The southernmost part of Africa has a rich mammalian fauna. Tasmania, the southernmost part of the Australian region, supports two monotremes, many marsupials, several native rodents, and several bats. On Tierra del Fuego, at the southern tip of South America, occur a bat, several rodents, a fox, otters, and a llama. The chiropteran family Vespertilionidae occurs almost everywhere there is land except in arctic areas, and the families Leporidae, Muridae, Sciuridae, Canidae, Mustelidae, and Felidae are native to all continents except Australia. All oceans, and all seas connected to the oceans, are inhabited by cetaceans; odontocetes also live in some large rivers and lakes.

Dispersal and Faunal Interchange

Animal Dispersal. Dispersal occurs when an individual or a population moves from its place of origin to a new area. The ability to disperse is as basic as the ability to reproduce, and as necessary to the survival of a species. A spacing of members of a population so that each individual can satisfy its environmental needs is critical to all organisms. Territoriality is one familiar means by which this spacing is insured, and the young of territorial species usually establish home ranges largely separate from those of other individuals, including their parents. The pressures exerted by reproduction and the necessity for the spacing of individuals create a tendency of populations to occupy ever-increasing areas, to colonize unoccupied areas, and to repopulate areas where they were previously extirpated. The more widespread a species, the less likely it is to be forced into extinction by local mortality, and as a result natural selection has usually favored those species that have broad distributions. A high adaptive premium is placed on dispersal ability. Udvardy (1969:12) has stated that "without evolved means of dispersal most animal populations would have succumbed, over a period of time, to the vicissitudes of the environment."

The ability of a population to expand into new areas depends on its innate dispersal ability (which is greater, for example, in fliers than in burrowers), on the breadth of environmental conditions it can tolerate, and on the presence of barriers. Barriers may be ecological, for example, environmental conditions under which a species cannot survive, or more simply physical, such as bodies of water, precipitous cliffs or mountains, or rough lava formations. If enough information were available, much of the story of zoogeography could be told by considering the way in which animal dispersal patterns have been modified by the location, effectiveness, and longevity of barriers.

Migration and Faunal Interchange.   Certain regions have apparently been major centers of origin of mammalian groups. Many families first appear in the fossil record in Eurasia, and North America seems also to have been the place of origin for many groups. The present mammalian faunas of regions such as Africa and South America were partly derived from mammalian migrations from northern continents. Despite uncertainty as to the place of origin of many mammalian groups (where a group first appears in the fossil record is generally taken as its place of origin), movements of mammals from place to place are in some cases well documented by the fossil record.

Simpson (1940) recognized several avenues of faunal interchange. The *corridor* is a pathway that offers relatively little resistance to mammalian migration and along which considerable faunal interchange would be expected to occur. Such a continuous corridor now exists across Eurasia; interchange of animals between Europe and Asia is highly probable and has apparently occurred frequently. A *filter route* allows passage of certain animals, but stops others (Fig. 19–1). Selective filtering has occurred at times along Beringia, the land bridge that has periodically connected Siberia and Alaska. When this bridge was present late in the Pleistocene, as an example, conditions were such that only animals adapted to cold climates could migrate between these two continents; mammals intolerant of cold conditions were denied use of this route. Mountain ranges, deserts, or tropical areas may also form filter routes. The third and most restrictive route is the *sweepstakes route*. This is a pathway that will probably not be crossed by large numbers of any given type of animal, but may be followed by an occasional individual. Such a pathway is that between New Guinea and Australia or between Africa and Madagascar. Dispersal via a sweepstakes route must occur by swimming or flying, or by such uncertain means as rafting from one land mass to another on floating vegetation or debris. The probability that an animal will follow a sweepstakes route is extremely low if the route is long, as, for example, from North America to Hawaii, but is increased if an animal is small and can cling to floating material, is aquatic, or can fly. (The only land mammals that reached Hawaii without the help of humans were bats). De-

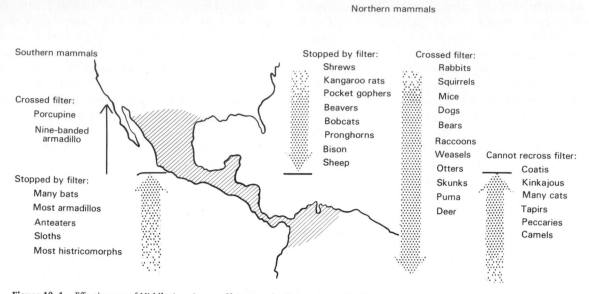

Figure 19-1. Effectiveness of Middle America as a filter route for Recent mammals. The cross-hatched area is the filter zone, and the success or failure of animals in crossing the barrier is shown. (After Simpson, 1965b)

spite the unlikelihood of a mammal's dispersal via a sweepstakes route, such dispersal has occurred and has been the means by which some mammals originally reached Australia and South America.

Mammals of the Zoogeographic Regions

The zoogeographical realms shown in Fig. 19–2, which are the basis for the organization of this discussion, were proposed by Wallace (1876) and have been widely used in discussions of zoogeography.

Palearctic Region. This region includes much of the northern part of the Old World and is the largest zoogeographic region. Included in this vast area are Europe, North Africa, Asia (except India, Pakistan, and southeastern Asia), and the Middle East. The climate is largely temperate, but contrasting conditions exist, from the intense heat of North Africa to the arctic cold of northern Siberia. Broad areas of coniferous forests, comparable in many ways to those of northern North America, are typical of

much of the northern Palearctic Region, and deserts are widespread in the south. The Palearctic is at present separated from the Ethiopian Region by deserts, from the Oriental Region by the Himalayas, and from the Nearctic by the Bering Strait.

The Palearctic mammalian fauna includes 35 families and resembles most strongly the Oriental fauna, with which it shares 60 percent of its families (Table 19–1). Because of repeated faunal interchange across the land bridge that periodically spanned the Bering Strait, the Palearctic shares nearly 55 percent of its mammalian families with the Nearctic. Many genera, and a few species, of the families Soricidae, Vespertilionidae, Muridae, Canidae, Ursidae, Mustelidae, Felidae, and Cervidae occur in both regions. Only one family, Seleviniidae, is endemic to the Palearctic Region. (A taxon is endemic to an area if it lives nowhere else.)

Nearctic Region. This area includes nearly all of the New World north of the tropical sections of Mexico, and contains habitats ranging from semitropical thorn forest to arctic tundra. The mammalian fauna includes 34 families, some of

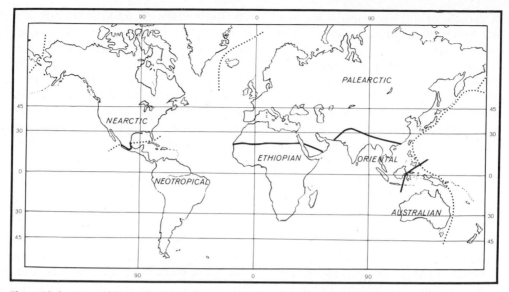

Figure 19–2. A map of the world showing the zoogeographic regions discussed in this chapter.

which are mostly tropical in distribution (for example, Emballonuridae, Desmodontidae, and Tayassuidae), together with some primarily boreal families (Dipodidae, Castoridae, and Ursidae). Only one Nearctic family (Aplodontidae) is endemic. The mammalian fauna of the Nearctic resembles most closely that of the Neotropical (Table 19–1).

Neotropical Region. This region features great climatic and biotic diversity and includes all of the New World from tropical Mexico south. Much of the area is tropical or subtropical, and broad areas are covered with spectacular evergreen rain forest. Tropical savanna and

grasslands occupy parts of the southern half of South America, and there are deserts in the south and along the western coast. The higher parts of the Andes support montane forests and alpine tundra. The South American part of the Neotropics was isolated from the rest of the world through most of the Cenozoic, but the Isthmus of Panama has provided a connection between South America and North America since the late Pliocene.

This region is second only to the Ethiopian Region in diversity of mammals. The Neotropical supports 46 families of mammals and has the largest number of endemic families (20). Espe-

Table 19–1: Comparison of the Mammals of the Faunal Regions

Region	No. of Families	No. of Endemic Families	Percentage of Families Also Found in					
			P	Nc	Nt	E	O	A
Palearctic (P)	35	1	—	55	28	48	60	30
Nearctic (Nc)	34	1	49	—	52	24	37	25
Neotropical (Nt)	46	20	37	84	—	22	35	25
Ethiopian (E)	50	14	69	38	24	—	72	40
Oriental (O)	43	4	74	52	33	62	—	40
Australian (A)	20	12	17	16	11	16	19	—

The family Hominidae and the families of marine mammals were discounted when the listings were made.

cially characteristic of the Neotropical are marsupials, bats (including three endemic families), primates (two endemic families), edentates (two endemic families), and histricomorph rodents (11 endemic or nearly endemic familes). Two species of the genus *Lama* live in South America and are the only New World representatives of the family Camelidae. (Wild Old World camelids occur only in the Gobi Desert of Mongolia.) Tapirs are restricted to the Neotropical and Oriental regions. The Neotropical mammalian fauna most strongly resembles that of the Nearctic, but it also shares one third of its families with the Oriental Region.

Ethiopian Region.   This region takes in Madagascar and Africa north to the Atlas Mountains, the Sahara, and the southern Arabian Peninsula. Deserts, tropical savannas, tropical forests, montane forests, and even alpine tundra are all represented, and the most extensive tropical savannas in the world occur in Africa.

The Ethiopian Region has the greatest number of mammalian families (50) of any faunal region, and next to the Neotropics, the greatest number of endemic families (14). The impressive array of ungulates that inhabits the savannas of Africa is unmatched elsewhere, and Africa is the last important stronghold of the families Equidae, Rhinocerotidae, Elephantidae, and Hippopotamidae. Although the only endemic artiodactylan family is Giraffidae, nearly all of the African genera of antelope (Bovidae) are endemic. The primitive lemuroid primates of Madagascar and the diverse group of cercopithecid primates of Africa are especially typical of the region, and two of the four genera of great apes live in Africa. Apart from South America, Africa is the only area with a fairly diverse histricomorph rodent fauna. Viverrid carnivores reach their greatest diversity in the Ethiopian Region, where 23 of the 25 genera are endemic. The Ethiopian mammalian fauna most closely resembles that of the Oriental Region.

Oriental Region.   Included in this region are India, Indochina, southern China, Malaya, the Philippine Islands, and the islands of Indonesia east to a line (imaginary and controversial) between Borneo and Celebes and between Java and Lombok. The area is dominated by tropical climates and once supported, before extensive land clearing by humans, broad areas of tropical forests. Deserts occur in the Pakistan area. The Oriental Region is partly isolated from the Palearctic by deserts in the west and by the Himalayas to the north and northeast.

The mammalian fauna of the Oriental Region includes 43 families and resembles most strongly that of the Palearctic area, with which it shares 74 percent of its families of mammals. Many (62 percent) of the Oriental families of mammals also occur in the Ethiopian Region. The most distinctive elements of the Oriental mammalian fauna are all of tropical affinities. Four families of primates occur in this region. Four families of mammals—Tupaiidae (tree shrews), Cynocephalidae (flying lemurs), Tarsiidae (tarsiers), and Platacanthomyidae (spiny dormice)—are endemic, and each occupies forested tropical areas. Some 15 percent of Oriental families occur elsewhere only in the Ethiopian Region (Lorisidae, Pongidae, Manidae, Elephantidae, Rhinocerotidae, and Tragulidae). The presence in both areas of rhinoceroses and elephants, great apes and lorises, and a diversity of viverrid carnivores makes the mammalian faunas of the Oriental and Ethiopian regions seem much alike, but between these areas there are some striking faunal differences. Lacking in the Oriental Region are lemuroid primates, the distinctive African histricomorph rodents, and the diverse assemblage of antelope so typical of African savannas.

Australian Region.   This region includes Australia, Tasmania, New Guinea, Celebes, and many of the small islands of Indonesia (New Zealand and the Pacific area are not included). In the area are islands of various size and degree of isolation. The island continent of Australia comes closer to New Guinea than to any other large island, but these land masses are separated by the Torres Strait, 160 kilometers wide. The northern part of the area, including New Guinea and parts of the eastern coast of Australia, are covered with tropical forest, but much of Australia is tropical savanna or desert. Some of the most arid deserts in the world occur in the interior of Australia.

The Australian Region is famous for its unusual mammalian fauna, and to the popular mind Australia itself is an area supporting marsupials almost exclusively. Actually, even Australia has nearly as many placental families (eight) as marsupial families (ten). Fifty percent of the families of the Australian Region are marsupials, and 60 percent (the monotremes and marsupials) are endemic. The mammals of The Australian Region have their closest affinities with those of the Ethiopian and Oriental regions.

Oceanic Region.   The oceans of the world compose the Oceanic Region. In this region live the whales and porpoises, most of the seals, the sea lions and walruses, and the inhabitants of isolated oceanic islands (usually bats, introduced murid rodents, and other mammals associated with humans). The large islands are included in the region with which their faunas have the most in common. Greenland, for instance, is included in the Nearctic Region, and Iceland is included in the Palearctic. The biogeography of pinnipeds is discussed later in this chapter (p. 352).

Historical Zoogeography of Mammals

Mammalian Evolution and Faunal Succession.   Since the beginning of their spectacular adaptive radiation in the Late Cretaceous and the Paleocene, mammals have followed an evolutionary pattern typical of most plant and animal groups. In general, they have become progressively more diverse and better able to completely and efficiently exploit the niches available to animals with the basic mammalian structural plan. Along with this trend toward full occupation of the environment has gone a tendency toward increasing specialization. Whereas many Paleocene mammals were generalists and could probably rather inefficiently utilize a wide variety of foods, relatively few living mammals have this mode of life. A modern herbivore, for example, typically eats only a certain type of plant material, but its teeth and digestive system are adapted to efficient utilization of the food. Behavioral adaptations favoring selective foraging have also developed. The "average" modern mammal, relative to its Paleocene or Eocene counterpart, is able to find its food with less expenditure of energy and to derive more energy from that food.

As mentioned in Chapter 18, biotic communities evolve just as do the interacting organisms that compose them. Whenever two or more organisms have attempted to play the same role in a community, to occupy the same ecological niche, the unstable situation that developed resulted in faunal change. One organism became master of the niche, and the others either moved away, evolved the ability to occupy a different niche, or became extinct. Natural selection favored those structural or behavioral modifications that allowed an animal to be the strong competitor, to most efficiently occupy its niche. Thus, a succession of mammalian faunas have occupied the earth, each better able than its predecessors to efficiently exploit the environmental possibilities of the times (Table 19–2). A number of factors, such as weather patterns and vegetation, have importantly "guided" the evolution of mammalian faunas. In addition, a factor of major importance has been the timing of migrations.

Faunal Stratification and Origins.   All large land masses support stratified faunas: not all of the animals have occupied the areas for the same length of time, nor have they all come from the same place. As an example, the mammalian fauna of Africa consists of animals representing families or genera that evolved elsewhere and dispersed to Africa, together with representatives of groups that apparently evolved on the African continent and have oc-

Table 19–2:   Ecological Replacement of Older Genera by Younger Genera in the Pleistocene Record

Older Genera	Younger Genera
Nannippus (three-toed horse)	*Equus* (modern horse)
Stegomastodon (mastodon)	*Mammuthus,* mammoth (elephant)
Capromeryx (pronghorn)	*Antilocapra* (pronghorn)
Hypolagus	*Lepus* (hare), *Sylvilagus* (rabbit)
Pliophenacomys	*Microtus* (meadow vole)
Arctodus	*Ursus* (brown and grizzly bears)

After Hibbard, Ray, et al., 1965.

cupied the area throughout much of the Cenozoic. Horses had their origin in North America in the Paleocene (Jepsen and Woodburne, 1969) and probably did not reach Africa until the Pleistocene. Today horses (zebras) still form a conspicuous part of the African scene but are completely absent from the New World. Old World monkeys (Cercopithecidae), on the other hand, probably originated in North Africa and have seemingly occupied Africa continuously since the Oligocene. The Old World monkeys clearly represent part of an early African faunal stratum, whereas horses are part of a late one.

Historical Zoogeography of Pinnipeds: A Case History.   The story of the origins and dispersal of the seals, sea lions, and walruses, pieced together from a spotty fossil record and from knowledge of the changing configurations of land masses, provides an instructive example of historical biogeography. This section is based on the work of Repenning, Ray, and Grigorescu (1979).

The family Otariidae (sea lions and fur seals) originated in the temperate waters of the North Pacific 10 million years ago, in the late Miocene. Early otariids persisted as a conservative evolutionary line until about 6 million years ago, when the lineage leading to the Alaskan fur seal *(Callorhinus)* appeared and occupied northern waters. The remaining otariids were adapted to temperate waters, and tropical and subtropical seas were a barrier to their dispersal. Some 3 million years ago, with the closing of the Central American seaway (the seaway separating North and South America), cool waters extended southward along the Pacific coast of the New World, and some of these temperate-water otariids spread southward. Their appearance in the Southern Hemisphere is documented by fossil beds in Peru. From these southern waters, the ancestors of the southern fur seal *(Arctocephalus)* moved southward entirely around Antarctica. In the North Pacific, the remaining ancestral otariids evolved into the highly successful sea lions 3 million years ago. In the early Pleistocene, perhaps with the cooling of marine waters, sea lions spread southward and now occupy the Australian Region and both sides of southern South America.

The family Odobenidae (the walruses) appeared 14 million years ago in the warm-temperate seas of the mid-northern latitudes of the North Pacific. By about 8 million years ago, odobenids were more diverse than otariids and had divided into two lines of descent: one lineage stayed in the temperate waters of the north, while the other moved southward and became adapted to tropical or subtropical waters. This second (warm-water) lineage, from which the modern walrus evolved, spread to the Caribbean via the Central American seaway 6 to 8 million years ago. With the subsequent closing of this seaway, and warming of Caribbean waters, walruses dispersed northward into the cold North Atlantic. The lineage that remained in the North Pacific became extinct 4 million years ago, and there were no walruses in this area until less than 1 million years ago, when the North Atlantic walrus moved through the

Arctic Ocean to the North Pacific. Today, walruses in both the North Pacific and the North Atlantic are restricted to cold northern waters.

The seals (Phocidae) had a North Atlantic origin in the Miocene 15 million years ago. By about 5 million years ago, the two modern phocid lineages had diverged: the Phocinae, the "northern seals," adapted to cooling northern seas and invaded even arctic waters; in contrast, the Monachinae, the "southern seals," retreated southward, entered the Pacific before the closure of the Central American seaway, and reached Antarctica, where several species are abundant today. From the phocine lineage evolved a series of cold-water species, including the ringed seal *(Phoca hispida),* which occupies the Arctic Ocean and has the most northerly distribution of any pinniped.

The phocine seal *(Phoca siberica)* of Lake Baikal in Russia probably gained access to this lake some 300,000 years ago by moving up rivers from the Arctic Ocean to lakes at the foot of the central Siberian ice sheet. With the retreat of the ice sheets and the drying of many Pleistocene lakes, this seal was isolated in Lake Baikal. The Antarctic species are not the only living seals derived from the monachine lineage; several species inhabit warm waters (the Black Sea, Mediterranean, Caribbean, and waters of the Hawaiian Islands). Among the Antarctic species is the Weddell seal *(Leptonychotes weddelli),* which occupies the icebound margins of Antarctica and perhaps represents the extreme in adaptation to severe polar conditions.

Clearly the distribution of each pinniped family has been strongly influenced by climatic conditions, water temperatures, and patterns of ocean currents. Seemingly each group has become progressively better able to cope with cold boreal or cold austral waters.

Continental Drift, Mammalian Evolution, and Dispersal

Continental Drift. Teaching in geology and paleontology in North America has been dominated for many years by the view that the positions of the continents and the intervening oceans are fixed, that they have remained immutable back through the vast sweep of geologic time. Because they accepted these tenets, most North American paleontologists were forced to rely on often tenuous intercontinental land bridges or sweepstakes dispersal to account for intercontinental movements of terrestrial animals. Within the last 20 years, however, our geologic, paleontologic, and biogeographic perspective has been drastically transformed by convincing evidence in favor of the theory of continental drift. As put by L.M. Gould (1973), acceptance of the theory of continental drift "has created the greatest revolution in man's thinking about his earth since the Copernican Revolution."

Since 1756, when the German minister Theodor Lilienthal noted that "facing coasts of many countries, though separated by the sea, have a congruent shape" (N. Calder, 1972:42), the possibility that continents have not been immovable has repeatedly been considered. The proposition that continents have drifted over the earth was formulated in detail by Wegener in 1915, but our modern views of continental drift are based more on the careful work of DuToit (1957). These researchers and others observed that the shapes of the eastern coastlines of the New World could be fit in jigsaw-puzzle fashion against the western coastlines of the Old World, and they took this evidence to indicate that the continents of the two hemispheres had originally formed one great land mass that had split apart.

The discovery by Colbert of the ancient reptile *Lystrosaurus* in Antarctica, on a "summer" day (in December) in 1969, put the capstone on the pyramid of evidence supporting continental drift. This Triassic nonaquatic reptile had previously been found on other southern continents; its distribution could be explained only by assuming that the continents had once been connected. Wegener's wild-eyed theory was vindicated—the weight of the evidence became too much for even the staunchest

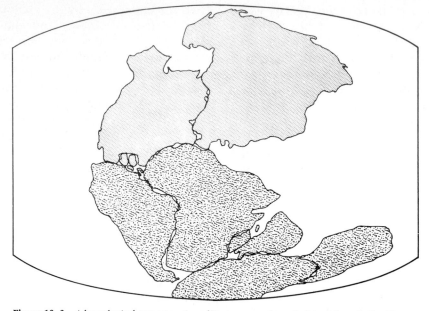

Figure 19-3. A hypothetical reconstruction of Pangaea near the end of the Paleozoic Era. The northern land masses (stippled) represent Laurasia, and the southern masses represent Gondwanaland. (From Colbert, 1973)

"antidrifters" to bear. (References on continental drift appear in the bibliography under Colbert, 1973; Dietz and Holden, 1970; and Menard, 1969.)

One present view of the history of the major land masses, based on an acceptance of continental drift, is that 200 million years ago there was a single great land mass, Pangaea, to use Wegener's term (Fig. 19-3). (An earlier cycle of continental drift, during the Paleozoic, is less well understood than the Mesozoic-Cenozoic cycle under consideration.) This supercontinent was divided by a series of rifts that by the end of the Triassic (180 million years ago) had split it into a northern land mass, Laurasia, and a southern series of land masses, collectively called Gondwana (Fig. 19–4). By the end of the Cretaceous (65 million years ago), South America had moved westward, well away from Africa, which was separated from Laurasia by a narrow sea. The eastern coast of North America and the western coast of Europe were presumably still in contact, but further drifting of the continents throughout the Cenozoic led to the existing arrangement of these land masses.

The force behind the slow, inexorable movement of continents is the spreading of the sea floor. The lithosphere of the earth—its crust, roughly 70 kilometers thick—is divided into a series of huge plates (Fig. 19–5). These largely rigid plates are constantly but slowly changed by the addition at one border of upwelling molten rock and destruction at the other border where the plates plunge down into the earth and again become part of its deep, molten core (Fig. 19–6). This vast system of crustal movement is called plate tectonics and is a major focal point of modern geologic research.

Continents float passively on the lithosphere and are carried along with the movements of the plates. Geologic and paleontologic evidence not only documents the divisions of the land masses and their movements away from

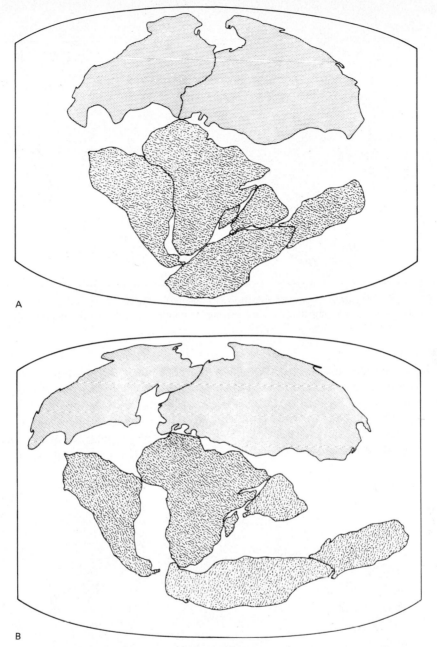

A

B

Figure 19-4. Pangaea as it may have been (A) in the Triassic and (B) in the Cretaceous. (From Colbert, 1973)

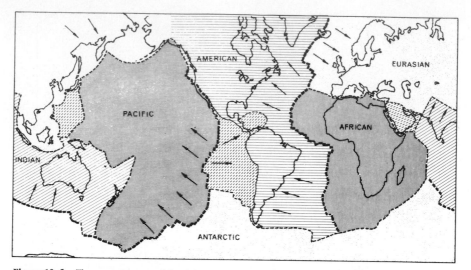

Figure 19–5.   The present extent of the major tectonic plates and the direction of their movement. Plate movement is influenced by upwelling of molten rock from deep within the earth along the rift lines between two plates (see Fig. 19–6). (From Colbert, 1973)

each other but also suggests that collisions have occurred between continents carried on adjacent plates. Thus India was carried northward on the Indian plate and collided in the Cenozoic with the Eurasian plate. The concept of continental drift and plate tectonics has given great unity to the geologists' view of the creation of major land forms: the major mountain chains of the world are formed in part by deformation of the earth's crust at the leading edges of moving plates; earthquake zones are concentrated along these lines of tension; and exten-

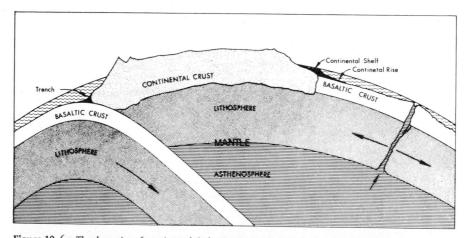

Figure 19–6.   The dynamics of continental drift. This figure, from Colbert (1973), is explained by him as follows: "Rifting between two tectonic plates (American and African) results in the welling up of molten rock from the depths, to form the midoceanic ridge (right). The westward drift of the American plate, carrying the continental block, caused a collision with the pacific plate, this latter being forced down to be destroyed in the mantle. The zone of collisions is marked by the pushing up of the Andes and the formation of a deep trench, with resultant deep earthquakes."

sive ocean trenches occur where the basaltic crust and the lithosphere of a plate plunge deep into the earth (Fig. 19–6).

The revolution in geologic thought is obvious, but to biologists also the shift from the static-continent to the dynamic plate concept is of tremendous importance. Just as continents have separated, collided, or drifted progressively farther apart, so too have terrestrial biotas been isolated or brought together, entire distribution patterns of marine biotas profoundly altered, and global ecological diversity shifted. To the evolutionary biologist, the movements of the earth's crust provide "the stage for all biological activity" (McKenna, 1972).

It is obvious that when one considers the biogeography of individual species or of entire biotas, one must take into account plate tectonics and continental drift. The beauty of these concepts is that they often provide explanations for a diverse array of biogeographic patterns that have long appeared inexplicable.

Continental Drift and Mammalian Diversity. As pointed out by B. Kurten (1969), the fact that mammals evolved during a span of

the earth's history when continents were moving apart may be a key to mammalian diversity. The Cenozoic radiation of mammals occupied a shorter span of time (65 million years) than that available to reptiles for their radiation during the age of reptiles (200 million years), but mammals have diversified to a much greater extent, as reflected by a greater number of orders (about 30 orders of mammals to 20 orders of reptiles). Kurten believes that the greater diversity of mammals is a result of continental drift. Mammals evolved on several land masses under conditions of isolation or semi-isolation, whereas reptiles developed before the continents had moved far apart and therefore developed under conditions allowing freer faunal interchange between evolving stocks.

A striking feature of mammalian evolution has been duplication of functional and, to some extent, structural types in separate groups. Examples of such convergent evolution are abundant. Members of several orders specialize in eating ants and termites (Fig. 19–7). The orders Marsupialia, Rodentia, Lagomorpha, Artiodactyla, and Perissodactyla all contain herbivorous,

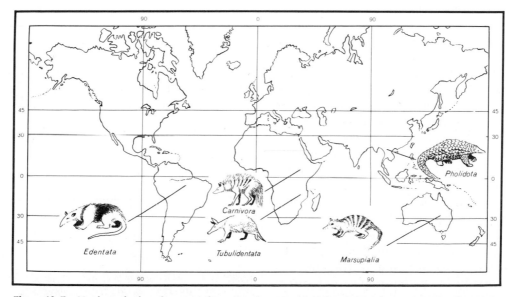

Figure 19-7. Members of at least five mammalian orders that occur in southern continents are adapted to eating ants and termites: Edentata (all members of the family Myrmecophagidae), Carnivora (the hyaenid *Proteles*), Tubulidentata (the aardvark *Orycteropus*), Marsupialia (the numbat *Myrmecobius*), and Pholidota (the pangolin *Manis*).

cursorial mammals that pursue basically similar modes of life. Small, terrestrial, insect-eating mammals have developed in at least four orders (Insectivora, Marsupialia, Edentata, and Rodentia). The greatest duplication has occurred in southern land masses, which have been longer and more completely isolated than have the Nearctic and Palearctic areas. Mammalian diversity, then, may be as much a result of the progressive Mesozoic and Cenozoic separation of the continents as of the structural and functional adaptability of the mammals themselves.

Nearctic and Palearctic Mammalian Faunas. The high percentage (roughly 55 percent) of Nearctic mammalian families that also occur in the Palearctic Region has long suggested Cenozoic avenues of dispersal between these regions. Although presently separated by a seaway across the now submerged Bering-Chukchi platform, Alaska and northeastern Asia are parts of the same tectonic plate and during much of the Cenozoic were connected by the Beringian land bridge. Although this bridge has allowed repeated faunal interchange between Asia and the Nearctic, it was not the key to the faunal similarities between Europe and the Nearctic in the Early Tertiary.

From the Cretaceous until roughly the end of the Eocene, a marine seaway, the Turgai Strait, stretched from the Mediterranean to the Arctic Ocean, separating Europe from Asia and interposing a barrier between the terrestrial faunas of these areas. Recent geologic evidence based on studies of plate tectonics points toward persistent connections between northern Europe and North America until the middle Eocene, 50 million years ago. These connections allowed dispersal between Europe and North America and account for Early Tertiary faunal similarities between the areas (McKenna, 1975a).

Since the middle Eocene, because of the drifting apart of Europe and North America, there has not been a continuous and direct North Atlantic land bridge. From the time of the severing of this dispersal route until the Turgai

Strait was drained in the Oligocene, Europe was isolated not only from North America but from Asia. Whereas the Early Tertiary was clearly a time when North America and Europe had broad faunal affinities, after the Eocene the only route of Palearctic-Nearctic biotic exchange was Beringia, and since that time North America has had its closest affinities with Asia.

The degree of faunal interchange between the Nearctic and Palearctic regions fluctuated during the Cenozoic, with peaks in the early Eocene, early Oligocene, late Miocene, and Pleistocene (Fig. 19–8). No faunal mingling seems to have occurred in the Recent, by which time the North Atlantic dispersal route was long since inundated and interchange across Beringia was barred by the Bering Strait. Because Beringia is at a high latitude, this dispersal route probably functioned as a filter barrier even before the climatic cooling of the Pleistocene. The rodent families Heteromyidae and Geomyidae evolved in the Middle Tertiary in response to the arid and semiarid conditions that developed in the southern part of the United States. These

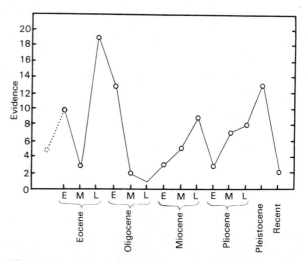

Figure 19-8. The intensity of interchange of land mammals between Eurasia and North America in most of the Cenozoic, as indicated by the numbers of closely related mammals on the two continents. (After Simpson, 1965)

families and those with southern distributions, such as the Phyllostomidae and Procyonidae, were apparently unable to reach or cross the Beringian land bridge.

The derivation of the Nearctic mammalian fauna is complex. There are a number of families now established in North America and widespread in Eurasia that had their origins in Europe or Asia and reached North America via Beringia or, in the Early Tertiary, across the North Atlantic route. The Talpidae and Soricidae, for example, first originated in Europe in the Eocene and appear in the North American fossil record in the Oligocene. Their route to North America must have been across the North Atlantic bridge, for these families do not appear in Asia until the Recent (Talpidae) or the Pliocene (Soricidae). The Felidae first appears in the Paleocene of Asia and occurs in both Europe and America in the Eocene. In this case, North America must have served as an immigration route to Europe, for at the time that felids appeared in Asia, the Turgai Strait separated this land mass from Europe. The Cervidae arose in Asia in the Oligocene and arrived in North America in the Miocene. This family clearly used the Beringian dispersal route. North America was also the source of some families that spread to the Old World. The Camelidae, Aplodontidae, Didelphidae, and Leporidae are examples. The primitive horse *Hyracotherium* died out at the end of the Eocene in the Old World, and all of the horses that appeared there after this time were emigrants from North America.

The Nearctic mammalian fauna has been augmented to a very limited extent by species of Neotropical origin. Except for some bats, members of only two families that originated in South America were able to become established in the Nearctic. A porcupine (Erethizontidae) has become widely established in the Nearctic, and during the Pleistocene large relatives of the armadillo—ground sloths and glyptodonts—were far more widespread in North America than the armadillo is now.

The Ethiopian Mammalian Fauna.   As mentioned in the discussion of the Palearctic Region, Ethiopian-Palearctic faunal interchange took place periodically in the Tertiary, but one-way transfers also seem to have occurred. The vertebrate fossil record indicates that, in the Eocene and again in the Oligocene, Africa received mammalian immigrants from Eurasia but there was no reciprocal movement of African mammals back to Eurasia. Such one-way dispersal was perhaps associated with the separation from Europe of land masses less than continent size (microplates) and their movement to Africa, thus "inoculating" Africa with northern mammals (Van Couvering and Van Couvering, 1975). Such Tertiary microplate movement in the western Mediterranean area has been described by a number of geologists. These microplates may well have functioned as classic "Noah's arks."

The present African mammalian fauna has clearly had a complex derivation. Tertiary immigrants from the north have contributed importantly, and adaptive radiation of some of these groups in Africa gave rise to a number of endemic orders and families. In the Late Tertiary, extensive faunal interchange with Eurasia reduced the distinctiveness of the African mammalian fauna. In the Pleistocene, however, the narrowing of the dispersal route between Africa and Eurasia and the development of extreme desert conditions throughout much of North Africa formed a strong filter barrier; the Ethiopian mammalian fauna again tended toward endemism. Africa, together with some parts of Asia, today serves as a refuge for a diverse Pleistocene world fauna that has disappeared elsewhere.

A high percentage of Palearctic families (48 percent) also occurs in the Ethiopian Region, a surprising situation in view of the present narrow and inhospitable single link between Africa and Eurasia at the Isthmus of Suez. (This link is now broken by the Suez Canal.) At various times during the Cenozoic, however, broad dispersal routes existed between North Africa and Eurasia. In the Paleocene, Africa was linked to Asia by Arabia and to Europe by the area that is now Turkey and Bulgaria; at least temporary links were seemingly present in the Oligocene, Miocene, and Pliocene epochs. Pleistocene and Recent connections have been much restricted and have been via the Isthmus of Suez, the breadth of which changed with fluctuations in sea level. Seemingly, then, there have been periodic opportunities since the Cretaceous for faunal exchange between Eurasia and Africa when equable climates did not exert the strong filtering effect typical of the inhospitable Recent route through Egypt and the Sahara.

South American Mammals.   The origins of the Neotropical mammalian fauna have long held the interest of distinguished scientists. Wallace (1876) was first to recognize the faunal interchanges that occurred between North and South America late in the Tertiary, and intensive paleontologic field work in the late nineteenth century provided a more complete understanding of these events. In 1893, von Zettel wrote that "there was thus accomplished, toward the end of the Pliocene, one of the most remarkable migrations of faunas that geology has been able to record." The classic works of Simpson (1950, 1965a, 1965b, 1969) did much to clarify our understanding of South American historical zoogeography, and recent field work and advances in our geologic knowledge have refined previously held views.

Throughout most of the Cenozoic, North America was isolated from South America by the Bolivar Trough Marine Barrier, a seaway through northwestern Colombia and southern Panama. In the late Pliocene, 3 million years ago, the emergence of the Panamanian land bridge provided a gateway for an intermingling of North and South American faunas. This classic natural experiment has been called the Great American Interchange. (The following discussion of this interchange is based largely on the work of L.G Marshall, Webb, et al., 1982.)

The mammalian participants in this interchange can be divided into two groups on the basis of the means and timing of dispersal. The first group was made up of "waif immigrants" that dispersed along island arcs in the late Miocene before the emergence of the Panamanian land bridge. This group includes the procyonids and murids, which moved into South America from North America, and the Megalonychidae and Mylodontidae (extinct ground sloths), which dispersed from South America to North America. The second group includes mammals that dispersed across the Panamanian land bridge at various times after its emergence. A diversity of mammals belong to this group, including North American taxa that immigrated to South America and South American taxa that entered North America (Table 19–3).

This spectacular reciprocal interchange of land mammals (discounting bats and manatees) was symmetrical: there were 32 families of land mammals in South America before the interchange, 39 families after it began, and then a late Pleistocene decline to the 35 families of today. In North America, in the same period, an equivalent number of families followed a comparable pattern of increase and decline. The unique outcome of this interchange was the rapid diversification of North American secondary immigrants (those that evolved from the initial immigrants) in South America. This radiation of North American immigrants resulted in a partial replacement of South American natives and an increase in mammalian diversity. Overall, the South American mammalian fauna was thus enriched but at the expense of some native taxa.

L.G. Marshall, Webb, et al. (1982) offer the following speculative explanations of the post-land-bridge mammalian history of South America. Documented orogenic (mountain-building) activity between 4.5 and 2.5 million years ago elevated the Andes from 2000 to 4000 meters. By providing a barrier to moist Pacific air masses, the newly elevated Andes caused the development of dry pampas, desert, and semidesert areas on the eastern (leeward) side of the mountains. The animals that occupied these

**Table 19-3:   Timing and Means of Dispersal of Families Involved in the
Great American Interchange**

Family	Time of Dispersal
North American Immigrants to South America	
Waif Immigrants Across Bolivar Trough Marine Barrier	
Procyonidae	Late Miocene
Muridae	
Immigrants Across Panamanian Land Bridge	
Mustelidae	Late Pliocene
Tayassuidae	
Canidae	
Felidae	
Ursidae	
Camelidae	Early Pleistocene
Cervidae	
Equidae	
Tapiridae	
Gomphotheriidae	
Heteromyidae	
Sciuridae	
Soricidae	Recent
Leporidae	
South American Immigrants to North America	
Waif Immigrants Across Bolivar Trough Marine Barrier	
Megalonychidae	Late Miocene
Mylodontidae	
Immigrants Across Panamanian Land Bridge	
Dasypodidae	
Glyptodontidae	Late Pliocene
Hydrochoeridae	
Erethizontidae	
Didelphidae	Early and middle Pleistocene
Megatheriidae	
Toxodontidae	Late Pleistocene
Callithricidae	
Cebidae	
Bradypodidae	Recent
Myrmecophagidae	
Dasyproctidae	
Echymyidae	

From L. G. Marshall, Webb, et al., 1982.

areas, which were previously savanna-woodland, retreated northward. This environmental change made new habitats available, and these were largely filled by mammals arising from radiations among North American groups (murids, canids, horses, llamas, javelinas). The greater ability of the North American mammals to occupy narrow niches, perhaps developed during their North American history, may have enabled them to occupy these newly opened habitats, with a corresponding reduction in the area available to some South American natives.

It is clear, then, that the South American mammalian fauna has a complex derivation. The unusually large number of endemic taxa is a reflection of the degree and duration of separation

Figure 19-9. Restorations of two South American notoungulates. (A) *Toxodon* (Toxodontidae), a Pleistocene form; (B) *Pyrotherium* (Pyrotheriidae), an Oligocene form. (After Patterson and Pascual, 1972)

of South America from other continents and of the fact that some native South American mammals of tropical affinities have been unable to disperse northward.

The Tertiary marsupial radiation in South America has been discussed previously (p. 53), but the extremely impressive ungulate radiation also deserves mention. The condylarth stock that reached South America very early in the Tertiary radiated rapidly, and by the end of the Paleocene, a diverse series of evolutionary lines were established. In isolation from North American ungulate stocks, the South American ungulates went their unique evolutionary ways. Although they clearly filled many of the same niches occupied by other lineages of ungulates in other parts of the world, many of the South American ungulates were anomalous-looking beasts unlike any ungulates elsewhere. Some had an arched profile with the snout lengthened into a proboscis, a low-slung body, and short limbs (Fig. 19-9B), and one can imagine that they moved ponderously.

These South American ungulates spanned a considerable size range. There were rat-size little ones and giants approaching the size of an elephant. Especially successful was the order Notoungulata, which included various herbivorous genera, one of the largest of which was *Toxodon,* a stubby-legged, rhinoceros-like beast some 3 meters in length (Fig. 19-9A). Another important order, Litopterna, included a number of smaller cursorial ungulates. One ad-

vanced Miocene genus *(Thoatherium)* had one-toed feet that not only were much more specialized than those of the contemporary North American horses but were even more specialized than the feet of present-day horses (Figs. 19-10 and 19-11).

These distinctive South American ungulates reached their peak of diversity and numbers in the Oligocene and Miocene, but they declined in the Pliocene and by the end of the Pleistocene only fossils remained. Patterson and Pascual (1972) stressed that the decline of the South American ungulates was not due to the invasion of South America by Nearctic ungulates and carnivores. Major faunal shifts occurred be-

Figure 19-10. A restoration of *Thoatherium* (Prototheriidae), a highly cursorial South American ungulate from the Miocene. (After Patterson and Pascual, 1972)

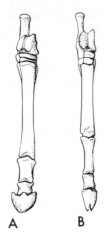

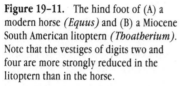

Figure 19-11.   The hind foot of (A) a modern horse *(Equus)* and (B) a Miocene South American litoptern *(Thoatherium).* Note that the vestiges of digits two and four are more strongly reduced in the litoptern than in the horse.

fore the emergence of the Panamanian land bridge. By this time, over half of the mammals occupying the adaptive zone of the large herbivores were not ungulates: of the ten families of large Pliocene herbivores, four were edentates and two were gigantic caviomorph rodents.

History of the Australian Mammals.   The native mammalian fauna of the Australian Region, and especially of Australia itself, is famous for its uniqueness. Many of the functional roles played elsewhere by carnivores or ungulates are pursued by marsupials, which are restricted either to Australia or to the Australian zoogeographical region.

For many years, the origin of the Australian marsupial fauna has been debated, but recent studies of continental drift and the occurrence of a fossil marsupial in Antarctica (M.O. Woodburne and Zinsmeister, 1982) strongly indicate a southern dispersal route. Marsupials originated in the New World, and by the Early Tertiary were present in both North and South America. Australia was situated far south, near

Antarctica, until as late as the Cretaceous or perhaps the Eocene. Marsupials probably entered Antarctica from South America in the Cretaceous and dispersed from there to Australia.

Although marsupials and eutherians both reached South America by the Early Tertiary, only marsupials entered Australia at that time. Why did eutherians fail to accompany the marsupials? Perhaps eutherians did not yet occur in the specific parts of South America or in the habitats from which marsupials dispersed to Antarctica and then to Australia, or perhaps marsupials were more abundant than placentals or had superior ability to cross water barriers. In any case, Australia drifted northward, away from Antarctica, in the early Tertiary and thus provided a refuge where, for a considerable time, marsupials underwent an adaptive radiation free from competition with terrestrial eutherians, the first of which probably reached Australia in the Pliocene.

Eight of the 20 families of Australian mammals are placentals. These groups arived in Australia at various times, and all can easily be assigned to families that occur in other areas. Some have undergone little change since their arrival in Australia.

These groups arrived in Australia at various times, and all can easily be assigned to families that occur in other areas. Some have undergone little change since their arrival in Australia.

Bats are a group that has remained nearly unchanged. Apparently, bats entered Australia at various times in the Tertiary, and of the 21 Australian genera of bats, only two are restricted to Australia. Interchanges of bats between New Guinea and Australia, which were perhaps frequent because bats can fly, have kept the Australian bats from differentiating markedly from those of the Oriental Region.

Rodents are abundant and diverse (about 13 genera) in Australia, and some species have undergone great specialization, but all of these rodents are clearly assignable to the widespread family Muridae. Simpson (1961) has divided the Australian murids into four groups, according to their history on this island continent: (1) the Rattus group consists of species that were introduced by Europeans and of species endemic to Australia that perhaps developed from pre-Pleistocene immigrants; (2) the old Papuan group contains genera that evolved in Australia from murid ancestors that arrived probably no later than the Pliocene; (3) the old Australians are a fairly diverse group of, in some cases, highly specialized rodents whose ancestors were the first rodents to reach Australia, probably in the Pliocene; (4) two semiaquatic murid genera; one of these apparently evolved in Australia from ancestors that came from New Guinea, and the other came recently from New Guinea.

The family Canidae is represented in Australia by the dingo (*Canis familiaris*), which was probably brought in by aborigines from New Guinea.

Perhaps the most remarkable feature of the Australian mammalian fauna is the presence of a marsupial assemblage that is fairly balanced, in the sense that it fills most of the terrestrial niches. Kangaroos and wallabies take the place of ungulates, dasyurids take the place of shrews and to some extent of rodents, phalangerids take the place of squirrels, and so on (Chapter 5). Placental mammals that reached Australia without any help from humans have in large part either filled adaptive zones that marsupials could not occupy, as in the case of the bats, or occupied niches that they could perhaps fill more effectively than could marsupials, as in the case of the murid rodents. Such recently introduced placentals as dingos and, more recently, rabbits and foxes have had the unfortunate effect of displacing native marsupials or preying heavily on them.

The Unusual Mammalian Fauna of Madagascar.

Islands long isolated from continents frequently have an unusual mammalian fauna. Such a fauna may be dominated by a group equally important nowhere else, as in the case of the marsupials of Australia, or may be extremely poor in mammals, as in the case of New Zealand, where the only native mammals are bats. Madagascar is an interesting example of a refugium supporting a primitive mammalian fauna with little ordinal diversity, many endemics, and a seemingly incomplete exploitation of habitats.

Madagascar is a large island 1600 kilometers in length, with a maximum width of 560 kilometers. It lies 420 kilometers east of the eastern coast of Africa. Madagascar had separated from Africa by the start of the Cenozoic and has been isolated from other land masses since that time. The island has supported six orders of mammals in Recent times (Table 19-4). Most of the mammals are endemic; the most highly diversified groups are the lemuroid primates, which probably arrived in the Eocene, and the tenrecid insectivores, which perhaps arrived as early as the Paleocene. The mammals of Madagascar reached there from Africa by island hopping and rafting. Many of the Malagasy mammals occupy niches filled elsewhere by mammals of different taxa (Table 19-5). This phenomenon is called complementarity by Darlington (1957:23). The only artiodactyl present before the arrival of humans was the now extinct hippopotamus *(Hippopotamus lemelii)*, and today

Table 19-4: Mammalian Fauna of the Panama Canal Zone and Madagascar

Order	Number of Species	
	Panama Canal Zone	Madagascar
Marsupialia	6	0
Insectivora	0	10
Chiroptera	40	12
Primates	5	10
Edentata	7	0
Lagomorpha	1	0
Rodentia	19	8
Carnivora	11	6
Perissodactyla	1	0
Artiodactyla[a]	3	1

[a]The artiodactyl from Madagascar, a hippopotamus, is now extinct.

After J. F. Eisenberg and Gould, 1970.

Table 19–5: Some Major Feeding Niches and the Mammals Occupying Them in Panama and Madagascar

	Anteaters		Primary Insectivore and Secondary Frugivore		Carnivore	
	Arboreal	*Terrestrial*	*Arboreal*	*Terrestrial*	*Arboreal*	*Terrestrial*
Panama (major genera only)	Edentata Myrmecophagidae *Cyclopes* *Tamandua*	Edentata Myrmecophagidae *Myrmecophaga*	Marsupialia Didelphidae *Marmosa* Primates Cebidae *Aotes* *Saguinus*	Edentata Dasypodidae *Cabassous* *Dasypus*	Carnivora Mustelidae *Eira* Felidae *Felis*	Carnivora Mustelidae *Mustela* *Galictus* Canidae *Urocyon*
Madagascar			Insectivora Tenrecidae *Echinops* Primates Lemuridae *Microcebus* *Cheirogaleus* Daubentoniidae *Daubentonia*	Insectivora Tenrecidae *Centetes* *Hemicentetes* *Oryzorictes* *Geogale*	Carnivora Viverridae *Cryptoprocta* *Galidia* *Fossa* *Viverricula*	

After J. F. Eisenberg and Gould. 1970.

the introduced river hog *(Potomochoerus)* is the only wild artiodactyl. The ungulate niche has largely gone unfilled, although a group of large Pleistocene lemurs, now extinct, may have been terrestrial herbivores.

The Island Syndrome.   Mammals isolated on islands typically face different selective pressures than do members of parental mainland stocks. On islands, competition is usually reduced or may be absent, predators are often absent or few kinds are present, and the flora may be depauperate. In some cases, as on some small desert islands off the eastern coast of Baja California, one or two species of rodents are the only mammalian inhabitants and just a fraction of the number of species of plants that occur on the mainland is present. Through time, island mammals tend to diverge from parental mainland stocks, but the pattern of divergence is not consistent for all species. Island mammals typically differ in size from mainland relatives, and Foster (1964) pointed out that, whereas some mammals become larger on islands, others become smaller. Island rodents and marsupials are generally larger than their mainland relatives; insectivores, lagomorphs, carnivores, and artiodactyls, however, are usually smaller. Examples are numerous. The *Peromyscus* inhabiting islands off the coast of British Columbia are unusually large, but the caribou that inhabited one of these islands (but is now extinct) was a dwarfed form. The gray fox *(Urocyon littoralis)* that lives on the Channel Islands off the coast of California is substantially smaller than the mainland *Urocyon cinereoargenteus,* and the elephant that lived on these islands in the Pleistocene was a dwarf. However, evolutionary patterns on islands are not completely consistent. Counter to the usual trend, not all island insectivores are small. Unusually large insectivores (solenodonts; p. 88) evolved on some of the islands of the West Indies, and the largest insectivore of all time *(Deinogalerix koenigswaldi)* lived on the Mediterranean island of Gargano. This insectivore was larger than a fox

and probably fed on rodents (Freudenthal, 1972).

The remarkable dwarfed Pleistocene mammals of the Mediterranean islands have been discussed by Sondaar (1977). In the Pleistocene, elephants *(Elephas)* and deer *(Cervus)* lived on many of these islands, and some islands supported hippopotami *(Hippopotamus)*. These mammals must have reached the islands by sweepstakes routes, for the extent to which they diverged morphologically from the mainland stocks and the fact that generally not all types occurred on an island suggests that access to the islands was across water. All of the three types listed above are known to be strong swimmers, and a single pregnant female could have founded a population on an island. Of special interest are the similar evolutionary trends exhibited by large mammals on a number of islands between which passage of terrestrial mammals would have been impossible. Elephants on the islands became strongly dwarfed relative to the parental mainland stock of *Elephas namadicus. Elephas falconeri* of Sicily, an example of extreme dwarfism, was roughly 1 meter high, some one quarter the size of its mainland progenitor, and relative to mainland elephants, *E. falconeri* had short distal segments of the limbs, cheek teeth with fewer enamel ridges, and a much lower skull (Fig. 19-12), with a reduction of the elaborate system of air sinuses. Short-leggedness was especially pronounced in the island deer, but the pig-size island hippopotami also became short-legged.

These patterns of parallel evolution probably resulted from similar selective pressures on the many isolated islands. No large predators were on the islands. Large size is an extremely effective adaptation to avoid predation, and without large predators great size was no longer of advantage. An unreliable food supply for herbivores may have favored smaller size, and in the absence of predators, overpopulation might have triggered periodic heavy mortality. Beds of deer bones found on the island of Crete are in-

Figure 19-12. Differences between the skull morphology of (A) the widespread mainland species *Elephas namadicus* and (B) the insular species *E. falconeri* from Crete; approximate skull lengths are 768 mm for *E. namadicus* and 288 mm for *E. falconeri.* (After Maglio, 1973)

terpreted by some paleontologists as evidence of mass mortality, and abnormalities of the bones suggest starvation as the cause of death. In the dwarf elephant *E. falconeri,* the reduction of the skull crest and the reduced number of enamel ridges on the molars were related to the general dwarfing (Maglio, 1973:94). The marked shortening of the limbs of the deer is thought by Sondaar (1977) to have been due to

two factors: the absence of predators and the consequent lack of need for speed, and the need for sturdy and well-braced limbs with which to negotiate mountainous terrain.

Just as mammals on islands are divergent structurally, some have changed behaviorally. As an example, desert woodrats *(Neotoma lepida)* on Danzante Island in the Gulf of California have very large home ranges, and males are

Table 19-6: Climatic Change in North America During the Cenozoic Era

Subdivision of the Cenozoic Era	Time Since Start of Each Division (millions of years)	Climate
Quaternary		
Holocene	0.012	Temperate; desert in southwestern U.S. and Mexico
Pleistocene	1.5	Episodic glaciation at northern latitudes
Tertiary		
Pliocene	5	Low equability; summer drought in
Miocene	23	west; deserts in southwestern U.S. and Mexico
Oligocene	33	Low equability; low mean annual temperatures at northern latitudes
Eocene	53	High equability; high mean annual
Paleocene	65	temperatures at northern latitudes

Partly after Wolfe, 1978.

resource-defense polygynists, whereas this species on the nearby mainland does not have these behaviors (Vaughan and Schwartz, 1980). Mammals that live on islands and have no mammalian predators often show little fear of humans. Blake (1887:49) found gray foxes on Santa Cruz Island to be remarkably tame. They commonly approached to within 1 meter of a person and regularly visited Blake's camp for scraps of food. These foxes even approached sleeping persons and pulled at their blankets

Climate and Mammalian Distribution. Mention has repeatedly been made in the foregoing chapters of the geologic epochs of the Cenozoic in relation to the evolutionary history of various mammalian families and orders. Cenozoic patterns of climatic change had a profound effect on the evolution, character, and distribution of plant communities, and the evolutionary patterns and distributions of mammals, in turn, were influenced by these floral changes. Although one cannot with assurance account for the myriad patterns of mammalian adaptation by recourse to Cenozoic climatic changes alone, there is no better point of beginning.

Although the Cenozoic has been a time of dramatic climatic and biotic change, it began with a long period of climatic equability (uni-formity, evenness). For 33 million years through the Paleocene and Eocene, over half of the duration of the Cenozoic (Table 19-6), climates were mild, and tropical and subtropical forests were widespread (Wolfe, 1978). Broad-leaved evergreen forests occupied the Puget Sound area of the Pacific Northwest, and the mean annual temperature fluctuation of 7 to 9 C° was half that of today. Northern California was tropical rain forest, and such forest extended some 30 degrees farther north than it does today. Fossil plants from the Eocene of Alaska (at 60 degrees north latitude) include fan palms and mangroves, plants indicative of a subtropical climate. At that time, the latitudinal temperature gradient along the Pacific coast of the United States was but half that of the present.

A climatic event of major importance (the terminal Eocene event) took place at the end of the Eocene. Within a short time (geologically speaking), perhaps 1 million years, the mean annual temperature declined sharply, by 12 to 13 C° at latitude 60 degrees in Alaska and by 10 to 11 C° at latitude 45 degrees in the Pacific Northwest. Equability also declined. The mean annual temperature range in western Oregon shifted from a mid-Eocene low of 3 to 5 C° to an

Oligocene high of about 25 C° (present range is 12 to 16 C°). Comparable changes seemingly occurred broadly over the Northern Hemisphere. Areas that previously had supported broad-leaved evergreen forest became dominated by broad-leaved deciduous forest, and tropical forests retreated southward.

Since the terminal Eocene event, although many climatic perturbations have occurred, a widespread trend has been toward increasing equability in the Northern Hemisphere (Wolfe, 1978; Wolfe and Tanai, 1979). Declines in the mean annual temperature range are documented for Alaska and eastern Asia, and during the middle Miocene in Alaska, this range was strikingly higher than it is today (26 to 27 versus 18 C°). Broad trends in mean annual temperature since the terminal Eocene event have differed according to latitude: at low latitudes, higher winter temperatures have caused an increase in the mean annual temperature; at middle latitudes (about 45 degrees), there has been no change; at high latitudes (as in Alaska), lower summer temperatures have led to a decrease in mean annual temperature and to a greater tendency for snowfields to last through the summer.

An important result of the increased latitudinal temperature gradients after the terminal Eocene event was that subtropical high-pressure cells increased in intensity and the western coasts of the continents developed patterns of summer drought. This Miocene-Pliocene trend toward seasonal aridity is well known (Chaney, 1944) and led to extensive deserts. The origins of such desert-adapted animals as kangaroo rats can be traced to the Miocene, when deserts or semideserts became widespread.

The Cenozoic climatic changes may be correlated with shifts in the inclination of the earth's rotational axis, a phenomenon proposed by Milankovitch (1938) as the cause of the recurring cycles of Pleistocene glaciation. Careful analyses of floral changes and of inferred climatic changes since the terminal Eocene event suggest that these changes are of the precise sort predicted by Milankovitch to occur under decreasing inclination of the earth's rotational axis. Assuming Cenozoic climatic changes to be the result of changes in inclination, the inclination decreased slowly from the Paleocene to the middle Eocene from about 10 to 5 degrees. A gradual late Eocene increase in inclination was followed at the end of the Eocene by a rapid increase to between 25 and 30 degrees. The inclination has decreased slowly since then to its present average of 23.5 degrees.

The Pleistocene was a time of pronounced climatic shifts, when periods of lowered temperatures alternated with periods of relative warmth. Temperatures during the cool periods may have been 4 to 8 C° below present temperatures, and temperatures in the intervening warm periods were probably higher than those now. Geologic and paleontologic evidence both indicate four major episodes of cool climates separated by three warm intervals. Accompanying the periods of cooling, which were apparently worldwide, were a number of spectacular environmental changes. Precipitation increased everywhere, and with increased snowfall, continental glaciers developed and pushed southward. At one time in the Pleistocene, over 25 percent of the land surface was covered with glaciers: Eurasia had 3,200,000 square miles of ice; the Nearctic ice sheet covered 4,500,000 square miles and during its greatest push southward reached what is now Kansas. Glaciers on Mount Kenya in Africa extended about 1700 meters below the present vestigial snow fields (at 4500 meters), and New Guinea and Madagascar had montane glaciers. The distributions of floras were changed. Boreal vegetational zones were pushed downward on mountains, and coniferous forests spread southward over areas that previously supported less boreal floras. Concurrently, tropical floras receded toward the equator and deserts became far more restricted than they are today.

The Pleistocene began about 1.5 million years ago and ended 12,000 years ago with the extinction in the Nearctic and Palearctic re-

Table 19–7:   Some Siberian Immigrants That Entered North America Across Beringia in the Pleistocene

Time of Interchange (millions of years ago)	Immigrant	Affinity
1.8	*Mammuthus,* mammoth elephant	Temperate to cold-temperate
	Synaptomys, bog lemming	Temperate to cold-temperate
	Microtus, vole	Temperate to cold-temperate
	Ondatra, muskrat	Temperate to cold-temperate
1.2	*Clethrionomys,* red-backed mouse	Cold-temperate
	Phenacomys, vole	Cold-temperate
	Synaptomys, bog lemming	Cold-temperate
	Microtus, vole	Temperate
	Bison, bison	Temperate
	Gulo, wolverine	Cold-temperate, arctic
	Smilodon, saber-tooth cat	Temperate
0.47	*Microtus pennsylvanicus,* meadow vole	Temperate
	M. montanus, montane vole	Temperate
	Ondatra, muskrat	Temperate
	Cervalces, extinct moose	Cold-temperate
	Rangifer, caribou	Arctic
	Oreamnos, mountain goat	Arctic alpine
	Ovibos, musk ox	Arctic
	Ovis, sheep	Temperate to arctic
	Alces, moose	Temperate to arctic
	Bos, yak	Arctic
	Saiga, Asian antelope	Cold-temperate
	Bootherium, extinct bovid	Cold-temperate
	Symbos, woodland musk ox	Cold-temperate
0.17	*Dicrostonyx,* collared lemming	Arctic
	Lemmus, lemming	Arctic
	Lagurus, vole	Cold-temperate
	Microtus, vole	Temperate
0.07	*Clethrionomys rutilis,* red-backed mouse	Arctic
	Microtus oeconomus, tundra vole	Arctic

Partly after Repenning, 1980.

gions of such common Pleistocene mammals as elephants, camels, woodland musk ox, ground sloths, horses, and the giant beaver. Following the retreat of the last continental glacier at the close of the Pleistocene was a warm, moist climatic phase (perhaps 8000 to 6000 years ago), followed by a warm, dry phase (about 6000 to 4000 years ago), when temperatures were higher than today and precipitation was lower.

Pleistocene episodes of faunal interchange between Siberia and North America were decisively affected by climate. This interchange was across the periodically emergent Beringia and

had a profound effect on the North American mammalian fauna. Arvicoline rodents have repeatedly used this route, and the following synopsis of these movements (based largely on the studies of Repenning, 1980, and Hoffmann, 1976, 1980) illustrates the importance of this corridor and its function as a filter bridge.

Some 1.8 million years ago, a buildup of Cordilleran ice fields (ice fields along the major north-south mountain chains of North America) caused a drop in sea level and left Beringia emergent. At this time, voles *(Microtus)* came from Siberia into North America. Fossil evidence indicates a pattern of differentiation between arvicoline faunas of different areas that provides clear-cut evidence that the Cordillera were a barrier separating two dispersal routes from Siberia, a maritime (coastal) route and an interior "ice-free corridor." The Beringian climate at this time was apparently temperate and allowed the passage of animals not adapted to arctic conditions.

Another wave of immigrants entered North America via Beringia about 1.2 million years ago, during a glacial episode when sea level dropped. Among this group were several arvicoline genera *(Microtus, Phenacomys, Clethrionomys)* that are present today in North America and *Neodon,* which survives as a relict in central Mexico. Typically, each wave of Siberian immigrants displaced those arvicolines already in North America, and some displaced types survived for at least a time in southern refugia. The Beringian climate of this time was probably cold-temperate. All the microtines that entered except *Neodon* were subarctic or coniferous forest species and followed the ice-free corridor, becoming established east of the Rocky Mountains.

The next exchange, 470,000 years ago, re-sulted in the establishment of the important living species *Microtus pennsylvanicus,* which appears suddenly in almost every appropriate fossil locality east of the Rocky Mountains, and *M. montanus.*

The two last interchanges, 175,000 and 70,000 years ago, allowed entry into North America of tundra or cold steppe species, including two lemmings *(Dicrostonyx, Lemmus)* and several voles *(Lagurus, Clethrionomys rutilis, Microtus oeconomus).* The last two species are still present in the North American arctic tundra.

The climate of Beringia probably shifted from cold-temperate at the start of the Pleistocene to arctic near its end and thus acted as an ever more rigorous filter bridge (Hoffmann, 1976, 1980). A review of the Pleistocene immigrants into North America is given in Table 19-7.

During glacial advances, the ranges of boreal mammals extended well south of present limits. Remains of the musk ox *(Ovibos),* arctic shrew *(Sorex arcticus),* and collared lemming *(Dicrostonyx)* have been found well south of their present northern ranges (Fig. 19-13). Two voles that today have separate ranges that ex-

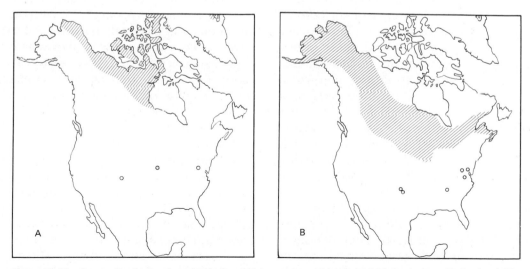

Figure 19-13.   Present distribution (cross-hatched) and Pleistocene record (circles) of (A) the musk ox *(Ovibos)* and (B) the arctic shrew *(Sorex arcticus).* (After Hibbard, Ray, et al., 1965)

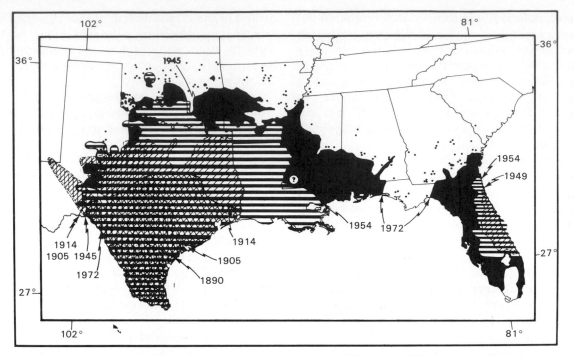

Figure 19-14.   Spread of the armadillo in the United States in the last 100 years. The dates and different patterns of hatching roughly indicate the spread at various times. Scattered single records during a 1972 survey are indicated by dots. (From Humphrey, 1974)

tend far north occurred together in the Pleistocene in the area that is now Virginia and Pennsylvania (Hibbard, Ray, et al., 1965). Abundant evidence verifies the occurrence of northern assemblages of mammals during the Pleistocene as far south as Kansas and Oklahoma. There were reciprocal northward movements of subtropical or desert mammals during interglacial times, as indicated by the fossil occurrence of such animals as the hog-nosed skunk *(Conepatus)* and jaguar *(Panthera onca)* far north of their present ranges. A fossil record of the jaguar, for example, is from northern Nebraska, 1300 kilometers north of the animal's present range.

One of the most common and obvious patterns of mammalian distribution—the occurrence of isolated or semi-isolated populations of northern mammals on mountain ranges at fairly low latitudes—is the result of Pleistocene

southward migrations of boreal faunas. During glacial advances, assemblages of boreal mammals were widespread in lowlands well south of their present ranges. Concurrent with the movements of these mammals northward during the retreat of cool climates were movements of boreal mammals into montane areas. Here, because of the effect of elevation on climate, cool refuges were available. Many of these montane populations have persisted in "boreal islands" far south of the northern stronghold of their closest relatives, and the zonation of mammalian distributions on some mountain ranges in the southwestern United States has perhaps resulted from Pleistocene faunal movements (Findley, 1969). The White Mountains of east central Arizona constitute a boreal island, where several species having northern affinities are isolated from other populations of their species. Populations of least chipmunks *(Eutamias minimus)* and red-backed mice *(Clethriono-mys gapperi)* are isolated in this way, and the water shrews *(Sorex palustris)* in the White Mountains are separated by 300 kilometers of inhospitable (for this shrew) habitat from the nearest other water shrew populations, which occur in northern New Mexico.

Some relict populations, left behind after northward and eastward retreats of boreal conditions, also occur in some moist lowland localities. One such isolated population of the southern bog lemming *(Synaptomys cooperi)* occurs in a restricted marshy area in western Kansas (Hibbard and Rinker, 1942); another popula-tion, which occupies an area of moist habitat only 100 meters wide and about 1.5 kilometers long, is at a fish hatchery in extreme southwest-ern Nebraska (J. K. Jones, 1964:220).

Humans have obviously reduced the ranges of many mammals, but some species are extend-ing their ranges northward today, perhaps in re-sponse to the present warm climatic cycle. The armadillo has extended its range from southern Texas into much of the central and southern United States in recent years (Fig. 19-14), and the cotton rat *(Sigmodon hispidus)* and op-posum *(Didelphis virginiana)* are also moving northward (E. R. Hall, 1958).

Chapter 20

Behavior

The behavior of any animal is of great interest because it is, so to speak, "the proof of the pudding." In the case of the pronghorn antelope, for example, great running speed became part of a unified functional system only because of a complex of behaviors that evolved in association with this ability. The formation of herds, sexual dimorphism, and systems of social behavior; the preference for open situations; the flashing of the white rump patch as a danger signal to other antelope; and the remarkable ability to detect enemies at a distance all allow the pronghorn antelope to utilize its great speed effectively to escape predators. How an animal uses its morphological and physiological equipment is of vital adaptive importance and forms the substance of behavior.

The behavior of mammals is of particular interest because of its flexibility and variability. Mammals learn much more rapidly than other vertebrates and can modify behavior on the basis of past experience. This ability, superimposed on a rich array of innate (instinctive and unlearned) responses, or behaviors, makes for complex behavioral patterns that often differ widely between species. Remarkably well-developed sense organs, coupled with a brain capable of rapid evaluation of complex sensory information, have enlarged the perceptual sphere of mammals and have facilitated the evolution of communication and rich social behavior.

The Ethological Approach

This chapter deals largely with ethology, the study of behavior in relation to structure and mode of life or, as put by Tinbergen (1963), "the biological study of behavior." One might suppose that behavior could be more readily observed and analyzed than could other aspects of biology and that detailed behavioral information on many species would have been assembled relatively early, but this is not the case. Indeed, little is known of the behavior of many animals that are well known morphologically. The study of mammalian ethology has reached the stage of synthesis (see, for example, Geist, 1974, and R.J. Jarman, 1974), and such excellent contributions as those of Ewer (1968, 1973), E.O. Wilson (1975), J.F. Eisenberg (1981), and J.F. Eisenberg and Kleinman (1983) offer compilations and analyses of our knowledge of mammalian behavior.

Although mammals are remarkable in their ability to learn and to profit from experience, built-in patterns of behavior form an important part of their behavioral repertoire. Such innate behavior is not individually variable within a species; it is unlearned and is a part of an individual's heritage that is shared with other members of the species. These built-in behaviors are best regarded as simple sequences of movement elicited by specific stimuli. The term *Erbkoordination,* coined by Lorenz (1950), seems especially appropriate and refers to simple but specific hereditary movement or pattern of coordination. Such behavior in canids or hyaenids seems to be the lowering of the head, rump, and tail—assuming the "submissive posture"—in response to the menacing jaws, high head, cocked ears, and high tail of a dominant individual. Behavioral patterns have evolved along with structural and physiological features and have equal adaptive importance.

Clearly, mammals are not completely unique behaviorally. They are set apart from other vertebrates by their superior ability to learn, remember, and innovate, but they resemble other vertebrates in their wide use of innate behaviors.

Nonsocial Behavior

A number of behaviors relate basically not to social interactions but to such vital activities as feeding and seeking or preparing shelter. Feeding behavior is highly variable and in some species is a nonsocial activity. In some species, the preparation of shelter involves remarkably complex nonsocial behavior.

Feeding Behavior. Herbivores utilize food that is unable to escape and therefore are spared some of the problems that carnivores face. Herbivores are generally equipped to utilize efficiently only specific types of vegetation, however, and must often face seasonal food shortages and cope with plant material that is difficult to digest. Specialized feeding behaviors, together with specializations of the dentition and digestive system, tend to maximize the return of energy from food relative to the energy expended in securing it.

Some of the most specialized foraging behaviors occur among rodents. Pocket gophers (Geomyidae), mole rats (Bathyergidae), and other fossorial rodents dig complex burrow systems, and part or all of their diet consists of underground parts of plants they encounter. Because of the tremendous energetic cost of burrowing, one would expect fossorial herbivores to have evolved behaviors favoring the most efficient system of finding food. The basic geometry of burrow spacing in one species of pocket gopher *(Thomomys bottae)* was found to be remarkably uniform both within one burrow system and from one system to another (Reichman, Whitham, and Ruffner, 1982). The

basic building unit of the system, as well as the distance between forks and the lengths of the branches, was uniform within and among systems and is analogous to the nodes and internodes of plants. These units can be combined to increase overall burrow length, but the uniform spacing is maintained (Fig. 20–1). Such consistent burrow spacing suggests intense selection in pocket gophers for precise, uniform, probably innate burrowing behavior.

The closely related kangaroo rats and pocket mice (Heteromyidae), by contrast, travel far from their burrows and gather seeds from the soil by using the long claws of the small forefeet. These rodents rapidly collect seeds in the cheek pouches but tend to gather seeds that are superior energetically to those randomly available in the soil. Kangaroo rats concentrate on the few widely spaced patches with many seeds, whereas pocket mice utilize the more abundant small patches each of which contains fewer seeds (Reichman, 1980; Reichman and Oberstein, 1977). Typically, many loads are taken to the burrow in an evening. Later, in the safety of the burrow, the animals become even more selective: of the seeds gathered, the rodents eat those richest in energy (Reichman, 1977). This style of foraging is not optimal in the sense that energy is expended to collect seeds that are not eaten, but it enables the animals to gather food rapidly and to minimize above-ground exposure to predators. This delayed-eating behavior departs from the pattern of many rodents, which make decisions as to suitable food and consume it as soon as it is found. Some kangaroo rats can apparently discriminate between dry seeds and those that are moist and would spoil in the burrow. Moist seeds are placed in shallow pits near the burrow, covered with soil, and allowed to dry before being transferred to underground caches (Shaw, 1934).

Hoarding behavior is highly developed in a number of mammals. Some rodents, some shrews, some carnivores, and the pika (a lagomorph) store food. Hoarding behaviors seem-

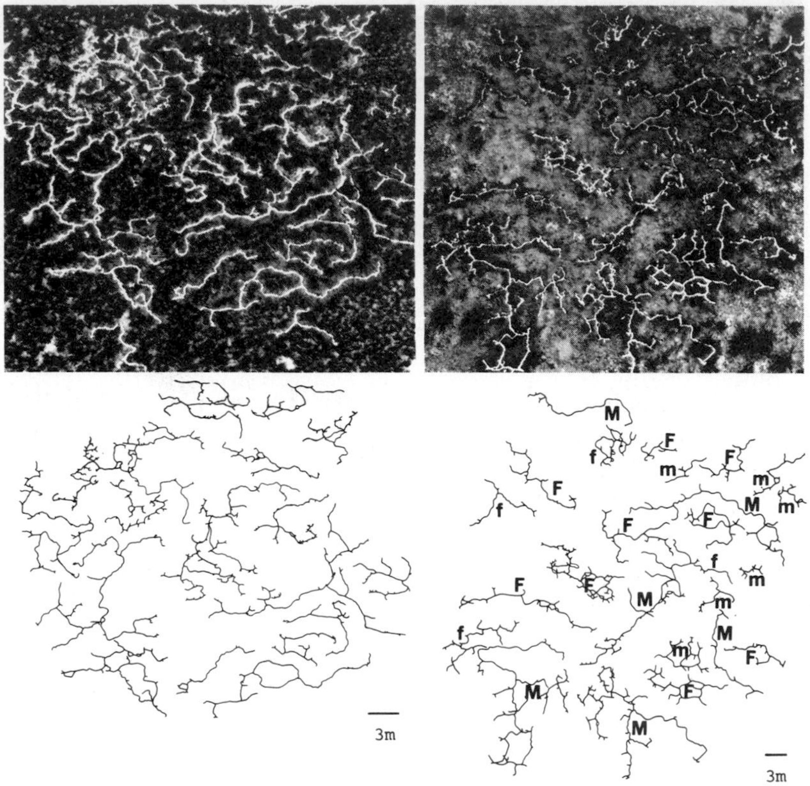

Figure 20-1. Aerial photographs and line drawings of pocket gopher burrows excavated at two sites near Cottonwood, Arizona. Burrow systems were marked with lime. The diagram on the right designates burrows of adult males (M), adult females (F), nonreproductive males (m), and nonreproductive females (f). (From Reichman, Whitham, and Ruffner, 1982, p. 688, by permission of the Ecological Society of America)

ingly evolved under selective pressures imposed by periodic food shortages or by seasonal difficulties in foraging. An example of the latter situation was reported by Kenagy (1973a) for the little pocket mouse *(Perognathus longimembris),* a tiny heteromyid that cannot forage during the cold part of the year because of the extremely high energetic cost of maintaining homeothermy. This rodent stays in its burrow continuously for up to five months during the winter, alternating between feeding on seeds it has stored and periods of dormancy. Many sciurid, cricetid, and heteromyid rodents store food in underground or protected caches. "Scatter hoarding" is practiced by some squirrels, which bury food at scattered points in the home range.

The caching behavior of the North American red squirrels *(Tamiasciurus hudsonicus)* is particularly notable. These squirrels depend for food on fir and pine seeds. Cones are cached in holes in large middens formed by the litter of cone fragments that accumulates beneath a squirrel's favorite feeding sites. The middens are frequently 6 to 10 meters in diameter, contain from two to ten bushels of cones, and are in shady situations where the moisture retained in the midden aids in preserving the green cones (Finley, 1969). Small numbers of cones are commonly cached in logs or pools of water. The cones are harvested in late summer and fall and are cut on occasion at the rate of 29 per minute (Shaw, 1936:348). The squirrels are such effective harvesters that one pine in northern California lost 93 percent of its 926 cones to them (Schubert, 1953). Seeds from the cached cones are eaten during the winter, and when snow is deep, access burrows through the snow into the midden are maintained.

The dextrous forefeet of rodents are often used in handling food. The bipedal squirrel-like feeding posture frees the hands, which grip and manipulate the food while it is gnawed by the incisors. This behavior is seemingly innate, for hand-reared rodents handled food in an essentially adult manner when they were still too poorly coordinated to maintain their balance (Ewer, 1968:32). The small forefeet of kangaroo rats and pocket mice skillfully sift seeds from the soil and put them in the cheek pouches. Even some rodents with forelimbs highly specialized for digging retain considerable manual dexterity. The plains pocket gopher *(Geomys bursarius),* which has powerfully built forelimbs and large forefeet equipped with long claws, holds food in typical rodent fashion when it is eating. This fossorial rodent has several specialized feeding behaviors that avoid the ingestion of soil and that demand considerable dexterity. Dirt or water is carefully stripped from leaves by the claws, and unusually wet or sandy food is often held in both forepaws and shaken rapidly (Vaughan, 1966b). Probably the importance of the rodent forelimbs for handling food has limited the extent to which the limbs have become specialized for digging or locomotion.

Predators have evolved behaviors that facilitate the pursuit, capture, and killing of prey, which in most cases have defensive or predator-avoidance strategies. Some behavioral patterns are common to a wide array of both placental and marsupial carnivores. The neck bite is such a behavior. This killing technique was studied in the house cat by Leyhausen (1956), who presented the predators with normal and headless rats and with rats with the head fastened to the tail end. The cats aimed their bites at any constriction in the body; with normal prey this results in the neck bite. Grabbing prey across the back and shaking it violently is another pattern shared by many carnivores. When hunting mice in grass, many carnivores, from house cats and African servals *(Felis serval)* to foxes and coyotes, freeze and listen intently to locate the mouse, then make a high leap and come down on the mouse with the feet (Fig. 20–2).

Most canids are solitary hunters and quite cursorial and capture prey by virtue of speed or, occasionally, endurance. Experience and learning are of great importance, and canids are highly adaptable. A coyote with access to a waterfowl marsh may learn to capture moulting ducks, while its relative in the desert patrols the perimeters of sand dunes for kangaroo rats. Canids have an extraordinary sense of smell, and prey is often detected initially by wind-borne scent. Adaptability and catholic tastes, rather than a specialized style of hunting, are the keys to the success of solitary canids.

The cats employ more specialized hunting and killing techniques than canids. Cats are not long-distance runners but usually depend on short rushes directed against surprised prey. The sudden rushes of lions seldom cover more than 50 to 100 meters, and leopards and smaller felids frequently make only several bounds to reach their prey. The cheetah, an exceptional felid, may chase an antelope several hundred meters at speeds up to 112 kilometers per hour!

In order to use the typical feline hunting technique effectively, a cat must get close to its prey. The stalking of prey by felids involves a series of beautifully coordinated behaviors described in detail by Leyhausen (1956). When prey is sighted, the cat crouches low to the ground and approaches, using the "slink-run" and taking advantage of every object offering concealment. As the distance to the prey is reduced, the cat moves more slowly and cautiously. At the last available cover, the cat stops and "ambushes." The body is held low and, just before the attack, the heels are raised from the ground and the weight is shifted forward, just as sprinters ready themselves in the last, tense instant before the sound of the gun. The brief rush to the prey ends in a spring; the forefeet clutch the animal, but the hind feet remain planted and steady the cat for the possible struggle. Some of the components of this total behavior pattern may be observed in kittens as they stalk an insect or a scrap of paper. The cat makes the kill not by belaboring the prey, as do many canids, but either by a powerful bite at the base of the skull or the neck, which crushes the back of the skull or some of the cervical vertebrae and

Figure 20–2. A coyote listening to a mouse, preparing to leap, and leaping on the mouse with all four feet. This same sequence of actions is used by many small to moderate-size carnivores. (After Murie, 1940)

the spinal cord, or, in the case of large prey, by strangulation. The shortening of the felid jaws is a specialization that contributes to the power of the bite. The tiger may kill prey as large as buffalo by gripping the throat and waiting for strangulation (Schaller, 1967). The cheetah depends on this style of killing.

Most cats are solitary, and cooperative effort in killing prey is rare. An exception is the African lion, the only truly social felid, which often hunts in groups in which there is some cooperation between individuals, with adult females doing most of the killing. The lion typically deals with large prey, often as heavy as or heavier than itself. Although cooperative effort improves success, a prey animal is typically killed by a single lion. Schaller (1972), who observed many kills, found that, when attacking prey the size of a zebra or wildebeest, the lion attempts to bring the prey to the ground by clutching the rump, hind legs, or shoulders with its forepaws and throwing the prey off balance. When the prey falls, the lion grabs with its jaws for the neck or nose and maintains a grip until the prey is suffocated. On occasion, a lion puts its mouth over the muzzle of the prey, surely a specific and effective means of strangulation. Schaller points out that, by centering the bite on the neck or nose, the lion immobilizes the horns, remains clear of thrashing hooves, and can easily keep the victim on the ground. This specialized killing behavior enables a single animal to kill large prey and, of great importance, reduces the risk of serious injury from the powerful prey.

Unusual behavior patterns enable some carnivores to break the exoskeletons of invertebrates and the shells of eggs. Some mongooses use the forefeet to throw objects against hard surfaces (Dücker, 1957; Ewer, 1968:48). Clomerid millipedes have an unusually hard exoskeletal armor, roll into a sphere when disturbed, and are thus invulnerable to many predators. The banded mongoose *(Mungos mungo),* however, smashes a millipede by using both front feet to throw the millipede between the hind legs against a rock (Eisner and David, 1967). Similarly, the spotted skunk *(Spilogale putorius)* breaks eggs by kicking them against rocks. (Van Gelder, 1953). The sea otter *(Enhydra lutris)* smashes the sturdy shells of mollusks by using a tool: the otter floats on its back with a flat stone on its chest, grasps the mollusk with its forepaws, and pounds it against the stone (Fisher, 1939). An individual was observed to pound mussels *(Mytilus)* on a stone 2237 times during a feeding period lasting 86 minutes (K.R.L. Hall and Schaller, 1964:290). These otters are clearly selective in their choice of stones and may use the same one repeatedly.

Shelter-Building Behavior. Many mammals have evolved elaborate shelter-building behaviors that aid them in maintaining homeostasis. The nests, burrows, or houses of mammals provide insulation that augments the animal's own pelage and reduces the rate of thermal conductance from the animal to the external environment or vice versa. Insulation from extremes of both heat and cold is important in reducing the energy expended in thermoregulation. The woodrat *(Neotoma)* collects a variety of materials with which it builds houses or improves the shelter provided by rock crevices or vegetation (Fig. 20–3). At the center of the woodrat house, or in a burrow beneath it, is a carefully constructed nest, which may be globular or cup-shaped and is formed of grasses and plant fibers. Many terrestrial rodents construct similar nests beneath logs or rocks or in burrows. Arboreal rodents frequently build nests in the branches of trees or in hollows in trees. Some nest-building behaviors are perhaps common to many rodents, but the choice of nesting site seems to be species-specific. For example, red tree mice *(Phenacomys longicaudus)* of the humid coastal belt of Oregon and California build their nests only in Douglas firs *(Pseudotsuga menziesii),* the needles of which provide the primary food of the mice (Benson and Borell, 1931).

Fossorial rodents follow rather complex patterns of movement when digging. Probably many of the specific components of the total digging sequence are innate behaviors. Pocket gophers (Geomyidae) use the forefeet to loosen the soil by powerful downward sweeps, and the hindlimbs kick the accumulated soil backward from beneath the animal. Pocket gophers periodically eject soil from a burrow entrance by pushing it with the chin and forelimbs. Careful studies of the pocket gopher by T.R. Kennerly (1971) have shown that the long, complex series of behavior patterns associated with mound building are basically innate but may be modified by learning. "Autoformulated releasers" probably play an important role in guiding mound building in rodents (and probably many other behavioral sequences in other mammals). Such a releaser is any alteration an animal makes to its perceptual environment that releases the animal's subsequent behavior. Thus, the mound of earth itself, and changes in the mound resulting from the pocket gopher's activity, release subsequent behaviors associated with mound building. The animal characteristically alternates direction in pushing soil from the burrow; it pushes a series of five to 20 loads to the right, a similar series to the left, and so on. The frequency distribution of directions of pushing soil indicates that efforts are mainly in three directions: either directly in front of the mouth of the burrow or at an angle of 90 degrees to either side. This results in the fan-shaped mound so typical of pocket gophers. That learning plays a part in burrowing and mound building is suggested by the fact that young animals are less successful in plugging the openings of burrows than are older animals.

Figure 20-3. (A) House of a Stephen's woodrat *(Neotoma stephensi)* at the base of a juniper tree *(Juniperus* sp.). Note the pile of sticks to the right of the tree and in the crotch between two trunks. (B) Nest of a Stephen's woodrat. This nest, about 180 mm in diameter and composed of grass and shredded juniper bark, was exposed when a woodrat house was dismantled. (D. M. Kuch)

The burrowing and mound-building behaviors of some African mole rats (Bathyergidae) differ markedly from those of pocket gophers. *Heliophobius,* for example, uses its incisors to excavate soil and pushes the dislodged soil in back of its body with its feet. The animal transports a load of soil to the surface by backing up against it with the rump and large hind feet. The forefeet push the animal backwards, and the head and upper incisors are braced against the

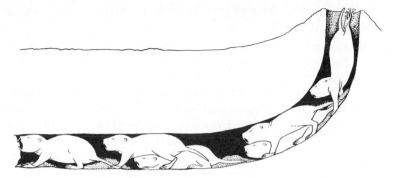

Figure 20–4. The digging chain of the naked mole rat *(Heterocephalus glaber)* of East Africa. (After Jarvis and Sale, 1971)

roof of the burrow to gain purchase. Unlike pocket gophers, which appear briefly at the surface each time they push a load of soil onto the mound, *Heliophobius* pushes a core of soil onto the mound without appearing on the surface.

Several kinds of Neotropical bats utilize leaf shelters, and among these the "leaf-tents" are of special interest. Honduran white bats *(Ectophylla alba),* for example, roost in groups of from one to six individuals beneath the large leaves of platanillo *(Heliconia).* The bats cut the side veins extending from the midrib of the leaf in such a way that the end of the leaf bends down and forms a tent. One colony of bats uses several of these shelters alternately.

dles other members and returns to a position in back of the digger. The digger is replaced by another mole rat periodically. The new digger often uses its incisors to pull the animal it is relieving back from its position, frequently to the accompaniment of loud squeaking by the latter. On occasion, the labor is divided differently among members of the chain, and each member in back of the digger passes soil back to the next. The well-coordinated, social digging behavior of these animals enables them to extend their burrow systems rapidly during the wet seasons when digging is easiest. As might be expected, burrow systems of colonies of naked mole rats are extensive and extremely complex.

Social Burrowing Behavior

Especially remarkable is the cooperative burrowing behavior of the naked mole rat. Jarvis and Sale (1971) described and photographed this "digging chain" (Fig. 20–4). The lead member of the chain chisels soil away with its protruding incisors and kicks it back to the second member, which takes the load and backs toward the mouth of the burrow. When it reaches the mouth of the burrow, it kicks the soil out of the burrow with a powerful sweep of the hind feet. The soil spews out as it would from a miniature volcano, and the mound takes the form of a small volcanic crater (Fig. 20–5). After kicking the soil out, this member of the chain strad-

Figure 20–5. The mound formed when naked mole rats kick soil from the mouth of their burrow. (From Jarvis and Sale, 1971)

Communication

Communication has often been broadly defined to include all interactions between animals that transmit information between them, but if all types of stimulus-reception sequences are regarded as communication, then essentially all behavior of one animal that can be perceived by another must be regarded as communication. For the purposes of discussion here, the definition of communication proposed by Otte (1974) will be used. Only a small segment of the multitude of stimuli received when an animal "views" its environment with its receptors is produced by other organisms and, through natural selection, has become modified to convey information. Only these stimulus-reception sequences involve communication.

Communication signals, then, are "behavioral, physiological, or morphological characteristics fashioned or maintained by natural selection because they convey information to other organisms" (Otte, 1974:385). "By far the greatest part of the whole system of communication seems to be devoted to the organization of social behavior" (Marler, 1965:584). This comment was applied to primates but may be equally valid for all mammals. Each type of communication—visual, olfactory, auditory, and tactile—will be considered separately, but it should be stressed that usually a complex of several kinds of communication signals passes between animals.

Visual Communication. Visual signals involving displays were perhaps derived in vertebrates from movements showing intention (to flee, for example), from displacement activities (seemingly inappropriate actions that typically occur when two opposing "desires," such as to escape or to attack, are in conflict), and from such autonomic responses as the erection of hair (Hinde, 1970:668). The evolution of displays is in the direction of reduced ambiguity. They have tended to become simplified, exaggerated, and stereotyped (repeated without variation). Highly developed facial muscula-

ture, the ability of the body and ears to assume a variety of postures, the control many mammals have over the local erection of hair, and large and conspicuous secondary sexual characteristics, such as horns, allow mammals a remarkable breadth of visual communication. Some visual signals are familiar: the dog wags its tail as a sign of friendship, and the cat arches its back, erects its fur, and raises its tail in a defensive threat. Such displays can be observed readily, and their functions or messages can often be recognized, but only a start toward understanding the broad area of visual communication in mammals has been made.

Facial expressions are of great importance in communication, and natural selection has often favored the development of distinctive facial markings that focus attention on the head. As described by Lorenz (1963), the facial expressions and ear postures of dogs signal degrees of aggressiveness or submissiveness. The posture of the head and the facial expression of many ungulates provide visual signals to other members of the herd or to territorial or sexual rivals (Fig. 20–6). An elk *(Cervus elaphus)* ready to run from danger elevates its nose and opens its mouth (McCullough, 1969); Grant's gazelle *(Gazella granti)* holds its head high, elevates its nose, and pricks its ears forward when challenging another male (R.D. Estes, 1967). The head of both these animals is conspicuous: the elk's because the dark brown head and neck contrast strongly with the pale body, and the gazelle's because of bold black patterns. The intricate facial expressions of primates are frequently made more obvious by distinctive and species-specific patterns of pelage coloration and by brightly colored skin.

The body is used for signaling in many species. This type of signaling is particularly well developed in ungulates that inhabit open areas and that gain an advantage from coordinated herd action. Grant's gazelle and Thompson's gazelle *(Gazella thompsonii)* of Africa, which have two warning displays (R.D. Estes, 1967), twitch the flank skin (conspicuously marked in

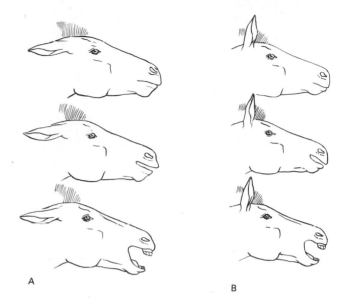

Figure 20-6. Facial expressions of horses: (A) three intensities of threat expressions; (B) three successive stages of greeting expressions. (After Trumler, 1959)

A

B

G. thompsonii) just as they begin to run from a predator that has entered the minimum flight distance (the minimum distance at which an approaching enemy causes the animals to run). The most effective display is a stiff-legged bounding gait, called "stotting," used as the gazelles begin to run. The conspicuousness of this display is enhanced by the erection and flaring of the hair of the white rump patch. In some monkeys and apes, the presentation of the hind quarters as if inviting copulation is a social gesture symbolic of friendship and is accepted by a brief "token" mounting (Heinroth-Berger, 1959). Kangaroos threaten one another by standing bipedally at their maximum height.

The effectiveness of visual signaling is heightened in many species of mammals by weapon automimicry in the form of striking (and to our eyes handsome) markings (Guthrie and Petocz, 1970). The ears are commonly used signaling devices in mammals, and—seemingly because of the proximity of the ears to the horns of artiodactyls, which are universally used in signaling—the ears in many species are marked or adorned with hair in such a way as to mimic the horns. This probably strengthens the visual

signal given by the horns, as well as making the posture of the ears extremely obvious. Several examples include the prong of the pronghorn antelope horn (Fig. 13–3), which gives a hooked effect that is mimicked by the black and hooked ear tip, the horns of the roan antelope (Fig. 20–7), which are mimicked by the tufted ears, and the marks inside the ears of the klipspringer (Fig. 13–26A), which give the impres-

Figure 20-7. The drooping ear tips of the roan antelope *(Hippotragus equinus)* are an example of automimicry. (Taken in Kenya by R. G. Bowker)

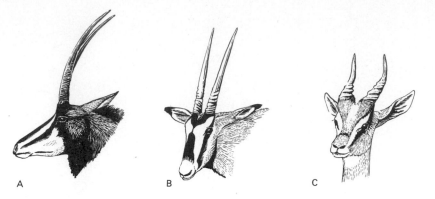

Figure 20–8. Facial markings of antelope. (A) Sable antelope *(Hippotragus niger)*; (B) oryx *(Oryx beisa)*; (C) young Grant's gazelle *(Gazella granti)*.

sion of a halo of short horns. Facial markings may also play a role in automimicry by accentuating the horns. In the sable antelope, oryx, and Thompson's gazelle, black markings create a design that extends the contours of the horns (Fig. 20–8).

Olfactory Communication. J.F. Eisenberg and Kleiman (1972:1) define olfactory communication as "the process whereby a chemical signal is generated by a presumptive sender and transmitted (generally through the air) to a presumptive receiver who by means of adequate receptors can identify, integrate, and respond (either behaviorally or physiologically) to the signal." A chemical signal that elicits a response in a conspecific receiver is known as a *pheromone,* whereas an *allomone* conveys a message to a receiver of a different species. Olfactory communication is effective because specific chemicals can convey very specific messages, and a scent mark on an object will persist and yield a message long after it is deposited. Because scent is released into the air and disperses rapidly, however, a receiver must have a sense of smell acute enough to find the source by detecting concentration gradients. Also, olfactory signals "broadcast" in the air are as available to a predator as to a conspecific. Pheromones of mammals have a variety of sources.

Urine and feces contain metabolic wastes that serve as chemical signals. Many kinds of mammals are highly specific in their choice of urination and defecation sites, and in some species a stereotyped routine is associated with urination and defecation. The dik-dik *(Madoqua kirki)*, a small, brush-dwelling, African bovid, deposits its feces in conspicuous piles at the borders of its territory, urinates on the piles, and makes scratch marks around them with its hoofs (Fig. 20–9). These obvious piles provide both olfactory and visual signals announcing territorial boundaries (Hendrichs and Hendrichs, 1971). All defecation and urination by the coyote can be regarded as scent marking and are most common in places where "intrusions" into a home range occur (M.C. Wells and Bekoff, 1981). Peters and Mech (1975) found that scent marking with urine by wolves was concentrated along the borders of pack territories. They present evidence that wolves of one pack respect the territorial boundaries of another pack and suggest that marking provides for the full use of available space and resources.

Urine and feces seemingly convey a considerable amount of information. Males of most species of mammals can recognize when a female is in estrus by the smell of her urine, and usually copulation will not be attempted until this time. In winter, the red fox *(Vulpes vulpes)* produces urine that contains several chemical compounds not present at other seasons (Jorgenson, Novotny, et. al., 1978). One compound is unique to the male. The red fox is one of the least social of the canids, mates in winter, and increases its scent marking at this time. Urine is probably especially important during the winter in olfactory communication between the sexes. It is used for communication by a bull elk maintaining a harem. Such a bull does not urinate normally; instead, the penis is extended and the animal squirts urine on the belly and the thick hair of the chest. The smell of the urine may have an important communication function (McCullough, 1969:82, 99, 110). By self-marking with his metabolic wastes, the bull is probably advertising his general physical condition. While he is in excellent condition, the urine advertises this information to rival bulls; when his condition declines, this is also conveyed to other bulls via the smell of the urine. This type of communication may avoid disruption of breeding activity and of the harem by delaying attempts at deposing the harem master until his exhaustion allows a fresh bull to replace him.

Scent urination in goats *(Capra hircus)* was studied by Coblentz (1976), who suggested that, in addition to advertising age dominance and physical condition to other males, the scent urination or self-marking behavior may have evolved because of its importance in male-female interactions. The behavior perhaps hastens estrus and synchronizes it with the period of peak physical condition in the male. Thus scent urination would increase the fitness of the dominant male by enabling him to make the most, reproductively, of the brief time of his peak condition.

Scent marking is used widely as a means of communication and is commonly an expression of dominance. In her careful consideration of scent marking, Ralls (1971:449) indicated that it is used by mammals "in any situation where they are both intolerant of and dominant to other members of their species." Ewer (1973:243), in her discussions of scent marking by carnivores, makes a similar point, and the work on European rabbits *(Oryctolagus cuniculus)* by Mykytowycz (1968) and on Mongolian gerbils *(Meriones unguiculatus)* by Thiessen, Owen, and Lindzey (1971) raises several major points: (1) the maturation and use of scent glands are controlled by gonadal hor-

Figure 20–9. A male dik-dik *(Madoqua kirki)* carefully smelling its dung pile (left) and then marking the pile by scratching the soil around it. (Taken in Kenya by M. H. Bowker)

mones produced at sexual maturity, (2) most scent marking is done by dominant males, and (3) scent marking is associated with the possession of a territory. Experimental verification of the relationship between social rank and scent marking has come from studies of house mice *(Mus musculus)*. Dominant males avidly marked the entire cage floor, whereas subordinate males voided urine in only a few places in the corners of a cage. Urination by the dominant male was regulated by interactions with another male: previously isolated dominant males immediately increased their urinary scent marking when caged with a subordinate male (Desjardins, Maruniak, and Bronson, 1973).

The signals provided by urine and feces may also be valuable to such refuging species as bats. To some bats, the urine and feces deposited in retreats may provide valuable olfactory signals that aid the animals in finding optimal roosts.

Also important as sources of pheromones are a variety of glands. Glands associated with the mouth, eyes, sex organs, anus, and skin are known to produce chemicals used in olfactory communication. Secretions from five locations on the body of the Australian honey glider *(Petaurus breviceps)* serve functions ranging from attracting newborn young in the case of the pouch gland of the female to contributing to a community odor within the social group in the cases of the frontal and sternal glands (Schultze-Westrum, 1965). Müller-Schwarze (1971) described a number of pathways of social odors in mule deer *(Odocioleus hemionus);* these are shown in Fig. 20–10.

Reproductive behavior in some, and perhaps most, terrestrial mammals is strongly influenced by the sense of smell. Experimental removal of the olfactory bulbs of laboratory rats impairs male reproductive behavior, and in male laboratory mice and hamsters *(Mesocricetus)* so treated, copulatory behavior was abolished. The vomeronasal organ has been regarded by many as playing an important sexual

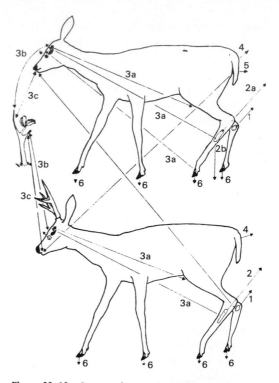

Figure 20–10. Sources of scents used in intraspecific communication and pathways of social odors in the mule deer *(Odocoileus hemionus)*. The scents of the following are transmitted through the air: tarsal organ (1), metatarsal gland (2a), tail (4), and urine (5). When the animal lies down, the metatarsal gland marks the ground (2b). The hind leg is rubbed against the forehead (3a), and the forehead is rubbed against twigs (3b). Marked objects are sniffed and licked (3c). The interdigital glands (6) deposit scent on the ground. (From Müller-Schwarze, 1971)

role in the male. After smelling the genital area and urine of a female, some mammals, notably perissodactyls, artiodactyls, and some carnivores, make a characteristic facial expression involving the upward curling of the upper lip and often the lifting of the head. This distinctive behavior, called *Flehmen* by K.M. Schneider (1930), has been thought by some to be important in activating the vomeronasal organ and in perceiving sexual pheromones (Fig. 20–11). This hypothesis has gained support from laboratory studies on the hamster *(Mesocricetus au-*

ratus) by Powers and Winans (1975), who found that destruction of the afferent nerves of the vomeronasal organ produced a disruption of copulatory behavior in one third of the altered animals. This procedure coupled with destruction of the afferent nerves of the olfactory bulbs completely eliminated copulatory behavior in all experimental animals. Powers and Winans suggest that input from both the vomeronasal organ and the olfactory bulbs is necessary for the arousal of sexual activity. Olfactory communication may also play an important role in the reproductive cycles of some primates. The smell of vaginal secretions of rhesus monkeys *(Macaca rhesus)* in estrus is sexually stimulat-

ing to males and promotes copulation (Michael, Keverne, and Bonsall, 1971).

Self-anointing, an unusual signaling behavior shared by a number of species of hedgehogs (Erinaceidae), involves the smearing of saliva over the quills and results in a pungent smell easily detectable by humans. This behavior has been studied in the European hedgehog *(Erinaceus europaeus)* by Brockie (1976). When removed from their nests, nestlings self-anoint; seemingly this aids the mother in recovering them. Self-anointing is a ritual that may be important primarily in spreading an individual's scent over the body (Poduschka; 1970), and face washing behavior may serve a similar function in many mammals (V.F. Eisenberg, 1981:362).

Some kinds of mammals are known to be able to discriminate between individuals of their species entirely by scent, and this ability is probably very widespread. Even in the primitive primate *Lemur fulvus,* this ability is well developed (Harrington, 1976). The mechanism for this recognition in the small Indian mongoose *Herpestes auropunctatus* has been studied by Gorman and coworkers (Gorman, 1976; Gorman, Nedwell, and Smith, 1974). The anal pockets of this mongoose produce acids used in scent marking objects within the animal's home range. The glands contain a series of six volatile carboxylic acids derived from bacterial decomposition of sebaceous and apocrine secretions in the pocket. The six acids occur in different relative amounts in different individuals, and animals recognize each other by scent on the basis of the unique carboxylic acid profiles. Bacterial production of these acids has been demonstrated in a number of mammals. Sebum and, to a limited extent, apocrine secretions are waterproofing agents deposited on the pelage of mammals. When metabolized by skin bacteria, these substances produce odor-yielding carboxylic acids. Gorman (1976) suggests that selection has favored the concentration of sebaceous and apocrine glands into discrete organs, where

Figure 20–11. (A) A female Grant's gazelle *(Gazella granti)* urinates when approached from the rear by a male. (B) The male performs Flehman behavior. (A taken in Kenya by J. E. Varnum; B taken in Kenya by K. P. Dial)

bacteria can produce the carboxylic scents used in olfactory signaling in relatively large amounts.

A reasonable summary statement, in my judgment, is that of J.F. Eisenberg and Kleiman (1972:24), who regard scent "as a means of exchanging information, orienting the movement of individuals, and integrating social and reproductive behavior."

Auditory Communication.   The sense of hearing in mammals is acute, and auditory communication is of great importance. Indeed, the sounds of some mammals that are rarely seen are commonly heard. Impressive choruses of howling coyotes may be heard nightly in some parts of the western United States where the animals themselves are only occasionally seen. The importance of nearly constant auditory communication to a herd animal is difficult for one to imagine. Virtually continuous noises made by the members of a herd integrate the group by keeping individuals apprised of each other's location. When he was very close to a herd of tule elk, McCullough (1969:71) observed that "there is a continuous array of sounds—foot bone creaking, stomach rumbling, teeth grinding, and others." The crunching of vegetation by McCullough was instantly distinguished by the elk as distinct from similar sounds made by their own feeding. In caribou *(Rangifer tarandus),* also, the creaking and snapping of foot bones can be heard for considerable distances and enable scattered members of a herd to remain in auditory contact (Kelsall, 1970).

Vocal communication is widely used by mammals. In humans, of course, this type of communication reaches its most complicated development, but even in other primates some type of "language" can be recognized. Vervet monkeys *(Cercopithecus aethiops)* give different alarm calls to announce different predators, and to each call the monkeys have a different response: they look into the trees in response to the leopard call, at the ground after the snake call, and at the sky after the eagle call (Seyfarth, Cheney, and Marler, 1980). Gunnison's prairie dog *(Cynomys gunnisoni)* gives alarm calls that are differentiated into local dialects (Slobodchikoff and Coast, 1980). The complexity of the call is related to the complexity of the habitat: More complex calls are given where shrubs, rocks, and tree stumps occur; less complex calls are given by prairie dogs living in open grassland. The Japanese macaque *(Macaca fuscata)* has a repertoire of some 25 sound signals (Mizuhara, 1957). The more basic sounds used by the rhesus monkey may be linked by a series of intermediate sounds, and one basic sound may grade independently into other calls (Rowell, 1962). This yields a remarkably complex and rich vocal repertoire. The functional importance of some sounds can be recognized. The quiet grunt, for example, is used by many primates to maintain contact with each other (Marler, 1965:568), and vocal sounds are used by many mammals to announce their position or to maintain or re-establish contact with one another. This seems to be the function of howling choruses of canids and the calls of young in a variety of species.

The functions of the varied sounds made by cetaceans are as yet not well understood, but many are clearly used in communication. Some "vocal" sounds may keep members of a social group aware of one another's position or signal aggression. Some tail slapping and loud splashing by dolphins may provide long-distance communication between scattered members of a foraging group.

Many mammals have vocalizations that seem to serve primarily as territorial advertisements. Among primates, for example, male howler monkeys *(Alouatta* spp.) of the Neotropics and the woolly lemur *(Indri)* of Madagascar make loud, resonant territorial calls. Probably no mammalian vocalizations are more impressive than those of the gibbons *(Hylobates* spp.) of southeastern Asia. The gibbon's territorial calls are vividly described by J.T. Marshall and Marshall (1976): "Each pair of gibbons daily advertises its territory by loud singing accompanied by gymnastics—a show of force. The female's great call dominates the half-hour morning bout. It is a brilliant theme lasting 20 seconds or more, repeated every 2 to 5 minutes. It swells in volume after soft opening notes, achieves a climax in pitch, intensity, or rapidity (at which time the gymnastics occur), then subsides. The male's shorter phrases, varying according to the species, either appear at appointed times during the great call, follow it as a coda, are interspersed between great calls (the female's opening notes command his silence during her aria), or are broadcast from his sleeping tree during a predawn chorus.'' Each species has a different song, the calls of the male and female differ, each individual has a slightly different set of calls, and a subadult will often join in with the female. Territorial songs by a family thus advertise the family's species and location, the sex and individual identity of each singer, and the presence of a subadult.

The vocal repertoire of some rodents consists basically of ultrasonic signals. Ultrasonics play an important role in the integration of the reproductive behavior of the laboratory rat (*Rattus norvegicus;* Barfield and Geyer, 1972): a 50-kilohertz call is associated with aggression and such aspects of sexual behavior as solicitation and mounting, and a 22-kilohertz signal is given by reproductively refractory or unreceptive individuals. Parent-young communication is also based on ultrasonics. Parents respond to the ultrasonic distress vocalizations of helpless young by bringing them back to the nest, and a decrease in the acoustical energy of the calls as the young grow older is associated with the development of homeothermy and the attending decrease in vulnerability to cold (Noirot, 1969). Some ultrasonic calls by rodents probably serve as territorial announcements (Sewell, 1968).

Vocalizations may facilitate individual recognition in many kinds of mammals. P. Brown (1976) found that female pallid bats (*Antrozous pallidus*) and their young recognize each other on the basis of distinctive vocal signatures. To the attentive human ear, the "who-

oop" call of the spotted hyena (*Crocuta crocuta*) differs from one individual to another and may well provide for individual recognition among the hyenas themselves. There are pronounced differences (involving the spacing of pulses and the addition of snorts) among the threat calls of elephant seal (*Mirounga angustirostra*) bulls. These differences give each bull a unique vocal signature that allows individual recognition among bulls competing for females on the breeding ground (Shipley, Hines, and Buchwald, 1981). Similarly, the extremely loud, gruntlike calls of male hippopotami may provide for individual recognition. After camping for several months on the western shore of Lake Baringo in Kenya, my wife and I became able to recognize differences among the calls of different hippopotami. Each bull called consistently from "his" segment of shallow water or shoreline, and the calls were seemingly used to announce the presence of a bull on its territory.

Relative to what remains to be learned, we know little about vocal communication in mammals under natural conditions. Here again is a promising area for research.

Tactile Communication.   The use of tactile communication by mammals is probably widespread. The sexual behavior of many mammals includes precopulatory activities by the male, such as laying the chin on the female's rump, nuzzling the genital area, or touching various other parts of her body. These behaviors are presumably sexually stimulating to the female or at least cause her to accept mounting by the male. Perhaps tactile stimuli are of greatest importance in connection with sexual activities in most mammals, but in primates they have assumed other roles.

Grooming of one individual by another is the most important form of tactile communication in primates (Sebeok, 1977), and the function far overshadows that of cleaning. In general, females groom most often and for longer periods than males, and males groom each other much less often than they groom females. Grooming typically serves as a form of appeasement, reduces social tension, and in many primates is important in establishing and maintaining social contact between individuals. Grooming is regularly practiced after aggressive encounters, when it dissipates tensions and reestablishes "amicable" social contact. Often one animal will seek contact with another of higher social rank by grooming it, and mothers will distract young being weaned by grooming them.

Mutual grooming occurs also in collared peccaries *(Dicotyles tajacu)* and is described by Sowls (1974). Animals rub their heads against each other's flanks and rump, which bears a much-used scent gland, in a ritual that seems to serve as a greeting ceremony but may also have a scent-marking component. The African dik-dik uses its nose to touch various parts of another's body. Often the female rejoining her young or a male rejoining a female will perform this behavior. Seemingly, this is a tactile reassurance and a reassertion of familiarity akin to mutual grooming (Bowker, 1977).

Observational Learning

An extremely important kind of "communication" in many mammals is observational learning. Young mammals, through repeated observation of their mother or of other adults, learn lessons vital to survival. These lessons often involve the choice and handling of food. Weigl and Hanson (1980) found that naive young red squirrels *(Tamiasciurus hudsonicus)* who had an experienced adult squirrel to observe learned feeding techniques twice as fast as did naive young without an adult to observe. These authors conclude that observational learning is crucial to the safe and rapid exploitation of new foods and habitats. Mother-to-young transfer of dietary preferences, which may be partly by observation, may be of great importance to many folivores forced by plant defensive chemicals to be highly selective feeders. Observational learning is probably important also in the early pol-

ishing of hunting techniques by a young predator and, to a social predator like the lion, could be important during much of an individual's early life. Gaudet and Fenton (1984) found that three species of bats learned a novel feeding behavior more rapidly by observing this behavior in other bats than by being experimentally conditioned as isolated individuals.

Solitary Mammals

Many mammals, including some marsupials, rodents, and insectivores and many carnivores, are essentially solitary and territorial. The home range of a pocket gopher (Geomyidae), for example, contains its burrow system, which, except in the breeding season, is occupied by a single animal and is apparently defended against interlopers. Although the burrow system may be extended and some sections may be plugged, most individuals probably occupy the same area throughout most of their lives. These animals may be colonial, but they are never social. Concentrations of burrow systems often occur locally, but each burrow system is exclusively "owned." Similarly, in the case of two species of ground squirrel (*Spermophilus mexicanus* and *S. armatus*), the colony consists of a number of solitary individuals living in close proximity (Balph and Stokes, 1963; Edwards, 1946).

Leyhausen (1964) regards some carnivores, such as various kinds of felids, as territorial but not truly solitary. Individuals occupying neighboring territories probably meet on occasion, gain some familiarity with one another, and establish a social order featuring mutual respect for one another's territorial rights. Bailey (1974) found that, in Idaho, female bobcats occupied virtually exclusive but relatively small ranges, whereas those of males were larger and overlapped ranges of other males and of females. Few interactions between adults were observed, and signaling by scent marking was the primary means by which an animal advertised its presence. For many solitary species, little is known of the brief periods of social life, but an awareness of the identities and positions of their nearest conspecific neighbors is probably a critical part of their perceptual world.

Social Behavior

Evolution of Social Patterns.   For many years, much of the literature on social behavior reflected a belief that there is some pervasive benefit derived from sociality, some automatic increase in survival and fitness resulting from a social life. In his thoughtful consideration of the evolution of social behavior, Alexander (1974) stressed that, to the contrary, social living has important disadvantages. Competition for food, mates, and space is heightened, and the conspicuousness gained is disadvantageous to groups of prey and groups of predators alike. An additional liability for social animals is the rapid spread of disease or parasites. Seasonal differences in the areas over which social groups of mangabey monkeys *(Cercocebus albigena)* disperse to feed are probably related to the need to avoid such danger (Freeland, 1980). In the wet season, when rains wash vegetation clean of fecal contamination each day, these monkeys occupy smaller areas than in the dry season, when fecal contamination remains for long periods.

Monogamy (the habit of having only one mate) occurs in a variety of mammals, from some bats and rodents to some artiodactyls. Our knowledge of this mating system was reviewed by Kleiman (1977), who recognized two types. Type 1, termed *facultative,* occurs when the population densities of a species are so low that only one member of the opposite sex is available for mating. Type 2, termed *obligate,* occurs when the carrying capacity of a habitat is so low that only a single female can occupy a home range and she cannot raise a litter without help from conspecifics. Monogamous mammals in general typically display little morphological or behavioral dimorphism, and the members of a pair interact infrequently except during the early stages of pair-bond formation and when rearing young. In obligate monogamy, sexual maturation in the young is delayed until after their association with the parents; older juveniles may help raise younger siblings; and the father often aids in the feeding, defense, and socializing of the young. Kleiman estimates that fewer than 3 percent of mammalian species are monogamous, although this system is spread throughout the mammalian orders. In some families of mammals—the marmosets (Cebidae) and dogs and foxes (Canidae), for example— monogamy is the major type of mating system. How long monogamous pair bonds last under natural conditions is difficult to determine, but the bonds clearly persist in some species through a number of reproductive seasons or as long as both members of the pair live.

Alexander (1974) discusses three broad advantages of sociality. (1) An individual's vulnerability to predation may be reduced by effective group defense or herd behavior. Defense of the group by dominant males is an important antipredator strategy of baboons, and a cohesive, running herd of ungulates presents a problem for predators. "The safety of the herd consists of the cohesive mass of animals running in an organized manner. The animals exposed are only those on the outside, and even these are protected by the number of flying hoofs and the ebbs and surges within the group. The vast array of movement has a disorienting effect on the observer's vision" (McCullough, 1969:72). Under a variety of situations, "hiding" within the herd is an effective means of escape; selection against the straggler or the individual who breaks from the herd is intense. The conspicuous "misanthrope" is the animal most easily singled out and killed by predators. (2) The cooperative effort of a predatory group (such as hyenas or wolves) may be effective in bringing down large prey that could not be killed by a single predator. With baboons, scattered but rich sources of food can be found more often by many searchers than by a single animal. (3) A paucity of safe nocturnal or diurnal retreats may have forced a partly social life on such animals as baboons and some bats.

After animal groups form, refinements in social behavior evolve. Alexander views these refinements as serving several functions. They may increase the advantages gained by group living. Such a behavior as the formation of a defensive ring of animals by musk oxen *(Ovibos moschatus)* tends to increase the invulnerability of the herd to predation by wolves. Further advantages may also be gained by groups of predators: the precise positions and spacing maintained by individuals of some foraging groups of cetaceans may increase the ability of the group to perceive and capture prey.

Most important, in Alexander's view, is the fact that the evolution of social behavior affects reproductive competition between group members and the reproductive performance of the population at large. The social system of the hamadryas baboon *(Papio hamadryas)* described by Kummer (1968b), for example, is based on the one-male unit. An adult male maintains a group of from one to several females, which are threatened or punished when they stray. This is a stable unit, and the male copulates with his females only, but by keeping his females with him constantly and by being aware as they come into estrus, he ensures his fitness. Young males are often unit followers, have opportunities to copulate with estrus females on occasion, and become familiar with the social behaviors that may later be used in gaining and maintaining their own unit. During the evolution of this system, the fitness of the socially integrated individuals was presumably greater than that of the individual who was solitary or did not learn the behavioral tactics associated with the social life.

Many ethologists agree that the evolution of social systems in animals is associated with increased fitness of individuals, but a social system in any group of animals is tested against the constraints imposed by a specific environment. Perhaps as strongly as any factor, the abundance and distribution of food limit evolutionary options. Indeed, the evolution of some social systems may have been influenced primarily by selective pressures imposed by the distribution of food in time and space. Using baboons as examples again, the one-male social unit of the hamadryas baboon seems well adapted to dry areas, where productivity of the habitat is low and food has a patchy distribution but is nowhere abundant. The savanna baboons, however, occupy a more productive area where food is scattered but a patch may provide abundant food; these animals forage in large social groups.

E.O. Wilson (1975:456) regards milk as the key to sociality in mammals. Mothers must invest considerably more energy in the care of postnatal young than in the growth and development of intrauterine young. Young are associated with their mother during much of their early lives, and during this time they are protected from a variety of dangers, including temperature extremes. Not only is avoidance of cold important when young are still poor thermoregulators, but in large ungulates protecting young from heat stress may demand specialized maternal care. Female giraffes, for example, hide their young where they are both concealed from predators and shaded from the mid-day sun. Before young giraffes have developed considerable body mass, they cannot withstand the heat load gained while foraging in the open and cannot afford the water loss associated with evaporative cooling (Langman, 1982). This mother-young group is the basic social unit in mammals, and even in species that are solitary as adults, the bond between young and mother is close. Because the care and nourishment of young demand from the females a tremendous amount of time and energy, the females "are the limiting resource in sexual selection" (E.O. Wilson, 1975:426). Males, on the other hand, typically invest little time and energy in the young and are thus free to make behavioral adjustments, such as polygynous breeding or the holding of harems, that increase their fitness.

Just as the distribution of food influences

the evolution of social behavior, so too does it affect the evolution of mating systems. Females tend to compete for food and, in the case of bats, for shelter, whereas males compete for females (Bradbury, 1977). By excluding other conspecific males from an area containing resources sought after by females, a male can gain exclusive access to females. If it is not feasible to protect an area and its resources, either a male can attach itself to a group of females and drive away other males (as in the harem system of some ungulates) or, if females do not assemble in communal groups, males and females can meet and copulate at a mating area advertised by the male. These strategies are considered in the next section.

Social Behavior and Mating Systems. The separation of social behavior and mating systems in discussions of mammalian behavior is difficult and artificial, and for this reason the two will be discussed together in this section. Patterns of social behavior in mammals are so diverse that broad summary statements are hard to frame. An additional problem is the incompleteness of our knowledge of the social behavior of many mammals. As might be expected, the large, more spectacular mammals and game species have been most thoroughly studied, but some of the most important groups of mammals remain poorly understood. Although rodents and bats together make up over half of the known species of mammals, we know relatively little of social behavior in these groups. This section gives a necessarily cursory overview of the sociobiology of mammals. There are excellent treatments of this subject by E.O. Wilson (1975), Eisenberg (1966, 1981), and Ewer (1968, 1973).

Social behavior has evolved in two marsupial families. In the Petauridae, several species are social to some extent, but sociality is perhaps best developed in the honey glider *(Petaurus breviceps)*, in which cohesive family units are dominated by males (p. 69). In the family Macropodidae, sociality is developed to varying degrees, and probably the most highly evolved marsupial social behavior occurs in the whiptail

wallaby *(Macropus parryi)*. The population studied by Kaufmann (1974b) consisted of subunits called mobs. The members of a mob occupy a home range to the near exclusion of members of other mobs, but the area is not defended. The social organization of a mob is loose, but some structure is provided by a rather flexible dominance hierarchy between the males that is maintained by nonviolent, ritualized fighting. (A dominance hierarchy is a fairly permanent social system based on dominance. Each individual recognizes its "position"; that is to say, it recognizes the animals that it can dominate and those that dominate it.) An estrous female is typically accompanied for from one to three days by her dominant-male consort, who has exclusive mating rights.

Among insectivores, sociality is rare. Eisenberg and Gould (1970) found that some species of the family Tenrecidae exhibit limited social behavior. In the most social tenrecid, *Hemicentetes semispinosus* (Fig. 7–11), several females and their young and a single male occupy the same den. Among members of the Soricidae, the largest family of insectivores, the greatest display of sociality occurs when a number of shrews occupy the same nest, presumably in the interest of lowering the energy cost of thermoregulation. Solitary lives and the use of olfactory and perhaps acoustical signals to bring the sexes together in the breeding season are characteristic of most insectivores.

Although our knowledge of the sociobiology of bats is extremely fragmentary, there is no doubt that within this group a wide array of social systems occur, and some species are known to have complex social behavior. A few species are completely solitary except during copulation and when the mother-young bond is maintained briefly. A more common pattern, typically of vespertilionid bats and some species in other families, involves the separation of the sexes at the time of parturition. Females form nursery colonies exclusive of males, with the sexes associating again when the young can fly and forage. Monogamous family groups are formed by a few bats in the families Emballon-

Figure 20-12. The hovering display of the Neotropical emballonurid bat *Saccopteryx bilineata*. (B. Dale, National Geographic Society)

uridae, Nycteridae, Rhinolophidae, Megadermatidae, and Vespertilionidae. Of great interest as examples of complicated social behavior are the several species of harem-forming bats that have been studied.

Thanks to the careful work of Bradbury and Emmons (1974), the elaborate social organization of some Neotropical bats is well known. The sociobiology of *Saccopteryx bilineata* can serve as a basis for comparing social behavior in several closely related members of the family Emballonuridae. This small insectivorous bat (Fig. 20–12) occurs widely in the Neotropics, where it lives by day in colonies in the buttress cavities at the bases of large tropical trees. Each colony is organized into a number of harems, and each harem and its territory are maintained as a discrete social unit by a single male, with each unit containing from one to eight adult females. The harem territories are from about 0.10 to 0.36 square meters in area, and females are regularly spaced from 5 to 8 centimeters apart. This individual distance seems inviolate, except by young, which up to the age of about two months can approach their mother. Territorial defense by males involves a remarkably

intricate series of displays. Frequently, males face each other across a common territorial boundary and give brief, high-frequency "barks." A mutual approach by two males often leads to a wavy pattern of boundary patrol. During the least frequent but most intense type of border dispute, one male charges across the boundary and strikes at the other with closed wings. The other male makes a brief flight and lands forcefully near the first, causing it to retreat in turn. A quite different display, called salting, seemingly includes both visual and olfactory signals. The male approaches another bat with forearm extended but with the digits flexed and the chiropatagium folded, opens the orifice of the wing gland, and shakes the wing the way a person would shake a salt cellar. Males salt both females in their own territory and another male's females across a territorial boundary, and will also salt another male across a boundary. Never did Bradbury and Emmons observe the antagonists sustain injury during agonistic displays.

In a different behavioral context, territorial males use a diverse repertoire of displays during interactions with females. The males are first to return to the colony sites at dawn, and when females and their young begin to return, the males begin singing. Each song is a long series of chirps, and some components of the song are audible to humans. As the females and young begin alighting in numbers in their respective territories, the males stop singing and turn to various displays, such as hovering (Fig. 20–12). The period of displaying and of sorting animals into their territories generally occupies several hours, after which the relative positions of the bats in the colony remain fairly constant. A summary of behavior during a 60-minute period is shown in Fig. 20–13. Females are highly aggressive toward one another and rigidly maintain their individual distances; this tendency is probably a major factor limiting harem size. Females vocalize frequently and spontaneously, and the vocalizations are highly variable.

There are suprisingly sharp contrasts be-

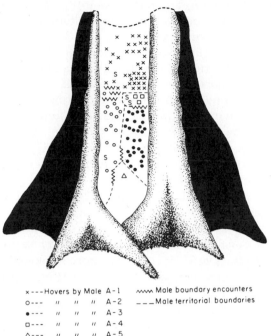

x ---Hovers by Male A-1   ᴧᴧᴧ Male boundary encounters
o---   "   "   " A-2   ___ Male territorial boundaries
●---   "   "   " A-3
□---   "   "   " A-4
△---   "   "   " A-5
s --- Salts

Figure 20–13. The occurrence of various behaviors (hovers, saltings, and boundary displays) by the adult males of a colony of *Saccopteryx bilineata* during a 60-minute observation period. (From Bradbury and Emmons, 1974)

tween the social behavior of *Saccopteryx bilineata* and that of two closely related species, *S. leptura* and *Rhynchonycteris naso*. *Saccopteryx leptura* occurs in small groups of twos and threes, males seem tolerant of one another, and perhaps this species is monogamous. The third species, *R. naso,* roughly resembles *S. leptura* in social structure.

Detailed observations of the behavior of a laboratory colony of *Carollia perspicillata* have been made by Porter (1978). This frugivorous phyllostomid bat is common and widespread in the Neotropics. The laboratory colony contained six males and four females and was maintained in a large flight cage. Early in the study, the animals established a stable pattern of roosting: the four females roosted in one corner of the cage and were guarded by one male, and the other five males roosted singly or in pairs else-

where. None of these five males interacted with females, nor did they fly within 1 meter of the females' corner. Two months after the observations began, two females bore young. Within six weeks, a male that had previously roosted with other males 5 meters from the harem began roosting alone within 3 meters of the females. Soon the interloper began roosting about 1 meter from the females and became aggressive toward the harem male. On the first day of confrontation, the males periodically approached each other, the harem male closest to the harem, and hung close to each other with necks extended until they were nearly nose to nose. Their mouths were partly open, and their wings were partly spread, but both maintained a frozen posture except for the rapid flicking of their tongues. The bats then rocked back and forth from foot to foot and occasionally struck at each other with their wings. After about a week of repeated confrontations, the nose-to-nose posture was used progressively less, wing shaking and rocking became more intense, and the interaction came to resemble a boxing match (Fig. 20–14.) The boxing activity was interrupted only briefly during the night by feeding or grooming.

Thirteen days after the initial confrontation, the interloper began roosting nearer the

Figure 20–14. Two male *Carollia perspicillata* performing the "boxing" ceremony. (F. L. Porter)

females while the harem male began leaving the harem more frequently than usual. On the seventeenth day, the new male displaced the original harem male and assumed control of the harem; he began roosting with the females and defended their immediate area. At this time, the former harem male began roosting 1.5 meters from the harem. Neither competing male was injured, but both seemed exhausted. During the week following the displacement, the boxing displays became less frequent and were finally abandoned.

The amazingly specialized hammer-head bat has been described previously, and its use of leks has been mentioned (p. 111). The raucous vocalizations of the male are used not only for attracting females but for establishing dominance ranking and for maintaining a small territory in the lek. Males able to occupy the sites in the lek most favored by females probably do most of the breeding, and males may take several years to establish themselves in these places.

Primates are primarily social mammals, and they display some of the most complex mammalian social systems. A major trend in primates is toward the abandonment of the sense of smell and the refinement of visual depth perception. The eyes have broadly overlapping visual fields, and stereoscopic vision allows the precise discrimination of distance necessary for rapid arboreal locomotion and for dextrous manual manipulation of objects. As might be expected, visual and auditory signals are of prime importance, manual dexterity in the form of mutual grooming is widely used in social situations, and olfaction is of relatively little importance, especially in higher primates. An impressive body of literature is available on primate social behavior, and only brief comments on this subject are appropriate here. (An excellent overview of the sociobiology of nonhuman primates is given by E.O. Wilson, 1975.)

Even in the most primitive primates, the Lemuridae, sociality is well developed, but relative to those of many other primates, lemurid social systems are rather simple. The mouse lemur *(Microcebus murinus)*, studied by Petter (1962) and R.D. Martin (1973), occurs on Madagascar in dispersed "population nuclei." The proportion of females to males in these nuclei is four to one, and as many as 15 females may occupy the same nest. Surplus males not accompanying groups of females often occupy nests on the periphery of the area. Although mouse lemurs occupy nests together, perhaps because nest sites are at a premium, there is no organized social life, and animals forage alone. This primate, therefore, must be regarded as a basically solitary creature that is flexible enough to be able to occupy nests communally.

The ring-tail lemur *(Lemur catta)*, studied by Jolly (1966, 1972), is far more advanced socially, being perhaps the most socially progressive of all lemuroid primates. This lemur lives in troops that range in size from ten to 20 or more animals. Adult males and females are equally represented in the troop, and their total numbers are usually equalled by the numbers of young. Troops occupy exclusive areas, and there is little intertroop contact. Social organization within a troop is based on dominance patterns. Females are dominant over males, a reversal of the usual primate system. A male-dominance hierarchy is established, and dominant males seem to be able to remain for long periods with a troop but (surprisingly) do not have first access to estrous females.

As befits their rather lowly position in the primate taxonomic scheme, lemurs make wide use of olfactory signals. Both sexes mark branches with secretions from the genitals, and, using the palms of the hands, males mark branches with other secretions. Scent glands occur on the chest and forearms of males. During aggressive confrontations, the tail is pulled between the forearms, annointed with scent, and then lifted high and waved to disperse the scent. Males indulge in "stink fights," which involve palmar marking, tail marking and waving, and often displacement of one animal by the other. The animals face each other when performing the scent marking, and the visual displays by the two animals, each using the conspicuously banded tail, seem to be mirror images of each other. The dominant animal moves forward while the other retreats. Vocalizations are also important during social interactions, and a variety of vocal signals are used.

Although a variety of types of social organization occurs among higher primates, most must learn to be responsive to a complex *social field*. An individual remains aware simultaneously of the attitudes and displays of a number of members of the social group and of the social ranks of these animals. Manipulation of the social field becomes an important aspect of the behavior of many primates, and even the ranking of an individual may depend in part upon its effectiveness as a manipulator. As a result of enduring social bonds between female baboons, two "friends" can put up an intimidating united front when one is threatened by a third individual. In some primate social systems, the rank of a female depends partly on her close association with a dominant male and on her ability to depend upon his help or protec-

tion during aggressive confrontations. Her status may abruptly decline if the male is deposed from his dominant position. In some baboon troops and in other primate societies in which group structure and cohesiveness are maintained by strong dominance patterns, the dominant male is the focal point of attention. An individual's behavior and the behavior of the entire social group are geared to the responses of this leader.

Savanna baboons (*Papio* spp.), which pursue an almost entirely terrestrial life, are often vulnerable to attack by a predator in places where immediate escape to trees is impossible. Food is frequently scattered, and a troop must forage over wide areas, thus increasing the chance of encounters with predators. A large and tightly organized social group has evolved in these baboons, perhaps largely in response to this pressure. These groups include from roughly a dozen to over 150 individuals. Each group occupies a largely exclusive home range; although home range boundaries are ususally respected, they are seemingly not defended. When a group is moving (Fig. 20–15), males

Figure 20–15. Positions of the members of a moving troop of baboons *(Papio anubis)*. (After K. R. L. Hall and DeVore, 1965)

Figure 20-16. The threat "yawn" of a male baboon *(Papio anubis)* displays the large canines. The eyes are closed during the threat, and the whitish lids are conspicuous. (I. DeVore)

quickly respond to threats from any quarter, and their united action provides the primary defense of the troop.

Sexual dimorphism is pronounced in baboons and enhances the male's intimidating appearance as well as his fighting ability. The male of the anubis baboon *(Papio anubis),* for example, is roughly twice as large as the female, is more powerfully built, and has comparatively huge canines. The long fur over the crown, neck, and shoulders accentuates the impression of size and ferocity (Fig. 20–16).

The mating pattern of baboons seems to have a consistent relationship to dominance ranking. Whereas subadult, juvenile, and less dominant males copulate with females in the early states of estrus, dominant males have exclusive rights to females during the period of maximal sexual swelling (the time when ovulation occurs). In some groups, only the highest ranking male copulated with females during the height of the swelling (Fig. 20–17), and in a group observed by DeVore (1965) in Kenya, not one dominant male attempted copulation until the swelling was at its peak.

A series of contrasts between the social behavior of savanna-dwelling baboons, such as *Papio cynocephalus,* and the hamadryas baboon *(Papio hamadryas),* a desert dweller, reflect differing suites of selective pressures. Group size, for example, differs markedly. Although savanna baboons must often seek scattered food patches, these patches are typically rich enough to support large social groups: In Amboseli National Park of Kenya, groups contained an average of 51 individuals (Altmann and Altmann, 1970). The hamadryas baboon, on the other hand, not only faces a scattered and unpredictable food supply, but food patches are seldom rich enough to support large groups of baboons. These foraging groups of baboons are small, from several animals to perhaps 20. In addition, social organization differs between these baboon species. Savanna baboons form only one type of social unit, the group, which includes many females, their young, and multiple mature males. The male-female pair bond is brief, lasting only a few hours or days, during the female's estrus period. The hamadryas baboon, in sharp contrast, has four levels of social organization (Kummer, 1968, 1984). The smallest and most tightly knit unit is the family group, consisting of a single mature male, one to several adult females, their young, and often a young adult bachelor male "follower." Several such family groups band together to form a less tightly knit foraging unit called the clan, and a number of these clans form a fairly stable traveling unit called a band, which usually includes some 60 baboons. Many bands tolerate each other in order to sleep in safety on the same cliff at night;

this loose aggregation is called a troop and contains several hundred baboons.

The complexity and subtlety of baboon behavior is illustrated by an examination of the ways male hamadryas baboons acquire mates. The male leader of a family group guards his females carefully: The male threatens or attacks a female that strays, and her response is to move close to him. The most direct and violent means of acquiring a mate involves a follower attacking and defeating a leader; the females of the deposed leader are divided among the oldest followers of the family units within the leader's clan. Some young adult followers, however, adopt an alternate, more subtle strategy. Such a follower is intensely interested in handling and caring for infants, and his first female is a small, pre-reproductive juvenile. After months of cautious maneuvering by the follower, he is allowed by the group leader and by the juvenile female's mother to care for the little female, and even to cradle her in his arms while they sleep. In time the bond between the follower and his young female is accepted; this pair bond is still respected by the family leader and other prime males of the clan when the female reaches sexual maturity. Until this time, the follower never tries to copulate with his young female. This system enables a young male to bypass compe-

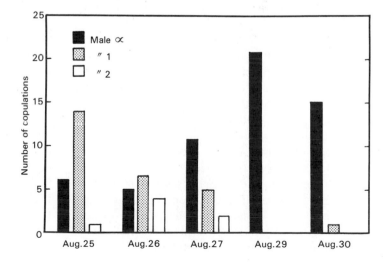

Figure 20–17. A comparison of the frequency of copulation by three male baboons with a female entering estrus. Note that the alpha (dominant) male copulated exclusively with the female on 29 August, at which time she was fully in estrus. (After K. R. L. Hall and DeVore, 1965)

tition with prime males, but he may lose some two years of his reproductive life while he waits for his female to be able to bear his young.

Most anthropoid primates retain a largely arboreal style of life, and among these species there is considerable diversity in social organization. The simple social system of the dusky titi monkey *(Callicebus moloch,* Cebidae) of South America occurs in several other cebids. The social unit of the titi is the closely knit family group, which includes a mated pair and one or two young. The pair is probably mated for life. Members of a family unit stay close together when foraging, and frequent confrontations between two neighboring units seem to confirm territorial boundaries. Despite their simple social organization, these animals use a broad array of signals, and their vocal repertoire is one of the richest known. Moynihan (1966) believes that, in the acoustically confusing forest community they occupy, where an extraordinary diversity of birds are calling and other cebids are vocalizing, the complex vocabulary of the titi facilitates precise, unambiguous, and "private" communication.

The social organization of a group of arboreal primates affects their foraging behavior. As an example, a consistent spacing is maintained by a group of capuchin monkeys *(Cebus nigrivittatus)* moving through the trees. The front and center positions are occupied by the dominant male and female and by individuals tolerated by them, while peripheral positions toward the rear are occupied by individuals of lower social rank (J.G. Robinson, 1981). The monkeys near the center of the group can afford to be less alert for predators than those monkeys with peripheral positions and can therefore eat fruit more rapidly.

Studies of the African red colobus monkey *(Colobus badius tephrosceles)* by Struhsaker (1976) provide evidence that a social system involving large multimale social groups has evolved in rain forest as well as savanna or desert environments. This colobus monkey is an arboreal leaf eater and lives in tropical forests. Its social units are large multimale groups that average 45 animals. These groups are not territorial, and home ranges overlap broadly, but spacing calls tend to keep neighboring groups apart.

The great apes (Pongidae) have social systems that display no radical departures from basic primate patterns, but there are some unique features. Groups of the mountain gorilla *(Gorilla gorilla beringei)* include from two to 30 animals. Social interplay between individuals is amiable, and assertions of dominance are low-keyed (Fossey, 1972; Schaller, 1963, 1965a, 1965b). Particularly notable is the age-graded male troop, with the nucleus of the group consisting of the dominant silver-backed male (ten years of age or older) and adult females and their young. Additional males, including less dominant silver-backs and black-backed males, attach themselves to the periphery of the group.

The chimpanzee *(Pan troglodytes)* has in recent years been the subject of considerable field observation by a number of workers, including Izawa (1970), Izawa and Itani (1966), Nishida and Kawanaka, (1972), Sugiyama (1973), and Van Lawick-Goodall (1968). The basic social unit of chimpanzees is an often dispersed group of from 30 to 80 animals that show considerable fidelity to a large home range. Particularly unusual is the looseness of the organization of the social group, with intricate patterns of establishment and dissolution of small parties. Highly evolved visual, tactile, and vocal communications are used. When a party of chimpanzees finds trees bearing fruit, their almost manic vocalizations and actions attract other parties to the bonanza.

Primates are clearly set apart from other mammals by the complexity of their social systems, by the closeness and permanence of their social bonds, both between opposite sexes and between members of the same sex, and by their highly developed vocal and visual communication. Much remains to be learned of primate sociobiology, and of special importance is the behavior of forest-dwelling species.

Social behavior is not widespread in the Lagomorpha. Pikas (*Ochotona* spp.) live in colonies, but individuals are solitary, and most rabbits are also solitary except during the breeding season. The European rabbit *(Oryctolagus cuniculus)* is an exception. Certain males have a number of females in their territories, and the females establish an order of dominance. Some social groups are dominant to others and maintain larger territories.

Widely divergent social systems occur in the Rodentia, from the solitary system of pocket gophers (Geomyidae) to the eusocial system of the naked mole rat (*Heterocephalus glaber,* p. 266). Social evolution in rodents has been influenced by their patterns of daily and seasonal activity, by the temporal and spatial distribution of food and suitable habitat, and by predation. Simple social systems seem to prevail. In solitary rodents, such as pocket gophers, kangaroo rats (Heteromyidae), and New World porcupines (Erethizontidae), each individual is solitary except briefly when the male and female are together during mating. In the beaver *(Castor canadensis),* a simple level of sociality occurs. A beaver colony is a family group in which mutual tolerance is usually the rule and apparently no social structure exists (Tevis, 1950). In wild colonies of Norway rats *(Rattus norvegicus),* a similar colonial system prevails, but a dominance hierarchy gives order to the system (Steiniger, 1950).

Arvicoline rodents display several types of social and mating systems, from strict monogamy for the prairie vole (*Microtus ochrogaster;* J.A. Thomas and Birney, 1979), to polygyny or promiscuity for the taiga vole (*M. xanthognathus;* Wolff, 1980) and meadow vole (*M. pennsylvanicus;* Madison, 1980). The ability of some arvicolines to be socially flexible is probably associated with their tendency toward population fluctuations. The social biology of the California vole *(M. californicus)* is comparatively well known through the work of Pearson (1960) and Lidicker (1973, 1976, 1979, 1980) and provides a model for what seems to be an intermediate system. A family group of California voles maintains and shares a runway system, to the exclusion of other individuals and families. The family group has from two to 12 (mean = 6) individuals, including one male and one or more females and their young. Monogamy is the rule at low population densities, but at high densities polygyny is common. A pair bond probably occurs, at least at low densities, and some paternal care has been observed in this and some other species of *Microtus*. The inflexibility of the male-female pair system was demonstrated by Lidicker (1976, 1979), who introduced several pairs into a small enclosure. The fighting and mortality that ensued usually reduced the population to a single pair, which reproduced successfully.

The Olympic marmot *(Marmota olympus),* which lives near timberline in Washington's Olympic Mountains, has evolved a system in which amiability between colony members avoids time-consuming aggression and allows individuals to make the most of the brief (40 to 70 days) growing season. Colonies of Olympic marmots are tightly organized social units. Each usually includes a male, two females, young of the year, and young of the two previous summers. Social contact is frequent and amiable, and during the early morning individuals enter occupied burrows and exchange ritualized greetings with burrow residents; there are no individual territories, with all areas occupied by a colony being equally accessible to all colony members. No clear or strongly enforced dominance hierarchy exists.

The black-tail prairie dog (*Cynomys ludovicianus*), which has been carefully studied by King (1955, 1959), is highly social. Prairie dogs formerly occupied many parts of the western United States, where they occurred in large "towns," often including over 1000 animals and covering many acres. (Now, lamentably, prairie dog towns are rare in many parts of the west, owing partly to intensive, government-sponsored poisoning campaigns.) The functional social units are *coteries,* which generally consist of an adult male, several adult females, and a group of young. No dominance hierarchy exists in the coterie. The paths, burrows, and food in the area held by a coterie are shared by its members, but hostility between coteries is the universal pattern.

Members of the coterie become familiar with each other in part by grooming, playing, and "kissing" behaviors. During kissing, the mouth is open and the incisors are bared. This behavior is seemingly a ritualized method of distinguishing between friend and foe. Faced with the threatening expression of open mouth and bared teeth, a trespasser retreats, while a fellow coterie member meets its "friend" and kisses. A two-syllable territorial call is used to proclaim ownership of territory. A repetitive, high-pitched yelp is a warning of danger. During the spring, when females are pregnant or lactating, the coterie system partially dissolves and some yearlings and adults establish themselves beyond the territorial limits of their coterie. The personnel of coteries thus may change, but the territory itself is stable.

An individual gains several advantages from this social system. Many eyes are watchful for danger, and many voices are ready to sound a warning. The effect of the animals' foraging is to keep vegetation low over a wide area and to provide terrestrial carnivores with little concealment. Perhaps equally effective in providing for long-term occupancy of an area, the animals are kept spaced so that overuse of food plants is generally avoided.

Our knowledge of the behavior of cetaceans is as yet extremely incomplete, but current evidence indicates that most species are social. Not only do some cetaceans travel and forage in social groups in which some consistent spatial organization is evident, but cooperative behavior is known. Several dolphins *(Delphinus delphis)* were observed lifting an injured individual to the surface where it could breathe, and females without young will help a mother tend its young. These and other behaviors have been regarded by Connor and Norris

(1982) as evidence that dolphins are reciprocal altruists and that the system of altruism is independent of genetic relatedness. The theory of reciprocal altruism, as formulated by Trivers (1971), suggests that an individual's apparent loss of fitness due to an altruistic act toward a partner is compensated for when the partner reciprocates with an altruistic act. Several aspects of the life history of dolphins that would be expected to be associated with reciprocal altruism are listed by Conner and Norris: (1) there is no dominance hierarchy, (2) life spans are long (40 to 70 years), allowing time for familiarity and practiced interactions among members of a social group, (3) the period of parental care of the young is long, (4) females cooperate in feeding young, (5) members of a social group are strongly interdependent for protection from predators. Some scientists believe that the tragic strandings of entire social groups of cetaceans on beaches may be due, at least in part, to repeated attempted altruistic acts toward a stranded member.

Cetaceans are known to have a wide array of vocalizations. Studies of several kinds of cetaceans indicate vocal repertoires of from 16 to 19 signals, plus other signals made by snapping the jaws shut or hitting the water with the tail (M.C. Caldwell and Caldwell, 1972; Dreher and Evans, 1964). The long, intricate, melodious songs of the humpback whale *(Megaptera)* are perhaps used for long-range communication in the open sea.

Although most carnivores are not highly social, highly organized social systems have evolved in some species. The lion "pride" and the spotted hyena "clan" are examples. Some diurnal viverrids are also social, with the packs of the banded mongoose *(Mungos mungo),* which may include 40 animals, representing the extreme.

Among the most highly social canids (the wolf, *Canis lupus;* the African hunting dog, *Lycaon pictus;* the dhole of southern Asia, *Cuon alpinus*), a dominance hierarchy gives structure to the pack. By cooperative hunting, these cursorial animals can kill large prey. The coyote *(Canis latrans)* is socially flexible (Bekoff and Wells, 1980). When carrion provides large, defensible patches of food, coyotes form small, cohesive packs, but when depending on small, dispersed prey, such as rodents, coyotes are solitary.

The dwarf mongoose *(Helogale parvula),* a common inhabitant of woodland and savanna in East Africa, illustrates a viverrid pattern of sociality. This small (350 grams) insectivore lives in packs averaging ten individuals, and termite mounds provide den sites and refuge from predators. Each pack includes a breeding pair (the alpha male and female), which are the dominant and usually oldest members of the pack. Subdominant males and females may copulate, but these females either do not conceive or abort in early pregnancy. Cooperative behavior among pack members is highly developed (Rood, 1979). Small young are guarded at the den by "babysitters" (but not by their parents) while the rest of the pack forages, and weaned young are fed insects by all pack members. A loud, repetitive alarm call is used to announce predators, and this call may be taken up by the entire pack. Mutual grooming and social contact between colony members are common.

Kruuk (1972) studied spotted hyenas from mid-1964 to early 1968 in the Ngorongoro Crater and the Serengeti Plains of Tanzania. His excellent work revealed that, contrary to popular opinion, hyenas kill much of their own food. Of particular interest, he observed a remarkable social system. The basic social unit of the spotted hyena is the clan, which may contain as many as 80 animals. The clan system is periodically disrupted in the Serengeti, where seasonal migrations of wildebeest and zebras result in drastic shifts in food supply. Each hyena clan defends a territory, the boundaries of which are maintained in part by systematic scent marking. Territorial disputes are often violent, and individuals are occasionally killed during border warfare. Females are larger than males and are dominant to them. A rather complex dominance

hierarchy exists within a clan, and strong bonds develop between females. At a kill, dominance is often not asserted and competition is based largely on the ability to eat extremely rapidly (this is aided by profuse salivation) rather than on fighting prowess or size (Fig. 20–18). Females tend their cubs at a central denning area, and here young receive early training (or practice) in the social rituals of the species.

Spotted hyenas offer an interesting example of intraspecific mimicry. The external genitalia of the female hyena mimic the male penis and scrotum: the female's clitoris is very large, resembles a penis, and is erectile, and two sacs filled with fibrous tissue form a false scrotum. So closely alike are the external genitalia of the two sexes that, beginning in Aristotle's time (384–322 B.C.), hyenas have been considered by many to be hermaphroditic.

When two hyenas meet, they typically go through a meeting ritual, part of which involves mutual examination of the external genitalia. Kruuk describes one such greeting: "When the male reaches the female, he quickly touches the side of her head with his muzzle, then lifts his leg, whereupon the female does the same, both

Figure 20–18. A group of spotted hyenas feeding on a lion-killed zebra. There are few indications of dominance rank among members of such groups. The young hyena on the right was able to eat side by side with the much larger adults. The birds are white-backed vultures *(Gyps bengalensis).* (Taken in Masai Mara Game Reserve, Kenya)

now showing full erection. They spend about 8 seconds sniffing each other's genitals, then separate, each walking away in opposite directions." Kruuk concludes that the intraspecific mimicry evolved because of its importance during meeting ceremonies. This ritual probably enables individuals to be close to each other briefly and to "identify" each other while attention is attracted to the genitals and to a course of action other than fighting. This cooling-off period perhaps allows aggressive tendencies to subside. The leg-lifting action seems to be an appeasement gesture, hence its initiation typically by a subordinate animal. The adaptive value of such behavior to a spotted hyena is perhaps heightened by its flexible social life. Although often social, hyenas may be solitary for varying lengths of time, and peaceful meeting behavior and recognition of individuals is frequently of importance.

The social system of the spotted hyena seems highly adaptive when considered in relation to the animal's environment. This predator is well adapted both for scavenging and for killing its own prey. Scavenging is largely a solitary enterprise, whereas success in killing large animals is greatly improved by group action. Hyenas therefore must have a social organization that favors social hunting and peaceful social feeding in addition to behaviors that enable individuals to safely meet one another after being separated. In predators with such powerful offensive weapons as the teeth of hyenas, control of aggression is of critical importance, and the primary function of the great array of scents, displays, and vocal signals is probably to restrict the use of these weapons (Fig. 20–19).

The social behavior of the brown hyena (Hyaena brunnea) of southern Africa has only recently become known through the work of Owens and Owens (1978, 1980) and is noteworthy in several ways. This hyena is an opportunist which scavenges, kills its own food, or eats melons as the occasion demands. Like the spotted hyena, the brown hyena forms clans, makes extensive use of scent marking, and for-

ages either singly or together with other clan members, but in several important ways the social systems of these hyenas differ. In contrast to the spotted hyena system, (1) the brown hyena clan has but one adult male, who is the dominant member; (2) vocalizations are not loud and are used for short-range communication; (3) when her cubs are two and a half months old, a mother takes them to a communal den, where, together with an assortment of other cubs of various ages, they are nursed indiscriminately by all of the clan's lactating females until they are 14 months old; (4) the females of the clan, whether or not they have cubs, together with the male relatives of the cubs, help to raise the cubs by carrying food to the communal den. Owens and Owens liken the communal den to a hyena kindergarten, where, in the course of riotous play each evening, young practice the ritualized fighting behaviors (such as neck biting) that they will use later in establishing rank in the social hierarchy. In contrast to the spotted hyena, there is no genital mimicry in the brown hyena.

Careful long-term field studies on the lion (Panthera leo) in Serengeti National Park of Tanzania by Schaller (1972) and by B.C.R. Ber-

Figure 20–19. Two spotted hyenas doing the "parallel walk," a threat behavior directed toward a third hyena. (From Kruuk, 1972)

Figure 20-20. Two female members of a lion pride nuzzle one another after having killed and fed on a zebra. (Taken in Amboseli National Park, Kenya)

tram (1973, 1975) have provided a fascinating picture of the social life of this animal. The social unit of the lion is the pride. This permanent and fairly stable social unit has a nucleus of from about three to 12 adult females. There is virtually no recruitment of outside females: all of the females are born and grow up within the pride. A pride probably lasts for many decades, and its female members are all closely related. When a young female reaches three years of age, she is either accepted as a member of the pride or driven from it. Rejected females, and males that are not attached to prides, become nomads, tend to follow the migratory movements of their prey, and constitute about 15 percent of the total lion population in the Serengeti. A pride is usually controlled by several adult males, which defend it against outside males. Each pride occupies a territory that is largely exclusive of those of other prides. Although the boundaries of the territory shift to some extent, essentially the same area is held by a pride year after year. All members of a pride are not together at all times. Individuals may hunt alone, or part of a pride may separate from the rest, but the members of a pride are familiar with each other, and social contacts are often peaceful or seemingly affectionate (Fig. 20–20). A member of the central sisterhood of the pride leads a stable, if at times violent, life, and her reproductive life is about 13 years.

A male, on the other hand, does not associate consistently with a single pride throughout life, and his reproductive life may be only two or three years. The life cycle of the male can be divided into several periods. Young males stay with their pride until they are three years of age, at which time they are either forced from the pride or leave it voluntarily. Often several males leave the pride together; these males may be brothers or—because the females of a pride are all grandmothers, mothers, sisters, or daughters—at least closely related. The young male outcasts become members of the nomadic population, and for the first time in their lives are unable to depend for food on the hunting prowess of the experienced females of their pride. A good deal of scavenging is typically done by these nomads. After roughly two years of nomadic life, these males are approaching the prime of life and are sexually mature and sufficiently formidable, in terms of strength and aggressiveness, to take over a pride. The pride they take over is virtually always not the one into which they were born. A pride lacking males may be taken over peacefully, or several males past the prime of life may be displaced and their pride taken over. On occasion, violent fighting accompanies the displacement, and because a group of males can successfully challenge males holding a pride whereas one or two males cannot, selection favors tight social ties between males. The new owners disrupt the life of the pride: pregnant females may abort, the cycle of females coming into estrus is interrupted, and the newcomer males may even kill cubs. After a few months, however, females again begin coming into estrus, they are bred by the new males, and males often help take care of the cubs.

After only two or three years, however, when the males are aging and perhaps some have died or have been injured, they are driven out by a new group of prime males. The displaced males, their reproductive life terminated, are again part of the nomadic segment of the population. Because they have become accustomed to being provided for by females, and because of their declining physical condition, their life expectancy is not great.

All the females in a pride tend to synchronize the bearing of young, and a cub can nurse from lactating females other than its mother. This communal care of young can perhaps be explained by the close genetic relationships between females of a pride and by the increased survival of cubs with familiar companions (B.C.R. Bertram, 1975).

In general, however, lions are inefficient reproductively. That is to say, an extremely low percentage of copulations result in offspring. In addition, the mortality rate of cubs is about 80 percent, and the killing of cubs by lions is an important cause of this high death rate. Bertram (1975) details the situation as follows: "Assuming that lions mate every 15 minutes for three days, that only one in five three-day mating periods results in cubs, that the mean size of litters is two and a half cubs and that the mortality among cubs is 80 percent, then a male must mate on the average some 3000 times for each of his offspring reared to the next generation." Because each copulation is so relatively unimportant and because the males of a pride are typically closely related and are thus genetically similar, there seems to be little pressure on the males to fight for the chance to copulate with an estrous female. This inefficiency and lack of

competition between males may be closely related to the rather unique mode of life of a top predator. The lion has no important predator except humans, and the size of the pride is perhaps controlled largely by periodic food shortages. The life span of lions is fairly long; reproductive inefficiency, therefore, does not prejudice the survival of the pride. Under these conditions, there is seemingly no strong selective pressure favoring highly efficient reproduction, but the critical factor may be the reduction of aggression in a pride to a level permitting the survival of at least some young. Bertram suggests that reproductive inefficiency and reduction of competition between males result in increased stability of the pride, with less frequent changes of the male guard and hence greater chances for the survival of cubs.

Wemmer (1979) has discussed evidence suggesting that some saber-tooth mammals that killed large prey lived in social groups. There would be little advantage to a solitary predator in killing large prey, for such prey could neither be defended against scavengers nor eaten before it spoiled. Also, the long period required for the full development of the sabers probably necessitated a longer period of mother-young association than occurs in living felids.

In the pinnipeds that are polygynous—the otariids, some phocids, and the walrus—the males are extremely vocal, are much larger than the females, and maintain breeding territories. In the California sea lion *(Zalophus californianus californianus),* studied by Richard S. Peterson and Bartholomew (1967), large bulls establish territories adjacent to the water at sites favored as hauling-out places by females, which arrive at the rookery a few days before they give birth. Nonterritorial bulls usually form aggregations apart from the breeding rookery. Because the same females do not continuously occupy a male's territory and because males make no effective effort to herd females into territories, the term "harem" cannot be applied to the females in a territory. Females enter estrus and copulation occurs roughly two weeks after par-

Figure 20-21. A boundary ceremony between two male California sea lions; the animals are confronting each other and "head shaking." (From Richard S. Peterson and Bartholomew, 1967)

turition. Fighting between males occurs during the establishment of territories, and males on established territories signal their possession by incessant barking. Little actual fighting occurs after territories are established, but a boundary ceremony between males on adjoining territories periodically reaffirms boundaries. These ceremonies involve an initial charge toward one another, followed by open-mouthed head shaking as the animals confront each other at close quarters (Fig. 20–21), and a final standoff in which the bulls stare obliquely at each other. The ceremony is so precisely ritualized that, should animals get uncomfortably close to one another, they adroitly avoid contact. Females are aggressive toward one another through much of the breeding season; again, however, injury is avoided by ritualized aggressive threats. Although males may be on territories in a rookery from May through August, each male maintains a territory for only one or two weeks; territories are thus occupied by a succession of males.

The northern fur seal *(Callorhinus ursinus)* has a breeding season that extends from June to December, and large numbers of animals assemble on the Pribilof Islands of the North Pacific (Fig. 20–22). Nonbreeding males form bachelor "cohorts" around the edges of the breeding grounds. Territorial males herd fe-

males into their territory and maintain fairly stable harems (Richard S. Peterson, 1965).

In contrast to the pinnipeds mentioned above, male elephant seals *(Mirounga angustirostris)* establish a social hierarchy on the breeding ground but are not territorial (Le Boeuf and Peterson, 1969). The highest ranking males stay close to breeding females, and breeding success is closely correlated with social rank. In one study area on Año Nuevo Island off the coast of California, four of the highest ranking males, which constituted but 6 percent of the 71 bulls in the area, copulated with 88 percent of the 120 females. At another study area on the same island, the alpha bull (the bull at the top of the hierarchy) maintained its rank throughout the breeding season and was involved in 73 percent of the observed copulations.

The social life of African elephants *(Loxodonta africana)* is known through the work of Laws and Parker (1968), Hendrichs and Hendrichs (1971), McKay (1973), and Douglas-Hamilton (1972, 1973). The elephant social system is structured at several levels. The first level is that of the family group, including an old matriarch and ten to 20 females and their offspring. Because of the long life span of elephants, the family unit generally includes grandmothers, mothers, sons, daughters, grand-

Figure 20–22.   Groups of northern fur seals *(Callorhinus ursinus)* on St. Paul Island, Bering Sea. Each group of females is attended by a male. (From Orr, 1976)

sons, and granddaughters. Tight social bonds between females may last 50 years or more. The second level of the social system is the kinship group, consisting of several family groups that remain in the same vicinity and on occasion mingle peaceably. Under some conditions, as during migration, many kinship groups may band together to form clans, containing on occasion 100 or more animals. The clan probably has no social cohesion at any level above the kinship group.

Males leave the family units when they become sexually mature and assemble in all-male groups in which dominance is established by ritualized fighting and sparring. Dominant males are temporarily attached to family units with females in estrus. Especially remarkable is the importance of cooperative and altruistic behavior within the family unit. A nursing elephant is allowed to suckle from any lactating female, young females approaching sexual maturity are solicitous of the well-being of small calves, and

the safety of a calf seems to be the concern of the entire family unit. When threatened, the adult members of the family form a united defensive phalanx.

Ungulate social behavior is of particular interest for several reasons. Many species are large, occupy open situations and can be observed easily, and have therefore been well studied. Because these open-country dwellers are probably the most highly social of all ungulates, we have a reasonably good understanding of ungulate social behavior. Further, a growing knowledge of the environments occupied by a variety of ungulates has provided a basis for a theoretical approach to the relationships between ecology and the evolution of social behavior, morphological features, and color patterns (R.D. Estes, 1974; Geist, 1974; R.J. Jarman, 1974).

Some ungulates form small, tightly knit herds of closely related individuals. An example is the extremely social collared peccary *(Tay-*

assu tajucu), which occupies a large range from the southwestern United States to Argentina. The social unit of this peccary is a herd of from 12 to 15 members, with equal numbers of males and females. Herd members remain close to one another, often resting with bodies touching or, on cold nights, huddling together. A dorsal scent gland is used for mutual scent marking and for marking the boundaries of a herd's territory. A remarkable degree of amiability and cooperation characterizes the peccary herd. All adults are tolerant of all young and do not dominate them, all lactating females cooperate in nursing all young, males do not compete for estrus females, and all adults cooperate in attacking small carnivores such as bobcats and coyotes (Byers, 1981). Regular periods of rambunctuous play involve all herd members and probably reinforce bonds of "friendship."

Many ungulates have breeding cycles that feature harems of females, each maintained by a dominant male. The area occupied by the females is the strongly defended territory, or in some cases, the male maintains a "breeding territory" that is defended even when females are not present. Breeding is done mostly by mature, vigorous, aggressive males. McCullough

(1969:99) reported that, in a herd of elk he studied, only 12 percent of the bulls—the largest individuals—did 84 percent of the observed copulating.

The rutting behavior of the elk is especially well known from the studies of Darling (1937), Graf (1955), McCullough (1969), Struhsaker (1967), and others. McCullough recognized four main categories of bulls during the breeding season. Primary bulls are powerful, mature individuals that shed the velvet from their antlers early and are the first to establish harems. Secondary bulls are large individuals that take over the harems by defeating the primary bulls as the latter become exhausted. Tertiary bulls assume control of the harems after the secondary bulls decline. Opportunist bulls are those whose only contact with cows is by chance. When a bull becomes exhausted through constantly keeping cows herded together, driving rival bulls away, and copulating, all while unable to obtain adequate food and rest (Figs. 20–23 and 20–24), it is beaten in a fight with a fresh bull, who takes over the harem from the deposed master.

Of particular interest are the ritualized social behaviors evident during the rut. Bulls ad-

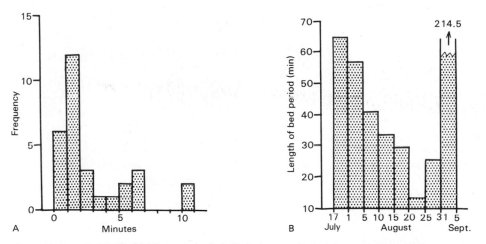

Figure 20–23. (A) Length of feeding periods of a bull elk *(Cervus elaphus)* at an intense rutting stage. (B) Changes in bedding (resting) periods of a bull elk as the rut progressed; this bull was defeated and gave up his harem on 31 August. (After McCullough, 1969)

Figure 20-24. A bull elk during the rut. (A) The bull bugling; the coat of this animal is caked with mud after wallowing. (B) The bull on the right (the same animal whose feeding and bedding periods are shown in Fig. 20-23) is in poor condition, has broken antlers, and is close to the end of his reign as harem master. (C) The same bull on 12 September, nearly two weeks after his defeat. Note the scars and broken antlers. (From McCullough, 1969)

vertise their location, vigor, and sexual readiness by several acts. The most obvious and characteristic of these is "bugling," a stirring, high-pitched call with considerable carrying quality. In addition, a bull frequently thrashes low vegetation and rakes the ground with his antlers while spurting urine onto his venter. Bulls also wallow in boggy places, gouge the soil with the antlers, and frequently rub the antlers and scrape the incisors against tree trunks. The master bull drives competitors away from the harem by a ritualized charge: with the head and neck extended, he jogs stiff-legged at the intruder while grinding the teeth and lifting the upper lip to display the upper canines. If the interloper does not retreat, the bulls face each other after a series of preliminary behaviors and smash their antlers together. Such an encounter typically occurs after the physical condition of the primary bull is on the wane and is a serious test of strength, with each contestant attempting to lunge forward and catch its adversary off balance. Such encounters may end indecisively, but if the master bull is clearly defeated, his reign as harem master is ended. In one case observed by McCullough (1969:89) the harem master was deposed one month after shedding the velvet and exhibited no further rutting behavior; he even ignored cows in estrus.

This mating system has costs as well as benefits for the male. Fighting success and reproductive success in British red deer (*Cervus elaphus*) are positively related: the bull that wins fights over harems most often benefits by being able to mate with the greatest number of females (Clutton-Brock, Albon, et al., 1979). But the costs are high. After a several-year study of fighting in a population of red deer, Clutton-Brock, Albon, et al. estimated that virtually all bulls are slightly injured during fighting at some time in life and from 20 to 30 percent are permanently injured. In addition, depletion of energy reserves by fighting during the rut may reduce bull survival rates through the food shortages of winter.

R.J. Jarman (1974) has given order to our view of bovid behavior by relating the sociobiology of bovids to their ecology. Jarman has shown that, in contrast to forests, grasslands (open woodlands, savannas, and plains) support the higher diversity and biomass of bovids. In grasslands the bulk of the food available to bovids is grass. Whereas a high percentage of each grass plant is edible, much less of a tree or shrub can be eaten. Grasslands therefore produce proportionally more food per growing season for bovids, but growing seasons are short. Although they produce less food for bovids, forests have moisture and plant productivity distributed more evenly throughout the year.

How do bovids respond to these different habitats and patterns of productivity? Grass provides great quantities of forage, but it offers less energy and protein than many small forbs. Because of the relatively high metabolic rate of small antelope and the attending need for relatively rapid tissue replacement, these animals have higher energy and protein needs than do the large species and tend to be selective feeders, utilizing forbs or only the most nutritious parts of grasses. The small antelope, therefore, because they depend on a food base that is far less abundant and more dispersed than grasses, do not attain the densities of the larger grass-eating bovids. During dry seasons, when the nutritional level of grasses is low, the larger bovids typically lose weight and vigor, and during the occasional prolonged drought may die (Hillman and Hillman, 1977).

Associated with the differences in density between large and small bovids are sharp differences in social organization. The five categories of Jarman (1974), as outlined by E.O. Wilson (1975), are shown in Table 20–1 and indicate the relationships between habitat, feeding style, and social organization in African bovids.

R.D. Estes (1974) has stressed a major structural and behavioral dichotomy within the family Bovidae. The ancestral bovids were probably small forest dwellers, perhaps resembling today's forest-dwelling duikers (Fig. 20–25). The expansion of grasslands in the Miocene and Pliocene of Eurasia and Africa set the stage for the movement of bovids into open grassland or savanna habitats. This major evolutionary step brought some bovids under the influence of new suites of selective forces and led to their structural and behavioral divergence from the persistently forest-dwelling species. Viewed today, the dichotomy is between, on the one hand, the forest-dwelling browsers that are generally small, cryptically marked, and simple-

Table 20–1:   Behavioral/Ecological Classification of Some African Bovids

Social Organization and Feeding Style	Body Size (kg)	Antipredator Behavior	Examples
Class A			
Solitary or in pairs or family groups; small, permanent home range; highly diversified diet, but selective feeders	1–20	Freeze, dash to cover and freeze, or lie down; do not outrun or counterattack predators	Dik-dik *(Madoqua)*, duiker *(Cephalophus)*
Class B			
Several female-offspring units associate; group size 1 to 12; permanent home range; males solitary; diversified diet	15–100	Similar to class A, but with some outrunning of predators for short distances	Reedbucks *(Redunca)*, vaal rhebuck *(Pelea)*, oribi *(Ourebia)*, lesser kudu *(Tragelaphus imberbis)*
Class C			
Larger herds, of six to hundreds of members; males have breeding territories; selective browsers and grazers	20–200	Diverse; hiding used in heavy cover, running used in open situations; communication of alarm behavior important	Kob, waterbuck, lechwe *(Kobus)*, gazelles *(Gazella)*, impala *(Aepyceros)*, greater kudu *(Tragelaphus strepsiceros)*
Class D			
During sedentary times, societies as in Class C species; gigantic herds develop during migration; feeders on variety of grasses; selective as to plant parts eaten	100–250	Run from large predators or mount unified counterattack on smaller predators	Wildebeest *(Connochaetes)*, hartebeest *(Alcelaphus)*, topi *(Damaliscus)*
Class E			
Large, stable herds of females and young with males organized into dominance hierarchies; herd size up to 2000; no coalescing of herds during migration; unselective grazers or browsers	200–700	Run from predators or mount unified counterattack even on large predators; groups respond to distress calls of young	Buffalo *(Syncerus caffer)*, beisa oryx *(Oryx beisa)*, gemsbok *(Oryx gazella)*, probably eland *(Taurotragus)*

Based on R.J. Jarman, 1974, and E.O. Wilson, 1975

horned and escape from predators by hiding and, on the other hand, the open-country grazers that are of medium or large size, are conspicuously marked, possess large and often complex horns, and use their speed in the open to escape predators. The behavioral dichotomy is also clearly delineated: the forest dwellers either are solitary or live in small family groups, and scent marking is the primary means of communication; the open-country grazers are typically highly gregarious and use primarily visual signals.

As noted and summarized in Table 20–1, the social organization of the classes of bovids

Figure 20-25. A red duiker *(Cephalophus natalensis)* in dense bush in southern Kenya.

line, comments on the social behavior of several species follow.

Although both the impala and the Uganda kob *(Kobus kob)* have somewhat similar social organizations (Class C in Table 20-1), their breeding behavior differs. The impala has been studied by M.V. Jarman (1970) and by R.J. Jarman and Jarman (1974). Dominant male impalas, constituting roughly one third of the population of adult males, maintain territories in the breeding season in the most favorable habitat. These territories form a mosaic of adjoining areas, and each dominant male defends his area against other males of comparable social status. Females and bachelor males occupy home ranges that typically include a number of territories. A territorial male attempts to round up females that enter this area and keep them within it. In bachelor herds, the hierarchy is based partly on age distinctions, with older, larger-horned animals dominating younger ones. Adult males of different ages are not always set apart by obvious clues, such as body and horn size, and among these animals a social hierarchy is established by frequent aggressive encounters featuring displays and fighting (Fig. 20-26). Males at or near the top of the bachelor hierarchy challenge territorial males, and repeated encounters between a territorial male and his challenger might span several weeks. A male holding a prime territory much frequented

recognized by Jarman and Wilson forms a progression from the small social units of the selective feeders of the bush and forest to the very large herds of the unselective feeders of the grassland. To flesh out the skeleton of this out-

Figure 20-26. Sparring between two pairs of male impala *(Aepyceros melampus)* that are members of a bachelor herd. (Taken in Tarangire National Park, Tanzania)

by impala is kept busy herding females, checking for females in estrus, and keeping bachelor and competitive adult males at a distance. These males become exhausted and lose their territory more quickly than do males holding less preferable areas. In areas with seasonal precipitation, the territorial system is abandoned during the dry season.

Both male and female Uganda kob return to traditional and permanent territorial breeding grounds (leks) to breed, and breeding occurs throughout the year (Buechner, 1962, 1963, 1974; Beuchner and Roth, 1974). Each lek is 200 meters in diameter and is defended by a single male through displays and ritualized fighting. The turnover of males on territories is high. Males on territories with the most traffic of females and invading males are generally displaced within ten days. As they enter estrus, the females move into the leks and pass freely through the territories of the males, who defer to one another as the females cross territorial boundaries. Both sexes show great fidelity for their home lek, and males are especially faithful to the vicinity of the lek. Over half of the mature males tagged by Buechner (1974) remained within 50 meters of the home lek from 90 to 100 percent of the time. The Kafue lechwe (Kobus leche kafuensis) is the only other antelope known to exhibit lekking behavior (Schuster, 1976).

The social behavior of the topi (Damaliscus korrigum) is flexible and of special interest because it reflects the different habitats of this antelope. In the woodland population of Serengeti National Park, each male defends a year-round territory of about 1 to 3 square kilometers, and each social group of females (two to eight animals) tends to stay within one male's territory. Each territory includes the year-round requirements of the females; thus a male is not merely defending a mating area. In the plains population of the Serengeti, by contrast, male territories are small and are used by females only during the wet season (the breeding season). On the plains, during the dry season,

males and females congregate in large herds, and the home ranges of these females are 50 times as large as those of woodland females. In this season, each plains male intermittently abandons the large herd, returns to his territory (sometimes a distance of 10 to 15 kilometers), and renews his scent marks (Duncan, 1975). Throughout much of the wet season, each male occupies his territory and tries to herd females into it. Although this small breeding territory is maintained all year—at considerable cost in energy—it is of reproductive importance for only part of the year.

The social organization of many ungulates is sufficiently flexible to respond to seasonal changes in resources. The herd size of the African buffalo shifts seasonally, from about 500 in the wet season to 100 or fewer in the dry season (Sinclair, 1974), and several other African ungulates, including Thompson's gazelles and common zebras, respond similarly to seasonal changes in food supply.

The advanced social organization of the African buffalo (Synceros caffer) has been studied by Sinclair (1970, 1974, 1977) in Tanzania. This nonterritorial animal forms herds of from 50 to 2000 animals. The size of a given herd is rather constant, at a mean size of about 350 animals. For the first three years of its life, a young buffalo tends to remain near its mother, and bonds between mothers and daughters seem closer than those between mothers and sons. In the third year of life, males begin to leave their mother, and when they are four and five years old they form subgroups within the herd. Older, adult males that remain with the herd establish a linear dominance hierarchy. The repeated sparring typical of immature males may result in the formation of this hierarchy. The less dominant males are driven from the herd and form small bachelor groups that remain separate from the mixed herds. Old males, over about ten years of age, leave the herd and become extremely sedentary; they are either solitary or form small social units (Fig. 20–27). The breeding is done largely by the dominant males of the

Figure 20-27. A small social unit of three old African buffalo bulls *(Synceros caffer)*. Old bulls remain separate from large herds and are therefore more vulnerable to predation by lions. (Taken in Amboseli National Park, Kenya)

herd, with the highest ranking ones having the greatest access to estrous females.

Especially remarkable is the way in which the herd functions as a tightly knit unit. An entire herd will rally to the defense of a young or adult animal in distress, and their formidable united front will discourage even the largest predators. A herd also moves and feeds as a closely massed unit. There is little attempt to maintain individual distance, and the bodies of herd members may on occasion be touching.

Geist (1974) points out that the widespread substitution by bovids of ritualized combat and aggressive displays for damaging physical contact has probably evolved under selection exerted by high densities and high diversities of predators. Bovids that attack and injure others invite damaging counterattack and are likely to be wounded, whereas those that use nondamaging, ritualized fighting are less likely to sustain injury. Because predators often concentrate on conspicuously wounded animals, selection by predators strongly favors the adherence by bovids to ritualized intraspecific

contests. This great development of ritualized combat in African bovids, which must face many diverse predator populations, contrasts with the more damaging encounters between members of northern species, such as bighorn sheep *(Ovis canadensis)*, that are under far less pressure from predators.

Degree of sexual dimorphism, which varies widely among ungulates, has been related to habitat and social behavior by some authors (Geist, 1974; R.J. Jarman and Jarman 1979). In general, where high-quality habitat is patchy, there is intense competition among dominant males for territories within these patches and for the females that occupy them. The impala, which lives under these conditions, is extremely dimorphic: males are one and a half times as large as females and have flamboyant horns, whereas females are hornless. In woodland habitats, which can support only low densities of ungulates, the inhabitants are selective feeders and a male and a female may mate for life and share a territory. An example of such a type is the dik-dik, in which sexual dimorphism

Figure 20-28. Two male oryx *(Oryx gazella)* maneuvering their horns into position for ritualized fighting. (Taken in Buffalo Spring Game Reserve, Kenya, by J. E. Varnum)

is slight. Dimorphism is also slight among species that occupy open grassland with a high carrying capacity. These ungulates occur in large herds, in which defense of females by males is not feasible, and males and females compete more or less on equal terms for food. African buffalo and wildebeest are examples. The resemblance between the male and female wildebeest is heightened by the female's tuft of hair that resembles the penis tuft of the male.

An interesting relationship exists between the degree of ritualization of fighting and the structure of the head weaponry. In elk, the branching pattern of the antlers is such that antlers of combatants interlock, reducing the chance that tines will inflict injury during tests of strength; injuries usually occur when the antlers are disengaged and one animal is attempting to escape. In the genus *Oryx,* the horns are long, unbranched, and pointed and extremely dangerous weapons. Fighting is highly stylized (Fig. 20-28), with the pushing occurring with the foreheads together but with little contact between the horns (Walther, 1958). This test of strength occurs within the limits of a pattern of behaviors that avoids injury from the dangerous horns. Encounters between male Grant's gazelles, which have quite dangerous horns, usually go no further than intimidation by a neck display, whereas in Thompson's gazelle, R.D. Estes (1967:189) found that "natural selection

has operated on horn configuration and fighting style to produce a relatively safe type of parry-thrust combat, thus obviating the need for a display substitute." In most cases, complementary specializations of head weapons and fighting style have ensured that opponents come together and measure each other's strength without physical damage. The pattern of rings around the horns of many African antelope may serve as "nonslip" devices that reduce the danger of injury to fighting males. Horns occur in females in about half of African antelope genera. The horns of females are more specialized as stabbing weapons than those of the males. Horns are present mostly in the females of the larger, heavier genera, enabling them to defend their young against some predators (Packer, 1983).

Eusociality in Mammals. Among the remarkably highly evolved social systems of insects are those in which only a single female colony member (the queen) breeds and the remainder of the colony is divided into social castes. (A caste is a group of morphologically distinctive individuals that perform specialized labor in a colony.) These *eusocial* insects typically have a long life span, overlapping generations, and cooperative care of eggs and young, and the reproductive female is the member of the colony least vulnerable to predation or accident. Insects with this system include some bees, some ants, social wasps, and termites. Only one vertebrate animal, the naked mole rat *(Heterocephalus glaber),* is known to be eusocial. Our knowledge of this remarkable animal is due to the excellent studies of Jarvis (1978, 1981) and Withers and Jarvis (1980).

Naked mole rats are small (25 to 50 grams), nearly hairless rodents that occupy the hot, dry parts of Kenya, Ethiopia, and Somalia. They live in large colonies of up to 40 individuals, with each colony occupying an extensive burrow system. The mole rats eat enlarged roots and tubers of plants adapted to long dry seasons, and most burrowing is done when the usually hard soils are made friable by rain. These animals have a low metabolic rate and extremely limited thermoregulatory ability (p. 463).

In many ways, the social system of the naked mole rat parallels the eusocial systems of insects. The colony is composed of three castes. "Frequent workers" are small (25 to 30 grams), nonbreeding individuals that burrow cooperatively (p. 381), forage, and build the communal nest. Members of this caste make many trips to the nest with food for the other castes. "Infrequent workers" are large (about 35 grams) relative to frequent workers and work at roughly half the rate of the latter. "Nonworkers" are the largest colony members (about 46 grams) and rarely work but do care for the young. They are brought food by the frequent workers, and when nonworkers sleep they are often joined by other mole rats. Their huddling together reduces the energy expended by the colony. The single breeding female performs no colony tasks, breeds with nonworker males, and produces from one to four very large litters (up to 24 young per litter) each year. Some individuals in each litter grow more rapidly than their siblings and become larger than the frequent workers; such individuals may replace infrequent workers or nonworkers that die. A laboratory colony studied by Jarvis had 16 frequent workers, nine infrequent workers, and eight nonworkers.

Suppression of reproduction in all but the single breeding female is probably under pheromonal control. Jarvis has shown that physical contact between the breeding female and other females is necessary for such suppression and suggests that the pheromone is carried in the urine and transmitted from the breeding female to others at the communal latrine. All males produce sperm, but the small frequent worker males have difficulty in copulating with the large breeding female. Females other than the single breeding female have ovaries that are seemingly quiescent; they contain many primordial and primary follicles but few more-mature follicles. Jarvis has found that breeding females are extremely difficult to capture,

suggesting that they are the individuals least vulnerable to predation.

Under what selective pressures would this unique social system have evolved? Jarvis hypothesizes that the demanding environment—with its high temperatures, sparse food supply, and long dry seasons during which burrowing is extremely difficult—have put a high adaptive premium on conserving energy. When there is cooperative digging and a division of labor among diggers, the energy cost per digger per unit of burrow length is probably reduced. A low metabolic rate and the huddling together of many resting animals conserve energy, favor longevity and a low rate of population turnover, and allow the reproductive output of one female to maintain colony size.

Threat and Appeasement.   Threat behaviors are among the most familiar activities of mammals. A dog lifts its upper lip to expose the length of its upper canines, a cat opens its mouth and hisses, and some rodents grind their teeth. These actions all signal a readiness to fight or to attack if the antagonist does not retreat or take other appropriate action. A threat is typical of a situation in which conflicting tendencies preclude either an immediate attack or a hasty retreat.

Threats can be simple, as in animals that merely open the mouth wide to display the teeth (Fig. 20–29), or complex, as in some horned artiodactyls in which both distinctive postures and movements are involved, but usually seem to advertise the most important offensive weapons. Visual threats may be made more impressive or startling by such audible threats as explosive hisses or growls. Strictly defensive threats are often used by animals that are under pressure from an aggressor but would gladly escape if possible. The cat's threat with the back arched and the side of the body confronting an aggressor is such a behavior, and the oblique stare used by bull sea lions on adjacent territories is seemingly a ritualized defensive threat. In an extreme defensive posture, many carnivores, such as cats and weasels, lie on the back with

Figure 20-29. Open-mouthed threat by a pouched "mouse" (*Dasyuroides byrnei,* Marsupialia). (J. Hudson)

the teeth and claws ready for action. The animal has retained its intention to defend itself while abandoning any inclination to attack.

Some mammals have carried this type of defensive behavior a step further by discarding the pretense of defense. Such an appeasement posture or behavior is a complete surrender and contains no elements likely to trigger an opponent's aggression. Complete vulnerability is emphasized, and the response on the part of the dominant animal is to cease its hostile activity. Wolves and many other mammals appease their dominant opponents by lying on the back with the vulnerable throat and underside unprotected. In several artiodactyls, lying down serves as appeasement (Burckhardt, 1958; Walther, 1966); and a subordinate black wildebeest *(Connochaetes gnou)* may roll on its side with its belly toward its superior and the side of its head on the ground (Ewer, 1968:177). In Grant's gazelle, a lowering of the head, the reverse of the high-headed threat posture, is adopted by a submissive animal (Walther, 1965). In some primates, the presenting of the rump as the female does prior to copulation is an appeasement gesture. The brightly colored skin on the rump of some Old World monkeys may serve in part to make the rump conspicuous and thus make "presenting" appeasement gestures more effective. A "grin" serves as an appeasement in some higher primates—perhaps this is akin to responding to the cowboy's demand to "smile when you say that, podner."

Appeasement behavior clearly serves several purposes. It allows an animal being defeated in a fight to avoid further injury and in many cases allows a subordinate animal to avoid a contest. In highly social species, threat and appeasement behaviors foster the peaceful perpetuation of a dominance hierarchy and allow animals to be close to one another with a minimum of energy wasted on aggressive interactions. Ritualized appeasement behavior may even be important in permitting a subordinate animal to seek social contact without risking attack (Schenkel, 1967).

Friendly Behavior. In many social species, patterns of friendly behavior, and often close bonds between individuals, help maintain social structure. Grooming is the most common type of friendly behavior. This may involve the grooming of infants by the mother, mutual grooming by adults, as in primates, or grooming as part of courtship behavior. Other friendly or recognition behaviors are smelling of the mouth and the anal-genital region in dogs, the mutual embrace of chimpanzees *(Pan troglodytes)* described by Goodall (1965:471), and the mutual pressing together by duiker antelope *(Cephalophus maxwelli)* of the maxillary scent gland (Ralls, 1971:446). The choral howling so characteristic of such social canids as wolves may keep members of a group apprised of the location of other members and may strengthen familiarity and bonds between individuals of the group. The nuzzling and tail wagging of wolves preparing for a hunt may create a common mood in preparation for a cooperative effort (Lorenz, 1963). Play featuring frequent body contact, as among members of a peccary herd, is perhaps of great importance in cementing friendship bonds.

Activity Rhythms

A striking aspect of animal behavior is the rhythmic, or cyclic, pattern of activity. Some species are active at night *(nocturnal)* and some during the day *(diurnal);* some are active primarily at dawn and dusk *(crepuscular).* The activity periods tend to be at regular intervals— the time of emergence of a particular species of bat may differ by no more than 2 or 3 minutes night after night. Animals also exhibit other kinds of cyclic behavior. The timing of reproduction is cyclic, and in some mammals, such as some rodents and bats, daily or seasonal shifts occur between highly active and torpid states. Migratory movements are also cyclic. Daily activity rhythms, those based on a 24-hour cycle, are termed *circadian rhythms* and are better understood than are other types of rhythms.

Circadian rhythms differ markedly from one species to another. Most mammals are nocturnal, but even in two nocturnal species there are contrasts between the patterns of activity (Fig. 20-30). In general, small mammals that are especially vulnerable to predation, such as rodents, tend to be nocturnal (chipmunks and ground squirrels are exceptions), whereas less vulnerable species, such as many ungulates, may be active during the day. The activity cycles of carnivores seem to be geared to the circadian cycles of their prey or to the period when hunting is most rewarding. Martens *(Martes americana)* frequently forage by day, when red squirrels *(Tamiasciurus hudsonicus)* are active, whereas coyotes *(Canis latrans)* usually hunt at dawn, dusk, and night, when rabbits and rodents are feeding.

Circadian rhythms are also influenced by interactions between species with similar environmental needs. In some cases, competition between species is reduced or eliminated because their activity cycles are out of phase. Two species of fishing bats *(Noctilio),* both of which

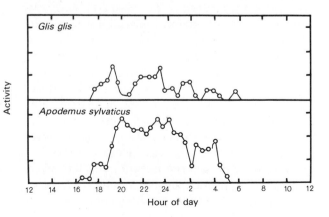

Figure 20-30. The autumn activity cycles of two nocturnal rodents. (After Eibl-Eibesfeldt, 1958)

feed over water, avoid interfering with one another partly by foraging at different times of the night (Hooper and Brown, 1968). Clearly, the circadian rhythm of an animal is part of its total adaptation to its particular mode of life and environment and has evolved just as have morphological characters.

The question of whether circadian cycles are endogenous (internally controlled) or exogenous (ultimately regulated by external stimuli) has occupied the attention of many biologists. Clearly, some strong endogenous control is present in many species. As an example, careful work on the flying squirrel *(Glaucomys volans)* by DeCoursey (1961) showed that, even under constant environmental conditions, including continuous darkness, flying squirrels maintained regular activity periods that deviated only ± 2 minutes from the mean value for activity periods under natural conditions. When a laboratory animal whose circadian cycle is not in phase with the natural 24-hour light-dark cycle is again exposed to normal day and night conditions, its cycle rapidly shifts and becomes "synchronized," that is, it becomes adjusted and locked—*entrained*—to the 24-hour cycle (V.G. Bruce, 1960). Circadian cycles, and other animal behaviors, are seemingly regulated by intricate and as yet poorly understood interactions between endogenous and exogenous factors.

As might be expected if circadian cycles are adaptive, they shift seasonally in some species (Fig. 20–31). Attending the seasonal changes in environmental temperatures are changing metabolic demands put on small mammals, and some shifts in circadian rhythms may allow the animals to avoid activity during times of most intense temperature stress. Studies by Wirtz (1971) have shown that, in the deserts of California, the antelope ground squirrel *Ammospermophilus leucurus)* is most active at midday in the winter, whereas in the summer the greatest activity is in the morning and late afternoon and the midday heat is avoided. The shift from nocturnal activity in the summer to diurnal activity in the winter by a bank vole (Fig. 20–31) probably results in a considerable saving of energy.

Activity cycles are also clearly geared to the basic metabolic demands of animals. In shrews, activity periods are distributed more or less evenly throughout the day, and smaller shrews have shorter and more frequent bursts of activity (Crowcroft, 1953). The high metabolic rate of the smaller shrews requires frequent periods of feeding and short intervals between feedings. Larger mammals, such a rabbits and rats, which have a much lower metabolic rate than small shrews, can meet their energy needs by feeding at dusk and at night, and some rabbits probably have only two feeding periods per 24-hour cycle.

Although less thoroughly studied than circadian rhythms, *circannian rhythms* play an equally prominent role in the lives of some mammals. Among vertebrates, such vital activities as breeding, migration, and hibernation are phased on an annual cycle, or a circannian rhythm. Pengelley (1967) defined a circannian rhythm as an endogenous cycle that has a length of approximately one year. Among mammals, such rhythms have been documented by Pengelley and Fisher (1963) for the golden-mantled ground squirrel *(Spermophilus lateralis),* by D.E. Davis (1967) for the woodchuck *(Marmota monax),* and by Heller and Poulson (1970) for two species of ground squirrels *(Spermophilus)* and four species of chipmunks *(Eutamias).*

Circannian rhythms are a major key to the survival of some temperate-zone and arctic mammals. In the words of Heller and Poulson (1970), these rhythms allow an organism "to anticipate, and thus prepare for, a future, annually occurring environmental condition such as cold weather, drought, food scarcity or optimal breeding time." The rhythm also ensures some flexibility of response to cyclic environments that may differ markedly from year to year. In addition, circannian rhythms enable "the organism to integrate a large number of en-

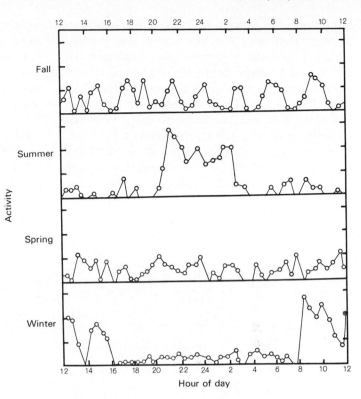

Figure 20-31. Seasonal changes in the daily activity cycles of a vole *(Clethrionomys glareolus)*. (After Eibl-Eibesfeldt, 1958)

vironmental cues and, through phasing the rhythm, respond most favorably to conserve energy and to ensure reproductive success.''

In hibernators in temperate regions, circannian rhythms make the animals sensitive to falling temperatures and declining body weight in the autumn so that the onset of hibernation may be hastened by unfavorable conditions or delayed by favorable temperatures and food supplies. In arctic areas, the extremely harsh environment and the sudden onset of winter, coupled with the brief time available for breeding and for putting on fat in preparation for hibernation, make flexibility nonadaptive. Here the adaptive premium shifts to a precise, inflexible, optimal schedule. Breeding at the optimal time each year, regardless of climatic conditions, probably ensures the greatest reproductive success, and precision in the onset of hibernation ensures maximum over-winter survival. Even in tropical areas, circannian

rhythms may be highly adaptive. In Kenya, for example, the African false vampire bat *(Cardioderma cor)* and the giant leaf-nosed bat *(Hipposideros commersoni)* become pregnant well before the onset of the late March-April rainy season, seemingly in anticipation of the burst of insect abundance that accompanies the rains. This pattern of breeding is probably controlled by a circannian rhythm.

The factors controlling these remarkable rhythms are as yet poorly understood. For arctic hibernators, Heller and Poulson (1970) suggest that photoperiod is the environmental factor of primary importance in phasing the underlying circannian rhythm. In temperate-zone hibernators, in contrast, although the timing of breeding seems inflexible, the animals are responsive to environmental conditions in the fall and may delay or hasten their entrance into hibernation. The situation is not entirely simple, however, for even temperate-zone hibernators that occupy the same area do not follow the same circannian rhythms. The activity of the golden-mantled ground squirrel, an inhabitant of mountains in the western United States, is controlled largely by an endogenous rhythm. This animal feeds into the fall and stores relatively nonperishable seeds in its burrow, and entrance into hibernation is relatively tightly scheduled. The Belding ground squirrel *(Spermophilus beldingi),* which often lives almost side by side with the golden-mantled ground squirrel, feeds on green material that decomposes quickly if stored underground. This squirrel feeds as long as possible in the fall, putting on more and more fat, and stores no food. Whereas the golden-mantled ground squirrel can perhaps afford greater rigidity in the timing of its hibernation because of the cushion of stored food in the hibernaculum, the Belding ground squirrel must depend entirely on food stored in the form of body fat and thus feeds as long as such activity is energetically feasible.

Chapter 21

Reproduction

Because of its primary importance to all life, reproduction is subserved by virtually every structural, physiological, and behavioral adaptation of an individual or a species. The unique mammalian pattern of reproduction must be of great antiquity: mammary glands, nourishment of the newborn or newly hatched with milk, and a close mother-young bond probably evolved together with the diphydont dentition in the Late Triassic, but the bearing of live young (viviparity) may have appeared later.

Of all the vertebrate classes, the reproductive pattern typical of mammals departs most from that of primitive vertebrates. Primitive, ancestral vertebrates were presumably egg layers, and this style of reproduction, or some fairly modest variation on this theme, is typical of all classes of vertebrates but the Mammalia. In all mammals but prototherians, young remain within the uterus during their embryonic and fetal life, and it is here that embryonic tissues and organs differentiate and the fetus grows. Nourishment and protection for the intrauterine young are provided by the mother, and under most conditions fetal survival rates are high. After birth, all young mammals are nourished by milk from the mother, and parental care, or in most cases maternal care, lasts until the young are reasonably capable of caring for themselves. The young of some mammals stay with their parents through an additional period in which they learn complex foraging and social behavior. In sharp contrast, in most nonmammalian vertebrates (birds are an exception), the young have little or no parental care after hatching or, in the case of ovoviviparous animals, after birth. In mammals, the combined effect of high fetal survival rate and extended postpartum care is an increase in reproduction efficiency in terms of energy expenditure per young reaching reproductive maturity. Most lower vertebrates lay great numbers of eggs at tremendous metabolic cost, and the success of the species depends on the survival of an extremely small percentage of young. For any given young of a lower vertebrate, survival is unlikely. In mammals, on the contrary, relatively few young are produced but the likelihood for survival of any given young is fairly high.

The Mammalian Placenta

One of the most distinctive and important structures associated with reproduction in therian mammals is the placenta. Differences among the major placental types have been used in distinguishing some of the higher taxonomic categories of mammals (subclasses and infraclasses), and some primary contrasts between reproductive patterns in mammals depend partly on placental differences.

A functional connection between embryo and uterus is necessary in animals in which the fetus develops within the uterus and in which nutrients for the fetus come directly from the uterus rather than from yolk stored in the ovum. This connecting structure, the placenta, allows for nutritional, respiratory, and excretory interchange of material by diffusion between the embryonic and maternal circulatory systems and consists of both embryonic and uterine tissues. The placenta also functions as a barrier that excludes from the embryonic circulation bacteria

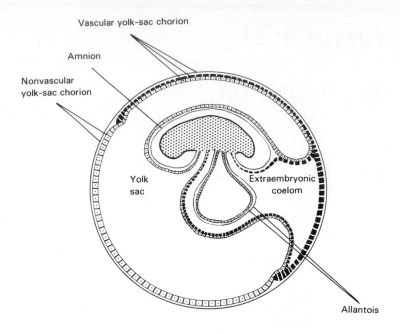

Figure 21-1. Embryo and extraembryonic membranes of a marsupial *(Didelphis)*. (After Torrey, 1971)

and many large molecules. In addition, the eutherian placenta produces certain food materials and synthesizes hormones important for the maintenance of pregnancy. Mammals are not unique in having a placenta, for certain fishes and reptiles establish placenta-like connections that allow diffusion of materials between the vascularized oviduct and the embryo. Among mammals, the major types of placentae differ sharply in structure and in the efficiency with which they facilitate the nourishment of the embryo.

Choriovitelline Placenta. This, the most primitive type of mammalian placenta, occurs in all marsupials except those of the family Peramelidae, the bandicoots. In marsupials with a choriovitelline placenta, the yolk sac is greatly enlarged to form a placenta. The blastocyst (an embryological state when the zygote has become a hollow ball of cells enclosing a fluid-filled cavity) does not actually implant itself deep in the uterine mucosa, as is the case in eutherians, but merely sinks into a shallow depression in the mucosa. The contact is strengthened by the wrinkling of the blastocyst wall that lies

against the uterus, and this wrinkling increases the absorptive surface of the blastocyst. The embryo is nourished largely by "uterine milk," a nutritive substance secreted by the uterine mucosa and absorbed by the blastocyst. The embryo also derives nourishment from limited diffusion of substances between the maternal blood in the eroded depression in the mucosa and the blood vessels within the large yolk sac of the blastocyst (Fig. 21–1).

Chorioallantoic Placenta. This type of placenta occurs in the bandicoots and in all eutherian mammals. Although similar to the eutherian chorioallantoic placenta in basic structure, the peramelid placenta achieves less effective transfer of substances between the fetal and maternal circulations. In peramelids, the allantois is fairly large and becomes highly vascularized; the blastocyst rests against the endometrium (the inner, mucosa layer of the uterus) on the side where the allantois contacts the chorion. At the point of contact with the blastocyst, the uterus is highly vascularized and the part of the chorion against the vascularized endometrium is more or less lost. At this point of approxima-

Table 21-1: Breeding Cycles of Several Marsupial Families

Family	Breeding Season	Polyestrous or Monestrous	Litter Size	Gestation Period (Days)	Suckling Period (Days)
Didelphidae	March–October	Either	2–25	13	70–80
Dasyuridae	April–December	M	3–12	8–34	49–150
Peramelidae	March–June	P	1–7	15	59
Phalangeridae	All year	1 litter/2 years or P	1–6	17–35	42–165
Macropodidae	All year	Either	1–2	24–43	64–270

Data on suckling period from Sharman, 1970.

tion of maternal bloodstream and allantois, exchange of materials occurs across the allantoic membranes. Because the peramelid allantois lacks villi and only its corrugations increase its absorptive surface, a limited surface area is available for exchanges of material between the maternal and fetal bloodstreams; supplementary nutrition is supplied by uterine milk. Probably due partly to the lack of villi and the resulting lack of absorptive efficiency of the allantois, the gestation period of peramelids is fairly short and the suckling period is much longer (Table 21–1).

In eutherian mammals, the chorioallantoic placenta reaches its most advanced condition with regard to facilitating rapid diffusion of materials between the fetal and uterine bloodstreams. In eutherians, the blastocyst first adheres to the uterus and then sinks into the endometrium. As implantation proceeds, chorionic villi grow rapidly and push farther into the endometrium as local breakdown of uterine tissue occurs. The resulting tissue "debris" is often called embryotroph; this nutritive substance is absorbed by the blastocyst and nourishes the embryo until the villi are fully developed and the embryonic vascular system becomes functional. In response to the presence of the blastocyst, the uterus becomes highly vascularized at the site of implantation. When the eutherian placenta is fully formed, the complex and highly vascularized villi provide a remarkably large surface area through which rapid interchange of materials between

the maternal and fetal circulations can occur (Fig. 21–2). The extent to which the villi increase the surface area available for diffusion is difficult to imagine but is suggested by the fact that the total length of the villi in the human placenta is roughly 48 kilometers (Bodemer, 1968).

Among eutherians, the degree to which the maternal and fetal bloodstreams are separated in the placenta varies widely. Lemurs, some ungulates (suids and equids), and cetaceans have an *epitheliochorial placenta,* in which the epithelium of the chorion is in contact with the uterine epithelium and the villi rest in pockets in the endometrium. Under these structural conditions, oxygen and nutrients must pass through the walls of the uterine blood vessels and through layers of connective tissue and epithelium before entering the fetal bloodstream. In ruminant artiodactyls, the uterine epithelium is eroded locally and there is contact between the chorionic ectoderm and the vascular uterine connective tissue. This is a *syndesmochorial placenta.* In carnivores, erosion of the endometrium is carried further and the epithelium of the chorion is in contact with the endothelial lining of the uterine capillaries. This is called an *endotheliochorial placenta.* Destruction of the endometrium in some mammals may involve even the endothelium of the uterine blood vessels, allowing blood sinuses to develop in the endometrium; the chorionic villi may then be in direct contact with maternal blood. This *hemochorial placenta* occurs in

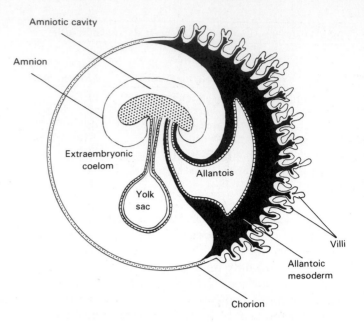

Figure 21-2. Embryo and extraembryonic membranes of a placental mammal. (After Balinsky, 1975)

some insectivores, bats, higher primates, and some rodents. In rabbits and some rodents, the destruction of placental tissue is so extreme that only the endothelial lining of the blood vessels in the villi separates the fetal blood from the surrounding maternal blood sinuses (Arey, 1974:147). In this case, a *hemoendothelial placenta* results.

The shape of the placenta is governed by the distribution of villi over the chorion. Several different distributions of villi occur in mammals. The lemurs, some artiodactyls, and perissodactyls have a *diffuse placenta,* which has a large surface area because the villi occur over the entire chorion. Ruminant artiodactyls have a *cotyledonary placenta* consisting of more or less evenly spaced groups of villi scattered over the mostly avillous chorion. Carnivores have a *zonary placenta,* in which a continuous band of villi encircles the equator of the chorion. In the *discoidal placenta,* villi occupy one or two disk-shaped areas on the chorion; this type occurs in insectivores, bats, some primates, rabbits, and rodents.

At birth, the fetal contribution to the placenta is always expelled as part of the "after-

birth," but the maternal part may or may not be lost at this time. In mammals with an epitheliochorial placenta, the villi pull out of the uterine pits in which they fit, none of the endometrium is pulled away, and no bleeding occurs at birth. This placenta is termed *nondeciduous.* In mammals with placentae allowing more intimate approximation of uterine and fetal bloodstreams, there is extensive erosion of the uterine tissue and extensive intermingling of uterine and chorionic tissue, and the uterine part of the placenta is torn away at birth, resulting in some bleeding. This type of placenta is *deciduous.* The hemorrhaging after birth is soon stopped by the collapse of the uterus, by contractions of the myometrium (the smooth muscle layer of the uterus), which tend to constrict the blood vessels, and by clotting of blood.

The rate at which substances move from the maternal to the fetal bloodstream in the placenta is of course increased when the number of interposed membrane barriers is reduced. Because of the difference between the number of such barriers in the human and the pig, for example, sodium is transferred 250 times more efficiently by the human placenta than by the pig placenta (Flexner, Crowie, et al., 1948). The remarkable absorptive ability of the allantoic placentae of such mammals as insectivores, bats, primates, rabbits, and rodents is due largely to the great surface area afforded by the complex system of villi, to the extensive erosion of uterine mucosa and the resulting development of blood sinuses into which the villi extend, and to the loss of nearly all of the membranes separating uterine from fetal blood.

The Estrus Cycle, Pregnancy, and Parturition

In mammals, reproduction is characterized by a series of cyclic events that are under nervous and hormonal control. As with many complex functions of the vertebrate body, the regulation of the reproductive cycle is maintained by environmental and social cues and by reciprocal controls between endocrine organs and their secretions. The events characterizing the stages in the mammalian reproductive cycle are well known, but details of the hormonal regulation of these events are not completely understood. The ovarian cycle results in the development of ova, their release from the ovary, and their passage into the uterus; the uterine cycle involves a series of cyclic changes in the uterus.

The ovarian cycle includes two major phases: (1) the growth of the follicle (the ovum, or egg cell, surrounded by specialized cells of the ovary) and release of the ovum and (2) the development of the corpus luteum (a glandular structure at the surface of the ovary) from the ruptured follicle. This cycle seems to be controlled largely by pituitary and ovarian hormones. The pituitary produces FSH (follicle-stimulating hormone) and LH (luteinizing hormone), which act together to stimulate growth of the follicle and initiate estrogen secretion by the ovary. Estrogen acts on the pituitary to stimulate increased production of LH, to initiate production of LTH (luteotropic hormone), and

to reduce secretion of FSH. Under the influence of interactions between these hormones, ovulation occurs (ova are released from follicles) and the corpus luteum forms from the ruptured follicle. Maintained by LTH from the pituitary, the corpus luteum produces progesterone, which sensitizes the uterus for implantation. If fertilization of ova does not occur, the corpora lutea regress and estrogen and progesterone production is reduced. The pituitary responds by resuming production of FSH, and another ovarian cycle is initiated.

This pattern, involving spontaneous ovulation, occurs widely among mammals, but some deviations are known. In the cat and some rodents, the follicles develop but ovulation does not occur until after copulation. In rabbits, the follicles do not develop fully before copulation and a long estrus may occur; ripening of the follicles and ovulation are initiated by copulation. In some rodents, pheromonal and tactile stimuli from the male initiate ovulation.

As the ovarian cycle proceeds, a series of cyclic changes occur in the uterus. Just before ovulation, the endometrium becomes thicker; this uterine stage is called the *proliferation phase*. In many mammals, a period of "heat," during which the female is receptive to the male, occurs at the end of the proliferation phase. This time of receptivity immediately prior to ovulation is termed *estrus*. After ovulation, the endometrium develops further and becomes highly vascularized; this is the *progestational phase* of the uterine cycle.

In most mammals, if the ova are not fertilized, the endometrium shrinks and the vascularization is reduced. In nonprimate mammals, no extensive bleeding occurs during the regression of the endometrium and the period of receptivity is short; in these animals, the uterine cycle is referred to as the *estrus cycle*. Some mammals have repeated estrus cycles during a year and are said to be *polyestrous;* others have a single estrus cycle per year and are *monestrous*.

The ovarian and uterine cycles in primates are different, to some extent, from those in other mammals. This primate cycle is called the *menstrual cycle.* In humans and in most other primates, considerable bleeding is typical of the time of endometrial breakdown, ovulation occurs at regular intervals throughout the year, and females may be receptive over an extended period. Even in such advanced primates as the gorilla *(Gorilla gorilla),* however, copulation is cyclic and occurs rarely except when the vulva of the female is swollen in association with ovulation (Nadler, 1975).

When copulation occurs, the sperm reach the oviducts in a matter of minutes in some species, and fertilization of the ova usually occurs within 24 hours of ovulation. The zygotes move down the oviducts, aided by contractions of the muscles of the oviducts, and usually reach the uterus and implant within a few days.

In eutherians, the delicate hormonal control of pregnancy is exerted by interactions between hormones produced by the pituitary, the ovary, and the uterus. In the early part of pregnancy, chorionic gonadotropin is of critical importance in preserving the corpora lutea and in preventing regression of the endometrium. This hormone is produced first by the trophoblast (the epithelial covering of the blastocyst that develops prior to implantation) during its implantation in the endometrium and then by the chorion, which develops from the trophoblast. During early pregnancy, the corpus luteum, because of its production of progesterone, is important in maintaining pregnancy by keeping the endometrium in a thickened and highly vascularized condition and by altering the ability of the myometrium to perform coordinated contractions that might expel the embryo. In some species, progesterone sensitizes the endometrium and increases the efficiency of blastocyst implantation.

The maintenance of pregnancy in many mammals is not entirely under the control of the corpus luteum. Instead, as pregnancy continues, hormones are produced progressively more by the placenta and less by the ovary. In humans, the placenta is thought to produce chorionic gonadotropin and estrogens and is probably the most important source of progesterone. During the latter stages of pregnancy, the placenta seems to be a nearly independent endocrine gland which, in humans at least, takes over the functions of the pituitary gland and the corpus luteum.

An important hormone of pregnancy, but one whose function is important mainly at parturition, is *relaxin.* This hormone, which causes relaxation of the pelvic ligaments and the pubic symphysis in preparation for parturition, is known to occur in a variety of mammals and may be universal among mammals. The concentration of relaxin in the bloodstream increases toward the end of pregnancy. In some mammals, such as pocket gophers (Geomyidae), the connective tissue joining the pubic bones is resorbed at puberty and a gap between the pubic bones remains during the rest of the animal's life (Hisaw, 1924). Relaxin may be produced by the uterus or by the placenta and in humans is known to be produced by the ovaries during pregnancy (Guyton, 1976). Passage of the very large fetus through the birth canal of the Brazilian free-tail bat *(Tadarida brasiliensis),* and perhaps other bats, is facilitated by an elastic interpubic ligament that can stretch to 15 times its original length (Crelin, 1969).

Birth is accomplished by rhythmic and powerful contractions of the uterine myometrium, aided by the abdominal muscles. Continued contractions force the placenta from the uterus and the vagina. These contractions are seemingly under the control of interacting hormones. *Oxytocin,* produced by the hypothalamus and stored in the posterior lobe of the pituitary, occurs in increasingly higher concentrations in the maternal bloodstream toward the end of pregnancy; oxytocin can initiate contractions of the uterus. Apparently the reduced concentrations of progesterone late in pregnancy are insufficient to block the effects of the increasing levels of oxytocin, with resulting contractions of the myometrium and parturition. The amount of estrogen also increases late in pregnancy, and the sensitizing of the uterus by estrogen shortly before parturition may allow oxytocin to initiate uterine contractions.

The newborn mammal is nourished by milk produced by the mother's mammary glands. Under the influence of estrogen and progesterone from the placenta, these glands undergo considerable growth during pregnancy. Milk secretion is stimulated and regulated by *prolactin,* produced by the anterior lobe of the pituitary. Prolactin is secreted in progressively larger amounts during the latter part of pregnancy and after parturition, and when the inhibition of prolactin by placental hormones is removed after birth, milk secretion begins. Milk production is partly under neural control and continues only as long as the suckling stimulus persists.

The Marsupial-Placental Dichotomy

The recognition of the marsupial-placental dichotomy is based on a number of biological differences; primary among these are the contrasting reproductive patterns. Whereas marsupials bear virtually embryonic young after a brief gestation period, many placentals bear highly precocious young and all placentals bear, after a relatively long gestation period, young that are anatomically complete. Why this difference? Lillegraven (1976) considered evidence bearing on this question, and the following remarks are based largely on his discussions.

Vertebrates have an extremely effective immune-response system to destroy invading foreign antigenic materials. During prolonged gestation in vertebrates lacking substantial stored energy reserves (such as yolk), an intimate contact between mother and fetus must ensure efficient physiological exchange, especially throughout the period of organogenesis and rapid growth. In this critical period, however, because the paternal antigens may be recognized as foreign by the maternal system, the fetus risks destruction during its extended "parasitism" of the mother. The mechanisms tending to avoid this immune response are not understood for most vertebrates, but for mammals some information is available (J.M. Anderson, 1972; Hughes, 1974).

In those marsupials that have been studied, the eggshell membranes are retained through the early two thirds of the gestation period. Although these membranes permit the passage of nutrients and probably of enzymes and antibodies, they interpose a barrier between the antigen-bearing parts of the embryo and the lymphocytes in the uterine fluid that initiate the immunological reaction; and immune-response rejection of the early fetus is thus avoided. Late in the gestation period, however, after the shedding of the shell membranes has occurred and the fetal and maternal tissues are in close contact, the stage is set for an immunological attack. The apparently precipitous birth of the rudimentary young after the short gestation period is seemingly an adaptation enabling marsupials to avoid this attack.

Although eutherian mammals lack shell membranes, the early stages of the zygote are separated from maternal tissues by the zona pellucida (a noncellular layer surrounding the zygote) and do not elicit an immune response. Later embryological stages are protected by the trophoblast and by the decidua (the uterine mucosa contacting the trophoblast) and its noncellular external secretion (Wynn, 1971). Adcock, Teasdale, et al. (1973) found that, in humans, chorionic gonadotropin blocks the action of maternal lymphocytes, protects the surface of the trophoblast, and allows the fetus to be accepted. Throughout the gestation, despite the close apposition of the uterine tissues and the fetus, at least one layer of trophoblast constantly provides a barrier between fetal and maternal tissues. Because of its remarkable capacity for isolating the fetus from maternal tissues that would initiate an immunological response, while at the same time not hindering efficient transport of essential materials, the trophoblast has allowed eutherians the luxury of a long gestation during which embryogenesis and rapid growth occur and an anatomically complete young is formed. Lillegraven (1976) states, "The 'invention' of the trophoblastic tissues by primeval eutherians was probably the single most important event in the history of the subclass."

This event may well have been important in allowing a structural diversity among eutherians (from forms with grasping hands to those with hoofs or wings to marine forms with flippers) that marsupials could never match. Because all marsupials must be born with forelimbs that enable them to immediately make their way to the marsupium, diversity of forelimb structure is impossible. Limitations imposed by the necessarily brief gestation period and long extrauterine development while attached to a nipple in the pouch have thus greatly restricted the adaptive options open to marsupials.

Major Reproductive Patterns

Although several deviations from the usual scheme occur, most mammals follow a similar reproductive pattern with regard to development of the embryo. After ovulation, the ovum passes down the oviduct, where it is fertilized. Early cell cleavages occur during the several days the zygote is passing down the oviduct, and by approximately the time the zygote enters the uterus, it has reached the blastocyst stage. After further enlargement, the blastocyst implants in the endometrium. Implantation occurs between the fifth and the fourteenth day after copulation, and the timing of implantation varies little within a species. After implantation, the embryo develops in the placenta a system of membranes and blood vessels that allows diffusion of nutrients and waste materials between the uterine and embryonic blood vessels.

In eutherian mammals, the length of time from fertilization to implantation is considerably shorter than the period between implantation and birth. Typically, fertilization occurs shortly after ovulation and the development of the embryo from fertilization to birth is an uninterrupted process. Perhaps in response to specialized activity cycles, some mammals have abandoned this usual pattern of continuous development. One departure involves a delay of ovulation and fertilization until long after copulation (delayed fertilization); another is typified by normal fertilization and early cell cleavages but also by an arresting of embryonic development at the blastocyst stage (delayed implantation); and another involves a long delay in the development of the blastocyst after it has implanted (delayed development).

Delayed Fertilization.   This pattern of development occurs in a number of bats inhabiting northern temperate regions. As early as 1879, Fries recognized that the males of some species of the families Rhinolophidae and Vespertilionidae could store viable sperm through the winter, long after spermatogenesis had ceased; later studies detailed the reproductive cycles of the females of these species (M.J. Guthrie, 1933; C.G. Hartman, 1933; Wimsatt, 1944, 1945). These remarkable reproductive cycles are seemingly adaptations to continuous or periodic winter dormancy and occur in a number of New World and Old World species in the genera *Rhinolophus, Myotis, Pipistrellus, Eptesicus, Nycticeius, Lasiurus, Plecotus, Miniopterus,* and *Antrozous.* Delayed fertilization may be the typical pattern in all but the tropical members of the family Vespertilionidae. Excellent papers by Wimsatt (1944, 1945) and by Pearson, Koford, and Pearson (1952) describe delayed fertilization as it occurs in vespertilionids, and the following remarks are based largely on those studies.

The reproductive cycle of the big-eared bat *(Plecotus townsendii)* in California follows a timetable similar to that of many temperate-zone vespertilionids. The testes descend into the scrotum in the spring. This migration is caused mostly by increased production of testosterone, which is cyclical in bats. The males become reproductively active in August. The testes begin to enlarge in the spring and are largest in September; spermatogenesis occurs mostly in late August and September. The testes regress and spermatogenesis ends before winter, but the caudal epididymides retain motile sperm through February and the accessory reproductive organs remain enlarged throughout the winter. Young males are not fertile in their first autumn. In the females, a single follicle enlarges in the autumn but remains in the ovary throughout the winter. A female may be inseminated repeatedly in the fall and winter, and males frequently copulate with hibernating females, although usually all females are inseminated by the end of November. The most typical vespertilionid pattern is for most copulation to occur before hibernation. The sperm are stored in the uterus, where they remain motile for at least 76 days. In *P. townsendii,* ovulation usually occurs in late February or March, either while the females are still at the winter roost or shortly after they leave. In many species inhabiting cold regions, ovulation occurs shortly after the females emerge from the hibernacula. Implantation is nearly always in the right horn of the uterus in *P. townsendii,* but ovulation occurs with equal frequency in each ovary. The gestation period in this species is highly variable, ranging from 56 to 100 days. This variation is probably due to regional differences in ambient temperatures and hence to the different body-temperature routines that occur in bats of widely separated colonies. Periodic torpor or low body temperatures after the beginning of gestation slow the development of the embryo.

Several features of this unique cycle are especially noteworthy. (1) The development of the male reproductive organs is out of phase; that is, the testes have regressed when the caudal epididymides and accessory organs are most enlarged and when breeding activity is at its peak. (2) Males retain viable sperm in the caudal epididymides long after spermatogenesis has ceased. (3) Females do not ovulate until long after they have been inseminated but are able to store viable sperm for several months.

(4) Because of differing metabolic routines in different individuals, the rate of development of the embryo is highly variable.

Delayed fertilization is seemingly a highly advantageous adaptation in mammals with long periods of dormancy. Spermatogenesis, enlargement of reproductive organs, and copulation require considerable energy. In species that practice delayed fertilization, these activities occur in the late summer and autumn, when males are in excellent condition and have abundant food, rather than in spring, when the animals are in their poorest condition and when food (insects) may not yet be abundant. Ovulation and zygote formation occur almost immediately upon emergence from dormancy, rather than being delayed until after males attain breeding condition and copulation occurs. The female can therefore channel more energy into nourishment of the embryo than would be available if copulation occurred immediately after hibernation. Perhaps the major advantage is that the time of parturition is hastened and thus young have the longest possible time to develop before the winter period of dormancy.

Delayed Development. Fleming (1971) showed that, in the Jamaican fruit bat *(Artibeus jamaicensis),* development of the blastocyst is delayed. A blastocyst conceived after the birth of young in July or August soon implants in the uterus, as in most mammals, but then becomes dormant, and further development is delayed until mid-November. This delayed development allows the young resulting from late summer matings to be born in early spring, when fruit is abundant.

Delayed Implantation. This deviation from the normal reproductive pattern occurs in a variety of mammals representing the orders Chiroptera, Edentata, Carnivora, and Artiodactyla (Tables 21–2 and 21–3). These mammals obviously do not share a common heritage; in addition, they occupy a wide variety of habitats and pursue differing modes of life. Delayed implantation in each group, therefore, has probably evolved separately and in response to differ-

Table 21–2.   Periods During Which Blastocysts Remain Dormant in Some Mammals with Obligate Delayed Implantation or Delayed Development

Species	Dormancy of Blastocyst (Months)
Order Chiroptera	
Equatorial fruit bat *(Eidolon belvum)*	3+
Jamaican fruit bat *(Artibeus jamaicensis)*	2.5
Order Edentata	
Nine-banded armadillo *(Dasypus novemcinctus)*	3.5–4.5
Order Carnivora	
Grizzly bear *(Ursus arctos)*	6+
Polar bear *(U. maritimus)*	8
River otter *(Lontra canadensis)*	9–11
Harbor seal *(Phoca vitulina)*	2–3
Gray seal *(Halichoerus grypus)*	5–6
Walrus *(Odobenus rosmarus)*	3–4
Order Artiodactyla	
Roe deer *(Capreolus capreolus)*	4–5

Data mostly from J.C. Daniel, 1970, data on *A. jamaicensis* from Fleming, 1971.

ent selective pressures. Delayed implantation is either *obligate* and constitutes a consistent part of the reproductive cycle or *facultative* and provides for a delay of implantation on occasions when an animal is nursing a large litter. A good discussion of delayed implantation is given by J.C. Daniel (1970).

In mammals with obligate delayed implantation, ovulation, fertilization, and early cleavages up to the blastocyst stage occur normally but further development of the blastocyst is arrested and it does not implant in the uterine endometrium. The blastocyst remains dormant in the uterus for periods from 12 days to 11 months. Little blastocyst growth occurs during dormancy, which begins generally when the embryo consists of from roughly 100 to 400 cells. The western spotted skunk *(Spilogale gracilis)* studied by Mead (1968) follows a reproductive pattern fairly typical of mammals with delayed implantation. Males become fertile in the summer, and copulation and fertilization of the ova occur in September. The zygote

undergoes normal cleavage but stops at the blastocyst stage; the blastocysts float freely in the uterus for 180 to 200 days. After implantation, the gestation period is about 30 days, and the young are usually born in May. During dormancy, each blastocyst is covered by a thick and durable *zona pellucida*. This general pattern of delayed implantation occurs in a number of carnivores, but the timing of the cycle varies from species to species (Table 21–3).

The adaptive advantage of delayed implantation is not understood for all species. In *Macrotus waterhousii* (Bradshaw, 1962; Wimsatt, 1969b) and in *Miniopterus schreibersii,* in which delayed implantation is known, this specialization may confer the same advantages as those resulting from delayed fertilization in temperate-zone vespertilionids.

Facultative delayed implantation occurs in some species in which the female is inseminated soon after the birth of a litter. This type of delay is known in some marsupials, some insectivores, and some rodents. In certain rodents

Table 21-3:   Reproductive Cycles of Some North American Mammals with Delayed Implantation

	Breeding Season	Time of Implantation	Length of Delay Period (Months)	Time of Parturition	Litter Size	Gestation Period (Months)
Long-tail weasel *(Mustela frenata)*	July	March	8	April–May	6–10	9
Ermine *(Mustela erminea)*	June–July	March	8.5–9	April–May	6–10	9.5–10
Mink *(Mustela vison)*	Feb–March	March	0–1	April–May	3–8	1.3–2.3
Marten *(Martes americana)*	July–August	March–April	8	May	2–3	9
Fisher *(Martes pennanti)*	March–April	Feb–March	11	March–April	2–4	11.5–12
Wolverine *(Gulo gulo)*	Spring (?)–Summer	Jan–Feb	5+	March–April	2–4	8–9+
Badger *(Taxidea taxus)*	July–Aug	Feb	6	March–April	2–3	8
Western spotted skunk *(Spilogale gracilis)*	Sept	April	6–7	May–June	4–7	8
Black bear *(Ursus americanus)*	June	Nov	6	Jan–Feb	1–4	7
Northern fur seal *(Callorhinus ursinus)*	Late July	Nov–Dec	3.5–4.5	Late July	1	12

Courtesy P.L. Wright

that have postpartum estrus, implantation of blastocysts is delayed when the female is suckling a large litter.

Our understanding of the factors controlling normal blastocyst development or dormancy in eutherian mammals is incomplete. Present evidence suggests that estrogen causes the uterine endometrium to form proteins essential for rapid growth of the blastocyst and that a deficiency of these proteins results in blastocyst dormancy (J.C. Daniel, 1970). One protein seemingly responsible for regulation of blastocyst differentiation and growth was named "blastokinin" (Krishnan and Daniel, 1967). Experimentally administered doses of estrogen or progesterone or both have been used in an attempt to initiate growth of dormant blastocyst in mammals with obligate delayed implantation. These procedures have not been successful in renewing growth of the blastocyst, indicating that in such animals some blocking of the action of estrogen in the endometrium must occur (J.C. Daniel, 1970). McLaren (1970) has proposed that, during lactation in mice *(Mus)*,

implantation is delayed by an initial inability of the blastocyst to "hatch" from the zona pellucida, which must be shed before implantation can occur.

Delayed implantation is an important part of the reproductive cycles of many marsupials. In most marsupials, the suckling of the young in the pouch inhibits estrus and ovulation, but in some kangaroos and wallabies (Macropodidae), a type of delay occurs that is termed *embryonic diapause* by Sharman (1970). In most macropodids for which embryonic diapause is known, the mother undergoes postpartum estrus; copulation occurs and the ovum is fertilized early in the life of the young she carries in the pouch. The suckling of the pouch young initiates neural and hormonal responses that arrest the activity of the corpus luteum and induce dormancy of the blastocyst; cell division in the blastocyst ceases, and it does not implant. In contrast to the dormant eutherian blastocyst, which consists of an inner cell mass that gives rise to the embryo and a hollow sphere of cells that gives rise to extraembryonic membranes, the marsupial blastocyst consists of only 70 to 100 cells that form a single spherical layer of cells of one type (protoderm). The marsupial blastocyst is surrounded by protective coverings consisting of an albumin layer and a shell membrane. When the young leaves the pouch, development of the corpus luteum and growth of the blastocyst resume, the blastocyst implants, and rapid growth of the embryo resumes.

In marsupials, the young suckle after they leave the pouch for a period roughly comparable to the suckling period in eutherian mammals of similar size, but the intrauterine period for the marsupial fetus is often short (Table 21–1). As a result, a newborn young may be attached to a nipple and suckling while a much older young is returning periodically to suckle from a separate nipple. In both marsupials and eutherians, the composition of the milk changes during pregnancy. In marsupials, the milk secreted early in the suckling period contains little or no fat, whereas milk secreted later in the period may contain as much as 20 percent fat. During double suckling in kangaroos, a remarkable thing occurs: separate mammary glands concurrently produce vastly different milks, although both glands are seemingly under the same hormonal influences. The gland supporting the pouch young produces milk containing little fat, and the gland supporting the advanced young produces milk with three times as much fat. The physiological basis for this remarkable arrangement is not known.

The available evidence suggests that, in marsupials, no extraovarian hormones are secreted during pregnancy. The placenta, important as an endocrine organ in eutherians, does not serve this function in marsupials. The reproductive physiology of marsupials is reviewed by Sharman (1970), who believes that the differences between marsupial and eutherian reproduction point to a separate evolution of viviparity in these two groups after they diverged from a common oviparous ancestral stock.

This view is taken to task effectively by Lillegraven (1976), who presents strong evidence in favor of the opposing view. Lillegraven suggests that the common ancestor of marsupials and placentals was viviparous and bore virtually embryonic young; he hypothesizes that viviparity did not evolve independently in marsupials and placentals. These groups of mammals seemingly diverged in the Early Cretaceous, and compelling paleontological and anatomical evidence indicates an origin from a common ancestral stock. Temporary closures of the eyes and ears in newborn marsupials guard against desiccation, and partial closure of the mouth ensures secure attachment to the nipple and immovable jaws in which the dentary-squamosal jaw joint (absent at birth) can develop. In placental mammals that bear precocious young, these closures have no function, yet they still occur briefly during intrauterine development and perhaps represent a developmental stage inherited from a common ancestor with marsupials. This ancestor may have borne rudimentary young after a brief gestation period, just as marsupials do today.

Broad Patterns: Size, Energy, and Reproduction

The enormous variation among mammals in size and life-history strategies is associated with great differences in reproductive performance. If one will abide exceptions and develop the picture in broad-brush style, such aspects of reproductive performance as litter size, frequency of litters, number of litters in a lifetime, rate of growth of young, and duration of the mother-young bond can be related to body size and metabolic rate. (For a detailed treatment of these relationships, see J.S. Eisenberg, 1981.)

In general, small mammals have a high metabolic rate and the rate progressively decreases as one ascends the size scale of mammals (p. 449). This broad pattern is clear, although the metabolic rate in certain species that pursue demanding styles of life (such as fossorial rodents, leaf-eating edentates, and porpoises inhabiting cold water) is higher or lower than what would be predicted on the basis of size. As a result of their high metabolic rate, small mammals need more energy per unit of body weight than do large mammals. In order to rebuild tissue at a higher rate, they must have a diet that is higher in protein, and they "wear out" more quickly and have shorter life spans.

For small mammals, then, the best reproductive investment is to have large and frequent litters, rapid development of young, and short periods of dependence for the young. Small mammals typically allocate a relatively high percentage of their lifetime energy budget to reproduction, but there is wide variation on this theme. Kangaroo rats live several years, have small litters, and may even fail to reproduce during a year when desert rainfall is extremely low. Among several species of Madagascan tenrecs, there is a reversal of the usual litter-size, body-size trend: in these mammals, litter size tends to increase with increasing body size (J.S. Eisenberg, 1981). The ultimate small-mammal strategy is that of semelparous species (species capable of having only one reproductive cycle in a lifetime). Males of the small Australian marsupial "mouse" *(Antechinus stuartii),* for example, go through one cycle of spermatogenesis at about 11 months of age and then die some four weeks later (Lee, Bradley, and Braithwaite, 1977). Females may live longer, but most also have only one reproductive cycle.

Large mammals, those above 10 kilograms in weight, generally have a single young per litter, the gestation period is long, and the mother-young bond is long-lasting. An extreme example is the African elephant *(Loxodonta africana):* the single young is born after a gestation period of nearly two years, and the tight mother-young bond may persist for ten years or more. Among large mammals, too, wide variation occurs. The carnivores typically have unexpectedly large litters, with wolves *(Canis lupus)* having up to six pups per litter and African lions *(Panthera leo)* up to five. The herbivores, in contrast, have small litters. Not only do most deer and antelope have one young per litter, but even the tiny (1.5 kilograms) Javan mouse deer *(Tragulus javanicus)* has a single young per birth.

Clearly low metabolic rates and long life spans are not entirely responsible for small litters in large mammals. Associated with large size are several important capabilities that favor small litters. Relative to small ones, large mammals are vulnerable to a smaller group of predators and can better defend their young; their motility is greater, and some can move great distances to find nutritious food. Also, long life favors complex social systems that allow for greater survival of adults and defense of young.

From the great diversity of mammalian reproductive strategies, then, a broad pattern emerges: small mammals face different physiological and life-history constraints than do large ones and have generally responded by evolving different reproductive patterns (Table 21–4).

Control of the Timing of Reproduction

A major factor affecting reproductive success in mammals is the precise timing of reproduction to coincide with favorable environmental conditions. This timing is mediated by interactions between environmental, behavioral, and physiological stimuli. These complex interactions defy simple explanation but have recently become better understood (Bronson, 1979).

Many environmental factors have some effect on the timing of reproduction in mammals, but photoperiod (the period of light during the daily light-dark cycle), temperature, energy, and nutrition are probably of prime importance. Seasonal breeding in many temperate-zone mammals is partly regulated by photoperiod (Elliott, 1976; Kenagy and Bartholomew, 1981), but some opportunistic species that respond to unpredictable food abundance are unaffected by photoperiod. The effects of temperature on reproduction are often difficult to interpret. The testes of male pocket mice *(Per-*

Table 21–4: Gestation Periods of Some Eutherian Mammals

Taxon	Common Name	Body Weight (g)	Gestation Period (Days)[a]	Litter Size[a]	Altricial or Precocial Young
Small Mammals					
Erinaceus europaeus	European hedgehog	1000	34–49	1–7	A
Sorex vagrans	Wandering shrew	5	20	5.2	A
Blarina brevicauda	Short-tail shrew	16	21–22	3–9	A
Lepus californicus	Black-tail jackrabbit	2870	41–47	2.3	P
Sciurus carolinensis	Eastern gray squirrel	425	44	2.7	A
Dipodomys merriami	Merriam's kangaroo rat	35	33	2	A
Cricetus cricetus	Hampster	150–250	20	11	A
Peromyscus truei	Pinyon mouse	27	26.2	3.4	A
Microtus pennsylvanicus	Eastern meadow vole	32	21	4	A
Jaculus orientalis	Jerboa	130–225	29	2	A
Large Mammals					
Lycaon pictus	African hunting dog	20,000	72–73	7	A
Canis latrans	Coyote	11,000	58–61	4	A
Panthera tigris	Tiger	134,000	103	1–4	A
Eumetopias jubatus	Northern sea lion	350,000	270	1	P
Tursiops truncatus	Bottlenose dolphin	155,000	360	1	P
Loxodonta africana	African elephant	2,766,000	660	1	P
Equus burchelli	Zebra	219,000	330–360	1	P
Diceros bicornis	Black rhinoceros	1,081,000	475	1	P
Phacochoerus aethiopicus	Warthog	65,000	171–175	1–4	P
Antilocapra americana	Pronghorn antelope	40,000	230–240	1–2	P

[a]Note that, relative to small mammals, large mammals have long gestation periods and small litters.
Data from J.F. Eisenberg, 1981.

ognathus formosus) exposed experimentally to high temperatures were small relative to those of mice exposed to low temperatures, and at high temperatures (35°C) reproduction in house mice was depressed (Pennycuik, 1969). This inverse relationship is unusual, however, for low temperatures usually inhibit testicular growth (J.R. Clarke and Kennedy, 1967; Kenagy, 1981).

Two prime factors affecting the timing of reproduction are energy and nutrition. Gestation, and especially lactation, demand large amounts of energy, and it is axiomatic that, in mammals faced with seasonal variations in food availability, breeding coincides with the time (or times) of food abundance. Unpredictable

food resources may cause irregular reproduction: in Nevadan deserts, when the annual seed crop fails to appear because precipitation has been scant, kangaroo rats (Dipodomys merriami) do not breed (Beatley, 1969). In the Great-Basin desert of southwestern Idaho, the Townsend ground squirrel (Spermophilus townsendi) has a single pulse of reproduction in early spring when green forage is available, but will suspend reproduction in response to inadequate food supply. As an example, green forage was scarce due to low precipitation when Townsend ground squirrels emerged from hibernation in 1977. Whereas the males entered reproductive condition, the females did not. There was no reproduction this year and the population density was reduced by more than one half (Smith and Johnson, 1985).

A specific nutritional factor has long been suspected of influencing reproduction in such herbivores as voles. Recently, such a factor has been demonstrated (Berger, Negus, et al., 1981; Sanders, Gardner, et al., 1981). Both male and female montane voles *(Microtus montanus)* become reproductive rapidly when they ingest 6-methoxybenzoxazolinone (6-MBOA), a compound derived from young, actively growing plants. This chemical cue allows the voles to initiate reproduction when the growing season has begun—when survival of the young of this short-lived rodent would be high—and synchronizes breeding with abundant plant growth in a fluctuating environment. Although still subject to experimental verification, there is evidence that 6-MBOA is a cue for reproduction in a number of vertebrates and may affect population dynamics and population cycles in arvicoline rodents.

Behavioral-physiological regulation of reproduction is clearly of great importance, and the importance of pheromones has been established by laboratory studies. Mammals deposit urine at various places in the course of their activities, and this urine serves as an individual's olfactory "signature" or "fingerprint" (Caroom and Bronson, 1971; R.B. Jones and Nowell, 1973a, 1973b) and provides information on species, sex, reproductive condition, and social status. Such olfactory cues can regulate reproduction by triggering endocrine responses or modifying behavior. In association with tactile stimuli, chemicals in male urine called *priming pheromones* regulate the reproductive maturation and timing of ovulation in female house mice and deer mice *(Peromyscus maniculatus)* by inducing a series of hormonal responses (Bronson, 1971; Bronson and Maruniak, 1975). Males are also sensitive to urinary cues: the male endocrine system responds to female urine by secreting hormones that Bronson (1979) hypothesizes increase the effect that the male's urine has on the female's reproductive system.

Among prepubertal females, female urine suppresses reproductive maturation and overrides the effect of male urine for a period during development. In the presence of a male, female puberty can occur at 25 days of age but may not occur until 50 days in the absence of a male. Tactile stimuli are important at various stages of the reproductive cycle, and domination of one mouse by another can drastically reduce the secretion of sex hormones in the subordinant.

These interplays between social cues and the endocrine system have been studied primarily in house mice, but similar interplays are known for a number of species. Female deer mice enter estrus in response to male urine (Whitten, 1956), and among several species of American arvicoline rodents, estrus occurs only in the presence of a male (MacFarlane and Taylor, 1982). Pheromonal initiation of estrus occurs also in the cuis *(Galea musteloides),* a South American caviid rodent (Rood and Weir, 1970; Weir, 1971; Weir and Rowlands, 1973). In this case, the tactile stimulation by the chin of the male on the rump of the female is augmented by scent from a large gland on the chin of the male. The importance of a pheromonal cue has been demonstrated for primates also: male rhesus monkeys *(Macaca rhesus)* become sexually active in response to vaginal secretions of an estrus female (Michael, Keverne, and Bonsall, 1971). Under natural conditions, scent urination in male goats *(Capra hirtus)* hastens estrus in females and synchronizes it with prime condition in the male (Coblentz, 1976).

Male-induced termination of pregnancy is a remarkable phenomenon mediated by social, olfactory, and endocrine factors. Such termination is initiated in the laboratory by the replacement of the original male by an unfamiliar male. In response to a new male, females of several species of voles will abort, enter estrus, and breed with the new male within a few days (Stehn and Jannett, 1981). The appearance of the new male induces estrus no matter the reproductive condition of the female. Male-induced abortion is to be expected in many arvicolines and could influence the population dynamics of wild populations.

Infanticide by males is an additional factor affecting the timing of reproduction in some mammals. Under some conditions, male langur monkeys *(Presbytis entellus)* kill young during a troop takeover (Hrdy, 1977), and infanticide is recorded among several other primates (Goodall, 1977; Rudran, 1973, 1979; Struhsaker, 1977). Infanticide in male collared lemmings *(Dicrostonyx groenlandicus)* is directed toward unfamiliar young, but males show little or no infanticidal behavior toward their own young (Mallory and Brooks, 1978). Male infanticide hastens the onset of estrus in the langur monkey and perhaps in other species; this alteration in the timing of reproduction seemingly increases the fitness of the infanticidal male. In the collared lemming, a female that becomes pregnant by an infanticidal male has a shorter-than-normal gestation period.

Reproductive Cycles and Life-History Strategies

Underlying the tremendous variations in mammalian reproductive cycles is a broad pattern. Mammals can be segregated into two groups, those that bear altricial young (helpless, naked young in which the eye and ear openings are covered by membranes and locomotion and thermoregulation are undeveloped) and those that bear precocial young (well-developed, fur-bearing young in which the eyes and ears are functional and thermoregulation and locomotion are well developed). Each of these patterns is typically associated with a different life-history strategy.

Mammals with altricial young live under unstable conditions with seasonal or unpredictable food abundance. They are small and subject to heavy predation pressure. Litters are large (often seven young or more), the young are born in a nest, and the gestation and suckling periods are short. The young grow rapidly

and reach sexual maturity early. Life spans are short, the brain size is relatively small, the mother-young bond is brief, and social behavior is simple. Estrus is short and frequently triggered by male-female interactions. Under favorable environmental conditions, breeding may occur repeatedly throughout the year. Their high reproductive rates allow these mammals to be reproductive opportunists and take advantage of even brief periods of food abundance. With such opportunism go high population turnover and population densities that are unstable seasonally and from year to year. An array of mammals, including tree shrews, insectivores, many kinds of rodents, and small carnivores, fit this pattern. Examples of altricial mammals are the European lemming *(Lemmus lemmus)* and the American montane vole *(Microtus montanus)*. Females of both these species can breed at a remarkably early age (15 and 21 days, respectively; Asdell, 1974; Kalela, 1961), and the gestation period is only some 21 days. The polyestrous females may have several litters during the summer growing season, but seldom will they survive to bear young during a second reproductive season. In extreme cases, reproduction may continue nearly all year in some arvicoline rodents (J.R. Baker and Ranson, 1933; Greenwald, 1956). Populations of lemmings and voles are famous for their dramatic fluctuations.

Mammals with precocial young, among which are the ungulates, cetaceans, primates, hyraxes, and some rodents, typically live in a fairly stable environment with a predictable food base. These mammals are often large and reach sexual maturity late, and some are not subject to intense predator pressure. The estrus cycle is long, ovulation is usually spontaneous, and gestation is prolonged. The usually single young is not born in a nest but commonly accompanies (or clings to) the female virtually from birth. Lactation is also prolonged, with an enduring mother-young social bond developing in many species. The brain of the precocious mammal is usually large, social behavior is complex, and an individual may spend its entire life as a member of a social group. These mammals have a low reproductive rate, but the survival rate of young is high because of the extended period of maternal care. Population stability, a low reproductive rate, low population turnover, and dependence on a stable environment make these animals vulnerable to habitat alteration by humans. An example of a precocious mammal is the South American woolly monkey *(Lagothrix lagotricha)*. This 5.4-kilogram primate mates first at eight years of age, has a long gestation (255 days) and a single, large (1 kilogram) young, and young are born at widely spaced intervals (1.5 to two years). Lactation lasts from nine to 12 months, the mother-young bond is tight, and social behavior is complex.

The length of the gestation period in mammals is, in general, positively correlated with body weight and with the degree of development of the newborn. The larger the mammal, the longer the gestation period, and for mammals of equal weight, the one with the heaviest neonate has the longest gestation (Huggett and Widdas, 1951). Elephant shrews, edentates, hystricomorphous rodents, cetaceans, and primates depart sharply from this trend, however. These departures are the basis for the following examination of the relationships between specific aspects of life-history strategies and reproductive patterns.

Elephant shrews have an unusually long gestation period and bear highly precocious young. This mouse-size (45 grams) *Macroscelides proboscideus,* for example, has a gestation period nearly two weeks longer (76 versus 63 days) than the 622 times larger (28,000 grams) wolf *(Canis lupus)*. Most species of elephant shrews do not use burrows or nests, rest on the surface of the ground, remain perpetually alert for predators, and escape by rapid bounds along carefully maintained trails. Survival of young elephant shrews depends on their precociously developed sensory and locomotor abilities and thus indirectly on their long gestation period.

Edentates, which have a very specialized style of life (p. 219), have a long gestation period, but there is no even progression toward longer gestation with increasing body size (Table 21–5). McNab (1979, 1980) has found that edentates have unusually low metabolic rates and has related this to their myrmecophagous (ant-eating) or folivorous (leaf-eating) habits. The case of the two-toed sloth *(Choloepus hoffmanni)* is of particular interest. This modest-size (9 kilograms) animal has a gestation period of 332 days, some 3.5 months longer than the American elk *(Cervus elaphus canadensis),* which is 22 times larger (200 kilograms). Both of these mammals are herbivorous, and both bear a single, precocious young at long intervals (12 months for the elk and 18 months for the sloth). Whereas the elk is highly motile, however, and ranges widely, feeds selectively on a wide variety of plants, and has relatively modest protein and energy needs per unit of weight because of its large size, the arboreal sloth is sedentary, occupies a small (1.96 hectares) home range, and feeds on the leaves of but a few kinds of trees. Not only are leaves often low in energy and protein, but they contain defensive secondary compounds (such as tannins and terpenoids) that retard digestion. This sloth became adapted to a dependable and ubiquitous but energetically marginal food by adopting a very low metabolic rate, heavy insulation, and a countercurrent heat exchange system in the limbs (p. 143). The reproductive pattern is also the result of chronic selection to reduce energy needs: the extemely long gestation period avoids the rapid mobilization of energy, allows for the development of a precocial young able to cling to its mother, and reduces the energetically costly lactation period.

Hystricomorphs, in striking contrast to other rodents, have an extremely long gestation period, and such is the case even among those that have a life style closely comparable to that of non-hystricomorphous rodents. For example, the hystricomorph *Ctenomys talarum* and the non-hystriocomorph pocket gopher *(Thomomys bottae)* are both fossorial herbivores, both

Table 21-5: Reproductive Data for Some Eutherian Mammals With Unusually Long or Short Gestation Periods Relative to Body Size

Taxon	Common Name	Body Weight (g)	Gestation Period (Days)	Litter Size	Altricial or Precocial Young
	Macroscelidea				
Macroscelides proboscideus	Elephant shrew	45	76	1–2	P
Elephantulus rufescens	Long-eared elephant shrew	58	56	1–2	P
	Edentata				
Dasypus novemcinctus	Long-nosed armadillo	4000	120	4	A
Bradypus variagatus	Three-toed sloth	4500	175	1	P
Choloepus hoffmanni	Two-toed sloth	9000	332	1	P
Myrmecophaga tridactyla	Giant anteater	32,000	190	1	A
	Hystricomorpha				
Ctenomys talarum	Tuco-tuco	150	100	4	A
Octodon degus	Degu	200	90	5	A
Galea musteloides	Cuis	400	52	3	P
Cavia aperea	Guinea pig	500	61	3	P
Chinchilla laniger	Chinchilla	500	111	2	P
Lagostomus maximus	Plains viscacha	3000	154	2	P
Erethizon dorsatum	North American porcupine	5000	217	1	P
Myocastor coypus	Coypu	6000	132	5	P
	Mysticeti				
Balaenoptera acutirostrata	Minke whale	5,854,540	300	1	P
B. borealis	Sei whale	12,201,286	330	1	P
Megaptera novaeangliae	Humpback whale	31,837,850	345	1	P
B. physalis	Fin whale	50,172,680	370	1	P
B. musculus	Blue whale	121,400,000	360	1	P
	Strepsirhine Primates				
Microcebus murinus	Mouse lemur	70	60	2	A
Galago senegalensis	Galgao	229	124	1	P
Nycticebus coucang	Slow loris	1230	193	1	P
Perodicticus potto	Potto	1348	193	1	P
Lemur catta	Lemur	2290	128	1	P
	Haplorhine Primates				
Cibuella pygmaea	Pygmy marmoset	145	145	2	P
Callithrix jacchus	Short-tusked marmoset	241	144	2	P
Pithecia pithecia	Saki monkey	1046	163	1	P
Lagothrix lagotricha	Woolly monkey	5400	225	1	P
Hylobates lar	Gibbon	6800	166	1	P
Theropithecus gelada	Gelada baboon	9830	170	1	P

Data from J.F. Eisenberg, 1981.

weigh roughly 150 grams, and both bear about four altricial young. The gestation period of *C. talarum,* however, is more than three times as long as that of *T. bottae* (100 versus 29 days). Why should large and small, altricial and precocial species of hystricomorphs (Table 21–5) have long gestation periods? Weir (1973) hypothesizes that this is a basic character that

evolved in hystricomorphs during their Early Tertiary isolation in South America and that severe climatic conditions at that time favored precocial young and long pregnancies.

Baleen whales stand out as remarkable exceptions to the general body-weight, gestation-period trend (Table 21–5). These whales, the largest animals of all time, have gestation periods similar to or shorter than those of camels or horses. Thus, the blue whale *(Balaenoptera musculus)*, 250 times heavier than the Bactrian camel *(Camelus bactrianus)*, has a shorter gestation period (360 versus 406 days). The reason for this amazing rate of fetal growth is unknown. One possibility is that, by being extremely efficient harvesters of plankton, the most abundant marine food source, and by spending "summers" in boreal or austral waters where plankton reaches peak productivity, female whales may be able to invest great amounts of energy in the fetus and still store the reservoir of blubber necessary for migration and early lactation (young are usually born in tropical waters). Another possibility is that carrying a fetus through more than one migratory cycle and continuing pregnancy through tropical fasting periods is unfeasible energetically; the very rapid development of a fetus may be more economical.

The primates as a group have long gestation periods, and as shown by Table 21–5, those of the small primitive primates are most remarkable. Although primate young are precocial, they depend on a long period of maternal care and indoctrination into an often complex social system. This is reflected by the typically long lactation period. The evolution of an early-primate reproductive pattern involving long gestation and precocial young may have been critical in setting the stage for the highly social lives of higher primates.

Litter Size and Reproductive "Seasons"

Because the metabolic cost of raising large and well-nourished litters is paid by a lowering of future reproduction, litter size represents the best reproductive investment for the environmental situation in which any population is living (G.C. Williams, 1967). This best investment may differ within a species from area to area (A.W. Spencer and Steinhoff 1968). Within a species, large litters occur at northern latitudes and high elevations, where severe winters and brief growing seasons limit the number of litters per year (Table 21–6). The general pattern is for the mammals of a boreal community to have a few large litters each year, for those of a less severe temperate area to have smaller but more frequent litters, and for those in tropical communities to have many small litters each year. Within any area, however, strategies typically differ from one species to another. Thus, in Panama, most rodents breed throughout the year but some species breed seasonally (Fleming,

Table 21-6: Geographic Variation in Deer Mouse *(Peromyscus maniculatus)* Litter Size as Indicated by Embryo Counts

Area	Elevation (m)	Number of Females	Mean Litter Size
Plains			
Larimer Co., Colorado	1550–1600	56	4.0
Mountains			
Plumas Co., California	1050–1500	96	4.6
Larimer Co., Colorado	1700–2000	37	4.4
Larimer Co., Colorado	2400–3350	47	5.4
Routt Co., Colorado	3200	111	5.6

Data for California from Jameson, 1953:48; for Larimer Co., Colorado, from A.W. Spencer and Steinhoff, 1968:283; for Routt Co., Colorado, from Vaughan, 1969:60.

1970), and in a subalpine community in Colorado, some rodents have several litters each summer and some have only one (Vaughan, 1969).

Litter size may be under many kinds of control. The strategy in a population of the California vole *(Microtus californicus)* was to produce the largest possible litters when conditions were favorable. Litters were largest in March, April, and May, when food was abundant, and smaller at other times (Krohne, 1981). Postpartum litter reduction may be important in mammals but is poorly understood. A careful experimental study by McClure (1981) demonstrated that, under food deprivation during lactation, female eastern woodrats *(Neotoma floridana)* practiced sex-biased litter reduction. The mothers actively discriminated against males by neglecting them, and during the course of lactation in 32 food-deprived females, the sex ratio of litters shifted in favor of females. Because males are polygynous, not every male has an equal chance of reproducing, whereas all females probably reproduce. When energy is at a premium and some young must die, the loss of male young reduces the female's fitness less than does the loss of female young.

In the view of Pianka (1976), optimal reproductive tactics involve maximizing an individual's reproductive value (the sum of all present plus future offspring) at every age. Reproductive effort, therefore, should vary inversely with residual reproductive value (expectation of offspring). The oldfield mouse *(Peromyscus polionotus)* follows these tactics (Dapson, 1979). In one population of this species, the mice lived about two years, reproduced at a moderate rate, and had few differences in reproductive pattern among the cohorts (segments of the population). A contrasting strategy was followed by another population of this same mouse: one cohort, made up of younger mice that had a chance of surviving to reproduce the following year, had a moderate reproductive effort, whereas another cohort, consisting of older animals with low

residual reproductive value (animals that would not survive to reproduce the following season), had litters in rapid succession and had double the number of young per female.

The timing of reproduction is a vital factor influencing reproductive success. During gestation, and especially during lactation, the mother's energy needs increase tremendously, and during the period when young are weaned and are becoming independent, their survival depends on adequate food. All phases of reproduction may be within the seasonal period of food abundance if it is long. If this period is short, however, gestation may occur at a stressful time, with lactation and weaning (or just weaning) occurring when food is most abundant.

Even in tropical regions, there is typically seasonality in food abundance, and mammalian reproductive patterns are responsive to this. In many species of Neotropical bats, the weaning of young coincides with peak food abundance just after the start of the rainy season (D.E. Wilson, 1979), but among phyllostomid bats there are three common reproductive patterns (Fig. 21–3). Similarly, there are several reproductive patterns among African bats (Table 21–7). The timing of reproduction in some species varies geographically. In Gabon, Africa, colonies of the bat *Hipposideros caffer* usually give birth in October in southern latitudes and in March north of the equator. The majority of tropical rain forest mammals breed seasonally (Bourliere, 1973). This is the case for rodents in Africa (Delany, 1971; Dubost, 1968; Rahm 1970) and the typical pattern for monkeys of the genus *Cercopithecus* (Fig. 21–4).

Boreal or temperate species reproduce in spring or summer, and for hibernators this period is barely long enough for the young to prepare for winter. In subalpine areas in the western United States, those young ground squirrels *(Spermophilus* spp.) unable to store enough fat during the brief summer die during hibernation. Kunz and Anthony (1982) found that, for two species of bats in Massachusetts, there is a

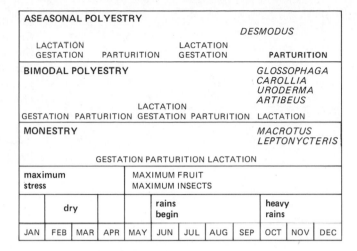

Figure 21-3. Three major reproductive patterns of Neotropical phyllostomid bats and the seasonal climatic changes affecting them. (From D. E. Wilson, 1979, p. 370, by permission of The Texas Tech Press)

high premium on rapid postnatal growth. Such growth is essential if fat depositon in the young is to be adequate for winter survival. In addition, females drop to about 85 percent of their postpartum weight during lactation, and the faster the young grow, the more time there is available for the female to recover fat stores before hibernation.

The reproductive cycles of small desert mammals are timed to take advantage of annual or semiannual bursts in the growth of ephemeral forbs (Beatley, 1976). Reichman and Van De Graff (1975) found that, in Arizona, both sexes of the kangaroo rat *Dipodomys merriami* ate more green plant material during the semiannual periods of plant growth than during the rest of the year and that there were surges in reproduction immediately following these pc-

Table 21-7:   Reproductive Patterns in Selected Paleotropical (African) Bats

Polyestrous: Yearlong Asynchronous Breeding	Polyestrous: Asynchronous Breeding for Part of Year	Diestrous: Two Synchronous Breeding Periods	Monestrous: One Synchronous Breeding Period
Rousettus aegyptiacus (P)	*Epomophorus wahlbergi* (P)	*Rousettus aegyptiacus* (P)	*Eidolon helvum*[a] (P)
R. lanosus (P)		*Myonycteris torquata* (P)	*Myonycteris torquata* (P)
Epomophorus labiatus (P)		*Taphozous mauritianus* (E)	*Rhinopoma hardwickei* (RM)
Epomops franqueti (P)		*Nycteris hispida* (N)	*Coelura afra* (E)
		N. thebaica (N)	*Taphozous nudiventris* (E)
		Cardioderma cor (MG)	*Hipposideros commersoni* (R)
		Rhinolophus landeri (R)	*H. cyclops* (R)
		Tadarida condylura (M)	*H. caffer* (R)
		Lavia frons (MG)	*Pipistrellus nanus* (V)
			Eptesicus somalicus (V)
			Tadarida pumila (M)

[a]This species has delayed fertilization; copulation and parturition seem not to be synchronous within a population.

Abbreviations: E, Emballonuridae; M. Molossidae; MG, Megadermatidae; N, Nycteridae; P, Pteropodidae; R, Rhinolophidae; RM, Rhinopomatidae; V, Vespertilionidae.

Data mostly from Kingdon, 1974; some data taken by T. J. O'Shea and T. A. Vaughan in Kenya.

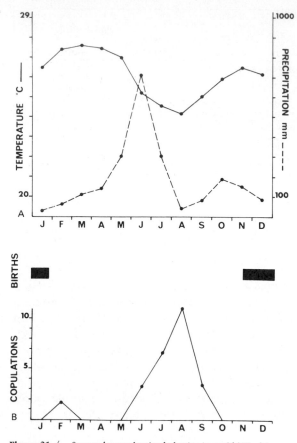

Figure 21-4. Seasonal reproductive behavior in an Old World monkey (Lowe's guenon, *Cercopithecus campbelli lowei*) inhabiting a tropical forest in Ivory Coast, Africa. (A) Mean monthly temperature and mean monthly precipitation. (B) Number of copulations per 100 hours of observation each month. (From Bourliere, 1973)

productively active, whereas only 14 percent of the males and none of the females sampled at the drier site were in reproductive condition. Perhaps reproduction is triggered in these kangaroo rats, as it is known to be in a vole, by a compound in the green leaves.

Postnatal Growth

As a broad rule, small mammals have higher postnatal growth rates than do large mammals. A small mammal will attain adult weight, or a given percentage thereof, at an earlier age than a large mammal. Lactation is roughly twice as costly energetically as gestation (Randolph, Randolph, et al., 1977), and because of the higher metabolic rate of the small mammal, it can presumably mobilize energy for lactation more rapidly. Among mammals of similar size, however, both metabolic rates and postnatal growth rates vary (Fig. 21–5).

Postnatal growth rates reflect the life history of a mammal. High rates have evolved under such demanding conditions as stressful environments and short seasons for preparing for hibernation. The need for high growth rates in young bats that must store fat for a long hibernation has been mentioned. Equally demanding conditions have selected for extremely rapid growth in some marine mammals, such as the elephant seal *(Mirounga leonina)*. This is a huge animal; males reach weights of over 3000 kilograms. An average female weighs 46 kilograms at birth, and this weight doubles by 11 days of age and quadruples by 21 days (Laws, 1953:33). The young Weddell seal *(Leptonychotes weddelli)* doubles its weight within two weeks after birth (G.C.L. Bertram, 1940:32). The rapid growth of pinnipeds is facilitated by the high-energy milk these animals produce (up to 53 percent fat). In addition, the suckling period is long, up to 1.5 years in the walrus *(Odobenus rosmarus)* and one year in some California sea lions *(Zalophus californianus;* Richard S. Peterson and Bartholomew,

riods. A similar relationship was found for female Arizona pocket mice *(Perognathus amplus)* at times of the year when these rodents were active. Van De Graaff and Balda (1973) found striking differences between the reproductive activity of two populations of *D. merriami* in Arizona at sites only 147 kilometers apart; one site had received three times more rain in the autumn than had the other and had vastly more green vegetation. At the wetter site, 90 percent of the adult kangaroo rats were re-

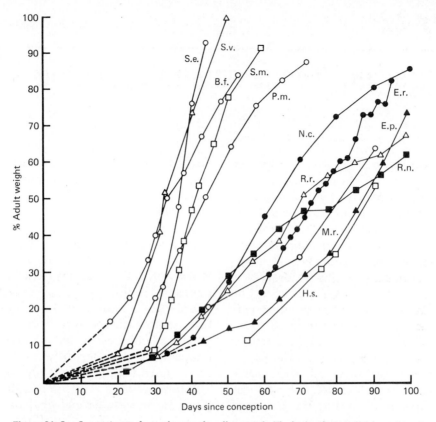

Figure 21–5. Comparisons of growth rates of small mammals. The broken lines indicate hypothetical intrauterine growth. Abbreviations: S.e., *Suncus etruscus* (Soricidae); S.v., *Sorex vagrans* (Soricidae); B.t., *Baiomys taylori* (Muridae); S.m., *Suncus murinus* (Soricidae); P.m., *Peromyscus maniculatus* (Muridae); N.c., *Neotoma cinerea* (Muridae); E.r., *Elephantulus rufescens* (Macroscelididae); E.p., *Eutamias panamintinus* (Sciuridae); M.r., *Marmosa robinsoni* (Didelphidae); H.s., *Hemicentetes semispinosus* (Tenrecidae); R.r., *Rattus rattus* (Muridae); R.n., *Rattus norvegicus* (Muridae). (From J. F. Eisenberg, The Mammalian Radiations, 1981, p. 306, by permission of The University of Chicago Press)

1967:44). The rapid growth allowed by rich milk and a long suckling period prepares the young for facing the stresses of winter at sea.

Low growth rates in small mammals may be associated with unusually low metabolic rates (as is probably the case with the tenrec, *Hemicentetes semispinosus;* Fig. 21–5) or with an adaptively long mother-young association. Consider the following case discussed by J.F. Eisenberg (1981:307). Relative to body weight of the adult, postnatal growth is more rapid in vespertilionid bats than in phyllostomid bats (Kleiman and Davis, 1978), and the former apparently have richer milk. Phyllostomids have a larger brain relative to body weight than do vespertilionids and relatively complex, less stereotyped foraging strategies. The low-energy milk is perhaps an adaptive feature associated with prolonged suckling, a mother-young bond of long duration, and a correspondingly long "training period" for the young. The Neotropical phyllostomids can afford a long suckling period because the young need not prepare for hibernation.

Chapter 22

Metabolism and Temperature Regulation

Some major barriers to mammalian distribution are easily recognized. Bodies of water, arid lands, or mountains may be absolute barriers to dispersal, depending on the environmental tolerances of the specific mammals. Equally limiting, however, are environmental temperatures. The distributions of some mammals—Neotropical sloths, for example—might be described most precisely by reference to the extremes of temperature and to the seasonal patterns of temperature change that can be tolerated. Air temperatures from −50 to 50°C may be encountered at various times and places on the earth, but at best mammals can survive body temperatures of approximately 45 to 0°C only and can be normally active only within the range of body temperatures between roughly 42 and perhaps 30°C. In mammals, interspecific differences in the ability to withstand temperature extremes occur even in closely related species, and it is not surprising that no one species is adapted to facing the full range of temperature extremes known for mammals as a group. Just as some mammals are adapted to a few food sources or to a restricted type of habitat, some can live only within a narrow range of temperatures. Knowledge of metabolism and temperature regulation in mammals is essential to understanding how these animals can adapt to the great variety of ecological settings they occupy.

Endothermy: Benefits and Costs

Most animals are *ectothermic* (their body temperature is regulated largely by heat gained from the environment rather than by metabolic heat), but mammals and birds are *endothermic* (their body temperature is controlled largely by metabolic activity and by precise regulation of heat exchange with the environment). Most endotherms maintain a high and fairly constant body temperature throughout life, but for this they pay an extremely high cost in energy. When not under thermal stress, a mammal expends five to ten times more energy for maintenance than a reptile of equal size and equal body temperature (38 to 40°C). At lower temperatures, the cost of maintaining a high body temperature rises abruptly: a mammal uses 33 times more energy than a reptile at 20°C and 100 times more at 10°C. A foraging mouse uses 20 to 30 times more energy than a foraging lizard of equal weight. In small mammals, such as most rodents, 80 to 90 percent of the total energy budget goes for thermoregulation. The costs of endothermy are clearly high. What are the benefits?

A variety of benefits have been recognized. Endotherms can be active under an imposing array of temperature extremes, ranging from intense desert heat to extreme arctic cold. They are freed from dependence on the sunlit part of the daily light-dark cycle and from becoming inactive during cold seasons. Most mammals have responded to this freedom by being nocturnal, and many are active through all seasons. Maximum oxygen-transport capability and high rates of enzymatic action are supported by constant high body temperatures. In their excellent treatment of endothermy and activity in vertebrates, Bennett and Ruben (1979) suggest that the primary advantage of endothermy is in greatly enhancing the ability to sustain high lev-

Table 22–1:   Maximum Aerobic Power Produced by the Activity of Small Vertebrates

Species	Body Weight (g)	Aerobic Power[a] (mW/g)
Fish		
Sockeye salmon *(Oncorhynchus nerka)*	55	3.9
Largemouth bass *(Micropterus salmoides)*	150	2.1
Amphibians		
Great Plains toad *(Bufo cognatus)*	40	8.5
Leopard frog *(Rana pipiens)*	38	2.8
Reptiles		
Green iguana *(Iguana iguana)*	800	2.8
Desert iguana *(Dipsosaurus dorsalis)*	35	11.2
Gopher snake *(Pituophis catenifer)*	548	2.9
Red-eared turtle *(Pseudemys scripta)*	305	5.8
Mammals		
Merriam's chipmunk *(Eutamias merriami)*	75	39
Norway rat *(Rattus norvegicus)*	286	27
House mouse *(Mus musculus)*	34	52
Fruit bat *(Pteropus gouldii)*	779	60

[a]After Bennett and Ruben, 1979.

els of activity. The maximum capacity of endotherms to use aerobic metabolism to produce power surpasses that of ectotherms by a factor of roughly 10 (Table 22–1). An example given by Bennett and Ruben of the relative ability of an ectotherm and an endotherm to increase oxygen consumption at high activity levels compares a 1-kilogram lizard and a 1-kilogram mammal. Resting and maximal rates of oxygen transport are 2 and 9 milliliters of oxygen per minute in the lizard and 9 and 54 milliliters of oxygen per minute in the mammal. In mammals (or in cynodont therapsids antecedent to mammals), endothermy probably evolved in the Late Triassic under selective pressure for sustained activity and stamina; as a costly by-product of these capabilities, the resting metabolic rate was raised.

Endothermy is the result of an increased metabolic rate in many organs of the body. These high metabolic rates are correlated with far higher concentrations of mitochondria in the cells of mammalian tissues than in those of lizards (Bennett, 1972) or other ectotherms.

There is no single endothermic pattern. Many large mammals have sufficient body mass, and therefore enough thermal inertia, to maintain their body temperature within narrow limits. For small mammals, however, in which the ratio of mass to surface area favors rapid heat exchange, wider fluctuations of body temperature are common. Indeed, for small mammals a regular pattern of body temperature change through the circadian cycle may be the rule rather than the exception. Abert's squirrel *(Sciurus aberti),* a pine forest dweller of the western United States, displays one such pattern. Body temperature in this squirrel ranges between 35.2 and 41.1°C, averaging 39.0°C; it is low (38.0°) when the squirrel rests and high (above 40°) when the squirrel forages on the ground (Golightly and Ohmart, 1978).

Selection has favored body temperatures and metabolic rates that save energy or facilitate the exploitation of a particular environment. There are many variations on the endothermic theme among mammals, and the survival of a species depends just as much on matching its thermoregulation pattern to its style of life and environment as it does on foraging efficiency or predator avoidance.

All animals that maintain a reasonably constant body temperature must balance heat gains and losses. Whereas lizards and many other ectotherms do this by practicing *behavioral thermoregulation,* mammals rely primarily on metabolic adjustments. A lizard will lie in the sun when its body temperature is below the preferred level and will seek shade when its temperature is above this level. Mammals use behavioral thermoregulation too, but their high resting metabolic rate and their activity under temperature extremes impose special thermoregulatory burdens. To maintain a constant body temperature when the ambient temperature is below body temperature, a mammal must balance heat produced by metabolism with heat lost to the environment. When the environmental temperature exceeds body temperature, metabolic heat and heat gained from the environment must be dissipated by some cooling device (such as evaporation). A major evolutionary trend in mammals is toward reducing the costs of these adjustments.

Some mammals, termed *heterotherms,* save energy by maintaining a constant body temperature at some times and allowing their temperature to fluctuate at other times. Many kinds of bats in temperate areas, for example, maintain a constant body temperature while foraging but allow their body temperature to approach ambient temperature when they rest.

Hair probably evolved in the earliest mammals or their forebears as a means of conserving energy. Because hair is such a poor thermal conductor, it greatly decreases the amount of body heat lost to the environment, by decreasing *thermal conductance* from the skin to the environment. (Thermal conductance is expressed as the metabolic cost, in cubic centimeters of oxygen per gram of body weight, for a given time interval per Celsius degree difference between body temperature and environmental temperature; Bartholomew, 1968.) The units of

this quantity are thus cm^3 $O_2/g/h/C°$. In some mammals in extremely cold environments, the difference between body temperature and environmental temperature may be 70 C° or more; because the rate of loss or gain of heat in a body is proportional to the difference between the body's temperature and that of the environment, it is essential that such mammals reduce heat loss by lowering their thermal conductance. Insulation, in the form of fur or blubber (or feathers in birds), is the primary means of reducing thermal conductance. Control of the blood supply to peripheral parts of the body is also important.

Each endotherm has a *thermal neutral zone* within which little or no energy is expended on temperature regulation. This zone is "the range of temperatures over which a homeotherm can vary its thermal conductance in an energetically inexpensive manner and on a short time scale" and maintain a constant body temperature (Gordon, Bartholomew, et al., 1977:371). Within this zone, the fluffing or compressing of the fur, local vascular changes, or shifts in posture suffice to maintain thermal homeostasis.

At the lower limit of the thermal neutral zone is the *lower critical temperature,* the point below which the balance between metabolic heat production and heat lost to the environment cannot be maintained by metabolically inexpensive variations in thermal conductance. Below the lower critical temperature, oxidative metabolism must be increased to keep the body temperature constant. Obviously, if a constant body temperature is to be maintained over a wide range of ambient temperatures, adjustments of both thermal conductance (through changes in insulation) and heat production (through metabolic changes) are necessary.

The *upper critical temperature* is the point above which a constant body temperature can be maintained only by increasing the metabolic work being done to above the resting level in order to dissipate heat. This temperature is far less variable than the lower critical temperature

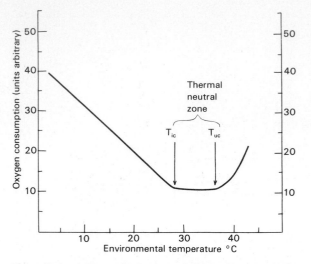

Figure 22-1. Oxygen consumption as a function of environmental temperature in a hypothetical mammal. Abbreviations: T_{lc}, lower critical temperature; T_{uc}, upper critical temperature. (After Bartholemew, 1968)

but is of great importance to desert mammals, which usually do not have access to drinking water and must strictly minimize water loss. Animals faced with temperatures above the upper critical temperature dissipate heat by evaporative cooling, which involves considerable water loss. Because such loss in desert species is extremely disadvantageous, these animals try to avoid temperatures above the upper critical temperature, often by spending the day in cooler shelters and being active on the surface only at night. Figure 22–1 summarizes the relationship between oxygen consumption and environmental temperatures in mammals.

Coping with Cold

Terrestrial Mammals. Many terrestrial mammals of temperate or boreal regions can remain active at ambient temperatures 30°C or more below body temperature. Major adaptations of these animals to cold include large size, improved insulation, high metabolic rate, regional heterothermy, and modified behavior.

Large size is a common and effective adaptation to cold. In general, large size favors heat conservation and small size favors heat dissipation: the larger the animal, the greater the volume or mass relative to surface area, and the smaller the animal, the greater the surface area relative to mass. Surface area is proportional to the square of body length, and volume is proportional to the cube of length. The surface-area-to-volume ratio, then, varies as the two-thirds power of the weight. The empirical relationship between body temperature (T_b), lower critical temperature (T_{lc}), and body weight (W) in mammals is represented by the expression $T_b - T_{lc} = 4 W^{0.25}$. "Because T_b is essentially independent of body weight in mammals, as weight decreases, T_{lc} approaches T_b" (Gordon, Bartholomew, et al., 1977:371). When foraging, small nocturnal mammals usually face temperatures below their lower critical temperature. This temperature for a 20-gram mouse is approximately 29°C, a temperature considerably higher than that usually encountered by nocturnal mammals.

Basal metabolic rate (that necessary for simply maintaining life in a resting organism in thermal neutrality), lower critical temperature, and thermal conductance all vary inversely with body size and are intimately related. Metabolic rate, as measured by oxygen consumption per gram of body weight per hour, rises so precipitously with decreasing body weight that a mammal weighing less than about 3 grams would be unable to eat sufficient food to sustain activity. Rates of oxygen consumption differ markedly even among small mammals (Fig. 22–2): the tiny masked shrew consumes oxygen at a rate over four times that of the larger deer mouse. Carrying the comparison further, the mouse consumes oxygen 10 times faster than the horse (H. A. Krebs, 1950).

Boreal mammals are generally larger than their ecological counterparts and close relatives of warmer areas. For example, the collared lemming *(Dicrostonyx groenlandicus)* of the Greenland tundra, a giant among nonaquatic ar-

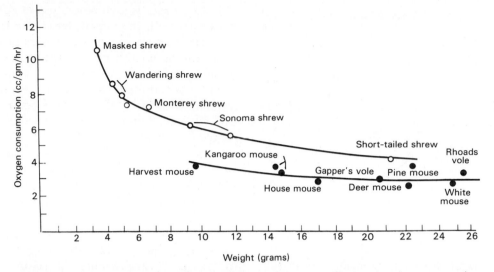

Figure 22-2. Oxygen consumption as a function of body weight in some small mammals. (After Pearson, 1948)

vicoline rodents, is more than twice as large (76 versus 32 grams) as its related eastern North American counterpart, the meadow vole (*Microtus pennsylvanicus*). The heaviest rabbits are those of the Arctic, Alaskan wolves are far heavier than those of Mexico, and the Alaskan moose is the heaviest member of its family. The uniformly large size of marine mammals (which will be discussed later) is an adaptation to living in cold water.

Effective insulation is an important feature of cold-adapted mammals. This insulation is so remarkably effective in some species that the thermal neutral zone may extend down to −40°C, as in the case of the arctic fox (*Alopex lagopus*). In many mammals active in the cold, the length of the woolly underfur and of the longer guard hairs varies seasonally; the summer pelage, which is acquired in spring, is short and has reduced insulating ability, but the winter coat, which replaces the summer pelage in autumn, is long and has great insulating ability. The hollow hair of some ungulates is remarkable insulation and allows winter activity under extreme conditions. Pronghorn antelope (*Antilocapra americana*) often forage on windswept plains at temperatures below 0°C. The low thermal conductance of hair is of particular importance for species that in winter must endure both extreme cold and food shortages. Subcutaneous fat may also reduce thermal conductance, but this is used primarily in marine mammals and is seldom important in land-dwellers.

High metabolic rates, especially high peak metabolic capacities, are typical of cold-dwellers. The high metabolic rates of voles and lemmings (Table 22-2) are generally regarded as an adaptation to activity through all seasons in temperate or boreal areas. The red fox and arctic fox, both of which forage through winter cold, have metabolic rates that are considerably higher than expected for mammals of their respective sizes (Table 22-3).

Under cold stress, many mammals practice regional heterothermy. Extremities, such as legs and ears, which are poorly insulated and dissipate heat rapidly, are allowed to become cool, thereby reducing heat loss by minimizing the temperature differential between these parts and the environment. This cooling of the extremities is due to vasoconstriction or to a countercurrent heat-exchange system (p. 143). In an Eskimo dog exposed to cold, the deep-body temperature was 38°C, the toe pads were 0°C,

Table 22-2: Resting Metabolic Rates of Selected Mammals

Taxon	Common Name	Body Weight (g)	Metabolic Rate (ml O₂/g/h)
	Monotremata		
Ornithorhynchus anatinus	Platypus	1200	0.46
Tachyglossus aculeatus	Echidna	4200	0.22
	Marsupialia		
Antechinus stuart	Marsupial "mouse"	37	1.53
Sminthopsis crassicaudata	Marsupial "mouse"	14	1.67
Sarcophilus harrisii	Tasmanian devil	6700	0.28
Trichosurus vulpecula	Brush-tail possum	1982	0.32
	Insectivora		
Sorex cinereus	Masked shrew	3–5	16.8
Blarina brevicauda	Short-tail shrew	14–18	5.3
Microsorex hoyi	Pygmy shrew	2.3–3.5	16.7
	Edentata		
Dasypus novemcinctus	Nine-banded armadillo	4000	0.20
Choloepus hoffmanni	Two-toed sloth	9000	0.19
Bradypus variegatus	Three-toed sloth	4500	0.18
	Rodentia		
Tamiasciurus hudsonicus	Red squirrel	190	1.7
Dipodomys merriami	Merriam's kangaroo rat	35	1.2
Heterocephalus glaber	Naked mole rat	32	0.4
Peromyscus eremicus	Cactus mouse	22	2.2
Microtus pennsylvanicus	Meadow vole	32	3.2
Dicrostonyx groenlandicus	Collared lemming	76	3.9
Capromys pilorides	Cuban hutia	4300	0.227
	Carnivora		
Canis familiaris	Dog	9666	0.34
Vulpes vulpes	Red fox	4440	0.55
Fennecus zerda	Fennec	1106	0.36
Ursus americanus	Black bear	77,270	0.36
Mustela rixosa	Least weasel	40	8.15
	Proboscidea		
Elephas maximus	Indian elephant	2,730,000	0.15
	Perissodactyla		
Equus caballus	Horse	260,000	0.25
	Artiodactyla		
Sus scrofa	Pig	75,000	0.11
Rangifer tarandus	Caribou	105,000	0.36
Oreamnos americana	Mountain goat	60,000	0.26
Ovis aries	Sheep	30,000	0.34

Data from J. F. Eisenberg, 1981; Jarvis, 1978; and Noll-Banholzer, 1979a.

and the tops of the feet 8°C (Irving, 1966). Regional heterothermy is a major cold adaptation of arctic ground squirrels *(Spermophilus parryii)*. By selective vasoconstriction in the limbs, the peripheral parts, such as toe pads, are stabilized at from 2 to 5°C, some 35 C° below body-core temperature (Baust and Brown, 1980).

Behavior plays a critical role in reducing cold stress. Many small mammals curl up so that

Table 22–3: Thermoregulation Data for Four Species of Canids

Species	Body Weight (g)	BMR[a] (ml O₂/g/h)	% of Expected BMR	Body Temperature (°C)	LCT[b] (°C)
Dog *(Canis familiaris)*	9666	0.342	99.7	38.0	24
Arctic fox *(Alopex lagopus)*	4700	0.526	128.1	37.9	−30
Red fox *(Vulpes vulpes)*	4440	0.550	132.0	38.7	8
Fennec *(Fennecus zerda)*	1106	0.358	60.7	38.1	24

[a]Basal metabolic rate.
[b]Lower critical temperature.
After Noll-Banholzer, 1979a.

their body form is nearly spherical. This minimizes the ratio of surface area to volume and tends to protect such lightly insulated parts as the face and feet. Nest-building behavior is important for small mammals: nests, usually built in protected places, provide insulation that greatly augments that provided by the pelage. Group thermoregulation, involving several animals huddling together, reduces the exposed surface area of each animal, is important in some boreal rodents (West, 1977), and is apparently widespread in mammals. Social behavior may be an important energy-conserving strategy. Taiga voles *(Microtus xanthognathus),* which are active through severe interior Alaskan winters, enhance over-winter survival by living in groups of from five to ten animals (Wolff and Lidicker, 1981). The communal nest

of these voles is always occupied by one or several animals, and nest temperature is thus kept well above ambient temperature (Fig. 22–3). Morton (1978) observed nest sharing in a small marsupial *(Sminthopsis crassicaudata)* and regarded it as an energy-saving behavior. The response of seeking shelter or a favorable place is also of importance. An animal foraging at ground level beneath a deep snowpack faces temperatures near 0°C, whereas ambient temperatures above the snow may be many degrees below zero. Wintering herds of deer or elk often prefer ridge tops or south-facing slopes, where nighttime cold is moderated by the early morning sun, and during Alaskan winters, moose abandon many basins into which cold air drains from surrounding mountains.

The long, slim body of some small mam-

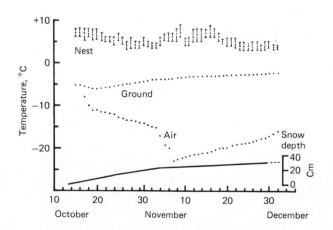

Figure 22–3. Daily mean air temperature, ground temperature, and temperature in the nest of the taiga vole *(Microtus xanthognathus)* from 15 October to 1 December 1977, at a site 155 km northwest of Fairbanks, Alaska. The daily ranges of nest temperature are shown by the vertical bars. Increasing ground temperatures after 20 October were due to progressively deeper snow cover. (From Wolff and Lidicker, 1981, by permission of Springer-Verlag, Inc.)

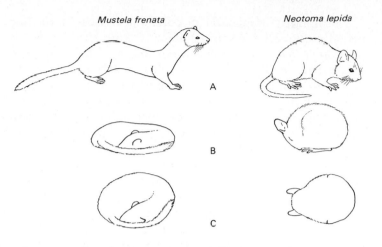

Mustela frenata   *Neotoma lepida*

A

B

C

Figure 22–4. Postures of weasels and woodrats: (A) under normal resting conditions; (B,C) under cold stress (B, lateral view; C, vertical view). (After J. H. Brown and Lasiewski, 1972)

mals seems maladaptive in terms of heat conservation. Long-tailed weasels *(Mustela frenata),* for example, pay a high energy price for being long and thin (J. H. Brown and Lasiewski, 1972). The surface-area-to-weight ratio of this weasel is some 15 percent greater than that of the woodrats *(Neotoma),* and under cold stress weasels cannot assume a spherical form by curling up (Fig. 22–4). As a result, cold-stressed weasels have a metabolic rate 50 to 100 percent greater than less slender mammals of comparable weight. The enhanced ability of the weasel to follow its prey into burrows probably more than compensates for the metabolic cost attending the elongate form.

Aquatic and Semiaquatic Mammals. Temperature regulation is a demanding problem to mammals that inhabit cold water. The rate of heat loss by an endotherm in water is some 10 to 100 times as great as the loss rate in air of the same temperature (Kanwisher and Sundnes, 1966:398). Arctic and antarctic waters are near 0°C year round, and northern lakes and rivers approach this temperature in winter. Consequently, a temperature differential of some 35°C between deep-body temperature and ambient temperature is usual in mammals swimming in these waters. Despite the thermal inhospitability of this environment, cold waters are permanantly inhabited by cetaceans, and

pinnipeds spend much of their lives in such waters.

In addition, the muskrat *(Ondatra zibethicus),* beaver *(Castor canadensis),* some shrews *(Sorex* spp.), some otters *(Lontra canadensis, Enhydra lutris),* and the mink *(Mustela vison)* spend considerable time in cold water. These semiaquatic mammals lose heat to the water most rapidly from the foot pads, the nose, and other bare surfaces; most of the body is insulated by a layer of air entrapped by the fur. Nonetheless, heat is lost far more rapidly during immersion in water than when the animal is in air. W. A. Calder (1969) found that, in two species of shrews *(Sorex palustris* and *S. cinereus)* and two species of mice *(Zapus princeps* and *Peromyscus maniculatus),* thermal conductance in the water when the fur had entrapped air was 4.5 times that in air; when the fur was wet to the skin, the conductance rose to nine times that in air. Calder also measured heat loss in the water shrew *(S. palustris),* the smallest homeothermic diver. The body temperature of water shrews with air entrapped in the fur dropped an average of 1.4°C in 30 seconds during dives beneath the surface of the water, whereas shrews with fur wet to the skin had a temperature drop of 4.5°C in the same time. The meticulous grooming and drying of the fur by a shrew after a dive is clearly highly adaptive.

Prolonged immersion in cold water is especially difficult for small mammals. Both the platypus *(Ornithorhynchus anatinus)* and the muskrat are small (about 1200 grams), and both swim for long periods. The platypus has a resting metabolic rate 35 percent below that of eutherian mammals of comparable size, but in the water its metabolic rate increases by a factor of 3.2 (Grant and Dawson, 1978). Heat loss from the extremities is reduced in water by countercurrent heat exchange. When a muskrat swims, vasoconstriction and countercurrent systems keep its limbs at ambient temperature in cool and cold water, but vasodilation in the limbs allows for heat dissipation at a water temperature of 30°C or above (Fish, 1979). Because of the very high rate of thermal conductance in the limbs of these two semiaquatic mammals, they would lose heat extremely rapidly when swimming if the limbs were kept near body-core temperature.

Some nearly permanent inhabitants of the sea, such as otariid seals, similarly use entrapped air as insulation, but many marine mammals (cetaceans, phocid seals, and walruses) lack insulative fur, and their bodies are in contact with water that may in extreme cases be 40°C below the deep-body temperature. How they maintain a constant body temperature under such demanding conditions is of considerable interest.

These marine mammals have a thick layer of subcutaneous blubber that forms an insulating envelope around the deep, vital parts of the body. A substantial amount of the weight of a marine mammal may be contributed by the blubber. For example, blubber constitutes 25 percent of the weight of the Weddell seal *(Leptonychotes weddelli;* W. C. Bruce, 1915). In the small (75 kilograms) harbor porpoise *(Phocoena),* 40 to 45 percent of the weight is blubber, and only 20 to 25 percent is muscle (Kanwisher and Sundnes, 1966:405). Studies of seals by Irving and Hart (1957) have shown that the skin temperature varies directly with the water temperature down to 0°C. The cooled surface of

the body and the thick blubber are seemingly an effective insulation, as indicated by the fact that the lower critical temperature of some seals is 0°C. The skin of seals has a well-developed vascular supply, and the temperature gradient between the deep parts of the body and the skin is controlled largely by changes in the blood supply to the skin.

The young of seals and polar bears face especially severe thermoregulatory problems. These young must face extreme cold but are far smaller than the adults and thus lack the heat-conserving advantages of large size. The baby harp seal *(Pagophilus groenlandicus)* is born on drifting ice in the North Atlantic in winter and must survive air temperatures of −20°C or below. The pup weighs but 11 kilograms, whereas its mother may weigh as much as 140 kilograms. The pup has long fur, which offers better insulation than the short fur of the adult, and thermogenic adipose tissue internally and along the back, which yields energy during shivering and helps maintain the body-core temperature (Blix, Grav, and Roland, 1979). Thermogenic adipose tissue ("brown fat") is specialized fat that releases energy rapidly when metabolized; it is used, for example, by shivering bats arousing from torpor. This adipose tissue is transformed to insulative blubber when the pup is several days old. Young harps seals use hypothermia (lowered body temperature) to conserve energy during extreme cold, wind, or rain. Northern fur seals *(Callorhinus ursinus)* born on islands in the North Pacific weigh only 5 to 6 kilograms at birth, whereas the adult female weighs roughly 55 kilograms. The insulation of the pups consists of dense fur and a layer of blubber 2 to 4 millimeters thick. Their tolerance to cold is largely due to a high metabolic rate supported by rich milk, shivering thermogenesis, and vasoconstriction in the skin (Blix, Miller, et al., 1979). These pups live close to their threshold of cold tolerance, and during cold rains many of the weaker young die. Polar bear cubs *(Thalarctos maritimus)* also face demanding conditions. They emerge from the natal den in April weighing 10 kilograms; the mother weighs about 170 kilograms. Because of their dense pelage and extremely high resting metabolic rate, the cubs have a lower critical temperature (-30°C), than the adults (Blix and Lentfer, 1979). Although cubs tolerate low air temperatures, icy water is a threat. Under experimental conditions, the deep-body temperature of cubs immersed in icy water dropped 11°C in 30 minutes. Their chance of surviving a comparable period in the icy sea would be slim.

Some of the most extreme thermal demands faced by endotherms are those met by cetaceans. Whales and porpoises live their entire lives in the water, and some species continuously occupy water at or near the freezing point. All cetaceans have insulating layers of blubber, but an extreme situation is faced by a small porpoise, which must maintain a deepbody temperature some 40°C higher than that of the sea, from which it is insulated by only 2 centimeters of blubber. An inflexible pattern of thermoregulation is inadequate even in inhabitants of the sea, which offers a relatively constant thermal environment. Some cetaceans migrate seasonally from cold waters to warm tropical seas. Because of the high thermal conductivity of water, skin temperatures generally equal water temperatures, and variations in water and skin temperatures of roughly 20 to 30°C may occur seasonally. The temperature of the body core, however, remains constant, and insulation requirements therefore may vary fivefold. Heat production by mammals varies tremendously as a result of changes in metabolic level. Metabolic activity and heat production increase roughly ten times in an animal going from a resting state to one of maximum exertion. It has been estimated that a cetacean at rest in cold water needs roughly 25 to 50 times the insulation it needs when swimming at high speed in tropical waters (Kanwisher and Sundnes, 1966:399).

Gigantic differences in the ability of cetaceans to keep warm result from differences in body size and in thickness of blubber. The biggest whale is 10,000 times as heavy as the smallest porpoise, has roughly a 10 times more advantageous mass-to-surface-area ratio with regard to heat retention, and has a much thicker shell of blubber. Because of these differences, the whale has approximately a 100-fold advantage over the small porpoise in its ability to keep warm. The very factors working in favor of heat retention in the large cetaceans, however, are obviously disadvantageous under conditions of great activity or warm waters. Because of the vast bulk of these animals, dissipation of heat is an acute problem. As an example of the slowness of heat dissipation in large cetaceans, the deep-muscle temperature of a dead and eviscerated rorqual *(Balaenoptera)* dropped only 1°C in 28 hours (Kanwisher and Leivestad, 1957). Clearly, cetaceans must have considerable "thermal versatility." How is this versatility achieved?

As is usual in considerations of biological problems, no single answer is appropriate, nor has sufficient research been done on the problem to suggest even most of the probable answers. Although much remains to be learned, several points seem well established. First, metabolic rates of cetaceans differ markedly from species to species. The small porpoises have much higher basal metabolic rates than do large whales, far higher, in fact, than what would be predicted on the basis of weight. The harbor porpoise *(Phocoena phocoena),* for example, metabolizes at about 1.6 times the predicted rate.

Second, blood flow through the well-developed vascular system in the flippers, dorsal fin, and flukes of cetaceans allows these structures to function effectively as heat dissipators under conditions of heat stress. The flow can apparently be shut down during cold stress, allowing for a minimum of heat loss from these surfaces.

Third, a remarkable series of vascular specializations allows for great variations in the thermal resistance offered by the blubber. A system of countercurrent heat exchange in the vascular network supplying the blubber minimizes heat loss to the blubber and skin, and hence to the environment. In cetaceans, a second venous system in the blubber bypasses the countercurrent system during heat stress and allows considerable heat loss to the environment when heat dissipation is of prime importance. Similar countercurrent and bypass systems occur in the flippers and fins. The extremities and much of the surface of the body can thus dissipate heat or can be maintained under an altered vascular supply that provides for maximal heat retention.

The longitudinal folds of blubber on the throat of rorquals *(Balaenoptera)* may function partly as a cooling device (Gilmore, 1961) by providing increased surface area for heat dissipation. The highly vascularized skin at the bottom of these grooves can be exposed to the water. Morrison (1962) found that these grooved anterior parts of the humpback whale *(Megaptera)* were slightly cooler than other parts of the body, suggesting their importance in heat dissipation.

The great quantities of blubber on large whales (up to 20 centimeters thick) are seemingly not primarily useful as insulation. Because of their size, these animals could probably maintain a constant deep-body temperature, without increased heat production, with much less insulation. These fat deposits may be useful primarily as food stores that can support an animal during periods of migration and fasting. The consumption of only half of a whale's blubber could support the animal at a basal metabolic rate for from four to six months (Parry, 1949).

Coping with Heat

Some of the most severe problems in thermoregulation are those faced by mammals living in hot regions. In many low-latitude deserts, daytime surface and air temperatures in the summer rise well above the body temperature of most mammals. Under such conditions, heat from the environment is absorbed while at the same time the animals are producing considerable metabolic heat. In order to maintain thermal homeostasis, these animals must avoid as much as possible the absorption of heat from the enviornment, dissipate such heat as is absorbed, and lose endogenous heat. Unless the body temperature is elevated, these heat transfers must occur against a thermal gradient, from the relatively cool animal to the relatively hot environment. Such heat transfers invariably involve evaporative cooling, a luxury that most desert organisms cannot afford since they live in a region where water is in critically short supply. Nonetheless, even extremely hot and arid deserts are occupied by mammals, and some kinds, notably rodents, are quite common in such areas. A variety of physiological, anatomical, and behavioral adaptations have allowed mammals to inhabit these seemingly inhospitable regions.

Avoidance of High Temperatures.   Most desert animals are never subjected to extremely high diurnal temperatures, nor are they able to survive them; their success is based on the ability to avoid extremely high temperatures rather than to cope with them. Perhaps the saving grace of the desert is the typically great daily and seasonal fluctuation in temperature. Temperatures frequently drop markedly at night, and winters are usually cool or cold. As a result, soil temperatures well below the surface are never high (Fig. 22-5), even in the summer, and to this refuge of coolness and relatively high humidity nearly all desert rodents retreat during the day. All but a very few desert rodents are strictly nocturnal, and all are more or less fossorial; these mammals are active above ground in the part of the circadian cycle when temperatures are lowest. The studies of McNab (1966) and MacMillen and Lee (1970) suggest that most fossorial and nocturnal desert rodents have metabolic rates below those predicted on the basis of body size. The low rates are associated with lowered metabolic heat production while the animals are in the humid burrows during the daytime summer heat. This lowered heat production probably obviates the use of wasteful (in terms of water loss) evaporative cooling in order to dissipate heat while the rodents are below ground.

Most desert carnivores are also largely nocturnal, some are fossorial, and all seek shelter during the hottest part of the day. Some, like the fennec *(Fennecus zerda)* of the intensely hot North African deserts have become physiologically adapted to the heat they must face in spite of being nocturnal. The foxlike fennec is small (1.1 kilograms) and has a very low metabolic rate (Table 22–3) and heart rate for its size. When inactive during the day in its relatively cool burrow, its heart rate and temperature drop, as does, presumably, its rate of heat production.

Various other means of avoiding daytime heat are used by mammals of the deserts of the southwestern United States. During the day, the white-throated woodrat *(Neotoma albigula)* remains in burrows insulated by piles of sticks and other debris. Frequently these houses are

Figure 22–6. A jackrabbit *(Lepus californicus)* on a hot summer day in the Kofa Mountains of Arizona. Note the dilation of the blood vessels in the ears.

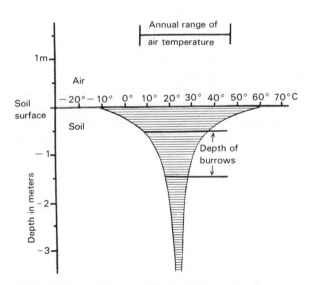

Figure 22–5. Annual range of subsoil temperatures in a desert in Arizona. Note that most rodent burrows are at depths at which heat or cold stress is never encountered. (After Misonne, 1959)

built in the shade of low-growing vegetation or under mesquite. In the burrow beneath a woodrat house near Yuma, Arizona, the temperature fluctuated only 2.5°C throughout a 24-hour cycle and never exceeded 34.4°C; temperatures outside the house varied 13.7°C and reached 41.5°C (J. H. Brown, 1968:24). Jackrabbits *(Lepus* spp.) stay in "forms" in the shade of bushes during the day. Steady or pulsating vasodilation of vessels in the huge ears of *L. californicus* (Fig. 22–6) allows for rapid dissipation of heat when the ambient temperature is below body temperature (R. W. Hill and Veghte, 1976). Bighorn sheep *(Ovis canadensis)* and javelinas *(Tayassu tajacu)* often take shelter in rock grottos or in the shade of steep rock outcrops, where for much of the day their body temperature is above air temperature and they can dissipate heat.

Terrestrial Thermoregulation in Pinnipeds. As might be expected, the very adaptations that enable a seal or sea lion to reduce heat dissipation in cold water make the animals unable to stand high temperatures. Whittow (1974) and his coworkers at the University of Hawaii studied the terrestrial thermal budget of California sea lions *(Zalophus californianus),* which inhabit some coasts of Mexico and the Galapagos Islands, where high temperatures occur regularly. At air temperatures of about 30°C, sea lions were unable to maintain a constant body temperature after the sea water had evaporated from their body surfaces. With continued exposure under experimental conditions, body temperature rose to slightly over 40°C. When the sea lions slept, their heat production dropped 24 percent, an obvious advantage for energy conservation and temperature regulation, but they were unable to dissipate sufficient heat in direct sunlight to avoid heat stress. Under stress, the animals fanned their flippers, which are known to sweat, thus increasing evaporative cooling. They also urinated and wet the underside of the body, thus further increasing evaporation. Under experimental conditions, however, these behaviors were inadequate and the animals were increasingly hyperthermic. Only when they were able to wet their bodies in the sea did body temperatures drop and stabilize.

Terrestrial heat production by a sea lion is dissipated approximately as follows: 2 percent is lost by respiratory evaporative cooling, 12 percent by evaporation from the skin, 52 percent by nonevaporative heat loss (conduction and convection) from the skin, and 15 percent by conduction from the parts of the body against the sand. Nineteen percent of the metabolic heat is stored, leading eventually to an elevation of body temperature. It is obvious why sea lions and other pinnipeds have difficulty staying out of the water for long periods on a warm day, and increased activity on land at night is understandable. Although their physiological makeup limits the amount of time they can spend hauled out on a warm or hot day, by choosing the windy side of an island and by basking at sites where spray from breaking waves repeatedly wets them and increases evaporative cooling, sea lions can considerably extend their resting time on land.

Mammals Unable to Avoid High Temperatures.   Some mammals have evolved physiological, anatomical, and behavioral strategies for tolerating long exposure to air temperatures higher than body temperatures. This section deals with these adaptations.

Large size itself is advantageous to mammals that must tolerate high temperatures. Because of the weight-to-surface-area ratio discussed earlier, the larger the animal, the greater its ability to withstand exposure to high temperatures due to a relatively reduced surface area for heat gain. Stated differently, large animals have greater thermal inertia than do small ones. Of additional importance, just as insulation in the form of thick pelage slows the loss of body heat in low ambient temperatures, fur slows the penetration of heat to the body surface when temperatures are high.

Studies of temperature regulation in the camel by K. Schmidt-Nielsen (1959) revealed a carefully regulated and highly adaptive circadian cycle of changes in body temperature. Camels in the Sahara in winter, when cool temperatures (from roughly 0 to 20°C) prevailed, had fairly constant body temperatures that varied between 36 and 38°C. The fluctuations in body temperature were not random, but followed the same pattern day after day, regardless of weather. In the summer, variations in body temperature were considerably greater; generally body temperature was between 34 and 35°C in the morning and reached a peak of approximately 40°C late in the day (Fig. 22–7). The camels were seemingly able to regulate their temperature but did so only above or below these extremes; when body temperatures reached 40.7°C, evaporative cooling in the form of sweating was used to dissipate heat, and body temperatures never exceeded 40.7°C. Thus, the camel accepted a heat load during the day that

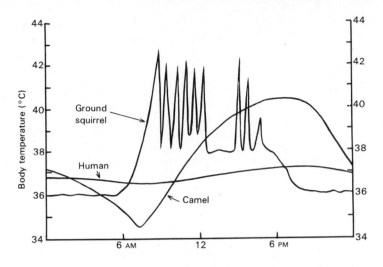

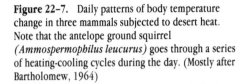

Figure 22-7. Daily patterns of body temperature change in three mammals subjected to desert heat. Note that the antelope ground squirrel *(Ammospermophilus leucurus)* goes through a series of heating-cooling cycles during the day. (Mostly after Bartholomew, 1964)

sharply elevated its temperature, but during the relative coolness of the desert night the heat stored during the day was passively dissipated and the body temperature was allowed to drop some (perhaps according to Schmidt-Nielsen, to a critical lower limit below which metabolism is disrupted). K. Schmidt-Nielsen (1964:44) estimated that it would require the expenditure of 5 liters of water for a camel to dissipate by evaporative cooling the heat load accepted during a hot day. In an animal that does not have frequent access to water, such daily water loss would lead to fairly rapid dehydration. An additional advantage of high body temperature during the day results from the narrowing of the gap between environmental and body temperature; the smaller this temperature differential, the lower the rate of heat flow from the environment to the body.

As an adaptation to intense heat, similar patterns of temperature fluctuation occur in the oryx *(Oryx),* the eland *(Taurotragus),* and the gazelle *(Gazella),* three African antelope that occur in desert or savanna areas. The African elephant *(Loxodonta africana)* also accepts a heat load in the day and then dissipates it at night (Hiley, 1975). The oryx frequently inhabits areas where no shade is available, and its

ability to withstand a diurnal heat load is exceptional. Under experimental conditions, the oryx could withstand exposure to an ambient temperature of 45°C for 12 hours (C. R. Taylor, 1969a:91). During this period, the body temperature rose above 45°C and was sustained at this level for up to eight hours with no injury to the animal. Hence, rather than gaining heat from the environment the oryx was actually losing heat. Such high body temperatures would kill most mammals fairly quickly, but circulatory specializations apparently allow the oryx to survive such extreme "overheating."

In the oryx, in gazelles, and probably in many other antelope, the brain, seemingly the most heat-sensitive organ, is provided in the three antelope named above with a specialized countercurrent cooling system of its own in the sinus cavernosus (C. R. Taylor, 1969a:92). The external carotid artery, on its way to the brain, divides into many branches in this sinus, and these branches are in close proximity to veins returning from the nasal passages (Fig. 22-8). This system is called a carotid rete. These veins carry relatively cool blood because evaporative cooling of the nasal mucosa cools the blood supplying these surfaces. Countercurrent heat exchange in the carotid rete assures that the

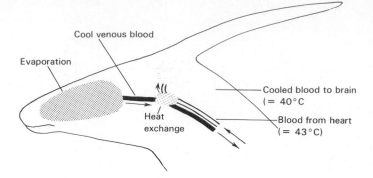

Figure 22-8. Countercurrent cooling system in the brain of a gazelle as proposed by C. R. Taylor and Lyman (1972).

blood supply of the brain is cooler than that of most of the rest of the body (Fig. 22–9). A similar countercurrent system involving a carotid rete occurs in the domestic sheep, goat, cat, and dog and is probably of much wider occurrence.

The reactions of ungulates to high temperatures may depend in part on their physiological condition: dehydrated animals typically respond differently to heat stress than do animals with unlimited access to water. This is shown well by the tiny African dik-diks (*Madoqua*

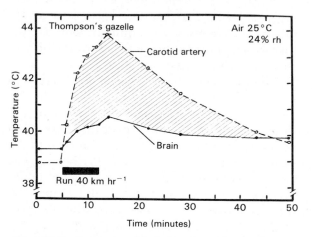

Figure 22-9. Temperature of the brain and carotid artery of a Thompson's gazelle *(Gazella thompsonii)* before, during, and after a run of 7 min at 40 km/h. The temperature of the carotid artery rose precipitously after the gazelle began to run and exceeded the brain temperature until 40 min after the run. (After C. R. Taylor and Lyman, 1972)

kirki and *M. guentheri*) studied by Maloiy (1973). At temperatures between 20 and 45°C, patterns of change in body temperature, respiration rate, and cutaneous evaporation differed between dehydrated and hydrated animals, with differences between the changes in body temperature being most striking (Fig. 22–10). Until the ambient temperature became extremely high, about 45°C, the dehydrated dik-dik's body temperature was above ambient temperature and the animal could lose some heat to its environment. At high temperatures, evaporative cooling, largely through panting, is the main route of heat loss. During panting at temperatures above the thermal neutral zone, the metabolic rate of the dik-dik drops to a level below that of thermoneutrality, thereby reducing heat production and increasing evaporative cooling (Maskrey and Hoppe, 1979). This lowered rate is probably due to a redistribution of blood similar to that in the domestic sheep. Many of the internal organs of a panting sheep receive a reduced blood supply, but blood flow to the diaphragm increases by a factor of 3 or 4 (Hales, 1973). (Thermoregulation in relation to water economy in dik-diks is discussed on p. 481.) A dehydrated oryx reduces its metabolic rate at high temperatures sufficiently to reduce evaporative water loss to 17 percent below that of an individual with free access to water.

Tolerance to high body temperatures may require specializations (as yet unknown) of var-

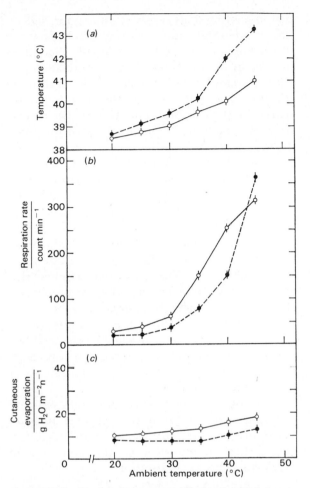

Figure 22–10. Physiological responses of hydrated and dehydrated dik-diks *(Madoqua kirki)* to different temperatures: (A) rectal temperature, (B) respiratory rate, (C) cutaneous evaporation. (From Maloiy, 1973)

ample, indicated a tolerance of an ambient temperature of 43°C for at least seven hours (K. Robinson and Lee, 1941). Panting utilizes evaporative cooling of the mouth, tongue, and probably most important, the nasal mucosa (K. Schmidt-Nielsen, Bretz, and Taylor, 1970). In the dog, and in many other mammals with a long snout and excellent olfactory ability, the turbinal bones of the nasal cavity are intricately rolled and provide a large surface area for olfactory epithelium in the nasal mucosa. This moist surface is ideal for evaporative dissipation of heat. The tongue is probably also important as a site of heat loss during panting, for the blood flow to it increases sharply at onset of panting and during heat stress increases six times over normal.

The resting respiratory rate of a dog is roughly 30 per minute, but this rate rises abruptly, with virtually no intermediate rate, to over 300 per minute during panting. The lateral nasal glands, which open some 2 centimeters inside the opening of each nostril, supply a major share of the water used in evaporative cooling during panting in dogs (Blatt, Taylor, and Habal, 1972). Under experimental conditions, the rate of secretion of one of these glands in a dog rose from no secretion at 10°C to 9.6 grams per hour at 50°C. Between 20 and 40 percent of the evaporative cooling during panting at high temperatures results from evaporation of the fluid from these glands. Because the glands are situated anterior to the turbinals, they tend to keep the nasal mucosa moist when air is drawn rapidly in through the nostrils during panting.

This system has some advantages over evaporative cooling by sweating. There is little loss of salt during panting, whereas salt loss during sweating (except probably in donkeys and camels) is always appreciable. In addition, adequate ventilation of evaporative surfaces always occurs during panting; in cool, still air, however, sweating is seemingly not equally efficient. One disadvantage of panting is that the increased activity increases metabolism, thereby contribut-

ious enzyme systems, such as those of the muscles. Many of the physiological reactions known to occur in animals not subject to heat stress do not proceed properly at elevated temperatures; also, many enzymes are unstable or denatured at high temperatures.

Typical panting, which involves rapid and shallow respiration, is used entirely for heat dissipation and is an effective aid to temperature regulation. Laboratory studies of dogs, for ex-ing more heat to be dissipated. Studies of respiratory frequency in dogs (Crawford, 1962) indicated that these animals pant at the resonant frequency of oscillation of the diaphragm (the natural frequency of vibration of this structure) and may therefore economize on energy output. Considering water loss relative to total body surface area of a mammal, the amounts of water lost in sweating and panting are probably similar. Both panting and sweating are obviously not effective means of cooling in high humidities.

Exertion and Heat Stress. During heavy exercise, mammals produce heat much faster than it can be dissipated. Indeed, for moderate-size mammals (5 to 200 kilograms), the most intense thermal stress encountered is during heavy exertion. C. R. Taylor (1974) calculated that the rate of excess heat production in the domestic dog during heavy exercise is ten times the highest possible heat gain the dog could face in the hottest desert. The rise in deep-body temperature during exertion in mammals is far more rapid than that in resting mammals in desert heat. Because the mammalian brain begins to function abnormally at temperatures only 4 to 5°C above resting body temperature, shielding the brain from high temperatures during exercise is vital.

Measurements of brain temperature during exertion have been made for two mammals, the domestic dog *(Canis familiaris)* and Thompson's gazelle *(Gazella thompsonii)*. During rapid running, this gazelle is able to keep its brain at a temperature 2.7°C below that of the blood in the carotid artery. During such exertion, Thompson's gazelle produces heat at 40 times the resting rate, but the brain is protected by the carotid rete (p. 460). With the rapid panting associated with exercise, a high rate of heat exchange in the carotid rete is maintained. Although the carotid rete in the dog is rudimentary, it is a highly effective heat exchanger during heavy exercise because of the acceleration of respiratory evaporative cooling (M. A. Baker and Chapman, 1977). In exercising dogs, respiratory evaporative cooling is roughly double that in panting dogs at rest (C. R. Taylor, 1981). This is seemingly due to a change in the pattern of breathing during exercise: at this time, dogs breathe in and out through the mouth and nose, whereas panting dogs at rest breathe in through the nose and out through the mouth. Future studies will probably demonstrate that carotid retia in a number of mammals cool the brain during exercise.

Rabbits lack the carotid rete, but their large ears (Fig. 22–6) dissipate heat. Most of the excess metabolic heat generated during exercise can be lost by the ears of the black-tail jackrabbit *(Lepus californicus)*. After exercise, the arteries in the pinnae are dilated and the pinna temperature is near 30°C (R. W. Hill, Christian, and Veghte, 1980). The experiments that yielded these data were run at air temperatures (below 20°C) at which dissipation of heat from the relatively warm ear to the cooler air would be rapid. The extreme reluctance of jackrabbits to run far on a summer day in the desert suggests that this cooling system has its limitations at high ambient temperatures.

During exercise, many mammals sweat and dissipate heat by evaporative cooling. Because sweating causes rapid water loss and is ineffective at high humidities, its use among wild mammals is probably limited.

Patterns of Thermoregulation

Fossorial Mammals. A number of families of rodents (Ctenomyidae, Octodontidae, Bathyergidae, Muridae Rhizomyidae, Spalacidae, Geomyidae) have members that are fossorial herbivores. These rodents are convergent in general body form and to some extent in style of life. Fossorial rodents are protected from surface environmental extremes and from most predators, and in some environments they have abundant below-ground food. All burrowers face one major problem, however: burrowing takes a great deal of energy. Vleck (1979) estimates that burrowing a given distance consumes 360 to 3400 times as much energy as moving the same distance on the surface. This outlay of energy produces metabolic heat, but closed and humid burrow systems provide poor conditions for evaporative or convective cooling. Fossorial rodents have low metabolic rates; according to McNab (1966), this is a means of minimizing heat stress.

An extreme and fascinating example of adaptation to fossorial life in a harsh environment is offered by the naked mole rat *(Heterocephalus glaber;* Bathyergidae).

This animal (Fig. 16–23) is highly social (Jarvis, 1978, 1981; Jarvis and Sale, 1971), living in colonies of 39 or more individuals in semidesert parts of East Africa. The workers (p. 417) average 32 grams in body weight. These rodents have a high rate of thermal conductance (because they are naked), a basal metabolic rate

less than 60 percent of the expected rate, and a low and labile body temperature. They have the poorest thermoregulatory ability of any known mammal. The narrow thermal neutral zone (31 to 34°C) is within the range of burrow temperatures recorded by Jarvis during the dry season. As shown in Fig. 22–11, the metabolic rate of mole rats increases, but only to modest levels, in response to temperatures below 30°C; the decline in the body temperature and metabolic rate of individual mole rats resting at ambient temperatures between 25 and 20°C indicates that these animals had given up their attempt at physiological thermoregulation.

Why did the naked mole rat evolve such an unusual thermoregulatory pattern? McNab (1966) hypothesized that it is of importance primarily in reducing the probability of overheating during burrowing in hot, humid bur-

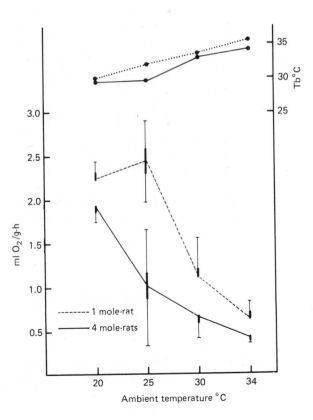

Figure 22-11. Differences in metabolic rates and body temperatures of naked mole rats *(Heterocephalus glaber),* resting alone or huddling together in groups of four, at different ambient temperatures. (After Jarvis, 1978, vol. 6, Bulletin of the Carnegie Museum of Natural History)

rows. Jarvis (1978), however, suggests that this animal's unique life-history strategy evolved under selective pressures arising from its harsh environment, which is characterized by high temperatures and limited food. She views the low metabolic rate and high thermal conductance rate as prerequisites for energy-saving behavioral thermoregulation. When heat-stressed, a mole rat can passively and rapidly unload heat from its naked body by moving to a cool section of the burrow and can become warm passively by seeking a warmer section. When resting, it can reduce heat dissipation and the cost of maintaining a preferred body temperature by huddling with other individuals (Fig. 22–11). The naked mole rat's behavior and physiological peculiarities, it would seem, are part of an integrated strategy furthering survival under demanding environmental conditions.

In fossorial mammals other than rodents, several typical adaptations are reduction of body mass, low basal metabolic rate, and increased thermal conductance (McNab, 1979). In addition, a low body temperature occurs among some mammals, such as armadillos (McNab, 1980) and hedgehogs (Shkolnik and Schmidt-Nielsen, 1976), that seek refuge in burrows. Vleck (1981) points out that the cost of burrowing increases with body size but the benefits do not. Because foraging by burrowing becomes less economical as body size increases, most mammals that use borrowing to find food are small.

Ground Squirrels. Hudson, Deavers, and Bradley (1972) compared thermoregulation in nine species of ground squirrels *(Spermophilus)* and found these animals to be physiologically remarkable in several ways. They have unusually low metabolic rates, down to 50 percent below that expected for mammals of their size, and, as an odd accompaniment, their thermal conductance can be up to 50 percent greater than expected. In other words, ground squirrels dissipate heat especially rapidly while producing heat at an unexpectedly low rate.

The maintenance of thermal stability by ground squirrels seems to depend largely on (1) their remarkable control over the rate of cutaneous heat loss, (2) their ability to raise their heat production sharply, and (3) their capacity to dissipate heat at high temperatures by evaporative cooling. Hudson and his coworkers found that, in the laboratory, an animal's thermal conductance and metabolic rate typically changed markedly in response to changes in ambient temperature. Ground squirrels clearly have a high level of vasomotor control of the skin: they can increase cutaneous blood flow when heat dissipation is necessary and decrease it when heat conservation is of advantage. A delicate balance between metabolic rate and conductance thus maintains a fairly constant body temperature, while at very high temperatures panting occurs (see Table 22–4 for changes in breathing rates) and heat is dissipated rapidly via both the respiratory tract and the skin.

Hudson, Deavers, and Bradley (1972) present an interesting and reasonable hypothesis: the high level of control of physical temperature regulation and the low basal metabolic rate of ground squirrels evolved as adaptations to hibernation, a strategy requiring rapid cooling and warming of the body. These same physiological abilities preadapted ground squirrels to life in seasonally hot areas by allowing the dissipation of heat when there is little disparity between body temperature and ambient temperature. Probably only minor quantitative changes in physiological performance fitted some species for the very high temperatures of deserts.

Especially remarkable in their ability to be active in the searing midday heat are the desert ground squirrels (Fig. 22–12). The physiological ecology of an antelope ground squirrel *(Ammospermophilus leucurus)* of the southwestern deserts of the United States is especially well known. This small (90 grams) squirrel is active even when temperatures in the sun reach 70°C and temperatures in the shade are above 41°C (Chappell and Bartholomew, 1981a). The thermal neutral zone of this animal is high (33

Table 22–4:   Highest Tolerated Ambient Temperatures for Ground Squirrels and Some Responses

Species Spermophilus	N	Highest Ambient Temp. Tolerated (°C)	Mean Breathing Rate at Highest Amb. Temp. Tolerated (min⁻¹)	Body Temp. (°C)	Amb. Temp. at Which 100% of Metabolic Heat is Evaporated (°C)
S. lateralis	4	<39	316	41.4	—
S. armatus	5	39	268	41.0	39.0
S. richardsoni	5	39	192	41.5	—
S. spilosoma	5	40	240	41.0	39.6
S. leucurus	7	41	—	41.3	41.0
S. townsendi	5	41	230	41.0	40.8
S. tereticaudus	7	46	—	41.2	—

From Hudson, Deavers, and Bradley, 1972.

to 41° C), and when air temperatures are within this zone, body temperature is maintained slightly above air temperature, allowing some dissipation of heat to the environment (Hudson, 1962). In intense heat, this squirrel can tolerate a body temperature of up to 43.6°C without sweating or panting. Periodically it retreats to its relatively cool burrow, presses its venter against the ground, and loses heat to the sub-strate (Fig. 22–7). Under laboratory conditions, an antelope ground squirrel was able to reduce its body temperature from 41 to 38°C within 3 minutes in a chamber with an ambient temperature of 25°C. In winter, antelope ground squirrels conserve substantial amounts of energy by brief episodes of nocturnal hypothermia, during which they allow their body temperature to drop to about 33°C (Chappell and Bartholomew, 1981b).

Marsupials.   A number of marsupials, including some of the larger kangaroos *(Macropus),* the quokka *(Setonix brachyurus),* a rabbit-size macropodid, and the Tasmanian devil *(Sarcophilus harrisii),* a raccoon-size dasyurid, are excellent temperature regulators in the face of heat stress. Panting is the primary means of evaporative cooling in marsupials (T. J. Dawson, 1973b). Under conditions of extreme heat stress, however, both sweating and licking occur. Sweating is important mostly in the large kangaroos, in which it is used in addition to panting to dissipate heat during sustained exertion (T. J. Dawson, 1973b; Dawson, Robertshaw, and Taylor, 1974). At high ambient temperatures, many marsupials resort to salivating heavily and licking the appendages. This may be an extremely effective means of heat loss, for vascular specializations of the forelimbs of the

Figure 22–12.   An antelope ground squirrel *(Ammospermophilus harrisi)* foraging on a July morning in the Kofa Mountains of Arizona, when the temperature was 40°C.

red kangaroo *(Megaleia rufa)*, and very probably those of other large kangaroos, provide for increased blood flow to the forelimbs when heat stress initiates the licking behavior (Needham, Dawson, and Hales, 1974).

An interesting example is the rat kangaroo *(Potorous tridactylus)*, studied by Hudson and Dawson (1975). Thermoregulation at high temperatures in this small and rather generalized macropodid involves a specialized system of evaporative cooling. This marsupial weighs about 1 kilogram, and its metabolic rate and body temperature (36°C) are low relative to those of eutherian mammals of similar size; thermal conductance from the well-furred body is low. (Studies by MacMillen and Nelson, 1969, and by T. J. Dawson and Hulbert, 1970, have shown that a number of marsupials have body temperatures equivalent to those of eutherians but have metabolic rates that are about two thirds those of placentals of comparable size.) During exercise, at ambient temperatures below body temperature, heat is dissipated by the rat kangaroo primarily by panting, but at ambient temperatures approaching and exceeding body temperature, the bare tail, which contributes 9.4 percent of the total surface area, is a major route for heat loss. Vasodilation in the skin of the tail allows for increased nonevaporative heat loss, and at temperatures near and above body temperature, profuse sweating of the tail, but not of the body, produces rapid evaporative cooling. Constant side-to-side movement of the tail further facilitates evaporation. The maximum rate of sweating in the tail is extremely high, amounting to 620 to 650 g/m^2/h, roughly double the highest measured rates in such eutherians as horses and cows.

Rat kangaroos are nocturnal, prefer areas of dense ground cover, and thus probably avoid intense daytime heat; therefore sweating from the tail may be of importance primarily during exertion when the animals are hopping. Because of the low metabolic rate, good insulation of the body is seemingly essential for heat conservation under cool conditions, whereas the evolu-

tion of the compensatory responses of the tail—delicate vasomotor control of the skin and profuse sweating—has provided the animal with a system for the rapid dissipation of heat at high temperatures or during exertion.

Behavioral Thermoregulation

Most discussions of temperature regulation in mammals center on physiological means of producing, conserving, and dissipating heat. Because physiological temperature regulation is extremely well developed in mammals, this is appropriate. Because this type of thermoregulation generally involves adjustments in evaporative cooling and metabolic rate, however, it is costly, both in energy and in water. As a result, selection has favored the evolution of behavioral control of body temperature as a "cheap" means of thermoregulation. While it is more or less important to many mammals, behavioral temperature regulation plays a central role in the lives of a few mammals.

Behavioral thermoregulation is a conspicuous part of the daily routine of some African rock-dwelling hyraxes. Field studies by Sale (1970) indicate the importance of behavior in adjusting to heat or coolness, and laboratory studies by Bartholomew and Rainy (1971) attest to the unusual system of body temperature regulation in the rock hyrax *(Heterohyrax brucei)*. The body temperature of normally active individuals varies from 35 to 37°C and is affected by ambient temperatures. The standard metabolic rate is some 20 percent below that expected on the basis of body weight, and the mean minimum heart rate (118 beats per minute) is 52 percent below the expected level.

Outside their nocturnal retreats, hyraxes adjust posture and location to exploit the environment to maintain an appropriate body temperature. I observed a colony of *H. brucei* on the Yatta Plateau in southern Kenya, in July 1973, and noted the following behavior. When the animals first emerge from their nocturnal re-

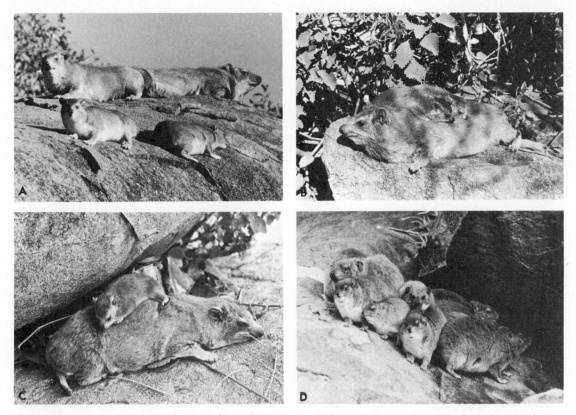

Figure 22-13. Hyraxes behaviorally thermoregulating. (A) Two *Procavia johnstoni* (above) and two *Heterohyrax brucei* (below) basking with their bodies broadside to the early morning sun. (B) A female *P. johnstoni* and her young basking in mottled light and shade in late morning as the ambient temperature is increasing. (C) A female *P. johnstoni* and her young stretched out on deeply shaded rock during the heat of the day. (D) Animals of both species huddled together on a cool day. (H. N. Hoeck)

treats in deep rock crevices, they avoid extensive contact between their ventral surfaces and the cool rock, turn broadside to the first rays of the sun, and bask while maintaining a semispherical body form, which presumably prevents excessive heat dissipation. As the air and the rock begin to warm, the hyraxes sprawl on the rock, presenting a large surface area to the sun (Fig. 22–13A). Bartholomew and Rainy (1971) found that the lowest body temperatures of hyraxes were reached shortly before sunrise, despite their huddling together during the night. The basking utilizes solar rather than metabolic heat to raise body temperature. After basking, the hyraxes generally feed.

As the temperature rises abruptly during the morning, the hyraxes often move first to the dappled shade beneath the sparse foliage of trees or bushes (Fig. 22–13B) and then, when ambient temperatures reach 30°C or above, to deep shade. During these hot times, the hyraxes often lie full length on the rock (Fig. 22–13C). At peak ambient temperatures during the afternoon, they tend to remain on deeply shaded rock. Before dark, they may again be sprawled out, but this time on warm rock in the open. Solar radiation and shaded sites seemingly provide means of passively absorbing and dissipating heat, respectively.

Of particular importance, is the fact that

the rocks on which the hyraxes live provide an auxiliary means of adjusting body temperature. The rock outcrops form massive heat sinks with vastly more thermal inertia than air. During the day, shaded or partially shaded rock heats relatively slowly and heat transfer from the ventral surfaces of the hyraxes to the rock probably allows for cooling; fully shaded rock remains cooler than the air even during the hottest part of the day and may enable the animals to unload heat passively. On the other hand, on a breezy evening just before dark, air temperatures decline rapidly but the surfaces of previously sunlit boulders are relatively warm and radiate heat absorbed during the day. The hyraxes then use the heat from the rock to compensate for body heat lost to the air. On cool, cloudy days, hyraxes huddle tightly together, thus reducing the surface area exposed to the air (Fig. 22–13D).

That hyraxes can regulate body temperature at very low metabolic cost is reflected by their casual approach to feeding. On a day when I had 12 hyraxes under observation for most of the morning, six did not feed until at least 3 hours after they had begun basking. Behavioral thermoregulation seemingly took precedence over feeding.

Hypothermia and Saving Energy

Hypothermia, the lowering of body temperature, is used by many mammals as a means of saving energy. Some use hypothermia seasonally, some use it daily, and others use both patterns. Adaptive hypothermia occurs in monotremes, some marsupials, and some members of the orders Insectivora, Chiroptera, Primates, Carnivora, and Rodentia.

Hibernation, a form of seasonal use of adaptive hypothermia, is defined by Watts, Øritsland et al. (1981) as "a specialized adaptive seasonal reduction in metabolism concurrent with the environmental pressures of food unavailability and low environmental temperatures." Although long denied status as a hibernator by

many biologists because of its modest use of hypothermia, the black bear *(Ursus americanus)* fits this definition. In many areas, this animal retires to a protected place into which it has carried insulative nesting material to lie on and remains there from October to April. A bear studied in the hibernation chamber of a laboratory used far less energy each day than an active bear (640 versus 1245 kilocalories), had a low heart rate (16 beats/minute), and an average body temperature of 35°C (Watts, Øritsland, et al., 1981). Rogers (1981) found that bears hibernating in Minnesota maintained body temperatures above 31°C, varying some 7°C from their normal temperature. A wild hibernating Alaskan black bear had a heart rate of 8 beats/minute when sleeping soundly in December, whereas the heart rate of active bears in summer is from 50 to 80 beats/minute.

A number of energy-saving physiological changes occur during adaptive hypothermia, or torpor. These include lowering of the heart rate, progressive vasoconstriction, suppression of shivering, reduction in breathing rate, and lowered oxygen consumption. These changes occur during entry into torpor and usually precede a decline in body temperature. Torpor in summer is called estivation, and winter torpor is hibernation.

A number of stimuli for hibernation have been studied, and it is clear that different animals respond to different stimuli or respond differently to the same stimulus. Some murid and heteromyid rodents become torpid rather quickly in response to low air temperatures and lack of food, but some, the golden hamster *(Mesocricetus auratus),* for example, require two to three months of cold "preparation" before becoming torpid. Mammals with circannian rhythms, such as the arctic ground squirrel *(Spermophilus parryii),* enter torpor in response to diminishing photoperiods and falling temperatures in the autumn (p. 421). Laboratory studies by Dawe and Spurrier (1969), in which the injection of blood from hibernating thirteen-lined ground squirrels *(Spermophilus tridecemlineatus)* into active animals was followed by hibernation by the latter, suggest that some "trigger substance" in the blood may initiate hibernation in squirrels.

Preparation for hibernation often (but not always) involves great increases in body weight. This gain ranges in sciurids from 80 percent of the fat-free weight in the golden-mantled ground squirrel *(Spermophilus lateralis)* to 30 percent of this weight in the yellow-pine chipmunk *(Eutamias amoenus;* Jameson and Mead, 1964).

Although small mammals can exploit small food items (such as seeds) that are usually unavailable to larger animals and have access to a nearly limitless array of retreats, small size is a liability energetically. The high metabolic rates of small mammals must be sustained by high intakes of food, and seasonal changes in food availability present severe problems. Winters in the north and dry seasons in the deserts and in many tropical areas are times of potential food shortage for small mammals, and these are also periods when temperature or lack of moisture may limit activity.

It is not surprising, therefore, that some small species have evolved means of surviving periods of food shortage and temperature stress and of using times of moderate temperatures and high food productivity to reproduce and to store food or fat. Many small (up to woodchuck size) mammals periodically conserve energy by allowing the body temperature to drop to near that of the environment. This is not a primitive feature, a manifestation of some ancestral inability to sustain a steady temperature at all times, but is instead a highly adaptive ability. An extreme example of this ability is shown by a small pocket mouse *(Perognathus parvus;* MacMillen, 1983a). When hibernating at ambient temperatures down to 2°C, this mouse maintains a body temperature about 1°C above ambient temperature. At ambient temperatures between 2 and −5°C, it increases its metabolic rate just enough to maintain its body temperature at 2°C. Thus, despite a body and brain temperature near freezing, this mouse makes precise adjustments in metabolic rate that maintain body temperature above lethal limits. Such adaptive hypothermia may well have been a factor important in furthering the success of the two largest mammalian orders, Rodentia and Chiroptera. Many small bats would be unable to forage only at night and fast throughout the day if they could not conserve energy in the day by hypothermia, and seasonally hostile areas would not be inhabited by some small rodents if these animals retained constant thermal homeostasis.

Entry into torpor by some small mammals seems to be triggered by lack of food, but in others entry is spontaneous. Several species of pocket mice and a kangaroo mouse *(Microdipodops pallidus)* enter torpor spontaneously (J. H. Brown and Bartholomew, 1969; A. R. French, 1977; Kenagy, 1973a; Meehan, 1976). When its food was rationed, the duration of torpor in the kangaroo mouse was inversely related to the number of seeds provided. Torpor in many small rodents (such as *Microdipodops,* some *Perognathus,* and some *Peromyscus*) is a circadian phenomenon: these animals are torpid by day and active and normothermic (the body temperature maintained by an active mammal) by night. In hibernators, these periods of torpor become progressively longer (up to five days), with intervening arousals. During the shallow circadian torpor of some rodents, body temperature stays above roughly 15°C, whereas during deep torpor in some rodents and some bats body temperature drops to between 1 and 5°C.

Species differ in their "critical" body temperature—the lowest body temperature they can withstand without arousing. This temperature is fairly high in some heteromyids that undergo shallow torpor (12°C in *Perognathus hispidus*) but as would be expected, low in mammals using deep torpor. Critical body temperature is close to freezing in some bats of the genus *Myotis,* 2.8°C in the golden-mantled ground squirrel, and 4°C in the European hedgehog *(Erinaceus europaeus).*

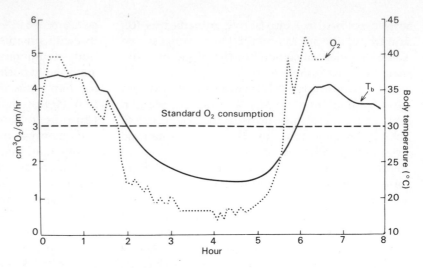

Figure 22-14. Pattern of changes in oxygen consumption and body temperature in the cactus mouse *(Peromyscus eremicus)* during entry into and arousal from torpor at an ambient temperature of 19.5°C. The standard consumption line is for active mice at an ambient temperature of 20°C. The cycle of torpor was initiated by deprivation of food and water. (After MacMillen, 1965)

Arousal from hibernation is energetically costly and is associated with the metabolism of energy-rich "brown fat" (Chaffee and Roberts, 1971) in some species and with shivering in others. The high cost of arousal suggests that the most advantageous strategy for a hibernator would involve continuous deep torpor, but surprisingly, periodic arousal during the hibernation period is the rule among those species studied. The broad pattern is one of progressively increasing periods of torpor through the early stages of hibernation and decreasing periods in the late stages. Among several rodents, the maximum periods of torpor were from 12 to 33 days, and a period of 80 days was recorded for the little brown bat *(Myotis lucifugus)*. The European hedgehog was found to be torpid 31 percent of the time at 10°C and 80 percent of the time at 4.5°C. (For an excellent review of torpor in mammals, see Hudson, 1973.)

Periods of torpor are seemingly characteristic of the life cycles of some rodents inhabiting hot regions. In the cactus mouse *(Peromyscus eremicus)*, an inhabitant of the deserts of the southwestern United States and northern Mexico, torpor may occur in both summer and winter (MacMillen, 1965). Cactus mice remain in their burrows for several weeks during the driest part of the summer, and laboratory animals entered torpor in the summer in response to a reduced food supply or, in some cases, to restricted access to water. In the winter, laboratory animals were torpid by day and active by night when their food was in short supply, and they were able to become torpid at ambient temperatures below 30°C (Fig. 22-14) but could not survive torpor bouts if body temperatures dropped below 15°C. The cactus mouse has a narrow thermal neutral zone (28 to 35°C) and a low basal metabolic rate for a mammal its size, features seemingly typical of small mammals able to enter torpor under moderate temperatures. According to MacMillen, the summer torpor in the cactus mouse is a device for reducing food and water consumption and for surviving periods of water and food shortage on the surface. The California pocket mouse *(Perognathus californicus),* an inhabitant of seasonally dry chaparral areas, undergoes a daily cycle of diurnal torpor in the laboratory when its food supply is reduced and maintains a delicate balance between food supply and metabolic economy (Tucker, 1962, 1965). Its periods of torpor are adjusted so that the shorter the food supply, the longer the daily torpor becomes. Both the cactus mouse and the California pocket mouse are adapted to moderately high-temperature torpor; neither can arouse, and both will die, when the body temperature goes below about 15°C.

Torpor in some species may strongly reduce competition for limited food between sympatric rodents during periods of food shortage, and MacMillen (1983a) regards torpor in *Peromyscus* of the southwestern United States as an important physiological response to water stress.

The Mohave ground squirrel *(Spermophilus mohavensis)* remains in its burrow from August to March, a period spanning both extremely hot and extremely cold seasons. Laboratory studies indicate that during this period the animals are intermittently torpid for periods of from several hours to several days (Bartholomew and Hudson, 1960, 1961). This squirrel is able to elevate its temperature from 20 to 30°C in from 20 to 35 minutes and even during its active period in spring and summer has an unusually variable body temperature (from 31 to 41.5°C).

Some moderate-size carnivores can avoid harsh winter conditions by entering hypothermia. Several American badgers *(Taxidea taxus)* studied in outdoor pens in Laramie, Wyoming, stayed underground for 70 consecutive winter days (Harlow, 1981). One badger entered torpor 30 times during this period. The heart rate of torpid badgers dropped from 55 to 25 beats/minute, and body temperature declined from 38 to 29°C. Harlow estimated that during torpor there was a 27 percent reduction in energy expenditure.

Temperature Regulation in Bats

Studies of a variety of bats have both clarified and complicated the picture of temperature regulation in these animals. Among different species, contrasting reactions to temperature changes occur, and within the Chiroptera most mammalian styles of temperature regulation are represented.

Seemingly the larger megachiropterans are homeotherms. Those that have been studied are able to maintain body temperature within fairly narrow limits (35 to 40°C) over a range of ambient temperatures from roughly 0 to 40°C. No diurnal torpor occurs in these bats, which are usually quite active during the day in their communal roots. Many pteropodid bats react to cold stress by shivering and by enveloping the body with the wings (Fig. 22–15), which serve as blankets that provide considerable insulation for the body (Bartholomew, Leitner, and Nelson, 1964). Many megachiropterans roost in trees where they are periodically exposed to direct sunlight.

Several devices for lowering body temperature were observed by Bartholomew and his coworkers in animals under heat stress. Vasodilation occurred in surfaces such as the scrotum, wing membranes, and ears; these naked surfaces are seemingly efficient heat dissipators. Other reactions to high temperatures were extension of the wings, fanning of the wings, and panting. Under intense heat stress, the animals salivated

Figure 22–15. An African epauletted bat (*Epomorphorus wahlbergi*, Pteropodidae) with its wing membranes enshrouding its body. (R. G. Bowker)

copiously and licked their bodies. Regular use of wing licking as an aid to evaporative cooling under natural conditions has been observed in southern Kenya by T. J. O'Shea (personal communication). On a number of hot afternoons, he observed roosting epauletted fruit bats *(Epomophorus wahlbergi)* partially spread their wings and lick the membranes, in which the engorged blood vessels could be seen clearly. These bats were hanging beneath the sparse canopy of an acacia tree. A different thermoregulatory pattern occurs in some megachiropterans; Bartholomew, Dawson, and Lasiewski (1970) have shown that some of the smaller species in New Guinea are capable of becoming torpid.

Compared with megachiropterans, microchiropteran bats are highly variable in their responses to temperature extremes. Among tropical or subtropical microchiropterans, two extreme patterns of response have been found. The Australian species *Macroderma gigas* (Megadermatidae), probably because of its large size, is able to maintain a stable body temperature in the face of ambient temperatures as low as 0°C, and many of the reactions to temperature extremes in this bat are similar to those of megachiropterans (Leitner and Nelson, 1967). At the other extreme, the Neotropical species *Desmodus rotundus,* the vampire bat, seems unable to regulate its body temperature (Lyman and Wimsatt, 1966). The inability of this species to dissipate heat causes death at ambient temperatures as low as 33°C. Nor can this species withstand low temperatures: after an initial short-lived attempt at maintaining the body temperature by increasing metabolism, the body temperature drops and body temperatures between 17 and 27°C are often lethal. Vampire bats have no ability to rewarm the body if the ambient temperature is not raised. The style of temperature regulation of most tropical bats probably lies somewhere between these extremes.

Many tropical microchiropteran bats from the Old World and from the Neotropics have similar circadian activity cycles. These bats are active at night and inactive during the day, and this activity cycle is reflected by their temperature cycle. In the Neotropical phyllostomid bats that have been studied, body temperatures are from 37 to 39°C during the night and 2 to 3°C lower during the day (Morrison and McNab, 1967). In general, these bats are able to maintain a high body temperature despite moderately low ambient temperatures; only one species cannot sustain its normal body temperature at ambient temperatures below 12°C. Some Old World members of the families Rhinopomatidae, Emballonuridae, Megadermatidae, and Rhinolophidae have responses to temperature extremes similar to those of phyllostomids (Kulzer, 1965). Broadly speaking, though, cold stress can usually be tolerated by tropical bats for only fairly short periods, after which the body temperature falls uncontrollably. Body temperatures below 20°C are often fatal.

The Molossidae, a widespread but largely tropical and subtropical group, seem to have a pattern of temperature regulation intermediate between that of the tropical families and that typical of microchiropterans inhabiting temperate areas. The strictly tropical molossids have some circadian variation in body temperature but cannot cope with low temperatures and die when the body temperature goes below 20°C. Molossids that inhabit temperate areas, however, can tolerate body temperatures as low as 10°C. In southern California in the winter, the western mastiff bat *(Eumops perotis)* becomes torpid during the day, when its body temperature drops to within 1 to 2°C of the ambient temperature at temperatures from 9 to 28°C. The metabolic rate is spontaneously elevated in the afternoon and evening, and the bats are active at night (Leitner, 1966).

Adaptive hypothermia, often involving (at different seasons) both daily torpor and hibernation, occurs in many vespertilionids of north temperate areas and seems to be the key to the survival of some species in cool or cold regions. During the summer, some microchiropterans of temperate zones undergo daily torpor *("Ta-*

gesschlaflethargie''; Eisentraut, 1934) at low ambient temperatures. Tremendous metabolic savings are realized by microchiropteran bats that become hypothermic at low temperatures. Under experimental conditions, the average metabolic rate of six *Myotis lucifugus* kept at an ambient temperature of 35°C was 33 times that of these bats when kept at 5°C (Henshaw, 1970:201). In addition, strikingly abrupt rises in metabolic rate occur during flight. The metabolic rate of the phyllostomid bat *Phyllostomus hastatus* was found by S. P. Thomas and Suthers (1972) to be about 30 times greater in a flying individual than in one resting at a body temperature of 36.5°C. It is obvious that, by evolving foraging strategies that allow the briefest possible periods of flight and, in bats capable of adaptive hyothermia, by lengthening the daily periods of hypothermia, bats in cool areas can balance energy input and output even under conditions that demand minimal energy turnover.

Some microchiropterans, such as the pallid bat *(Antrozous pallidus),* are able to roost in summer in situations allowing the utmost energy savings. In many hot desert and semidesert areas, pallid bats roost in deep crevices in cliffs, and these cliffs function as tremendous heat sinks. In one area in Arizona where these bats were studied, temperatures deep within crevices remained close to 30°C throughout the circadian cycles on summer days. The temperature gradient in several crevices late on a July afternoon was from 31° deep in the crevices to 46° at their entrances. By adjusting their position in the thermal gradient of a crevice, the bats were able to unload heat to the substrate after nocturnal activity and lower body temperature early in the day, maintain a low body temperature and low metabolic rate until late afternoon, and passively raise the body temperature prior to emergence by moving toward the relatively hot mouth of the crevice (Vaughan and O'Shea, 1976). In early spring, when days and nights were cool, the pallid bats in the study area preferred to roost in shallow, south-facing poorly insulated crevices, in which temperatures often rose from about 10°C in the early morning to 32°C just before dark. Here at sundown the bats could passively attain a body temperature high enough to allow flight.

Some bats that are homeothermic in summer abandon this pattern well before winter. Fat deposition is known to occur in the late summer or early fall in some species of vespertilionids that hibernate (W. W. Baker, Marshall, and Baker, 1968; Ewing, Studier, and O'Farrell, 1970; Krzanowski, 1961; Weber and Findley, 1970), and three species of *Myotis* studied by M. J. O'Farrell and Studier (1970) became hypothermic during this time. In *M. thysanodes,* the metabolic rate for homeothermic individuals at an ambient temperature of 20.5°C is 6.93 cm^3 O$_2$/g/h; the rate drops to 0.59 cm^3 O$_2$/g/h in hypothermic bats. This drop results in the saving of 2.81 kilocalories per day as a bat becomes hypothermic. Fat is deposited in preparation for hibernation at the rate of 0.17 grams per day in the period of maximum fat accumulation, which requires 1.60 kilocalories per day. This energy is available primarily because of the late summer-autumn shift to daily hypothermia (Ewing, Studier, and O'Farrell, 1970; Krzanowski, 1961). Some birds accumulate fat in preparation for migration by greatly increasing food intake, but this increased intake does not occur during the period of fat deposition in *M. lucifugus* and *M. thysanodes.*

Winter hibernation in bats differs from short-term torpor largely in the length of dormancy and in the levels to which the metabolic rate and temperature drop. The duration of hibernation for bats differs widely among species and within a species, depending on the area. In the northeastern United States, *M. lucifugus* remains in hibernation for six or seven months, from September or October to April or May (W. H. Davis and Hitchcock, 1965). Periods of hibernation for bats in warmer areas are probably considerably shorter. Ewing, Studier, and O'Farrell (1970) used amount of fat accumulated by bats in the autumn as a basis for estimating duration of hibernation and found the following for several New Mexican bats: *M. lucifugus,* 165 days; *M. yumanensis,* 192 days; *M. thysanodes,* 163 days. Because no allowance was made for the metabolic drain occasioned by intermittent periods of activity, these estimates are probably too high. Hock (1951) estimated that the metabolic rate of hibernating bats at ambient temperatures not much above freezing was 0.1 cm^3 O_2/g/h, which is 257 to 385 times lower than the rate for a flying bat estimated by Studier and Howell (1969). At ambient temperatures near 5°C, bats in deep hibernation maintain a body temperature about 1°C above ambient temperature. These bats are responsive to certain stimuli and will begin arousal when handled or when subjected to unusual air movement. As a defense against freezing to death, bats spontaneously raise the metabolic rate at dangerously low ambient temperatures (below roughly 5°C) and either arouse fully or regulate body temperature and remain in hibernation.

Energy Costs of Locomotion: Advantages of Large Size

The scaling of size in mammals is a fascinating but complex topic. What are the advantages gained by small or large size? At this point, let us consider size in relation to the energy costs of locomotion. A 10-kilogram coyote and a 400-kilogram horse have similar top speeds, but at such speeds how do these animals compare with regard to the expenditure of energy?

Some time ago, A. V. Hill (1950) made a series of predictions as to how energy use during running would change with the size of the runner. (Energy consumption by the muscles of a running animal is generally regarded as the result of the transformation of chemical energy to mechanical energy.) Hill reasoned that, although large and small runners could often reach similar top speeds, the rates of work and energy use at these speeds would be higher in the small runners. His logic was that, whether an animal is small or large, each gram of muscle performs the same amount of work and consumes the same amount of energy during a stride, but the short legs of the small animal have to take many strides to cover the same distance covered in one stride by a large animal. When large and small mammals run at the same speed, then, the small ones should have the higher stride rates and should consume more energy per unit of body weight.

These proposed relationships have been studied experimentally by C. R. Taylor and his associates (Fedak, Heglund, and Taylor, 1982; Heglund, Fedak, et al., 1982; Taylor, Heglund, and Maloiy, 1982; Taylor, Schmidt-Nielsen, and Raab, 1970), who used mammals ranging in weight from 21 grams (a house mouse, *Mus musculus*) to 254 kilograms (African cattle, *Bos indicus*).

Table 22–5:   Speed, Stride Frequency, and Metabolic Energy Consumer at the Trot-Gallop Transition by Three Size Classes of Animals

Body Mass (kg)	Speed at Trot-Gallop Transition (m/s)	Stride Frequency at Trot-Gallop Transition (Stride/s)	Energy Used (J/kg/stride)
0.01	0.51	8.54	5.59
1.0	1.53	4.48	5.00
100	4.61	2.35	5.53

Values calculated by C. R. Taylor, Heglund, and Maloiy (1982) from equations given by N. C. Heglund, Taylor, and McMahon (1974).
From C. R. Taylor, Heglund, and Maloiy, 1982.

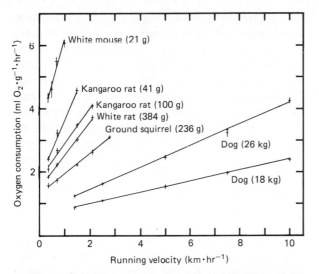

Figure 22-16. Oxygen consumption by mammals of different sizes at various running speeds. (From C. R. Taylor, Schmidt-Nielsen, and Raab, 1970, p. 1105, by permission of the American Journal of Physiology)

force is generated and dissipated more rapidly as an animal runs faster. With increasing speed, more muscle fibers that have a rapid contraction-relaxation cycle are brought into play. Each cycle uses a unit of energy, and the increase in the cost of rapid locomotion is perhaps due to the increased use of rapid-cycling muscle fibers.

Muscular force must be generated and dissipated more rapidly in small mammals than in larger ones because, at comparable speeds, small mammals have higher stride rates. The muscles of small mammals contain "faster" fibers that have rapid contraction-relaxation cy-

Several important relationships were demonstrated by these studies:

1. At the trot-gallop transition speed, the amount of energy used per stride per gram of muscle is nearly constant over a wide range of body sizes (Table 22–5). Mouse, baboon, and horse all expend nearly the same amount of energy per gram of muscle during a stride.

2. The metabolic cost of muscle action in running animals increases linearly with speed. As shown in Fig. 22–16, the amount of energy a mammal expends increases as running speed increases.

3. The mass-specific (per gram) use of energy by a running animal decreases as a function of weight (Fig. 22–17), varying as the -0.3 power of body mass. Thus, when a chipmunk and a horse are running at the same speed, each gram of chipmunk uses 15 times more energy than each gram of horse.

Some probable explanations of these relationships are available. Seemingly, muscular

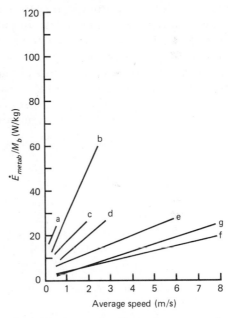

Figure 22-17. Mass-specific use of energy (use of energy per unit of weight) at various speeds by animals of different sizes. The notation E_{metab}/M_b (W/kg) = metabolic energy consumed in watts per kilogram of body weight. Key: a, 42-g painted quail; b, 100-g chipmunk; c, 1.2-kg guinea fowl; d, 4.3-kg turkey; e, 5.0-dog; f, 103-kg ostrich; g, 107-kg horse. (A watt is a unit of power equal to about $\frac{1}{746}$ of an English horsepower.) (From Fedak, Heglund, and Taylor, 1982, p. 38, Journal of Experimental Biology)

cles and use energy at a high rate. The decrease in the use of "fast" muscle fibers accompanying increased body size may partially account for the mass-specific decrease in the cost of running in large mammals.

Fedak, Heglund, and Taylor (1982) found that, during high-speed running by large mammals, the energy expended by the muscles was not sufficient to provide the work necessary for the total kinetic energy (energy of motion) developed. The authors concluded that the storage of energy by muscles and tendons and its release by elastic recoil provide a significant part of the total kinetic energy. Whereas in large mammals elastic recoil is of considerable importance in locomotion, its contribution to locomotor efficiency in small mammals is unknown.

These relationships provide at least a partial explanation for the scaling of size in cursorial mammals. Artiodactyls, perissodactyls, and cursorial carnivores generally weigh at least 10 kilograms and thus expend less energy per gram of body weight during running than smaller mammals. The mechanical problems associated with the support and propulsion of great weight may set upper limits on the sizes of runners, but some quite heavy mammals (such as 500-kilogram horses) are rapid and enduring runners. The swiftest cursors, however, generally weigh from about 50 to 125 kilograms (antelope, for example), and the cheetah, the fastest of all runners, weighs between 50 and 65 kilograms.

Bipedal locomotion, involving leaping or bounding, has evolved independently many times in the class Mammalia. Questions as to the relative energy cost of bipedal versus quadrupedal locomotion have inspired much controversy. Some researchers have claimed that, over a considerable range of hopping speeds in bipeds (such as kangaroos), there is no increase in energy cost attending increasing speed. Recent work (such as that by S. D. Thompson, MacMillen, et al., 1980) indicates, however, that no such energy savings is associated with bipedal hopping. The present consensus is that hopping bipeds and running quadrupeds have similar weight-specific expenditures of energy. Although hopping is no less costly than running, it may be of importance in some small mammals as a means of avoiding predators.

Chapter 23

Water Regulation

Roughly 35 percent of the earth's land surface is desert, where water is the primary limiting factor for plant and animal life. These desert areas are characterized by intense daytime heat in the summer, intense solar radiation by day and maximal heat loss by night, resulting in great daily changes in temperature (commonly up to 30°C), extremely low humidity through most of the year, and small amounts of precipitation, often at irregular intervals. To an animal abroad on the desert on a summer day, the searing dry winds and the radiation and reflection of heat from the hot, pale-colored soil add to the harshness of the environment. Few equally hostile environments occur on earth, and to the casual observer the desert gives the impression of overwhelming sterility. This impression is deceptive, however, for in reality the desert supports a great variety of life. Most mammals who live on the desert remain hidden in shelters by day and forage in the relative cool of night.

The abundance of mammalian life on the desert and the severity of this environment are well described by E. R. Hall (1946:1). In the morning, "scores of burrow openings around sandy dunes attest the density of population of small mammals—a density equaled in few other habitats—and inspection discloses that, in nearly every burrow, a short distance back from the entrance the occupant has snugly packed a plug of moist sand to shut him away from the dangers of day. Before a person's curiosity is half satisfied about the burrows and the dozens of stories told by the tracks, the sun is up—and with it the wind, the wind that obliterates every telltale mark and burrow opening, leaving only smooth sand in their places. Little by little the

heat returns." Each desert mammal has evolved means for coping with the extreme conditions of this environment, none of which is more acute than the lack of water.

Water is absolutely essential to life; to all mammals, life depends on the maintenance of an internal *water balance* within fairly narrow limits. (Water balance results when intake—through drinking, eating, and production of metabolic water—equals output—through lactation, evaporation from skin and lungs, defecation, and urination.) Most mammals are under intense discomfort when water loss reduces their body weight by as little as 10 or 15 percent, and death occurs in many mammals when such loss reduces the body weight by 20 percent. Loss of water occurs rapidly on the desert: water loss in a human on a hot summer day in the southwestern deserts of the United States has been recorded as 1.41 percent of body weight per hour; comparable figures for the donkey and dog are 1.24 and 2.62, respectively (K. Schmidt-Nielsen, 1964:27). Deprived of drinking water, a human or a dog can survive only a day or two of exposure to the desert in the summer. Nonetheless, some small desert rodents are independent of drinking-water availability and must satisfy their water needs by utilizing water in their food and, more important, metabolic water. Similarly, some large mammals of desert areas must maintain water balance with only occasional access to drinking water. Although much remains to be learned about mammalian adaptations for water conservation in arid environments, excellent studies have provided a solid base of knowledge.

A number of solutions to the problem of

maintaining water balance without drinking water are used by desert mammals. These solutions depend on seasonal weather patterns, size of the animal, timing of activity cycles, diet, and a variety of behavioral, structural, and physiological features. The following discussions do not cover the subject of water conservation in mammals exhaustively but do consider the adaptations that permit some mammals to maintain water balance in dry environments.

Mammals Dependent on "Wet" Food

A number of mammals that occupy deserts or semiarid areas are no better adapted to surviving without considerable moisture in the diet than are mammals of fairly moist areas. Even in some areas with fairly high precipitation, small mammals do not have regular access to drinking water and, as in the case of some desert rodents, satisfy their water requirements by eating moist food.

Succulent plants provide water for some desert rodents, such as the white-throated woodrat (*Neotoma albigula;* Fig 23–1), which occupies the hot deserts of the southwestern United States and northern and central Mexico. Paradoxically, this rodent needs large amounts of water, which it obtains largely from cactus. MacMillen (1964a) found that the desert wood-

Figure 23–1. A white-throated woodrat *(Neotoma albigula)* in the Kofa Mountains of Arizona. This woodrat occurs widely in the deserts of the southwestern United States.

rat *(N. lepida)* and the cactus mouse *(Peromyscus eremicus)* also utilize large quantities of cactus *(Opuntia)* as a source of both food and water. These mammals have evolved the ability to cope metabolically with oxalic acid, a compound abundant in cactus and toxic to some mammals (K. Schmidt-Nielsen, 1964:146). The ability to obtain water from cacti and to deal with oxalic acid is not limited to the rodents mentioned above, all of which belong to the family Muridae, but also occurs in the rodent family Geomyidae, the pocket gophers. The northern pocket gopher *(Thomomys talpoides),* inhabiting fairly dry short-grass prairies of Colorado, obtains water by eating prickly pear cactus (Vaughan, 1967). During the dry midwinter period, this plant constituted 79 percent of the diet of these gophers in one area. Anyone who has had contact with prickly pear cacti must admit that the behavioral ability of small mammals to cope with the spiny armor of these plants is as impressive as their physiological ability to deal with oxalic acid.

Other plants provide water for rodents inhabiting arid regions. Juniper leaves *(Juniperus* spp.) are the primary source of water for Stephen's woodrat *(Neotoma stephensi)* in Arizona (Vaughan, 1982), and other woodrats include juniper in their diet. The desert ground squirrels *(Spermophilus* and *Ammospermophilus)* also obtain moisture from their food, which is largely green vegetation and insects, and such foods provide water seasonally for many other desert rodents.

Some desert rodents obtain water from succulent plants that contain high salt concentrations; these animals have kidneys that are able to produce highly concentrated urine (urine that has little water relative to the contained solutes). The North African sand rat *(Psammomys obesus,* Muridae) is such an animal. This small rodent obtains water from the fleshy leaves of halophytic plants (plants that grow in salty soil), which grow along dry river beds in the desert (K. Schmidt-Nielsen, 1964:183). These leaves are 80 to 90 percent water but contain higher concentrations of salt than seawater does and also have large amounts of oxalic acid. In order to utilize this water source, the sand rat must produce urine with extremely high concentrations of salt and must be able to metabolize large quantities of oxalic acid. An Australian hopping mouse, *Notomys cervinus* (Muridae), has a remarkably well-developed ability to concentrate salts in its urine and probably also uses the succulent but highly saline leaves of halophytic plants as a water source (MacMillen and Lee, 1969). Similarly, a South American desert-dwelling rodent, *Eligmodontia typus,* can eat the saline leaves of halophytes and maintain water balance by producing concentrated urine (Mares, 1977). The tammar wallaby is known to be able to drink seawater (Kinnear, Purohit, and Main, 1968). This ability occurs in some populations of the western harvest mouse *(Reithrodonotomys megalotis),* a rodent that commonly inhabits semiarid or arid areas and can live in salt marshes, areas regarded as "physiological deserts" because of the physiological problems of utilizing water from the highly saline sap of the plants or from seawater. Kenagy (1973b) found that the kangaroo rat *(Dipodomys microps)* uses the moist leaves of saltbush *(Atriplex canescens)* for food and avoids the salt-rich peripheral tissues of the leaves by trimming them off with the incisors.

Most deserts support a number of carnivorous and insectivorous mammals whose moisture requirements are seemingly met by the water in their food. The grasshopper mouse *(Onychomys),* a small rodent widely distributed in the deserts and semiarid sections of the western United States and Mexico, is almost exclusively insectivorous at some times of the year. This mouse has thrived in the laboratory on an entirely meat diet, with no drinking water (K. Schmidt-Nielsen, 1964:185). In the North American deserts, kit foxes *(Vulpes macrotis),* badgers *(Taxidea taxus),* and coyotes *(Canis latrans)* must also be able to derive sufficient water from their meat diet, for these animals often live in areas remote from drinking water. Schmidt-Nielsen (1964:126) found that the desert hedgehog *(Hemiechinus auritus,* an insectivore) and the fennec *(Fennecus zerda,* a fox), both inhabitants of North African deserts, could get adequate water from a predominantly carnivorous diet, as could the mulgara *(Dasycercus cristicauda),* an Australian dasyurid marsupial (K. Schmidt-Nielsen and Newsome, 1962). The fennec can maintain water balance for at least 100 days on a diet of mice and no drinking water. This small animal has an unusually low rate of evaporative water loss and equals water-independent desert rodents in its ability to concentrate urine (Noll-Banholzer, 1979b; Table 23–1), a capacity which enables it to excrete the large amounts of urea yielded by its high-protein diet.

Table 23–1: Relative Urine-Concentrating Ability of Some Mammals as Indicated by Osmotic Concentration of Urine from Dehydrated Animals

Species	Common Name	Urine Osmolality (mOsm/liter)	Diet
Dipodomys merriami	Merriam's kangaroo rat	3165	Granivore
Peromyscus crinitus	Canyon mouse	3047	Omnivore
Perognathus longimembris	Little pocket mouse	1675	Granivore
Onychomys torridus	Grasshopper mouse	2733	Insectivore
Neotoma lepida	Desert woodrat	2436	Herbivore
Lepus californicus	Black-tail jackrabbit	3600	Herbivore
Fennecus zerda	Fennec	4022	Carnivore
Canis familiaris	Dog	2608	Carnivore
Felis sylvestris	Cat	3118	Carnivore

Data from MacMillen, 1972, K.A. Nagy, Shoemaker, and Costa, 1976, and Noll-Banholzer, 1979a, 1979b.

Figure 23-2. Gemsbok *(Oryx gazella)* in the Kalahari Desert of Africa. These animals are able to go for long periods without drinking. (F. C. Eloff)

Few large ungulates inhabit barren deserts where no drinking water or cover is available. One notable exception is the oryx, or gemsbok *(Oryx;* Fig. 23–2), a large antelope that occurs in arid and semiarid sections of Africa and even penetrates the borders of the Sahara. The oryx has an amazing ability to withstand intense desert heat (p. 459); perhaps more remarkable is the animal's lack of dependence on drinking water. Careful studies by C. R. Taylor (1969a) showed that the water needs of the oryx are probably satisfied by its food, which consists of leaves of grasses and shrubs that by day may contain as little as 1 percent water. After nightfall, as the temperature drops and the humidity rises, these parched leaves absorb moisture from the air and probably contain approximately 30 percent water during much of the night (Fig. 23–3). By feeding at night, therefore, the oryx can manage a nightly intake of some 5 liters of water with its forage. This is at best a minimal amount of water for a 200-kilogram mammal living in shelterless desert and is sufficient for the oryx only because of a combination of mechanisms that favor water conservation.

Some of the major means for reducing water loss in a dehydrated oryx include (1) voluntary hyperthermia (body temperature above the normal level) and the sparing use of evapo-

rative cooling except under extreme conditions, when panting but not sweating is used; (2) reduction of evaporative water loss by lowering of the metabolic rate at high ambient temperatures during the day; (3) reduced permeability of the skin, resulting in reduced water loss by diffusion; (4) reduced respiratory rate and greater extraction of oxygen from inspired

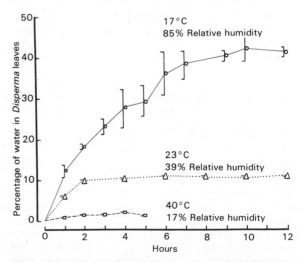

Figure 23-3. Water in *Disperma* leaves at various humidities and temperatures. The hygroscopic leaves may be an important water source for Grant's gazelles *(Gazella granti)* in some areas. (After Taylor, 1968a)

air during the night, and hence reduced water loss via expired air; (5) lowering of metabolic rate and respiratory water loss by more than 30 percent during the cool night. Taylor also studied the eland *(Taurotragus)* and found that its diet of moist acacia leaves, together with physiological specializations to reduce water loss similar to those in the oryx, allowed this largest of African antelope to be independent of drinking water availability.

Even between quite similar ungulates, strategies for conserving water may differ, as indicated by studies of Grant's gazelle *(Gazella granti)* and Thompson's gazelle *(G. thompsoni)* by C. R. Taylor (1968b, 1972). These antelope occur together widely in East Africa, but the range of Grant's gazelle extends into the harsh deserts of northern Kenya while Thompson's gazelle is restricted to less arid areas. Unexpectedly, however, under experimental conditions involving peak temperatures of about 40°C, Grant's gazelle had a rate of pulmocutaneous water loss about one third higher per kilogram of body weight than that of Thompson's gazelle. This apparent paradox was resolved when Taylor (1972) considered the performance of these antelope under extreme heat. At air temperatures above 42°C, Thompson's gazelle used panting to increase evaporative cooling; at an air temperature of 45°C, it maintained a body temperature of 42.5°C. Grant's gazelle, by contrast, did not resort to evaporative cooling but allowed its body temperature to rise to 46°C and was thus able to dissipate heat to the air. It is obvious that, under desert conditions, with limited access to drinking water and intense heat at midday, the temperature regulation strategy of Grant's gazelle would be highly adaptive. As in the case of the oryx, the major source of water for desert-dwelling Grant's gazelles may be leaves that absorb water at night.

In many large mammals, water stress probably has a pronounced effect on how much water is expended through evaporative cooling. Two African antelope delayed their use of evaporative cooling when dehydrated. Whereas hydrated eland *(Taurotragus oryx)* began to

Table 23-2: Maximal Levels of Urine Concentration in Dehydrated East African Mammals

Species	Urine Osmolality (mOsm/kg H_2O)
Dik-dik antelope	4300
One-hump camel	3100
Oryx	2900
Fat-tail sheep	2900
African goat	2800
Impala antelope	2300
Donkey	1500
Zebu cattle	1400

From Maloiy, 1973.

sweat at a skin temperature of 31.3°C, dehydrated eland began to sweat at 35.7°C. In hartebeest *(Alcelaphus buselaphus)*, panting started at a skin temperature of 33.4°C in hydrated animals but was triggered by a deep-body temperature of 39.5°C in water-stressed animals (Finch and Robertshaw, 1979). Both of these animals underwent voluntary hyperthermia during the day.

The social behavior of a mammal may markedly influence its pattern of water regulation. Kirk's dik-dik *(Madoqua kirki)* is a sedentary, territorial antelope that lives in brushy, often hot and dry areas of East Africa. Because each pair or family group has a rigidly defined, exclusive area, extensive daily movements to and from sources of water would involve prohibitively frequent aggressive interactions with other dik-diks. Each individual thus relies on water from plants within its own territory. This constraint, intensified in the dry seasons by high temperatures and reduced plant moisture levels, provides strong selective pressures which have been effective: under stress from heat and restricted access to water, the dik-dik has the most highly concentrated urine of any ungulate studied (Table 23-2). Under these conditions, the savings due to decreased urine volume and increased urine osmolality average about 20 to 40 grams of water per day per animal (Maloiy, 1973).

Considerations of temperature regulation in a number of ungulate species have led C. R.

Taylor (1970) to conclude that there are two major systems of temperature regulation in ungulates. Small ungulates under heat stress use hyperthermia, allowing body temperature to rise to 46°C in extreme cases, avoid sweating, and often avoid panting. The strategy of Grant's gazelle exemplifies this system. Large ungulates, on the other hand, have a relatively low surface-area-to-mass ratio and can afford to sweat. The correlation between body size and degree of use of cutaneous water loss proposed by Robertshaw and Taylor (1969) seems valid; under heat stress, the eland and the African buffalo *(Syncerus caffer)* dissipate heat by sweating, whereas the dik-dik and Thompson's gazelle dissipate heat from the respiratory tract by panting.

Hyperthermia during running may be an important adaptation in some mammals. C. R. Taylor, Schmidt-Nielsen, et al. (1971) found that the body temperature of a running African hunting dog *(Lycaon pictus)* is higher than that of the domestic dog (41.2 versus 39.2°C) and the percentage of running-produced heat lost by respiratory evaporation is lower (25.1 versus 49.7 percent). Taylor and his coworkers suggest that the greatly reduced pulmonary evaporation rate of the hunting dog may allow this animal to increase the distance it can chase prey. Lest the student assume that, for the physiologist, the course of research is all sweetness and light, I offer the following quotation from their paper: "The hunting dog is a quite intractable animal, and its odor alone makes it unsuitable for domestication. Nevertheless, at some inconvenience to our surroundings, a male hunting dog was hand-reared from birth and trained to run a treadmill while wearing a mask."

Periodic Drinkers

In many arid or semiarid regions, scattered water holes or widely separated rivers offer water to mammals that can move long distances. The extensive grasslands of Africa form such an area, as did the North American Great Plains before the coming of settlers. Most large mammals in such areas probably drink every day or two in hot weather and seemingly are unable to survive for long periods without drinking. A few ungulates, however, occupy an intermediate position with regard to water needs. Although they can go for moderate periods without drinking, these mammals are not independent of drinking water availability, as are some desert rodents, and must drink water periodically. Such a mammal is the camel.

Our present knowledge of the water metabolism of the camel is largely a result of the work of K. Schmidt-Nielsen, Houpt, and Jarnum (1956, 1957). Their work, done in the northwestern Sahara on local domestic camels *(Camelus dromedarius),* substantiated the popular idea that camels can tolerate long periods without drinking water, but more important, Schmidt-Nielsen and his group explained the adaptations allowing this tolerance. The ability of their experimental animals to tolerate dehydration was remarkable. One camel went without water for 17 days in the winter on a diet of dry food; during this period, it lost 16.2 percent of its body weight. In some areas, camels that foraged on native vegetation in the winter were never watered. Two camels kept without water for seven days in the heat of the summer lost slightly over 25 percent of their body weight. All of these animals drank tremendous amounts of water after their periods of dehydration, and none showed ill effects.

The camel economizes on water in several ways. Its body temperature drops sharply at night and then rises slowly during the heat of the day. It is able to tolerate considerable hyperthermia, and typically the day is largely over before the animal's body temperature rises to levels at which evaporative cooling, in the form of sweating, must occur. Thus, relative to humans under similar conditions, for example, very little moisture is expended each day in cooling the camel (Fig. 23–4). As in the oryx, excess heat gained by day is lost passively at

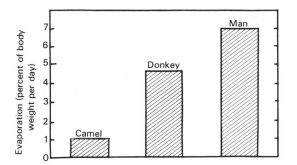

Figure 23-4. Water used for temperature regulation in subjects exposed to the Sahara sun in June. The camel and donkey were tested during periods when they were deprived of water. (After K. Schmidt-Nielsen, Houpt, and Jarnum, 1956)

night. Further water saving results from the modest ability of the kidneys to concentrate urine and from the absorption of water from fecal material. Despite these important water-saving devices, however, the camel loses water steadily through evaporation and in the urine and feces, and perhaps most remarkable is its ability to tolerate tremendous losses of weight (up to 27 percent of body weight) during intervals of dehydration.

Apparently the proportions of water lost from various parts of the body differ in humans and camels. When a person on the desert has lost water equal to about 12 percent of body weight, the blood plasma loses considerable water and becomes viscous; as a result, the heart has difficulty moving the blood and the rate of blood circulation decreases. This leads to a marked reduction in the rate of dissipation of metabolic heat, to a sudden rise in body temperature, and to rapid death. Even in a camel that has been dehydrated to the extent of losing 20 percent of its body weight, however, close to the normal plasma water content is retained but large amounts of water are lost from interstitial fluids and from intracellular protoplasm. In a camel deprived of water for eight days, the loss of volume due to water loss was 38 percent for interstitial fluids and 24 percent for intracellular water. However, only 10 percent reduction in plasma water occurred (K. Schmidt-

Nielsen, 1964:65). Although the camel becomes strikingly dehydrated and emaciated during periods without water, the blood apparently retains its fluidity and its ability to contribute to heat dissipation without straining the circulatory system. The donkey *(Equus asinus)* was also studied by Schmidt-Nielsen (1964:81) and proved to be as capable of tolerating weight loss due to dehydration as the camel, but the donkey had a rate of water expenditure 2.5 times that of the camel and could therefore not be independent of water for intervals of more than a few days.

Bedouin goats *(Capra hircus)* are not only remarkably tolerant of dehydration but can recoup their water losses amazingly rapidly. These small (25 kilograms), black goats are driven by nomadic peoples through dry, sparsely vegetated parts of North Africa and the Middle East (Fig. 23-5). The goats can survive a 25 to 30 percent loss of body weight by dehydration, and within 2 minutes are able to drink from 30 to 40 percent of their body weight in water (Choshniak and Shkolnik, 1977). The capacious rumen stores water for a considerable time after drinking and releases it slowly to the body tissues. The camel and the donkey (Table 23-3) also recover body weight quickly by drinking. A camel that had lost 25 percent of its body weight recovered its original weight within 10 minutes by drinking, and a

Figure 23-5. Bedouin goats *(Capra hircus)* belonging to nomadic Gashgai tribesmen near Gach Saran, southern Iran. These small goats can survive on scant forage and infrequent access to water. (R. P. Vaughan)

Table 23-3:   Water Consumption by a Donkey Deprived of Water in June on a Desert

Days Without Water	Initial Body Weight (kg)	Dehydr. Weight (kg)	Weight Loss (kg)	% of Body Weight	Water Consumed (liters)		Water Consumed as % of Dehydr. Body Wt.	
					Within 5 min	Within 2 h	Within 5 min	Within 2 h
4	107.0	74.5	32.5	30.4	20.5	23.5	27.2	31.5
4	104.1	73.0	31.1	29.9	20.3	26.9	27.8	36.8

Data from K. Schmidt-Nielsen, 1964.

similarly dehydrated donkey drank back its weight in 2 minutes. The amount of water drunk in a short time is almost unbelievable: in a few minutes, a dehydrated male camel studied by Schmidt-Nielsen drank approximately 104 liters of water.

Heat Exchange Systems and Water Conservation

The observation by Schmidt-Nielsen that kangaroo rats exhale air that is cooler than body temperature led to studies showing that the nasal passages of many mammals serve as heat exchange systems (Langman, Maloiy, et al., 1979; K. Schmidt-Nielsen, Bretz, and Taylor, 1970; K. Schmidt-Nielsen, Schroter, and Shkolnik 1982). These systems result in a significant reduction of pulmonary water loss.

The nasal passages of rodents function as heat exchangers with alternating flow in opposite directions in a single tube rather than steady flow in opposite directions in adjacent tubes (as in retia mirabilia). Inhaled air that is below body temperature cools the moist nasal mucosa, which is further cooled when water evaporates from it. Inhaled air then becomes warmed and saturated with water in the lungs. During expiration, this humid, relatively warm air passes back through the narrow nasal passages and over the cool mucosa. The expired air is cooled and thus moisture condenses out onto the mucosa. This moisture is subsequently absorbed

back into the animal's system. This pattern, repeated with every respiratory cycle, results in expired air that is far below body temperature and substantially below the temperature of the inhaled (or ambient) air. Although the exhaled air is saturated with water, it is far cooler than the air in the lungs and thus contains considerably less water. A kangaroo rat with a body temperature near 38°C, resting in air at 30°C and 25 percent relative humidity, exhaled air that was 27°C. In this case, 54 percent of the water used to humidify the inhaled air was recovered from the exhaled air by condensation on the mucosa. Ambient temperature influences the rate of water recovery; the cooler the inhaled air is the more the nasal passages are cooled; consequently the more the air being exhaled is cooled and thus the more water recovered. At an ambient temperature of 15°C and 25 percent relative humidity, the kangaroo rat mentioned above would salvage up to 88 percent of the water used to humidify the inhaled air. MacMillen (1972) has shown that, in rodents in general (irrespective of the habitat occupied), evaporative water loss is directly related to ambient temperature. By being nocturnal, rodents are active when temperatures are lowest and thus evaporative water losses are minimal.

Similar heat exchange systems provide for reduced evaporative water loss in the giraffe *(Giraffa camelopardalis)* and the camel. Studies by Langman, Maloiy, et al. (1979) showed that, in ambient air of 21°C, the giraffe exhaled air that was 28°C, some 20°C below body temperature. This yielded a recovery of 56 percent of the water used to saturate the air in the lungs, a savings amounting to 1.5 to 2 liters per day, roughly one fifth of the giraffe's daily water requirement.

Although the heat exchange systems of the giraffe and of rodents markedly reduce evaporative water loss, these animals exhale air that is saturated with water. A dehydrated camel, however, not only cools exhaled air but also desaturates it. This is due to a hygroscopic (water-absorbent) layer of dried mucous and cellular debris that coats the nasal passages. This layer absorbs water rapidly during exhalation, and as a result the exhaled air is dried. K. Schmidt-Nielsen (1981) estimated that, if a camel in an air temperature of 28°C at night exhaled air that was at this same temperature but at only 75 percent relative humidity, it would be losing only one third as much water by pulmonary evaporation as it would be forming as a by-product of metabolism. These estimates are not intended to be full accounts of the camel's water intake and loss, but they do illustrate the effectiveness of the coupling in the camel of a heat exchange system and hygroscopic moisture exchange.

Water-Independent Desert Rodents

Many rodents that inhabit deserts, seasonally dry chaparral, or woodlands must survive for extended periods without access to preformed water (water already in the form of H_2O, in contrast to water formed as a by-product of such chemical reactions as the oxidation of starches). Among the rodent members of these habitats are species that primarily eat seeds. These rodents, some of which are saltatorial and can move rapidly over considerable distances in search of concentrations of seeds, occupy even the most barren and inhospitable deserts of the world. They represent the families Heteromyidae, Dipodidae, and Muridae. These "water-independent" rodents share two life-history features: they are nocturnal and they are fossorial. The

importance of these features will be considered later.

As a basis for further discussion, the routes of water intake and loss in water-independent rodents should be reviewed. Sources of water include succulent foods, metabolic water (water released as a by-product of metabolism; seeds high in carbohydrates have high yields of metabolic water), and drinking water. Water is lost by lactation, defecation, urination, and pulmocutaneous evaporation. Seed-eating desert rodents do not have access to drinking water and eat little succulent food; their major water source is metabolic water. Water lost in lactation is important periodically to females (p. 492). Urinary water loss is typically reduced in these rodents by the concentration of urine, and water is absorbed from fecal material. The primary channel for water loss is pulmocutaneous evaporation. Such loss may account for 90 percent of total water loss, and MacMillen and Grubbs (1976) and MacMillen (1972) demonstrated that there is no difference in the rate of such water loss between desert and nondesert rodents. Because water-independent rodents do not sweat and may have reduced cutaneous water loss, pulmonary evaporation is of greatest importance.

Water intake in kangaroo rats and other seed-eating desert rodents can be accounted for fairly easily. Many seeds are high in carbohydrates, which yield large amounts of water when they are oxidized. For example, for every 100 grams of dry barley metabolized, 53.7 grams of water is produced. This may be augmented by preformed water in the food: seeds in the soil or on the surface at night absorb moisture, and seeds stored in nests in burrows may contain as much as 20 percent water (Morton and MacMillen, 1982).

An important factor in reducing water loss is the ability of the kidneys of water-independent rodents to concentrate urine. The kangaroo rat's urine is roughly five times as concentrated as that of a human; in excreting comparable

Table 23-4:   Amount of Water Lost in Feces of Kangaroo
Rat (*Dipodomys merriami*) and White Rat (*Rattus norvegicus*)

	Feces (g dry matter/ 100 g food)	Water (mg/ g dry fecal matter)	Water Lost with Feces (g/100 g barley eaten)
Kangaroo rat	3.04	834	2.53
White rat	6.04	2246	13.6

After K. Schmidt-Nielsen, 1964.

amounts of urea, the kangaroo rat uses one fifth as much water as do humans. The concentration of dissolved compounds in the urine of kangaroo rats may be roughly twice that of seawater, and in the laboratory these animals have maintained water balance by drinking seawater. The urine of humans, on the other hand, has a concentration of dissolved compounds lower than that of seawater; when a human drinks seawater, the excretion of the dissolved salts requires the withdrawal of water from body tissues, resulting in dehydration.

Fecal water loss is also low in the kangaroo rat and other desert rodents. In eliminating wastes from comparable amounts of food, the laboratory white rat *(Rattus)* used five times as much water to form feces as did the kangaroo rat (Table 23-4). This saving in the kangaroo rat is due partly to more complete utilization of food and to a reduced amount of fecal material, but it is also the result of an exceptional ability to withdraw water from the contents of the intestine.

Of central importance to desert rodents, then, is the balance between water lost via pulmonary evaporation (the principal route of water loss) and water gained from the metabolism of food (the major source of water). This balance as it relates to ambient temperature and body size has been studied by MacMillen and his associates (MacMillen, 1972; MacMillen and Christopher, 1975; MacMillen and Grubbs,

1976; MacMillen and Hinds, 1983a, 1983b). The following discussion is based on these studies.

As discussed previously, pulmonary water loss is directly related to ambient temperature: the lower the ambient temperature, the less water rodents lose from exhaled air. This relationship probably holds for many mammals; despite the use of both heat exchange and hygroscopic moisture exchange, the camel (p. 485) resting in warm air (38°C) would not reduce evaporative water loss as well as the kangaroo rat resting in cool air (15°C) and using only nasal heat exchange. Desert seed-eating rodents are nocturnal; they are active in the coolest part of the daily cycle, when pulmonary water loss is lowest. Except in summer, temperatures in the desert drop dramatically at night. For much of the year, foraging desert rodents are faced with temperatures below their thermal neutral zone and must raise their metabolic rate accordingly. As metabolic rate rises, so does production of metabolic water. To desert rodents, therefore, the relationship between evaporative water loss, production of metabolic water, and ambient temperature is critically important. By means of careful analyses of data on simultaneous measurements of evaporative water loss and oxygen consumption (an indicator of the level of production of metabolic water), MacMillen and Grubbs (1976) showed that, at ambient temperatures below 16.6°C, many rodents produce more water metabolically than they lose by evaporation. At even lower temperatures, metabolic water production still further exceeds evaporative water loss (Fig. 23-6). There is, however, considerable interspecific variation in the ability to limit evaporative water loss. Merriam's kangaroo rat, for example, far surpasses average performance: metabolic water production by this species equals evaporative water loss at an ambient temperature of 17°C (MacMillen and Hinds, 1983a).

If the ability of desert rodents to maintain water balance is strongly affected by temperature, they would be expected to have little need

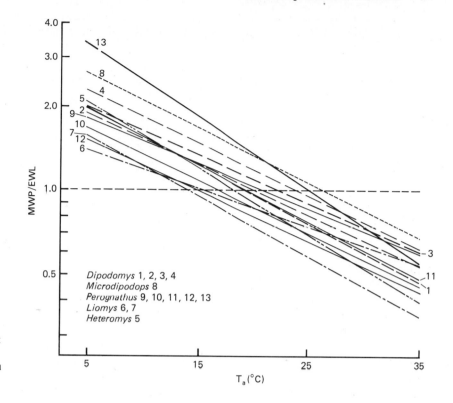

Figure 23-6. Relationship between metabolic water production and evaporative water loss (MWP/EWL) and ambient temperature (T_a) among 13 species of heteromyid rodents. The horizontal line shows the temperature at which MWP = EWL. (From MacMillen and Hinds, 1983a, p. 159, by permission of the Ecological Society of America)

for conserving water by such means as concentrating urine during favorable (cool) seasons but a greater need in unfavorable (warm or hot) seasons. MacMillen has found this to be the case. The rodents of a California desert concentrated their urine more in hot months than in cool months (Table 23–5). The ability to adjust renal function in response to environmental conditions is probably of general occurrence among water-independent rodents. The Australian hopping mouse *(Notomys alexis)* resembles the rodents studied by MacMillen in that it uses its extraordinary ability to concentrate urine only when facing water stress (Hewitt, Wheldrake, and Baudinette, 1981). Similarly, the spring mouse *(Acomys cahirinus)* of North Africa and the Middle East increases urine concentration rapidly in response to water restriction (Daily and Haines, 1981). MacMillen and Grubbs (1976) conclude that nocturnality provides desert rodents with a favorable relation-

ship between evaporative water loss and metabolic water production. Given this relationship, water balance in these rodents is maintained simply by the choice of an appropriate diet and by proper control of urinary and fecal water loss.

By being fossorial, water-independent rodents vastly reduce evaporative water loss. Desert rodents spend a major share of their time in burrows; some species are active on the surface for as little as 1 hour each night (Kenagy, 1973a). These rodents commonly plug the mouth of their burrows, and in such burrows, relative humidity is high. Evaporative water loss markedly decreases as the humidity of inhaled air increases and is affected more by humidity changes in small mammals than in large ones (D.P. Christian, 1978). Temperatures in burrows are moderate, even when surface temperatures are high (Fig. 22–5). For most of their lives, therefore, desert rodents live under con-

Table 23-5:   Seasonal Changes in Urine-Concentrating Ability of Rodents in the Desert Near Joshua Tree, California

Species (Diet)	Mean Urine: Plasma Osmotic Pressure Ratio $\pm$ S.D.[a]				
	Dec 1970 (0°C)[b]	Mar 1971 (4°C)[b]	June 1971 (17°C)[b]	Sept 1971 (20°C)[b]	Dec 1971 (0°C)[b]
Dipodomys merriami (granivore)	2.7 $\pm$ 0.6 (16)[c]	5.7 $\pm$ 1.7 (10)	11.4 $\pm$ 1.5 (10)	9.3 $\pm$ 3.1 (15)	4.8 $\pm$ 1.7 (2)
Perognathus fallax (granivore)	2.5 (1)	5.4 (1)	9.3 $\pm$ 1.5 (2)	7.9 $\pm$ 2.7 (6)	—
Perognathus longimembris (granivore)	—	4.4 $\pm$ 1.7 (8)	13.9 $\pm$ 0.9 (6)	3.7 $\pm$ 0.2 (2)	—
Neotoma lepida (herbivore)	3.7 $\pm$ 1.2 (2)	—	—	4.5 $\pm$ 0.9 (3)	—
Onychomys torridus (insectivore)	10.1 $\pm$ 0.4 (2)	—	8.4 (1)	10.3 (1)	9.3 (1)
Peromyscus crinitus (omnivore)	10.1 $\pm$ 0.9 (5)	7.7 $\pm$ 1.2 (3)	11.6 $\pm$ 1.2 (5)	12.5 $\pm$ 1.9 (9)	9.4 $\pm$ 3.3 (4)

[a]All values are from field-collected samples.
[b]Mean minimum temperature for month.
[c]Values in parentheses are sample size.
After MacMillen and Christopher, 1975.

ditions of high humidities and moderate temperatures and are well insulated from the rigors of the desert environment.

The importance to desert rodents of storing seeds in subterranean caches has probably been underestimated. Morton and MacMillen (1982) considered the hypothetical water budget of a kangaroo rat on a diet of desert annual seeds stored in a burrow at 76 percent relative humidity and estimated that 28 percent of the animal's daily water requirement could be supplied in the dry season by the fairly high water content (17.2 percent by weight) of the seeds. They concluded that, for some desert-dwelling rodents, the high water content of stored seeds may be important in tipping the delicate dry-season water balance in favor of survival.

The ability to be independent of preformed water is by no means unique to a few rodents. Each major desert region in the world is well populated by rodents, and seemingly each of these regions supports some highly specialized species adapted to a dry diet. Such specializations occur in at least some members of the

North American heteromyid genera *Perognathus, Dipodomys,* and *Microdipodops.* Probably all species of heteromyids that are restricted to deserts can survive on dry diet; not all inhabit deserts, however, and not all are equally well adapted to dry conditions. For example, *Dipodomys agilis,* an inhabitant of coastal southern California and northwestern Baja California, occurs largely in coastal sage scrub, where conditions of extreme aridity occur only in midsummer. This kangaroo rat is marginal in its ability to subsist on a dry diet and cannot survive indefinitely without preformed water (Carpenter, 1966; MacMillen, 1964a).

Among desert rodents of the Old World, two jerboas (*Jaculus jaculus* and *Dipus aegypticus*) and several gerbils of two genera (*Meriones* and *Gerbillus*) can live on dry food, and some surpass kangaroo rats in their ability to concentrate urine (Table 23-6). Adaptations to a dry diet have clearly evolved independently in several rodent families (Heteromyidae, Dipodidae, Muridae). Striking convergent evolution in these families has led to saltatorial ad-

Table 23–6:   Concentration of Urea in Urine, Water Loss from Lungs and Skin, and Percentage of Water in Feces of Desert Rodents

Species	Max. Urea Conc. in Urine (mmol/liter)	Pulmocutaneous H_2O Loss (mg/cm^{3}O$_2$)	Amt. of H_2O in Feces (%)
Sciuridae			
Ammospermophilus leucurus (North America)	2860	0.53	
Heteromyidae			
Dipodomys merriami (North America)	3840	0.54	45.2
Dipodomys spectabilis (North America)	2710	0.57	
Cricetidae			
Gerbillus gerbillus (Egypt)	3410		
Dipodidae			
Jaculus jaculus (Northern Africa)	4320		
Muridae			
Notomys alexis (Australia)	5430	0.91	48.8
Notomys cervinus (Australia)	3140	0.76	51.8
Leggadina hermannsburgensis (Australia)	3920	1.15	50.4

Data from MacMillen and Lee, 1967.

aptations in some members of each family as well as to similar specializations favoring water conservation. The most concentrated urine yet measured is that of a murid Australian hopping mouse *(Notomys alexis)*. This rodent occupies some of the most arid deserts in the world, regions where ten years may pass between rains. This species and two other desert-inhabiting murids of Australia were studied by Macmillen and Lee (1967, 1969), who found that all three species could live on dry seeds with no drinking water. Relative to the kangaroo rat, these murids had higher rates of pulmocutaneous and fecal water loss, but in general they had a greater ability to concentrate urine (Table 23–6).

Two species of spiny mice *(Acomys,* Muridae) studied by Shkolnik and Borut (1969) in the desert of Israel are remarkable in their unusual pattern of adaptation to arid conditions. These animals have highly specialized kidneys that can concentrate urine to a greater degree than can the kangaroo rat kidney, but the spiny mice have an evaporative water loss two to three times as great as that in Merriam's kangaroo rat. Probably primarily due to high water loss through the skin, spiny mice are unable to subsist on a diet of dry seeds. Apparently the high cutaneous water loss is important as a means of dissipating heat in a hot climate, and the great ability of the kidney to concentrate urine, coupled with a diet high in land snails (which have a high water content), compensates for the extravagant use of water in thermoregulation. The relationship between an omnivorous diet and modest water-regulatory ability is demonstrated by members of the genus *Peromyscus,* which are nearly ubiquitous in North America. The desert-dwelling *Peromyscus* maintain water balance by including succulent foods in their diet and by conserving water by becoming torpid during especially dry and stressful times (MacMillen, 1983a).

Water Regulatory Efficiency in Heteromyid Rodents

The following remarks are based on the work of MacMillen and Hinds (1983a), whose findings contribute importantly to our understanding of the adaptive importance of water regulatory ability in heteromyids.

From a Neotropical origin in the Early Tertiary, heteromyids radiated in association with a flora of the southwestern United States and northern Mexico that became adapted to progressively greater aridity and a more seasonal rainfall. Primary productivity under desert conditions was low, and heteromyids exploited seeds, often those of short-lived annuals. In a single packet, seeds provided both energy and water (from the oxidation of carbohydrates). Physiological adaptations that reduced heteromyid energy use included lowered metabolic rates and the use of torpor by the smaller species. Specialized water metabolism enabled these rodents to survive with no water other than that from food.

Although under laboratory conditions most species of kangaroo rats *(Dipodomys)* and all species of pocket mice *(Perognathus;* Fig. 23–7) remain healthy on a dry-seed diet, the urine of pocket mice has roughly half the osmotic concentration of that of kangaroo rats. Pocket mice, then, should be in a more favorable state of water balance than kangaroo rats. MacMillen and Hinds tested this expectation by measuring the primary route of water input (MWP; metabolic water production) and output (EWL; evaporative water loss) in *Dipodomys merriami* (36 grams) and *Perognathus longimembris* (8.0 grams) fed millet seeds and subjected to a wide temperature range. The measurements demonstrate (Fig. 23-8) that in both species the ratio MWP:EWL (an expression of the state of

water balance) is invariably related to ambient temperature. The smaller *P. longimembris* is always in a more favorable state of water balance than *D. merriami,* as indicated by the more dilute urine of the former. As an indication of the greater water regulatory efficiency of *P. longimembris* at higher temperatures, the ambient temperatures at which MWP = EWL is some 10°C higher in *P. longimembris* than in *D. merriami.* MacMillen and Hinds regard urine concentration, the traditional criterion of water regulatory efficiency in desert rodents, to be inappropriate and propose that a more meaningful indicator for granivorous rodents is the ambient temperature at which MPW = EWL.

The demonstrated relationship between metabolism and evaporative water loss led MacMillen and Hinds to hypothesize that water regulatory efficiency in heteromyids should vary inversely with body weight. They confirmed this hypothesis by experiments on five genera and 13 species of heteromyids (Fig. 23–6) which led to their evaluation of the adaptive significance of water regulatory efficiency in heteromyids.

As outlined previously, these authors believe that, under selective pressures associated with increasing aridity, heteromyids became granivorous, improved water regulatory ability, reduced energy needs by reducing metabolic rate, and became smaller. At a weight threshold of 35 to 40 grams, a divergence in locomotor

Figure 23-7. A rock pocket mouse *(Perognathus intermedius)* in the Kofa Mountains of Arizona. This heteromyid is widely distributed in the deserts of the southwestern United States.

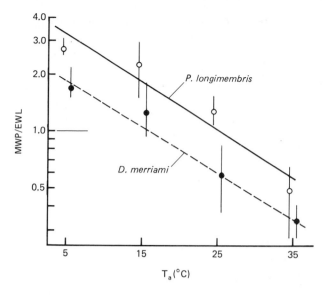

Figure 23–8. Relationship between metabolic water production and evaporative water loss (MWP/EWL) in *Dipodomys merriami* and *Perognathus longimembris* while on a diet of dry millet seeds. An extension of horizontal line at MWP/EWL = 1.0 indicates the temperature at which MWP = EWL. Note that this equality occurs at a much higher temperature in *P. longimembris* (about 25°C) than in *D. merriami* (about 17°C). (From MacMillen and Hinds, 1983a, by permission of the American Institute of Biological Sciences)

style occurred: the larger *Dipodomys* (more than about 40 grams) became bipedal hoppers and the smaller *Perognathus* (less than 40 grams) maintained quadrupedal locomotion. MacMillen and Hinds propose that the ecological importance of the observed mass-related differences in water regulatory efficiency is in favoring survival in the small species that have limited locomotor ability. Whereas bipedal hopping provides kangaroo rats with a rapid means of locomotion, and one that allows erratic escape from predators in open situations, the slower quadrupedal locomotion of *Perognathus* leaves these animals more vulnerable to predators on open ground and restricts the foraging area. Thus *Dipodomys* can forage widely and selectively and can choose seeds high in carbohydrates (seeds that have high yields of metabolic water), and the water regulatory ability of *Dipodomys* became fixed at an intermediate level. *Perognathus,* in contrast, must forage less widely and eat a variety of seeds, some of which, because of relatively low carbohydrate content yield relatively less metabolic water. Such a diet demands a high degree of water regulatory ability. Such basic differences

in water regulatory ability may have an important influence on the organization of desert rodent communities.

Water Balance in Bats

The kidneys of some bats are specialized to concentrate urine, but these animals are seemingly not independent of drinking water availability. Carpenter (1969) found that two desert-dwelling insectivorous bats produced concentrated urine, but their need for water was increased by high evaporative water losses during flight and when they were not torpid. He estimated that the bats lost 3.09 percent of their body weight through evaporation per hour of flight. After a careful consideration of the bats' water budgets, Carpenter concluded that they were not independent of drinking water but that their ability to fly long distances to drink water enabled them to maintain water balance in desert areas. Urine concentrations in the little brown bat *(Myotis lucifugus)* reach peak levels during high evaporative cooling and directly after feeding, but by drinking after feeding the bats avoid water stress (Geluso and Studier, 1979). A marine-fish- and crustacean-eating bat *(Pizonyx vivesi,* Vespertilionidae) that inhabits desert islands and desert coasts of the Gulf of California has the ability to concentrate urine to the extent that it can utilize seawater as a water source (Carpenter, 1968). Because of high evaporative water losses, particularly during flight, the water gained from this bat's food probably is not sufficient to meet its water requirements, and presumably it must drink seawater.

Lactation and Water Balance

The loss of water by a female during lactation is substantial, and it is not surprising that some mammals recycle such water. Cat breeders have long been familiar with behaviors only recently discussed in the scientific literature. The lactating domestic cat *(Felis sylvestris)* stimulates urination and defecation in her young by licking their genital areas. She then ingests the urine and feces. This keeps the nest sanitary and recycles to the mother much of the water lost in the milk. Similar behavior occurs in several Australian murid rodents, the dingo *(Canis familiaris),* and two species of kangaroos (Baverstock and Green, 1975). These species recovered about 30 percent of the water lost in milk. A laboratory study indicated that most of the water lost by female house mice *(Mus musculus)* during lactation, with the exception of that dissipated by evaporative water loss from the young, was recovered in a similar manner by the mother (Baverstock, Watts, and Spencer, 1979). Behaviors that result in recycling water lost in milk are probably widespread among mammals that bear altricial young or have limited access to water.

Ecological Considerations

The importance of knowledge of interspecific differences in water metabolism to an understanding of the roles of mammals in the ecosystems of arid regions has only recently been appreciated. The ecological displacement of some species of rodents results in part from their different means of satisfying water requirements. In a study of a semidesert rodent fauna in California, MacMillen (1964a) found that water metabolism differed significantly from species to species. He also found that each of the six members of a desert rodent community exhibited different temporal patterns of changes in urine concentration and concluded that these patterns reflected both the level of renal water conservation and the solute load imposed by the diet (MacMillen, 1972). Seasonal shifts in diet by desert rodents may be part of a strategy that favors both water balance and optimal energy return. In some deserts, the patterns of seasonal

Table 23-7: Twelve-Hour Body Weight Loss in Bats Caged Singly and in Groups of Four[a]

Species	Roost Ambient Temp. (°C) Mean	Range	Roost Rel. Humidity (%) Mean	Range	Percent Wt. Loss in Singly Caged Individuals Mean	Range	Percent Wt. Loss in Group-Caged Individuals Mean	Range
Myotis lucifugus (site 1)	26.8	15.6–31.1	23	18–31	10.5 (5)[b]	7.7–12.8	9.9 (4)	7.7–11.5
M. lucifugus (site 2)	26.1	15.6–30.4	32	27–40	10.9 (7)	8.8–13.0	11.2 (5)	9.0–11.5
M. thysanodes	26.8	15.6–31.1	23	18–31	15.8 (8)	9.0–21.8	10.9 (3)	10.1–11.5
M. velifer	22.0	20.7–23.3	64	53–96	8.2 (4)	5.6–9.8	8.4 (2)	8.1–8.8

[a]Time period was from 8 A.M. to 8 P.M.
[b]Sample size.
From Studier, Proctor, and Howell, 1970.

Table 23-8: Average Percentage of Body Weight Lost by Bats Deprived of Water Until Stressed

Species	Sample Size	Weight Lost (% of body wt.)
Myotis lucifugus	12	32.3
M. thysanodes	8	31.7
M. yumanensis	3	31.6
M. velifer	4	22.8

From Studier, Proctor, and Howell, 1970.

dormancy in small mammals may favor water economy by enabling these animals to avoid periods of intense heat and low humidity.

Some desert mammals, such as the jackrabbit *(Lepus californicus),* are not fossorial, do not have access to drinking water, and in summer must face temperatures as high as 50°C. This jackrabbit responds in several ways to chronic water restriction. Basal metabolism drops by 30 percent; evaporative water loss is reduced 50 to 70 percent, partly due to limited use of evaporative cooling; fairly concentrated urine (Table 23-1) and very dry feces are produced; and the animals undergo voluntary hyperthermia to 41°C (K.A. Nagy, Shoemaker, and Costa, 1976; Shoemaker, Nagy, and Costa, 1976). Despite these responses, this jackrabbit's diet probably must contain at least 68 percent water for the animal to maintain water balance in summer. The water balance problems of this animal must be taken into account in any consideration of diet and habitat preferences.

Pulmocutaneous water loss is high in some bats exposed to moderately high ambient temperatures, and the choice of roosting sites as well as the geographic distributions of some bats may be limited by their inability to avoid daily dehydration. Clustering during roosting, a common behavior in some bats, reduced pulmocutaneous water loss markedly in one species *(Myotis thysanodes)* studied by Studier, Proctor, and Howell (1970), as shown by Table 23-7. It was further found that several species of *Myotis* can tolerate considerable weight loss caused largely by pulmocutaneous water loss (Table 23-8); because of high rates of such loss, however, two species were unable to survive two days without access to water.

Chapter 24

Acoustical Orientation and Specializations of the Ear

Because the mammals most familiar to us depend largely on vision for perceiving their environment, it is surprising to note that about 20 percent of the known species of mammals use acoustical orientation as their primary means, or at least as an important secondary means, of "viewing" their surroundings. Most bats, some members of the orders Insectivora and Carnivora, and probably all odontocete cetaceans use acoustical orientation. Future research may demonstrate that the use of such orientation is even more widespread among mammals.

Echolocation in Bats

Foraging insectivorous bats and bats flying in deep caverns face seemingly insurmountable problems: they must perceive tiny prey and obstacles in nearly complete darkness. The remarkable nocturnal performance of bats indicates that these problems of perception have been effectively solved. Bats are able to capture insects with great efficiency and speed in darkness. The pursuit and capture of an insect take a mustached bat (*Pteronotus parnellii;* Fig. 24–1) but 0.25 to 0.33 of a second (Novick, 1970). Bats deep in caverns, flying in complete darkness, can detect the walls of the cavern as well as avoid collisions with hundreds of other bats that are circling and maneuvering abruptly.

Ultrasonic pulses are emitted by the bat, and the echoes of these pulses reflected by ob-

jects allow the bat to acoustically "see" in the dark. Even the "nature" of objects can be analyzed to some extent.

When considering echolocation in bats, we are dealing almost entirely with members of the suborder Microchiroptera, all of which use echolocation. Most members of the suborder Megachiroptera use visual perception instead of echolocation, but how this serves at low levels of illumination is not known. Among megachiropterans, only members of the genus *Rousettus* are known to echolocate; their technique is unique, however, in that the pulses are audible and are made by tongue clicking (Kulzer, 1956, 1958; Mohres and Kulzer, 1956; Novick, 1958a).

Although microchiropterans clearly depend on echolocation for perceiving their environment in darkness, the eyes are present in all bats and sight has by no means been totally abandoned. Many phyllostomids (leaf-nosed bats), for example, have large eyes and obviously make use of them. Olfaction may be of great importance for detecting food in bats that eat fruit, nectar, pollen, or small vertebrates. In species with well-developed visual and olfactory capabilities, echolocation is perhaps primarily used for perceiving nearby obstacles.

An accurate picture of the bat's use of acoustical orientation was long in emerging. As early as 1793, Lazaro Spallanzani performed experiments that suggested that bats use acoustical rather than visual perception when avoiding

Figure 24-1. (A) A mustached bat *(Pteronotus parnellii)* in flight. (B) Close-up of the face, showing the lips formed into a "megaphone" during echolocation. (A by T. Strickler; B by O. W. Henson)

obstacles and when feeding. Not until the early 1940s, however, was the use of echolocation by bats conclusively demonstrated by the careful laboratory experiments of Griffin and Galambos (1940, 1941) and by the observations of Dijkgraaf (1943, 1946). Continued research and the use of refined electronic equipment have contributed to our present detailed, but as yet incomplete, knowledge of echolocation in bats.

Evolution of Echolocation.   E. Gould (1970, 1971) hypothesized that the sonar pulses of bats may have been derived originally from vocalizations that established or maintained spacing or contact between bats. The repetitive communication sounds used by infant bats, and similar pulses perhaps used originally during flight to maintain adequate spacing of foraging individuals, may have secondarily become important in connection with detecting prey and avoiding obstacles. According to E. Gould (1971:311), "The prominence with which continuous, graded signals pervade the lives of such social and nocturnal mammals as bats suggests that echolocation is an inextricable and integral part of a communication system." This author suggests that some of the vocalizations used by early bats may have been inherited from their insectivore ancestors, in which auditory communication was perhaps as important as it has been shown to be in some living insectivores (J.F. Eisenberg and Gould, 1970; E. Gould, 1969).

Adaptive Importance of Echolocation.   Probably ever since their appearance in the Paleocene or Late Cretaceous, bats have "owned" the aerial adaptive zone during the nocturnal segment of the diel cycle. No birds can match the nocturnal insect-catching ability of bats, nor have birds exploited nocturnal fruit eating or nectar feeding. In many tropical areas, the most important flying predators of small vertebrates are seemingly bats, not birds. Bats and birds, the only two flying vertebrates of the Cenozoic, may have divided the diel cycle early in the era: birds dominated the aerial zone during daylight, and bats dominated it during darkness. In trop-

ical areas, which characteristically support a great diversity of both bats and birds, the changing of the guard at sunset is dramatic. Bird activity and singing suddenly begin to diminish as the sun disappears; at the same time, bats begin to appear and become increasingly more evident as darkness descends. When the twilight glow in the west is gone, most birds are inactive and silent, but the cries of bats regularly penetrate the cacophonous blending of frog and insect noises. This chiropteran domination of the nocturnal air is seemingly due largely to their ability to use echolocation. The perfection of highly maneuverable flight and echolocation in bats may have occurred more or less concurrently during their early evolution, and together these abilities were probably responsible for the spectacular nocturnal success of bats.

Production of Ultrasonic Pulses.   The echolocation pulses used by microchiropterans are produced by the larynx. The cricothyroid muscles (muscles that tense the vocal cords) were shown by Novick (1955) to be essential for normal emission of pulses. These are mostly of frequencies well above those audible to the human ear, which can detect frequencies up to roughly 20 kilohertz. There are humanly audible components (the so-called ticklaute) in the vocalizations of many bats. The pulses, however, are only the low-intensity components of cries with higher harmonics (frequencies that are integral multiples of the fundamental frequency).

The pulses emitted by bats are often of high intensity (great loudness) and are emitted through either the mouth or the nose. In most vespertilionids, molossids, and noctilionids, the pulses are of the highest intensities recorded for bats and come from the mouth, which is kept open during flight. The rhinolophids (horseshoe bats) emit high-intensity pulses through the nostrils. Members of the families Nycteridae, Megadermatidae, and Phyllostomidae, on the other hand, produce relatively low-intensity pulses, which is why these groups are called the whispering bats (Griffin, 1958:232). Whereas the bats that emit high-intensity pulses catch flying insects, the whispering bats feed largely on fruit, nectar, small vertebrates, insects on the ground, vegetation, or combinations of these foods. Some authors (for example, Griffin, 1958:251; Novick, 1970:39) have suggested that these soft pulses are well adapted to close-range perception of surfaces, such as rocks or tree trunks, or of complex tropical environments with interlacing vines and branches and stratified foliage. Perhaps an advantage is gained by not receiving echoes from distant objects and thereby limiting the complexity of information from an already complex perceptual situation.

Just how loud are the pulses emitted by bats? Inasmuch as we cannot hear them, their loudness must be expressed in terms of an energy unit called the dyne (1 dyne is the force necessary to accelerate 1 gram of mass 1 centimeter per second per second). Under ideal circumstances, the threshold of human hearing is roughly 0.0002 dyne per square centimeter. When recorded 5 centimeters from a bat's mouth, the least intense pulses of whispering bats are approximately 1 dyne per square centimeter, whereas the loudest pulses of other bats are near 200 dynes per square centimeter. Such loud pulses are comparable to the painfully intense noise made by nearby jet engines.

The strange faces of bats, always a source of amazement to those unfamiliar with these animals, may have an important function in connection with echolocation. Mohres (1953) showed that the horseshoe-like structure surrounding the nostrils of a rhinolophid (Fig. 8–27) serves as a diminutive megaphone to concentrate the short-wavelength pulses emitted by these bats into a beam; these 80- to 100-kilohertz pulses have wavelengths of only 3 or 4 millimeters. In addition, because the nostrils are situated almost exactly 0.5 wavelength apart, the pulses emitted through the nostrils undergo interference and reinforcement of a sort that tends to beam the pulses. The facial patterns of many bats (Fig. 24–2) may well function similarly to direct pulses in such a way that some species can scan their environment with a concentrated beam of ultrasonic pulses much as we probe the darkness with a flashlight beam.

Sensitivity to Outgoing Pulses. A number of structural and physiological specializations reduce the sensitivity of bats to their own outgoing pulses. Vibrations are transmitted mechanically in mammals by the ear ossicles (malleus, incus, and stapes) from the tympanic membrane to the oval window of the inner ear. Two muscles of the mammalian middle ear dampen the ability of the ossicles to transmit vibrations when an animal is subjected to unusually loud sounds or when it is vocalizing. These muscles—the tensor tympani, which change the tension on the tympanic membrane, and the stapedius, which changes the angle at which the stapes contacts the oval window—are extremely well developed in bats, and their contraction reduces the bat's sensitivity to pulses. Jen and Suga (1976), using highly sophisticated electronic equipment, found that action potentials of the cricothyroid laryngeal muscles were followed 3 milliseconds (1 millisecond is 1/1000 of a second) later by action potentials of the middle ear muscles. This coordination of laryngeal and middle ear muscles ensures the contraction of the latter just prior to vocalization and attenuates (weakens in intensity) by some 25 percent the auditory self-stimulation (Suga and Shimozawa, 1974). Of additional importance, neural attenuation of the direct reception of sonar sounds occurs in the brain. Nerve impulses arising from direct reception and passing from the cochlea to the inferior colliculus of the brain are attenuated by the neurons of the lateral lemniscus of the brain. This change, plus that effected by the middle ear muscles, attenuates the neural events by 40 percent. Suga and Shimozawa suggest that similar attenuating mechanisms occur in humans and keep the sounds of our own speech from becoming disturbingly loud. Direct reception in bats is doubtless reduced also by the beaming of sounds by the lips and the complex noses and nose leaves (Fig. 24–2).

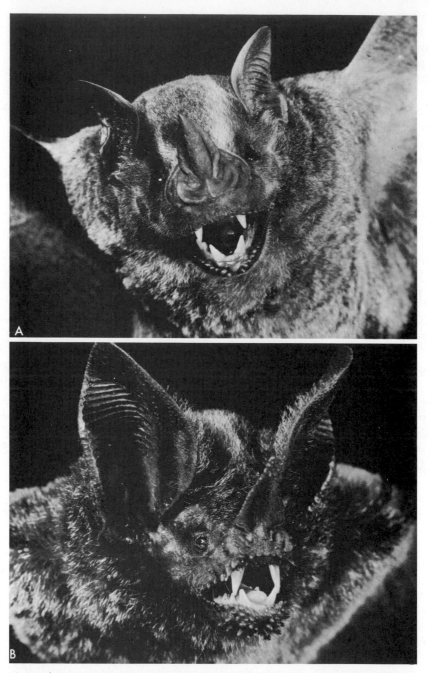

Figure 24–2. Faces of phyllostomatid bats: (A) Jamaican fruit-eating bat *(Artibeus jamaicensis)*; (B) fringe lipped bat *(Trachops cirrhosis)*, a carnivorous-omnivorous species. (N. Smythe and F. Bonaccorso)

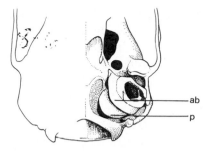

Figure 24–3. Ventral view of the posterior part of the skull of a bat *(Myotis volans)*, showing how loosely the periotic bone (*p*) and auditory bulla (*ab*) are attached to the skull.

An additional structural refinement is that the bones housing the middle and inner ear are insulated from the rest of the skull. This bony otic capsule does not contact other bones of the skull (Fig. 24–3) and is insulated from the skull by blood-filled sinuses or fatty tissue. During the emission of pulses, the conduction of vibrations from the larynx and the respiratory passages to the inner ear is thus greatly reduced.

The Advantage of High Frequencies. Because high frequencies are more severely attenuated by air than are low frequencies, one might wonder why echolocation in bats is based almost entirely on high frequencies. Interference from such sounds as insect chirps and frog calls is doubtless partially avoided by the use of high frequencies, but this factor may not be of primary importance. Perhaps of greater importance is the relationship between prey size and the wavelength of echolocation pulses. The higher the frequency of a sound, the shorter its wavelength. Frequencies of roughly 30 kilohertz have a wavelength of approximately 11.5 millimeters, roughly the size of a small moth; this balance between prey size and wavelength is ideal, for objects approximately the size of one wavelength reflect sound particularly well. Some species of bats are able to detect wires with as small a diameter as 0.08 millimeter (1/ 30 of a wavelength), but in general the wave-

lengths of the pulses emitted by bats are in the range that is most efficient for the detection of small- to medium-size insects.

Different Styles of Echolocation and Their Adaptive Importance. Adaptive radiation in foraging styles among insectivorous bats has been accompanied by the evolution of a variety of echolocation signal types. The following discussion considers how different echolocation strategies enable bats to exploit contrasting modes of foraging. This discussion draws heavily from the work of Simmons, Howell, and Suga (1975) and Simmons, Fenton, and O'Farrell (1979).

The frequencies of bat-produced ultrasonic pulses range from approximately 210 to 20 kilohertz, and the pulses of microchiropteran bats have both frequency-modulated (FM) and constant-frequency (CF) components (Fig. 24–4). The FM signals are typically brief, from 0.5 to 10 milliseconds, but cover a broad bandwidth (frequency range), usually sweeping through at least one octave. The CF signals may be either short (less than 10 milliseconds) or long (10 to over 50 milliseconds). The echo of the FM signal best conveys one type of information to the bat, and the CF signal best conveys another. Targets make complex modifications in the spectrum of the echoes of broad-

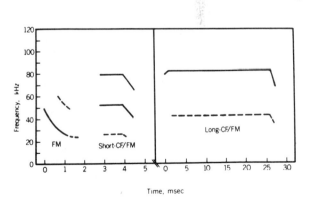

Figure 24–4. Sound spectrograms showing the three major types of echolocation signals used by microchiropteran bats. (From Simmons, 1974)

band FM signals; these signals are best suited to detecting the size and shape, surface details, and range of a target. Constant-frequency signals, on the other hand, have a narrow bandwidth, and because the amount of information an echo conveys is proportional to the bandwidth of the signal, CF signals are poor for conveying details about the characteristics of the target. If the target is roughly as long as or longer than the wavelength of the CF signal, however, this signal will indicate the presence of the target. The long CF signal is of importance because it allows the bat to determine whether the target is approaching or receding and at what rate. This discrimination is based on the Doppler shift, discussed later.

Each pattern of orientation sounds provides a different means of obtaining information, and each exploits in a somewhat different way the information-bearing echoes of FM and CF sounds. The hearing of FM bats is most acute through the sweep of frequencies of their FM signals. In the mormoopid, which uses CF/FM signals, the greatest acuity is at about the level of the CF component. The hearing of rhinolophids is precisely tuned to the frequency of the long CF component; the greater the duration of the CF component, the sharper the tuning of the auditory system.

The echolocation performance—the degree of accuracy with which the system allows a bat to assess its immediate environment—has been studied in a number of species. Simmons (1973) tested the range-detecting ability of one species each from the families Vespertilionidae, Phyllostomidae, Mormoopidae, and Rhinolophidae. All of the bats could discriminate range differences of from 1 to 2 centimeters, and this remarkable performance was apparently achieved by cross-correlation of the transmitted pulse with the returning echo. The essential variable from which the bats estimate distance is the time it takes for a pulse to reach a target and the echo to return. By comparing the relative target-ranging abilities of the bats in relation to the bandwidths of their echolocation pulses,

Simmons determined that the FM components of the pulses are used for target ranging. Laboratory studies on a limited number of species have further demonstrated the ability to discriminate differences in target size, shape, and direction. Simmons, Lavender, et al. (1974) used *Eptesicus fuscus* in laboratory experiments to determine this bat's ability to detect shape details of a target. The bats could discriminate differences of less than 1 millimeter in the depth of small holes in a target. The holes modify the spectrum of the echo from the target by absorbing sound energy at certain frequencies in the bandwidth of the bat's FM sweep; with changes in hole depth, the absorption peaks are shifted to different frequencies. Seemingly, bats can use this ability to associate features of the sonar echo spectrum with target shape when selecting appropriate prey from among the variety of nocturnal insects they perceive.

Foraging in Open Spaces. Some narrow-winged, rapid-flying bats, such as most molossids, hunt in open areas, often high above the ground. Insects are perceived against a completely uncluttered background and pursued through space devoid of obstacles. The Brazilian free-tail bat *(Tadarida braziliensis),* an example of such a forager, uses intense CF or slightly FM pulses when searching for insects; then, as an insect is perceived and pursued, the echolocation signals become progressively more strongly frequency-modulated. As the bat nears its target, the FM sweep is from about 50 to 25 kilohertz. The rate of emission and duration of the signal also change during pursuit, from 10 to 100 or 150 per second and from 10 to 0.5 millisecond. Pulse duration under a variety of conditions is shown in Table 24–1.

What kinds of information does the bat receive from echoes of these signals? Short signals, whether CF or narrow-band (spanning a narrow band of frequencies), yield echoes that enable bats to detect the presence of prey but do not carry sufficient information to allow the bat to recognize shape details or to capture. Echoes from FM (broad-band) signals, however,

Table 24-1: Pulse Duration Under a Variety of Conditions in Bats

Bat	Pulse Duration (ms)	Conditions	References
Megaderma lyra	0.54–1.2 (limits)	Approaching goal or obstacle	Mohres and Neuweiler, 1966
	0.72–1.8 (limits)	Searching in laboratory	
Rhinolophus ferrumequinum	50–65	Takeoff and flight in lab	Schnitzler, 1968
	Decreasing to 10	Approaching landing	
Rhinolophus euryale	30–45	Takeoff and flight in lab	Schnitzler, 1968
	Decreasing to 7	Approaching landing	
Noctilio leporinus	7.4 (5.9–9.4)	Searching flight in outdoor cage	Suthers, 1965
	14.3 (11.1–16.7)	Searching flight in wild	
	13.8 (10.7–16.0)	Searching flight in wild	
	Decreasing to 1	End of prey location	
Pteronotus parnellii	14–26	Searching flight in lab	Novick and Vaisnys, 1964
	Rising from 20–21 to 28–37	Detecting fruit fly	
	Decreasing to 6.8	End of insect pursuit	
Pteronotus psilotis	2.9–4.8	Searching flight in lab	Novick, 1965
	Decreasing linearly to 0.6–1.0	Fruit fly pursuit and capture	
Vampyrum spectrum	1.5–1.8	Searching flight	Bradbury, 1970
	0.5–1.5	Approach	
	About 0.5	Terminal portion of flight	
Myotis lucifugus	2–3	Searching flight in lab	Griffin, 1962
	Decreasing to 0.3–0.5	End of insect pursuit	
Plecotus townsendii	2–5	Searching flight in lab	A.D. Grinnell, 1963b
	Decreasing to 0.3–0.5	End of insect pursuit	
Eptesicus fuscus	2–4	Flying in laboratory	Griffin, 1958
	10–15	Flying in open at 10 m altitude	
	0.25–0.5	Late terminal phase	
Lasiurus borealis	2.4–3.0	Pursuit of mealworms in lab	F.A. Webster and Brazier, 1968
	0.3–0.5	Late terminal phase	

After Novick, 1971.

carry detailed information on the size, shape, and exact position of the target. Decisions as to whether or not to attack and information guiding the final, often erratic pursuit of the prey are based on these FM signals. By changing to FM signals and increasing the bandwidth of the FM sweep during pursuit, the bat progressively improves its ability to recognize the suitability of prey and its position. The use of a high signal repetition rate (the "terminal buzz") during the final stages of pursuit allows the bat to continuously monitor the position of an insect being approached rapidly. The terminal buzz is used by a wide variety of bats during the final phase of insect pursuit.

Foraging in the Presence of Clutter.   The bats most familiar to casual observers in temperate areas are those that pursue prey in the presence of obstacles. Such bats, including most members of the large family Vespertilionidae, forage in the proximity of vegetation or other obstacles and must avoid collisions with this clutter when foraging for insects. The familiar North American big brown bat *(Eptesicus fuscus)* is of this type and produces a variety of orientation sounds. When searching for prey, it emits short, multiple-harmonic (therefore broad-band) FM plus CF signals. As prey is approached, the CF portion is left out and the FM part is progressively shortened to 0.5 millisecond and the repetition rate increases from roughly 10 to 200 per second. As with the Brazilian free-tail bat, these sounds are intense.

In terms of function, the short CF signal serves in detecting prey and the FM portion yields a "picture" of background obstacles and a precise image of the prey during pursuit. This echolocation pattern, involving the constant monitoring of obstacles while searching for and pursuing insects, is typical of many insectivorous bats.

Foraging in Dense Clutter.   A number of tropical bats, as well as some temperate species, glean prey from leaves or the irregular surface of the ground, or pursue insects between the interlacing branches of dense vegetation. Whether it be the spear-nosed bat *(Phyllostomus hastatus)* foraging in a Neotropical forest or the pallid bat *(Antrozous pallidus)* catching prey on a rocky desert hillside, these bats must discriminate between their potential food and its complex background.

The spear-nosed bat accomplishes this by using complex echolocation signals. These sounds, from 0.5 to 1.0 millisecond in duration, contain four harmonics and span a wide range of frequencies (25 to 65 kilohertz). This broad bandwidth facilitates good resolution of both the target and its complex surroundings. Many bats use such multiple-harmonic, broad-band signals during the final stages of insect pursuit, and even the Brazilian free-tail bat, discussed previously as an example of a bat that pursues prey in situations free of clutter, will use a three-harmonic, broad-band signal when in a confined space.

A second strategy used by some bats foraging in dense clutter is to concentrate on tracking the movement of the prey and, in some cases, of its wing beats. Laboratory studies have demonstrated that some members of the family Rhinolophidae and the mormoopid *Pteronotus parnellii* use the long CF component of their pulse for analyzing the velocity of targets relative to their own velocity. This ability is based on the Doppler shifts in the echoes from targets approaching or moving away from a source (Simmons, 1974). For a stationary bat with a fixed target, or for a bat keeping precise pace with a target, the echo returning from the target has the same frequency as the original pulse; if the prey is moving away from the bat, the echo is Doppler-shifted to a lower frequency, whereas the shift is to a higher frequency if the prey is moving toward the bat. Bats using long CF/FM pulses bring the frequency of the echo to that to which their ears are tuned by altering the frequency of emitted pulses. This remarkable compensatory response occurs within fractions of a second, often in response to erratic movement by a target (Fig. 24–5). The longer the CF component of the pulse, the greater the sensitivity of the bat to target velocities. Computed velocity resolutions listed by Simmons, Howell, and Suga (1975) indicate that the European *Rhinolophus ferrumequinum,* with its very long CF pulse (up to 60 milliseconds), can perceive relative target velocities of less than 0.04 meter per second; the moderately long-pulsed *P. parnellii* (with pulses up to 28 milliseconds) can perceive velocities as low as 0.10 meter per second. These bats, then, concentrate on the trajectory of prey while remaining aware of background clutter. This ability to track the movement of prey with precision enables some CF bats to intercept rather than pursue prey, thus reducing the energy expended on foraging.

Detection of Prey by Fishing Bats.   The Neotropical fishing bat *Noctilio leporinus* locates prey by detecting ripples caused by small fish or by recognizing parts of the fish that break the

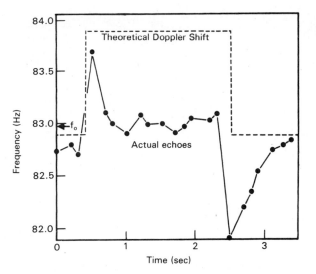

Figure 24–5. Doppler-compensating response of *Rhinolophus ferrumequinum* to a 2.2-s Doppler shift in artificial echoes. Simmons explains this illustration as follows: "The magnitude of the Doppler shift, 1.00 kHz upward (based upon a transmitted frequency of 83.00 kHz), corresponds to an approaching target velocity of 2.0 m/s. The dotted line shows the frequencies to be expected in echoes if the bat continued to emit CF signals at a frequency of 82.90 kHz, while the data points and solid line indicate the frequencies actually returned to the bat, showing that the bat decreased its transmitted frequencies to compensate for the simulated Doppler shift." (From Simmons, 1974)

water surface. During attacks on prey, the sequence is from short CF/FM to FM signals and only one harmonic is used. This fairly simple, single-harmonic pattern is seemingly well adapted to this bat's unique foraging style. Of central importance is the detection of a disturbance on a fairly uniform background, a problem similar to that facing a bat flying high above obstacles.

Jacks of All Echolocation Trades. Simmons, Fenton, and O'Farrell (1978) stress that some bats are extremely adaptable in their use of echolocation. This is the case with both the spear-nosed bat (Phyllostomidae) and the Brazilian free-tail bat (Molossidae). Each of these bats uses an array of signals that spans nearly the entire diversity of echolocation types known for bats. Such flexibility may well occur widely among phyllostomid and molossid bats.

Complexities and Speculations. Novick (1971) suggested that three types of evaluation might be made by a long CF/FM bat during the approach phase: (1) an evaluation of the relative target movement on the basis of an assessment of the frequency of the CF echo; (2) an evaluation of the distance of the target and the speed with which it is being approached on the basis of the interval between the end of the emission of a pulse and the end of its echo (the shorter this time, the closer the target); (3) an evaluation of the size and direction of travel of the target on the basis of assessment of the FM echo.

Some long CF/FM bats, adapted to detecting moving prey at long distances and to precise tracking of prey, enhance their ability to scan their surroundings by coordinating body, head, and ear movements. As an example, when the African species *Hipposideros commersoni* hangs from an acacia branch and scans for prey, its pendent body revolves continuously back and forth through an arc of approximately 180 degrees; the head is in constant motion up and down and the tips of the ears vibrate forward and backward. Such body and head movements enable the bat to use its strongly beamed pulses to meticulously scan the space surrounding its perch. But why the ear movements? Simmons, Howell, and Suga (1975) judged that, by moving the direction the ears aim, these long CF/FM bats scan the vertical plane. The movements of the ears, so rapid as to be perceived by an observer as a vibration or fluttering, are out of phase: the tip of one ear moves forward, toward the target, while the other moves backward, away from the target. Further, the ear movements are in approximate synchrony with the pulses being emitted by the bat. These ear movements, alternately toward and away from the target, may heighten the Doppler shift of echoes from moving targets and improve the bat's discrimination of movement. Even without a complete understanding of functional relationships, however, one cannot help but wonder at the elaborate neural coordination of ster-

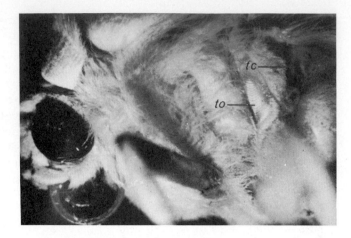

Figure 24-6. The tympanic cavity (*tc*) and timbal organ (*to*) of an arctiid moth. (K. D. Roeder)

eotyped body, head, and ear movements with pulse emission rate.

Bats Versus Moths. Studies by Roeder and his colleagues (Dunning, 1968; Dunning and Roeder, 1965; Roeder, 1965; Roeder and Treat, 1961) have revealed a remarkable series of adaptations by certain moths in response to predation by bats. Nocturnal moths of the families Noctuidae, Ctenuchidae, Geometridae, and Arctiidae have an ear on each side of the rear part of the thorax (Fig. 24–6). Each ear is a small cavity within which is a transparent membrane; the ears are sensitive to a wide range of frequencies and allow the insects to detect the ultrasonic pulses of foraging bats. Upon detecting the approach of a bat, the moths alter their level flight and adopt various erratic flight patterns (Fig. 24–7). Some members of these families of moths have carried the business of evading bats to an even greater extreme and have a noise-

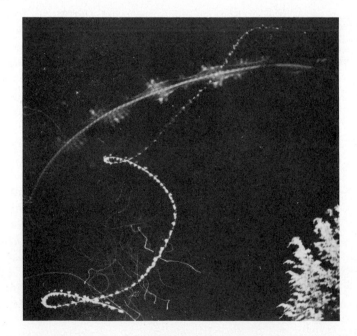

Figure 24-7. The evasive tactics of a moth approached by a bat. (F. A. Webster)

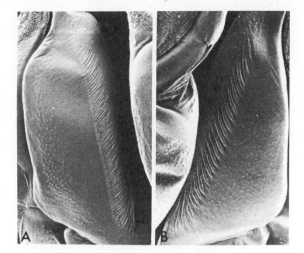

Figure 24-8. Scanning electron micrographs of the microtymbals of two species of moths: (A) *Haploa clymene* and (B) *Pyrrharctia isabella*. Contraction of a muscle that inserts on the concave inner surface of the corrugated episternite causes the microtymbals (the corrugations) to buckle in sequence, and the buckling of each microtymbal produces a signal lasting less than 1 ms and containing ultrasonic frequencies. The number and shape of the microtymbals determine the number and acoustical qualities of the clicks of each sequence. (From Fenton and Roeder, 1974)

making organ on each side of the thorax (Fig. 24-8). When the moths are disturbed, these organs produce trains of clicks with prominent ultrasonic components. Under laboratory conditions, flying bats about to capture mealworms tossed into the air regularly turned away from their targets when confronted with recorded trains of moth-produced pulses (Table 24-2). These pulses apparently protect moths from bats; probably the moths are announcing their identity as bad-tasting prey. Some arctiid moths, and perhaps some noctuids, are unpalatable to bats, and their identification by a bat having previous experience with them might be aided by the moth's ultrasonic clicks. Dunning (1968) found that captive *Myotis lucifugus* avoided three species of arctiid moths when the moths were sounding off but bit into them when they were quiet. Two of the moths were rejected when tasted, but the third, presumably a palatable form mimicking the sounds of the unpalatable species, was eaten.

Recent evidence suggests, however, that some moths jam a bat's echolocation system by

Table 24-2: Responses of Bats to Simultaneous Presentation of Mealworms and Bat or Moth Sounds

Bat Number	Number of Tosses	Number of Contacts (%)	Dodges (%)	Attempts (%)
Bat sounds presented				
3	67	88	7	5
4	150	79	8	12
5	92	65	27	5
Total	309	77	14	8
Moth sounds presented				
3	95	14	86	3
4	249	14	83	3
5	121	11	87	2
Total	465	13	85	2
No sounds presented				
3	141	99	0	1
4	373	98	0	1
5	167	97	1	1
Total	681	98	1	1

After Dunning and Roeder, 1965.

producing sounds that resemble the terminal buzz of a bat closing in on an insect (Fullard, Fenton, and Simmons, 1979).

Investigations of the strategies bats use to hear and feed on moths have shown that predator-prey interactions between these animals are complex, with considerable fine-tuning of adaptations favoring prey detection, predator avoidance, and inconspicuousness on the part of both predator and prey (Fenton and Fullard, 1981). In moth-bat confrontations, the relative times or distances at which each animal detects the other is critical. Bats using loud signals in the 20- to 50-kilohertz range can be detected by some moths at distances up to 40 meters, whereas bats probably detect moths at no more than roughly 20 meters. Neither bats nor moths do consistently well, however, for some bats detect insects at ranges of only 1 meter, and experiments using an African slit-faced bat (*Nycteris macrotis*) showed that moths cannot detect this bat at distances greater than 0.2 meter. The fact that ears are present in more than 95 percent of the moths of Ontario and in more than 85 percent of those in southern Africa suggests that sound detection is important.

Moths have apparently become tuned to bat signals. In most areas, the echolocation frequencies most used by bats are those to which the ears of moths are most highly sensitive. Most bats of North America have echolocation signals at frequencies between 15 and 60 kilohertz, and the ears of moths of this region are tuned to these frequencies. Since some bats obviously eat moths, however, we might well ask, then, how such bats avoid the early warning systems of moths.

One major bat strategy is to use frequencies to which moths are relatively insensitive. Bats that use echolocation signals of low intensity, and those that have high-frequency signals, can escape detection except at close range. In the Ethiopian, Oriental, and Australasian tropics, roughly one third of the species of insectivorous bats use signals of one or both of these types and thereby presumably partially avoid

detection by moths. Fenton and his associates have found that, in many cases, bats with such signals tend to specialize on moths. Each of three moth-eating rhinolophids were found to use very high frequencies (100, 139, and 210 kilohertz).

Another strategy that allows bats to remain inconspicuous to moths is that of partially abandoning echolocation. Some bats, such as some megadermatids, do not use echolocation for detecting prey but instead listen for sounds made by the prey while remaining silent themselves.

Bat-moth predator-prey interactions provide an interesting example of evolutionary gamesmanship. Echolocation is a major key to the great success of bats, allowing them to perceive and capture insects in darkness. Bat echolocation signals provide moths with a means of detecting foraging bats, however, and moths have developed antipredator behaviors that depend on this ability. A third countermove is seemingly in progress: the great variety of hunting styles and echolocation behavior among insectivorous bats may be a response to moth predator-avoidance behavior (Fenton and Fullard, 1981).

Echolocation in Cetaceans

Just as bats must cope with darkness, cetaceans frequently must be able to perceive their underwater environment under conditions that render vision difficult, if not impossible. In some waters inhabited by cetaceans, suspended material such as soil particles or plankton limits visibility to a few meters or even a few centimeters. Water transmits light poorly, and even under ideal conditions visibility under water is limited. Also, some cetaceans forage at considerable depths, where there is not a trace of light. It is not surprising, then, that some cetaceans have developed echolocation.

Probably all odontocetes (toothed whales and dolphins) use echolocation for avoiding obstacles and for detecting prey. Mysticetes (the baleen whales), however, are not known to echolocate. A tremendous variety of sounds, some having a fascinating musical quality, are made by both mysticetes and odontocetes; the wailing, creaking, and squealing noises of cetaceans have become commonplace to sailors operating sonar equipment at sea. Payne (1970) has recorded the remarkable "songs" of the humpback whale *(Megaptera)*. Some of these underwater sounds have a communication function (see, for example, C.W. Clark and Clark, 1980). The following discussion primarily considers pulses used by cetaceans in connection with echolocation.

Insulation of Ear from Skull. Each ear of a cetacean functions as a separate hydrophone, allowing the animal to localize a sound source by discriminating (as we do) between the times the sound is received by each ear. The pressure that sound transmitted through water exerts on the bones of the entire skull causes vibrations to be transmitted by the skull. When the bone that houses the middle and inner ear is attached rigidly to the skull, as it is in most mammals, vibrations from water are transmitted through the bones of the skull and reach the ear from various directions. As a consequence, when a mammal with this type of skull is submerged, it is unable to localize accurately the source of a sound. Because sound localization is of great importance to cetaceans that use echolocation, these animals have evolved several structural features that insulate the bone surrounding the middle and inner ear (tympanic bulla) from the rest of the skull.

First, the tympanic bullae are not fused to the skull in any cetacean, and in the specialized porpoises and dolphins, the bullae are separated by an appreciable gap from adjacent bones of the skull. In addition, the bullae are insulated by an extensive system of air sinuses unique to cetaceans. These air sinuses surround the bullae and extend forward into the enlarged pterygoid fossae (Fig. 15–6), and each sinus is connected by the eustachian tube to the cavity of the middle ear. The sinuses are filled with an oil-mu-

cous emulsion, foamed with air, and are surrounded by fibrous connective tissue and venous networks. These sinuses can apparently retain air even when subjected to pressures of 100 atmospheres, pressures higher than those to which cetaceans are subjected during deep dives. Purves (1966:360) used a foam in which gelatin was substituted for mucous and oil and found that, at such pressures, the air became dispersed in the mixture as tiny bubbles. The foam in the air sinuses apparently forms a layer around the bullae that retains remarkably constant sound-reflecting and sound-insulating qualities through a wide range of pressures.

Norris (1964, 1968) regards the extremely thin back part of the lower jaw of delphinids as an acoustical window. He holds that sound passes into the skin and blubber overlying the dentary, through the thin part of this bone, which at its thinnest may be only 0.1 millimeter thick, to the intramandibular fat body, which leads directly to the wall of the auditory bulla, into which the sound presumably passes. Weight is given to the Norris hypothesis by experiments done by Yanagisawa, Sata, et al. (1966), who found that the jaw is the most acoustically sensitive area of the dolphin's head.

Echolocation Patterns. Since the account by Schevill and Lawrence (1949) of the underwater noises made by the white whale *(Delphinapterus)*, considerable research has been done on the vocalizations of cetaceans. Much research has dealt with one of the common dolphins, *Tursiops truncatus* (Kellogg, Kohler, and Morris, 1953; Lilly, 1962, 1963; Norris, Prescott, et al., 1961; Schevill and Lawrence, 1953; F.G. Wood, 1959). *Tursiops* is able to detect obstacles and recognize food by means of echolocation, and, in the use of short pulses, its sonar system resembles that of bats. *Tursiops* is capable of producing a great variety of sounds, but of primary importance for echolocation are the trains of clicks that it emits. The clicks are audible to humans but cover a wide spectrum of frequencies, some of which are ultrasonic. The rate of click emission varies among odontocetes: the killer whale *(Orcinus orca)* has a slow repetition rate of from six to 18 clicks per second (Schevill and Watkins, 1966), whereas the Amazon dolphin *(Inia geoffrensis)* emits click trains at rates of from 30 to 80 per second (D.K. Caldwell, Prescott, and Caldwell, 1966). The pulse rate rises as a porpoise approaches a target, and *T. truncatus* can distinguish between a piece of fish and a substitute water-filled capsule with a similar shape (Norris, Prescott, et al., 1961), or even between sheets of different thicknesses of the same metal (W.E. Evans and Powell, 1967).

Where do the cetacean clicks and buzzes originate? Experimental observations using ultrasonic beams indicate that these sounds are produced by vibrations of the nasal plugs (R.S. Mackay and Liaw, 1981). These plugs are muscular valves at the blowhole that close the nares. How cetaceans produce whistles is not definitely known.

The sperm whale *(Physeter catodon)* presents an unusually interesting case of echolocation among cetaceans because it uses a unique mode of foraging that is probably made possible by echo ranging. (No proof that sperm whales echolocate is available, but this ability has been inferred on the basis of data from other odontocetes.) The click of a sperm whale is known to consist of a series of pulses (Backus and Schevill, 1966). The click lasts roughly 24 milliseconds and is composed of up to nine separate pulses. These vary in duration from 2 to 0.1 millisecond, and the interpulse intervals are 2 to 4 milliseconds. The clicks are repeated at rates of from less than one click per second to 40 per second. Of particular interest is the fact that the sperm whale feeds largely on squid that it takes at depths (down to at least 1000 meters) at which prey is scarce and light is virtually absent. It appears, therefore, that the sperm whale is able to utilize deep-sea foraging largely because it is able to use echo scanning to locate food under conditions that require efficient long-range echolocation. Backus and Schevill (1966:525) estimated that sperm whales, by using echo scanning, can detect prey up to 400 meters away.

Some small cetaceans that inhabit turbid water have tiny eyes and presumably are dependent on echolocation. One of the most highly specialized of these is the blind river dolphin *(Platanista gangetica),* an inhabitant of the muddy and murky waters of the Ganges, Indus, and Brahmaputra river systems of India and Pakistan. This unusual dolphin habitually swims on its side with the ventralmost flipper either touching the bottom or moving within 2 or 3 centimeters of it (Fig. 15–17). *Platanista* has greatly reduced eyes that are barely visible externally. The lens is absent, but the retina apparently retains the ability to detect light, although doubtless no image can be formed. The tiny eye opening is surrounded by a sphincter muscle, and another muscle opens the sphincter. Blind river dolphins in captivity continuously produced series of pulses at rates of from 20 to 50 per second, primarily in the frequency range between 15 and 60 kilohertz, and the animals have a remarkable ability to direct their pulses into a narrow beam (Herald, Brownell, et al., 1969).

Studies by W.E. Evans, Sutherland, and Beil (1964) have demonstrated that the shape of the skull of *Platanista* affects the directional beaming of pulses (which are emitted through the blowhole). The pulses are reflected by the concave front of the skull and focused by the "melon," a lens-shaped fatty structure that gives a domed profile to the forehead of many odontocetes. The skull of *Platanista* is modified by large flanges from the maxillary bones (Fig. 24–9). These prominent flanges, rounded on the outside but with an intricate, radial pattern of latticework on the inside, probably serve as acoustical baffles that, with the melon, concentrate the pulses into a narrow beam (Herald, Brownell, et al., 1969). Observations of swimming dolphins and reception of these pulses with a hydrophone and amplifier system showed that they are indeed effectively beamed. When a dolphin's snout was directed as little as 10 degrees on either side of the receiver, the intensity

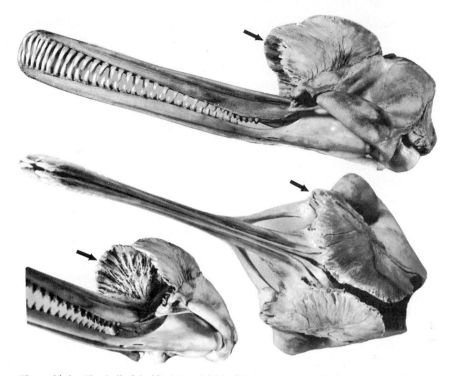

Figure 24–9. The skull of the blind river dolphin *(Platanista gangetica),* showing the highly developed extensions of the maxillary bones. (From Herald, Brownell, et al., 1969)

of the pulses dropped markedly, and a far greater drop occurred when the angle was greater than 40 degrees. As the dolphin swims on its side, it moves its head constantly in a sweeping action close to the bottom. One is tempted to speculate that the dolphin is scanning the area ahead with a beam of pulses, using a system similar in some ways to that of the horseshoe bat, which also uses beamed pulses. Such scanning in the dolphin might serve effectively both in determining bottom contours and in finding food.

Echolocation by Insectivores, Pinnipeds, and Other Mammals

Echolocation is known to occur in four orders of mammals (Insectivora, Chiroptera, Odontoceti, and Carnivora) and may have evolved independently in each of these groups. Highly suggestive evidence also points toward the occurrence of echolocation in some other orders.

The high-pitched sounds made by insectivores when they explore unfamiliar surroundings or objects (Crowcroft, 1957; Komarek, 1932; Reed, 1944; Swinhoe, 1870) are apparently used for echolocation. In a series of carefully controlled laboratory experiments, E. Gould, Negus, and Novick (1964) demonstrated that three species of *Sorex* could echolocate. These shrews searched around an elevated disk, found a lower platform, and jumped to it, all without the use of tactile, visual, or olfactory senses. While the shrews searched their environment, they emitted pulses with frequencies between 30 and 60 kilohertz; the pulse duration was from 5 to above 33 milliseconds. The familiar short-tail shrew of the eastern United States, *Blarina brevicauda,* produced similar pulses, and all shrews studied produced pulses with the larynx. Laboratory trials revealed that tenrecs (Tenrecidae) from Madagascar could also echolocate (E. Gould, 1965). These primitive insectivores produced pulses by clicking the tongue, and the pulses were of frequencies audible to humans (from 5 to 17 kilohertz). Careful experimental work by Buchler (1976) demonstrated the use of echolocation by the wandering shrew *(Sorex vagrans).*

Pinnipeds produce a variety of underwater sounds (Ray and Schevill, 1965; Schevill and Watkins, 1965), some of which are of a type that might be used for echolocation. The Weddell seal *(Leptonychotes weddelli)* of Antarctic waters, for example, emits chirps with frequencies up to 30 kilohertz. These chirps are produced regularly as the seals swim between air holes, and Watkins and Schevill (1968) suggested that they may have a navigational value. The California sea lion *(Zalophus californianus)* is probably able to echolocate (W.E. Evans and Haugen, 1963; Poulter, 1963; Schusterman and Feinstein, 1965) and makes extensive use of vocal signals for communication both in the water and on land. As in the case of the insectivores, our knowledge of echolocation in pinnipeds is limited.

Diverse lines of evidence indicate that some form of echolocation may occur in still other mammals. Flying lemurs (order Dermoptera) move in a series of jumps interspersed with pauses and emit series of pulses (Burton, 1949; Tate, 1947); these could be used in echolocation. *Antechinus,* a small, nocturnal, tiny-eyed Australian marsupial that is convergent toward shrews, may possibly echolocate (E. Gould, Negus, and Novick, 1964). Two rodents, the fat dormouse (*Glis,* Gliridae) and the golden hamster (*Mesocricetus,* Muridae), are able to locate perches without the use of tactile, visual, or olfactory senses and can presumably echolocate (Kahmann and Ostermann, 1951). Young mice are known to emit ultrasonic pulses (Zippelius and Schleidt, 1956), but the pulses are important in communication (p. 389), and their use for echolocation has not been demonstrated. Griffin (1958:297) discusses the ability of some blind people to echolocate. One is tempted to speculate that a latent ability to echolocate is common to many mammals.

The Specialized Ear of the Kangaroo Rat

One of the most remarkable cases of ear specialization is found in the kangaroo rat *(Dipodomys).* In this animal, the enormous auditory bullae, which have a total volume greater than that of the braincase, were a source of wonder to mammalogists for years. In recent decades, however, partly due to the studies of D.B. Webster (1961; 1963; 1966; Webster, Ackermann, and Longa, 1968; Webster and Stack, 1968), the functional importance of this and of other auditory specializations of the kangaroo rat has been clarified.

Before discussing the inflated bullae, let us consider some other important specializations. First, the malleus lacks the anterior and lateral ligaments, that in most mammals brace this bone, and therefore rotates unusually freely. In addition, the manubrium of the malleus, which rests against the tympanic membrane, is greatly lengthened; it thus serves as a lever arm that transforms the relatively weak vibrations of the tympanic membrane into more powerful movements transmitted to the incus and stapes. Further, the tympanic membrane is exceptionally large, and the footplate of the stapes, which rests against the oval window of the inner ear, is small. This forms a piston system of sorts. Because force per unit area (pressure) is increased in proportion to the difference in surface areas of the two structures, relatively low pressure on the large tympanic membrane is transformed into relatively high pressure on the fluid in the inner ear via the small footplate of the stapes. There is, then, great amplification of pressure by the combined means of the piston system and the long lever arm of the malleus. The degree of this amplification is expressed by the *transformer ratio,* which in Merriam's kangaroo rat is extremely high, about 97:1. (This ratio in humans is roughly 18:1.)

The inflation of the auditory bullae of kangaroo rats (Fig. 16–13) results in a great increase in the volume of the air-filled chambers surrounding the middle ear. This increase in volume reduces the resistance the enclosed air offers to the compression caused by inward movement of the tympanic membrane. Consequently, the damping effect on vibrations of the tympanic membrane is diminished. The advantage gained during transmission of low-frequency sounds is especially great, for these sounds cause relatively great movements of the membrane. Laboratory experiments in which the bullae of kangaroo rats were filled with plasticine demonstrated that the transmission ability of the tympano-ossicular system in individuals lacking use of the enlarged middle ear chambers is seriously lowered. Both experimental and control animals, however, were most sensitive to low-intensity sounds with frequencies between 1 and 3 kilohertz.

In an attempt to simulate natural predator-prey confrontations, Webster tested the reactions of kangaroo rats to two predators adapted to hunting in darkness. He used the sidewinder rattlesnake *(Crotalus cerastes),* which detects its prey by using heat-sensitive facial pits, and the barn owl *(Tyto alba),* which can use hearing to locate prey. By using delicate recording equipment, Webster found that the wings of owls produce low-frequency (below 3 kilohertz) whirring sounds as the birds swoop toward prey, and a rattlesnake produces a short burst of low-frequency sound (perhaps when scales rub against the substrate) just before it strikes. Under experimental conditions, when a kangaroo rat heard these low-intensity, low-frequency sounds produced by attacking predators, it made a sudden vertical leap and avoided capture. In contrast, individuals with artificially reduced middle ear volumes, and presumably reduced ability to hear faint sounds, could not evade capture. Clearly, the beautifully coordinated specializations of the tympano-ossicular system and the auditory bullae are highly adaptive: in darkness, kangaroo rats can detect the faint sounds of attacking predators, make evasive leaps, and go on to reproduce their sharp-eared kind.

Kangaroo rats are not alone in having enlarged bullae. Many other rodents, including other heteromyids, some South American hystricomorphs, gerbils (Muridae), jerboas (Dipodidae), the springhaas (Pedetidae), and Australian hopping mice (*Notomys,* Muridae), have this specialization, as do elephant shrews (Macroscelididae). Whether or not the adaptive importance of enlarged bullae is the same among all these mammals is not known. Many kinds of mammals that live in deserts have auditory bullae that are larger than those of relatives occupying cooler and more humid areas. Sound is known to be absorbed to different degrees under different conditions of temperature and humidity (Knudsen, 1931, 1935), and the enlargement of the bullae may compensate for the poor sound-carrying qualities of warm, dry desert air.

Chapter 25

Human Impact on Mammals

We humans have long been interested in our fellow mammals and have long exploited them. More than a million years ago, *Australopithecus* was killing and eating baboons and antelope, and the use of mammals for food remains characteristic of most cultures today. Many kinds of mammals have been domesticated, and some are taught to work for their owners. The trained Indian elephant lifts and drags teak logs in the tropical forests of Ceylon, where the periodically saturated soil limits the usefulness of vehicles; dogs help some African hunters capture antelope and other game; trained rhesus monkeys pick coconuts from tall trees and drop them to their masters; and even the vicious and rebellious camel has been trained. The raising of various kinds of mammals—from homozygous strains of mice for medical research to stocky beef cattle—is an important enterprise today. The very distribution of early humans was probably influenced by their ability to kill their fellow mammals, for the skins and furs of mammals may have enabled primitive humans, probably endowed with hopelessly inadequate insulation, to penetrate cool or cold regions.

Wild mammals are still under pressure from humans in nearly all parts of the world. In many primitive areas, the inhabitants hunt year around and either are basically hunters and partly depend on mammals for food or at least hunt to supplement limited food supplies. In the so-called developed countries, many people hunt for sport during regulated seasons. In the United States, fur trapping has long been important. Not only was the exploration of the western United States largely accomplished by trappers seeking new territory, but in more recent years trapping has been an important source of

seasonal income to some. In many sections of the United States, trapping provides extra income for people who have irregular employment. When prices for a particular type of fur are high, these people may put intense pressure on this mammal. The bobcat, for example, is trapped heavily in the West when its fur is valuable, with the obvious result that, at least locally, its populations have been decimated. In my judgment, this destruction of many thousands of carnivores each year for the sake of furthering high fashion in the United States and abroad is impossible to justify.

Hunting remains important in many parts of the United States, and the sale of ammunition, firearms, and other equipment associated with hunting is big business. At its best, hunting provides an opportunity for people to get away from their work in the city and to experience the sights, smells, and sounds of primitive country. At its worst, however, hunting results in the unnecessary killing of many nongame species of birds and mammals, in considerable loss of livestock, and in human deaths.

Wild animals can be costly. Pocket gophers, rabbits, meadow voles, ground squirrels, and even deer and elk may damage crops or rangeland, and efforts to combat these losses are frequently expensive. In addition, the United States federal government supports considerable research on mammals and finances the local control of some species. In 1971, federal animal control programs cost United States taxpayers over $8 million. Unfortunately, control programs using poisons against coyotes have frequently resulted in tremendous losses of other carnivores, such as gray foxes, kit foxes, and badgers. Many biologists deplore these losses and regard such programs as a misuse of federal monies.

The Impact of Humans

The long-term exploitation of mammals by humans has had a devastating impact. In the last 400 years, some 36 species of mammals have become extinct, and today over 100 species are threatened by extinction. Some species were disposed of remarkably summarily. Steller's sea cow of the Bering Sea was pushed to extinction only 27 years after its first discovery by white hunters. Sea otters, which were hunted along the Pacific Coast of North America at least as early as 1786, were killed for their valuable fur; probably more than 200,000 were killed between 1786 and 1868 (Evermann, 1923:524). By 1900, these animals were rare over much of their range, and they were seemingly lucky to have survived until protected by legislation in the early 1900s. Not so lucky was the grizzly bear in California. In the 1890s, grizzlies still persisted in the San Gabriel Mountains near Los Angeles, but the last known southern California grizzly was killed in 1916, and the last verified occurrence in California was in 1922, in the foothills of the Sierra Nevadas in central California (J. Grinnell, Dixon, and Linsdale, 1937:93). Only about 60 years were required to bring the grizzly in California from fair abundance to total extirpation. In Mexico, a population of grizzlies that survived in a small mountain range in central Chihuahua in 1957 was probably wiped out by 1963, the very year when funds were raised by the World Wildlife Fund to set aside a refuge for the animals.

Mammals in the Eastern Hemisphere have fared no better. The quagga, a zebra that inhabited southern Africa, was extirpated in the wild about 1860, and another type of zebra was exterminated by roughly 1910. The Arabian oryx, well on its way to extinction, has been hunted in recent years with machine guns mounted on jeeps—surely a manly style of hunting. The black rhinoceros of East Africa has been extirpated over broad areas, and its survival in the wild seems unlikely.

The impact of humans has nowhere been more strongly felt than by some marine mammals. Through persistent hunting, the blue whale, the largest animal of all times, has been reduced to a total population of probably no more than several hundred individuals. Other human activities have left their mark: pesticides have seemingly caused declines in bat populations in some areas, a freeway in California has transected the unusually small range of the handsome Morro Bay kangaroo rat, and the propellers of power boats have injured or killed many members of a declining population of Florida manatees.

Habitat destruction is a major problem over large areas of the Eastern Hemisphere. During parts of 1973 and 1974, I lived at Bushwhackers Safari Camp, a group of thatch-roofed cabins situated well away from any village in southern Kenya. In May 1973, Bushwhackers was surrounded by continuous tropical deciduous woodland, which supported not only a rich assemblage of smaller mammals but also rhinoceroses, giraffes, hippopotami, and buffalo. At night, hyenas whooped as they patrolled the banks of the nearby Athi River, lions were heard occasionally, and waterbuck, bushbuck, dikdik, and lesser kudu were seen regularly near camp.

During my stay there was rapid change, however. The land surrounding the camp became progressively more heavily settled by Kamba families, whose shambas were scattered throughout the woodland. (Shambas are small, several-acre plots, cleared by cutting and burning vegetation, where families live, plant corn, and often keep goats or cattle.) Although the practice was illegal, many families set snares for game in the thorn fences surrounding the shambas or in game trails in the woodland. By August 1974, the rhinoceroses, giraffes, and buffalo were nearly gone, and during even a short hike through the bush one could find the remains of wild animals snared and butchered for food.

I returned to Bushwhackers briefly in November 1982 and viewed a scene becoming common over an increasingly broad section of Africa. The area was given over to shambas and was criss-crossed by foot trails and roads. Cattle and goats grazed the few remaining areas of bush, and clearings that had formerly supported dense grass were barren and trampled. The western sky in the evening was streaked with smoke from fires burning the vegetation cleared from land to be used for shambas. At night we heard no hyenas; in the morning the sandbars along the river bore no tracks made during the night by hippopotami or rhinoceroses or buffalo.

I place no blame on the Kamba families. These people are trying to improve their lot, or in some cases simply to survive: they plant corn in new land and exploit the wildlife that remains in an effort to supplement their monotonous diet. Nor can blame be placed on the poor people who are having the same effect on the land in such places as Mexico, other parts of Africa, and southeastern Asia. In all these areas, however, the clearing of land and the use of subsistence agriculture, while providing only a marginal living for the people, are rapidly restricting many types of wildlife habitat to islands surrounded by land under pressure from the heavy hand of man.

On the positive side, the national park systems in Kenya, Tanzania, Zambia, and a number of other African countries deserve high praise for their efforts to maintain areas in which wildlife is protected. The African game rangers in many of the national parks and game reserves are doing a dedicated job in controlling poaching and the intrusion of domestic livestock. Through tourism, these natural areas bring large amounts of revenue to many countries, a situation that helps reinforce the countries' growing regard for wildlife.

When a country has economic problems, however, wildlife protection programs often suffer. In Kenya in 1983, the governmental support of national parks and reserves was insufficient to allow for adequate patrolling of boundaries. In Tsavo West National Park, in April 1983, thousands of Masai cattle crossed a section of the western boundary each day and foraged inside the park in an area already heavily used by wild ungulates.

It is obvious that the fate of wildlife depends on the persistence of appropriate habitat, which is being destroyed over broad areas at an ever-increasing rate by ever-expanding human populations. In many developing countries, where the focal point of the inhabitants' existence is the day-to-day search for food and fuel, the pressure on land and wildlife is intense. This pressure becomes vastly more acute when revolution, struggles for independence, or strife between political factions is associated with reigns of lawlessness.

The situation in Angola, Africa, eloquently described by Huntley (1976), provides a tragic case in point. Here, in Quissama National Park, for example, hundreds of red buffalo (*Syncerus caffer nanus*), members of a primitive race of African buffalo, have been lawlessly slaughtered with machine guns by soldiers after meat for themselves and their supporters, and in Luando District Nature Reserve, the future of the giant sable antelope has been uncertain since the Conservation Department personnel fled after repeated threats on their lives by undisciplined bands of liberation soldiers. During an aerial survey of Quissama National Park in November 1974, only bleaching skeletons and a network of jeep tracks were seen on the plains where red buffalo, eland, roan antelope, and reedbuck had once been abundant. As seems typically to be the case, white people were the major exploiters. Their attitude is embodied in the statement of an incorrigible white poacher charged with killing four elephants: "If the blacks are to be given Angola, I'm not leaving them any ivory."

The Angolan wildlife tragedy, unhappily not a unique and isolated case, is summed up by Huntley: "The dark cloud that has moved over Angola will surely pass, but in its wake little of a once-rich fauna and flora will remain. We cannot expect that many of the already threatened species—giant sable, gorilla, chimpanzee, manatees, and so on—will survive, but we can only hope that the small nuclei that escape the ho-locaust will suffice to repopulate the national parks. As hard as the realities of the situation are to bear, those who are devoted to the conservation of the country's magnificent wildlife heritage must face the truth and speak it, that the world may record, if not learn from, its mistakes.''

We have clearly reached the eleventh hour. If we do not learn from such mistakes, the wildlife of the world will pay a devastating price.

But it is impossible to lament the decimation of wildlife without recognizing the terrible human tragedies that accompany such upheavals as that occurring in Angola. Man's inhumanity to man must be of prime concern. For many centuries, history has been dominated by the overriding inability of the human race to develop social systems that allow people to live together and countries to coexist peacefully. Some politicians have suggested (with little justification, it would seem) that fear of our modern instruments of war would hasten development of peaceful ways of solving our social problems.

The need for the human race to set its own house in order is basic both to our own survival and to the perpetuation of the biological richness of the world. Clearly, our ability to solve social problems and the future of wildlife are tightly linked. People not under the pressures and stresses occasioned by high populations and limited resources, and at peace with one another, can work toward saving the biotas of the world—fearful people with empty stomachs make poor conservationists. On the success of the politicians, scientists, and teachers of the world in halting the rise in human populations and stopping strife between peoples hinges the survival of our biotic heritage.

A New Conservation Ethic

Ecologists and conservationists have argued that each element of a biota plays an essential part in the ecosystem it occupies, and that the loss of even a seemingly insignificant species might tip the delicate biotic balance. We do not begin to know the degree of pressure that most ecosystems can tolerate before collapse. Biologists have stressed practical problems: the pollution of water should be avoided not only because of the potential for serious public health problems but also because the ecosystem of a stream might be drastically altered and species of fish from which we derive pleasure or some monetary return might disappear. Range managers have emphasized economic problems: unwise grazing practices alter a grassland ecosystem to the point where its economic importance is reduced; in concrete terms, the weight that each head of cattle gains per day may be reduced to a point at which ranchers can no longer realize a profit. But can we, with due respect for honesty, justify conservation with only these kinds of arguments?

In his excellent discussion entitled *The Conservation of Non-resources,* Ehrenfeld (1976) points out that attempts to justify the conservation of many species on the basis of their economic importance are unjustifiable scientifically. The conservation doctrines lose force if a species is destroyed without the disruption of its ecosystem. If the ecosystem is destroyed following the loss of the species, however, not only is it too late to save the day, but the cause-and-effect relationship involved can never be proved and may not even be hypothesized.

Perhaps we must look to a new conservation ethic. Few species can be proved essential to the survival of their ecosystems or to have great economic value; nonetheless, each species forms a part of a biological richness developed over millions of years, and each is worthy of perpetuation at least in part because of what Ehrenfeld terms its ''natural art value.'' Two moving presentations of this view are quoted by Ehrenfeld. In his book *Ulendo: Travels of a Naturalist in and out of Africa,* Carr (1964) states, ''It would be cause for world fury if the Egyptians should quarry the pyramids, or the French should loose urchins to throw stones in the Louvre. It would be the same if the Americans dammed the Valley of the Colorado. A reverence for original landscape is one of the humanities. It was the first humanity. Reckoned in terms of human nerves and juices, there is no difference in the value of a work of art and a work of nature. There is this difference though. . . . Any art might somehow, some day be replaced—the full symphony of the savanna landscape never.''

Regarding specific nonresource species, in this case small primates called lion tamarins (three species of the genus *Leontideus*), Coimbra-Filho, Magnanini, and Mittermeier (1975) write: ''In purely economic terms, it really doesn't matter if three Brazilian monkeys vanish into extinction. Although they can be (and previously were) used as laboratory animals in biomedical research, other far more abundant species from other parts of South America serve equally well or better in laboratories. Lion tamarins can be effectively exhibited in zoos, but it is doubtful that the majority of zoo-goers would miss them. No, it seems that the main reason for trying to save them and other animals like them is that the disappearance of any species represents a great esthetic loss for the entire world. It can perhaps be compared to the destruction of a great work of art by a famous painter or sculptor, except that, unlike a man-made work of art, the evolution of a single species is a process that takes many millions of years and can never again be duplicated.''

Although effort expended on the perpetuation of some species can be justified on the basis of economic importance, I would regard this natural art value of many species as their greatest importance. A more basic argument would insist that a species should be preserved because of a reverence for its vast evolutionary history and that each species has a right to play out its evolutionary role; we have no moral right to

set ourselves up as the instrument of their destruction. Certainly, present attitudes must change if our biotas are to survive, and an as yet unformulated rationale or conservation ethic may unite enough of humankind behind the conservation cause to turn the tide.

If the present decimation of the earth's wildlife is to be curtailed, at least the following steps must be taken:

1. Human population growth must be halted and the need for space and resources stabilized.
2. Large tracts of land undisturbed by humans must be maintained for wildlife.
3. Our exploitation of many species (whales are notable examples) must be drastically reduced.
4. A broad understanding of the meaning of such biotic interactions as predation must underlie an interest in preserving balanced faunas.
5. Control of animals threatening crops and livestock must be local, with no attempt to exterminate a species over wide areas with little respect to the damage it is doing.
6. The use of biocides must be carefully controlled.
7. We must accept some types of economic losses and inconvenience caused by wildlife, and must feel that these are more than compensated for by our enjoyment of a rich and balanced biota.

Because the survival of the world's wildlife and indeed our very own survival are in our hands, one cannot help but fervently hope that the word *sapiens* (meaning wise) becomes a justly earned part of the name *Homo sapiens*.

Bibliography

Note: This list includes, in addition to publications cited in the text, a few additional important references by cited authors.

Abbott, J.H., and A.E. von Doenhoff, 1949. Theory of Wing Sections. McGraw-Hill, New York.

Abbott, K.D., 1971. Water economy of the canyon mouse, *Peromyscus crinitus stephensi.* Comp. Biochem. Physiol. 38A:37.

Ables, E.D., 1969. Home range studies of red foxes *(Vulpes vulpes).* J. Mamm. 50:108.

Adcock, E.W., F. Teasdale, C.S. August, S. Cox, G. Meschia, F.C. Battaglia, and M.A. Naughton, 1973. Human chorionic gonadotropin: Its possible role in maternal lymphocyte suppression. Science 181:845.

Alcorn, S.M., S.E. McGregor, G.D. Butler Jr., and E.B. Kurtz Jr., 1959. Pollination requirements of the saguaro *(Carnegiea gigantea).* Cactus Succ. J. Amer. 31:39.

Alexander, R.D., 1974. The evolution of social behavior. Ann. Rev. Ecol. System. 5:325.

Allee, W.C., A.E. Emerson, O. Park, T. Park, and K.P. Schmidt, 1949. Principles of Animal Ecology. W.B. Saunders, Philadelphia.

Allen, G.M., 1940. The mammals of China and Mongolia, part 2. Amer. Mus. Nat. Hist., New York.

Allen, J.A., 1924. Carnivora collected by the American Museum Congo Expedition. Bull. Amer. Mus. Nat. Hist. 47:73.

Allin, E.F., 1975. Evolution of the mammalian middle ear. J. Morph. 147:403.

Allison, J.J., and H.M. Cook, 1981. Oxalate degradation by microbes of the large bowel of herbivores: The effect of dietary oxalate. Science 212:675.

Altenbach, J.S., 1977. Functional morphology of two bats: *Leptonycteris* and *Eptesicus.* Amer. Soc. Mamm., Spec. Publ. No. 5.

Altman, P.L., and D.S. Dittmer, 1964. Biology Data Book. Fed. Amer. Soc. Exp. Biol. Washington, D.C.

Altmann, S.A., and J. Altmann, 1970. Baboon ecology: African field research. University of Chicago Press, Chicago. 220pp.

Amoroso, E.C., and J.H. Matthews, 1951. The growth of the grey seal *(Halichoerus grypus)* from birth to weaning. J. Anat. 85:426.

Andersen, D.C., and J.A. MacMahon, 1981. Population dynamics and bioenergetics of a fossorial herbivore, *Thomomys talpoides* (Rodentia: Geomyidae), in a spruce-fir sere. Ecol. Monogr. 51:179.

Andersen, H.T. (ed.), 1969. The Biology of Marine Mammals. Academic Press, New York.

Anderson, J.M., 1972. Nature's Transplant: The Transplantation Immunology of Viviparity. Appleton-Century-Crofts, New York.

Anderson, J.W., 1954. The production of ultrasonic sounds by laboratory rats and other mammals. Science 119:808.

Anderson, S., 1967a. Primates, *in* Recent Mammals of the World (S. Anderson and J.K. Jones Jr., eds.). Ronald Press, New York.

————, 1967b. Introduction to the Rodents, *in* Recent Mammals of the World (S. Anderson and J.K. Jones Jr., eds.). Ronald Press, New York.

———— and J.K. Jones, Jr. (eds.), 1984. Orders and Families of Recent Mammals of the World. Wiley, New York.

Andersson, A., 1969. Communication in the lesser bushbaby *(Galago senegalensis moholi).* Unpublished M.S. thesis, University of Witwatersrand.

Annison, E.F., and D. Lewis, 1959. Metabolism in the Rumen. Methuen, London.

Ansell, W.F.H., 1971. Order Artiodactyla, *in* The Mammals of Africa: An Identification Manual (J. Meester and H.W. Setzer, eds.). Smithsonian Institution Press, Washington, D.C.

Archer, M., and J.A.W. Kirsch, 1977. The case for the Thylacomyidae and Myrmecobiidae, Gill, 1872, or why are marsupial families so extended? Proc. Linn. Soc. New South Wales 102:18.

Arey, L.B., 1974. Developmental Anatomy, rev. 7th edn. W.B. Saunders, Philadelphia.

Arieli, R., and A. Ar, 1981a. Heart rate responses of the mole rat *(Spalax ehrenbergi)* in hypercapnic, hypoxic, and cold conditions. Physiol. Zool. 54:14.

————, ————, 1981b. Blood capillary density in heart and skeletal muscles of the fossorial mole rat. Physiol. Zool. 54:22.

Armitage, K.B., 1962. Social behaviour of a colony of the yellow-bellied marmot *(Marmota flaviventris)*. Anim. Behav. 10:319.

Armstrong, R.B., C.D. Ianuzzo, and T.H. Kunz, 1977. Histochemical and biochemical properties of flight muscle fibers in the little brown bat, *Myotis lucifugus.* J. Comp. Physiol. (B) 119:141.

Asdell, S.A., 1964. Patterns of Mammalian Reproduction, 2nd edn. Cornell University Press, Ithaca, N.Y.

Augee, M.L., and E.J.M. Ealey, 1968. Torpor in the echidna, *Tachyglossus aculeatus.* J. Mamm. 49:446.

Aumann, G.D., 1965. Microtine abundance and soil sodium levels. J. Mamm. 46:594.

Axelrod, D.I., and H.P. Bailey, 1968. Cretaceous dinosaur extinction. Evolution 22:595.

Backhouse, K.M., 1954. The grey seal. Univ. Durham Coll. Med. Gaz. 48(2):9.

Backus, R.H., and W.E. Schevill, 1966. *Physeter* clicks, *in* Whales, Dolphins and Porpoises (K.S. Norris, ed.). University of California Press, Berkeley.

Bailey, T.N., 1974. Social organization in a bobcat population. J. Wildl. Manage. 38:435.

Baker, H.G., 1961. The adaptation of flowering plants to nocturnal and crepuscular pollinators. Quart. Rev. Biol. 36:64.

——, 1973. Evolutionary relationships between flowering plants and animals in American and African tropical forests, Tropical Forest Ecosystems in Africa and South America: A Comprehensive Review (B.J. Meggers, E.S. Ayensu, and W.D. Duckworth, eds.). Smithsonian Institution Press, Washington, D.C.

—— and B.J. Harris, 1957. The pollination of *Parkia* by bats and its attendant evolutionary problems. Evolution 11:449.

Baker, J.R., and R.M. Ranson, 1933. Factors affecting the breeding of the field mouse *(Microtus agrestis)*. Proc. Roy. Soc. London 113B:486.

Baker, M.A., and L.W. Chapman, 1977. Rapid brain cooling in exercising dogs. Science 195:781.

Baker, R.H., 1971. Nutritional strategies of myomorph rodents in North American grasslands. J. Mamm. 52:800.

Baker, R.J., 1967. Karyotypes of bats of the family Phyllostomatidae and their taxonomic implications. Southwestern Nat. 12:407.

Baker, W.W., S.G. Marshall, and V.B. Baker, 1968. Autumn fat deposition in the evening bat *(Nycticieus humeralis)*. J. Mamm. 49:314.

Balinsky, B.I., 1975. An Introduction to Embryology, 4th edn. W.B. Saunders, Philadelphia.

Balph, D.E., and A.W. Stokes, 1963. On the ethology of a population of Uinta ground squirrels. Amer. Midl. Nat. 69:106.

Banfield, A.W.F., 1954. Preliminary investigation of the barren ground caribou. Part II. Life history, ecology and utilization. Can. Wildl. Serv., Wildl. Manage. Bull. ser. 1., no. 10B.

Barash, D.P., 1973. The social biology of the Olympic marmot. Anim. Behav. Monogr. 6:172.

——, 1974. The evolution of marmot societies: A general theory. Science 185:415.

——, 1976. Social behavior and individual differences in free-living Alpine marmots *(Marmota marmota)*. Anim. Behav. 24:27.

Barfield, R.J., and L.A. Geyer, 1972. Sexual behavior: Ultrasonic postejaculatory song of the male rat. Science 176:1349.

Barghusen, H.R., and J.A. Hopson, 1970. Dentary-squamosal joint and the origin of mammals. Science 168:573.

Bartholomew, G.A., 1964. Symposia of the Society for Experimental Biology 18:7. Academic Press, New York.

Bartholomew, G.A., 1968. Body temperature and energy metabolism, *in* Animal Function: Principles and Adaptations (M.S. Gordon, G.A. Bartholomew, A.D. Grinnell, C.B. Jorgensen, and F.N. White, eds.). Macmillan, New York.

—— and T.J. Cade, 1957. Temperature regulation, hibernation, and estivation in the little pocket mouse, *Preognathus longimembris*. J. Mamm. 38:60.

—— and N.E. Collias, 1962. The role of vocalization in the social behavior of the northern elephant seal. Anim. Behav. 10:7.

——, W.R. Dawson, and R.C. Lasiewski, 1970. Thermoregulation and heterothermy in some of the smaller flying foxes (Megachiroptera) of New Guinea. Z. Vergl. Physiol. 70:196.

—— and J.W. Hudson, 1960. Aestivation in the Mohave ground squirrel. *Citellus mohavensis*. Bull. Mus. Comp. Zool. Harvard 124:353.

——, ——, 1961. Desert ground squirrels. Sci. Amer. 205(5):107.

————, ————, 1962. Hibernation, estivation, temperature regulation, evaporative water loss, and heart rate of the pigmy opossum, *Cercatetus nanus*. Physiol. Zool. 29:26.

————, P. Leitner, and J.E. Nelson, 1964. Body temperature, oxygen consumption, and heart rate in three species of Australian flying foxes. Physiol. Zool. 37:179.

———— and M. Rainy, 1971. Regulation of body temperature in the rock hyrax, *Heterohyrax brucei*. J. Mamm. 52:81.

Bateman, G.C., and T.A. Vaughan, 1974. Nightly activities of mormoopid bats. J. Mamm. 55:45.

Bateman, J.A., 1959. Laboratory studies of the golden mole and the mole rat. Afr. Wildlife 13.

Bateson, G., 1966. Problems in cetacean and other mammalian communication, *in* Whales, Dolphins and Porpoises (K.S. Norris, ed.). University of California Press, Berkeley.

———— and B. Gilbert, 1966. Whaler's Cove dolphin community: An interim report. The Oceanic Institute, Makpuu Point, Waimanalo, Oahu, Hawaii.

Batzli, G.A., and F.A. Pitelka, 1970. Influence of meadow mouse populations on California grassland. Ecology 51:1027.

————, ————, 1971. Condition and diet of cycling populations of the California vole, *Microtus californicus*. J. Mamm. 52:141.

Baust, J.G., and R.T. Brown, 1980. Heterothermy and cold acclimation in the arctic ground squirrel, *Citellus undulatus*. Comp. Biochem. Physiol. 67:447.

Baverstock, P.R., and B. Green, 1975. Water recycling in lactation. Science 187:657.

————, C.H.S. Watts, and L. Spencer, 1979. Water balance of small, lactating rodents: The total water balance picture of the mother-young unit. Comp. Biochem. Physiol. 63:247.

Beatley, J.C., 1969. Dependence of desert rodents on winter annuals and precipitation. Ecology 50:721.

————, 1976. Rainfall and fluctuating plant populations in relation to distributions and numbers of desert rodents in southern Nevada. Oecologia 24:21.

Bee, J.W., and E.R. Hall, 1956. Mammals of Northern Alaska. University of Kansas Publ., Mus. Nat. Hist., Misc. Publ. No. 8.

Beer, J.R., 1961. Seasonal reproduction in the meadow vole. J. Mamm. 42:483.

————, R. Lukens, and D. Olson, 1954. Small mammal populations on the islands of Basswood Lake, Minn. Ecology 35:437.

———— and R.K. Meyer, 1951. Seasonal changes in the endocrine organs and behavior patterns of the muskrat. J. Mamm. 32:173.

Bekoff, M., and M.C. Wells, 1980. The social ecology of coyotes. Sci. Amer. 242:130.

Bel'kovich, V.M., and A.V. Yablokov, 1963. Marine animals "share experience" with designers. Nauka Zhizn' 30:61.

Bell, R.H.V., 1970. The use of the herb layer by grazing ungulates in the Serengeti, *in* Animal Populations in Relation to Their Food Resources (A. Watson, ed.). Blackwell Scientific, Oxford.

————, 1971. A grazing ecosystem in the Serengeti. Sci. Amer. 225(1):86.

Bennett, A.F., 1972. A comparison of activities of metabolic enzymes in lizards and rats. Comp. Biochem. Physiol. B42:637.

———— and J.A. Ruben, 1979. Endothermy and activity in vertebrates. Science 206:649.

Benson, S.B., 1933. Concealing coloration among some desert rodents of the southwestern United States. University of California Publ. Zool. 40:1.

———— and A.E. Borell, 1931. Notes on the life history of the red tree mouse. *Phenacomys longicaudus*. J. Mamm. 12:226.

Berg, L.S., 1950. Natural regions of the U.S.S.R. Macmillan, New York.

Berger, P.J., N.G. Negus, E.H. Sanders, and P.D. Gardner, 1981. Chemical triggering of reproduction *Microtus montanus*. Science 214:69.

Bertram, B.C.R., 1973. Lion population regulation. E. Afr. Wildl. J. 11:215.

————, 1975. The social system of lions. Sci. Amer. 232:54.

————, 1975. Social factors influencing reproduction in wild lions. J. Zool. (London) 177:463.

Bertram, G.C.L., 1940. The biology of the Weddell and crabeater seals, with a study of the comparative behavior of the Pinnipedia. Brit. Mus. (Nat. Hist.) Sci. Repts. Brit. Graham Land Exped. 1934–1937, 1:1.

Berzin, A.A., 1971. The sperm whale. Pacific Sci. Res. Inst. Fisheries Oceanogr., Trans. 1972, Israel Program for Scientific Trans., no. 600707, Jerusalem.

Black, H.L., 1974. A North Temperate bat community: Structure and prey populations. J. Mamm. 55:138.

Blair, W.F., 1939. Some observed effects of stream-valley flooding on mammalian populations in eastern Oklahoma. J. Mamm. 20:304.

————, 1951. Evolutionary significance of geographic variation in population density. Texas J. Sci. 1:53.

Blake, E.W. Jr., 1887. The coast fox. West. Amer. Sci. 3:49.

Blatt, C.M., C.R. Taylor, and M.B. Habal, 1972. Thermal panting in dogs: The lateral nasal gland, a source of water for evaporative cooling. Science 177:804.

Blix, A.S., H.J. Grav, and K. Ronald, 1979. Some aspects of temperature regulation in newborn harp seal pups. Amer. J. Physiol. 236:R188.

———— and J.W. Lentfer, 1979. Modes of thermal protection in polar bear cubs: At birth and on emergence from the den. Amer. J. Physiol. 236:R67.

————, L.K. Miller, M.C. Keyes, H.J. Grav, and R. Elsner, 1979. Newborn northern fur seas (Callorhinus ursinus): Do they suffer from cold? Amer. J. Physiol. 236:R322.

Bloedel, P., 1955. Hunting methods of fish-eating bats, particularly Noctilio leporinus. J. Mamm. 36:390.

Bodemer, C.W., 1968. Modern embryology. Holt, Rinehart and Winston, New York.

Bodenheimer, F.S., 1949. Problems of vole populations in the Middle East: Report on the population dynamics of the Levant vole (Microtus guentheri D.). Azriel Print. Works, Jerusalem.

Bond, R.M., 1945. Range rodents and plant succession. Trans. N. Amer. Wildl. Conf. 10:229.

Boonstra, R., 1978. Effect of adult Townsend voles (Microtus townsendii) on survival of young. Ecology 59:242.

———— and C.J. Krebs, 1979. Viability of large- and small-sized adults in fluctuating vole populations. Ecology 60:567.

Bourliere, F., 1956. The natural history of mammals. Knopf, New York.

————, 1973. Comparative ecology of rain forest mammals, in Tropical Forest Ecosystems in Africa and South America: A Comparative Review (B.J. Meggers, E.S. Ayensu, and W.D. Duckworth, eds.). Smithsonian Institution Press, Washington, D.C.

Boving, B.G., 1959. Implantation. Ann. N.Y. Acad. Sci. 75:700.

Bowers, M.A., and H.D. Smith, 1979. Differential habitat utilization by sexes of the deer mouse, Peromyscus maniculatus. Ecology 60:869.

Bowker, M., 1977. Behavior of Kirk's dik-dik, Rhynchotragus kirki. Unpublished PhD thesis, Northern Arizona University, Flagstaff.

Bradbury, J.W., 1970. Target discrimination by echolocating bat Vampyrum spectrum. J. Exp. Zool. 173:23.

————, 1977. Social organization and communication, in Biology of Bats (W. Wimsatt, ed.). Vol. 3. Academic Press, New York.

———— and L.H. Emmons, 1974. Social organization of some Trinidad bats. I: Emballonuridae. Z. Tierpsychol. 36:137.

Bradford, D.F., 1976. Space utilization by rodents in Adenostoma chaparral. J. Mamm. 57:576.

Bradshaw, G.V.R., 1962. Reproductive cycle of the California leaf-nosed bat, Macrotus californicus. Science 136:645.

Bramble, D.M., 1982. Mammalian cranial kinesis: Shock absorption in lagomorphs. ASM Abstracts, 62d Annual Meeting of the American Soc. of Mammalogists, Abstract 90.

Breadon, G., 1932. The flying fox in the Punjab. J. Bombay Nat. Hist. Soc. 35:670.

Broadbooks, H.E., 1970. Home ranges and territorial behavior of the yellow-pine chipmunk, Eutamias amoenus. J. Mamm. 51:310.

Brockie, R., 1976. Self-anointing by wild hedgehogs, Erinaceus europaeus. Anim. Behav. 24:68.

Brodie, P.F., 1975. Cetacean energetics: An overview of intraspecific size variation. Ecology 56:152.

Bromley, P.T., 1969. Territoriality in pronghorn bucks on the National Bison Range, Moiese, Montana. J. Mamm. 50:81.

Bronson, F.H., 1971. Rodent pheromones. Biol. Reprod. 1:344.

————, 1979. The reproductive ecology of the house mouse. Quast. Rev. Biol. 54:265.

———— and J.A. Maruniak, 1975. Male-induced puberty in female mice: Evidence for a synergistic action of social cues. Biol. Reprod. 13:94.

Brosset, A., 1968. Permutation du cycle chez Hipposideros caffer au voisinage de l'equateur. Biologica Gabonica 4.

Brown, J.H., 1968. Adaptation to environmental temperature in two species of woodrats, Neotoma cinerea and N. albigula. Misc. Publ. Mus. Zool. Univ. Mich. 135:1.

———— and G.A. Bartholomew, 1969. Periodicity and energetics of torpor in the kangaroo mouse, *Microdipodops pallidus.* Ecology 50:705.

———— and D.W. Davidson, 1977. Competition between seed-eating rodents and ants in desert ecosystems. Science 196:880.

————, ————, and O.J. Reichman, 1979. An experimental study of competition between seed-eating desert rodents and ants. Amer. Zool. 19:1129.

———— and R.C. Lasiewski, 1972. Metabolism of weasels: The cost of being long and thin. Ecology 53:939.

Brown, L.N., 1967. Ecological distribution of six species of shrews and comparison of sampling methods in the central Rocky Mountains. J. Mamm. 48:617.

Brown, P., 1976. Vocal communication in the pallid bat, *Antrozous pallidus.* Z. Tierpsychol. 41:34.

Bruce, V.G., 1960. Environmental entrainment of circadian rhythms. Cold Spr. Harb. Symp. Quant. Biol. 25:29.

Bruce, W.C., 1915. Measurements and weights of antarctic seals, *in* Report on the Scientific Results of the Voyage of S.Y. "Scotia" During the Years 1902, 1903, and 1904 . . . Edinburgh, Scottish Oceanogr. Lab.

Bryant, J.P., 1981. Phytochemical deterrence of snowshoe hare browsing by adventitious shoots of four Alaskan trees. Science 213:889.

Buchler, E.R., 1976. The use of echolocation by the wandering shrew. Anim. Behav. 24:858.

Buckner, C.H., 1969. Some aspects of the population ecology of the common shrew, *Sorex araneus,* near Oxford, England, J. Mamm. 50:326.

Buechner, H.K., 1961. Territorial behavior in Uganda kob. Science 133:698.

————, 1963. Territoriality as a behavioral adaptation to environment in the Uganda kob. Proc. XVI Int. Congr. Zool. 3:59.

————, 1974. Implications of social behavior in the management of the Uganda kob, *in* The Behavior of Ungulates and Its Relation to Management (V. Geist and F.R. Walther, eds.). International Union for Conservation of Nature and National Resources, Morges, Switzerland.

———— and H.D. Roth, 1974. The lek system of the Uganda kob. Amer. Zool. 14:145.

Bugge, J., 1974. The cephalic arteries of hystricomorph rodents, *in* The Biology of Hystricomorph Rodents (I.W. Rowlands and B. Weir, eds.). Symp. Zool. Soc. London. Academic Press, New York.

Bullard, E., 1969. The origin of the oceans. Sci. Amer. 221:66.

Burckhardt, D., 1958. Kindliches Verhalten als Ausdrucksvewegung im Fortpflanzungszeremoniell einiger Wiederkauer. Rev. Suisse Zool. 65:311.

Burrell, H., 1927. The platypus. Angus and Robertson, Sydney.

Burt, W.H., 1940. Territorial behavior and populations of some small mammals in southern Michigan. Misc. Publ. Mus. Zool., Univ. Michigan 45:1.

————, 1943. Territoriality and home range concepts as applied to mammals. J. Mamm. 24:346.

Burton, M., 1949. Wildlife of the World. Long Acre, London.

Busnel, R.G. (ed.), 1966. Animal Sonar Systems, Biology and Bionics, Tomes I and II. Laboratoire de Physiologie Acoustique, Paris.

Butler, P.M., 1969. *In* Fossil Vertebrates of Africa (L.S.B. Leakey, ed.), Vol. 1. Academic Press, New York.

————, 1972. Some functional aspects of molar evolution. Evolution 26:474.

———— and Z. Kielan-Jaworowska, 1973. Is *Deltatheridium* a marsupial? Nature 245:105.

Byers, J.A., 1981. Peaceable peccaries. Nat. Hist. 90:61.

Cadenat, J., 1959. Rapport sur les petits cétacés ouestafricains. Bull. Inst. Franc. Afrique Noire 21:1167.

Calaby, J.H., 1971. The current status of Australian Macropodidae. Australian Zool. 16:17.

Calder, N., 1972. The restless earth: A report on the new geology. Viking Press, New York.

Calder, W.A., 1969. Temperature relations and underwater endurance of the smallest homeothermic diver, the water shrew. Comp. Biochem. Physiol. 30:1075.

Caldwell, D.K., J.H. Prescott, and M.C. Caldwell, 1966. Bull. So. California Acad. Sci. 65:245.

Caldwell, M.C., 1972. Behavior of marine mammals: Sense and communication, *in* Mammals of the Sea: Biology and Medicine (S.H. Ridgway, ed.). Charles C Thomas, Springfield, Ill.

———— and D.K. Caldwell, 1970. Further studies on audible vocalizations of the Amazon freshwater dolphin, *Inia geoffrensis.* Contrib. Sci. 187:1.

Camp, C.L., and N.S. Smith, 1942. Phylogeny and

functions of the digital ligaments of the horse. Univ. California, Mem. 13:69.

Campbell, C.B.G., 1966. Taxonomic status of tree shrews. Science 153:436.

Carleton, M.D., 1984. Introduction to rodents, *in* Orders and Families of Recent Mammals of the World (S. Anderson and J.K. Jones Jr., eds.). Wiley, New York.

———— and G.G. Musser, 1984. Muroid rodents, *in* Orders and Families of Recent Mammals of the World (S. Anderson and J.K. Jones Jr., eds.). Wiley, New York.

Caroom, D., and F.H. Bronson, 1971. Responsiveness of female mice to preputial attractant: Effects of sexual experience and ovarian hormones. Physiol. Behav. 7:659.

Carpenter, R.E., 1966. A comparison of thermoregulation and water metabolism in the kangaroo rats *Dipodomys agilis* and *Dipodomys merriami*. Univ. California Publ. Zool. 78:1.

————, 1968. Salt and water metabolism in the marine fish-eating bat, *Pizonyx vivesi*. Comp. Biochem. Physiol. 24:951.

————, 1969. Structure and function of the kidney and the water balance of desert bats. Physiol. Zool. 42:288.

———— and J.B. Graham, 1967. Physiological responses to temperature in the long-nosed bat, *Leptonycteris sanborni*. Comp. Biochem. Physiol. 22:709.

Carr, A., 1974. Ulendo: Travels of a naturalist in and out of Africa. Knopf, New York.

Cates, R.G., and D.F. Rhoades, 1977. Patterns in the production of antiherbivore chemical defenses in plant communities. Biochem. Systematics and Ecol. 5:185.

Chaffee, R.R.J., and J.C. Roberts, 1971. Temperature acclimation in birds and mammals. Ann. Rev. Physiol. 33:155.

Chaney, R.W., 1944. Summary and conclusions, *in* Pliocene Floras of California and Oregon (R.W. Chaney, ed.). Carnegie Inst., Washington, D.C. Publ. 553.

Chappell, M.A., 1978. Behavioral factors in altitudinal zonation of chipmunks *(Eutamias)*. Ecology 59:565.

———— and G.A. Bartholomew, 1981a. Standard operative temperatures and thermal energetics of the antelope ground squirrel, *Ammospermophilus leucurus*. Physiol. Zool. 54:81.

————, ————, 1981b. Activity and thermoregulation of the antelope ground squirrel *Ammospermophilus leucurus* in winter and summer. Physiol. Zool. 54:215.

Chapskiy, K.K., 1936. The walrus of the Kara Sea. Trans. Arct. Inst., Leningrad, Tom. 67 (in Russian; resumé in English).

Chitty, D., 1954. Tuberculosis among wild voles with a discussion of other pathological conditions among certain mammals and birds. Ecology 35:227.

————, 1958. Self-regulation of numbers through changes in viability. Cold Spr. Harb. Symp. Quant. Biol. 22:277.

————, 1960. Population processes in the vole and their relevance to general theory. Can. J. Zool. 38:99.

———— and H. Chitty, 1962. Population trends among the voles at Lake Vyrnwy, 1932–1960. J. Anim. Ecol. 35:313.

Choshniak, I., and A. Shkolnik, 1977. Rapid rehydration in the black Bedouin goats: Red blood cells fragility and the role of the rumen. Comp. Biochem. Physiol. 56A:581.

Christian, D.P., 1978. Effects of humidity and body size on evaporative water loss in three desert rodents. Comp. Biochem. Physiol. 60:425.

Christian, J.J., 1950. The adreno-pituitary system and population cycles in mammals. J. Mamm. 31:247.

————, 1954. The relation of the adrenal cortex to population size in rodents. Doctoral dissertation, Johns Hopkins School of Hygiene and Public Health, Baltimore.

————, 1959a. Control of population growth in rodents by interplay between population density and endocrine physiology. Wildl. Dis. 1:1.

————, 1959b. The roles of endocrine and behavioral factors in the growth of mammalian populations, *in* Comparative Endocrinology (A. Gorbman, ed.). Columbia Univ. Symposium, New York.

————, 1963. Endocrine adaptive mechanisms and the physiologic regulation of population growth, *in* Physiological Mammalogy (W.V. Mayer and R.G. Van Gelder, eds.). Academic Press, New York.

———— and D.E. Davis, 1956. The relationship between adrenal weight and population status in Norway rats. J. Mamm. 37:475.

————, ————, 1966. Adrenal glands in female voles *(Microtus pennsylvanicus)* as related to reproduction and population size. J. Mamm. 47:1.

—————, V. Flyger, and D.E. Davis, 1960. Factors in mass mortality of a herd of sika deer. Chesapeake Sci. 1:79.

Clark, A.B., 1978. Sex ratio and local resource competition in a prosimian primate. Science 201:163.

Clark, C.W., and J.M. Clark, 1980. Sound playback experiments with southern right whales *(Eubalaena australis)*. Science 202:663.

Clark, W.E.L., 1971. The antecedents of man. Quadrangle Books, New York.

Clarke, J.R., and J.P. Kennedy, 1967. Effect of light and temperature upon gonad activity in the vole *(Microtus agrestis)*. Gen. Comp. Endocrinol. 8:474.

Clarke, M.R., 1979. The head of the sperm whale. Sci. Amer. 240:128.

Clemens, W.A., 1968. Origin and early evolution of marsupials. Evolution 22:1.

—————, 1970. Mesozoic mammalian evolution, *in* Annual Review of Ecology and Systematics (R.F. Johnston, ed.), Vol. I. Annual Reviews, Inc., Palo Alto, Calif.

—————, 1979a. Marsupialia, *in* Mesozoic Mammals: The First Two-Thirds of Mammalian History (J.A. Lillegraven, Z. Kielan-Jaworoska, and W.A. Clemens, eds.). University of California Press, Berkeley.

—————, 1979b. Notes on the Monotremata, *in* Mesozoic Mammals: The First Two-Thirds of Mammalian History (J.A. Lillegraven, Z. Kielan-Jaworoska, and W.A. Clemens, eds.). University of California Press, Berkeley.

————— and Z. Kielan-Jaworowska, 1979. Multituberculates, *in* Mesozoic Mammals: The First Two-Thirds of Mammalian History (J.A. Lillegraven, Z. Kielan-Jaworoska, and W.A. Clemens, eds.). University of California Press, Berkeley.

————— and L.G. Marshall, 1976. Fossilium Catalogus: American and European Marsupialia. W. Junk (The Hague), Pars. 123:1.

————— and M. Plane, 1974. Mid-Tertiary Thylacoleonidae (Marsupialia, Mammalia). J. Paleont. 48(4):652.

Clough, G.C., 1965. Lemmings and population problems. Amer. Sci. 53:199.

Clutton-Brock, T.H., S.D. Albon, R.M. Gibson, and F.E. Guinness, 1979. The logical stag: Adaptive aspects of fighting in red deer *(Cervus elaphus)*. Anim. Behav. 27:211.

Coblentz, B.E., 1976. Functions of scent-urination in ungulates with special reference to feral goats *(Capra hircus* L.). Amer. Nat. 110:549.

Coe, M.J., 1967. Preliminary notes on the spring hare *Pedetes surdaster larvalis* in East Africa. E. Afr. Wildl. J. 5.

Coimbra-Filho, A.F., A. Magananini, and R.A. Mittermeier, 1975. Vanishing gold: Last chance for Brazil's lion tamarins. Animal Kingdom, Dec., p. 20.

Colbert, E.H., 1948. The mammal-like reptile *Lycaenops*. Bull. Amer. Mus. Nat. Hist. 89:353.

—————, 1949. The ancestors of mammals. Sci. Amer. 180:40.

—————, 1962. The weights of dinosaurs. Amer. Mus. Novitates 2076:1.

—————, 1973. Wandering Lands and Animals. Dutton, New York.

—————, 1980. Evolution of the Vertebrates, 2nd edn. Wiley, New York.

—————, 1982. Personal communication.

Cole, R.W., 1970. Pharyngeal and lingual adaptations in the beaver. J. Mamm. 51:424.

Conaway, C.H., 1958. Maintenance, reproduction and growth of the least shrew in captivity. J. Mamm. 39:507.

Conley, W.H., 1971. Behavior, demography and competition in *Microtus longicaudus* and *M. mexicanus*. Ph.D. thesis, Texas Tech University, Lubbock.

Conner, R.C., and K.S. Norris, 1982. Are dolphins reciprocal altruists? Amer. Nat. 119:358.

Cott, H.B., 1966. Adaptive Coloration in Animals. Methuen, London.

Coulombe, H.N., 1970. The role of succulent halophytes in water balances of salt marsh rodents. Oecologia (Berl.) 4:223.

Cowan, I.M., 1947. The timber wolf in the Rocky Mountain national parks of Canada. Can. J. Res. 25:139.

—————, 1956. Life and times of the coast black-tailed deer, *in* The Deer of North America. (W.P. Taylor, ed.). Stackpole Company, Harrisburg, Pa.

Crawford, E.C. Jr., 1962. Mechanical aspects of panting in dogs. J. Appl. Physiol. 17:249.

Crelin, E.S., 1969. Interpubic ligament: Elasticity in pregnant free-tailed bat. Science 164:81.

Crisler, L., 1956. Observations of wolves hunting caribou. J. Mamm. 37:337.

—————, 1958. Arctic Wild. Harper, New York.

Crompton, A.W., 1971. The origin of the tribosphenic molar, *in* Early Mammals. (D.M. Kermack and K.A. Kermack, eds.). J. Linn. Soc. (London) Zool., 50:(1):165–180.

————, 1972. The evolution of the jaw articulation in cynodonts, *in* Studies in Vertebrate Evolution (K.A. Joysey and T.S. Kemp, eds.). Oliver and Boyd, Edinburgh.

————, 1974. The dentitions and relationships of the southern African mammals *Erythrotherium parringtoni* and *Megazostrodon rudnerae*. Bull. Brit. Mus. Nat. Hist. (Geol.). 24:399.

———— and K. Hiiemae, 1969. How mammalian molar teeth work. Discovery 5(1):23.

————, ————, 1970. Molar occlusion and mandibular movements during occlusion in the American opossum, *Didelphus marsupialis*. L.J. Linn. Soc. (London) Zool. 49:21.

———— and F.A. Jenkins, Jr., 1968. Molar occlusion in Late Triassic mammals. Biol. Rev. 43:427.

————, ————, 1973. Mammals from Reptiles: A Review of Mammalian Origins. Ann. Rev. Earth Planetary Sci., Vol. I, Annual Reviews, Palo Alto.

————, ————, 1979. Origin of mammals, *in* Mesozoic Mammals: The First Two-Thirds of Mammalian History (J.A. Lillegraven, Z. Kielan-Jaworoska, and W.A. Clemens, eds.). University of California Press, Berkeley.

————, C.R. Taylor, and J.A. Jagger, 1978. Evolution of homeothermy in mammals. Nature 272:333.

Crowcroft, P., 1953. The daily cycle of activity in British shrews. Proc. Zool. Soc. London 123:715.

————, 1957. The Life of the Shrew. Max Reinhart, London.

————, 1966. Mice All Over. Foulis, London.

Curry-Lindahl, K., 1962. The irruption of the Norway lemmings in Sweden during 1960. J. Mamm. 43:171.

Dagg, A.I., 1962. The role of the neck in the movements of the giraffe. J. Mamm. 43:88.

Daily, C.S., and H.B. Haines, 1981. Evaporative water loss and water turnover in chronically and acutely water-restricted spiny mice *(Acomys cahirinus)*. Comp. Biochem. Physiol. 68A:349.

Dalquest, W.W., 1948. The mammals of Washington. University of Kansas Publ., Mus. Nat. Hist. 2:1.

Daniel, J.C. Jr., 1970. Dormant embryos of mammals. Bio. Sci. 20(7):411.

Daniel, M.J., 1979. The New Zealand short-tailed bat, *Mystacina tuberculata:* A review of present knowledge. New Zealand J. Zool. 6:357.

Dapson, R.A., 1979. Phenologic influences on cohort-specific reproductive strategies in mice *(Peromyscus polionotus)*. Ecology 60:1125.

Darling, F.F., 1937. A Herd of Red Deer. Oxford University Press, London.

Darlington, P.J., 1957. Zoogeography: The Geographical Distribution of Animals. Wiley, New York.

Dasmann, R.F., and A.S. Mossman, 1962. Population studies of impala in Southern Rhodesia. J. Mamm. 43:375.

David, A. 1968. Can young bats communicate with their parents at a distance? J. Bombay Nat. Hist. Soc. 65:210.

Davidson, D.W., J.H. Brown, and R.S. Inouye, 1980. Competition and the structure of granivore communities. BioScience 30:233.

Davis, D.E., 1951. The relation between level of population and pregnancy of Norway rats. Ecology 32:459.

————, 1967. The annual rhythm of fat deposition in woodchucks *(Marmota monax)*. Physiol. Zool. 40:391.

Davis, R.B., C.F. Herreid Jr., and H.L. Short, 1962. Mexican free-tailed bats in Texas. Ecol. Monogr. 32:311.

Davis W.H., 1970. Hibernation: Ecology and physiological ecology, *in* Biology of Bats (W.A. Wimsatt, ed.). Academic Press, New York.

———— and H.B. Hitchcock, 1965. Biology and migration of the bat, *Myotis lucifugus,* in New England. J. Mamm. 46:296.

———— and W.Z. Likicker Jr., 1956. Winter range of the red bat. J. Mamm. 37:280.

Dawe, A.R., and W.A. Spurrier, 1969. Hibernation induced in ground squirrels by blood transfusion. Science 163:298.

Dawson, M.R., 1958. Later Tertiary Leporidae of North America. University of Kansas Paleont. Contributions, Vertebrata, Art. 6.

————, 1967a. Fossil history of the families of Recent mammals, *in* Recent Mammals of the World (S. Anderson and J.K. Jones Jr., eds.). Ronald Press, New York.

————, 1967b. Lagomorph history and stratigraphic record. Essays in Paleontology and Stratigraphy, Raymond C. Moore Commemorative Volume,

University of Kansas, Department of Geology Spec. Publ. 2.

————— and L. Krishtalka, 1984. Fossil history of the families of Recent mammals, *in* Orders and Families of Recent Mammals of the World (S. Anderson and J.K. Jones Jr., eds.). Wiley, New York.

—————, E.-K. Li, and T. Qi, In Press. Eocene ctenodactyloid rodents (Mammalia) of eastern and central Asia. Bull. Carnegie Mus. Nat. Hist.

Dawson, T.J., 1973a. Primitive mammals, *in* Comparative Physiology of Thermoregulation, Vol. III (C.G. Whittow, ed.). Academic Press, New York.

—————, 1973b. Thermoregulatory responses of the arid zone kangaroos *Megaleia rufa* and *Macropus robustus*. Comp. Biochem. Physiol. 46:153.

————— and A.J. Hulbert, 1970. Standard metabolism, body temperature, and surface areas of Australian marsupials. Amer. J. Physiol. 218:1233.

—————, D. Robertshaw, and C.R. Taylor, 1974. Sweating in the red kangaroo: An evaporative cooling mechanism during exercise but not heat. Amer. J. Physiol. 227:494.

DeCoursey, P., 1960. Phase control of activity in a rodent. Cold Spr. Harb. Symp. Quant. Biol. 25:49.

—————, 1961. Effect of light on the circadian activity rhythm of the flying squirrel, *Glaucomys volans*. Z. Vergl. Physiol. 44:331.

Deevey, E.S. Jr., 1947. Life tables for natural populations of animals. Quart. Rev. Biol. 22:283.

Degerbøl, M., and P. Freuchen, 1935. Mammals, Vol. 2, *in* Report of the fifth Thule expedition, 1921–1924. Copenhagen, Nordisk Forlag. Part I. Systematic notes, by Degerbøl, Part II. Field notes and biological observations, by Freuchen.

Delany, M.J., 1971. The biology of small rodents in Mayanja Forest, Uganda. J. Zool. London 165:85.

Desjardins, C., J.A. Maruniak, and F.H. Bronson, 1973. Social rank in house mice: Differentiation revealed by ultraviolet visualization of urinary marking patterns. Science 182:939.

DeVore, I. (ed.), 1965. Primate Behavior. Holt, Rinehart and Winston, New York.

————— and K.R.L. Hall, 1965. Baboon ecology, *in* Primate Behavior (I. DeVore, ed.). Holt, Rinehart and Winston, New York.

Dietz, R.S., and J.C. Holden, 1970. The breakup of Pangaea. Sci. Amer. 223(4):30.

Dijkgraaf, S., 1943. Over een merkwaardige functie wan den gehoorzin bij vleermuizen. Versla-
gen Nederlandsche Akademie Wetenschappen Afd. Naturkunde 52:622.

—————, 1946. Die Sinneswelt der Fledermäuse. Experientia 2:438.

—————, 1960. Spallanzani's unpublished experiments on the sensory basis of object perception in bats. Isia 51:9.

Dobzhansky, T., 1950. Mendelian populations and their evolution. Amer. Nat. 84:401.

Douglas-Hamilton, I., 1972. On the ecology and behavior of the African elephant: The elephants of Lake Manyara. Ph.D. thesis, Oxford University, Oxford.

—————, 1973. On the ecology and behavior of the Lake Manyara elephants. E. Afr. Wildl. J. 11:401.

Dreher, J.J., 1966. Cetacean communication: Small-group experiments, *in* Whales, Dolphins and Porpoises (K.S. Norris, ed.). University of California Press, Berkeley.

————— and W.E. Evans, 1964. Cetacean communication, *in* Marine Bio-acoustics (W.N. Tavolga, ed.). Pergamon Press, Oxford.

Dryden, G.L., 1975. Establishment and maintenance of shrew colonies. Zool. Yearbook 15:12.

—————, 1976. Personal communication.

Dubost, G., 1968. Aperçu sur le rythme annuel de reproduction des muridés du nord-est du Gabon. Biol. Gabonica 4:227.

Dücker, G., 1957. Fard- und Helligkeitssehen und Instinkte bei Viverriden und Feliden. Zool. Beitr. (Berl.) 3:25.

Duncan, P., 1975. Topi and their food supply. Ph.D. dissertation, University of Nairobi, Kenya.

Dunning, D.C., 1968. Warning sounds of moths. Z. Tierpsychol. 25:129.

————— and K.D. Roeder, 1965. Moth sounds and the insect-catching behavior of bats. Science 147:173.

Durrell, G.M., 1954. The Bafut Beagles. Rupert Hart-Davies, London.

DuToit, A.L., 1957. Our Wandering Continents. Oliver and Boyd, Edinburgh.

Eadie, W.R. 1952. Shrew predation and vole populations on a localized area. J. Mamm. 33:185.

Edwards, R.L., 1946. Some notes on the life history of the Mexican ground squirrel in Texas. J. Mamm. 27:105.

Egorov, Y.E., 1966. Relationships between American mink and otter in Bashkiria, *in* Acclimatization of Animals in the U.S.S.R., Jerusalem.

Ehrenfeld, D.W., 1976. The conservation of non-resources. Amer. Sci. 64:648.

Eibl-Eibesfeldt, I., 1958. Das Verhalten der Nagetiere. Handb. Zool. 8:1.

Einarsen, A.S., 1948. The Pronghorn Antelope. Wildlife Management Institute, Washington, D.C.

Eisenberg, J.F., 1966. The social organization of mammals. Handb. Zool. (Berl.) 10:1.

Eisenberg, J.F., 1981. The Mammalian Radiations: An Analysis of Trends in Evolution, Adaptation, and Behavior. University of Chicago Press.

————— and E. Gould, 1970. The tenrecs: A study in mammalian behavior and evolution. Smithson. Contrib. Zool. 27:1.

————— and D.G. Kleiman, 1972. Olfactory communication in mammals. Ann. Rev. Ecol. System. 3:1.

————— and D.G. Kleiman, (eds.), 1983. Advances in the Study of Mammalian Behavior. Spec. Publ. No. 7, American Society of Mammalogy.

Eisentraut, M., 1934. Der Winterschlaf der Fledermäuse mit besonderer Berücksichtigung der Warmeregulation. Z. Morphol. Oekol Tiere 29:231.

—————, 1957. Aus dem Leben der Fledermäuse und Flughunde. Jena, Veb. Gustav Fischer Verlag.

—————, 1960. Heat regulation in primitive mammals and in tropical species. Bull. Mus. Comp. Zool. Harvard 124:31.

Eisner, T., and J.A. Davis, 1967. Mongoose throwing and smashing millipedes. Science 155:577.

Ellerman, J.R., 1940. The Families and Genera of Living Rodents, vol. I. British Museum of Natural History.

—————, 1941. *Ibid,* vol. II.

—————, 1949. *Ibid,* vol. III.

Elliot, P.W., 1969. Dynamics and regulation of a *Clethrionomys* population in central Alberta. Ph.D. Thesis, University of Alberta.

Elliott, V.A., 1976. Circadian rhythms and photoperiodic time measurement in mammals. Fed. Proc. 25:2339.

Eloff, F.C., 1967. Personal communication.

Eloff, G., 1958. The functional and structural degeneration of the eye in the African rodent moles *Cryptomys bigalkei* and *Bathyergus maritimus*. S. Afr. J. Sci., 54.

Else, P.L., and A.J. Hulbert, 1981. Comparison of the "mammal machine" and the "reptile machine": Energy production. Amer. J. Physiol. 240:R3.

Elsner, R., 1965. Hvalradets Skrifter Norske Videnskaps-Akad. Oslo 48:24.

—————, 1969. Cardiovascular adjustments to diving, *in* The Biology of Marine Mammals (H.T. Andersen, ed.). Academic Press, New York.

Elton, C., 1942. Voles, Mice and Lemmings. Clarendon Press, Oxford.

Emerson, S.B., and L. Radinsky, 1980. Functional analysis of sabertooth cranial morphology. Paleobiol. 6:295.

Emmons, L.H., 1981. Morphological, ecological, and behavioral adaptations for arboreal browsing in *Dactylomys dactylinus* (Rodentia, Echymyidae). J. Mamm. 62:183.

Ende, P. Van den, 1973. Predator-prey interactions in continuous culture. Science 18:562.

Erickson, A.B., 1944. Helminth infections in relation to population fluctuations in snowshoe hares. J. Wildl. Manage. 8:134.

Erlinge, S., 1972. Interspecific relations between otter *Lutra lutra* and mink *Mustela vison* in Sweden. Oikes 23:327.

—————, 1974. Distribution, territoriality and numbers of the weasel *Mustela nivalis* in relation to prey abundance. Oikos 25:308.

—————, 1975. Feeding habits of the weasel *Mustela nivalis* in relation to prey abundance. Oikos 26:378.

Errington, P.L., 1937. What is the meaning of predation? Smithsonian Rep. for 1936, p. 243.

—————, 1943. An analysis of mink predation upon muskrats in north-central United States. Agric. Exp. Sta. Iowa State Coll. Res. Bull. 320:797.

—————, 1946. Predation and vertebrate populations. Quart. Rev. Biol. 21:144, 221.

—————, 1957. Of populations, cycles and unknowns. Cold. Spr. Harb. Symp. Quant. Biol. 22:287.

—————, 1963. Muskrat Populations. Iowa State University Press, Ames.

—————, 1967. Of Predation and Life. Iowa State University Press, Ames.

Espenshade, E.B. (ed.), 1971. Goode's World Atlas. 13th ed. Rand-McNally, Chicago.

Estes, J.A., and J.F. Palmisano, 1974. Sea otters: Their role in structuring nearshore communities. Science 185:1058.

Estes, R.D., 1966. Behavior and life history of the wildebeest *(Connochaetes taurinus* Burchell). Nature, Lond. 212:999.

————, 1967. The comparative behavior of Grant's and Thompson's gazelles. J. Mamm. 48:189.

————, 1974. Social organization of the African Bovidae, *in* The Behavior of Ungulates and Its Relation to Management (V. Geist and F.R. Walther, eds.). International Union for Conservation of Nature and Natural Resources, Morges, Switzerland.

———— and J. Goddard, 1970. Prey selection and hunting behavior of the African wild dog. J. Wildl. Manage. 31(1):52.

Evans, F.G., 1942. The osteology and relationships of elephant shrews (Macroscelididae). Bull. Amer. Mus. Nat. Hist. 80(4):85.

Evans, W.E., and J. Bastian, 1969. Marine mammal communication: Social and ecological factors, *in* The Biology of Marine Mammals (H.T. Andersen, ed.). Academic Press, New York.

———— and R.M. Haugen, 1963. An experimental study of the echolocation ability of a California sea lion, *Zalophus californianus* (Lesson). Bull. So. Calif. Acad. Sci. 62:165.

———— and B.A. Powell, 1967. Proc. symp. bionic models of animal sonar systems. Frascati, Italy, 1966, p. 363. Labor. d'Acoustique Animal, Jouy-en-Josas, France.

————, W.W. Sutherland, and R.G. Beil, 1964. *in* Marine Bioacoustics, vol. 1 (W.N. Tavolga, ed.). Pergamon Press, Oxford.

Evermann, B.W., 1923. The conservation of marine life of the Pacific. Sci. Mon. 16:521.

Evernden, J.F., D.E. Savage, G.H. Curtis, and G.T. Jones, 1964. Potassium-argon dates and the Cenozoic mammalian chronology of North America. Amer. J. Sci. 262:145.

Ewer, R.F., 1968. Ethology of Mammals. Plenum Press, New York.

————, 1973. The Carnivores. Cornell University Press, Ithaca, N.Y.

Ewing, W.G., E.H. Studier, and M.J. O'Farrell, 1970. Autumn fat deposition and gross body composition in three species of *Myotis*. Comp. Biochem. Physiol. 36:119.

Faegri, K., and L. Van Der Pijl, 1966. The Principles of Pollination Ecology. Pergamon Press, New York.

Farentinos, R.C., P.J. Capretta, R.E. Kepner, and V.M. Littlefield, 1981. Selective herbivory in tassel-eared squirrels: Role of monoterpenes in ponderosa pines chosen as feeding trees. Science 213:1273.

Fedak, M.A., N.C. Heglund, and C.R. Taylor, 1982. Energetics and mechanics of terrestrial locomotion. II: Kinetic energy changes of the limbs and body as a function of speed and body size in birds and mammals. J. Exp. Biol. 97:23.

Feder, M.E., 1981. A cold look at paleophysiology. Paleobiology 7:144.

Feeny, P.P., 1975. Biochemical coevolution between plants and their insect herbivores, *in* Coevolution of Plants and Animals (P.H. Raven , ed.). University of Texas Press, Austin.

Fenton, M.B., and J.H. Fullard, 1981. Moth hearing and feeding strategies of bats. Amer. Sci. 69:266.

Fenton, M.B., and K.D. Roeder, 1974. The microtymbals of some Arctiidae. J. Lepidopt. Soc. 28:205.

————, D.W. Thomas, and R. Sasseen, 1981. *Nycteris grandis* (Nycteridae): An African carnivorous bat. J. Zool. London 194:461.

Fields, R.W., 1957. Histricomorph rodents from late Miocene of Colombia, South America. University of California Publ. Geol. Sci. 32:273.

Finch, V.A., and D. Robertshaw, 1979. Effect of dehydration on thermoregulation in eland and hartebeest. Amer. J. Physiol. 237:R192.

Findley, J.S., 1967. Insectivores and dermopterans, *in* Recent Mammals of the World (S. Anderson and J.K. Jones, eds.). Ronald Press, New York.

————, 1969. Biogeography of Southwestern boreal and desert mammals, *in* Contributions in Mammalogy (J.K. Jones Jr., ed.). University of Kansas, Mus. Nat. Hist., Misc. Publ. No. 51.

———— and D.E. Wilson, 1974. Observations on the Neotropical disk-winged bat, *Thyroptera tricolor* Spix. J. Mamm. 55:562.

Fink, B.D., 1959. Observation of porpoise predation on a school of Pacific sardines. California Fish and Game 45(3):216.

Finley, R.B. Jr., 1969. Cone caches and middens of *Tamasciurus* in the Rocky Mountain region, *in* Contributions in Mammalogy (J.K. Jones Jr., ed.). University of Kansas, Mus. Nat. Hist., Misc. Publ. No. 51.

Fischer, G.A., J.C. Davis, F. Iverson, and F.P. Cronmiller, 1944. The winter range of the Interstate

Deer Herd. U.S. Dept. Agric., Forest Serv., Region 5:1. (mimeo).

Fish, F.F., 1979. Thermoregulation in the musk-rat *(Ondatra zibethicus):* The use of regional heterothermia. Comp. Biochem. Physiol. 64:391.

Fisher, E.M., 1939. Habits of the southern sea otter. J. Mamm. 20:21.

Fitch, H.S., 1948. Ecology of the California ground squirrel on grazing lands. Amer. Midl. Nat. 39:513.

————, R. Goodrum, and C. Newman, 1952. The armadillo in the southeastern United States. J. Mamm. 33:21.

Fitzgerald, B.M., 1972. The role of weasel predation in cyclic population changes in the montane vole *(Microtus montanus)*. Ph.D. thesis, University of California, Berkeley.

————, 1977. Weasel predation on a cyclic population of the montane vole *(Microtus montanus)* in California. J. Anim. Ecol. 46:367.

Fleay, D., 1944. We Breed the Platypus. Robertson and Mullens, Melbourne.

Fleming, T.H., 1970. Notes on the rodent faunas of two Panamanian forests. J. Mamm. 51:473.

————, 1971. *Artibeus jamaicensis:* Delayed embryonic development in a Neotropical bat. Science 171:402.

———— and E.R. Heithaus, 1981. Frugivorous bats, seed shadows, and the structure of tropical forests. Biotropica 13:45.

————, E.T. Hooper, and D.E. Wilson, 1972. Three Central American bat communities: Structure, reproductive cycles, and movement patterns. Ecology 53:555.

Flessa, K.W., 1975. Area, continental drift and mammalian diversity. Paleobiology 1:189.

Flexner, L.B., D.B. Crowie, L.M. Hellman, W.S. Wilde, and G.J. Vosburgh, 1948. The permeability of the human placenta to sodium in normal and abnormal pregnancies and the supply of sodium to the human fetus as determined with radioactive sodium. Amer. J. Obst. Gyn. 55:469.

Fogden, A., 1974. A preliminary field study of the Western tarsier, *Tarsius bancanus* Horsefield, *in* Prosimian Biology (R.D. Martin, G.A. Doyle, and A.C. Walker, eds.). University of Pittsburgh Press, Pittsburgh.

Fons, P.R., 1974. Méthodes de capture et d'élevage de la Pachyure étrusque *Suncus etruscus* (Savi,

1822) (Insectivora, Soricidae). Z. Saugetierk. 39:204.

Forman, G.L., 1971. Comparative morphological and histochemical studies of gastrointestinal tracts of selected North American bats. University of Kansas Sci. Bull. 49:591.

————, 1973. Studies of gastric morphology in North American Chiroptera (Emballonuridae, Noctilionidae, and Phyllostomatidae). J. Mamm. 54:909.

Formozov, A.N., 1946. The covering of snow as an integral factor of the environment and its importance in the ecology of mammals and birds. Material for Fauna and Flora of the USSR, New Series Zool. 5:1.

————, 1966. Adaptive modifications of behavior in mammals of the Eurasian steppes. J. Mamm. 47(2):208.

Fossey, D., 1972. Living with mountain gorillas, *in* The Marvels of Animal Behavior (P.R. Marler, ed.). National Geographic Society, Washington, D.C.

Foster, J.B., 1964. Evolution of mammals on islands. Nature 202:234.

Fox, R.C., 1964. The adductor muscles of the jaw in some primitive reptiles. University of Kansas. Publ., Mus. Nat. Hist. 12:657.

Fox, R.F., 1971. Marsupial mammals from the early Campanian Milk River Formation, Alberta, Canada, *in* Early Mammals (D.M. Kermack and K.A. Kermack, eds.). Academic Press, New York.

Fraser, F.C., and P.E. Purves, 1955. The blow of whales. Nature 176:1221.

————, ————, 1960a. Anatomy and function of the cetacean ear. Proc. Roy. Soc. (London), B 152:62.

————, ————, 1960b. Hearing in cetaceans. Bull. Brit. Mus. Nat. Hist. Zool. 7:1.

Freeland, W.J., 1980. Mangabey *(Cercocebus albigena)* movement patterns in relation to food availability and fecal contamination. Ecology 61:1297.

———— and D.H. Janzen, 1974. Strategies in herbivory by mammals: The role of plant secondary compounds. Amer. Nat 108:269.

French, A.R., 1977. Periodicity of recurrent hypothermia during hibernation in the pocket mouse, *Perognathus longimembris*. J. Comp. Physiol. 115:87.

French, N.R., W.E. Grant, W. Grodzinski, and D.M. Swift, 1976. Small mammal energetics in grassland ecosystems. Ecol. Monogr. 46:201.

Freudenthal, M., 1972. *Deinogalerix koenig-*

swaldi nov. gen., Nov. Spec.: A giant insectivore from the Neogene of Italy, Scripta Geologica, Leiden 14:1.

Fries, S., 1879. Uber die Fortpflanzung der einheimischen Chiropteren. Zool. Anzeiger, Bd. 2:355.

Frith, H.J., and S.H. Calaby, 1969. Kangaroos. F.W. Cheshire, Melbourne.

Fullard, J.H., M.B. Fenton, and J.A. Simmons, 1979. Jamming bat echolocation: The clicks of arctiid moths. Canadian J. Zool. 57:647.

Fuller, W.A., 1967. Ecologie hivernale des lemmings et fluctuations de leurs populations. Terre Vie 114:97.

————, 1969. Changes in numbers of three species of small rodents near Great Slave Lake, N.W.T., Canada, 1964–1967, and their significance for general population theory. Ann. Zool. Fennici 6:113.

Gaines, M.S., and C.J. Krebs, 1971. Genetic changes in fluctuating vole populations. Evolution 25:702.

————, A.V. Vivas, and C.L. Baker, 1979. An experimental analysis of dispersal in fluctuating vole populations. Demographic perameters. Ecology 60:814.

Galambos, R., and D.R. Griffin, 1942. Obstacle avoidance by flying bats: The cries of bats. J. Exp. Zool. 89:475.

Garsd, A., and W.E. Howard, 1981. A 19-year study of microtine population fluctuations using time-series analysis. J. Mamm. 62:930.

Gaudet, C.L., and M.B. Fenton, 1984. Observational learning in three species of insectivorous bats (Chiroptera). Anim. Behav. 32:385.

Gawn, R.L.W., 1948. Aspects of the locomotion of whales. Nature 161:44.

Geist, V., 1971. Mountain-sheep: A study in behavior and evolution. University of Chicago Press, Chicago.

————, 1974. On the relationship of ecology and behavior in the evolution of ungulates: Theoretical considerations, *in* The Behavior of Ungulates and Its Relation to Management (V. Geist and F.R. Walther, eds.). International Union for Conservation of Nature and Nature Resources, Morges, Switzerland.

Geluso, K.N., and E.H. Studier, 1979. Diurnal fluctuation in urine concentration in the little brown bat, *Myotis lucifugus,* in a natural roost. Comp. Biochem. Physiol. 62:471.

Genelly, R.E., 1965. Ecology of the common mole-rate *(Cryptomys hottentotus)* in Rhodesia. J. Mamm. 46:647.

Gill, E.D., 1955a. The problem of extinction, with special reference to the Australian marsupials. Evolution 9:87.

————, 1955b. The Australian "arid period." Australian J. Sci. 17:204.

————, 1957. The stratigraphical occurrence and paleontology of some Australian Tertiary marsupials. Mem. Nat. Mus., Victoria 21:135.

Gilmore, R.M., 1961. Whales, porpoises, and the U.S. Navy. Norsk Hvalfangst-tid. 3:1.

Gingerich, P.D., N.A. Wells, P.E. Russell, and S.M. Ibrahim Shah, 1983. Origin of whales in epicontinental remnant seas: New evidence from the Early Eocene of Pakistan. Science 220:403.

Glander, K.E., 1977. Poison in a monkey's Garden of Eden. Nat. Hist. 86:35.

Golightly, R.T. Jr., and R.D. Ohmart, 1978. Heterothermy in free-ranging Abert's squirrels *(Sciurus aberti).* Ecology 59:897.

Goodall, J., 1965. Chimpanzees of the Gombe stream reserve, *in* Primate Behavior (I. DeVore, ed.). Holt, Rinehart and Winston, New York.

————, 1977. Infant killing and cannibalism in free-living chimpanzees. Folia Primat. 28:258.

Goodman, D., 1975. The theory of diversity-stability relationships in ecology. Quart. Rev. Biol. 50:237.

Goodwin, G.G., 1954. Mammals of the air, land, and waters of the world, *in* The Animal Kingdom (F. Drimmer, ed.). Doubleday, Garden City.

Gordon, M.S., G.A. Bartholomew, A.D. Grinnell, C.B. Jorgensen, and F.N. White, 1977. Animal Physiology: Principles and Adaptations (3d edn.). Macmillan, New York.

Gorman, M.L., 1976. A mechanism for individual recognition by odor in *Herpestes auropunctatus* (Carnivora, Viverridae). Anim. Behav. 24:141.

————, D.B. Nedwell, and R.M. Smith, 1974. An analysis of the anal scent pockets of *Herpestes auropunctatus* (Carnivora: Viverridae). J. Zool., London 172:389.

Gould, E., 1955. The feeding efficiency of insectivorous bats. J. Mamm. 36:399.

————, 1965. Evidence for echolocation in the Tenrecidae of Madagascar. Proc. Amer. Phil. Soc. 109(6):352.

————, 1969. Communication in three genera of shrews (Soricidae): *Suncus, Blarina,* and *Cryptotis.* Comm. Behav. Biol., Part A 3(1):11.

————, 1970. Echolocation and communication of bats, *in* About Bats (B.H. Slaughter and D.W. Walton, eds.). Southern Methodist University Press, Dallas.

————, 1971. Studies of maternal-infant communication and development of vocalization in the bats *Myotis* and *Eptesicus.* Comm. Behav. Biol., Part A 5(5):263.

———— and J.F. Eisenberg, 1966. Notes on the biology of the Tenrecidae. J. Mamm 47:660.

————, N.C. Negus, and A. Novick, 1964. Evidence for echolocation in shrews. J. Exp. Zool. 156:19.

Gould, L.M., 1973. Foreword, *in* E.H. Colbert, Wandering Lands and Animals. Dutton, New York.

Graf, W., 1955. The Roosevelt elk. Port Angeles, Wash.: Port Angeles Evening News.

Grand, T.I., and R. Lorenz, 1968. Functional analysis of the hip joint in *Tarsius bancanus* (Horsefield, 1821) and *Tarsius syrichta* (Linnaeus, 1758). Folia Primat. 9:161.

Grant, T.R., and T.J. Dawson, 1978. Temperature regulation in the platypus, *Ornithorhynchus anatinus:* Production and loss of heat in air and water. Physiol. Zool. 51:315.

Grant-Taylor, T.L., and T.A. Rafter, 1963. New Zealand natural radiocarbon measurements I-V. Radiocarbon 5:118.

Green, R.G., and C.L. Larson, 1938. A description of shock disease in the snowshoe hare. Amer. J. Hyg. 28:190.

————, ————, and J.F. Bell, 1939. Shock disease as the cause of the periodic decimation of the snowshoe hare. Amer. J. Hyg. 30B:83.

Greenwald, G.S., 1956. The reproductive cycle of the field mouse, *Microtus californicus.* J. Mamm. 37:213.

————, 1957. Reproduction in a coastal California population of the field mouse, *Microtus californicus.* Univ. California Publ. Zool. 54:421.

Griffin, D.R., 1951. Audible and ultrasonic sounds of bats. Experientia 7:448.

————, 1953. Bat sounds under natural conditions with evidence for echolocation of insect prey. J. Exp. Zool. 36:399.

————, 1958. Listening in the Dark. Yale University Press, New Haven, Conn.

————, 1962. Comparative studies of the orientation sounds of bats. Symp. Zool. Soc. London 7:61.

————, 1970. Migrations and homing of bats, *in* Biology of Bats (W.A. Wimsatt, ed.). Academic Press, New York.

————, D. Dunning, D.A. Cahlander, and F.A. Webster, 1962. Correlated orientation sounds and ear movements of horseshoe bats (part I). Nature 196:1185.

———— and R. Galambos, 1940. Obstacle avoidance by flying bats. Anat. Rec. 78:95.

————, ————, 1941. The sensory basis of obstacle avoidance by flying bats. J. Exp. Zool. 86:481.

———— and H.B. Hitchcock, 1965. Probable 24-year longevity records for *Myotis lucifugus.* J. Mamm. 46:332.

———— and A. Novick, 1955. Acoustic orientation of Neotropical bats. J. Exp. Zool. 130:251.

————, F.A. Webster, and C.R. Michael, 1960. The echolocation of flying insects by bats. Anim. Behav. 8:141.

Grinnell, A.D., 1963a. The neurophysiology of audition in bats: Intensity and frequency parameters. J. Physiol. 167:38.

————, 1963b. The neurophysiology of audition in bats: Temporal parameters. J. Physiol. 167:67.

Grinnell, J., 1914a. An account of the mammals and birds of the lower Colorado Valley with especial reference to the distributional problems presented. Univ. California Publ. Zool. 12:51.

————, 1914b. Barriers to distribution as regards to birds and mammals. Amer. Nat. 48:248.

————, 1914c. The Colorado River as a hindrance to the dispersal of species. Univ. California Publ. Zool. 12:100.

————, 1922. A geographical study of the kangaroo rats of California. Univ. California Publ. Zool. 24:1.

————, 1926. Geography and evolution in the pocket gopher. Univ. California Chron. 30:429.

————, 1933. Review of the Recent mammal fauna of California. Univ. California Publ. Zool. 40:71.

————, J.S. Dixon, and J.M. Linsdale, 1937. Fur-Bearing Mammals of California, 2 vols. University of California Press, Berkeley.

———— and T.I. Storer, 1924. Animal Life in

the Yosemite. University of California Press, Berkeley.

Grummon, R.A., and A. Novick, 1963. Obstacle avoidance in the bat *Macrotus mexicanus*. Physiol. Zool. 36:361.

Guilday, J.E., 1958. The prehistoric distribution of the opossum. J. Mamm. 39:39.

Guthrie, M.J., 1933. The reproductive cycles of some cave bats. J. Mamm. 14:199.

Guthrie, R.D., and R.G. Petocz, 1970. Weapon automimicry among mammals. Amer. Nat. 104:585.

Guyton, A.C., 1976. Textbook of Medical Physiology, 5th edn. W.B. Saunders, Philadelphia.

Hales, J.R.S., 1973. Effects of heat stress on blood flow in respiratory and non-respiratory muscles in the sheep. Pflugers Arch. Ges. Physiol. 345:123.

Hall, E.R., 1946. Mammals of Nevada. University of California Press, Berkeley.

————, 1951. American weasels. Univ. Kansas Publ., Mus. Nat. Hist. 4:1.

————, 1958. Introduction, Part II, *in* Zoogeography (C.L. Hubbs, ed.). American Association for the Advancement of Science, Washington, D.C.

———— and W.W. Dalquest, 1963. The mammals of Veracruz. Univ. Kansas Publ., Mus. Nat. Hist. 14:165.

———— and K.R. Kelson, 1959. The Mammals of North America. Ronald Press, New York.

Hall, K.R.L., 1965. Behaviour and ecology of the Wild Patas Monkeys, *Erythrocebus patas,* in Uganda. J. Zool. Soc. London 148:15.

————, 1968. Behaviour and ecology of the Wild Patas monkey, *in* Primates: Studies in Adaptation and Variability (P.C. Jay, ed.). Holt, Rinehart and Winston, New York.

———— and I. DeVore, 1965. Baboon social behavior, *in* Primate Behavior (I. DeVore, ed.). Holt, Rinehart and Winston, New York.

———— and G.B. Schaller, 1964. Tool-using behavior of the California sea otter. J. Mamm. 45:287.

Hamilton, W.J. Jr., 1937. The biology of microtine cycles. J. Agric. Res. 54:779.

————, 1939. American Mammals. McGraw-Hill, New York.

Hamilton, W.J. III, 1962. Reproductive adaptations of the red tree mouse. J. Mamm. 43:486.

Hanover, J.W., 1966. Genetics of terpenes. I:

Gene control of monoterpene levels in *Pinus monticola.* Dougl. Heredity 21:73.

————, 1971. Genetics of terpenes. II: Genetic variances in interrelationships of monoterpene concentrations in *Pinus monticola.* Heredity 27:237.

Hansen, R.M., 1962. Movements and survival of *Thomomys talpoides* in a mima-mound habitat. Ecology 43:151.

————, 1978. Shasta ground sloth food habits, Rampart Cave, Arizona. Paleobiology 4:302.

———— and A.L. Ward, 1966. Some relations of pocket gophers to rangelands on Grand Mesa, Colorado. Colo. Agric. Exp. Sta., Tech. bull., 88.

Hanson, D.D., 1971. The food habits and energy dynamics of *Neotoma stephensi*. MS thesis, Northern Arizona Univ., Flagstaff.

Hardy, R., 1945. The influence of types of soil upon the local distribution of some mammals in southwestern Utah. Ecol. Monogr. 15:71.

Harlow, H.V., 1981. Torpor and other physiological adaptations of the badger *(Taxidea taxus)* to cold environments. Physiol. Zool. 54:267.

Harrington, J.E., 1976. Discrimination between individuals by scent in *Lemur fulvus*. Anim. Behav. 24:207.

Hart, F.M., and J.A. King, 1966. Distress vocalizations of young in two subspecies of *Peromyscus*. J. Mamm. 47:287.

Hart, J.S., 1956. Seasonal changes in insulation of the fur. Can. J. Zool. 34:53.

Hart, R.T., 1932. The vertebral columns of ricochetal rodents. Bull. Amer. Mus. Nat. Hist. 58:599.

————, 1934. The pangolins and aard-varks collected by the American Museum Congo Expedition. Bull. Amer. Mus. Nat. Hist. 66:643.

————, 1936. Hyraxes collected by the American Museum Congo Expedition. Bull. Amer. Mus. Nat. Hist. 72:117.

Hartenberger, J.J., 1975. Nouvelles decouvertes de rongeurs dans le Deseadien (Oligocene Inferieur) de Salla Luribay (Bolivia). C.R. Acad. Sci. Paris, ser. D 280:427.

Hartman, C.G., 1933. On the survival of spermatozoa in the female genital tract of the bat. Quart. Rev. Biol. 8:185.

Hartman, D.S., 1979. Ecology and behavior of the manatee *(Trichechus manatus)* in Florida. Spec. Publ., Amer. Soc. Mammal. No. 5.

Harvey, M.J., and R.W. Barbour, 1965. Home ranges of *Microtus ochrogaster* as determined by a modified minimum area method. J. Mamm. 46:398.

Hatt, R.T., 1932. The vertebral column of ricochetal rodents. Bull. Amer. Mus. Nat. Hist. 63:599.

————, 1934. The pangolins and aard-varks collected by the American Museum Congo expedition. Bull. Amer. Mus. Nat. Hist. 67:643.

————, 1936. Hyraxes collected by the American Museum Congo expedition. Bull. Amer. Mus. Nat. Hist. 67:643.

Hayman, D.L., 1977. Chromosome number—constancy and variation, *in* The Biology of Marsupials (B. Stonehouse and D. Gilmore, eds.). Macmillan, London.

Hayward, J.S., and C.P. Lyman, 1967. Non-shivering heat production during arousal from hibernation and evidence for the contribution of brown fat, *in* Mammalian Hibernation, vol. III (K.C. Fisher et al., eds.). Oliver & Boyd, London.

————, ————, and C.R. Taylor, 1965. The possible role of brown fat as a source of heat during arousal from hibernation. Ann. N.Y. Acad. Sci. 131:441.

Hecht, M.K., 1975. The morphology and relationships of the largest known terrestrial lizard, *Megalania prisca* Owen, from the Pleistocene of Australia. Proc. Roy. Soc. Victoria 87:239.

Hediger, H., 1950. Gefangenschaftsgeburt ein afrikanischen Springhasen. Zool. Gart. Leipzig 17(5).

Heezen, B.C., 1957. Whales entangled in deep-sea cables. Deep Sea Res. 4:105.

Heglund, H.C., M.A. Fedak, C.R. Taylor, and G.A. Cavagna, 1982. Energetics and mechanics of terrestrial locomotion. IV: Total mechanical energy changes as a function of speed and body size in birds and mammals. J. Exp. Biol. 97:57.

Heglund, N.C., C.R. Taylor, and T.A. McMahon, 1974. Scaling stride frequency and gait to animal size: Mice to horses. Science 186:1112.

Heim de Balsac, H., 1954. Un genre inédit et inattendu de mammifera (Insectivore Tenrecidae) d'Afrique Occidentale. Compt. Rend. Acad. Sci., Paris 239.

Heinroth-Berger, K., 1959. Beobachtungen an handaufgezogenen Mantelpavianen *(Papio hamadryas* L.), Z. Tierpsychol. 16:706.

Heller, H.C., and T.L. Poulson, 1970. Circannian rhythms. II: Endogenous and exogenous factors controlling reproduction and hibernation in chipmunks

(Eutamias) and ground squirrels *(Spermophilus)*. Comp. Biochem. Physiol. 33:357.

Hellwing, S., 1971. Maintenance and reproduction in the white-toothed shrew, *Crocidura russula monacha* Thomas, in captivity. Z. Saügetierk 36:103.

Hendrichs, H., and U. Hendrichs, 1971. Dikdik und Elephanten. R. Piper, Munich.

Henshaw, R.E., 1970. Thermoregulation in bats, *in* About Bats (B.H. Slaughter and D.W. Walton, eds.). Southern Methodist University Press, Dallas.

———— and G.E. Folk Jr., 1966. Relation of thermoregulation to seasonally changing microclimate in two species of bats *(Myotis lucifugus* and *M. sodalis)*. Physiol. Zool. 39:223.

Henson, O.W. Jr., 1961. Some morphological and functional aspects of certain structures of the middle ear in bats and insectivores. Univ. Kansas Sci. Bull. 42:151.

————, 1965. The activity and function of the middle-ear muscles in echo-locating bats. J. Physiol. 180:871.

Heppes, J.B., 1958. The white rhinoceros in Uganda. Afr. Wildlife 12:273.

Herald, E.S., R.L. Brownell Jr., F.L. Frye, E.J. Morris, W.E. Evans, and A.B. Scott, 1969. Blind river dolphin: First side-swimming cetacean. Science 166:1408.

Herman, L.H., 1980. Cognitive characteristics in dolphins, *in* Cetacean Behavior (L.H. Herman, ed.). Wiley, New York.

Herman, L.M., M.F. Peacock, M.P. Yunker, and C.J. Madsen, 1975. Bottlenosed dolphin: Double-slit pupil yields equivalent aerial and underwater diurnal acuity. Science 189:650.

Hermanson, J.W., 1981. Functional morphology of the clavicle in the pallid bat, *Antrozous pallidus*. J. Mamm. 62:801.

———— and J.S. Altenbach, 1981. Functional anatomy of the primary downstroke muscles in the pallid bat, *Antrozous pallidus*. J. Mamm. 62:795.

Herreid, C.F. II, 1963. Temperature regulation and metabolism in Mexican free-tail bats. Science 142:1573.

————, 1967. Temperature regulation, temperature preference and tolerance, and metabolism of young and adult free-tail bats. Physiol. Zool. 40:1.

Hershkovitz, P., 1969. The Recent mammals of the Neotropical Region: A zoogeographic and ecological review. Quart. Rev. Biol. 44:1.

————, 1972. The Recent mammals of the

Neotropical Region: A zoogeographic and ecological review, *in* Evolution, Mammals, and Southern Continents (A. Keast, F.C. Erk, and B. Glass, eds.). State University of New York Press, Albany.

Hertel, A., 1969. Hydrodynamics of swimming and wave-riding dolphins, *in* The Biology of Marine Mammals (H.T. Andersen, ed.). Academic Press, New York.

Hesse, R., W.C. Allee, and K.P. Schmidt, 1951. Ecological Animal Geography, 2d edn. Wiley, New York.

Hewitt, S., J.F. Wheldrake, and R.V. Baudinette, 1981. Water balance and renal function in the Australian desert rodent *Notomys alexis:* The effect of diet on water-turnover rate, glomerular-filtration rate, renal-plasm flow and renal blood flow. Comp. Biochem. Physiol. 68A:405.

Hibbard, C.W., D.E. Ray, D.E. Savage, D.W. Taylor, and J.E. Guilday, 1965. Quaternary mammals of North America, *in* The Quarternary of the United States (H.E. Wright Jr. and D.G. Frey, eds.). Princeton University Press, Princeton, New Jersey.

————— and G.C. Rinker, 1942. A new bog-lemming *(Synaptomys)* from Meade County, Kansas. Univ. Kansas Sci. Bull. 28:25.

Hiiemae, K., 1967. Masticatory function in mammals. J. Dent. Res. 46(2):883.

Hilborn, R., 1975. Similarities in dispersal tendency among siblings in four species of voles. Ecology 56:1221.

Hildebrand, M., 1959. Motions of the running cheetah and horse. J. Mamm. 40:481.

—————, 1960. How animals run. Sci. Amer. 202(5):148.

—————, 1962. Walking, running, and jumping. Amer. Zool. 2:151.

—————, 1965. Symmetrical gaits of horses. Science 150:701.

—————, 1974. Analysis of Vertebrate Structure. Wiley, New York.

Hiley, P., 1975. How the elephant keeps its cool. Nat. Hist. 84:34.

Hill, A.V., 1950. The dimensions of animals and their muscular dynamics. Sci. Progr. 38:209.

Hill, J.E., 1974. A new family, genus and species of bat (Mammalia: Chiroptera) from Thailand. Bull. Brit. Mus. Nat. Hist. 27:301.

————— and T.D. Carter, 1941. The mammals of Angola, Africa. Bull. Amer. Mus. Nat. Hist. 78:1.

————— and S.E. Smith, 1981. *Crasseonycteris thonglongyai.* Mamm. Species 160:1.

Hill, R.W., D.P. Christian, and J.H. Veghte, 1980. Pinna temperature in exercising jackrabbits. J. Mamm. 61:30.

————— and J.H. Veghte, 1976. Jackrabbit ears: Surface temperatures and vascular responses. Science 194:436.

Hill, W.C.O., and J. Meester, 1971. Suborder Prosimii, Infraorder Lorisiformes, part 3.2, *in* The Mammals of Africa: An Identification Manual (J. Meester and H.W. Setzer, eds.). Smithsonian Institution Press, Washington, D.C.

Hillman, J.C., and A.K.K. Hillman, 1977. Mortality of wildebeest in Nairobi National Park during the drought of 1973–1974. E. Afr. Wildl. J. 15:1.

Hinde, R.A., 1970. Animal Behavior: A Synthesis of Ethology and Comparative Psychology, 2d edn. McGraw-Hill, New York.

Hisaw, F.L., 1924. The absorption of the public symphysis of the pocket gopher, *Geomys bursarius* (Shaw). Amer. Nat. 58:93.

Hock, R.V., 1951. The metabolic rates and body temperatures of bats. Biol. Bull. 101:289.

Hoeck, H.N., 1982. Population dynamics, dispersal and genetic isolation in two species of hyrax *(Heterohyrax brucei* and *Procavia johnstoni)* on habitat islands in the Serengeti. Z. Tierpsychol. 59:177.

—————, H. Klein, and P. Hoeck, 1982. Flexible social organization in hyrax. Z. Tierpsychol. 59:265.

Hoese, H.D., 1971. Dolphin feeding out of water in a salt marsh. J. Mamm. 52:222.

Hoffmann, R.S., 1958. The role of reproduction and mortality in population fluctuations of voles *(Microtus)*. Ecol. Monogr. 28:79.

—————, 1976. An ecological and zoogeographical analysis of animal migration across the Bering land bridge during the Quaternary period, *in* Beringia in Cenozoic (V.L. Kontrimavichus, ed.). Vladivostok: Academy of Science (in Russian).

—————, 1980. Of mice and men: Beringian dispersal and the ice-free corridor. Canadian J. Anthropol. 1:51.

Holling, C.S., 1959. The components of predation as revealed by a study of small mammal predation of the European pine sawfly. Can. Entomol. 91:293.

————, 1961. Principles of insect predation. Ann. Rev. Entomol. 6:163.

Hooper, E.T., 1952. A systematic review of the harvest mice (genus *Reithrodontomys*) of Latin America. Misc. Publ. Mus. Zool., Univ. Michigan 77:1.

————, 1968. Anatomy of middle-ear walls and cavities in nine species of microtine rodents. Univ. Michigan Occ. Papers 657:1.

———— and J.H. Brown, 1968. Foraging and breeding in two sympatric species of Neotropical bats, genus *Noctilio*. J. Mamm. 49:310.

Hopkins, D.M., 1959. Cenozoic history of the Bering land bridge. Science 129:1519.

Hopson, J.A., 1966. The origin of the mammalian middle ear. Amer. Zool. 6:437.

————, 1970. The classification of non-therian mammals. J. Mamm. 51:1.

———— and A.W. Crompton, 1969. Origin of mammals, *in* Evolutionary Biology (T. Dobzhansky, ed.). Appleton-Century-Crofts, New York.

Hornocker, M.G., 1970a. The American lion. Nat. Hist. 79:40.

————, 1970b. An analysis of mountain lion predation upon mule deer and elk in the Idaho Primitive Area. Wildl. Monogr. No. 21.

Horst, R., 1969. Observations on the structure and function of the kidney of the vampire bat *(Desmodus rotundus murinus), in* Physiological Systems in Semiarid Environments (C.C. Hoff and M.L. Riedesel, eds.). Univ. New Mexico Press, Albuquerque.

Houlihan, R.T., 1963. The relationship of population density to endocrine and metabolic changes in the California vole, *Microtus californicus*. Univ. Calif. Publ. Zool. 65:327.

Howe, J.G., W.E. Grant, and L.J. Folse, 1982. Effects of grazing by *Sigmodon hispidus* on the regrowth of annual rye-grass *(Lolium perenne)*. J. Mamm. 63:176.

Howell, A.B., 1930. Aquatic Mammals. Charles C Thomas, Springfield, Ill.

————, 1944. Speed in Animals. University of Chicago Press, Chicago.

Howell, D.J., 1974a. Acoustic behavior and feeding in glossophagine bats. J. Mamm. 55:293.

————, 1974b. Bats and pollen: Physiological aspects of the syndrome of chiropterophily. Comp. Biochem. Physiol. 48A:263.

———— and D. Burch, 1974. Food habits of some Costa Rican bats. Rev. Biol. Trop. 21:281.

———— and J. Pylka, 1976. Why bats hang upside down: A biomechanical hypothesis. J. Theor. Biol., 69:625.

Hrdy, S.B., 1977. The Langurs of Abu: Female and Male Strategies of Reproduction. Harvard University Press, Cambridge, Mass.

Hudson, J.W., 1962. The role of water in the biology of the antelope ground squirrel. Univ. Calif. Publ. Zool. 64:1.

————, 1965. Temperature regulation and torpidity in the pigmy mouse, *Baiomys taylori*. Physiol. Zool. 38:243.

————, 1973. Torpidity in mammals, *in* Comparative Physiology of Thermoregulation, vol. III. Academic Press, New York.

————, 1974. The estrous cycle, reproduction, growth, and development of temperature regulation in the pigmy mouse, *Baiomys taylori*. J. Mamm. 55:572.

———— and T.J. Dawson, 1975. Role of sweating from the tail in the thermal balance of the rat-kangaroo *Potorous tridactylus*. Aust. J. Zool. 23:453.

————, D.R. Deavers, and S.R. Bradley, 1972. A comparative study of temperature regulation in ground squirrels with special reference to the desert species. Symp. Zool. Soc. London 31:191.

Huey, R.B., 1969. Winter diet of the Peruvian desert fox. Ecology 50:1089.

Hugget, A.St.G., and W.F. Widdas, 1951. The relationship between mammalian foetal weight and conception age. J. Physiol. 114:306.

Hughes, R.L., 1974. Morphological studies on implantation in marsupials. J. Reprod. Fertil. 39:173.

Hult, R., 1982. Another function of echolocation for bottlenose dolphins *(Tursiops truncatus)*. Cetology 47:1.

Humboldt, A. von, and A. Bonpland, 1852–53. Personal Narrative of Travels to the Equinoctial Regions of America During the Years 1799–1804, 3 vols. Henry G. Bohn, London.

Humphrey, S.R., 1974. Zoogeography of the nine-banded armadillo *(Dasypus novemcinctus)* in the United States. BioScience 24:457.

————, 1975. Nursery roosts and community diversity in Nearctic bats. J. Mamm. 56:321.

Huntley, B., 1976. Angola, a situation report. Afr. Wildlife 30(1):10.

Hurley, P.M., 1968. The confirmation of continental drift. Sci. Amer. 218(4):52.

Hutchinson, G.E., 1957. Concluding remarks. Cold Spr. Harb. Symp. Quant. Biol. 22:415.

Iler, R.K., 1979. The chemistry of silica. Wiley, New York.

Ingles, L.G., 1949. Ground water and snow as factors affecting the seasonal distribution of pocket gophers. *Thomomys monticola.* J. Mamm. 30:343.

Irving, L., 1966. Adaptations to cold. Sci. Amer. 214(1):94.

————, 1969. Temperature regulation in marine mammals, *in* The Biology of Marine Mammals (H.T. Andersen, ed.). Academic Press, New York.

———— and J.S. Hart, 1957. The metabolism and insulation of seals as bare-skinned mammals in cold water. Can. J. Zool. 35:497.

————, H. Krog, and M. Monson, 1955. The metabolism of some Alaskan animals in winter and summer. Physiol. Zool. 28:173.

Izawa, K., 1970. Unit groups of chimpanzees and their nomadism in the savannah woodland. Primates 11:1.

———— and J. Itani, 1966. Chimpanzees in Kasakati Basin, Tanganyika. I: Ecological study in the rainy season 1963–1964. Kyoto Univ. Afr. Stud. 1:73.

Jaeger, E.C., 1950. The coyote as a seed distributor. J. Mamm. 31:452.

Jameson, E.W. Jr., 1952. Food of deer mice, *Peromyscus maniculatus* and *P. boylei,* in the northern Sierra Nevada, California. J. Mamm. 33:50.

————, 1953. Reproduction of deer mice *(Peromyscus maniculatus* and *P. boylei)* in the Sierra Nevada, California. J. Mamm. 34:44.

———— and R.A. Mead, 1964. Seasonal changes in body fat, water and basic weight in *Citellus lateralis, Eutamias speciosus and E. amoenus.* J. Mamm. 45:359.

Jarman, M.V., 1970. Attachment to home area in impala. E. Afr. Wildl. J. 8:198.

Jarman, R.J., 1974. The social organization of antelope in relation to their ecology. Behavior 58:215.

———— and M.V. Jarman, 1974. Impala behavior and its relevance to management, *in* The Behavior of Ungulates and Its Relation to Management (V. Geist and F.R. Walther, eds.). International Union for Conservation of Nature and Natural Resources, Morges, Switzerland.

————, ————, 1979. The dynamics of ungulate social organization, *in* Serengeti, Dynamics of an Ecosystem. (A.R.E. Sinclair and M. Norton-Griffiths, eds.). University of Chicago Press, Chicago.

Jarvis, J.U.M., 1973. The structure of a population of mole-rats, *Tachyoryctes splendens* (Rodentia: Rhizomyidae). J. Zool., London 171:1.

————, 1978. Energetics of survival in *Heterocephalus glaber* (Rüppell), the naked mole-rat (Rodentia: Bathyergidae). Bull. Carnegie Mus. Nat. Hist. 6:81.

————, 1981. Eusociality in a mammal: Cooperative breeding in the naked mole rat. Science 212:571.

———— and J.B. Sale, 1971. Burrowing and burrow patterns of East African mole-rats *Tachyoryctes, Heliophobius* and *Heterocephalus.* J. Zool., London 163:451.

Jay, P.C. (ed.), 1968. Primates, Studies in Adaptation and Variability. Holt, Rhinehart, and Winston, New York.

Jen, P.H.-S., and N. Suga, 1976. Coordinated activities of middle-ear and laryngeal muscles in echolocating bats. Science 191:950.

Jenkins, F.A. Jr., 1970. Limb movement in a monotreme *(Tachyglossus aculeatus):* A cineradiographic analysis. Science 168:1473.

————, 1971. Limb posture and locomotion in the Virginia opossum *(Didelphis marsupialis)* and in other non-cursorial mammals. J. Zool., London 165:303.

————, and A.W. Crompton, 1979. Triconodonta, *in* Mesozoic Mammals: The First Two-Thirds of Mammalian History (J.A. Lillegraven, Z. Kielan-Jaworoska, and W.A. Clemens, eds.). University of California Press, Berkeley.

————, A.W. Crompton, and T. Downs, 1983. Mesozoic mammals from Arizona: New evidence on mammalian evolution. Science 222:1233.

———— and F.R. Parrington, 1976. The postcranial skeletons of the Triassic mammals *Eozostrodon, Megazostrodon,* and *Erythrotherium.* Phil. Trans. Roy. Soc. London, B (Biol. Sci.), v. 273, p. 387.

Jenkins, H.O., 1948. A population study of the meadow mice *(Microtus)* in three Sierra Nevada meadows. Proc. Calif. Acad. Sci., ser. 4, 26:43.

Jepsen, G.L., 1966. Early Eocene bat from Wyoming. Science 154:1333.

————, 1970. Bat origins and evolution, *in* Biology of Bats (W.A. Wimsatt, ed.). Academic Press, New York.

———— and M.O. Woodburne, 1969. Paleocene hyrocothere from Polecat Bench Formation, Wyoming. Science 164:543.

Johannessen, C.L., and J.A. Harder, 1960. Sustained swimming speeds of dolphins. Science 132:1550.

Johansen, K., and J. Krog, 1959. Diurnal Body temperature variations in the birch mouse, *Sicista betulina.* Amer. J. Physiol. 196:1200.

Johnson, D.R., 1961. The food habits of rodents on rangelands of southern Idaho. Ecology 42:407.

————, 1964. Effects of range treatment with 2,4-D on food habits of rodents. Ecology 45:241.

Jolly, A., 1966. Lemur behavior: A Madagascar field study. University of Chicago Press, Chicago.

————, 1972. Troop continuity and troop spacing in *Propithecus verreauxi* and *Lemur catta* at Berenty (Madagascar). Folia Primat. 17:335.

Jones, C., 1967. Growth, development, and wing loading in the evening bat, *Nycticeius humeralis* (Rafinesque). J. Mamm. 48:1.

Jones, F.W., 1923. The Mammals of South Australia. Part I: The Monotremes and Carnivorous Marsupials. Government Printer, Adelaide.

————, 1924. The Mammals of South America. Part II: The Bandicoots and Herbivorous Marsupials. Government Printer, Adelaide.

————, 1925. The Mammals of South Australia. Part III: The Monodelphia. Government Printer, Adelaide.

Jones, J.K. Jr., 1964. Distribution and taxonomy of mammals of Nebraska. Univ. Kansas Publ., Mus. Nat. Hist. 16:1.

———— and R.R. Johnson, 1967. Sirenians, *in* Recent Mammals of the World (S. Anderson and J.K. Jones Jr. eds.). Ronald Press, New York.

Jones, R., 1968. The biographical background to the arrival of man in Australia and Tasmania. Arch. Phys. Anthrop. Oceania, III 3:186.

Jones, R.B., and N.W. Nowell, 1973a. Aversive effects of the urine of a male mouse upon the investigatory behavior of its defeated opponent. Anim. Behav. 21:707.

————, ————, 1973b. The effect of urine on the investigatory behavior of male albino mice. Physiol. Behav. 11:35.

Jorgenson, J.W., N. Novotny, M. Carmack, G.B. Copland, and S.R. Wilson, 1978. Chemical scent constituents in the urine of the red fox *(Vulpes vulpes)* during the winter season. Science 199:796.

Kahmann, H., and K. Ostermann, 1951. Wahrnehmen and Hervorbringen hoher Tone bei kleiner Saugetieren. Experientia 7:268.

Kalela, O., 1957, Regulation of reproduction rate in subarctic populations of the vole *Clethrionomys rufocanus* (Sund.). Ann. Acad. Sci. Fennicae, Ser. A 4(34):1.

————, 1961. Seasonal change of habitat in the Norwegian lemming *Lemmus lemmus* L. Ann. Acad. Sci. Fennicae, Ser. A 4(55):1.

————, 1962. On the fluctuations in the numbers of arctic and boreal small rodents as a problem of production biology. Ann. Acad. Sci. Fennicae, Ser. A 4(66):1.

Kangas, E., 1949. On the damage to the forests caused by the moose and its significance in the economy of the forest. Eripainos: Suomen Riista, vol. 4, p. 62 (English summary, p. 88).

Kanwisher, J., and H. Leivestad, 1957. Thermal regulation in whales. Norsk Hvalfangst-tid. 1:1.

———— and G. Sundnes, 1966. Thermal regulation in ceteceans, *in* Whales, Dolphins and Porpoises (K.S. Norris, ed.). University of California Press, Berkeley.

Kasuya, T., 1973, Systematic consideration of Recent toothed whales based on the morphology of the tympano-periotic bone. Sci. Rep. Whales Res. Inst. 25:1.

Kaufmann, J.H., 1974a. Habitat use and social organization of nine sympatric species of macropod marsupials. J. Mamm. 55:66.

————, 1974b. Social ethology of the whiptail wallaby. *Macropus parryi,* in northeastern New South Wales. Anim. Behav. 22:281.

Keast, A., 1972, Australian mammals: Zoogeography and evolution, *in* Evolution, Mammals, and Southern Continents (A. Keast, F.C. Erk, and B. Glass, eds.). State University of New York Press, Albany.

Keen, R., and H.B. Hitchcock, 1980. Survival and longevity of the little brown bat *(Myotis lucifugus)* in southeastern Ontario. J. Mamm 61:1.

Kellogg, W.N., 1961. Porpoises and sonar. University of Chicago Press, Chicago.

————, R. Kohler, and H.N. Morris, 1953. Porpoise sounds as sonar signals. Science 117:239.

Kelsall, J.P., 1970. Migration of the barren-ground caribou. Nat. Hist. 79:98.

Kenagy, G.J., 1972. Saltbush leaves: Excision of hypersaline tissues by a kangaroo rat. Science 178:1094.

————, 1973a. Daily and seasonal patterns of activity and energetics in a heteromyid rodent community. Ecology 54:1201.

————, 1973b. Adaptations for leaf eating in the Great Basin kangaroo rat, *Dipodomys microps*. Oecologia 12:383.

————, 1981. Effect of day length, temperature, and endogenous control on annual rhythms of reproduction and hibernation in chipmunks (*Eutamias* spp.). J. Comp. Physiol. 141:369.

————, and G.A. Bartholomew, 1981. Effects of day length, temperature, and green food on testicular development in a desert pocket mouse *Perognathus formosus*. Physiol. Zool. 54:62.

Kendeigh, S.C., 1961. Animal Ecology. Prentice-Hall, Englewood Cliffs, N.J.

Kennerly, T.E. Jr., 1964. Microenvironmental conditions of the pocket gopher burrow. Texas J. Sci. 14(4):397.

Kennerly, T.R., 1971. Personal communication.

Kermack, K.A., 1963. The cranial structure of triconodonts. Phil. Trans. Roy. Soc. London, Ser. B 246:83.

———— and Z. Kielan-Jaworowska, 1971. Therian and non-therian mammals, *in* Early Mammals (D.M. Kermack and K.A. Kermack, eds.). Academic Press, London.

———— and F. Musset, 1958. The jaw articulation of the docodonta and the classification of Mesozoic mammals. Proc. Roy. Soc. London, B 149:204.

Kevan, P.G., 1975. Sun-tracking solar furnaces in high arctic flowers: Significance for pollination and insects. Science 189:723.

Kielan-Jaworowska, Z., 1969. Preliminary data on the Upper Cretaceous eutherian mammals from Bayn Dzak Gobi Desert. Palaeontol. Polon. 19:171.

————, 1974. Migrations of the Multituberculata and the Late Cretaceous connections between Asia and North America. So. African Mus. Ann. 64:231.

————, 1975. Late Cretaceous mammals and dinosaurs from the Gobi Desert. Amer. Sci. 63:150.

————, T.M. Bown, and J.A. Lillegraven, 1979. Eutheria, *in* Mesozoic Mammals: The First Two-Thirds of Mammalian History. (J.A. Lillegraven, Z. Kielan-Jaworoska, and W.A. Clemens, eds.). University of California Press, Berkeley.

Kiltie, R.A., 1981. The function of interlocking canines in rain forest peccaries *(Tayassuidae)*. J. Mamm., 62:459.

King, J.A., 1955. Social behavior, social organization and population dynamics in a black-tail prairie dog town in the Black Hills of South Dakota. Contrib. Lab. Vert. Biol. Univ. Michigan. 67:1.

————, 1959. The social behavior of prairie dogs. Sci. Amer. 201(4):128.

———— (ed.), 1968. Biology of *Peromyscus* (Rodentia). Spec. Publ. No. 2, Amer. Soc. Mammal.

Kingdon, J., 1971. East African mammals, vol. I. Academic Press, New York.

————, 1974a. East African mammals, vol. IIA. Academic Press, New York.

————, 1974b. East African mammals, vol. IIB, Academic Press, New York.

Kinnear, J.E., A Cockson, P. Christensen, and A.R. Main, 1979. The nutritional biology of the ruminants and ruminant-like mammals: A new approach. Comp. Biochem. Physiol. 64:357.

————, K.G. Purohit, and A.R. Main, 1968. The ability of the tammar wallaby (*Macropus eugenii*, Marsupialia) to drink seawater. Comp. Biochem. Physiol. 25:761.

Kirsch, J.A.W., 1977. The six-percent solution: Second thoughts on the adaptedness of the marsupialia. Amer. Sci. 65:276.

Kitchen, D.W., 1974. Social behavior and ecology of the pronghorn. Wildl. Monogr. 38:1.

Kleiman, D.G., 1977. Monogamy in mammals. Quart. Rev. Biol. 52:39.

———— and T.M. Davis, 1978. Ontogeny and maternal care, *in* Biology of Bats of the New World Family Phyllostomatidae, part III (R.J. Baker, J.K. Jones Jr., D.C. Carter, eds.). Spec. Publ. Mus., no. 16. Texas Tech Press, Lubbock.

Klingel, H., 1967. Soziale Organisation und Verhalten freilebender Steppenzebras. Z. Tierpsychol. 24:580.

Klingener, D., 1964. The comparative myology of four dipodoid rodents (Genera *Zapus, Napaeoza-*

pus, Sicista and *Jaculus*). Misc. Publ. Mus. Zool., Univ. Michigan 124:1.

Knappe, H., 1964, Zur Funktion des Jacobsonschen Organs *(Organon vomeronasale Jacobsoni)*. Zool. Gart. (Leipzig) 28:188.

Knudsen, V.O., 1931. The effect of humidity upon the absorption of sound in a room, and determination of the coefficients of absorption of sound in air. J. Acoustical Soc. Am. 3:126.

————, 1935. Atmospheric acoustics and the weather. Sci. Mon. 40:485.

Komarek, E.V., 1932. Notes on mammals of Menominee Indian Reservation, Wisconsin. J. Mamm. 13:203.

Kooyman, G.L., 1968. An analysis of some behavioral and physiological characteristics related to diving in the Weddell seal. Anarctic Res. Ser. 11:227.

————, 1975a. A comparison between day and night diving in the Weddell seal. J. Mamm. 56:563.

————, 1975b. Physiology of freely diving Weddell seals. Rapp. P.-V. Reun. Cons. Int. Explor. Mer. 169:441.

———— and H.T. Andersen, 1969. Deep diving, *in* The Biology of Marine Mammals (H.T. Andersen, ed.). Academic Press, New York.

———— and W.B. Campbell, 1972. Heart rates in freely diving Weddell seals. Comp. Biochem. Physiol. 43:31.

————, R.L. Gentry, and D.L. Urquhart, 1976. Northern fur seal diving behavior: A new approach to its study. Science 193:411.

————, D.H. Kerem, W.B. Campbell, and J.J. Wright, 1971. Pulmonary function in freely diving Weddell seal, *Leptonychotes weddelli*. Resp. Physiol. 12:271.

Koshkina, T.V., 1965. Population density and its importance in regulating the abundance of the red vole (Russian translated by W.A. Fuller). Bull. Moscow Soc. Nat. Biol. 70:5.

———— and A.S. Kholansky, 1962. Reproduction of the Norwegian lemming *(Lemmus lemmus* L.) on the Kola Penninsula (Russian translated by W.A. Fuller). Zool. Zh. 41:604.

Kramer, M.O., 1960. J. Am. Soc. Naval Engrs, p. 25.

Krassilov, V., 1973, Mesozoic plants and the problem of angiosperm ancestry. Lethaia 6:163.

Krebs, C.J., 1963. Lemming cycle at Baker Lake, Canada during 1959–62. Science 146:1559.

————, 1964a. Cyclic variation in skull-body regressions of lemmings. Can. J. Zool. 42:631.

————, 1964b. The lemming cycle at Baker Lake, Northwest Territories, during 1959–62. Arctic Inst. N. Amer. Tech. Paper No. 15.

————, 1966. Demographic changes in fluctuating populations of *Microtus californicus*. Ecol. Monogr. 36:239.

————, 1970. *Microtus* population biology: Behavioral changes associated with the population cycle in *M. ochrogaster* and *M. pennsylvanicus*. Ecology 51:34.

———— and K.T. DeLong, 1965. A *Microtus* population with supplemental food. J. Mamm. 46:566.

————, M.S. Gains, B.L. Keller, J.H. Myers and R.H. Tamarin, 1973. Population cycles in small rodents. Science 179:35.

————, B.L. Keller, and R. Tamarin, 1969. *Microtus* population biology: Demographic changes in fluctuating populations of *M. ochrogaster* and *M. pennsylvanicus* in southern Indiana. Ecology 50:587.

———— and J.H. Myers, 1974. Population cycles in small mammals, *in* Advances in Ecological Research (MacFadyen, ed.). Academic Press, New York.

Krebs, H.A., 1950. Body size and tissue metabolism. Biochem. Biophys. Acta 4:249.

Krishnan, R.S., and J.C. Daniel, 1967. "Blastokinin": An inducer and regulator of blastocyst development in the rabbit uterus. Science 158:490.

Krohne, D.T., 1981. Intraspecific litter size variation in *Microtus californicus:* Variation within populations. J. Mamm. 62:29.

Kruger, L., 1966. Specialized features of the cetacean brain, *in* Whales, Dolphins and Porpoises (K.S. Norris, ed.). University of California Press, Berkeley.

Krumrey, W.A., and I.O. Buss, 1968. Age estimation, growth, and relationships between body dimensions of the female African elephant. J. Mamm. 49:22.

Kruuk, H., 1966. Clan-system and feeding habits of spotted hyaenas (*Crocuta crocuta* Erxleben). Nature, Lond. 209:1257.

————, 1970. Interactions between populations of spotted hyaenas *(Crocuta crocuta)* and their prey species, *in* Animal Populations in Relation to Their Food Resources (A. Watson, ed.). Blackwell Scientific Publications, Oxford.

————, 1972. The spotted hyena: A study of

predation and social behavior. University of Chicago Press, Chicago.

───── and H. Van Lawick, 1968. Hyenas, the hunters nobody knows. National Geographic, 134(1):44.

───── and W.A. Sands, 1972. The aardwolf (*Proteles cristatus* Sparrman, 1783) as predator of termites. E. Afr. Wildl. J. 10:211.

Krzanowski, A., 1960. Investigations of flights of Polish bats, mainly *Myotis myotis*. Acta Theriol. 4:175.

─────, 1961. Weight dynamics of bats wintering in the cave at Pulway (Poland). Acta Theriol. 4:249.

─────, 1964. Three long flights by bats. J. Mamm. 45:152.

Kühme, W., 1965. Freilandstudien zur Soziologie des Hyäenenhundes (*Lycaon pictus lupinus* Thomas 1902). Z. Tierpsychol. 225:495.

─────, 1966. Beobachtungen zur Soziologie des Löwens in der Serengeti-Steppe Ostafrikas. Z. Saugetierk. 31:205.

Kulzer, E., 1956. Flughunde erzeugen Oreintierung durch Zungenschlag. Naturwiss. 43:117.

─────, 1958. Untersuchungen über die Biologie von Flughunden der Gattung *Rousettus* Gray. Z. Morph. Ökol Biere. 47:374.

─────, 1960. Physiologische und morphologische Untersuchungen über die Erzeugung der Orientierungslaute von Flughunden der Gattung *Rousettus*. Z. Vergl. Physiol. 43:231.

─────, 1961. Über die Biologie der Nil-Flughunde (*Rousettes aegyptiacus*). Natur Volk. 91:219.

─────, 1963. Temperaturregulation bei Flughunden der Gattung *Rousettus* Gray. Z. Vergl. Physiol. 46:595.

─────, 1965. Temperaturregulation bein Fledermäusen (Chiroptera) aus berschiedenen Klimazonen. Z. Vergl. Physiol. 50:1.

Kummer, H., 1968a. Two variations in the social organization of baboons, *in* Primates: Studies in Adaptation and Variability (P.C. Jay, ed.). Holt, Rinehart and Winston, New York.

─────, 1968b. Social organization of hamadryas baboons. University of Chicago Press, Chicago.

───── 1984. From laboratory to desert and back: A social system of hamadryas baboons. Anim. Behav. 32:965.

Kunz, T.H., 1971. Ecology of the cave bat. *Myotis velifer*, in southcentral Kansas and northwestern Oklahoma. Unpublished Ph.D. dissertation, University of Kansas, Lawrence.

─────, 1973a. Population studies of the cave bat (*Myotis velifer*): Reproduction, growth, and development, Occas. Papers Mus. Nat. Hist., Univ. Kansas 15:1.

─────, 1973b. Resource utilization: Temporal and spatial components of bat activity in central Iowa. J. Mamm. 54:14.

─────, 1974. Feeding ecology of a temperate insectivorous bat (*Myotis velifer*). Ecology 55:693.

───── and E.L.P. Anthony, 1982. Age estimation and post-natal growth in the bat *Myotis lucifugus*. J. Mamm 63:23.

Kurten, B., 1969. Continental drift and evolution. Sci. Amer. 220(3):54.

Kurten, L., and U. Schmidt, 1982. Thermoreception in the common vampire bat (*Desmodus rotundus*). J. Comp. Physiol. 146:223.

Lacher, T.E. Jr., 1979. The comparative social behavior of *Kerodon rupestris* and *Galea spixii* in the xeric caatinga of northeastern Brazil. Thesis, University of Pittsburgh.

Lack, D., 1948. The significance of litter size. J. Anim. Ecol. 17:45.

─────, 1954a. The Natural Regulation of Animal Numbers. Oxford University Press, London.

─────, 1954b. Cyclic mortality. J. Wildl. Manage. 18:25.

─────, 1966. Population Studies of Birds. Clarendon Press, Oxford.

Lackey, J.A., 1967. Growth and development of *Dipodomys stephensi*. J. Mamm. 48:624.

Lamprey, H.F., G. Halevy, and S. Makacha, 1974. Interactions between Acacia, bruchid seed beetles and large herbivores, E. Afr. Wildl. J. 12:81.

Landry, S.O., 1957. The interrelationships of the New World and Old World histricomorph rodents. Univ. California Publ. Zool. 56:1.

Lang, H., and J.P. Chapin, 1917. The American Museum Congo Expedition Collection of Bats III: Field notes. Bull. Amer. Mus. Nat. Hist. 37.

Lang, T.G., 1966. Hydrodynamic analysis of cetacean performance, *in* Whales, Dolphins and Porpoises (K.S. Norris, ed.). University of California Press, Berkeley.

Langman, V.A., 1982. Giraffe youngsters need a little bit of maternal love. Smithsonian 12:95.

Langman, V.A., G.M.O. Maloiy, K. Schmidt-Nielsen, and R.C. Schroter, 1979. Nasal heat exchange in the giraffe and other large mammals. Resp. Physiol. 37:325.

Langworthy, M., and R. Horst, 1971. Reproductive behavior in a captive colony of *Molossus ater.* Paper presented at Second Southwestern Symposium on Bat Research, Univ. New Mexico, Albuquerque.

Lavocat, R., 1962. Réflexions sur l'origine et la structure du groupe des rongeurs, *in* Problèmes Actuels de Paleontologie. Colloq. Int. Cent. Nat. Rech. Sci., Paris.

————, 1973. Les rongeurs du Miocène d'Afrique Orientale. I: Miocene inférieur. Trav. Mém. Inst. E.P.A.E. Monpellier 1:1.

————, 1974. What is an hystricomorph? *in* The Biology of Hystricomorph Rodents (I.W. Rowlands and B. Weir, eds.). Symp. Zool. Soc. London. Academic Press, New York.

————, 1976. Rongeurs caviomorphes de l'Oligocene de Bolivia. II: Rongeurs du Bassin Deseadien de Salla-Luribay. Paleovertebrata 7:15.

————, 1980. The implications of rodent paleontology and biogeography to the geographical sources and origin of platyrrhine primates, *in* Evolutionary Biology of New World Monkeys and Contintal Drift (R.L. Ciochon and A.B. Chiarelli, eds.). Plenum, New York.

Lawhon, D.K., and M.S. Hafner, 1981. Tactile discriminatory ability and foraging strategies in kangaroo rats and pocket mice (Rodentia: Heteromyidae). Oecologia 50:303.

Lawlor, T.E., 1973. Aerodynamic characteristics of some Neotropical bats. J. Mamm. 54:71.

Lawrence, B., and A. Novick, 1963. Behavior as a taxonomic clue: Relationships of *Lissonycteris* (Chiroptera). Mus. Comp. Zool. 184:1.

Laws, R.M., 1953. The elephant seal (*Mirounga leonina,* Linn.) I: Growth and age. Falkland Is. Depend. Surv. Sci. Repts. 8:1.

————, 1970, Elephants as agents of habitat and landscape change in East Africa. Oikos 21:1.

———— and I.S.C. Parker, 1968. Recent studies on elephant populations in East Africa Symp. Zool. Soc. London 21:319.

Layne, J.N., 1958. Observations on freshwater dolphins in the upper Amazon. J. Mamm. 39:1.

————, 1965. Observations on marine mammals in Florida waters. Bull. Florida State Mus. 9:131.

————, 1968. Ontogeny, *in* Biology of *Peromyscus* (Rodentia) (J.A. King, ed.). Spec. Publ. No. 2, Amer. Soc. Mamm.

Lear, J., 1970. The bones on Coalsack Bluff: A story of drifting continents. Sat. Rev. 53(6):46.

Le Boeuf, B.J., and R.S. Peterson, 1969. Social status and mating activity in elephant seals. Science 163:91.

Lechleitner, R.R., 1958a. Certain aspects of behavior of the black-tailed jackrabbit. Amer. Midl. Nat. 60:145.

————, 1958b. Movements, density and mortality in a black-tailed jackrabbit population. J. Wildl. Manage. 22:371.

————, 1959, Sex ratio, age classes and reproduction of the black-tailed jackrabbit. J. Mamm. 40:63.

————, J.V. Tileston, and L. Kartman, 1962. Die-off of a Gunnison's prairie dog colony in central Colorado. I: Ecological observations and description of the epizootic. Zoonoses Res. 1:185.

Lee, A.K., 1963. The adaptations to arid environments in wood rats of the genus *Neotoma.* Univ. Calif. Publ. Zool. 64:57.

Lee, A.K., A.J. Bradley, and R.W. Braithwaite, 1977. Corticosterone levels and male mortality in *Antechinus stuarti, in* The Biology of Marsupials. (B. Stonehouse and D. Gilmore, eds.). University Park Press, Baltimore.

Leitner, P., 1966. Body temperature, oxygen consumption, heart rate and shivering in the California mastiff bat, *Eumops perotis.* Comp. Biochem. Physiol. 19:431.

———— and J.E. Nelson, 1967. Body temperature, oxygen consumption and heart rate in the Australian false vampire bat, *Macroderma gigas.* Comp. Biochem. Physiol. 21:65.

Lenfant, C., 1969, Physiological properties of blood marine mammals, *in* The Biology of Marine Mammals (H.T. Andersen, ed.). Academic Press, New York.

Leopold, A.S., T. Riney, R. McCain, and L. Tevis Jr., 1951. The jawbone deer herd. California Div. Fish and Game, Game Bull. 4:1.

————, L.K. Sowls, and D.L. Spencer, 1947. A survey of overpopulated deer ranged in the United States, J. Wildl. Manage. 11:162.

Lettvin, J.Y., E.R. Gruberg, R.M. Rose, and G. Plotkin, 1982. Dolphins and the bends. Science 216:650.

Leyhausen, P., 1956. Verhaltensstudien an Katzen. Z. Tierpsychol. 2:1.

————, 1964. The communal organization of solitary animals. Symp. Zool. Soc. London 14:249.

Lidicker, W.Z. Jr., 1968. A phylogeny of New Guinea rodent genera based on phallic morphology. J. Mamm. 49:609.

————, 1973. Regulation of numbers in an island population of the California vole: A problem in community dynamics. Ecol. Monogr. 43:271.

————, 1975. The role of dispersal in the demography of small mammals, *in* Small Mammals: Their Productivity and Population Dynamics. International Biol. Prog., vol. 5. (F.B. Golley, K. Petrusewicz, and L. Ryzkowski, eds.). Cambridge University Press, Cambridge.

————, 1976. Experimental manipulation of the timing of reproduction in the California vole. Res. Pop. Ecol. 18:14.

————, 1979. Analysis of two freely-growing enclosed populations of the California vole. J. Mamm. 60:447.

————, 1980. The social biology of the California vole. The Biologist 62:46.

———— and P.K. Anderson, 1962. Colonization of an island by *Microtus californicus,* analyzed on the basis of runway transects. J. Anim. Ecol. 31:503.

Lillegraven, J.A., 1969. Latest Cretaceous mammals of upper part of Edmonton Formation of Alberta, Canada, and review of marsupial-placental dichotomy in mammalian evolution. Univ. Kansas Paleontol. Contrib. 50:1.

————, 1974. Biogeographical considerations of the marsupial-placental dichotomy, *in* Annual Review of Ecology and Systematics, vol. 5 (R.F. Johnston, ed.). Annual Reviews, Inc., Palo Alto, Calif.

————, 1975. Biological considerations of the marsupial-placental dichotomy. Evolution 29:707.

————, 1979a. Introduction, *in* Mesozoic Mammals: The First Two-Thirds of Mammalian History (J.A. Lillegraven, Z. Kielan-Jaworoska, and W.A. Clemens, eds.). University of California Press, Berkeley.

————, 1979b. Reproduction in Mesozoic mammals, *in* Mesozoic Mammals: The First Two-Thirds of Mammalian History (J.A. Lillegraven, Z. Kielan-Jaworoska, and W.A. Clemens, eds.). University of California Press, Berkeley.

————, Z. Kielan-Jaworowska, and W.A. Clemens, 1979. Mesozoic Mammals: The First Two-Thirds of Mammalian History. University of California Press, Berkeley.

————, M.J. Kraus, and T.M. Bown, 1979. Paleogeography of the world of the Mesozoic, *in* Mesozoic Mammals: The First Two-Thirds of Mammalian History. (J.A. Lillegraven, Z. Kielan-Jaworoska, and W.A. Clemens, eds.). University of California Press, Berkeley.

Lilly, J.C., 1961. Man and Dolphin. Doubleday, New York.

————, 1962. Vocal behavior of the bottle-nosed dolphin. Proc. Amer. Phil. Soc. 106:520.

————, 1963. Distress call of the bottle-nosed dolphin. Stimuli and evoked behavioral responses. Science 139:116.

————, 1967. Mind of the Dolphin: A Nonhuman Intelligence. Doubleday, New York.

Lindlöf, B., E. Lindström, and A. Pehrson, 1974. Nutrient content in relation to food preferred by mountain hare. J. Wildl. Manage. 38:875.

Linsdale, J.M., 1946. The California Ground Squirrel. University of California Press, Berkeley.

Linzey, D.W., and A.V. Linzey, 1967. Maturational and seasonal molts in the golden mouse, *Ochrotomys nuttalli.* J. Mamm. 48:236.

Litchfield, C., A.J. Greenberg, D.K. Caldwell, M.C. Caldwell, J.C. Sipos, and R.G. Ackman, 1975. Comparative lipid patterns in acoustical and non-acoustical fatty tissues of dolphins, porpoises and toothed whales. Comp. Biochem. Physiol. 508:591.

Lockie, J.D., 1959. Estimation of the food of foxes. J. Wildl. Manage. 23:224.

————, 1966. Territory in small carnivores. Symp. Zool. Soc. London 18:143.

Long, A., R.M. Hansen, and P.S. Martin, 1974. Extinction of the Shasta ground sloth. Geol. Soc. Amer. Bull. 85:1843.

Lorenz, K., 1950. The comparative method of studying innate behavior patterns. Symp. Soc. Exp. Biol. 4:229.

————, 1963. Das sogenannte Böse, G. Borotha-Schoeler, Vienna. (English version, 1966. On Aggression. Methuen, London.)

Louch, C.D., 1958. Adrenocortical activity in two meadow vole populations. J. Mamm. 39:109.

Luckens, M.M., and W.H. Davis, 1964. Bats: Sensitivity to DDT. Science 146:948.

Lund, R.D., and J.S. Lund, 1965. The visual system of the mole, *Talpa europaea.* Exp. Neurol. 13:302.

Lyman, C.P., 1954. Activity, food consumption and hoarding in hibernators. J. Mamm. 35:545.

————, 1970. Thermoregulation and metabolism in bats, *in* Biology of Bats, vol. 1 (W.A. Wimsatt, ed.). Academic Press, New York.

———— and W.A. Wimsatt, 1966. Temperature regulation in the vampire bat, *Desmodus rotundus.* Physiol. Zool. 39:101.

MacArthur, R.H., 1955. Fluctuations of animal populations and a measure of community stability. Ecology 36:533.

———— and E.R. Pianka, 1966. On optimal use of a patchy environment. Amer. Nat. 100:603.

MacFarlane, J.D., and J.M. Taylor, 1982. Nature of estrus and ovulation in *Microtus townsendi* (Bachman). J. Mamm. 63:104.

MacKay, M.R., 1970. Lepidoptera in Cretaceous amber. Science 167:379.

Mackay, R.S., 1982. Dolphins and the bends. Science 216:650.

———— and H.M. Liaw, 1981. Dolphin vocalization mechanisms. Science 212:676.

MacLulich, D.A., 1937. Fluctuations in the numbers of the varying hare *(Lepus americanus).* Univ. Toronto Studies, Biol. Ser. No. 43.

MacMillen, R.E., 1964a. Population ecology, water relations, and social behavior of a southern California semidesert rodent fauna. Univ. Calif. Publ. Zool. 71:1.

————, 1964b. Water economy and salt balance in the western harvest mouse, *Reithrodontomys megalotis.* Physiol. Zool. 37(1):45.

————, 1965. Aestivation in the cactus mouse, *Peromyscus eremicus.* Comp. Biochem. Physiol. 16:227.

————, 1972. Water economy of nocturnal desert rodents, *in* Comparative Physiology of Desert Animals (G.M.O. Maloiy, ed.). Symp. Zool. Soc. London. Academic Press, New York.

————, 1983a. Water regulations in *Peromyscus.* J. Mamm. 64:38.

————, 1983b. The adaptive physiology of heteromyid rodents. Great Basin Nat. in press.

———— and E.A. Christopher, 1975. The water relations of two populations of noncaptive desert rodents, *in* Environmental Physiology of Desert Organisms (N.F. Hadley, ed.). Dowden, Hutchinson and Ross, Stroudsburg, Pa.

———— and D.E. Grubbs, 1976. The effects of temperature on water metabolism in rodents, *in* Progress in Animal Biometerology, vol. 1 (D.H. Johnson, ed.). Swetz and Zeitlinger, Lisse, The Netherlands.

———— and D.S. Hinds, 1983a. Water regulatory efficiency in heteromyid rodents: A model and its application. Ecology 64:152.

————, ————, 1983b. Adaptive significance of water regulatory efficiency in heteromyid rodents. BioScience 33:333.

———— and A.K. Lee, 1967. Australian desert mice: Independence of exogenous water. Science 158:383.

————, 1969. Water metabolism of Australian hopping mice. Comp. Biochem. Physiol. 28:493.

————, 1970. Energy metabolism and pulmocutaneous water loss of Australian hopping mice. Comp. Biochem. Physiol. 35:355.

———— and J.E. Nelson, 1969. Bioenergetics and body size in dasyurid marsupials. Amer. J. Physiol. 217:1246.

Madison, D.M., 1980. A review of the social biology of *Microtus pennsylvanicus.* The Biologist 62:20.

Maglio, V.J., 1973. Origin and evolution of the Elephantidae. Trans. Amer. Phil. Soc. 63:1.

Maher, W.J., 1967. Predation by weasels on a winter population of lemmings, Banks Island, Northwest Territories. Can. Field Nat. 81:248.

————, 1970. The pomarine jaeger as a brown lemming predator in northern Alaska. Wilson Bull. 82:130.

Mallory, F.F., and R.J. Brooks, 1978. Infanticide and other reproductive strategies in the collared lemming *(Dicrostonyx groenlandicus).* Nature 273:144.

Maloiy, G.M.O., 1973. The water metabolism of a small East African antelope: The dik-dik. Proc. Roy. Soc. London, B. 184:167.

Mares, M.A., 1977. Water economy and salt balance in a South American desert rodent *Eligmontia typus.* Comp. Biochem. Physiol. 56A:325.

Marler, P.R., 1965. Communication in monkeys and apes, *in* Primate Behavior (I. DeVore, ed.). Holt, Rinehart and Winston, New York.

———— and W.J. Hamilton III, 1966. Mechanisms of animal behavior. Wiley, New York.

Marshall, J.T. Jr., and E.R. Marshall, 1976. Gibbons and their territorial songs. Science 193:235.

Marshall, L.G., 1972. A study of the peramelid tarsus. Aust. Mammal. 1:67.

——, 1974. Why kangaroos hop. Nature, Lond. 248:174.

——, 1976. Evolution of the Thylacosmilidae, extinct saber-tooth marsupials of South America. PaleoBios. 23:1.

——, 1977. A new species of *Lycopsis* (Borhyaenidae: Marsupialia) from the La Venta fauna (Miocene) of Colombia. South America. J. Paleont. 51:633.

——, 1978. Evolution of the Borhyaenidae, extinct South American predaceous marsupials. Univ. Calif. Publ. Geol. Sci. 117:1.

——, 1981. The families and genera of the Marsupialia. Fieldiana, Geology, New Series 8:1.

——, 1982. Evolution of South American marsupialia, *in* Mammalian Biology in South America, (M.A. Mares and H.H. Genoways, eds.). Pymatuning Symposia in Ecology, Univ. Pittsburgh, vol. 6.

——, 1984. Monotremes and marsupials, *in* Orders and Families of Recent Mammals of the World (S.A. Anderson and J.K. Jones Jr., eds.). Wiley, New York.

——, S.D. Webb, J.J. Sepkoski Jr., and D.M. Raup, 1982. Mammalian evolution and the Great American Interchange. Science 215:1351.

—— and G.J. Weisenberger, 1971. A new dwarf shrew locality for Arizona. Plateau 43:132.

Martin, E.P., 1956. A population study of the prairie vole *(Microtus ochrogaster)* in northeastern Kansas. Univ. Kansas Publ. Mus. Nat. Hist. 8:361.

Martin, L.G., 1980. Functional morphology and the evolution of cats. Trans. Nebraska Acad. Sci. 8:141.

Martin, P., 1971. Movements and activities of the mountain beaver *(Aplodontia rufa)*. J. Mamm. 52:717.

Martin, R.D., 1973. A review of the behavior and ecology of the lesser mouse lemur *(Microcebus murinus,* J.F. Miller 1777), *in* Comparative Ecology and Behavior of Primates (R.P. Michael and J.H. Crook, eds). Academic Press, New York.

—— and S.K. Bearder, 1979. Radio bush baby. Nat. Hist. 88:77.

Martinson, D.L., 1968. Temporal patterns in the home ranges of chipmunks. J. Mamm. 49:83.

——, 1969. Energetics and activity patterns of short-tailed shrews *(Blarina)* on restricted diets. Ecology 50:505.

Maskrey, M., and P.P. Hoppe, 1979. Thermoregulation and oxygen consumption in Kirk's dik-dik *(Madoqua kirkii)* at ambient temperatures of 10–45°C. Comp. Biochem. Physiol. 62:827.

Matschie, P., 1899. Beitrage zur Kenmtnis von *Hypsignathus monstrosus* Allen. Sitz. Ber. Ges. Naturf. Freunde, Berlin.

Matthew, W.D., 1910. The phylogeny of the Felidae. Bull. Amer. Mus. Nat. Hist. 28:289.

——, 1915. Climate and evolution. New York Acad. Sci. Ann. 24:171.

Maynard Smith, J., 1976. Evolution and theory of games. Amer. Sci. 64:41.

Mayr, E., 1942. Systematics and the Origin of Species. Columbia University Press, New York.

——, 1963. Animal Species and Evolution. Harvard University Press, Cambridge, Mass.

McCabe, T.T., and B.D. Blanchard, 1950. Three species of *Peromyscus*. Rood Associates, Santa Barbara, Calif.

McCarley, H., 1959. The effect of flooding on a marked population of *Peromyscus*. J. Mamm. 40:57.

McClure, P.A., 1981. Sex-biased litter reduction in food-restricted woodrats *(Neotoma floridana)*. Science 211:1058.

McCullough, D.R., 1969. The tule elk, its history, behavior, and ecology. Univ. Calif. Publ. Zool. 88:1.

McKay, G.M., 1973. The ecology and behavior of the Asiatic elephant in southeastern Ceylon. Smithsonian Contrib. Zool. 125:1.

McKenna, M.C., 1972. Possible biological consequences of plate tectonics. BioScience 22:519.

——, 1975a. Fossil mammals and Early Eocene North Atlantic land continuity. Ann. Missouri Bot. Gard. 62:335.

——, 1975b. Toward a phylogenetic classification of the mammalia, *in* Phylogeny of the Primates (W.P. Luckett and F.S. Szalay, eds.). Plenum Press, New York.

McLaren, A., 1970. The fate of the zona pellucida in mice. J. Embryol. Exp. Morph. 23:1.

McLean, D.C., 1944. The prong-horned antelope in California. Bureau Game Cons., Calif. Div. Fish Game, San Francisco 30(4):221.

McNab, B.K., 1966. The metabolism of fossorial rodents: A study of convergence. Ecology 47: 712.

————, 1979. The influence of body size on the energetics and distribution of fossorial and burrowing mammals. Ecology 60:1010.

————, 1980. Energetics and the limits to a temperature distribution in armadillos. J. Mamm. 61:606.

————, 1982. Evolutionary alternatives in the physiological ecology of bats, in Ecology of Bats (T.H. Kunz, ed.). Plenum, New York.

McNaughton, S.J., 1976, Serengeti migratory wildebeest: Facilitation of energy flow by grazing. Science 191:92.

————, 1979. Grazing as an optimization process: Grass-ungulate relationships in the Serengeti. Amer. Nat. 113:691.

————, 1979. Grassland-herbivore dynamics, in Serengeti: Dynamics of an Ecosystem (A.R.E. Sinclair and M. Norton-Griffiths, eds.). University of Chicago Press, Chicago.

————, J.L. Tarrants, M.M. McNaughton, and R.H. Davis, 1985. Silica as a defense against herbivory and a growth promoter in African grasses. Ecology 66:528.

Mead, R.A., 1968. Reproduction in western forms of the spotted skunk (genus *Spilogale*). J. Mamm. 49:373.

Mech, L.D., 1966. The wolves of Isle Royale, U.S. Nat. Park Serv., Fauna ser. 7.

Meehan, T.E., 1976. The occurrence, energetic significance and initiation of spontaneous torpor in the Great Basin pocket mouse *(Perognathus parvus)*. Ph.D. dissertation, University of California, Irvine.

Menaker, M., 1961. The free-running period of the bat clock: Seasonal variations at low body temperature, J. Cell Comp. Physiol. 57:81.

Menard, H.W., 1969. The deep-ocean floor. Sci. Amer. 221:126.

Merriam, C.H., 1894. Laws of temperature control of the geographic distribution of terrestrial animals and plants. Nat. Geogr., 6:229.

————, 1899. Life zones and crop zones of the United States. Bull. U.S. Biol. Surv. 10:1.

Merrilees, D., 1968. Man the destroyer: Late Quaternary changes in the Australian marsupial fauna. J. Roy. Soc. West. Aust. 51:1.

Michael, R.P., E.B. Keverne, and R.W. Bonsall, 1971. Pheromones: Isolation of male sex attractants from a female primate. Science 172:964.

Milankovitch, M., 1938. Astronomische Mittel zur Erforschung der erdgeschichtlichen Klimate. Handbuch der Geophysik 9:593.

Miller, G.J., 1969. Man and Smilodon: A preliminary report on their possible coexistence at Rancho La Brea. Los Angeles Co. Mus. Contrib. Sci. 163:1.

Miller, G.S. Jr., 1907. The families and genera of bats. Bull. U.S. Nat. Mus. 57:1.

Miller, R.S., 1964. Ecology and distribution of pocket gophers (Geomyidae) in Colorado. Ecology 45:256.

————, 1967. Pattern and process in competition. Adv. Ecol. Res. 4:1.

————, 1969. Competition and species diversity. Brookhaven Symp. Biol. 22:63.

Misonne, X., 1959, Analyse zoogéographique des mammifères de l'Iran. Bruxelles, Inst. Royal Sci. Nat. Belgique, Mémoires, 2me sér. 59.

Mizuhara, H., 1957. The Japanese Monkey: Its Social Structure. San-ichi-syobo, Kyota (in Japanese).

Mobius, K., 1877. Die Auster und die Austernwirtschaft. Berlin. (Transl., 1880, The oyster and oyster culture.) Rept. U.S. Fish. Comm. 1880:683.

Mohr, E., 1941, Schwanzverlust und Schwanzregeneration bei Nagetieren. Zool. Anzeiger 135:49.

Mohres, F.P., 1953. Über die Ultraschallorientierung der Hufeisennasen (Chiroptera—Rhinolophidae). Z. Vergl. Physiol. 34:547.

————, 1966. Communicative characters of sonar signals in bats, in Animal Sonar Systems, Biology and Bionics, tome II (R.G. Busnel, ed.). Laboratoire de Physiologie Acoustique, Paris.

———— and E. Kulzer, 1956. Über die Orientierung der Flughunde (Chiroptera—Pteropodidae). Z. Vergl. Physiol. 38:1.

———— and G. Neuweiler, 1966. Ultrasonic orientation in megadermid bats, in Animal Sonar Systems, Biology and Bionics, tome I (R.G. Busnel, ed.). Laboratoire de Physiologie Acoustique, Paris.

Mooser, O., and W.W. Dalquest, 1975. Pleistocene mammals from Aguascalientes, Central Mexico. J. Mamm. 56:781.

Morhardt, J.E., 1970. Body temperatures of white-footed mice (*Peromyscus* sp.) during daily torpor. Comp. Biochem. Physiol. 33:423.

———— and D.M. Gates, 1974. Energy-exchange analysis of the Belding ground squirrel and its habitat. Ecol. Monogr. 44:17.

Morrison, P., 1959. Body temperatures in some Australian mammals. I: Chiroptera. Biol. Bull. 116:484.

————, 1962. Body temperatures in some Australian mammals. III: Cetacea (Megaptera). Biol. Bull. 123:154.

———— and B.K. McNab, 1962. Daily torpor in a Brazilian murine opossum *(Marmosa)*. Comp. Biochem. Physiol. 6:57.

————, 1967. Temperature regulation in some Brazilian phyllostomid bats. Comp. Biochem. Physiol. 21:207.

———— and F.A. Ryser, 1952. Weight and body temperature in mammals. Science 116:231.

————, ————, and A.R. Dawe, 1959. Studies on the physiology of the masked shrew *Sorex cinereus*. Physiol. Zool. 32:256.

Morton, M.L., and P.W. Sherman, 1978. Effects of a spring snowstorm on behavior, reproduction, and survival of Belding's ground squirrels. Canadian J. Zool. 56:2578.

Morton, S.R., 1978a. An ecological study of *Sminthopsis crassicaudata* (Marsupialia: Dasyuridae). II. Behavior and Social organization. Australian Wildl. Res. 5:163.

————, 1978b. Torpor and nest sharing in free-living *Sminthopsis crassicaudata* (Marsupialia) and *Mus musculus* (Rodentia). J. Mamm. 59:569.

Morton, S.R., and R.E. MacMillen, 1982. Seeds as sources of preformed water for desert-dwelling granivores. J. Arid Environ. 5:61–67.

Moynihan, M.H., 1966. Communication in the titi monkey, *Callicebus*. J. Zool., London 150:77.

Müller, F., 1969. Verhaltnis von Körperentwicklung und Cerebralisation in Ontogenese und Phylogenese der Sänger. Versuch einer Libersicht des Problems. Verh. naturf. Ges., Basel 80:1.

Muller, J., 1970. Palynological evidence on early differentiation of angiosperms. Biol Rev. 45:417.

Müller-Schwarze, D., 1971. Pheromones in the black-tailed deer *(Odocoileus hemionus columbianus)*. Anim. Behav. 19:141.

Murdoch, H.W., and A. Oaten, 1975. Predation and population stability, *in* Advances in Ecological Research (A. MacFadyen, ed.). Academic Press, New York.

Murie, A., 1940. Ecology of the coyote in the Yellowstone, U.S. Dept. Int. Nat. Park Serv., Fauna ser. 4.

————, 1944, The wolves of Mount McKinley, U.S. Dept. Int. Nat. Park Serv., Fauna ser. 5.

Myers, G.T., and T.A. Vaughan, 1964. Food habits of the plains pocket gopher in eastern Colorado. J. Mamm. 45:588.

Myers, J.H., and C.J. Krebs, 1971. Genetic, behavioral, and reproductive attributes of dispersing field voles, *Microtus pennsylvanicus* and *Microtus ochrogaster*. Ecol. Monogr. 41:53.

Mykytowycz, R., 1968. Territorial marking by rabbits. Sci. Amer. 218:116.

Nadler, R.D., 1975. Sexual cyclicity in captive lowland gorillas. Science 189:813.

Nagy, K.A., V.H. Shoemaker, and W.R. Costa, 1976. Water, electrolyte, and nitrogen budgets of jackrabbits *(Lepus californicus)* in the Mojave Desert. Physiol. Zool. 49:351.

Nagy, J.G., H.W. Steinhoff, and G.M. Ward, 1964. Effects of essential oils of sagebrush on deer rumen microbial function. J. Wildl. Mgmt. 28:785.

———— and R.P. Tengerdy, 1967. Antibacterial action of essential oils of *Artemisia* as an ecological factor. II: Antibacterial action of the volatile oils of *Artemisia tridentata* (big sagebrush) on bacteria of the rumen of the mule deer. Appl Microbiol. 16:441.

Needham, A.D., T.J. Dawson, and J.R.S. Hales, 1974. Forelimb blood flow and saliva spreading in the thermoregulation of the red kangaroo, *Megaleia rufa*. Comp. Biochem. Physiol. 49:555.

Negus, N.C., and A.J. Pinter, 1965. Litter sizes of *Microtus montanus* in the laboratory. J. Mamm. 46(3):434.

————, ————, 1966. Reproductive responses of *Microtus montanus* to plants and plant extracts in the diet. J. Mamm. 47(4):596.

Nel, J.A.J., 1978. Habitat heterogeneity and changes in small mammal community structure and resource utilization in the southern Kalahari. Bull, Carnegie Mus. Nat. Hist. 6:118.

Nelson, J.E., 1965a. Behavior of Australian Pteropodidae (Megachiroptera). Anim. Behav. 8:544.

————, 1965b. Movements of Australian flying foxes. Aust. J. Zool. 13:53.

Neumann, C.A., 1965/1966. Geo-marine Technol. 2:1; as cited by Ridgway, 1966.

Nicholson, P.J., 1963. Wombats. Timbertop Magazine, Geelong Grammar School, no. 8; p. 28.

Nishida, T., and K. Kawanaka, 1972. Interunit-group relationships among wild chimpanzees of the Mahali Mountains. Kyoto Univ. Afr. Stud. 7:131.

Noirot, E., 1969. Sound analysis of ultrasonic distress calls of mouse pups as a function of their age. Anim. Behav. 17:340.

Noll-Banholzer, U., 1979a. Body temperature, oxygen consumption, evaporative water loss and heart rate in the fennec. Comp. Biochem. Physiol. 62:585.

————, 1979b. Water balance and kidney structure of the fennec. Comp. Biochem. Physiol. 62:593.

Norberg, U.M., 1969. An arrangement giving a stiff leading edge to the hand wing in bats. J. Mamm. 50:766.

————, 1972. Bat wing structures important for aerodynamics and rigidity. Z. Morph. Tiere 73:45.

————, 1981. Allometry of bat wings and legs and comparison with bird wings. Philos. Trans. Royal Soc. London (B) 292:359.

Norris, K.S., 1964. Some problems in echolocation in cetaceans, in Marine Bio-acoustics (W.N. Tavolga, ed.). Pergamon Press, Oxford.

———— (ed.), 1966. Whales; Dolphins and Porpoises. University of California Press, Berkeley.

————, 1968. The evolution of acoustic mechanisms in odontocete cetaceans. Peabody Museum Centenary Celebration Volume, Yale University, New Haven, Conn.

————, 1969. The echolocation of marine mammals, in The Biology of Marine Mammals (H.T. Andersen, ed.). Academic Press, New York.

————, H.A. Baldwin, and D.J. Samson, 1965. Deep Sea Res. 12:505.

———— and T.P. Dohl, 1980. The structure and functions of cetacean schools, in Cetacean Behavior: Mechanisms and Processes (L.M. Herman, ed.). Wiley, New York.

———— and G.W. Harvey, 1972. A theory for the function of the spermaceti organ of the sperm whale (Physeter catodon), in Animal Orientation and Navigation, (S.R. Galler, K. Schmidt-Koenig, G.J. Jacobs, and R.E. Belleville, eds.). NASA Spec. Publ. 262.

———— and B. Møhl, 1983. Can odontocetes debilitate prey with sound? Amer. Nat. 122:85.

————, A. Prescott, D.V. Asa-Doran, and P. Perkins, 1961. An experimental demonstration of echolocation behavior in the porpoise, Tursiops truncatus (Montagu). Biol. Bull. 120:163.

———— and J.H. Prescott, 1961. Observations on Pacific cetaceans of California and Mexican waters. Univ. Calif. Publ. Zool. 63:291.

Novick, A., 1955. Laryngeal muscles of the bat and production of ultrasonic sounds. Amer. J. Physiol. 183:648.

————, 1958a. Orientation in paleotropical bats. II: Megachiroptera. J. Exp. Zool. 137:443.

————, 1958b. Orientation in paleotropical bats. I: Microchiroptera. J. Exp. Zool. 138:81.

————, 1962. Orientation in neotropical bats. I: Natalidae and Emballonuridae. J. Mamm. 43:449.

————, 1963a. Orientation in neotropical bats. II: Phyllostomatidae and Desmodontidae. J. Mamm. 44:44.

————, 1963b. Pulse duration in the echolocation of insects by the bat, Pteronotus. Ergeb. Biol. 261:26.

————, 1965. Echolocation of flying insects by the bat Chilonycteris psilotis. Biol. Bull. 128:297.

————, 1970. Echolocation in bats. Nat. Hist. 79(3):32.

————, 1971. Echolocation in bats: Some aspects of pulse design. Amer. Sci. 59(2):198.

———— and D.R. Griffin, 1961. Laryngeal mechanisms in bats for production of orientation sounds. J. Exp. Zool. 148:125.

———— and J.R. Vaisnys, 1964. Echolocation of flying insects by the bat Chilonycteris parnellii. Biol. Bull. 127:478.

Odum, E.P., 1971. Fundamentals of Ecology, 3d edn. W.B. Saunders, Philadelphia.

O'Farrell, M.J., and E.H. Studier, 1970. Fall metabolism in relation to ambient temperatures in three species of Myotis. Comp. Biochem. Physiol. 35:697.

O'Farrell, T.P., 1965. Home range and ecology of snowshoe hares in interior Alaska. J. Mamm. 46:406.

O'Gara, B.W., and G. Matson, 1975. Growth and casting of horns by pronghorns and exfoliation of horns by bovids. J. Mamm. 56:829.

————, R.F. Moy, and G.D. Bear, 1971. The annual testicular cycle and horn casting in the pronghorn (Antilocapra americana). J. Mamm. 52:537.

Orr, R.T., 1940. The rabbits of California. Occ. Papers California Acad. Sci. 19:1.

————, 1976. Vertebrate Biology, 4th ed. W.B. Saunders, Philadelphia.

Osborn, H.F., 1936–42. Proboscidea: A monograph of the discovery, evolution, migration, and extinction of the mastodonts and elephants of the world. Vol. 1. Moeritheroidea, Deinotheroidea, Mastodontoidea, Vol. 2. Stegodontoidea, Elephantoidea. Amer. Mus. Nat. Hist., New York. 1675 pp. (Although the taxonomic schemes and the phylogenetic patterns presented in this paper have been seriously questioned, the figures and discussions of structure are useful.)

O'Shea, T.J., 1982. Aspects of underwater sound communication in the West Indian manatee, *Trichechus manatus.* ASM Abstracts, 62d Annual Meeting of the American Society of Mammalogists, Abstract 89.

Otte, D., 1974. Effects and functions in the evolution of signaling systems. Ann. Rev. Ecol. System. 5:385.

Owens, D.D., and M.J. Owens, 1980. Hyenas of the Kalahari. Nat. Hist. 89:44.

Owens, M.J., and D.D. Owens, 1978. Feeding ecology and its influence on social organization in brown hyenas *(Hyaena brunnea)* of the Central Kalahari Desert. E. African Wildl. J. 16:112.

Packer, C., 1983. Sexual dimorphism: The horns of African antelopes. Science 221:1191.

Paine, R.T., 1969. A note on trophic complexity and community stability. Amer. Nat. 103:91.

Parker, P., 1977. An ecological comparison of marsupial and placental patterns of reproduction, *in* The Biology of Marsupials (B. Stonehouse and D. Gilmore, eds.). Macmillan, London.

Parrington, F.R., 1971. On the Upper Triassic mammals. Phil. Trans. Roy. Soc. London, B 261:231.

Parry, D.A., 1949. The structure of whale blubber and its thermal properties. Quart. J. Microbiol. Sci. 90:13.

Patterson, B., and R. Pascual, 1968. The fossil mammal fauna of South America. Quart. Rev. Biol. 43:409.

————, ————, 1972. The fossil mammal fauna of South America, *in* Evolution, Mammals, and Southern Continents (A. Keast, F.C. Erk, and B. Glass, eds.). State University of New York Press, Albany.

Payne, R.S., 1961. The acoustical location of prey by the barn owl *(Tyto alba).* Amer. Zool. 1:379.

————, 1970. Songs of the humpback whale. An LP Record by CRM Records, Del Mar, Ca.

———— and S. McVay, 1971. Songs of the humpback whales. Science 173:585.

Pearson, O.P., 1942. On the cause and nature of a poisonous action produced by the bite of a shrew *(Blarina brevicauda)* J. Mamm. 23:159.

————, 1947. The rate of metabolism of some small mammals. Ecology 28:127.

————, 1948. Metabolism of small mammals with remarks on the lower limit of mammalian size. Science 108:44.

————, "1959" 1960. Biology of the subterranean rodents, *Ctenomys,* in Peru. Mem. del Museo de Hist. Nat. "Javier Prado" 9:1.

————, 1963. History of two local outbreaks of feral house mice. Ecology 44:540.

————, 1964. Carnivore-mouse predation: An example of its intensity and bioenergetics. J. Mamm. 45:177.

————, 1966. The prey of carnivores during one cycle of mouse abundance. J. Anim. Ecol. 35:217.

————, 1971. Additional measurements of the impact of carnivores on California voles *(Microtus californicus).* J. Mamm. 52:41.

————, M.R. Koford and A.K. Pearson, 1952. Reproduction of the lump-nosed bat *(Corynorhinus rafinesquei)* in California. J. Mamm. 33:273.

————, 1960. Habits of *Microtus californicus* revealed by automatic photographic records. Ecol. Monogr. 30:231.

Pengelley, E.T., 1967. The relation of external conditions to the onset and termination of hibernation and estivation, *in* Mammalian Hibernation, vol. III (K.C. Fisher et al., eds.). Oliver and Boyd, London.

————, 1968. Interrelationships of circannian rhythms in the ground squirrel *Citellus lateralis.* Comp. Biochem. Physiol. 24:915.

———— and K.C. Fisher, 1963. The effect of temperature and photoperiod on the yearly hibernating behavior of captive golden-mantled ground squirrels *(Citellus lateralis tescorum).* Can. J. Zool. 41:1103.

Pennycuik, P.R., 1969. Reproductive performance and body weights of mice maintained for 12 generations at 34°C. Aust. J. Biol. Sci. 22:667.

Pequegnat, W.E., 1951. The biota of the Santa Ana Mountains, J. Entomol. Zool. 42:1.

Perkins, J., 1945. Biology at Little America III, the west base of the United States Antarctic service expedition 1939–1941. Proc. Amer. Phil. Soc. 89:270.

Perry, J.S., 1954. Some observations on growth and tusk weight in male and female African elephants. Proc. Zool. Soc. London 124:97.

Peters, R.P., and L.D. Mech, 1975. Scent-marking in wolves. Sci. Amer. 63:628.

Peterson, Randolph S., 1955. North American Moose. University of Toronto Press, Toronto.

Peterson, Richard S., 1965. Behavior of the northern fur seal. D.Sc. Thesis. Johns Hopkins University, Baltimore.

————— and G.A. Bartholomew, 1967. The natural history and behavior of the California sea lion. Amer. Soc. Mamm., Spec. Publ. No. 1.

Petter, J.J., 1962a. Ecological and behavioral studies of Madagascar lemurs in the field. Ann. N.Y. Acad. Sci. 102:267.

—————, 1962b. Recherches sur l'écologie et l'éthologie des lémuriens malgaches. Mém. Mus. Nat. Hist. Paris, ser. A. (Zool.), 27:1.

—————, 1965. The lemurs of Madagascar, in Primate Behavior (I. DeVore, ed.). Holt, Rinehart and Winston, New York.

Pianka, E.R., 1966. Convexity, desert lizards and spatial heterogeneity. Ecology 47:1055.

—————, 1976. Natural selection of optimal reproductive strategies. Amer. Zool. 16:775.

Pieper, R.D., 1964. Production and chemical composition of arctic tundra vegetation and their relation to the lemming cycle. Ph.D. thesis, University of California, Berkeley.

Pitelka, F.A., 1957a. Some characterisics of microtine cycles in the arctic. Eighteenth Ann. Biol. Coll., Oregon State College, p. 73.

—————, 1957b. Some aspects of population structure in the short-term cycle of the brown lemming in northern Alaska. Cold Spr. Harb. Symp. Quant. Biol. 22:237.

—————, 1958. Some aspects of population structure in the short-term cycle of the brown lemming in northern Alaska. Cold Spr. Harb. Symp. Quant. Biol. 22:237.

—————, 1964. The nutrient-recovery hypothesis for Arctic microtine cycles. I: Introduction, in Grazing in Terrestrial and Marine Environments (P.J. Crisp, ed.). Brit. Ecol. Soc. Symp. No. 4. Blackwell, Oxford.

—————, P.Q. Tomich, and G.W. Treichel, 1955. Ecological relations of jaegers and owls as lemming predators near Barrow, Alaska. Ecol. Monogr. 25:85.

Pivorunas, A., 1979. The feeding mechanisms of baleen whales. Amer. Scient. 67:432.

Poduschka, W., 1970. Das Selbstbespeicheln der Igel. Film CT 1320 der Bundesstaatl. Vienna: Hptst. wiss. Kinemat.

Pond, C.M., 1977, The significance of lactation in the evolution of mammals. Evolution 31:177.

Poole, E.L., 1936. Relative wing ratios of bats and birds. J. Mamm. 17:412.

Porter, F.L., 1978. Roosting patterns and social behavior in captive Carollia perspicillata. J. Mamm. 59:627.

Poulter, T.C., 1963. Sonar signals of the sea lion. Science 139:753.

Pournelle, G.H., 1968. Classification, biology, and description of the venom apparatus of insectivores of the genera Solenodon, Neomys, and Blarina, in Venomous Animals and Their Venoms (W. Bucherl, E.A. Buckley, and V. Deulofeu, eds.). Academic Press, New York.

Powers, J.B., and S.S. Winans, 1975. Vomeronasal organ: Critical role in mediating sexual behavior of the male hamster. Science 187:961.

Prakash, I., 1959. Foods of the Indian false vampire. J. Mamm 40:545.

Price, M.V., 1978. The role of microhabitat in structuring desert rodent communities. Ecology 59:910.

Pridham, J., 1965. Enzyme Chemistry of Phenolic Compounds. MacMillan, New York.

Pucek, M., 1968. Chemistry and pharmacology of insectivore venoms, in Venomous Animals and Their Venoms (W. Bucherl, E.A. Buckley, and V. Deulofeu, eds.). Academic Press, New York.

Pulliam, H.R., 1974. On the theory of optimal diets. Amer. Nat. 108:59.

Purves, P.E., 1966. Anatomy and physiology of the outer and middle-ear in cetaceans, in Whales, Dolphins and Porpoises (K.S. Norris, ed.). University of California Press, Berkeley.

Quilliam, T.A., 1966. The problem of vision in ecology of Talpa europaea. Exp. Eye Res. 5:63.

Rahm, U., 1970. Note sur la reproduction des sciuridés et muridés dans forêt équatoriale au Congo. Rev. Suisse Zool. 77:635.

Ralls, K., 1971. Mammalian scent marking. Science 171:443.

Randolph, P.A., T.C. Randolph, K. Mattingly, and M.M. Foster, 1977. Energy costs of reproduction in the cotton rat, Sigmodon hispidus. Ecology 58:31.

Ratcliff, H.M., 1941. Winter range conditions in

Rocky Mountain National Park. Trans. Sixth Amer. Wildl. Conf., p. 132.

Ratcliffe, F.N., 1932. Notes on the fruit bat of Australia. J. Anim. Ecol. 1:32.

Rathbun, G., 1973. The golden-rumped elephant shrew, A.W.L.F. News 8(3).

———, 1976. The ecology and social structure of the elephant shrews *Rhynchocyon chrysopygus* and *Elephantulus rufescens*. Ph.D. thesis, University of Nairobi, Kenya.

———, 1979. The social structure and ecology of the elephant shrews. Z. Tierpsychol. 20:1.

Rausch, R., 1950. Observations on a cyclic decline of lemmings *(Lemmus)* on the Arctic coast of Alaska during the spring of 1949. Arctic 3:166.

Ray, C., and W.E. Schevill, 1965. The noisy underwater world of the Weddell seal. Animal Kingdom, New York Zool. Soc. 68:34.

Redman, J.P., and J.A. Sealander, 1958. Home ranges of deer mice in southern Arkansas, J. Mamm. 39:390.

Reed, C.A., 1944. Behavior of a shrew mole in captivity. J. Mamm. 25:196.

———, 1951, Locomotion and appendicular anatomy in three soricoid insectivores. Amer. Midl. Nat. 45:513.

Reeder, W.G., and R.B. Cowles, 1951. Aspects of thermoregulation in bats. J. Mamm. 32:389.

Reese, J.B., and H. Haines, 1978. Effects of dehydration on metabolic rate and fluid distribution in the jackrabbit, *Lepus californicus*. Physiol. Zool. 51:155.

Regal, P.J., 1977. Ecology and evolution of flowering plant dominance. Science 196:622.

Reichman, O.J., 1975. Relation of desert rodent diets to available resources. J. Mamm. 56:731.

———, 1977. Optimization of diet through food preferences by heteromyid rodents. Ecology 58:454.

———, 1980. Factors influencing foraging in desert rodents, *in* Foraging Behavior: Ethological and Psychological Approaches (A.C. Kamil and T.D. Sargent, eds.). Garland STPM Press, New York.

———, and D. Oberstein, 1977. Selection of seed distribution types by *Dipodomys merriami* and *Perognathus amplus*. Ecology, 58:636.

——— and K.M. Van De Graaff, 1973. Seasonal activity and reproductive patterns of five species of Sonoran Desert rodents. Amer. Midl. Nat. 90:118.

———, ———, 1975. Association between ingestion of green vegetation and desert rodent reproduction. J. Mamm. 56:503.

———, T.G. Whitham, and G.A. Ruffner, 1982. Adaptive geometry of burrow spacing in two pocket gopher populations. Ecology 63:687.

Reid, R.T., 1970. The future role of ruminants in animal production, *in* Physiology of Digestion and Metabolism in the Ruminant (A.T. Phillipson, ed.). Oriel Press, Newcastle-upon-Tyne, England.

Reig, O.A., 1970, Ecological note on the fossorial octodont rodent *Spalacopus cyanus* (Molina). J. Mamm. 51:592.

Repenning, C.A., 1980. Faunal exchanges between Siberia and North America. Can. J. Anthropol. 1:37.

———, C.E. Ray, and D. Grigorescu, 1979. Pinniped biogeography, *in* Historical Biogeography, Plate Tectonics, and the Changing Environment (J. Gray and A.J. Boucot, eds.). Oregon State University Press, Corvallis.

Reysenbach de Haan, F.W., 1966. Listening underwater: Thoughts on sound and cetacean hearing, *in* Whales, Dolphins and Porpoises (K.S. Norris, ed.). University of California Press, Berkeley

Rhoades, D.F., 1979. Evolution of plant chemical defenses against herbivores, *in* Herbivores: Their Interactions with Plant Secondary Metabolites (G.A. Rosenthal and D.H. Janzen, eds.). Academic Press, New York.

Rice, D.W., 1967. Cetaceans, *in* Recent Mammals of the World (S. Anderson and J.K. Jones Jr., eds.). Ronald Press, New York.

——— and A.A. Wolman, 1971. The life history and ecology of the gray whale *(Eschrichtius robustus)*. Spec. Publ. No. 3, Amer. Soc. Mammal.

Ricklefs, R.E., 1973. Ecology, Chiron Press, Newton, Mass.

Ride, W.D.L., 1970. A Guide to the Mammals of Australia. Oxford University Press, New York.

Ridgway, S.H., 1966, Proc., Third Ann. Conf. Biol. Sonar Diving Mammals, p. 151. Stanford Res. Inst., Menlo Park, California.

———, B.L. Scronce, and J. Kanwisher, 1969. Respiration and deep diving in the bottle-nosed dolphin. Science 166:1651.

Ridgway, S.H., and R. Howard, 1979. Dolphin lung collapse and intramuscular circulation during free diving: Evidence from nitrogen washout. Science 206:1182.

————, ————, 1982. Dolphins and the bends. Science 216:651.

Rinker, G.C., 1954. The comparative myology of the mammalian genera *Sigmodon, Oryzomys, Neotoma,* and *Peromyscus* (Cricetinae), with remarks on their intergeneric relationships. Misc. Publ. Mus. Zool., Univ. Michigan 83:1.

Robertshaw, D., and C.R. Taylor, 1969. A comparison of sweat gland activity in eight species of East African bovids. J. Physiol., London 203:135.

Robinson, J.G., 1981. Spatial structure in foraging groups of wedge-capped capuchin monkeys *(Cebus nigrivittatus).* Anim. Behav. 29:1036.

Robinson, K., and D.H.K. Lee, 1941. Reactions of the cat to hot atmospheres. Proc. Roy Soc. Queensland, 53:159. Reactions of the dog to hot atmospheres, 53:171.

Robinson, P., C. Black, and M. Dawson, 1964. Late Eocene multituberculates and other mammals from Wyoming. Science 145:809.

Rodin, L.E., N.I. Bazilevich, and N.N. Rozov, 1975. Productivity of the world's ecosystems, *in* Productivity of World Ecosystems. NAS, Washington, D.C.

Roeder, K.D., 1965. Moths and ultrasound. Sci. Amer. 212(4):94.

———— and A.E. Treat, 1961. The detection and evasion of bats by moths. Amer. Sci. 49(2):135.

Rogers, L., 1981. A bear in its lair. Nat. Hist. 90:64.

Romer, A.S., 1966. Vertebrate Paleontology, 3d edn. University of Chicago Press, Chicago.

————, 1968. Notes and Comments on Vertebrate Paleontology. University of Chicago Press, Chicago.

————, 1969. Cynodont reptile with incipient mammalian jaw articulation. Science 166:881.

———— and Parsons, T.S., 1977. The Vertebrate Body. W.B. Saunders, Philadelphia.

Rood, J.P., 1958. Habits of the short-tailed shrew in captivity. J. Mamm. 39:499.

————, 1970a. Notes on the behavior of the pygmy armadillo. J. Mamm. 51:179.

————, 1970b. Ecology and social behavior of the desert cavy *(Microcavia australis).* Amer. Midl. Nat. 83:415.

————, 1972. Ecological and behavioral comparisons of three genera of Argentine cavies. Anim. Behav. Monogr. 5:1.

————, 1979. The social life of dwarf mongooses in the Serengeti. African Wildl. Leadership Found. 14:2.

———— and B.J. Weir, 1970. Reproduction in female wild guinea pigs. J. Reprod. Fert. 23:393.

Rose, K.D., 1982. Skeleton of *Diacodexis,* oldest known artiodactyl. Science 216:621.

———— and E.L. Simons, 1977. Dental function in the Plagiomenidae: Origin and relationships of the mammalian order Dermoptera. Univ. Michigan Contr. Mus. Paleont. 24:221.

Rosenzweig, M.R., D.A. Riley, and K. Krech, 1955. Evidence for echolocation in the rat. Science 12:600.

Rosevear, D.R., 1969. The Rodents of West Africa. British Museum Natural History, London.

Rowan, W., 1950. Winter habits and numbers of timber wolves, J. Mamm. 31:167.

Rowell, T.E., 1962. Agonistic noises of the rhesus monkey *(Macaca mulatta).* Symp. Zool. Soc. London 8:91.

Rudran, R., 1973. Adult male replacement in one-male troops of purple-faced langurs *(Presbytis senex senex)* and its effect on population structure. Folia Primat. 19:166.

————, 1979. Demography and social mobility in a red howler monkey *(A. seniculus)* population, *in* Vertebrate Ecology in the Northern Neotropics (J.F. Eisenberg, ed.). Smithsonian Institution Press, Washington, D.C.

Ruff, F.J., 1938. Trapping deer on the Pisgah National Game Preserve, North Carolina. J. Wildl. Manage. 2:151.

Rutherford, W.H., 1953. Effects of a summer flash flood upon a beaver population. J. Mamm. 34:261.

Sale, J.B., 1970. The behavior of the resting rock hyrax in relation to its environment. Zool. Afr. 5:87.

Sanders, E.H., P.D. Gardner, P.J. Berger, and N.C. Negus, 1981. 6-Methoxybenzoxazalinone: A plant derivative that stimulates reproduction in *Microtus montanus.* Science 214:67.

Sargent, A.B., and D.W. Warner, 1972. Movements and denning habits of a badger. J. Mamm. 53:207.

Saunders, J.K. Jr., 1963. Movements and activities of the lynx in Newfoundland. J. Wildl. Manage. 27:390.

Savage, J.M., 1974. The isthmiam link and the evolution of Neotropical mammals. Contrib. Sci., Nat. Hist. Mus., Los Angeles County 260:1.

Schaeffer, B., 1947. Notes on the origin and function of the artiodactyl tarsus. Amer. Mus. Novitates 1356:1.

Schaller, G.B., 1963. The Mountain Gorilla: Ecology and Behavior. University of Chicago Press, Chicago.

————, 1964. The Year of the Gorilla. University of Chicago Press, Chicago.

————, 1965a. The behavior of the mountain gorilla, in Primate Behavior: Field Studies of Monkeys and Apes (I. De Vore, ed.). Holt, Rinehart and Winston, New York.

————, 1965b. The Year of the Gorilla. Ballantine Books, New York.

————, 1967. The Deer and the Tiger: A Study of Wildlife in India. University of Chicago Press, Chicago.

————, 1972. The Serengeti Lion: A Study of Predator-Prey Relations. University of Chicago Press, Chicago.

Scheffer, V.B., 1958. Seals, Sea Lions and Walruses. Stanford University Press, Stanford, Calif.

———— and J.W. Slipp, 1944. The harbor seal in Washington state. Amer. Midl. Nat. 32:373.

Schenkel, R., 1966a. Play, exploration and territoriality in the wild lion. Symp. Zool. Soc. London 18:11.

————, 1966b. On sociology and behaviour in impala (Aepyceros melampus suara Matschie). Z. Säugetierk. 31:177.

————, 1967. Submission: Its features and functions in the wolf and dog. Amer. Zool. 7:319.

Schevill, W.E., and B. Lawrence, 1949. Underwater listening to the white porpoise, Delphinapterus leucas. Science 109:143.

————, ————, 1953. Auditory response of a bottle nosed porpoise, Tursiops truncatus, to frequencies above 100 kc. J. Exp. Zool. 124:147.

———— and C. Ray, 1965, The Weddell seal at home. Animal Kingdom, N.Y. Zool. Soc. 68:151.

———— and W.A. Watkins, 1965. Underwater calls of Leptonychotes (Weddell seal). Zoologica, N.Y. Zool. Soc. 50:45.

————, ————, 1966. Sound structure and directionality in Orcinus (killer whale). Zoologica, N.Y. Zool. Soc. 51(2):71.

————, ————, and C. Ray, 1963. Underwater sounds of pinnipeds. Science 141:50.

————, ————, ————, 1966. Analysis of underwater Odobenus calls with remarks on the development and function of the pharyngeal pouches. Zoologica, N.Y. Zool. Soc. 51(3):103.

Schmidt-Nielsen, B., and K. Schmidt-Nielsen, 1950a. Do kangaroo rats thrive when drinking sea water? Amer. J. Physiol. 160:291.

————, ————, 1950b. Evaporative water loss in desert rodents in their natural habitat. Ecology 31:75.

————, ————, 1951. A complete account of water metabolism in kangaroo rats and experimental verification. J. Cell Comp. Physiol. 38:165.

————, ————, J.T. Houpt, and S.A. Jarnum, 1956. Water balance of the camel. Amer. J. Physiol. 185:185.

————, ————, ————, ————, 1957. Urea excretion in the camel. Amer. J. Physiol. 188:477.

Schmidt-Nielsen, K., 1959. The physiology of the camel. Sci. Amer. 201:140.

————, 1964. Desert Animals: Physiological Problems of Heat and Water. Oxford University Press, New York.

————, 1972. Recent advances in the comparative physiology of desert animals, in Comparative Physiology of Desert Animals (G.M.O. Maloiy, ed.). Academic Press, New York.

————, 1981. Countercurrent systems in animals. Sci. Amer. 244:118.

————, W.L. Bretz, and C.R. Taylor, 1970. Panting in dogs: Unidirectional air flow over evaporative surfaces. Science 169:1102.

————, F.R. Hainsworth, and D.E. Murrish, 1970. Counter-current heat exchange in the respiratory passages: Effects on water and heat balance. Resp. Physiol., 9.263.

———— and A.E. Newsome, 1962. Water balance in the mulgara (Dasycercus cristicauda), a carnivorous desert marsupial. Aust. J. Biol. Sci. 15:683.

———— and B. Schmidt-Nielsen, 1952. Water metabolism of desert mammals. Physiol. Rev. 32:135.

————, ————, 1953. The desert rat, Sci. Amer. 189(1):73.

————, ————, 1954. Heat regulation in small and large desert animals, in Biology of Deserts

(J.L. Cloudsley-Thompson, ed.). Institute of Biology, London.

————, ————, T.A. Houpt, and S.A. Jarnum, 1956. The question of water storage in the stomach of the camel. Mammalia 20:1.

————, ————, ————, ————, 1957. Body temperature of the camel and its relation to water economy. Amer. J. Physiol. 188:103.

————, R.C. Schroter, and A. Shkolnik, 1981. Desaturation of exhaled air in camels. Proc. Royal Soc. London, B211:305.

Schneider, K.M., 1930. Das Flehmen. Zool. Gart. (Leipzig) 3:183; 4:349; 5:200, 287.

Schneider, R.H. Jurg Kugn, and G. Kelemen, 1967. De Larynx der Hypsignathus monstrosus Allen 1861. Ein Unifum in der Morphologie des Kehlkopfes. Z. Wiss. Zool.

Schnitzler, H.U., 1968. Echoortung bei der Ortungslaute der Hufeisen-Fledermause (Chiroptera-Rhinolophidae) in verschiedenen Orientierungssituationen. Z. Vergl. Physiol. 57:376.

Schoener, T.W., 1971. Theory of feeding strategies. Ann. Rev. Ecol. System. 2:369.

Scholander, P.F., 1940, Hvalradets Skrifter Norske Videnskaps-Akad, Oslo 22:1.

————, 1955. Evolution of climatic adaptation in homeotherms. Evolution 9:15.

————, R. Hock, V. Walters, and L. Irving, 1950a. Adaptation to cold in arctic and tropical mammals and birds in relation to body temperature, insulation, and basal metabolic rate. Biol. Bull. 99:259.

————, ————, ————, F. Johnson, and L. Irving, 1950b. Heat regulation in some arctic and tropical mammals and birds. Biol. Bull. 99:237.

———— and J. Krog, 1957. Countercurrent and vascular heat exchange; sloths. J. Appl. Physiol. 10:404.

———— and W.E. Schevill, 1955. Countercurrent and vascular heat exchange; whales. J. Appl. Physiol. 8:279.

————, V. Walters, R. Hock, and L. Irving, 1950. Body insulation of some arctic and tropical mammals and birds. Biol. Bull. 99:225.

Schubert, G.H., 1953, Ponderosa pine cone cutting by squirrels. J. Forestry 51:202.

Schultz, A.M., 1964. The nutrient-recovery hypothesis for Arctic microtine cycles. II: Ecosystem variables in relation to Arctic microtine cycles, in Grazing in Terrestrial and Marine Environments (P.J. Crisp, ed.). Brit. Ecol. Soc. Symp. No. 4. Blackwell, Oxford.

————, 1965. The tundra as a homeostatic system. Presented at A.A.A.S. Meeting, Dec. 1965, (Mimeo.)

————, 1969. A study of an ecosystem: The Arctic tundra, in The Ecosystem Concept in Natural Resource Management (G.M. Van Dyne, ed.). Academic Press, New York.

Schultz, C.B., M.R. Schultz, and L.D. Martin, 1970. A new tribe of saber-toothed cats (Barbourofelini) from the Pliocene of North America. Bull. Univ. Nebraska State Mus. 9:1.

Schultze-Westrum. T., 1965. Nochweis differenzierter Duftstoffe beim Gleitbeutler Petaurus breviceps papuanus Thomas (Marsupialia, Phalangeridae). Naturwiss 51(9):226.

Schuster, R.H., 1976. Lekking behavior in Kafue lechwe. Science 192:1240.

Schusterman, T.J., and S.N. Feinstein, 1965. Shaping and discriminative control of underwater click vocalization in a California sea lion. Science 150:1743.

Schwartz, C.C., J.G. Nagy, and W.L. Regelin, 1980. Juniper oil yield, terpenoid concentration, and antimicrobial effects on deer. J. Wildl. Mgmt. 44:107.

————, W.L. Regelin, and J.G. Nagy, 1980. Deer preference for juniper forage and volatile oil treated food. J. Wildl. Mgmt. 44:114.

Sclater, W.L., and P.L. Sclater, 1899. The Geography of Mammals. Kegan, Paul, Trench, Trubner, London.

Sebeok, T.A., (ed.), 1977. How Animals Communicate. Indiana University Press, Bloomington, London.

Selander, R.K., 1966. Sexual dimorphism and differential niche utilization in birds. Condor 68:113.

Selye, H., 1955. Stress and Disease. Science 122:625.

———— and J.B. Collip, 1936. Fundamental factors in the interpretation of stimuli influencing the endocrime glands. Endocrin. 20:667.

Semeonoff, R., and F. W. Robertson, 1968. A biochemical and ecological study of plasma esterase polymorphism in natural populations of the field vole, Microtus agrestis. Biochem. Genet. 1:205.

Sewell, G.D., 1968. Ultrasound in rodents. Nature 217:682.

Seyfarth, R.M., D.L. Cheney, and P. Marler, 1980. Vervet monkey alarm calls: Semantic communications in a free-ranging primate. Anim. Behav. 28:1070.

Sharman, G.B., 1970. Reproductive physiology of marsupials. Science 167:1221.

Shaw, W.T., 1934. The ability of the giant kangaroo rat as a harvester and storer of seeds. J. Mamm. 15:275.

————, 1936. Moisture and its relation to the cone-storing habit of the western pine squirrel. J. Mamm. 17:337.

Sheldrick, D., 1972. Death of the Tsavo elephant. Sat. Rev., Sept. 30, p. 29.

————, 1973. The Tsavo Story. Collins and Harvill Press, London.

Shipley, K., M. Hines, and J.S. Buchwald, 1981. Individual differences in threat calls of northern elephant seal bulls. Anim. Behav. 29:12.

Shirer, H.W., and H.S. Fitch, 1970. Comparison from radiotracking of movements and denning habits of the raccoon, striped skunk, and opossum in northeastern Kansas. J. Mamm. 51:491.

Shkolnik, A., and A. Borut, 1969. Temperature and water relations in two species of spiny mice (*Acomys*). J. Mamm. 50:245.

———— and K. Schmidt-Nielsen, 1976. Temperature regulation in hedgehogs from temperate and desert environments. Physiol. Zool. 49:56.

Shoemaker, V.H., K.A. Nagy, and W.R. Costa, 1976. Energy utilization and temperature regulation by jackrabbits (*Lepus californicus*) in the Mojave Desert. Physiol. Zool. 49:364.

Short H.L., 1966. Effects of cellulose levels on the apparent digestibility of seeds eaten by mule deer. J. Wildl. Manage. 30:163.

————, D.R. Dietz, and E.E. Remmenga, 1966. Selected nutrients in mule deer plants. Ecology 47:222.

Shortridge, G.C., 1934. The mammals of Southwest Africa, 2 vols. Heinemann, London.

Sige, B., 1971. Anatomie du membre anterieur chez un chiroptere molosside (*Tadarida* sp.) due Stampien de Cereste (Alpes-de-Haute-Provence). Palaeovert. 4:1.

Siivonen, L., 1954. Features of short-term fluctuations. J. Wildl. Manage. 18:38.

Silankove, N., H. Tagari, and A. Shkolnik, 1980. Gross energy digestion and urea recycling in the desert black Bedouin goat. Comp. Biochem. Physiol., 67:215.

Silverman, H.G., and M.H. Dunbar, 1980. Aggres-

sive tusk use by the narwhal (*Monodon monoceros*). Nature 284:57.

Simmons, J.A., 1971. Echolocation in bats: Signal processing of echoes for target range. Science, 171:925.

————, 1973. The resolution of target range by echolocating bats. J. Acoust. Soc. Amer. 54:157.

————, 1974. Response of the Doppler echolocation system in the bat, *Rhinolophus ferrumequinum*. J. Acoust. Sco. Amer. 56:672.

————, D.J. Howell, and N. Suga, 1975. Information content of of bat sonar echoes. Amer. Sci. 63:204.

————, M.B. Fenton, and M.J. O'Farrell, 1979. Echolocation and pursuit of prey by bats. Science 203:16.

————, W.A. Lavender, B.A. Lavender, C.A. Doroshow, S.W. Kiefer, R. Livingston, A.C. Scallet, and D.E. Crowley, 1974. Target structure and echo spectral discrimination by echolocating bats. Science 186:1130.

Simpson, G.G., 1937. Skull structure of the multituberculata. Bull. Amer. Mus. Nat. Hist. 73: 727.

————, 1940. Mammals and land bridges. J. Wash. Acad. Sci. 30:137.

————, 1943. Mammals and the nature of continents. Amer. J. Sci. 24:1.

————, 1944. Tempo and mode in evolution. Columbia University Press, New York.

————, 1945. The principles of classification and a classification of mammals. Bull. Amer. Mus. Nat. Hist. 85:1.

————, 1947. Evolution, interchange, and resemblance of North American and Eurasian Cenozoic mammalian faunas. Evolution 1:218.

————, 1950. History of the fauna of Latin America. Amer. Sci. 38:361.

————, 1951. Horses. Oxford University Press, New York.

————, 1952. Probabilities of dispersal in geologic time. Bull. Amer. Mus. Nat. Hist. 99.163.

————, 1953. The Major Features of Evolution. Columbia University Press, New York.

————, 1959. Mesozoic mammals and the polyphyletic origin of mammals. Evolution 13:405.

————, 1961. Historic zoogeography of Australian mammals. Evolution 15:431.

————, 1965a. Attending Marvels: A Patagonian Journal. Time, New York.

—————, 1965b. The Geography of Evolution. Chilton Books, Philadelphia.

—————, 1969. South American mammals, *in* Biogeography and Ecology in South America (E.J. Fihkan et al., eds.). Mono. Biol., 19. W. Junk, The Hague.

—————, 1970a. Additions to knowledge of the Argyrolagidae (Mammalia, Marsupialia) from the late Cenozoic of Argentina. Breviora, Mus. Comp. Zool. 361:1.

—————, 1970b. The Argyrolagidae, extinct South American Marsupials. Bull. Mus. Comp. Zool. 139:1.

Sinclair, A.R.E., 1970. Studies of the ecology of the East African buffalo. Ph.D. thesis, Oxford University, Oxford.

—————, 1974. The social organization of the East African buffalo, *in* The Behavior of Ungulates and Its Relation to Management (V. Geist and F.R. Walther, eds.). International Union for Conservation of Nature and Natural Resources, Morges, Switzerland.

—————, 1977 The African buffalo. University of Chicago Press, Chicago.

Slaughter, B.H., 1968. Earliest known marsupials. Science 162(3850):254.

————— and D.W. Walton (eds.), 1970. About Bats. Southern Methodist University Press, Dallas.

Slijper, E.J., 1936. Die Cetacean, vergleichend-anatomisch und systematisch. Capita Zool. 7:1.

—————, 1962. Whales. Basic Books, New York.

—————, 1979 Whales. Cornell University Press, Ithaca, N.Y.

Slobodchikoff, C., 1978. Experimental studies of predation on tenebrionid beetles by skunks. Behavior 66:313.

Slobodchikoff, C.N., and R. Coast, 1980. Dialects in the alarm calls of prairie dogs. Behav. Ecol. Sociobiol. 7:49.

Smith, G.W., and D.R. Johnson, 1985. Demography of a Townsend ground squirrel population in southwestern Idaho. Ecology 66:171.

Smith, H.M., 1960. Evolution of Chordate Structure. Holt, Rinehart, and Winston, New York.

Smith, J.D., 1972. Systematics of the chiropteran family Mormoopidae. Univ. Kansas Publ., Mus. Nat. Hist., Misc. Publ. No. 56, p. 1.

—————, G. Richter, and G. Storch, 1979. Wie Fledermäuse sich einmal ernährt haben. Ernährungsbiologie 79:482.

Smith, P.W., 1965. Recent adjustments in animals' ranges, *in* The Quaternary of the United States (H.E. Wright Jr. and D.G. Frey, eds.). Princeton University Press, Princeton, N.J.

Smith, R.B., 1971. Seasonal activities and ecology of terrestrial vertebrates in a Neotropical monsoon environment. M.S. thesis, Northern Arizona University, Flagstaff.

Smith, R.H., 1975. Nitrogen metabolism in the rumen and the comparative and nutritive value of nitrogen compounds entering the duodenum, *in* Digestion and Metabolism in the Ruminant. (I.W. McDonald and A.C.I. Warner, eds.). Univ. New England, Armidale, New South Wales, Australia.

Smythe, N., 1970. On the existence of "pursuit invitation" signals in mammals. Amer. Nat. 104:491.

—————, 1978. The natural history of the Central American agouti *(Dasyprocta punctata)*. Smithsonian Contrib. Zool. 257:1.

Soholt, L.S., 1973. Consumption of primary production by a population of kangaroo rats *(Dipodomys merriami)* in the Mojave Desert. Ecol. Monogr. 43:358.

Sondaar, P.Y., 1977. Insularity and Its Effects on Mammal Evolution. NATO Advanced Study Institute, Plenum Press, New York.

Sorenson, M.W., and C.H. Conaway, 1968. The social and reproductive behavior of *Tupaia montana* in captivity. J. Mamm. 49:502.

Southern, H.N., 1964. Handbook of British Mammals. Blackwell, Oxford.

Sowls, L.K., 1974. Social behavior of the collared peccary *Dicotyles tajacu, in* The Behavior of Ungulates and Its Relation to Management (V. Geist and F.R. Walther, eds.). International Union for Conservation of Nature and Natural Resources, Morges, Switzerland.

Spencer, A.W., and H.W. Steinhoff, 1968. An explanation of geographic variation in litter size. J. Mamm. 49:281.

Spencer, D.A., 1958a. Preliminary investigations on the northwestern *Microtus* irruption. U. S. Fish and Wildl. Serv., Denver Wildl. Res. Lab. Spec. Report.

—————, 1958b. Biological and control aspects, *in* the Oregon Meadow Mouse Irruption of 1957–1958. Federal Cooperative Extension service, Oregon State College, Corvallis.

Spitz, F., 1963. Estude des densites de population de *Microtus arvalis*. Pall. A Saint-Michal-en-L'Hern (Vendu). Mammalia 27:497.

Sprankel, H., 1965. Untersuchungen an *Tarsius.*

I: Morphologie des Schwanzes nebst ethologischen Bemerkungen. Folia Primat. 3:153.

States, J.S., 1980. Ecological Studies of Hypogenous Fungi. Progress Report, BFR Project #4.

Stehn, R.A., and F.J. Jannett Jr., 1981. Male-induced abortion in various microtine rodents. J. Mamm. 62:369.

Steiniger, F., 1950. Beiträge zur Soziologie und sonstigen Biologie der Wanerratte. Z. Tierpsychol. 7:356.

Stenlund, M.H., 1955. A Field Study of the Timber Wolf *(Canis lupus)* on the Superior National Forest, Minnesota. Minn. Dept. Cons. Tech. Bull. 4.

Stephenson, A.B., 1969. Temperatures within a beaver lodge in winter. J. Mamm. 50:134.

Sterling, I., 1969. Ecology of the Weddell seal in McMurdo Sound, Antarctica. Ecology 50:573.

Stevenson-Hamilton, J., 1947. Wildlife in South Africa. Cassell, London.

Stirton, R.A., R.H. Tedford, and M.O. Woodburne, 1967. A new Tertiary formation and fauna from the Tirari Desert, South Australia. Rec. South Aust. Mus. 15:427.

Stock, C., 1949. Rancho La Brea: A record of Pleistocene life in California. L.A. County Mus., Sci. Ser., no. 13, p.1.

Stones, R.C., and J.E. Wiebers, 1965. A review of temperature regulation in bats (Chiroptera). Amer. Midl. Nat. 74:155.

————, ————, 1967. Temperature regulation in the little brown bat, *Myotis lucifugus, in* Mammalian Hibernation, vol. III (K.C. Fisher, A.R. Dawe, C.P. Lyman, E. Schonbaum, and F.E. South Jr., eds.). Oliver & Boyd and American Elsevier, New York.

Storer, R.W., 1955. Weight, wing area, and skeletal proportions in three accipiters. Acta 11th Cong. Int. Ornithol., p. 278.

Storer, T.I., and R.L. Usinger, 1965. General Zoology. McGraw-Hill, New York.

Strickler, T.L., 1980. Downstroke muscle histochemistry in two bats, *in* Proc. Fifth Internat. Bat Research Conf. (D.E. Wilson and A.L. Gardner, eds.). Texas Tech Press, Lubbock.

Struhsaker, T.T., 1967. Behavior of elk *(Cervus canadensis)* during the rut Z. Tierpsychol. 24(1):80.

————, 1976. The Red Colobus Monkey. University of Chicago Press, Chicago.

————, 1977. Infanticide and social organization in the redtail monkey *(Cercopithecus ascan-* *ius schmidti)* in the Kibale Forest, Uganda. A. Tierpsychol. 4:75.

Studier, E.H., and D.J. Howell, 1969. Heart rate of female big brown bats in flight. J. Mamm. 50:842.

————, J.W. Proctor, and D.J. Howell, 1970. Diurnal body weight loss and tolerance of weight loss in five species of *Myotis*. J. Mamm. 51:302.

Suga, N., and T. Shimozawa, 1974. Site of neural attenuation of responses to self-vocalized sounds in echolocating bats. Science 177:1211.

Sugiyama, Y., 1973. Social organization of wild chimpanzees, *in* Behavioral Regulators of Behavior in Primates (C.R. Carpenter ed.). Bucknell University Press, Lewisburg, Pa.

Sussman, R.W., and P.H. Raven, 1978. Pollination by lemurs and marsupials: An archaic coevolutionary system. Science 200:731.

Suthers, R.A., 1965. Acoustic orientation by fish-catching bats. J. Exp. Zool. 158:319.

————, 1967. Comparative echolocation by fishing bats, J. Mamm. 48:79.

Swank, W.G., 1958. The Mule Deer in Arizona Chaparral. Arizona Game and Fish Dept., Wildl. Bull. No. 3.

Sweeney, R.C.H., 1956. Notes on *Manis temmincki*. Ann. Mag. Nat. Hist., London.

Swinhoe, R., 1870. On the mammals of Hainan. Proc. Zool. Soc. London, 1870:224.

Szalay, F.S., 1968. The beginnings of primates. Evolution 22:19.

————, 1969. Origin and evolution of function of the mesonychid condylarth feeding mechanism. Evolution 23(4):703.

Taber, R.D., and R.F. Dasmann, 1957. The dynamics of three natural populations of deer *Odocoileus hemionus columbianus*. Ecology 38:233.

Talbot, L.M., and M.H. Talbot, 1963. The wildebeest in western Masailand. Wildl. Monogr., Chestertown, no. 12, p. 1.

Talmage, R.V., and G.D. Buchanan, 1954. The armadillo *(Dasypus novemcinctus):* A review of its natural history, ecology, anatomy and reproductive physiology. The Rice Institute Pamphlet, Monograph in Biology, vol. 41.

Tappen, N.C., 1960. Problems of distributions and adaptations of the African monkeys. Cur. Anthrop. 1:91.

Tate, G.H.H., 1933. A systematic revision of the marsupial genus *Marmosa*. Bull. Amer. Mus. Nat. Hist. 66:1.

————, 1947. Mammals of eastern Asia. Macmillan, New York.

———— and R. Archbold, 1937. Results of the Archbold Expeditions. 16: Some marsupials of New Guinea and Celebes. Bull. Amer. Mus. Nat. Hist. 73:331.

Tavolga, M.C., 1966. Behavior of the bottlenose dolphin *(Tursiops truncatus)*: Social interaction in a captive colony, *in* Whales, Dolphins and Porpoises (K.S. Norris, ed.). University of California Press, Berkeley.

Taylor, C.R., 1968a. Hygroscopic food: A source of water for desert antelopes? Nature 219:181.

————, 1968b. The minimum water requirements of some East African bovids. Symp. Zool. Soc. London 21:195.

————, 1969a. The eland and the oryx. Sci. Amer. 220:89.

————, 1969b. Metabolism, respiratory changes, and water balance of an antelope, the eland. Amer. J. Physiol. 217(1):317.

————, 1970. Dehydration and heat: Effect on temperature regulation of East African ungulates. Amer. J. Physiol. 219:1136.

————, 1972. The desert gazelle: A paradox resolved, *in* Comparative Physiology of Desert Animals (G.M.O. Maloiy, ed.). Symp. Zool. Soc. London, 31. Academic Press, New York.

————, 1974. Exercise and Thermoregulation, *in* Environmental Physiology (D. Robertshaw, ed.). Internat. Rev. Science, Environ. Physiol., Butterworth, London.

————, 1977. Excercise and environmental heat loads: different mechanisms for solving different problems, *in* Environmental Physiology (D. Robertshaw, ed.). Internat. Rev. Physiol., University Park Press, Baltimore.

————, N.C. Heglund, and G.M.O. Maloiy, 1982. Energetics and mechanics of terrestrial locomotion. I: Metabolic energy consumption as a function of speed and body size in birds and mammals. J. Exp. Biol. 97:1.

———— and C.P. Lyman, 1967. A comparative study of the environmental physiology of an East African antelope, the eland, and the Hereford steer. Physiol. Zool. 40(3):280.

————, ————, 1972. Heat storage in running antelopes: Independence of brain and body temperatures. Amer. J. Physiol. 222:114.

————, K. Schmidt-Nielsen, R. Dmi'el, and M. Fedak, 1971. Effect of hyperthermia on heat balance during running in the African hunting dog. Amer. J. Physiol. 220:823.

————, ————, and J.L. Raab, 1970. Scaling of energetic cost of running to body size in mammals. Amer. J. Physiol. 219:1104.

————, C.A. Spinage, and C.P. Lyman, 1969. Water relations of the waterbuck, an East African antelope. Amer. J. Physiol. 217(2):630.

Taylor, W.T., and R.J. Weber, 1951. Functional Mammalian Anatomy. D. Van Nostrand, New York.

Tedford, R.H., 1967. The fossil Macropodidae from Lake Menindee, New South Wales. Univ. Calif. Publ. Geol. Sci. 64:1.

————, 1974. Marsupials and the new paleogeography, *in* Paleogeographic Provinces and Provinciality (C.A. Ross, ed.). Soc. Econ. Paleontol. Mineral., publ. 21.

————, M.R. Banks, N.R. Kemp, I. McDougall, and F.L. Sutherland, 1975. Recognition of the oldest known fossil marsupials from Australia. Nature, Lond. 255:141.

Tevis, L., 1950. Summer behavior of a family of beavers in New York State. J. Mamm. 31:40.

Thiessen, D.D., K. Owen, and G. Lindzey, 1971. Mechanisms of territorial marking in the male and female Mongolian gerbils *(Meriones unguiculatus)*. J. Comp. Physiol. Psychol. 77:38.

————, F.E. Regnier, M. Rice, M. Goodwin, N. Isaacks, and N. Lawson, 1974. Identification of a ventral scent marking pheromone in the male Mongolian gerbil *(Meriones unguiculatus)*. Science 184:83.

Thomas, J.A., and E.C. Birney, 1979. Parental care and mating system of the prairie vole, *Microtus ochrogaster*. Behav. Ecol. and Sociobiol. 5:171.

Thomas, S.P., and R.A. Suthers, 1972. The physiology and energetics of bat flight. J. Exp. Physiol. 57:317.

Thompson, D.Q., 1955. The role of feed and cover in population fluctuations of the brown lemming at Point Barrow, Alaska. Trans. North Amer. Wildl. Conf. 20:166.

Thompson, S.D., R.E. MacMillen, E.M. Burke, and C.R. Taylor, 1980. The energetic cost of bipedal hopping in small mammals. Nature 287:223.

Thorington, R.W., Jr., and S. Anderson, 1984. Primates, *in* Orders and Families of Recent Mammals of the World. (S. Anderson and J.K. Jones Jr., eds.). Wiley, New York.

Tinbergen, N., 1963. On aims and methods of ethology. Z. Tierpsychol. 20:410.

Tinkle, D.W., and I.G. Patterson, 1965. A study

of hibernating populations of *Myotis velifer* in northwest Texas. J. Mamm. 46:612.

To, L.P., and R.H. Tamarin, 1977. The relation of population density and adrenal gland weight in cycling and non-cycling voles *(Microtus)*. Ecology 58:928.

Tomich, P.Q., 1962. The annual cycle of the California ground squirrel, *Citellus beecheyi*. Univ. Calif. Publ. Zool. 65:213.

Torrey, T.W., 1971. Morphogenesis of the Vertebrates. Wiley, New York.

Townsend, M.T., 1935. Studies on some small mammals of central New York. Roosevelt Wildl. Ann. 4:1.

Trivers, R.L., 1971. The evolution of reciprocal altruism. Quart. Rev. Biol. 46:35.

Troughton, E., 1947. Furred Animals of Australia. Charles Scribner's Sons, New York.

Trumler, E., 1959. Das "Rossighkeitsgesicht" und ähnliches Ausdrucksverhalten bei Einhufern. Z. Tierpsychol. 16:478.

Tucker, V.A., 1962. Diurnal torpidity in the California pocket mouse. Science 136:380.

————, 1965. The relation between the torpor cycle and heat exchange in the California pocket mouse *Perognathus californicus*. J. Cell Comp. Physiol. 65:405.

Turner, B.N., 1971. The annual cycle of aggression in male *Microtus pennsylvanicus*, and its relation to population parameters. M.S. thesis, University of North Dakota.

Tuttle, M., 1981. Bat predation and the evolution of frog vocalizations in the Neotropics. Science 214:677.

Udvardy, M.D.F., 1969. Dynamic Zoogeography. Van Nostrand Reinhold, New York.

Vachrameev, V.A., and M.A. Akhmet'yev, 1972. The development of floras on the boundary of the Late Cretaceous and the Paleocene (on data from the study of leaf remains), *in* The Development and Replacement of the Organic World on the Boundary of the Mesozoic and Cenozoic (V.N. Shimansiy and A.N. Solov'yev, eds.). (Conf. April, 1972.) Abstr. pap. methodol. mater., Moscow: Akad. Nauk SSSR, Moscow Soc. Natur. (In Russian.)

Valentine, J.W., and E.M. Moores, 1970. Plate-tectonic regulation of faunal diversity and sea level: A model. Nature 228:657.

Van Couvering, J.A., and J.A.H. Van Couvering, 1975. African isolation and the Tethys seaway. Sixth Congress Regional Committee on Mediterranean Neogene Stratigraphy, Bratislava, p. 363.

Van De Graaff, K., and R.P. Balda, 1973. Importance of green vegetation for reproduction in the kangaroo rat, *Dipodomys merriami merriami*. J. Mamm. 54:509.

Van den Ende, P., 1973. Predator-prey interactions in continuous culture. Science 181:562.

Van Deusen, H.M., 1967. Personal communication.

————, 1969. Results of the Archbold Expeditions. No. 90. Notes on the echidnas (Mammalia, Tachyglossidae) of New Guinea. Amer. Mus. Novitates 2383:1.

————, 1971. *Zaglossus,* New Guinea's egg-laying anteater. Fauna 1:12.

———— and J.K. Jones Jr., 1967. Marsupials, *in* Recent Mammals of the World (S. Anderson and J.K. Jones Jr., eds.). Ronald Press, New York.

Van Gelder, R.G., 1953. The egg-opening technique of a spotted skunk. J. Mamm. 34:255.

Van Lawick-Goodall, J. 1968, A preliminary report on expressive movements and communication in the Gombe Stream Chimpanzees, *in* Primates: Studies in Adaptation and Variability (P.C. Jay, ed.). Holt, Rinehart and Winston, New York.

Van Valen, L., 1965a. The earliest primates. Science 150:743.

————, 1965b. Treeshrews, primates and fossils. Evolution 19:137.

————, 1966. Deltatheiridia, a new order of mammals. Bull. Amer. Mus. Nat. Hist. 132:1.

———— and R.E. Sloan, 1966. The extinction of the multituberculates. Syst. Zool. 15:261.

Vaughan, T.A., 1954. Mammals of the San Gabriel mountains of California. Univ. Kansas Publ., Mus. Nat. Hist. 7:513.

————, 1959. Functional morphology of three bats: *Eumops, Myotis, Macrotus*. Univ. Kansas Publ, Mus. Nat. Hist. 12:1.

————, 1961. Vertebrates inhabiting pocket gopher burrows in Colorado. J. Mamm., 42:171.

————, 1966a. Morphology and flight characteristics of molossid bats. J. Mamm. 47:249.

————, 1966b. Food-handling and grooming behaviors in the plains pocket gopher. J. Mamm. 47:132.

————, 1967. Food habits of the northern pocket gopher on shortgrass prairie. Amer. Midl. Nat. 77:176.

————, 1969. Reproduction and population

densities in a montane small mammal fauna, *in* Contributions in Mammalogy (J.K. Jones Jr., ed.). Misc. Publ., Mus. Nat. Hist., University of Kansas, No. 51.

————, 1970a. Adaptations for flight in bats, *in* About Bats (B.H. Slaughter and D.W. Walton, eds). Southern Methodist University Press, Dallas.

————, 1970b. The skeletal system. The muscular system. Flight patterns and aerodynamics, *in* Biology of Bats (W.A. Wimsatt, ed.). Academic Press, New York.

————, 1974. Resource allocation in some sympatric, subalpine rodents. J. Mamm. 55:764.

————, 1976. Nocturnal behavior of the African false vampire bat *(Cardioderma cor)*. J. Mamm. 57:227.

————, 1977. Foraging behavior of the giant leaf-nosed bat. E. Afr. Wildl. J. 15.

————, 1980. Woodrats and picturesque junipers, *in* Aspects of Vertebrate History: Essays in Honor of Edwin Harris Colbert (L.L. Jacobs, ed.). Mus. Northern Arizona Press, Flagstaff.

————, 1982. Stephen's woodrat, a dietary specialist. J. Mamm. 63:53.

————, 1985. Reproduction in Stephen's woodrat: The wages of folivory. J. Mamm. 66:429.

———— and G.C. Bateman, 1970. Functional morphology of the forelimbs of mormoopid bats. J. Mamm. 51:217.

———— and M.M. Bateman, 1980. The molossid wing: Some adaptations for rapid flight, *in* Proceedings of the Fifth International Bat Research Conference (D. Wilson and A. Gardner, eds.). Texas Tech Press, Lubbock.

———— and T.J. O'Shea, 1976. Roosting ecology of the pallid bat, *Antrozous pallidus*. J. Mamm. 57:19.

———— and S.T. Schwartz, 1980. Behavioral ecology of an insular woodrat. J. Mamm. 61:205.

———— and W.P. Weil, 1980. The importance of arthropods in the diet of *Zapus princeps* in a subalpine habit. J. Mamm. 61:122.

Vessey-FitzGerald, D.F., 1960. Grazing succession among East African game animals. J. Mamm. 41:161.

Villa-R., B., 1966. Los Murciélagos de Mexico. Inst. Biol., UNAM.

———— and E.L. Cockrum, 1962. Migration in the guano bat, *Tadarida brasiliensis mexicana*. J. Mamm. 43:43.

Vleck, D., 1979. The energy cost of burrowing by the pocket gopher *Thomomys bottae*. Physiol. Zool. 52:122.

————, 1981. Burrow structure and foraging costs in the fossorial rodent, *Thomomys bottae*. Oecologia 49:391.

Vogel, V.B., and geb. El-Kareh, 1969. Vergleichende Untersuchungen über den Wasserhaushalt von Fledermäusen *(Rhinopoma, Rhinolophus* und *Myotis)*. Z. Vergl. Physiol. 64:324.

Vogl, R.J., 1973. Ecology of the knobcone pine in the Santa Ana Mountains, California. Ecol. Monogr. 43:125.

Walker, E.P., 1968. Mammals of the World, 2d edn., 2 vols. Johns Hopkins Press, Baltimore.

Walker, E.P., et al., 1975. Mammals of the World, 3d edn. (J.L. Paradiso, ed.). Johns Hopkins University Press, Baltimore.

Wallace, A.R., 1876. The geographical distribution of animals, 2 vols. Harper, New York. Reprinted by Hafner, New York.

Walther, F., 1958. Zum Kampf- und Paarungsverhalten einiger Antilopen. Z. Tierpsychol. 15:340.

————, 1965. Verhaltensstudien an der Grantgazell (*Gazella granti* Brooke, 1872) im Ngorongoro-Krater. Z. Tierpsychol. 22:167.

————, 1966. Zum Liegeverhalten des Weissschwanzgnus (*Connochaetes gnou* Zimmerman, 1780). Z. Säugetierk. 31:1.

Ward, A.L., and J.O. Keith, 1962. Feeding habits of pocket gophers on mountain grasslands. Ecology 43:744.

Watkins, W.A., and W.E. Schevill, 1968. Underwater playback of their own sounds to *Leptonychotes* (Weddell seals). J. Mamm. 49:287.

Watson, A., 1958. The behavior, breeding and food-ecology of the snowy owl *Nyctea scandiaca*. Ibis 99:419.

Watson, R.M., 1968. Report on aerial photographic studies of vegetation carried out in the Tsavo area of Kenya. typescript. (Cited by Laws, 1970).

Watts, P.D., N.A. Øritsland, C. Jonkel, and K. Ronald, 1981. Mammalian hibernation and the oxygen consumption of a denning black bear *(Ursus americanus)*. Comp. Biochem. Physiol. 69a:121.

Weber, N.S., and J.S. Findley, 1970. Warm-season changes in fat content of *Eptesicus fuscus*. J. Mamm. 51:160.

Webster, D.B., 1961. The ear apparatus of the kangaroo rat, *Dipodomys*. Amer. J. Anat. 108:23.

————, 1962. A function of the enlarged middle-ear cavities of the kangaroo rat, *Dipodomys*. Physiol. Zool. 35:248.

————, 1963. A case of parallel evolution of the ear apparatus. Anat. Rec. 145:297.

————, 1966. Ear structure and function in modern mammals. Amer. Zool. 6:451.

————, R.F. Ackermann, and G.C. Longa, 1968. Central auditory system of the kangaroo rat, *Dipodomys merriami*. J. Comp. Neurol. 133(4):477.

———— and C.R. Stack, 1968. Comparative histochemical investigation of the organ of Corti in the kangaroo rat, gerbil and guinea pig. J. Morph. 129(4)413.

Webster, F.A., and O.G. Brazier, 1968. Experimental Studies on Echolocation Mechanisms in Bats. Aerospace Medical Research Laboratories, Wright-Patternson Air Force Base, Ohio.

———— and D.R. Griffin, 1962. The role of flight membranes in insect capture by bats. Anim. Behav. 10:332.

Wegener, A., 1915. Die Entstehung der Kontinente und Ozeane. Sammlung Vieweg. No. 23, Brunswick. (Translation, 1924. The Origin of Continents and Oceans. Methuen, London.)

————, 1966. The Origin of Continents and Oceans. (Translated from the fourth edition by John Birum.) Dover Publications, New York.

Wegge, P., 1975. Reproduction and early calf mortality in Norwegian red deer. J. Wildl. Manage. 39:92.

Weichert, C.K., 1965. Anatomy of the Chordates, McGraw-Hill, New York.

Weigl, P.D., and E.V. Hanson, 1980. Observational learning and the feeding behavior of the red squirrel *(Tamiasciurus hudsonicus):* The ontogeny of optimization. Ecology 61:213.

Weil, W.L.P., 1968. Food habits of the western jumping mouse in north-central Colorado. M.S. thesis, Colorado State University, Fort Collins.

Weir, B.J., 1971. The evocation of oestrus in the cuis, *Galea musteloides*. J. Reprod. Fert. 26:405.

————, 1973. The role of the male in the evocation of estrus in the cuis, *Galea musteloides*. J. Reprod. Fert. Suppl. 19:419.

———— 1974. The tuco-tuco and plains viscacha, *in* The Biology of the Histricomorph Rodents (I.W. Rowlands and B.J. Weir, eds.). Academic Press, New York.

———— and I.W. Rowlands, 1973. Reproductive strategies of mammals. Ann. Rev. Ecol. System. 4:139.

Wells, M.C., and M. Bekoff, 1981. An observational study of scent marking in coyotes, *Canis latrans*. Anim. Behav. 29:332.

Wells, P.V., and C.D. Jorgensen, 1964. Pleistocene wood rat middens and climatic change in the Mojave Desert: A record of juniper woodlands. Science 143:1171.

Wemmer, C., 1979. Social organization and the saber-tooth syndrome in carnivorous mammals. Sangetierk. Mitteil. 27:127.

West, S.D., 1977. Midwinter aggregation in the northern red-backed vole *(Clethrionomys rutilus)*. Can. J. Zool. 55:1404.

Wetzel, R.M., 1977. The Chacoan peccary, *Catagonus wagneri* (Rusconi). Bull. Carnegie Mus. Nat. Hist. 3:1.

Wetzel, R.M., R.E. Dubos, R.L. Martin, and P. Myers, 1975. *Catagonus,* an "extinct" peccary alive in Paraguay. Science 189:379.

Wharton, C.H., 1950. Notes on the life history of the flying lemur, *Cynocephalus volans*. J. Mamm. 31:269.

Whitaker, J.O. Jr., 1963. Food, habitat and parasites of the woodland jumping mouse in central New York. J. Mamm. 44:316.

White, A.C., 1948. The Call of the Bushveld. A.C. White, Bloemfontein, S. Africa.

Whittaker, R.H., 1970. Communities and Ecosystems. Macmillan, New York.

Whitten, W.K., 1956. Modifications of the oestrus cycle of the mouse by external stimuli associated with the male. J. Endocrinol. 13:399.

Whittow, G.C., 1974. Sun, sand, and sea lions. Nat. Hist. 83:56.

Wickler, W. Von, and D. Uhrig, 1969. Verhalten und okologische Nische der Gelbflügelfledermaus, *Lavia frons* (Geoffroy) (Chiroptera, Megadermatidae). Z. Tierpsychol. 26:726.

Williams, G.C., 1967. Natural selection, the costs of reproduction, and a refinement of Lack's principle. Amer. Nat. 100:687.

Williams, T.C., L.C. Ireland, and J.M. Williams, 1973. High altitude flights of the free-tailed bat, *Tadarida brasiliensis,* observed with radar. J. Mamm. 54:807.

Wilson, D.E., 1979. Reproductive patterns, *in* Biology of Bats of the New World Family Phyllostomatidae, part III. (R.J. Baker, J.K. Jones Jr., and D.C. Carter, eds.). Spec. Publ. Mus., no. 16. Texas Tech Press, Lubbock.

————— and J.S. Findley, 1970. Reproductive cycle of a Neotropical insectivorous bat, *Myotis nigricans*. Nature 225:1155.

Wilson, E.O., 1975. Sociobiology: The New Synthesis. University of Chicago Press, Chicago.

Wilson, R.W., 1960. Early Miocene rodents and insectivores from northeastern Colorado. Univ. Kansas Paleont. Cont., Vertebrata, Art. 7. p. 1.

—————, 1972. Evolution and extinction in Early Tertiary rodents. Proc. Int. Geol. Congr., 24(sec. 7):217.

Wimsatt, W.A., 1944. Further studies on the survival of spermatozoa in the female reproductive tract of the bat. Anat. Rec. 88:193.

—————, 1945. Notes on breeding behavior, pregnancy, and parturition in some vespertilionid bats of eastern United States. J. Mamm. 26:23.

—————, 1969a. Transient behavior, nocturnal activity patterns, and feeding efficiency of vampire bats *(Desmodus rotundus)* under natural conditions. J. Mamm. 50:223.

—————, 1969b. Some interrelations of reproduction and hibernation mammals. Symp. Soc. Exp. Biol. 23:511.

————— (ed.), 1970. Biology of Bats. Academic Press, New York.

————— and B. Villa-R., 1970. Locomotor adaptations in the disc-winged bat. Amer. J. Anat. 129:89.

Winge, H., 1941. The Interrelationships of the Mammalian Genera, vol. 1. C.A. Reitzels Forlag, Copenhagen. (Danish translated by E. Deichmann and G.M. Allen.)

Wirtz, W.O. II, 1968. Reproduction, growth and development, and juvenile mortality in the Hawaiian monk seal. J. Mamm. 49:229.

—————, 1971. Personal communication.

Wislocki, G.B., 1942. Studies on the growth of deer antlers. I: On the structure and histogenesis of the antlers of the Virginia deer *(Odocoileus virginianus borealis)*. Amer. J. Anat. 71:371.

—————, 1943. Studies on growth of deer antlers, *in* Essays in Biology. University of California Press, Berkeley.

—————, J.C. Aub, and C.M. Waldo, 1947. The effects of gonadectomy and the administration of testosterone proprionate on the growth of antlers in male and female deer. Endocrin. 40:202.

Withers, P.C., and J.U.M. Jarvis, 1980. The effect of huddling on the thermoregulation and oxygen consumption for the naked mole-rat. Comp. Biochem. Physiol. 66:215.

Wolfe, J.A., 1978. A paleobotanical interpretation of Tertiary climates in the Northern Hemisphere. Amer. Scient. 66:694.

————— and T. Tanai, 1979. The Miocene Seldovia Point flora from the Kenai Group, Alaska. U.S.G.S. Prof. Paper 1105.

Wolff, J.O., 1980. Social organization of the taiga vole *(Microtus xanthognathus)*. The Biologist 62:34.

————— and W.Z. Lidicker, 1981. Communal winter nesting and food sharing in taiga voles. Behav. Ecol. Sociobiol. 9:237.

Wood, A.E., 1935. Evolution and relationship of the heteromyid rodents with new forms from the Tertiary of western North America. Ann. Carnegie Mus. 24:73.

—————, 1937. Parallel radiation among the geomyoid rodents. J. Mamm. 18:171.

—————, 1955. A revised classification of the rodents. J. Mamm. 36:165.

—————, 1959. Are these rodent suborders? Syst. Zool. 7:169.

—————, 1965. Grades and clades among rodents. Evolution 19:115.

—————, 1974. The evolution of the Old World and New World hystricomorphs, *in* The Biology of Hystricomorph Rodents (I.W. Rowlands and B. Weir, eds.). Symp. Zool. Soc. London. Academic Press, New York.

—————, 1975. The problem of the hystricognathous rodents. Univ. Mich. Papers Paleontol. 12:75.

—————, 1980. The origin of caviomorph rodents from a source in Middle America: A clue to the area of the platyrrhine primates, *in* Evolutionary Biology of the New World Monkeys and Continental Drift (R.L. Ciochon and A.b. Chiarelli, eds.). Plenum, New York.

Wood, F.G. Jr., 1959. Underwater sound production and concurrent behavior of captive porpoises, *Tursiops truncatus* and *Stenella plagiodon*. Bull. Mar. Sci. Gulf Caribbean 3:120.

Woodard, T.N., R.J. Gutierrez, and W.H. Rutherford, 1974. Bighorn lamb production, survival, and mortality in south-central Colorado. J. Wildl. Manage. 38:771.

Woodburne, M.E., and R.H. Tedford, 1975. The first Tertiary monotreme from Australia. Amer. Mus. Novitates 2588:1.

Woodburne, M.O., and W.J. Zinsmeister, 1982. Fossil land mammal from Antarctica. Science 218:284.

Wunder, B.A., 1970. Temperature regulation and the effects of water restriction on Merriam's chipmunk. Comp. Biochem. Physiol. 33:385.

Wynn, R.M., 1971. Immunological implications of comparative placental ultrastructure, *in* The Biology of the Blastocyst. (R.J. Blandau, ed.). University of Chicago Press, Chicago.

Wynne-Edwards, V.C. 1959. The control of population density through social behavior: A hypothesis. Ibis 101:436.

————, 1960. The overfishing principle applied to natural populations and their food resources, and a theory of natural conservation. Proc. Int. Orn. Congr. 12:790.

————, 1962. Animal dispersion in relation to social behavior. Hafner, New York.

Yanagisawa, K., G. Sata, M. Nomoto, Y. Katsuki, E. Ibezono, A.D. Grinnell, and T.H. Bullock, 1966. Fed. Proc., Physiol. p. 1539 (abstr.).

Yoakum, J., 1958. Seasonal food habits of the Oregon pronghorn antelope (*Antilocapra americana oregona* Bailey). Inter. Antelope Conf. Trans. 9:47.

Young, J.Z., 1957. The Life of Mammals. Clarendon Press, Oxford.

————, 1975. The Life of Mammals, 2d edn. Clarendon Press, Oxford.

Young, R.A., 1976. Fat, energy, and mammalian survival. Amer. Zool. 16:699.

Young, S.P., and H.H.T. Jackson, 1951. The Clever Coyote. Wildlife Management Institute, Washington, D.C.

Zimmerman, E.G., 1965. A comparison of habitat and food of two species of *Microtus*. J. Mamm., 46:605.

Zippelius, H., and W.M. Schleidt, 1956. Ultraschalllaute bei jungen Mausen. Naturwisse., 21:1.

Zittel, K.A. von, 1893. Handbuck der Paleontologie. Vol. 4 (Mammalia). Ouldenberg, Munich.

Zoeger, J., J.R. Dunn, and M. Fuller, 1981. Magnetic material in the head of the Pacific dolphin. Science, 213:892.

Index